P9-CND-669

INTRODUCTION TO
DIFFERENTIAL EQUATIONS
AND DYNAMICAL SYSTEMS

Richard E. Williamson
Dartmouth College

The McGraw-Hill Companies, Inc.
New York   St. Louis   San Francisco   Auckland   Bogotá   Caracas
Lisbon   London   Madrid   Mexico City   Milan   Montreal   New Delhi
San Juan   Singapore   Sydney   Tokyo   Toronto

McGraw·Hill

A Division of The **McGraw·Hill** Companies

INTRODUCTION TO DIFFERENTIAL EQUATIONS AND DYNAMICAL SYSTEMS
Copyright © 1997 by The McGraw-Hill Companies, Inc. All rights reserved. Printed in the United States of America. Except as permitted under the United States Copyright Act of 1976, no part of this publication may be reproduced or distributed in any form or by any means, or stored in a data base or retrieval system, without the prior written permission of the Publisher.

This book is printed on acid-free paper.

1 2 3 4 5 6 7 8 9 SEM SEM 9 0 3 2 1 0 9 8 7

ISBN 0-07-070594-1

This book was set in Palatino by York Graphic Services, Inc.

The editors were Jack Shira and Alice Goehring; the cover and interior designer was Elizabeth Williamson; the cover production artist was Ikka Valli, Paloma Design; the production supervisor was Louis Swaim.

Quebecor Printing Semline, Inc., was the printer; Quebecor Printing Book Press, Inc., was the binder.

Cover photo by John Still, © Photonica.

Library of Congress Cataloging-in-Publication Data
Williamson, Richard E.
    Introduction to differential equations and dynamical systems/
    Richard E. Williamson.
        p.   cm.
    Includes index.
    ISBN 0-07-070594-1 (acid-free paper)
    1. Differential equations.   2. Differentiable dynamical systems.
    I. Title
    QA371.LW548   1996   96-25715
    515'.35—dc20   CIP

INTERNATIONAL EDITION
Copyright 1997. Exclusive rights by The McGraw-Hill Companies, Inc. for manufacture and export. This book cannot be re-exported from the country to which it is consigned by McGraw-Hill. The International Edition is not available in North America.

When ordering this title, use ISBN 0-07-114856-6

ABOUT THE AUTHOR

Richard E. Williamson is a Professor of Mathematics at Dartmouth College. He received his Ph.D. from the University of Pennsylvania, specializing in real and complex analysis. He has also worked in mathematical economic theory. Professor Williamson is the coauthor with Hale Trotter of the text *Multivariable Mathematics*.

CONTENTS

PREFACE

This book is for use in a first course in differential equations for students in pure and applied mathematics, the physical sciences, or engineering. One goal of the text is to integrate many of the classic ideas of differential equations and the more modern study of dynamical systems into an introductory course in differential equations. Chapters 2, 4, 5, and 6 cover many of the ideas fundamental to the study of dynamical systems and illustrate the ideas with concrete examples. Another goal is to provide a variety of relevant applications of differential equations within the framework of a traditional course. The approach is conservative in that all the important techniques of a traditional course are included and flexible in that various mixtures of applications with the traditional topic coverage are possible in constructing a specific course of study.

FEATURES OF THE BOOK

Dynamical Systems. In addition to coverage of the traditional topics of an introductory differential equations course, the book thoroughly discusses dynamical systems. This approach lets the instructor offer a more modern course and serves as a natural bridge to the use of computers in this course.

Integration of Computers and Numerical Methods. The presence of computers and graphic-numeric software affects the teaching and learning of our subject in at least two related ways. One way provides immediate pleasure we can all gain from appreciating the exquisite detail in the complicated graphs, trajectories, and vector fields that are so important for our understanding of our subject. In another way, the geometric and numeric insight we can gain from computing enables us to think concretely about concepts that were considered too abstract for an introductory course until computers became widely available.

Sections on numerical methods are included throughout the text and display some schematic descriptions of what simple computer code to implement the methods might look like. My own preference over the years has been to give students my own working programs written along these lines because they let students see in detail how the computations are carried out. However, because such programs are not easily transferable to other computing environments, at appropriate places throughout the text references are given to the Web address http://math.dartmouth.edu/~rewn/ where programs written in the platform-independent language Java can be downloaded for individual use. Commercially available computer software such as MAPLE, DERIVE, and Mathematica are other possibilities, and most of the recursive numerical work can be completed using programmable hand calculators. Individual exercises are marked with the icon 🖮 if they are appropriate for solving with the help of a computer or calculator. And of course software and hand calculators can be used for symbolic work.

Applications. This book includes a variety of differential equations applicable to engineering and the physical sciences, along with some applications to population dynamics. Extended applications are in Chapters 2, 4, 6, and 10. The needs and tastes of the students and the instructor will determine the selection of applications used in the course; many of these applications require computer or calculators, usually accompanied also by some paper and pencil work.

Exercises. Basic material is treated repeatedly, but with variations, in the exercises, and the more difficult exercises are broken down into carefully graduated steps. Starred exercises are somewhat more demanding and usually less fundamental to an introductory course; these exercises may also be the basis for group projects.

ORDER OF TOPICS

The order of the chapters follows a natural sequence for the study of differential equations. In particular, the treatment of second-order equations with non-constant coefficients is placed closer to the chapters on partial differential equations, which is where most of the related material on special functions is really used. This arrangement also allows time for early emphasis on the study of higher-dimensional systems, along with their various applications. However, Green's functions for second-order constant-coefficient operators appear in Chapter 3, making it possible to deal routinely with piecewise-continuous forcing functions without resorting to Laplace transforms.

The core Chapters 1, 3, and 5 are organized fairly traditionally, with optional sections and subsections marked accordingly. Chapters 2, 4, and 6 consist of independent sections with an applied flavor that can be chosen to suit a variety of tastes. A typical course would cover at most two or three of the sections in these even-numbered chapters; some courses would cover only one. The section in each of Chapters 2, 4, and 6 devoted to numerical methods has exercises keyed to the applications in the same chapter, making it natural, for example, to follow Section 5 on the inverse-square law and planetary motion in Chapter 6 by the section in the same chapter on numerical solution of systems.

SUPPLEMENTS

Instructor's Manual. A manual with detailed solutions for all the exercises has been prepared by Jeanne Albert, Allan Gunter, and Randy Zounes and is available to adopters of the textbook.

Student's Solutions Manual. A manual with worked solutions for selected exercises in the text is available for purchase by the student.

ACKNOWLEDGMENTS

I am grateful to my Dartmouth colleagues Jeanne Albert, Dennis Desormier, Allan Gunter, Dennis Healy, David Hemmer, John Lamperti, Tim Olson, Florin Pop, Reese Prosser, and Ben Tilly for their helpful suggestions, based on their teaching experience with preliminary versions of the book. Gregory Fredericks of Lewis and Clark College, John Palmer of the University of Arizona, Craig Tracy of the University of California at Davis, and Randy Zounes of Cornell University were also very helpful. Detailed written comments provided by the following reviewers were very important:

Bruce Berndt, University of Illinois, Urbana-Champaign

Paul Blanchard, Boston University

Stewart Davidson, Georgia Southern University

Victor Gummersheimer, Southeast Missouri State University

Robert Hunt, Humboldt State University

Joyce Longman, Villanova University

Carol Jean Martin, Dodge City Community College

Gary Meisters, University of Nebraska

Craig Tracy, University of California, Davis

Dr. R. Lee Van de Wetering, San Diego State University

Horace Wiser, Washington State University

My association with the publisher has been a pleasure throughout the production process. Jack Shira and Maggie Rogers provided careful oversight for the entire project. Alice Goehring's editing was superb, and Elizabeth Williamson created a beautiful design. My thanks go as well to the rest of the staff at McGraw-Hill who worked on the project.

Corrections and suggestions will be gratefully received by the author at rewn@dartmouth.edu or by regular mail at the Dartmouth Mathematics Department.

Richard E. Williamson
1997

Syllabus Suggestions

The listings by chapter and section are by no means exhaustive but display variety in emphasis to give some feeling for the book's flexibility. Sections or subsections not listed can of course be selected from the Contents at an instructor's discretion. Note in particular that either or both of Chapter 8 and Chapter 9 can be covered at any point after Chapter 3.

Chapter Number	Section Number		
	Basic Course	Dynamics Emphasis	Special Functions
1	1–3, 5	1–3, 5	1–3, 5
2	1	4–6	
3	1–3	1–3	1–3
4	1	2, 5, 7	
5	1, 2	1, 2	1, 2
6	1	5–8	
7	1	1, 2	
8	1, 2		
9	1, 2, 5		1–7
10	1–4		1–4
11			1–3

Introductory Survey

Isaac Newton's invention of what we now call differential equations has remained a cornerstone of pure and applied science, and the subject is still generating new ideas. One important force in current developments is the power of numerical computation and computer graphics to clarify ideas and in particular to display chaotic behavior. Indeed, a principal aim of this book is to incorporate these methods into the study of differential equations in a useful way at an introductory level. Reading the following remarks and examples will give you some idea of how the subject is organized and what some of the applications are.

Mathematically speaking, a **dynamical system** is a description of possible evolutions over time of the points in some space $\mathscr{S}$ called the **state space** of the system. In applied mathematics the points in a state space are often identified with possible pairs of positions and velocities of some physical configuration,

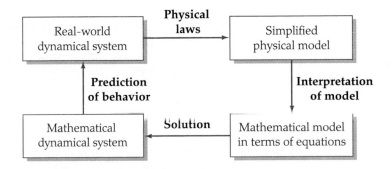

1

for example, a machine with springs and levers. If time t is measured continuously over an interval of the real numbers, the possible evolutions of a system will often be determined by a **differential equation** with t as an independent variable, that is, by an equation relating values of a function to corresponding values of one or more of its derivatives. (Formulating a differential equation is often our first purely mathematical step in dealing with a system, so it's sometimes natural to think of this equation itself as representing the system.) It is the interplay, shown schematically on page 1, between differential equation and dynamical system that is the central theme of this book, but specialists may find themselves focusing on this diagram at particular points. For example, a theoretical physicist might work across the top, a pure mathematician across the bottom, and an applied mathematician or scientist along the vertical directions. Equations with no designated time variable will also play a part in what follows, though dynamical models predominate in the applications.

The ideas sketched above apply also if time is measured in discrete jumps rather than continuously; we may still be dealing with a dynamical system, one in which a differential equation is replaced by a discrete recurrence relation as in the Newton, Euler, and Poincaré methods of Chapters 2, 4, and 6.

Since the examples that follow are primarily intended to give an intuitive feeling for what the subject is like, each one is tied to a particular application. Nevertheless, the equations are arranged according to a well-established mathematical classification. Such classifications are useful in part because different applications often lead to the same mathematical problem. Indeed, one of our aims is to show how a single mathematical idea can be used in several ways. Classification is useful also for predicting in general terms what sort of solution, if any, can be expected for a given type of problem. All these introductory examples are treated in more depth at appropriate points later on, and many of the details have been left out here in order not to overshadow general features. We begin with an example that appears prominently in many calculus courses.

---

e x a m p l e 1

Suppose $v = v(t)$ represents the velocity as a function of time t of an object of mass m falling in the vicinity of the earth's surface. Taking into account the pull of gravity on the object as well as the effect of air resistance will lead in some circumstances to considering a relationship of the form

$$m\frac{dv}{dt} = -kv + mg,$$

where g is a constant representing the acceleration of gravity and k is a constant determined by the amount of air resistance encountered at time t. If air resistance is negligible, we can assume $k = 0$, in which case we are just left with

$$\frac{dv}{dt} = g.$$

Integrating both sides with respect to t gives

$$v(t) = gt + v(0).$$

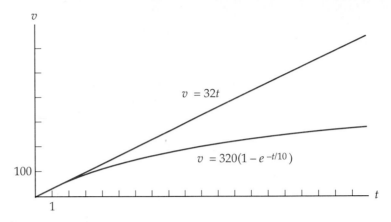

Figure IS.1 Effect of air resistance on the velocity of a falling object.

It follows that the graph of v as a function of t, shown in Figure IS.1 for the choices $m = 1$, $g = 32$, and $v(0) = v_0 = 0$, is a line with slope g. This outcome predicts in particular that the velocity will continue to increase at a steady rate until the object hits the surface of the earth, which is just what the differential equation $dv/dt = g$ says; beyond that we get a formula for velocity as a function of time. In case air resistance is taken into account, with resistance force proportional to velocity, we'll see in Chapter 1 that the velocity is expressible in terms of the interesting parameters g, k, m, and $v(0) = v_0$ in the form

$$v(t) = \frac{g}{k} + \left(v_0 - \frac{g}{k}\right)e^{-kt/m}.$$

The graph of a particular one of these solutions is also shown in Figure IS.1 for the choices $m = 1$, $g = 32$, $k = 0.1$, and $v(0) = 0$. (Note that if $v_0 = g/k$, the velocity is constant.)

example 2

A differential equation of the form

$$L\frac{d^2Q}{dt^2} + R\frac{dQ}{dt} + \frac{1}{C}Q = E(t)$$

can be used to predict the time variation of electric charge $Q = Q(t)$ in a simple electric circuit when $E(t)$ represents an externally applied voltage. One of the many applications for such a circuit is to produce an enhanced response to the applied voltage in the form of a more widely oscillating charge $Q(t)$. Here is a specific example:

$$\frac{d^2Q}{dt^2} + 0.3\frac{dQ}{dt} + Q = \cos t.$$

One particular function that satisfies this relation is $Q(t) = \frac{10}{3}\sin t$, and this function is called the *response* corresponding to the *input* $E(t) = \cos t$. By the choice of the constants L, R, and C the circuit has been "tuned" to produce the amplified response shown in Figure IS.2.

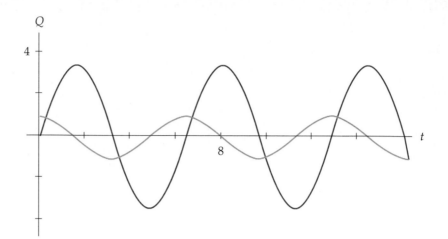

Figure IS.2 Input and amplified response from a circuit.

example 3

A horizontal beam of uniform cross section and constant linear density ρ will sag under its own weight. Let $y(x)$ represent the shape of the beam at distance x from one end. It can be shown that under these assumptions the function $y(x)$ nearly satisfies the **Euler beam equation**

$$EI\frac{d^4y}{dx^4} = -\rho.$$

The constant factors E and I depend, respectively, on the elasticity and cross-sectional configuration of the beam; their product $R = EI$ is a measure of a beam's rigidity, or resistance to bending. The shape that a beam actually takes depends on how it is supported at its two ends. Figure IS.3 shows the outline of the lower edge of a beam that is simply resting on a support at the left end and is *cantilevered* at the right end, that is, maintained horizontally, say by embedding it in a wall. The axis scaling is chosen to give an exaggerated view of the vertical deflection; if distance were measured in feet, the maximum deflection for such a 20-foot beam might be a fraction of an inch. Notice that the differen-

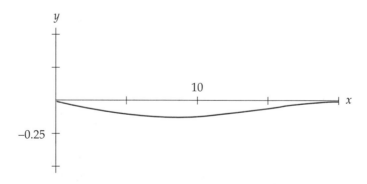

Figure IS.3 Beam profile; simply supported left end and cantilevered right end.

tial equation $EI\, d^4y/dx^4 = -\rho$ contains no information about what happens at the ends of the beam; this information has to be added by supplementary *boundary conditions* or *end conditions* on $y(x)$, in this case $y(0) = 0$, $y''(0) = 0$, and $y(20) = 0$, $y'(20) = 0$. For example, $y''(0) = 0$ requires flatness at the left end, and $y'(20) = 0$ requires the beam to be horizontal at the right end. The general solution of the differential equation contains four arbitrary constants of integration, which are then determined by the four end conditions. The resulting solution is described by the formula

$$y(x) = \frac{\rho}{EI}\left(-\frac{x^4}{24} + \frac{5x^3}{4} - \frac{500x}{3}\right).$$

Of course, for the picture some assumption had to be made about the relation between the product EI and the linear density ρ; in this example $EI = 7000\rho$.

Linearity of a Differential Equation. The differential equations in the three examples above have an important feature in common in that they are all **linear;** recognizing linearity when you see it is just a matter of checking that an equation has degree one in dependence on one or more unknown functions $y = y(x)$ and their derivatives. Thus the equation

$$\frac{d^2y}{dx^2} + 0.3x\frac{dy}{dx} + x^2y = x^2, \qquad \text{or} \qquad y'' + 0.3xy' + x^2y = x^2,$$

is linear, since it is a polynomial of degree one in the symbols y, $dy/dt = y'$, and $d^2y/dx^2 = y''$. (Recall that $d^2y/dx^2 \neq (dy/dx)^2$!) Likewise, the differential equation

$$\frac{dy}{dx} + x^2y = x^3$$

is linear since it is of first degree in both dy/dx and y. Many significant problems can be stated in terms of linear equations, and there are standard methods for providing complete solutions to lots of them, solutions that can in principle often be expressed in a single simple formula.

To say that an equation is **nonlinear** (i.e., not linear) is to say that its dependence on the functions, and their derivatives, that the equation determines is not of first degree. Thus $y'' + yy' = 0$ is nonlinear, since the term yy' has degree two as a function of the two variables y and y'.

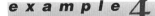

The differential equation

$$\frac{d^2y}{dt^2} + \tfrac{1}{5}\frac{dy}{dt} - y + y^3 = \tfrac{3}{10}\cos t$$

is nonlinear precisely because of the presence of the term y^3, which is of degree 3 in y. This equation is an example of a **Duffing equation** driven by a periodic oscillation with period 2π. Figure IS.4 shows a computer simulation for $0 \le t \le 600$ of an evolving solution satisfying $y(0) = y'(0) = 0$. The challenge posed by this rather irregular graph is to find some "order" hidden within its apparent "chaos." That issue is addressed concretely in Section 7C of Chapter 4.

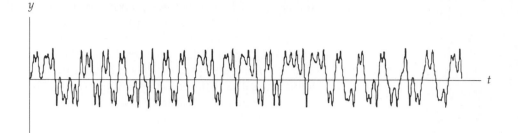

Figure IS.4 Approximate solution to the Duffing equation for $0 \leq t \leq 500$.

Nonlinear equations are important because they often provide much better models for physical behavior than do the simpler linear ones. Even though solution formulas are more difficult, or impossible, to find for nonlinear equations, the ability of computers to provide numerical or graphical solutions directly often allows us to bypass the need for a formula.

Order of a Differential Equation. The **order** of a differential equation is just the order of the highest-order derivative that occurs in the equation. Thus the orders of the equations in the previous three examples are, respectively, one, two, and four, while in the next two examples they are one and two. For linear equations, the number of arbitrary constants in a general solution formula is always the same as the order of the equation. (Unfortunately the corresponding situation for nonlinear equations isn't always so clear-cut. In particular a nonlinear equation may not have any solutions!)

In formal terms, a single n**th-order linear differential equation** of the type we've seen so far can always be written

$$a_n(x)\frac{d^n y}{dx^n} + a_{n-1}(x)\frac{d^{n-1}y}{dx^{n-1}} + \cdots + a_1(x)\frac{dy}{dx} + a_0(x)y = f(x),$$

where the functions $a_n(x), \ldots, a_0(x), f(x)$ all play a role in restricting the possible choices for solutions $y = y(x)$. Of course, we often use t, for time, as the independent variable instead of a space variable x, and a wide variety of other letters for the possible solutions instead of y.

e x a m p l e 5

The nonlinear first-order differential equation

$$\frac{dP}{dt} = kP\left(1 - \frac{P}{L}\right), \qquad \text{where } k, L \text{ are positive constants,}$$

is an example of what is called a **logistic equation** and can be used to model the growth of certain populations that have inherent limits to their ability to increase. The general solution has the form

$$P(t) = \frac{P_0 L}{(L - P_0)e^{-kt} + P_0},$$

where P_0 is the size of the population at time $t = 0$. The graph of this solution typically looks like the one in Figure IS.5.

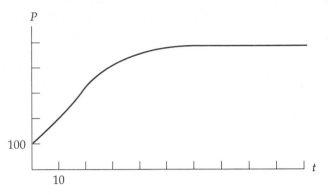

Figure IS.5 Limited population $P(t)$ with $P(0) = 100$, $L = 500$, and $k = 0.1$.

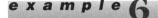 **e x a m p l e 6**   The position of a pendulum oscillating back and forth in a plane can be measured by the angle $\theta = \theta(t)$ that it makes with a vertical line at time t. A discussion based on elementary physical principles shows that $\theta(t)$ satisfies the second-order nonlinear differential equation

$$\frac{d^2\theta}{dt^2} = -\frac{g}{l}\sin\theta,$$

where l is the effective length of the pendulum and g is the acceleration of gravity. See Figure IS.6a. Unlike all the previous examples, there are no interesting solutions to this equation expressible in terms of the elementary functions of calculus. (The exceptional solution $\theta(t) = 0$ is called a *trivial* solution.) Our approach in this book is to use numerical approximations to the true solutions of this equation. These approximations can in principle be made to any desired degree of accuracy and can easily be converted to graphical form. A typical graph of $\theta(t)$ is shown in Figure IS.6b.

Higher-Dimensional Equations. Each of the next two examples deals with a **system** of differential equations that imposes constraints on more than one function. A system is **linear** if each equation in it is linear, and the **order** of a

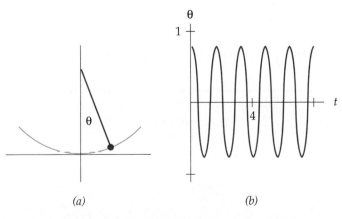

(a)   (b)

Figure IS.6 (a) Pendulum arc. (b) $\theta = \theta(t)$.

system is the maximum of the orders of the equations in the system. The number of real-valued functions determined by a system is called its **dimension.**

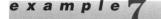

Populations of size $P_1(t)$ and $P_2(t)$ may each depend on the presence of the other for growth over time t. In the absence of the other species, each one might dwindle away according to some empirically discovered rule such as

$$\frac{dP}{dt} = -0.2P,$$

in which the negative right-hand side guarantees that $P(t)$ will decrease as time t increases. Such a decline is shown graphically in Figure IS.7a. The general formula for the solution is $P(t) = P(0)e^{-0.2t}$. On the other hand, the contribution of each population to the growth of the other can sometimes be described by a pair of linear equations, making up a first-order system of differential equations to be satisfied by two functions $P_1(t)$ and $P_2(t)$:

$$\frac{dP_1}{dt} = -0.2P_1 + 0.5P_2, \qquad \frac{dP_2}{dt} = 0.1P_1 - 0.2P_2,$$

in which the growth rate of each population is augmented by an amount proportional to the size of the other population. The resulting behavior of the two populations over time can also be expressed in terms of exponential functions, and a typical example appears in Figure IS.7b.

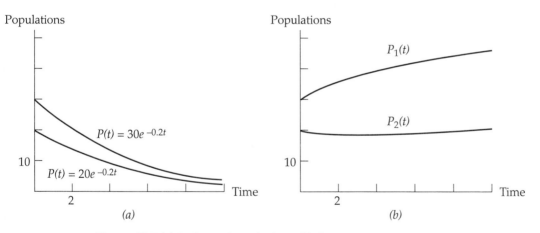

Figure IS.7 (a) Independent decline. (b) Cooperative increase.

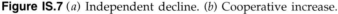

One of the earliest, and still one of the most significant, applications of differential equations is to the description of planetary motions as determined by the fundamental ideas of Newton's mechanics. Relative to a plane rectangular coordinate system centered at the center of mass of a star, a solitary planet has position $(x(t), y(t))$ at time t such that the second-order system of nonlinear equations

$$\frac{d^2x}{dt^2} = -\frac{G(M + m)x}{(x^2 + y^2)^{3/2}}, \qquad \frac{d^2y}{dt^2} = -\frac{G(M + m)y}{(x^2 + y^2)^{3/2}}$$

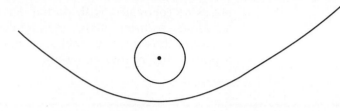

Figure IS.8 Closed planetary orbit and visitor's path.

is satisfied, where M is the mass of the star, m is the mass of the planet, and G is a constant associated with the inverse-square law of gravitational attraction. Rather than plot solution values $x(t)$ and $y(t)$ separately as functions of t, what is often done in examples like this is to plot the **trajectory,** or **orbit,** of a solution, that is, the graph representing the path actually followed by a planet. These paths can be described as ellipses for a single planet, though in the case of one-time visitors to a star the path will take the shape of a hyperbola. Two trajectories are shown in Figure IS.8.

example 9

Example 8 has to do with what is called the *two-body problem*, the two bodies typically being thought of as our sun and a single planet, or the earth and its moon, each pair moving in a fixed plane. The fact is that each of the three bodies just mentioned exerts some influence on the others. Taking all these effects into account in the *three-body problem* leads to a system of nine second-order nonlinear equations, since it takes three equations to determine the motion of each of the three bodies in 3-dimensional space. (For the sun and nine planets, we would have 30 such equations.) In cases of any practical significance, such systems must be solved by approximate numerical methods, and it's helpful in organizing the arithmetic to write position vectors, say, $\mathbf{x}_1 = (x_1, x_2, x_3)$, $\mathbf{x}_2 = (x_4, x_5, x_6)$, $\mathbf{x}_3 = (x_7, x_8, x_9)$ for each of the three bodies. In terms of these vectors, the nine equations of motion for the three-body problem can be written fairly briefly as follows:

$$\frac{d^2\mathbf{x}_1}{dt^2} = \frac{Gm_2}{r_{21}^3}(\mathbf{x}_2 - \mathbf{x}_1) + \frac{Gm_3}{r_{31}^3}(\mathbf{x}_3 - \mathbf{x}_1),$$

$$\frac{d^2\mathbf{x}_2}{dt^2} = \frac{Gm_1}{r_{12}^3}(\mathbf{x}_1 - \mathbf{x}_2) + \frac{Gm_3}{r_{32}^3}(\mathbf{x}_3 - \mathbf{x}_2),$$

$$\frac{d^2\mathbf{x}_3}{dt^2} = \frac{Gm_1}{r_{13}^3}(\mathbf{x}_1 - \mathbf{x}_3) + \frac{Gm_2}{r_{23}^3}(\mathbf{x}_2 - \mathbf{x}_3),$$

where r_{jk} stands for the distance between $\mathbf{x}_j$ and $\mathbf{x}_k$; that is, $r_{jk} = |\mathbf{x}_j - \mathbf{x}_k|$. Starting with given positions and velocities for the three bodies, it is routine to predict with good accuracy where they will be, and what their velocities are, at future times.

Partial Derivatives. So far the differential equations we've considered have as their solutions only functions of a single 1-dimensional variable, for example, t for time or x for distance. Differential equations of that kind are called **ordinary** differential equations, because they impose conditions only on ordinary derivatives as contrasted with partial derivatives. A differential equation whose solutions are differentiated in more than one variable (e.g., time and distance, or two or three space variables, or time together with space variables) is called a **partial differential equation.**

---

e x a m p l e 10

The linear partial differential equation

$$\frac{\partial u}{\partial t} = \frac{\partial^2 u}{\partial x^2}$$

is a **heat equation** in which $u = u(x, t)$ represents the temperature at time t and position x in a 1-dimensional heat-conducting medium such as a thin, heat-insulated wire. Generally speaking, partial differential equations are harder than ordinary ones to interpret directly. In this example, however, there is a direct qualitative interpretation that predicts something interesting about the solutions $u(x, t)$. Note that if $u(x, t)$ is concave up as a function of x for some value of t and some interval of x values, then $\partial^2 u(x, t)/\partial^2 t > 0$ on the interval. But the differential equation then implies that $\partial u(x, t)/\partial t > 0$; in other words, $u(x, t)$ is an increasing function of t. A careful look at the solution graph in Figure IS.9 will bear this interpretation out. Similarly, you can see that where $u(x, t)$ is concave down as a function of x, it is also decreasing as a function of t.

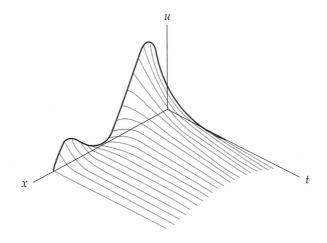

Figure IS.9 Temperature $u(x, t)$ as a function of discrete position x and continuously increasing time t.

EXERCISES

For each of the differential equations or systems of equations in Exercises 1 through 25, state (i) the order, (ii) the dimension, (iii) whether it is linear or nonlinear, and (iv) whether it is an ordinary differential equation (ODE) or a partial differential equation (PDE).

1. Equation for the height $h = h(t)$ of fluid in a vertical tank of uniform cross-sectional area A with fluid flowing from a hole of area a in the side of the tank:

$$\frac{dh}{dt} = -\frac{ka}{A}\sqrt{h}.$$

Here k is a constant depending on the properties of the fluid.

2. Equation for the position $s = s(t)$ of a falling object at time t, measured down from initial position s_0:

$$\frac{ds}{dt} = \sqrt{v_0^2 + 2g(s - s_0)},$$

where v_0 is the initial velocity and g is the acceleration of gravity.

3. Newton's law of cooling for the surface temperature $u = u(t)$ of an object in an environment maintained at temperature u_0:

$$\frac{du}{dt} = k(u - u_0), \qquad k = \text{const.}$$

4. The harmonic oscillator equation $m\, d^2y/dt^2 = -\pi r^2 \rho y$ for the vertical motion of a cylindrical buoy of mass m and radius r, floating in water of density ρ.

5. Van der Pol equation $d^2y/dt^2 - \epsilon(1 - y^2)\, dy/dt + \delta y = 0$ describing the behavior of periodic electrical discharges. (Note that setting $\epsilon = 0$ and $\delta = \pi r^2 \rho$ yields the previous equation.)

6. Damped harmonic oscillator equation with externally applied periodic force:

$$\frac{d^2y}{dt^2} + k\frac{dx}{dt} + hy = \cos t.$$

7. Catenary equation for the shape of a chain suspended between two supports:

$$y'' = \sqrt{1 + (y')^2}.$$

8. Equation for the velocity $v(t)$ of an object falling near the surface of the earth in a medium that offers high frictional resistance at high velocities: $dv/dt = -v - \frac{1}{2}v^2 + g$.

9. Damped oscillation of a pendulum: $d^2\theta/dt^2 = -0.2\, d\theta/dt - 5\sin\theta$, amplitude unrestricted.

10. Small, damped oscillation of a pendulum: $d^2\theta/dt^2 = -0.2\, d\theta/dt - 5\theta$. Here the term containing $\sin\theta$ in the preceding differential equation has been modified by the approximation $\sin\theta \approx \theta$, which may be good enough for some practical purposes if θ is small enough.

11. Relativistic differential equation for the speed v of a rocket in linear motion; v is regarded as a function of its decreasing mass m as fuel is burned:

$$m\frac{dv}{dm} + v_e\left(1 - \frac{v^2}{c^2}\right) = 0.$$

Here the constant v_e is the velocity of the expelled exhaust measured relative to the rocket, and c is the speed of light.

12. Setting $c = \infty$ in the previous (relativistic) problem yields the nonrelativistic differential equation appropriate for relating speed and mass of a rocket at lower speeds:

$$\frac{dv}{dm} = -\frac{v_e}{m}.$$

13. **Binet's equation** for the reciprocal $u = 1/r$ of the distance r from planet to star in polar coordinates: $d^2u/d\theta^2 + u = G/H^2$, G, H positive constants.

14. Relativistic equation for reciprocal $u = 1/r$ of the distance r from planet to star in polar coordinates: $d^2u/d\theta^2 + u = G/H^2 + (3G/c^2)u^2$, G, H constant. The constant c is the velocity of light. (Note that setting $c = \infty$ yields the previous equation.)

15. Equation for the shape of a wave moving with uniform speed in a narrow channel: $d^3y/dx^3 = \sigma y\, dy/dx$, σ constant.

16. Equation for the shape of a uniform beam subject to lengthwise compression:

$$EI\frac{d^4y}{dx^4} + F\frac{d^2y}{dx^2} = 0;$$

F is the compression force applied at one end, and both E and I are constant.

17. Bessel equation of order p:

$$x^2\frac{d^2y}{dx^2} + x\frac{dy}{dx} + (x^2 - p^2)y = 0.$$

18. System of equations satisfied by the displacements from equilibrium $x_1 = x_1(t)$ and $x_2 = x_2(t)$ of two equal masses m linked by springs of equal strength k to fixed supports:

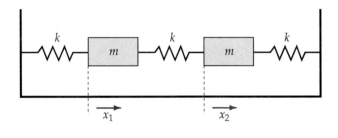

$$m\frac{d^2x_1}{dt^2} = -2kx_1 + kx_2, \qquad m\frac{d^2x_2}{dt^2} = kx_1 - 2kx_2.$$

19. Lotka-Volterra equations for the sizes of interacting parasite and host populations:

$$\frac{dP}{dt} = (a - bP)H, \qquad \frac{dH}{dt} = (cH - d)P, \qquad a, b, c, d \text{ const.}$$

20. Lotka-Volterra equations for the sizes H, K, P of three populations interacting over time:

$$\frac{dP}{dt} = (a - bP)H, \qquad \frac{dK}{dt} = K(P + H - e), \qquad \frac{dH}{dt} = (cH - d)P.$$

Here a, b, c, d, e are usually constant.

21. Lorenz equations derived by approximating partial differential equations for fluid flow:

$$\dot{x} = \sigma(y - x), \qquad \dot{y} = \rho x - y - xz, \qquad \dot{z} = -\beta z + xy.$$

Here σ, ρ, and β are constants. The specific choices $\sigma = 10$, $\rho = 28$, $\beta = \frac{8}{3}$ yield a system with solutions that appear to exhibit chaotic behavior.

22. One-dimensional equation for wave displacement $u(x, t)$:

$$\frac{\partial^2 u}{\partial t^2} = a^2 \frac{\partial^2 u}{\partial x^2},$$

where a determines the constant velocity of the wave motion.

23. Telegraph equation governing the long-distance flow of current in a wire:

$$A\frac{\partial^2 u}{\partial t^2} + B\frac{\partial u}{\partial t} + Cu = \frac{\partial^2 u}{\partial x^2},$$

where A, B, and C are constants. (Note that the choices $A = 1/a^2$ and $B = C = 0$ yield the previous equation.)

24. Korteweg-deVries equation governing a wave profile $u(x, t)$ in a narrow channel:

$$\frac{\partial u}{\partial t} + \frac{\partial u}{\partial x} + \beta u \frac{\partial u}{\partial x} + \frac{\partial^3 u}{\partial x^3} = 0, \qquad \beta \text{ const.}$$

25. Vector wave equation for small vibrations of a stretched string in three dimensions:

$$\rho(s)\frac{\partial^2 \mathbf{x}}{\partial t^2} = F\frac{\partial^2 \mathbf{x}}{\partial s^2},$$

where $\mathbf{x} = \mathbf{x}(s, t)$ is the position vector of a point on the string s units from one end at time t, $\rho(s)$ is the string density as a function of distance from the same end, and F is the constant tension force on the string.

1

First-Order Equations

1 SOLUTIONS AND INITIAL VALUES

It has been known for many years that formulating and interpreting differential equations often leads to profound scientific insights. In both pure and applied mathematics it is possible to establish equations relating an important function $y(x)$ to one or more of its derivatives, and then to use the equation to get information about $y(x)$. Such an equation relating a function and its derivatives is called a **differential equation.** Simple examples of differential equations are

$$y'(x) + y(x) = x \qquad \text{and} \qquad y'(x) = \sqrt{1 + y(x)^6}.$$

Using the alternative Leibnitz notation, and suppressing (x), the two equations look like

$$\frac{dy}{dx} + y = x \qquad \text{and} \qquad \frac{dy}{dx} = \sqrt{1 + y^6}.$$

Both these differential equations turn out to have infinitely many solutions, and since this is the typical state of affairs, it will be important to try to describe the multitude of solutions in a convenient and comprehensible way. This may not always be possible, but if we have some formula that contains as a special case all particular solutions $y(x)$, then it's customary to refer to that formula as the **general solution.** For example, we'll see that the first equation above has for its general solution on $-\infty < x < \infty$ the family of functions

$$y(x) = x - 1 + Ce^{-x}, \qquad c \text{ const.}$$

It turns out that by assigning an appropriate value to the constant C, we can pick a particular one of these solutions whose graph passes through any given

14

point (x_0, y_0) we choose. The second differential equation, $y' = \sqrt{1 + y^6}$, is more problematic in that a general solution can't be expressed in terms of the elementary functions of calculus or for that matter in terms of any other well-known class of functions. The standard response to that difficulty is to study the solutions by working directly with the differential equation; in particular we'll develop numerical methods in Chapter 2, Section 6 for displaying the properties of particular solutions. (If we're thorough enough in our investigations, we may find that in the process we've added to our supply of "well-known" functions.) The **order** of a differential equation is naturally defined to be the same as that of the highest-order derivative that appears effectively in the equation. Thus our two examples $y' + y = x$ and $y' = \sqrt{1 + y^6}$ are within the scope of the present chapter, namely, equations of order one. Equations such as $y'' + y = x$ and $y'' = \sqrt{1 + y^6}$ are of order two and will be treated in detail in Chapter 3.

1A   *Elementary Solution Formulas*

Equations are usually meant to be solved or interpreted in some way, and differential equations are no exception. A **solution** of a differential equation is a function $y = y(x)$, which, when substituted along with its derivatives into the differential equation, satisfies the equation for all x in some specified interval called the **domain** of the solution. In practice, there are sometimes elementary formulas for solutions, although by no means always. Sometimes solutions are defined in an **implicit** manner. For example, the equation $x^2 + y^2 = 1$ determines two different differentiable functions of x, $y = \sqrt{1 - x^2}$ and $y = -\sqrt{1 - x^2}$, both of which solve the differential equation $y' = -x/y$ for $-1 < x < 1$. In the following examples we will simply verify that certain formulas provide solutions; derivation of the solutions will be discussed later.

example 1

We'll verify that the differential equation
$$y' + y = x$$
has the solution $y(x) = x - 1$ for $-\infty < x < \infty$. We have $y'(x) = 1$, so
$$y' + y = 1 + (x - 1) = x, \qquad \text{for all real } x.$$
More generally, the same differential equation has the solution
$$y = x - 1 + ce^{-x}$$
for each choice of the constant c, for then $y' = 1 - ce^{-x}$, so
$$y' + y = (1 - ce^{-x}) + (x - 1 + ce - x) = x.$$
Thus there are actually infinitely many distinct solutions, and the solution $y(x) = x - 1$ can be obtained from the general formula above by setting $c = 0$. Figure 1.1a shows the graphs of the solutions corresponding to $c = 0$, $c = 0.15$, and $c = 0.25$. Notice that these graphs converge together as x increases, since the term ce^{-x} by which they differ tends to 0 as x tends to infinity.

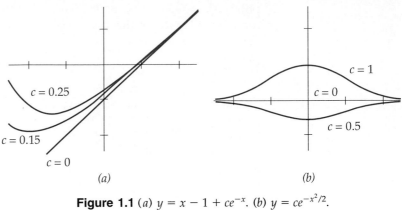

Figure 1.1 (a) $y = x - 1 + ce^{-x}$. (b) $y = ce^{-x^2/2}$.

e x a m p l e 2

The differential equation

$$y' + xy = 0$$

has the solutions $y = Ce^{-x^2/2}$, one for each value of C, since we can check that

$$y' + xy = (-Cxe^{-x^2/2}) + x(Ce^{-x^2/2}) = 0.$$

Figure 1.1b shows the solution corresponding to $C = -0.5$ and $C = 1$. For $C = 0$ we get $y = 0$ identically for all x, and the graph of this solution coincides with the x axis.

e x a m p l e 3

Suppose $f(x)$ defines a continuous function for $-\infty < x < \infty$. The differential equation $y' = f(x)$ has

$$y = \int_0^x f(t)\, dt, \qquad \text{or more generally} \qquad \int_0^x f(t)\, dt + C,$$

as solutions, one for each fixed C. In this example the number C distinguishes the various solutions from one another by making a uniform vertical shift in the graph. Figure 1.2a shows some examples for the case in which $f(x) =$

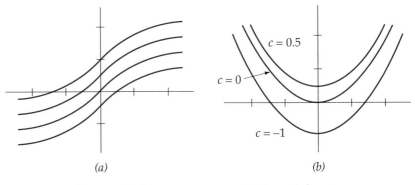

Figure 1.2 (a) $y = \arctan x + C$. (b) $y = \frac{1}{2}x^2 + C$.

$1/(1 + x^2)$. Choosing $C = 0$ gives $y = \int_0^x (1 + t^2)^{-1} \, dt = \arctan x$; if $C = -1$ then $y = \int_0^x (1 + t^2)^{-1} \, dt - 1 = \arctan x - 1$.

If $f(x) = (1 + x^2)^{-1}$ were replaced by some complicated function that did not have an indefinite integral in terms of an elementary formula, we could still find an approximation to the values of a solution by numerical approximation with Simpson's rule over successive intervals. (Recall that studying and computing a function like $\arctan x$ for the first time is also a lot of work.) On the other hand, arctan is already well known and has many nice properties, so a solution that can be expressed in terms of it is often to be preferred over a strictly numerical computation.

1B Initial Values and Solution Families

A typical first-order differential equation has infinitely many solutions on some interval, and each of these solutions has a graph passing through infinitely many points. To distinguish one solution from another in a convenient way, we can pick a point x_0 in the common domain of several solutions. Then let $y_0 = y(x_0)$, where $y(x)$ is some particular solution, and label that solution with the pair (x_0, y_0). Geometrically, what we just did was pick a point (x_0, y_0) on the given solution curve and then use that point to specify the solution. The choice of a point (x_0, y_0) is called an **initial condition** and is often expressed in the form $y(x_0) = y_0$. The question as to when such a condition specifies one and only one solution is taken up in Section 2C and in more detail in Chapter 10. In the meantime we'll examine each example with no prior assumptions about uniqueness.

example 4

The very simple differential equation

$$\frac{dy}{dx} = x$$

has solutions

$$y = \int_0^x t \, dt + C = \tfrac{1}{2}x^2 + C.$$

To pass through the point $(x_0, y_0) = (-1, 1)$, the solution graph must satisfy

$$1 = \tfrac{1}{2}(-1)^2 + C.$$

Thus $C = \tfrac{1}{2}$, and the desired solution is $y = (x^2 + 1)/2$; this and three other solutions are shown in Figure 1.2b.

example 5

We saw in Example 2 that the differential equation

$$\frac{dy}{dx} = -xy$$

has solutions

$$y = Ce^{-x^2/2}.$$

For its graph to pass through the point $(x_0, y_0) = (0, 2)$, a solution must satisfy
$$2 = Ce^0 = C.$$

Thus $C = 2$, and the particular solution selected by the condition $(x_0, y_0) = (0, 2)$ is $y = 2e^{-x^2/2}$.

---

e x a m p l e 6

The formula $y = mx$ describes a family of lines through the origin of the xy plane, each line with a different slope m. Differentiating the equation throughout with respect to x gives

$$\frac{dy}{dx} = m.$$

But $m = y/x$ if $x \neq 0$, so the differential equation satisfied by the family of lines is

$$x\frac{dy}{dx} = y.$$

It so happens that the constant C in Examples 4 and 5 is more than just a label; it represents the coordinate of the point where the corresponding solution crosses the y axis. In Example 6, m is a slope. Recognizing such interpretations for constants is not always possible, but it's worth doing when you can, because it adds some meaning that may turn out to be useful in thinking about a problem.

Trying to single out from all the solutions of a differential equation one that satisfies a given initial condition is called an **initial-value problem.** Thus an initial-value problem for what we'll call a **normal-form** first-order differential equation $y' = F(x, y)$ on an interval $a < x < b$ makes two requirements of a solution $y = y(x)$:

(i) $\qquad\qquad y'(x) = F(x, y(x)) \qquad$ for $a < x < b.$

(ii) $\qquad\qquad y(x_0) = y_0.$

For many of the most important differential equations the collection of all solutions is large enough that a quite arbitrary initial condition can be met; in practice this is often done by first identifying a set of solutions depending on one or more parameters, for example, a constant of integration C, and then choosing C so that the initial condition is satisfied. For a first-order equation a set of solutions will typically depend on just one parameter, and the term **1-parameter family** is used to describe such a set. In the previous two examples, the solution families $y = \frac{1}{2}x^2 + c$ and $y = ce^{-t^2/2}$ are 1-parameter families, with the parameters denoted by c. By reviewing how we arrived at them, it's easy to see that these two 1-parameter families describe *all* solutions of their respective differential equations. The next example is exceptional in that it has a nice 1-parameter family of solutions that fails to provide all solutions.

---

e x a m p l e 7

Suppose an object is dropped near the surface of the earth in such a way that you can ignore air resistance and slight variations in g, the acceleration of gravity. The object's displacement $s = s(t)$ from the dropping point where

$s(0) = 0$, measured as a function of time, satisfies a first-order differential equation with a simple 1-parameter family of solutions:

$$\frac{ds}{dt} = \sqrt{2gs}, \qquad s(t) = \left(\sqrt{\frac{g}{2}} \, t + c \right)^2.$$

(It's left as an exercise to show that the equation follows from the usual assumption that $d^2s/dt^2 = -g$.) Notice that both the differential equation and the solutions contain significant information about the object's motion. Since ds/dt is the object's velocity, the differential equation shows that velocity is completely determined by the distance s from the dropping point, a fact not immediately obvious from the solution formula. On the other hand, since $s(t)$ is the actual displacement as a function of time t, the parameter c has the significance that $0 = s(0) = c^2$; hence this initial condition requires that $c = 0$, which singles out the particular solution $s(t) = (\sqrt{g/2}t + 0)^2 = \frac{1}{2}gt^2$.

---

example 8

Example 7 is atypical in that the initial-value problem

$$\frac{ds}{dt} = \sqrt{2gs}, \qquad s(0) = 0$$

has an additional solution, namely, $s(t) = 0$ for all t, aside from the conventional one, $s(t) = \frac{1}{2}gt^2$. In cases like this we may need to resort to the physical interpretation that we place on the solution to help us decide which one to accept. Since we're talking about a falling object, the constant solution $s(t) = 0$ would seem to be ruled out, so we accept the other one. If you insist on having an interpretation for the constant solution, you could imagine it as describing what happens if you don't release the object but instead continue to hold on to it. The matter of nonuniqueness for the solution of an initial-value problem is taken up in more detail in Section 2C.

Examples 6 and 7 show that even if we have a solution formula for a physical problem, a more complete story may still be had by concentrating directly on the associated differential equation itself. But what if we start with a family of functions; can we find a differential equation satisfied by all members of the family? In practice we can best proceed as follows:

1. Differentiate the equations that define the members of the family with respect to a chosen independent variable to produce a second 1-parameter family of equations containing derivatives of the dependent variable.
2. Eliminate the parameter using, if necessary, both families of equations to get a differential equation independent of the parameter.

---

example 9

The equation $x + cy = 0$, with parameter c, describes a family of lines through the xy plane, each line with a different slope $-1/c$. Think of y as a function of x, and differentiate with respect to x to get

$$1 + c \frac{dy}{dx} = 0.$$

But $c = -x/y$ from the original equation, so the desired differential equation is

$$1 + \left(\frac{-x}{y}\right)\frac{dy}{dx} = 0, \quad \text{or} \quad x\frac{dy}{dx} = y, \quad \text{or} \quad \frac{dy}{dx} = \frac{y}{x}.$$

Satisfying an initial condition $y(x_0) = y_0$ can be done by solving $x_0 + cy_0 = 0$ for c, and this works just fine unless $x_0 = 0$; in that case the only initial condition we can satisfy is $y(0) = 0$.

---

e x a m p l e 10

The 1-parameter family of curves determined by the implicit relation

$$x^2 + c^2y^2 = 1, \quad c \neq 0,$$

consists of ellipses, all passing through the four points $(x, y) = (\pm 1, 0)$, $(0, \pm 1/c)$. If we want to regard x as an independent variable, the implicitly defined functions that could serve as solutions for a differential equation in y' are

$$y(x) = \frac{1}{c}\sqrt{1 - x^2}, \quad c \neq 0.$$

However, it's easier to work with the implicit relation directly. Using implicit differentiation with respect to x, we get $2x + 2c^2yy' = 0$, or $y' = -x/(c^2y)$. From the original quadratic equation we find $1/c^2 = y^2/(1 - x^2)$, so substitution into $y' = -x/(c^2y)$ gives

$$y' = -\frac{x}{y}\frac{y^2}{1 - x^2}, \quad \text{or} \quad y' = \frac{xy}{x^2 - 1}.$$

Note that the family of ellipses is defined only for $-1 \leq x \leq 1$ and that the vertical tangents at $(0, \pm 1)$ correspond to points at which the right side of the differential equation is undefined. (This differential equation also has solutions valid for $|x| > 1$; methods for finding them are developed in Sections 3 and 5.)

EXERCISES

1. For each of the following differential equations, carry out the details of checking to see whether the given function is a solution on the specified interval.

   (a) $y' + y = 0$; $y = ce^{-x}$, $-\infty < x < \infty$, c constant.

   (b) $y' = 2y$; $y = ce^{2x}$, $-\infty < x < \infty$, c constant.

   (c) $dy/dt + 2ty = 0$, $y = be^{-t^2}$, $-\infty < t < \infty$, b constant.

   (d) $y' = y + x$; $y = ce^{-x} - x - 1$, $-\infty < x < \infty$, c constant.

   (e) $x\, du/dx = 1$; $u = \ln(cx)$, $0 < x$, c positive, constant.

   (f) $dy/dx = 1 + y^2$; $y = \tan(x - c)$, $-\pi/2 + c < x < \pi/2 + c$, c constant.

   (g) $dz/dx = 2\sqrt{|z|}$, $z = x|x|$, $-\infty < x < \infty$.

   (h) $\dfrac{dy}{dx} = 3x^2y + 1$; $y = e^{x^3}\displaystyle\int_0^x e^{-t^3}\, dt$, $-\infty < x < \infty$.

(i) $\dfrac{dy}{dx} = \sqrt{1 + x^3};\ y = \displaystyle\int_0^x \sqrt{1 + t^3}\, dt + C,\ -1 < x < \infty,\ C$ constant.

(j) $\dfrac{dy}{dx} = 2x\sqrt{1 + x^6};\ y = \displaystyle\int_0^{x^2} \sqrt{1 + t^3}\, dt,\ -\infty < x < \infty.$

(k) $d^2y/dt^2 + 4y = 0,\ y = c_1 \cos 2t + c_2 \sin 2t,\ -\infty < t < \infty,\ c_1, c_2$ constant.

(l) $d^2y/dt^2 - 4y = 0;\ y = c_1 e^{2t} + c_2 e^{-2t},\ -\infty < t < \infty,\ c_1, c_2$ constant.

(m) $dy/dt = \sqrt{y};\ y = 0,\ -\infty < t < \infty.$

(n) $\dfrac{dy}{dt} = \sqrt{y};\ y = \begin{cases} 0, & t < 0, \\ t^2/4, & 0 \le t, \end{cases} \quad -\infty < t < \infty.$

(o) $y' = (y \ln y)/x;\ y = e^{cx},\ x > 0,\ c$ constant.

2. (a) The differential equation $(dx/dt)^2 + t^2 = 0$ has no real-valued solution $x = x(t)$ on a given interval $a < t < b$; explain why.

(b) The differential equation $(dx/dt)^2 + x^2 = 0$ has only one real-valued solution $x = x(t)$ on a given interval $a < t < b$; explain why. What is the unique solution?

3. For each of the following 1-parameter families of functions, find the particular solution that satisfies the associated initial condition; then sketch the graph of that solution. Note that some solutions are implicitly defined.

(a) $y = ce^{3x};\ y(0) = 7.$

(b) $y = ke^{-7x};\ y(0) = 7.$

(c) $x = \ln(ct);\ x(1) = -2.$

(d) $x = c \ln t;\ x(2) = -1.$

(e) $y = k \cos x;\ y(\pi) = 2.$

(f) $y = \tan(cx);\ y(\pi) - 1.$

(g) $x^2 + y^2 = c^2 > 0;\ y(1) = -1.$

(h) $x^2 - c^2 y^2 = 1,\ c \neq 0;\ y(2) = 1.$

(i) $y^2 = cx;\ y(3) = -3.$

(j) $x \ln y = k;\ y(1) = e.$

4. For each of the 1-parameter solution formulas in the previous exercise, find a first-order differential equation satisfied by all members of the family.

5. It doesn't necessarily follow that because some members of a 1-parameter family satisfy a differential equation then all members do. By substitution into the differential equation, find all values of the parameter r such that the corresponding function is a solution.

(a) $x^2 y'' + 5xy' + 4y = 0;$
       $y = x^r.$

(b) $y'' + 5y' + 4y = 0;\ y = e^{rx}.$

(c) $y' = (y \ln y)/x;\ y = re^x.$

(d) $y' = 3y;\ y = re^{rx}.$

6. The definition of the exponential function is sometimes introduced along with the **organic growth law.** According to this law, the size $P(t)$ of a population at time t satisfies

$$\frac{dP}{dt} = kP, \qquad 0 < k = \text{const.}$$

This equation is satisfied by $P(t) = Ce^{kt}$, where C is a constant that would normally be positive given the interpretation of P as a population size.

(a) Show that $C = P(0).$

(b) If $P(0) = 2$ and $P(2) = 6$, find k.

(c) Show that a population satisfying the organic growth law increases by the factor $R > 1$ in any given time interval of length $t_R = (\ln R)/k$.

7. The **radioactive decay equation** is

$$\frac{dQ}{dt} = -kQ, \qquad 0 < k = \text{const.}$$

This equation is satisfied by the quantity $Q(t)$ of a radioactive element remaining after time t, starting with some initial amount $Q(0)$. The differential equation has solutions of the form $Q(t) = Ce^{-kt}$, where C is constant.

(a) Show that $C = Q(0)$.

(b) If $Q(0) = 4$ and $Q(1) = 1$, find k.

(c) Show that a sample of radioactive element is decreased by multiplying by a factor r, with $0 < r < 1$, in any given time interval of length $t_r = -\ln r/k$.

8. Find a first-order differential equation satisfied by $y = y(x)$ and such that:

(a) The slope of the graph of $y(x)$ at (x, y) equals the distance from (x, y) to $(0, 0)$.

(b) The tangent line to the graph of $y(x)$ at (x, y) contains the point $(x + y, x + y)$.

(c) The rate of change of $y(x)$ with respect to x is proportional to x and inversely proportional to y.

9. Let g be the acceleration of gravity near the surface of the earth. ($g \approx 32.2$ feet per second per second.) By Newton's law an object falling with negligible air resistance has acceleration

$$\frac{dv}{dt} = g,$$

where $v = v(t)$ is the velocity of the object at time t. Use integration to derive the following relations.

(a) $v(t) = gt + v_0$, where v_0 is the velocity at time 0.

(b) $s(t) = \frac{1}{2}gt^2 + v_0 t + s_0$, where $s = s(t)$ is the distance at time t of the object from the reference point $s = 0$.

10. An object dropped near the earth's surface falls a distance $s(t) = \frac{1}{2}gt^2$ in time t. In particular, $s_0 = s(0) = 0$ and $v_0 = s'(0) = 0$.

(a) Show that $s = s(t)$ satisfies the first-order differential equation

$$\frac{ds}{dt} = \sqrt{2gs}.$$

(b) Show that the differential equation in part (a) has as solutions each member of the 1-parameter family

$$s = \left(\sqrt{\frac{g}{2}} t + c \right)^2 = \frac{1}{2}gt^2 + \sqrt{2g}ct + c^2.$$

(c) Show that the motions described in part (b) are quite special in that they all satisfy $v(0)^2 = 2gs(0)$.

11. The general solution to the falling-object problem treated in the two previous problems is

$$s = \tfrac{1}{2}gt^2 + v_0 t + s_0,$$

where v_0 is initial velocity and s_0 is initial displacement.

(a) Show that $s = s(t)$ satisfies the first-order differential equation

$$\frac{ds}{dt} = \sqrt{v_0^2 + 2g(s - s_0)}.$$

[*Hint:* Solve for t in terms of both s and ds/dt.]

(b) Show that the expression $v_0^2 + 2g(s - s_0)$ under the radical is always nonnegative, given our assumptions on s. [*Hint:* When does that expression reach its minimum as a function of t?]

12. **Flow of liquid from a tank.** A cylindrical tank with cross-sectional area A has an outlet hole in its side near the bottom. If $h = h(t)$ is the height of an ideal fluid above the outlet at time t and a is the area of the outlet hole, then $V(t)$, the remaining fluid volume at time t, satisfies **Torricelli's equation**

$$\frac{dV}{dt} = -a\sqrt{2gh}.$$

An intuitive justification for the equation is to note that it depends on having the outlet velocity equal to the free-fall velocity of a drop of fluid from height h, as derived in Exercise 11; thus $-dV/dt$ equals area a times outlet velocity $\sqrt{2gh}$. (A thoroughly scientific justification depends on principles of fluid mechanics.) Thus for an ideal fluid, the equation takes the form

$$\frac{dh}{dt} = -\frac{a}{A}\sqrt{2gh}.$$

(a) By analogy with the results of Example 7, show that the Torricelli equation has a solution of the form $h(t) = (bt + c)^2$. [*Hint:* Determine what the constants b and c must be.]

(b) Use your answer to part (a) to find out how long it would take for the fluid height above the outlet to drop from h_0 to 0. In particular, estimate how long it would take to empty a full cylindrical tank with diameter 10 feet and height 20 feet and a circular outlet at the bottom with diameter 6 inches.

2   DIRECTION FIELDS

Just as the graph of a function can be said to be a complete geometric representation of the function, so the direction field of a first-order differential equation, described below represents the differential equation. This fundamental idea can be used to get information about solutions without going to the trouble of finding formulas for solutions.

24   **CHAPTER 1**   First-Order Equations

2A   *Definition and Examples*

Since y' can be interpreted as a slope, a differential equation written in the **normal form**

$$y' = F(x, y)$$

assigns slope $F(x, y)$ to the point (x, y) in the xy plane. Such an assignment is called a **direction field** or **slope field.** Note that if $y(x)$ is a solution to the differential equation, then

$$y'(x) = F(x, y(x)).$$

But since $y'(x)$ is the slope of the tangent to the graph of the solution at the point with coordinates $(x, y(x))$, the number $F(x, y(x))$ is necessarily also equal to that slope. The slope then specifies a direction along the solution curve $y = y(x)$. From one point of view, we can think of starting with the solution $y = y(x)$ and locating a short segment of tangent line at each point of the graph of $y = y(x)$; some of these are shown in Figure 1.3a. In practice we draw a segment of slope $F(x, y)$ at selected points (x, y). In this way, we get a sketch of the direction field associated with the differential equation $y' = F(x, y)$. If the sketch is skillfully made, it is often possible to get a good idea of what the graphs of solutions $y = y(x)$ should look like; the principle is that a solution graph should be tangent to the segment located at each point that the graph passes through. Figure 1.3b shows such a sketch of a direction field, along with the graphs of three curves that are tangent to the segments in the direction field. The direction field is associated with the differential equation

$$y' = x - xy.$$

Thus the function F in the general discussion is $F(x, y) = x - xy$. In particular, $F(1, \frac{1}{2}) = \frac{1}{2}$ so that the segment through $(x, y) = (1, \frac{1}{2})$ has slope $\frac{1}{2}$. Notice particularly that there are apparently infinitely many possible curves that could be drawn tangent to the direction field. Of these infinitely many curves, Figure 1.3b shows just four; others could easily be drawn in.

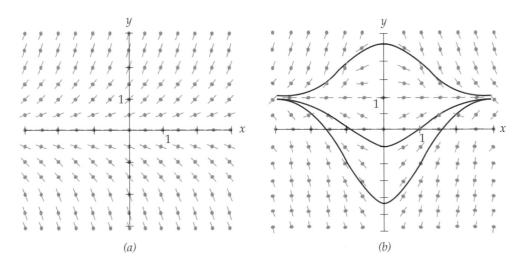

(a)   (b)

Figure 1.3 (a) Slope $F(x, y)$ at (x, y). (b) Slope $x - xy$ at (x, y).

e x a m p l e 1   The differential equation

$$\frac{dy}{dx} = \frac{y^2}{x^2 + y^2}$$

defines a direction field for all (x, y) except $(0, 0)$. To sketch the field, we can first make a table of slopes. In this example, we will restrict ourselves to the first quadrant, because replacing x by $-x$ or y by $-y$ leaves the slope unchanged. Table 1 lists the array of slopes.

Looking at this table doesn't by itself convey much understanding. It's only when the information is displayed graphically that we begin to understand something about the differential equation. These points (x, y), plus a few more, are plotted in Figure 1.4 along with segments having the assigned slopes. The pattern that emerges is a good guide to the general shape of the solution graphs. Though it can't be shown in the picture, it is worth noticing also what

Table 1: Direction Field Slopes for $dy/dx = y^2/(x^2 + y^2)$

(x, y)	dy/dx	(x, y)	dy/dx
$(0, 1)$	1	$(2, 3)$	$\frac{9}{13}$
$(0, 2)$	1	$(3, 0)$	0
$(0, 3)$	1	$(3, 1)$	$\frac{1}{10}$
$(1, 1)$	$\frac{1}{2}$	$(3, 3)$	$\frac{1}{2}$
$(1, 2)$	$\frac{4}{5}$	$(4, 0)$	0
$(1, 3)$	$\frac{9}{10}$	$(4, 1)$	$\frac{1}{17}$
$(2, 0)$	0	$(4, 2)$	$\frac{1}{5}$
$(2, 1)$	$\frac{1}{5}$	$(4, 3)$	$\frac{9}{25}$
$(2, 2)$	$\frac{1}{2}$	$(4, 4)$	$\frac{1}{2}$

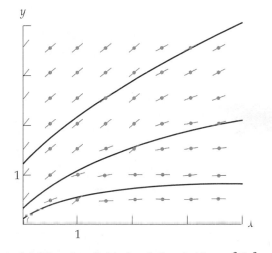

Figure 1.4 Direction field sketch for $dy/dx = y^2/(x^2 + y^2)$.

happens as $x \to \infty$ and $y \to \infty$. If $x \to \infty$ and y is held fixed, then the slopes tend to zero. If $y \to \infty$ with x held fixed, then the slopes tend to 1. Three solution curves are drawn in, showing their inclination to be consistent with the direction field. There is in fact an "elementary" solution formula for this differential equation, computable by one of the methods of Section 3C. But if what we want is simply a general picture of the behavior of solution graphs, we may be satisfied with a sketch of the direction field.

2B   Isoclines

Sketching a direction field requires a certain amount of judgment about how much information to include in the sketch. As with a caricature of a person's face, the trick is to include just enough information to allow the viewer to recognize what is being represented. Too much detail may obscure the essential features, and too little may fail to show a recognizable pattern. One way to organize a sketch is to begin by sketching some **isoclines,** that is, curves along which the slopes of the field are equal; this allows us to draw in parallel segments of the field along each isocline curve, making sure to include enough isoclines to ensure a good sketch of the field as a whole. Stated in terms of the function F in the differential equation $y' = F(x, y)$, an isocline is just an implicitly defined level curve $F(x, y) = m$ at level m. An isocline is *not* in general the graph of a solution; hence, the isocline curves themselves should not figure prominently, if at all, in the final sketch of a direction field.

example 2

The differential equation $y' = x$ has a very simple direction field, since the slopes are equal on vertical lines, that is, lines determined by equations $x = m$, where m is constant. These lines are the isoclines of the direction field and are indicated by the vertical dotted lines in Figure 1.5a. When $m = -1$, the slopes of the field are all -1; when $m = 0$, the slopes are all zero; when $m = 1$, the

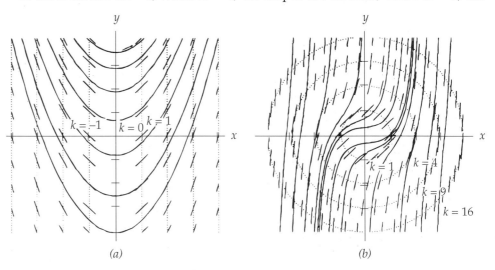

(a)   (b)

Figure 1.5 (a) $y' = x$. (b) $y' = x^2 + y^2$.

slopes are all 1; and so on. These are easily sketched in by hand, as in Figure 1.5a, once a single slope segment has been put in place, because the other segments on the same isocline are all parallel to the first one. Given a completed sketch of the direction field, it's an easy matter to sketch a few solution curves, which happen to be parabolas with equations of the form $y = \frac{1}{2}x^2 + c$, with c constant.

In Example 2 the ultimate aim was to sketch the graphs of the solutions to $y' = x$. Since these graphs were simply parabolas, one might argue that they could have been sketched without the aid of the direction field. However, making even moderately accurate sketches of parabolas is not as easy as you might think, and the direction field is definitely a help in doing the job by hand. In the next example, there are no available formulas for the solution graphs, so we rely more heavily on the direction field.

example 3

The differential equation $y' = x^2 + y^2$ has no elementary formula that describes its solutions. Furthermore, listing the slopes associated with a rectangular array of points requires computing many different numbers, even though noticing that the slopes are the same at (x, y) and $(-x, -y)$ is of some help. More effective is first to sketch the isocline curves $x^2 + y^2 = m$. Since these curves are circles of radius $\sqrt{m}$, it's natural to pick values such as $m = 1, 4, 9, 16$ as in Figure 1.5b. Then sketch in segments of slope m around the corresponding circle. Finally, sketch in the solution graphs, giving them at each point a tangent direction consistent with the nearby slope segments.

If a direction field and its isoclines are both hard to picture, there may be no alternative to the brute force method of plotting many segments. This is best done using computer graphics, and here is a description of how the underlying program could be constructed. Let's assume that we want to plot equal segments of slope $F(x, y)$ centered on points (x, y) at the corners of a square grid on a rectangular region R. Let s be the horizontal and vertical distances between points. A vector of length 1 parallel to a line with slope $F(x, y)$ is

$$\left(\frac{1}{\sqrt{1 + F(x, y)^2}}, \frac{F(x, y)}{\sqrt{1 + F(x, y)^2}} \right).$$

What we do is both add to and subtract from (x, y) a small multiple of this vector to find the endpoints of our segment centered at (x, y). Taking the small multiplier to be $s/3$, the process could be formalized as follows:

```
FOR x = x₁ TO x₂ BY STEP s
FOR y = y₁ TO y₂ BY STEP s
  LET d = (1 + F(x, y)²)^(1/2)
  PLOT LINE: (x − s/(3d), y − sF(x, y)/(3d)) TO (x + s/(3d), y + sF(x, y)/(3d))
NEXT x
NEXT y
```

Routines like this were used to plot some of the pictures in this book, and others like it have been incorporated into a number of widely available plotting routines. From the Web site http://math.dartmouth.edu/~rewn/ you can download a small application, called an Applet, named DIRFLD that will plot direction fields for differential equations of the form $y' = f(x, y)$ and then plot solution graphs through designated points. It is first necessary to specify the function $f(x, y)$ along with parameters determining the extent of the field as well as the spacing between points where the slopes are to be evaluated.

EXERCISES

1. For each of the following differential equations, sketch the associated direction field at points (x, y) with $-2 \le x \le 2$ and $-2 \le y \le 2$. Include enough points in your sketch so that a pattern becomes apparent. Then sketch in a few solution curves.

   (a) $y' = x$.

   (b) $y' = y$.

   (c) $y' = -1$.

   (d) $y' = 0$.

   (e) $y' = -y$.

   (f) $y' = -2x$.

2. For each of the following differential equations, sketch the associated direction field at points (x, y) for which the differential equation is defined and for which $-2 \le x \le 2$ and $-2 \le y \le 2$. Include enough points in your sketch so that a pattern becomes apparent. You may want to use isoclines to help in sketching the field. Then sketch in a few solution curves.

   (a) $y' = 2x^2 + y^2$.

   (b) $y' = x^2 + y$.

   (c) $y' = x - y$.

   (d) $y' = \sqrt{x^2 + y^2}$.

   (e) $y' = x/(1 + y^2)$.

   (f) $y' = \sqrt{1 - x^2 - y^2}$.

3. For each of the following initial-value problems, sketch enough of the associated direction field to allow you to make a sketch of the solution for increasing x. Then sketch the solution.

   (a) $y' = y + 2x$, $y(0) = 1$.

   (b) $y' = (x^2 + y^2)^{-1}$, $y(1) = 0$.

   (c) $y' = xy$, $y(0) = 1$.

   (d) $y' = xy$, $y(0) = 0$.

   (e) $y' = \cos y$, $y(0) = 1$.

   (f) $y' = xy + y^2$, $y(-1) = -1$.

4. Consider the differential equation

$$y' = \begin{cases} 0, & x < 0, \\ x, & x \ge 0, \end{cases}$$

   (a) Sketch the direction field for $-2 \le x \le 2$.

   (b) Find a two-piece formula for the solution to the differential equation satisfying the initial condition $y(0) = 1$.

5. There is no elementary formula for solutions of the differential equation $y' = \sqrt{1 + y^4}$; use its direction field to sketch solutions passing through the points $(-1, 0)$, $(0, 1)$, and $(1, 1)$.

6. There is no elementary formula for solutions of the differential equation $y' = \sqrt{1 + y^3}$; use its direction field to sketch solutions passing through the points $(-1, 0)$, $(0, 1)$, and $(1, 1)$. What about $(0, -2)$?

7. There is no elementary formula for solutions of $y' = 1/\sqrt{1 + y^4}$; use its direction field to sketch solutions passing through the points $(-1, 0)$, $(0, 1)$, and $(1, 1)$.

8. There is an elementary formula, namely, $y + \sqrt{1 + y^2} = Ce^x$, $C =$ const, satisfied by solutions of the differential equation $y' = \sqrt{1 + y^2}$, but it isn't very helpful for making a rough sketch of the solution satisfying $y(0) = 1$.

   (a) Use the direction field to sketch this solution.

   (b) Use implicit differentiation to verify that functions $y = y(x)$ that satisfy the elementary solution formula given above also satisfy the differential equation $y' = \sqrt{1 + y^2}$.

   (c) Show that solutions of the differential equation $y' = \sqrt{1 + y^2}$ also satisfy the *second-order* differential equation $y'' = y$. What can you conclude about concavity of the solution graphs of the first-order differential equation?

9. Define

$$F(y) = \begin{cases} 0, & y \le 0, \\ \sqrt{y}, & y > 0. \end{cases}$$

   (a) Sketch the direction field of the differential equation $y' = F(y)$ for $-2 \le x \le 2$, $-2 \le y \le 2$.

   (b) Verify that there are two distinct solutions of $y' = F(y)$ passing through each point of the form $(x_0, 0)$. [*Hint:* Consider $y = (x - x_0)^2/4$ for $x > x_0$.]

10. Differentiate each of the following differential equations throughout with respect to x to determine those regions where their solution graphs are concave up (where $y'' > 0$) or concave down (where $y'' < 0$). Use the information about concavity as an aid in sketching some typical solution graphs.

    (a) $y' = y^2$.

    (b) $y' = \sqrt{1 + y^4}$.

    (c) $y' = x + y$.

    (d) $y' = xy$.

    (e) $y' = \sin y$.

    (f) $y' = xy^2$.

    (g) $y' = \sin x$.

    (h) $y' = y/x$.

    (i) $y' = x/y$.

11. Show that if $F(x, y)$ has continuous partial derivatives $F_x(x, y)$ and $F_y(x, y)$ with respect to x and y in some region R of the xy plane, then solutions of the first-order equation $y' = F(x, y)$ also satisfy the second-order equation $y'' = F_x(x, y) + F_y(x, y)F(x, y)$.

12. Show that if a, b, c, d are constants with $ad - bc \ne 0$, the differential equation

$$y' = \frac{ax + by}{cx + dy}$$

    has lines through the origin for its isoclines. Show by example what can go wrong if $ad - bc = 0$.

13. (a) Show that every isocline of the differential equation $y' = y/x$ is itself the graph of a solution of the differential equation.

    (b) Show that no isocline of the differential equation $y' = x$ is a solution graph of the differential equation.

    (c) Find every isocline of the differential equation $y' = x + y$, and determine which, if any, is a solution graph of the differential equation.

(d) Find every isocline of the differential equation $y' = x/y$, and determine which, if any, is a solution graph of the differential equation.

14. Note that if $y = y(x)$ is a solution of $y' = f(x, y)$, then the graph of $y(x)$ is increasing at points (x, y) where $f(x, y) > 0$ and decreasing where $f(x, y) < 0$. Indicate by a sketch, or in some other way, the regions of the xy plane where solutions of the following differential equations are increasing and the regions where solutions are decreasing.

(a) $y' = x - y$.

(b) $y' = e^{\sin xy}$.

(c) $y' = x^3 - y$.

(d) $y' = x^2 - y^2$.

(e) $y' = 1 - x^2 - y^2$.

(f) $y' = x(x^2 - y)$.

15. Bugs in pursuit. Four identical bugs are on a flat table, each moving at the same constant speed. Use xy coordinates on the table, and locate bugs 1 through 4 initially in respective quadrants 1 through 4, each at one of the points $(\pm 1, \pm 1)$. Bug 1 always heads directly toward bug 2, bug 2 toward bug 3, bug 3 toward bug 4, and bug 4 toward bug 1, so their paths are mutually congruent.

(a) Use the symmetry of the paths to show that at any moment the bugs are at the corners of a square, and in particular, if bug 1 is at (x, y), then 2, 3, and 4 are, respectively, at $(-y, x)$, $(-x, -y)$, and $(y, -x)$.

(b) Using the notation of part (a), show that the paths of bugs 1 and 3 satisfy the differential equation

$$\frac{dy}{dx} = \frac{y - x}{x + y},$$

unless $x + y = 0$. What is the analogous differential equation for bugs 2 and 4?

(c) Sketch the direction field for the differential equation of bug 1, and use it to sketch his path starting at $(1, 1)$. *Note:* This differential equation has an elementary solution formula derivable by the methods of Section 3C.

16. Use a computer graphics routine to display the direction field for each of the following differential equations. Then sketch in a few solution curves.

(a) $y' = x^3 + y^3$, $-2 \le x \le 2$, $-2 \le y \le 2$.

(b) $y' = x^2 - y^2$, $-2 \le x \le 2$, $-2 \le y \le 2$.

(c) $y' = e^{-x^2 - y^2}$, $-2 \le x \le 2$, $-2 \le y \le 2$.

(d) $y' = \sin(x + y)$, $-4 \le x \le 4$, $-4 \le y \le 4$.

(e) $y' = \ln(1 + x^4 + y^4)$, $-2 \le x \le 2$, $-2 \le y \le 2$.

(f) $y' = \sqrt{1 + x^2 + y^4}$, $-4 \le x \le 4$, $-4 \le y \le 4$.

Orthogonal Trajectories. A curve that is perpendicular to each member of a given family of curves is called an **orthogonal trajectory** to the family. If the given family of curves satisfies a first-order differential equation $y' = f(x, y)$, then an orthogonal trajectory will quite generally be the graph of a function $y(x)$ that satisfies $y' = -1/f(x, y)$; since the product of the slopes of the two fields is -1 at each point, this just says that the direction field of the orthogonal trajectories is everywhere perpendicular to that of the given family.

17. For each of the following 1-parameter families of functions, find a first-order differential equation $y' = f(x, y)$ satisfied by the members of the family. Sketch in the most convenient order (i) the direction field of the orthogonal trajectory equation $y' = -1/f(x, y)$, (ii) a few graphs of orthogonal trajectories, and (iii) some graphs of members of the given family.

   (a) $y = cx$.

   (b) $y = x + c$.

   (c) $xy = c$.

   (d) $x^2 + y^2 = c^2$.

   (e) $y = cx^3$.

   (f) $y = (x + c)^2$.

   (g) $y = c \sin x$.

   (h) $y = \cos x + c$.

   (i) $y = c$.

   (j) $y = x^2 + c$.

   (k) $y = c/x$.

   (l) $y = c/x^2$.

18. A point P in the xy plane starts at the origin and moves up along the positive y axis, dragging behind it (as if attached by a string of length $a > 0$) another point Q that starts at $(a, 0)$ on the positive x axis. Thus the distance between the two points remains constantly equal to a, and the tangent line to the path of the trailing point Q always passes through the leading point P on the y axis. The path $y = y(x)$ of the following point Q in the first quadrant is called a **tractrix.**

   (a) Show that the tractrix satisfies the differential equation $y' = -\sqrt{a^2 - x^2}/x$. [*Hint:* Draw a picture.]

   (b) Show that the tractrix is concave up by showing that $y'' = a^2/(x^2\sqrt{a^2 - x^2}) > 0$.

   (c) Sketch the direction field of the differential equation for $0 < x < 1$, assuming $a = 1$.

   (d) Use the results of parts (b) and (c) to sketch the tractrix that starts at $(0, 1)$.

   (e) Integrating both sides of the differential equation with respect to x and assuming $y(a) = 0$ shows that a tractrix is the graph of

   $$y(x) = a \ln \left(\frac{a + \sqrt{a^2 - x^2}}{x} \right) + \sqrt{a^2 - x^2}.$$

   Verify this by substitution into the differential equation.

   (f) Show that the length of the tractrix over $x_0 \le x \le a$ is $a \ln (a/x_0)$. [*Hint:* Use the differential equation.]

2C Continuity, Existence, and Uniqueness

Even though the sketches that we make of direction fields are necessarily incomplete for practical reasons, the associated direction field is in principle a complete geometric representation of a differential equation $y' = F(x, y)$; for this reason the underlying concept is fundamental to understanding first-order differential equations. In relying on a sketch of a direction field for drawing solution graphs for $y' = F(x, y)$, we assume that the slope segments don't vary discontinuously from point to point. In particular we assume that $F(x, y)$ is a continuous function of the point (x, y). (Imagine how much trouble you'd have sketching solution curves if the field of slopes varied *discontinuously*, say in some random fashion.) Indeed it seems reasonable to expect that if the way $F(x, y)$ varies were sufficiently irregular, the corresponding differential equation might not have any solutions at all. In fact it can be proved, though we will not do so, that just continuity of $F(x, y)$ in a region R is enough to guarantee the existence of at least one solution whose graph contains a given point of R. Theorem 2.1 below guarantees that if $F(x, y)$ is not only continuous in a region R of the xy plane but that if also $\partial F / \partial y$ is continuous there, then $y' = F(x, y)$ has a solution graph passing through every point of R. Furthermore, there will be only one solution passing through each point; put slightly differently, the assertion that there is *at most* one solution through a point says that two different solution graphs will never cross each other. Referring back to Figure 1.5, we can conclude that the xy regions shown there are completely filled by non-overlapping solution graphs. From what we know about the parabolas $y = \frac{1}{2}x^2 + c$, this result seems fairly clear in the first picture but is not so clear just from looking at the second one.

■ 2.1 THEOREM

Assume that both $F(x, y)$ and its partial derivative $\partial F(x, y) / \partial y$ are continuous in the rectangle $R = \{(x, y): a < x < b, c < y < d\}$ with edges parallel to the axes. Then the differential equation $y' = F(x, y)$ has a unique solution passing through each point of R.

Note. The rectangle R referred to in the theorem might be infinite either horizontally or vertically; in particular R might be the entire xy plane.

For the proof of Theorem 2.1 refer to one of the many theoretical treatments of the subject. As we proceed, our solution techniques will provide proofs for important special cases. And Appendix C on existence of solutions shows how the general problem can be approached.

---

e x a m p l e **1**

Applying the theorem to the differential equation $y' = x^2 + y^2$ in Example 3 of Section 2B, we first note that $F(x, y) = x^2 + y^2$ and its partial derivative $\partial F(x, y) / \partial y = 2y$ are both continuous for all (x, y). The conclusion of Theorem 2.1 asserts that there is a unique solution graph through each point (x_0, y_0) and that no two of these graphs ever cross each other. (If they were to cross, then the uniqueness requirement would be violated upon following the graphs from the crossing point to some point at which the graphs became distinct.) Another way to look at the theorem's implication is that there is a unique solution $y = y(x)$ satisfying an arbitrary initial condition $y(x_0) = y_0$. Some of the solution graphs are shown in Figure 1.5*b*.

---

e x a m p l e **2**

Some condition stronger than just continuity of $F(x, y)$ is needed to guarantee the uniqueness of solutions described in Theorem 2.1. As to how uniqueness can fail, consider the following direction field. (This one is investigated in more detail in the exercises.) Let

$$F(x, y) = \begin{cases} \sqrt{y}, & \text{if } y > 0, \\ 0, & \text{if } y \leq 0. \end{cases}$$

Clearly $F(x, y)$ is continuous at all points (x, y) for which $y > 0$ and also for which $y < 0$. At points $(x, 0)$ we also have continuity, because $\lim_{y \to 0} \sqrt{y} = 0$. However, $\partial F(x, y)/\partial y = \frac{1}{2}y^{-1/2}$ when $y > 0$ and fails to exist when $y = 0$. Hence Theorem 2.1 fails to apply to the differential equation $y' = F(x, y)$. You can verify directly that the differential equation has two distinct solutions satisfying the initial condition $y(0) = 0$; one of them is $y_1(x) = \frac{1}{4}x^2$ for $x \geq 0$, and the other one is $y_2(x) = 0$ for all x.

e x a m p l e 3

Theorem 2.1 does apply to the differential equation $y' = \sqrt{y}$ of Example 2 as long as we restrict consideration to points (x, y) for which $y > 0$. In that case the partial derivative $\partial \sqrt{y}/\partial y = \frac{1}{2}y^{-1/2}$ is continuous, and we can conclude that there is a unique solution whose graph contains a specified point (x_0, y_0) with $y_0 > 0$.

e x a m p l e 4

If $F(x, y) = F(x)$ is independent of y and is continuous as a function of x on some interval $a < x < b$, then Theorem 2.1 applies to the differential equation $y' = F(x)$ at all points (x, y) of the infinite vertical strip containing the interval. The reason is that $\partial F(x)/\partial y$ is identically zero, and so of course is continuous. For example, the field determined in Example 2 of Section 2B by the function $F(x) = x$ is of this type. This example is typical of those that are continuous but independent of y in that the existence of a unique solution through each point can be inferred without using Theorem 2.1 by using both versions of the fundamental theorem of calculus as follows. Suppose $y = y(x)$ is some solution of the initial-value problem $y' = F(x)$, $y(x_0) = y_0$. We can integrate both sides of the differential equation between x_0 and x using one version of the fundamental theorem to get

$$y(x) - y(x_0) = \int_{x_0}^{x} y'(t)\, dt = \int_{x_0}^{x} F(t)\, dt.$$

Since $y(x_0) = y_0$, it follows that $y(x)$ must be related to $F(x)$ by the equation

$$y(x) = y_0 + \int_{x_0}^{x} F(t)\, dt.$$

This establishes the uniqueness of the solution, because if $y(x)$ satisfies the differential equation, then $y(x)$ must be given by the previous formula. Regardless of where it came from, the same formula provides the existence of a solution to the initial-value problem: Apply the other version of the fundamental theorem in differentiating both sides with respect to x to get $y'(x) = F(x)$, thus showing that the differential equation is satisfied; then set $x = x_0$ to show that the initial condition $y(x_0) = y_0$ is met. As a special case, the differential equation $y' = x$ of Example 2 of Section 2B, with initial condition $y(0) = c$, yields the unique solution $y = \frac{1}{2}x^2 + c$.

EXERCISES

1. Each of the following differential equations determines a direction field on a set S in the xy plane. Give a precise description, verbally or pictorially, of S in each case, and describe also the subset R of S where Theorem 2.1 guarantees a unique solution through each point of R.

   (a) $y' = xy^{4/3}$.   **(f)** $y' = x/y$.

   (b) $y' = xy^{1/3}$.   **(g)** $y' = \sin(x - y)$.

   (c) $y' = \sqrt{xy}$.   **(h)** $y' = (1 + y^2)/(1 - y^2)$.

   (d) $y' = \sqrt{x} + x^3y^4$.   **(i)** $y' = e^{\sin(xy)}$.

   (e) $y' = y/x$.

2. Does the differential equation

$$y' = \begin{cases} 0, & x < 0, \\ x, & 0 \le x, \end{cases}$$

   have a unique solution with its graph passing through an arbitrary point of the xy plane? Explain.

3. Define for all (x, y) the continuous function $K(x, y) = y^{2/3}$.

   (a) Sketch the direction field of the differential equation $y' = K(x, y)$ on the rectangle $-2 \le x \le 2$, $-2 < y < 2$.

   (b) Verify that there are *three* distinct solution curves through each point $(x_0, 0)$ and defined for all x. [*Hint:* Consider $y = (x - x_0)^3/27$ for $x \ge x_0$ and $y = 0$ for $x < x_0$.]

   (c) Explain in detail why this example doesn't contradict Theorem 2.1.

4. Define for all (x, y) the continuous function

$$F(x, y) = \begin{cases} \sqrt{y}, & \text{if } y > 0, \\ 0, & \text{if } y \le 0. \end{cases}$$

   (a) Sketch the direction field of the differential equation $y' = F(x, y)$ on the rectangle $-2 \le x \le 2$, $-2 < y < 2$.

   (b) Verify that there are two distinct solution curves through each point $(x_0, 0)$ and defined for all x. [*Hint:* Consider $y = (x - x_0)^2/4$ for $x > x_0$.]

   (c) Explain in detail why this example doesn't contradict Theorem 2.1.

5. A differential equation with the same essential features as the one in the previous exercise arises in mechanics.

   (a) A body dropped from rest under constant gravitational acceleration g moves distance $s = gt^2/2$ in time t. Show that this $s(t)$ satisfies the differential equation $ds/dt = \sqrt{2gs}$ with initial condition $s(0) = 0$.

   (b) Show that the initial-value problem in part (a) has a solution other than $s = gt^2/2$.

6. Define for all (x, y) the continuous function

$$G(x, y) = \begin{cases} x\sqrt{y}, & \text{if } y \ge 0, \\ 0, & \text{if } y < 0. \end{cases}$$

(a) Sketch the direction field of the differential equation $y' = G(x, y)$ on the rectangle $-2 \le x \le 2$, $-2 < y < 2$.

(b) Verify that there are two distinct solution curves through the point $(0, 0)$ and defined for all x. [*Hint:* Consider $y = x^4/16$ for $x > 0$.]

(c) Explain in detail why this example doesn't contradict Theorem 2.1.

7. (a) Verify that the differential equation $y' = \sqrt{1 - y^2}$ has solutions $y = \sin(x + c)$, where c is constant. Then sketch a solution graph passing through $(0, 0)$.

(b) Sketch the solution graph asked for in part (a) by relying entirely on a preliminary sketch of the direction field for $y' = \sqrt{1 - y^2}$, following the convention that $\sqrt{1 - y^2} \ge 0$. Show in particular that there is a solution graph through $(0, 0)$ given by

$$y(x) = \begin{cases} \sin x, & |x| \le \pi/2, \\ 1, & x > \pi/2, \\ -1, & x < -\pi/2. \end{cases}$$

(c) Note that $y(x) = 1$ is a solution with $y(\pi/2) = 1$. Is there a conflict between your results in parts (a) and (b) and Theorem 2.1? Explain.

8. Define for all (x, y) the continuous function

$$Q(x, y) = \begin{cases} \sqrt{1 - y^2}, & \text{if } -1 \le y \le 1, \\ 0, & \text{otherwise.} \end{cases}$$

(a) Sketch the direction field of the differential equation $y' = Q(x, y)$ on the rectangle $-\pi \le x \le \pi$, $-2 < y < 2$.

(b) Verify that there are two distinct solution curves through each point $(x_0, 1)$ and $(x_0, -1)$ and defined for all x. [*Hint:* Consider $y = \sin(x + c)$ for $-\pi/2 - c \le \pi/2 - c$; then extend the solution to have appropriate constant values.]

(c) Explain in detail why this example doesn't contradict Theorem 2.1.

*9. The right side of the differential equation $y' = (xy)^{3/2}$ is defined and continuous for all (x, y) in the closed quadrant $Q = \{(x, y): x \ge 0, y \ge 0\}$.

(a) Determine the open set R of points (x_0, y_0) for which Theorem 2.1 guarantees the existence and uniqueness of a solution to the initial-value problem $y(x_0) = y_0$.

(b) Define the function $F(x, y)$ for all (x, y) by

$$F(x, y) = \begin{cases} (xy)^{3/2}, & \text{for } (x, y) \text{ in } Q, \\ 0, & \text{for } (x, y) \text{ not in } Q. \end{cases}$$

Show that by applying Theorem 2.1 to the extended differential equation $y' = F(x, y)$ you can conclude that the initial-value problem $y(x_0) = y_0$ has a unique solution for all (x_0, y_0) in Q.

10. Suppose $y' = F(x, y)$ is a differential equation that meets the assumptions of Theorem 2.1 for all (x, y). Can it happen that the graphs of two different solutions of this differential equation have a point (x_0, y_0) in common? Explain your answer.

*11. The region on which Theorem 2.1 guarantees a unique solution graph containing a given point need not be a rectangle. Define the continuous function $F(x, y)$ for all (x, y) by

$$F(x, y) = \begin{cases} \sqrt{1 - x^2 - y^2}, & \text{for } x^2 + y^2 < 1, \\ 0, & \text{for } x^2 + y^2 \geq 1. \end{cases}$$

(a) Show that Theorem 2.1 can be applied to the open disc $x^2 + y^2 < 1$ so as to ensure the existence of a unique solution to $y' = F(x, y)$ with its graph passing through each point of the disc.

(b) Draw a conclusion analogous to that in part (a) for the open region $x^2 + y^2 > 1$.

(c) Can Theorem 2.1 be applied to $y' = F(x, y)$ on the entire xy plane? Explain.

(d) Use a sketch of the direction field of $y' = F(x, y)$ to sketch the graph of the solution to the differential equation on the interval $-2 \leq x \leq 2$ and that satisfies the initial condition $y(-2) = 0$. (For solution existence, see Peano's theorem, Theorem C.3, in Appendix C.)

12. Let $a > 0$ be constant in $y' = a(1 + y^2)$.

(a) Verify that $y = \tan(ax)$ is a solution of the differential equation for $-\pi/(2a) < x < \pi/(2a)$, but for no larger interval containing $x = 0$.

(b) Make a sketch of the direction field for $y' = 1 + y^2$, and use it to account for the behavior of the solution $y = \tan x$.

13. Define $P(x)$ for all x by

$$P(x) = \begin{cases} 2x, & x \leq 0, \\ x, & x > 0. \end{cases}$$

(a) Sketch the graph of $P(x)$ for $-2 \leq x \leq 2$.

(b) Sketch the direction field of the differential equation $y' = P(x)$.

(c) Find a solution to $y' = P(x)$ satisfying $y(-1) = 0$, and sketch the graph of the solution along with your sketch of the direction field.

(d) Is the solution you found in part (c) the only one satisfying both the differential equation and the initial condition? Give reasons for your answer.

*14. It might be supposed that nonuniqueness for our prime example, $y' = \sqrt{y}$, is equivalent to the failure of $\sqrt{y}$ to have a derivative at $y = 0$. This example shows that the matter is subtler than that. Let

$$y' = \begin{cases} -\sqrt{y}, & y \geq 0, \\ 0, & y < 0. \end{cases}$$

Show that the initial-value problem with $y(0) = 0$ has only the trivial solution $y(x) = 0$ for $0 \leq x$. In thinking about this it helps to sketch the direction field. Suppose there is a solution with $y(x_1) > 0$ for some $x_1 > 0$. Show that then $y(x) > 0$ for $x < x_1$ by Theorem 2.1 and hence that $y(0) > 0$. Use a similar argument if $y(x_1) < 0$.

3   INTEGRATION

Broadly speaking, integration in some form or other is the basis for all solution methods for differential equations. This section is devoted to finding elementary solution formulas for some first-order differential equations that can be analyzed by fairly direct application of formal integration techniques. In practice this means one or more of the following: (i) having the solution already stored in your brain as a well-known formula; (ii) looking the solution up in a table; (iii) finding the solution using a symbolic calculator; (iv) transforming the equation by a substitution so that (i), (ii), or (iii) applies.

3A $y' = f(x)$

A differential equation of the simple form $y' = f(x)$ can in principle be solved on any interval $a \le x \le b$ over which f is continuous. We can write the general solution either as

$$y(x) = \int f(x)\, dx + C \qquad \text{or} \qquad y(x) = \int_a^x f(u)\, du + C.$$

Differentiating either of these last two equations with respect to x shows that we have found a whole family of solutions. Solution families of equations of this simple form always have graphs consisting of parallel curves with constant vertical distance between points on two given graphs. One way to account for this parallelism is to observe that the slope specified by the direction field of $dy/dx = f(x)$ is the same for each fixed value of x, regardless of the vertical displacement. Examples are shown in Figure 1.2.

---

e x a m p l e 1

The differential equation $y' = x$ has general solution

$$y(x) = \int x\, dx + C = \tfrac{1}{2}x^2 + C,$$

valid for all real x. Similarly the differential equation $y' = \sin(1 + x^2)$ has general solution

$$y(x) = \int_0^x \sin(1 + u^2)\, du + C.$$

An important difference between this and the solution to $y' = x$ is that here we can't find an explicit elementary solution formula. Nevertheless, we can get quite a bit of information about the solution by using numerical integration to produce an accurate table of values. And the derivatives of $y(x)$ are readily available: since we know that $y'(x) = \sin(1 + x^2)$, it will follow that $y'' = 2x \cos(1 + x^2)$, and so on, for higher-order derivatives.

When the independent variable represents time, we denote it by t rather than x, and we can interpret an equation of the form $y' = v(t)$ as a condition that the position $y(t)$ at time t of a 1-dimensional motion has **velocity** $v(t)$. If $v(t)$ is differentiable, then $v' = a(t)$ is called the **acceleration** of the motion at time t. The absolute value $|v(t)|$ is called **speed.**

e x a m p l e 2

Near the surface of the earth, the acceleration due to gravity for a falling body has a nearly constant value $g = 32.2$ feet per second per second. Thus velocity satisfies the equation $v' = g$. Integration with respect to t gives $v = gt + C_1$. The constant C_1 is determined by observing the velocity v_0 at some time t_0. If $t_0 = 0$ we find

$$v(t) = gt + v_0.$$

Since position is related to velocity by $y' = v$, we can write a first-order equation for $y(t)$ in the form

$$y' = gt + v_0.$$

Integration gives $y = gt^2/2 + v_0 t + C_2$. If $y(0) = y_0$, as measured down from some reference point above ground level, then the constant C_2 must be y_0, so

$$y(t) = \tfrac{1}{2}gt^2 + v_0 t + y_0.$$

For example, if we were simply to drop a stone down a deep well, we would have $v_0 = 0$. Since it might be natural to measure distance down from the top of the well, we could set $y_0 = 0$, with the result

$$y(t) = \tfrac{1}{2}gt^2 \approx 16.1t^2 \text{ feet.}$$

If it took 4 seconds for the stone to hit the bottom of the well, we would know that it was about $16.1 \cdot 4^2 = 257.6$ feet deep.

　　If instead we measure up from the surface of the earth, the acceleration of gravity would act to decrease the velocity, so we would replace g by $-g$ in the preceding derivation. Thus a projectile thrown straight up from the surface of the earth ($y_0 = 0$) with velocity $v_0 = 100$ feet per second would have altitude in feet at t seconds given by

$$y(t) = -16.1t^2 + 100t.$$

The maximum altitude $y_{\max}$ occurs when the velocity $v = y'$ is zero. Hence we find $t_{\max}$ from

$$y'(t) = -32.2t + 100 = 0$$

to be $t_{\max} \approx 3.1$ seconds. Then $y_{\max} \approx 465$ feet at $t = 3.1$. Air resistance is assumed negligible, but is taken into account in Chapter 2, Section 3, and Chapter 4, Section 1.

　　In Example 2 we solved a second-order equation, $y'' = g$, by solving two successive first-order equations to get the solution

$$y(t) = \tfrac{1}{2}gt^2 + v_0 t + y_0;$$

this is the position at time t of an object with initial position y_0 and initial velocity v_0 that is being influenced by a constant gravitational acceleration g. The assumption was that distance was measured downward from some point

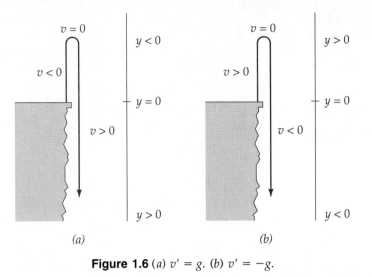

Figure 1.6 (*a*) $v' = g$. (*b*) $v' = -g$.

above the attracting body, say from the edge of a cliff or from the top of a deep well. Whether we choose to measure distance up from below or down from above is an arbitrary choice that we make. If, however, we follow the convention $y'' = g > 0$, note that since $y'' = (y')' = g$, the effect of g is at all times to increase velocity $v = dy/dx$. In particular, if v were to begin with negative values resulting from an initial upward motion (y would be decreasing here), then v would immediately begin to increase toward zero and positive values of y. (See Figure 1.6*a*.) In many applications it seems more natural to think of distance y as measured positively away from the attracting body. In that case the acceleration g acts to decrease velocity v, so the appropriate relation to start with is $v' = -g$. (See Figure 1.6*b*.) It's important to realize that the choice between using g or $-g$ for the acceleration is not dictated by the physical situation but rather by the chosen orientation of the coordinate line used for measuring distance.

---

e x a m p l e **3**

A projectile is fired straight up from the surface of the earth ($y_0 = 0$) with velocity $v_0 = 1000$ feet per second. If we choose to measure distance up from the surface of the earth, then at time $t \geq 0$ we'd have $v'(t) = -g$, since gravity acts to decrease velocity. Successive integrations yield

$$v(t) = y'(t) = -gt + v_0$$
$$\approx -32.2t + 1000$$

and

$$y(t) = -\tfrac{1}{2}gt^2 + v_0 t$$
$$\approx -16.1t^2 + 1000t.$$

Maximum height is reached when $v = 0$, at about $t_{max} \approx 1000/32.2 \approx 31$ seconds. The maximum height is about $y_{max} \approx -16.1(t_{max})^2 + 1000t_{max} \approx 15{,}528$ feet.

EXERCISES

1. Solve each differential equation, and find the particular solution that satisfies the associated initial condition.

   (a) $y' = x(1 - x)$, $y(0) = 1$.

   (b) $ds/dt = (t + 1)^2$, $s(1) = 2$.

   (c) $y' = x/(1 - x^2)$, $y(0) = 1$.

   (d) $du/dv = v^2 + 1$, $u(-1) = 1$.

   (e) $y'' = \sin x$, $y(0) = 1$, $y'(0) = 1$.

   (f) $y''' = 1$, $y(0) = y'(0) = y''(0) = 0$.

   (g) $dz/dt = te^t$, $z(0) = 1$.

   (h) $y' = \arctan x$, $y(0) = 0$.

   (i) $d^2x/dt^2 = e^t$, $x(0) = 1$, $dx/dt(0) = 0$.

   (j) $y'''' = x$, $y(0) = y''(0) = 0$, $y'(0) = y'''(0) = 1$.

Direction Fields. A differential equation of the special form $y' = f(x)$ has isoclines that are lines parallel to the y axis; thus to sketch the direction field you need to determine only one slope on each such line, making all slope segments centered on that line parallel to the first one. Most of the following differential equations can't be solved by integration in terms of elementary functions, but that doesn't matter, since you are asked to sketch the direction field for each one and then use the field to sketch a few solution graphs.

2. $y' = x^3$.

3. $y' = \sqrt{x^2 + 1}$.

4. $y' = 1/(1 + x^4)$.

5. $y' = \sqrt[3]{1 - x^3}$.

6. $y' = e^{-x^2}$.

7. $y' = x/(1 + x^4)$.

8. A projectile is fired up from ground level with an initial velocity of 5000 feet per second. What is the maximum altitude attained, and how long does it take to get there?

9. A weight is dropped from 5000 feet above ground. How long does it take to reach the ground, and with what velocity does it hit?

10. Suppose the objects described in the previous two exercises are sent on their way at the same time and are aimed directly at each other.

    (a) How long after release do they meet, and at what height above ground?

    (b) What initial velocity should the projectile be given so that the two objects meet 2500 feet above ground?

11. A block of wood is kicked across a smooth ice surface with initial velocity 20 feet per second. Friction with the ice produces a constant negative acceleration a. If the block stops sliding after 4 seconds, what is the value of a? How far does the block slide?

12. A projectile is fired straight up from the surface of the earth with initial velocity v_0 feet per second and with no other force but gravity acting on it. How large must v_0 be for the projectile to achieve a height of 100 feet? [*Hint:* Solve two equations for v_0.]

13. Estimate the velocity with which an object should be thrown down from a bridge if it is to have velocity 100 feet per second when it hits the water 100 feet below.

14. An object is thrown straight up from the surface of the earth with initial velocity v_0. Show that if air resistance is neglected, the maximum height is attained at $t_{max} = v_0/g$ and is $h_{max} = \frac{1}{2}v_0^2/g$.

15. Assuming a constant negative acceleration a for the application of an automobile's braking system, find a if the car comes to a stop within 100 feet from an initial speed of 100 feet per second.

16. A certain automobile's braking system provides an approximately constant *deceleration* of 30 feet per second per second on an asphalt pavement. Assume the brakes are applied when the car is going 60 miles per hour.

    (a) About how long will it take the car to stop?

    (b) Estimate the car's stopping distance.

    (c) If the pavement is wet, it is observed that the car stops in 3.5 seconds from an initial speed of 60 miles per hour. Estimate the deceleration rate and stopping distance under these conditions.

17. A spherical ball of ice with uniform density ρ melts so that its mass decreases at a rate proportional to its surface area, with proportionality constant $k > 0$. Show that if the radius of the ball is initially r_0, then it will be all melted after time $\rho r_0/k$. (Recall that a ball of radius r has volume $\frac{4}{3}\pi r^3$ and surface area $(d/dr)(\frac{4}{3}\pi r^3) = 4\pi r^2$.)

18. The estimate $g \approx 32.2$ feet per second per second that is often used for the acceleration of gravity near the earth's surface is related to Newton's **inverse-square law** by the equation

$$g = \frac{GM}{R^2},$$

where M and R are, respectively, the mass and mean radius of the earth, and G is a universal gravitational constant. To estimate the value g_m for the corresponding acceleration near the surface of the earth's moon, we use the estimates that the moon's mass and radius, respectively, are about 0.012 and 0.273 times that of the earth; hence follows the estimate

$$g_m \approx g\frac{0.012}{(0.273)^2} \approx 5.18 \text{ ft/s}^2.$$

    (a) Estimate the minimum initial velocity v_0 required for a jump to lift a person 6 inches off the earth's surface.

    (b) How high a jump would the velocity found in part (a) produce on the moon?

19. A ball is dropped 100 feet above the surface of a large planet and hits the surface about 1.5 seconds later.

    (a) Estimate the acceleration of gravity near the surface of the planet.

    (b) Estimate the impact velocity of the ball at the surface of the planet.

20. Show that if the acceleration of gravity at the surface of a planet is g_p, then the initial vertical velocity required for a high jumper to reach height h is $\sqrt{2hg_p}$.

21. A projectile is fired straight up from the surface of an airless planet with an initial velocity of 200 feet per second and is observed to reach a maximum height of 300 feet.

    (a) Estimate the acceleration of gravity on the planet.

    (b) If the acceleration of gravity on the planet were the same as that on earth, how high would the projectile have risen?

22. A stone is tossed up at the edge of a vertical cliff so that it just misses the edge on its way back down. Neglecting air resistance, show that the stone has the same speed, that is, the absolute value of velocity, when it passes the edge as it had from the initial toss.

23. A projectile is fired straight up with initial velocity v_0 from the surface of an airless planet where the acceleration of gravity is g_p. Show that the maximum height attained is $\frac{1}{2}v_0^2/g_p$.

24. A projectile is fired straight *down* with initial velocity v_0 from height s_0 above the surface of an airless planet where the acceleration of gravity is g_p. Show that the speed at impact is $\sqrt{v_0^2 + 2g_p s_0}$. Note that the initial conditions are different in the next problem, but the result is the same.

25. A projectile is fired straight *up* with initial velocity v_0 from height s_0 above the surface of an airless planet where the acceleration of gravity is g_p. Show that the speed at impact is $\sqrt{v_0^2 + 2g_p s_0}$. Note that the initial conditions are different in the previous problem, but the result is the same.

26. An object dropped near the surface of the earth under the influence of gravity alone falls distance $s = \frac{1}{2}gt^2$ in time t, where g is the acceleration of gravity.

    (a) Show that $s = s(t)$ satisfies the first-order differential equation

$$\frac{ds}{dt} = \sqrt{2gs}.$$

    (b) Verify that the initial-value problem $ds/dt = \sqrt{2gs}$, $s(0) = 0$, has the solution $s = \frac{1}{2}gt^2$. What other solution does it have?

27. Sometimes the area between the graph of $y = f(u)$ and the u axis over the interval from $u = a$ to $u = x$ is the same as the length of the graph over the same interval for all x. Verify that this is so if $f(u) = \cosh u = (e^u + e^{-u})/2$. Then find a first-order differential operation satisfied by all such functions $y = f(u)$.

Euler Beam Equation. Example 3 of the Introductory Survey contains Figure IS.3 and some background relevant to the next six exercises.

28. Suppose a uniform horizontal beam has profile shape $y = y(x)$, with x measured from one end. For a fairly rigid beam with uniform loading, $y(x)$ satisfies a 4th-order differential equation $y'''' = -P$, where $P > 0$ is a constant depending on the characteristics of the beam. If the ends of the beam are simply supported at the same level at $x = 0$ and $x = L$, then $y(0) =$

$y(L) = 0$. The extended beam also behaves as if its profile had an inflection point at each support, so that $y''(0) = y''(L) = 0$. (Figure IS.3 in the Introductory Survey shows only the left end simply supported.)

(a) Solve the differential equation by four successive integrations, and use the boundary conditions to show that the vertical deflection at point x, namely, $-y(x)$, is

$$\frac{P}{24}(x^4 - 2Lx^3 + L^3x), \qquad 0 \le x \le L.$$

(b) Where does the maximum amount of vertical deflection from the level at the ends occur, and how much is the maximum deflection?

(c) It's reasonable to expect that the beam's profile will be concave up between the supports; show that this is true.

(d) Explain how the value of P can be computed for a particular beam by making appropriate measurements.

29. Suppose a uniform horizontal beam has profile shape $y = y(x)$, with x measured from one end. For a fairly rigid beam with uniform loading, $y(x)$ typically satisfies a 4th-order differential equation $y'''' = -P$, where $P > 0$ is a constant depending on the characteristics of the beam. If the right end of the beam is embedded in a wall at $x = 0$ (the technical term is *cantilevered*) and simply supported at $x = L$, then $y(0) = y(L) = 0$ and $y''(0) = y'(L) = 0$. (Figure IS.3 in the Introductory Survey shows this situation.)

(a) Solve the differential equation by four successive integrations, and use the boundary conditions to show that the shape is described by the graph of

$$y(x) = -\frac{P}{48}(2x^4 - 3Lx^3 + L^3x).$$

(b) Show that the graph of $y(x)$ has an inflection on $0 < x < L$. What is the maximum vertical deflection?

(c) Make a sketch that shows the qualitative features of the graph of $y(x)$ for $0 \le x \le L$.

30. Let a horizontal beam have a profile described by the graph of $y = y(x)$. If the ends of the beam are horizontally embedded at the same height in vertical walls (i.e., cantilevered), then $y(x)$ satisfies conditions of the form $y(0) = y(L) = 0$ and $y'(0) = y'(L) = 0$. The **Euler beam equation** for $y(x)$ has the form $y'''' = -P$, $P > 0$.

(a) Show that the solution of the above Euler equation boundary-value problem is

$$y(x) = -\frac{P}{24}(x^4 - 2Lx^3 + L^2x^2).$$

(b) Show that the beam's maximum deflection occurs at the center, and find the amount of deflection there.

(c) Find the horizontal coordinates of the inflection points of $y(x)$.

(d) Make a sketch on a scale that shows the qualitative features of the graph of $y(x)$ for $0 \le x \le L$.

*31. A horizontal beam with profile graph $y = y(x)$ is embedded at both ends in such a way that that $y'(0) = y'(L) = 0$, but at different levels, so that $y(0) = 0$ and $y(L) = a$. Here L is the distance between the vertical supports. The previous exercise treats just the case $a = 0$.

(a) Solve the Euler beam equation $y'''' = -P$ with these boundary conditions.

(b) Show that the graph of $y(x)$ has a horizontal tangent strictly between 0 and L if and only if $PL^4 > 72|a|$.

(c) Make a sketch on a scale that shows the qualitative features of the graph of $y(x)$ for the two cases $PL^4 > 72|a|$ and $PL^4 < 72|a|$.

32. If one end of a beam profile $y = y(x)$ is embedded at $x = L$ so that $y(L) = y'(L) = 0$ and the other end is left hanging free at $x = 0$, the appropriate boundary conditions at the free end turn out to be $y''(0) = y'''(0) = 0$.

(a) Show that if $y(x)$ satisfies the Euler beam equation $y'''' = -P$, then the maximum deflection at the free end is $PL^4/8$.

(b) Make a sketch that shows the qualitative features of the graph of $y(x)$ for $0 \leq x \leq L$.

33. Boundary problems don't always have solutions, and there is sometimes a physical reason for this. Try to solve $y'''' = -R$, where R is a positive constant, subject to the boundary conditions $y(0) = y''(0) = 0$ and $y''(L) = y'''(L) = 0$.

(a) Explain in purely mathematical terms why there is no solution.

(b) Taking into account the interpretation of the boundary conditions in the previous exercise, explain in purely physical terms why there is no solution.

3B   Variables Separable

A differential equation that can be written in the form

$$g(y) \frac{dy}{dx} = f(x)$$

can in principle always be solved by integration. Letting $y = y(x)$ be a solution of the equation, we'll try to find an equation relating x and y. We can try to integrate both sides of the differential equation with respect to x, getting

$$\int g(y)\frac{dy}{dx}\, dx = \int f(x)\, dx. \tag{1}$$

Now suppose we can find indefinite integrals G and F such that $F' = f$ and $G' = g$. By the chain rule,

$$\frac{dG(y)}{dx} = G'(y) \frac{dy}{dx}$$

$$= g(y) \frac{dy}{dx},$$

so on the left side we have

$$\int g(y) \frac{dy}{dx} \, dx = \int \frac{d}{dx} G(y) \, dx$$

$$= G(y) + C_1.$$

The right side of the differential equation can be written

$$\int f(x) \, dx = F(x) + C_2.$$

Hence $G(y) + C_2 = F(x) + C_2$, or

$$G(y) = F(x) + c,$$

where $c = C_2 - C_1$. If we can now solve the equation for y in terms of x, we have a formula for solutions $y = y(x, c)$.

The process outlined is usually called **separation of variables,** because it involves getting x's on one side of the equation and y's on the other. Changing variables in the integral allows us to drop the dx's on the left side of equation (1). The resulting formal equation,

$$\int g(y) \, dy = \int f(x) \, dx,$$

still leads to the same solution: $G(y) = F(X) + c$, where $G' = g$ and $F' = f$. Equation (1) is sometimes written in **differential form**

$$g(y) \, dy = f(x) \, dx,$$

which can be interpreted as either

$$g(y) \frac{dy}{dx} = f(x) \qquad \text{or} \qquad g(y) = f(x) \frac{dx}{dy}.$$

Unfortunately, these separated forms are not always attainable for a first-order differential equation.

---

e x a m p l e 1

In biological studies it is often observed that the size $P(t)$ of a growing bacteria population is proportional to the rate of change $(dP/dt)(t)$. Expressing this proportionality in the form

$$\frac{dP}{dt} = kP, \qquad k \text{ const,} \tag{2}$$

gives us a first-order differential equation for P. Experience with the exponential function allows us to guess one solution. If we let

$$P(t) = Ce^{kt}, \qquad C \text{ const,}$$

we see that

$$\frac{dP}{dt}(t) = kCe^{kt} = kP(t)$$

for all real numbers t; in other words, $P = Ce^{kt}$ is a solution of the differential equation. Since we guessed this solution, the possibility remains that there are others. Under the assumption that the population size $P(t)$ is never zero, we

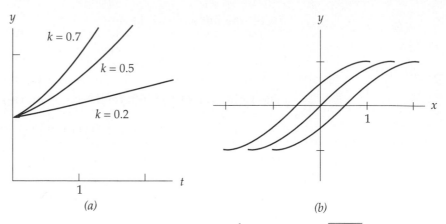

Figure 1.7 (a) $dP/dt = kP$, $P(0) = 1$; $P(t) = e^{kt}$. (b) $dy/dx = \sqrt{1 - y^2}$; $y = \sin(x + c)$.

can proceed as follows to separate variables. Divide equation (2) by P, getting

$$\frac{1}{P} \frac{dP}{dt} = k.$$

Now integrate both sides with respect to t:

$$\int \frac{1}{P} \frac{dP}{dt} \, dt = \int k \, dt.$$

The integral on the left is $\ln |P| + c_1$, and the integral on the right is $kt + c_2$. Hence we can lump the constants of integration together and write

$$\ln |P| = kt + c.$$

Since $e^{\ln z} = z$, we can apply the exponential function to both sides to get

$$|P(t)| = e^{kt+c}$$
$$= e^c e^{kt} = Ce^{kt},$$

where C is a necessarily positive constant: $C = e^c$. But $P(t)$ is differentiable, and since it must then be continuous, we must have either

$$P(t) = Ce^{kt} \quad \text{or} \quad P(t) = -Ce^{kt}.$$

Thus the constant can be either positive or negative in theory. If $P(t)$ does in fact represent a population size, then we use the first formula with $P(0) = C > 0$. Then $P(t) = P(0)e^{kt}$. Graphs of solutions satisfying $P(0) = 1$ are shown in Figure 1.7a for $k = 0.2, 0.5,$ and 0.7. Note that this picture doesn't contradict the uniqueness result of Section 2C, because the different values of k correspond to different differential equations rather than different solutions to the same equation.

---

e x a m p l e **2**

The differential equation $dy/dx = \sqrt{1 - y^2}$ in differential form is

$$\frac{dy}{\sqrt{1 - y^2}} = dx.$$

Integrating both sides gives $\arcsin y = x + c$. We can use any branch of the multiple-valued inverse-sine function we please, so assume we've picked the *principal branch,* for which arcsin 0 = 0; thus $-\pi/2 \le \arcsin y \le \pi/2$. Once a choice is made for the constant c, we then have the restriction $-\pi/2 \le x + c \le \pi/2$ on the values of x. Solving for y, we find $y = \sin(x + c)$.

Graphs of the solutions are shown in Figure 1.7b for $c = -1, 0,$ and 1. Note that the graphs are parallel in the x direction, because the direction field is independent of x. Note also that when, for example, $c = 0$, the solution graph comes to an abrupt end at $x = \pm\pi/2$. This apparent anomaly can be resolved in a geometrically natural way by extending the graph to the right with constant value 1 and to the left with value -1. See Exercise 16 at the end of this subsection.

Flow of Liquid from a Tank. Recall that free-fall position and velocity with initial conditions $y(0) = 0$ and $v(0) = 0$ are, respectively, $y(t) = \frac{1}{2}gt^2$ and $v(t) = gt$. We can solve the first equation for t to get $t = \sqrt{2y/g}$. Now set $gt = g\sqrt{2y/g}$ in the second equation to express **free-fall velocity in terms of distance:** $v = \sqrt{2gy}$. This last equation will be used to motivate the following discussion.

Suppose a tank has an outlet hole in its side near the bottom. Let $h = h(t)$ be the height of an ideal fluid above the outlet at time t, let $A(h)$ be the cross-sectional area of the tank at level h, and let a be the area of the outlet hole. Then $V(t)$, the remaining fluid volume at time t, satisfies **Torricelli's equation**

$$\frac{dV}{dt} = -a\sqrt{2gh}.$$

An intuitive justification for the equation is to note that it depends on having the outlet velocity equal to the free-fall velocity $\sqrt{2gh}$ of a drop of fluid from height h; thus dV/dt should be the negative of *outlet area a* times *outlet velocity* $\sqrt{2gh}$. (A more detailed justification depends on principles of fluid mechanics.) For an ordinary nonideal fluid, the equation takes the form

$$\frac{dV}{dt} = -ka\sqrt{2gh},$$

where $0 < k < 1$, the **flow constant** k depending on properties of the fluid such as viscosity and on the shape of the hole. For a tank with cross-sectional area $A(h)$ at height h, the volume V up to height h is given by an integral, and dV/dt by its derivative:

$$V(h) = \int_0^h A(y)\, dy, \qquad \text{so} \qquad \frac{dV}{dt} = A(h)\frac{dh}{dt}.$$

Thus Torricelli's equation appears as a differential equation to be satisfied by h:

$$A(h)\frac{dh}{dt} = -ka\sqrt{2gh}.$$

Note that the outlet area a need not always be constant but can be made to vary as a function of height h or time t. (For time dependence, imagine the gradual opening or closing of a valve.)

e x a m p l e 3

A tank in the shape of a circular cone has radius $r_0 = 5$ feet and vertical height $h_0 = 10$ feet. Hence the radius r at height h satisfies $r/h = \frac{5}{10}$, or $r = h/2$. (See Figure 1.8a.) The cross-sectional area at height h is $A(h) = \pi r^2 = \pi h^2/4$. A circu-

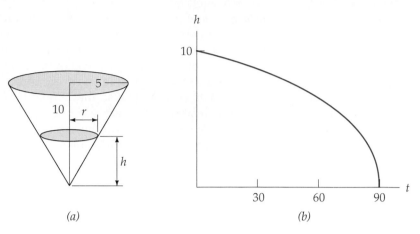

Figure 1.8 (a) $r/h = \frac{5}{10}$. (b) $h(t) = (316.23 - 3.5t)^{0.4}$.

lar outlet at the bottom of diameter 6 inches has area $a = \pi/16$ square feet. Suppose we want to use Torricelli's equation to find out how the fluid level varies as a function of time, starting with the tank full. Torricelli says

$$\frac{\pi h^2}{4} \frac{dh}{dt} = -\frac{k\pi}{16} \sqrt{2gh} \qquad \text{or} \qquad h^{3/2}\, dh = -k \sqrt{\frac{g}{8}}\, dt.$$

Integrate the separated equation with initial condition $h(0) = 10$ to get

$$\tfrac{2}{5}h^{5/2} = -k\sqrt{\frac{g}{8}}\, t + \tfrac{2}{5}10^{5/2} \qquad \text{or} \qquad h(t) = \left(10^{5/2} - 5k\sqrt{\frac{g}{32}}\, t\right)^{2/5}.$$

If we were dealing with an ideal fluid, we would take $k = 1$. The approximation $g \approx 32$ feet per second per second then yields $h(t) = (10^{5/2} - 5t)^{2/5}$. Under our assumptions we conclude that the tank would empty ($h = 0$) in time $t = 20\sqrt{10} \approx 63$ seconds. The empirical factor k could be determined by making an actual measurement of the time it takes to empty the tank; if this time turned out to be 90 seconds, we would solve $10^{5/2} - 5k \cdot 90 = 0$ to get $k \approx 0.7$; with this value of k, the practical numerical formula for $h(t)$ becomes $h = (316.23 - 3.5t)^{0.4}$. The graph is shown in Figure 1.8b.

---

example 4

Suppose the conical tank in Example 3 is modified by the addition of a valve at the bottom. With the tank full and the valve closed, someone begins to open the valve at the rate of $\frac{1}{100}$ square foot per second until it reaches its fully open area of $\pi/16$ square feet. What is the height of fluid in the tank at the time the valve becomes fully open? To find out, express the area a of the outlet as a function of t. Since the rate of increase of a is $\frac{1}{100}$ and $a(0) = 0$, we find $a(t) = t/100$; this holds until $a(t) = t/100 = \pi/16$, or until $t = 25\pi/4 \approx 19.6$ seconds. For this example we'll assume that $g = 32$ and that the flow constant k is 1. Torricelli's equation becomes

$$\frac{\pi h^2}{4} \frac{dh}{dt} = -\frac{8t}{100} h^{1/2} \qquad \text{or} \qquad h^{3/2}\, dh = -\frac{8t}{25\pi}\, dt.$$

Integrate the separated equation with initial condition $h(0) = 10$ to get

$$\tfrac{2}{5}h^{5/2} = -\frac{4t^2}{25\pi} + \tfrac{2}{5}10^{5/2} \qquad \text{or} \qquad h(t) = \left(10^{5/2} - \frac{2t^2}{5\pi}\right)^{2/5}.$$

At time $t = 25\pi/4$ seconds we find $h = (10^{5/2} - 125\pi/16)^{2/5} \approx 9.7$ feet when the valve becomes fully open. We can find subsequent behavior of $h(t)$ by an analysis like the one in Example 3.

EXERCISES

Solve each of the following equations by separating the variables. Then determine the constant of integration so that the given initial condition is satisfied and sketch the graph of that particular solution.

1. $y \, dy/dx = x^2$; $y = 1$ when $x = 2$.

2. $y' = x^2(1 + y^2)$; $y = 0$ when $x = 0$.

3. $dx + e^x y^2 \, dy = 0$; $y = 1$ when $x = 0$.

4. $s \, dt + t \, ds = 0$; $s = 1$ when $t = 1$.

5. $\cos u \, du - \sin v \, dv = 0$; $(u, v) = (\pi/2, \pi/2)$.

6. $y\sqrt{1 - x^2} \, dy = dx$; $(x, y) = (0, \pi)$.

7. Verify the indefinite integration formulas below. Also sketch the graph of the integrand and of the integral as functions of u for $u > 0$ and for $u < 0$, using the choice $\alpha = \tfrac{1}{2}$ for part (b).

(a) $\displaystyle\int (1/u) \, du = \ln |u| + C.$

(b) $\displaystyle\int |u|^\alpha \, du = u|u|^\alpha/(\alpha + 1) + C$, if $\alpha \neq -1$.

 Direction Fields. A differential equation of the special form $dy/dx = f(y)$ has isoclines that are lines parallel to the x axis; thus to sketch the direction field you need to determine only one slope on each such line, making all slope segments centered on that line parallel to the first one. All the following differential equations can be written in separated form but still are not necessarily solvable by integration in terms of elementary functions. Sketch the direction field for each one, and then use the field to sketch a few solution graphs.

8. $dy/dx = y^3$.

9. $dy/dx = \sqrt{y^2 + 1}$.

10. $dy/dx = 1/(1 + y^4)$.

11. $dy/dx = \sqrt[3]{1 - y^3}$.

12. $dy/dx = \sqrt{y}$, $y \geq 0$.

13. $dy/dx = y/(1 + y^4)$.

14. Newton's law of cooling asserts that the rate of change of the surface temperature $u(t)$ of an object is proportional to the difference between $u(t)$ and the temperature u_0 of the surrounding medium. Thus

$$\frac{du}{dt} = n(u - u_0),$$

where the constant of proportionality n is negative if $u > u_0$, so du/dt remains negative. Solve this equation under the assumption that u_0 is constant. Then sketch the graph of $u = u(t)$ under the assumption that $n = -0.5$, $u_0 = 10$, and $u(0)$, the initial temperature of the object, is 20.

15. In Example 2 of the text, the differential equation $dy/dx = \sqrt{1 - y^2}$ is shown to have the solution family $y = \sin(x + c)$, with the domain restrictions $-\pi/2 - c \leq x \leq \pi/2 - c$.

   (a) Show that if $y = \sin(x + c)$ is extended to have value 1 for $x > \pi/2 - c$ and -1 for $x < -\pi/2 - c$, then the extended solution is valid for all x.

   (b) Explain why the extension described in part (a) is consistent with the direction field of the differential equation.

16. In Example 3 of the text we arrived at the approximate formula $h(t) = (10^{5/2} - 5t)^{2/5}$ for the height of fluid in a certain conical tank at time t. Find the corresponding volume of fluid in the tank in two ways:

   (a) Use the formula $\frac{1}{3}\pi r^2 h$ for the volume of a right circular cone.

   (b) Integrate Torricelli's equation for dV/dt.

17. A water tank with constant cross-sectional area has an outlet of area a square feet at the bottom through which water is allowed to run freely, with flow constant $k = 1$. Water flows into the tank at a constant rate of q cubic feet per second. Show that if the height of the tank is less than $\frac{1}{64}q^2/a^2$, there will eventually be an overflow; otherwise the water level will steadily approach a fixed level. (Assume the acceleration of gravity is 32 feet per second per second.) [*Hint:* Interpret an appropriate modification of Torricelli's equation.]

18. A full water tank with constant cross-sectional area A and bottom outlet of area a empties in t_1 seconds. Assuming flow constant $k = 1$ and $g = 32$ feet per second per second, find the height of the tank.

19. Use Torricelli's equation

$$\frac{dV}{dt} = -ka\sqrt{2gh}$$

to do the following.

   (a) Show that if a tank has constant cross-sectional area A, then $h(t)$ satisfies

$$\frac{dh}{dt} = -\frac{ka\sqrt{2g}}{A}h^{1/2}.$$

   (b) Solve the differential equation.

   (c) Use your answer to part (b) to find out how long it would take for the fluid above the outlet to drop from initial height $h = h_0$ to $h = 0$. In particular, estimate how long that would take if the tank is cylindrical with diameter 10 feet, height 20 feet, the outlet is circular with diameter 6 inches, the constant k is 1, and the tank is initially full.

   (d) To estimate the value of the constant k empirically, you could measure the time it takes for the tank described in part (c) to empty and compute k from that information. Suppose the time it takes to empty the tank is 10 minutes; estimate k.

20. A tank 60 feet high and with cross-sectional area 100 square feet is to have its level lowered from full to half-full at the uniform rate of 1 foot per second. This is to be done by gradually increasing the scaled outlet area $ka(t)$ as a function of time over the required interval.

 (a) Find a formula for $ka(t)$. In particular, estimate $ka(0)$ and $ka(30)$.

 (b) Would it be feasible to empty the tank by continuing the method described above? Explain.

21. A full water tank h_0 feet high has constant cross-sectional area A and a bottom outlet of area a. Assuming flow constant $k = 1$ and $g = 32$ feet per second per second, show that the time it takes to empty the tank is $\frac{1}{4}A\sqrt{h_0}/a$ seconds.

22. A water tank of height h_0 feet and cross-sectional area A has two outlets of equal area a, one at the bottom and one l feet higher.

 (a) Show that the time in seconds required for the tank to drain from full to level l is

$$\frac{A}{12al}\,(h_0^{3/2} - l^{3/2} - (h_0 - l)^{3/2}).$$

 Assume flow constant $k = 1$ and $g = 32$ feet per second per second. [*Hint:* Find an appropriate modification of Torricelli's equation and solve it for time t.]

 (b) Show that as $l \to 0$ the result of part (a) tends to one-half the result of part (a) of the previous exercise. Why does this make sense physically?

23. An upright hemispherical bowl of radius 1 foot has an outlet hole near the bottom of radius 1 inch.

 (a) Express the upper surface area A of fluid in the bowl in terms of the height of the fluid.

 (b) Assuming flow constant $k = \frac{1}{2}$, about how long would it take for the full bowl to empty under the influence of gravity?

24. The problem here is to design a 1 cubic foot container, 4 feet high, with a circular cross section and an outlet at the bottom, that empties under the influence of gravity in 4 hours and in such a way that the fluid level drops at a constant rate. Assume throughout that the acceleration of gravity is 32 feet per second per second and that the outlet has flow constant $k = 1$.

 (a) Use Torricelli's equation to show that the radius of the container at height h above the bottom should have the form $r = p\sqrt[4]{h}$, where p is constant.

 (b) Show that the total volume requirement makes $p = \frac{1}{4}\sqrt{3/\pi}$ in part (a).

 (c) Find the area of the outlet if the container is to go from full to empty in exactly 4 hours.

 (d) Find the radius at the top of the container, and make a sketch of its profile as viewed from the side.

 (e) Explain how the container could be made to serve as a clock that measures time throughout a 4-hour period.

25. Imagine a balloon that starts with zero radius and gets inflated at a constant rate of C cubic units per second. Express the radius of the balloon as a function of time.

26. (a) Velocity $v = v(t)$ of a falling object subject only to a constant negative acceleration of gravity satisfies $dv/dt = -g$. Under the additional assumption that $v_0 = v(0) > 0$, show that the velocity is zero at time $t_1 = v_0/g$.

(b) A retarding force due to air resistance can be included in the equation in part (a) to give

$$\frac{dv}{dt} = -g - kv,$$

where k is a positive constant. Show that if $v_0 = v(0) > 0$, the velocity is zero at time

$$t_2 = \frac{1}{k} \ln\left(1 + \frac{kv_0}{g}\right).$$

(c) Give an intuitive physical argument to show that the times t_1 and t_2 in parts (a) and (b) should satisfy $t_1 > t_2$.

(d) Give a precise mathematical argument to show that the times t_1 and t_2 in parts (a) and (b) satisfy $t_1 > t_2$. [*Hint:* Is $x > \ln(1 + x)$ for $x > 0$?]

(e) Compute directly the right-hand limit

$$\lim_{k \to 0+} \frac{1}{k} \ln\left(1 + \frac{kv_0}{g}\right).$$

Is your result reasonable on physical grounds?

27. A particular parameter in a 1-parameter family of solutions can sometimes be advantageously replaced by another symbol. Here is an example.

(a) Solve $y' = -y^{3/2}$ by separating variables.

(b) Show that the equation $y' = -y^{3/2}$ has a family of solutions $y = 4/(x + c)^2$, each solution of the family being valid on intervals not containing a specified point $x = c$. Show, however, that this family fails to contain the constant solution $y = 0$.

(c) Replace the parameter c in part (b) by $1/C$, $C \neq 0$, and multiply numerator and denominator by C^2. Show that the resulting family contains all the solutions described in part (b) if we allow the possibility $C = 0$.

28. Consider the differential equation $dz/dt = -z^3$ with initial condition $z(0) = z_0$.

(a) Find a formula for the solution of the initial-value problem.

(b) Sketch the direction field of the equation.

(c) Sketch graphs for $t \geq 0$ of the solutions such that $z_0 = 2, 1, -1,$ and -2.

Differential Inequalities. Inequalities such as $y' \geq 0$ can often be used to determine qualitative properties of a function without necessarily specifying the function with as much precision as a differential equation may do. For example, $y' > 0$ on the interval $0 \leq x \leq 1$ implies only that $y(x)$ is increasing on the interval, while $y'' > 0$ implies that the graph of $y(x)$ is concave up there. Beyond that, we can sometimes integrate both sides of an inequality in such a way as to maintain inequality.

29. What conclusion can you draw about a function $y = y(x)$ under each of the following assumptions?

(a) $y'(x) \geq 1$ for $x \geq 0$, with $y(0) = 0$.

(b) $y''(x) \geq 1$ for $x \geq 0$, with $y(0) = y'(0) = 0$.

(c) $(e^{-x}y(x))' \geq 0$ for all real x, with $y(0) = 1$.

(d) $x \leq y''(x) \leq 2x$ for all real x, with $y(0) = 0$ and $y'(0) = 1$.

(e) $y'(x) \geq y(x) > 0$ for all x, with $y(0) = 1$.

(f) $y''(x) \leq 0$ for all x with $y(0) = 0$, $y'(0) = 1$.

30. (a) Solve the differential equation $y' = \sqrt{1 - y^2}$ by separation of variables, and express y in terms of x and a constant of integration. Then sketch solution graphs passing through $(0, -1)$, $(0, 0)$, and $(0, 1)$.

(b) Sketch the solution graphs asked for in part (a) by relying entirely on a preliminary sketch of the direction field for $y' = \sqrt{1 - y^2}$. Note that **singular solutions** $y(x) = 1$ and $y(x) = -1$ also satisfy the equation but are not contained in the solutions of part (a).

31. (a) Solve the differential equation $y' = 1 - y^2$ by separation of variables, and express y in terms of x and a constant of integration. Then sketch solution graphs passing through $(0, -2)$, $(0, 2)$, and $(0, 0)$.

(b) Sketch the solution graphs asked for in part (a) by relying entirely on a sketch of the direction field for $y' = 1 - y^2$.

(c) The given differential equation has the obvious **singular solutions** $y(x) = 1$ and $y(x) = -1$. Is either of these contained in the solution formula you derived in part (a)? Explain why, or why not.

3C   Transformations (Optional)

When the variables do not separate in a first-order equation, it may still be possible to change variables in such a way that the new equation separates. After solving the new equation, then change variables back again. An appropriate substitution will be suggested by the form of the differential equation, and in that regard the choice involves some educated guessing. We'll consider two types in detail.

LINEAR SUBSTITUTION. Suppose we have an equation that can be written in the form

$$\frac{dy}{dx} = f(ax + by + c).$$

If we let $u = ax + by + c$, then

$$\frac{du}{dx} = a + b\frac{dy}{dx}, \qquad \text{so} \qquad \frac{dy}{dx} = \frac{1}{b}\frac{du}{dx} - \frac{a}{b}.$$

The given equation becomes

$$\frac{1}{b}\frac{du}{dx} - \frac{a}{b} = f(u),$$

so the variables u and x can be separated.

---

example 1

For instance, in

$$\frac{dy}{dx} = (x + y)^2,$$

we let $u = x + y$, so $dy/dx = du/dx - 1$. Then

$$\frac{du}{dx} = u^2 + 1 \qquad \text{or} \qquad \frac{du}{1 + u^2} = dx.$$

Integrating, we find

$$\arctan u = x + c,$$

so $\arctan (x + y) = x + c,$ or $y = -x + \tan (x + c).$

Some solution curves are shown in Figure 1.9.

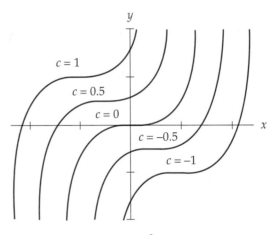

Figure 1.9 $dy/dx = (x + y)^2$; $y = -x + \tan (x + c)$.

QUOTIENT SUBSTITUTION. Suppose we have an equation that can be written in the form

$$\frac{dy}{dx} = g\left(\frac{y}{x}\right).$$

If we let $u = y/x$, then

$$\frac{du}{dx} = \frac{x\,dy/dx - y}{x^2} = \frac{1}{x}\frac{dy}{dx} - \frac{y}{x^2}.$$

We solve for dy/dx to get

$$\frac{dy}{dx} = x\frac{du}{dx} + \frac{y}{x} = x\frac{du}{dx} + u.$$

The given equation becomes

$$x \frac{du}{dx} = g(u) - u,$$

so the variables can be separated.

---

e x a m p l e 2

For instance, instead of

$$x^2 \frac{dy}{dx} = x^2 + xy + y^2,$$

we can write

$$\frac{dy}{dx} = 1 + \frac{y}{x} + \left(\frac{y}{x}\right)^2.$$

The function $g(u)$ is evidently $g(u) = 1 + u + u^2$. Letting $u = y/x$, we transform the given equation to

$$x \frac{du}{dx} = (1 + u + u^2) - u = 1 + u^2.$$

In separated form, we find

$$\frac{du}{1 + u^2} = \frac{dx}{x}.$$

We can integrate directly to get

$$\arctan u = \ln|x| + c.$$

Then put $u = y/x$ back again:

$$\arctan \frac{y}{x} = \ln|x| + c.$$

To simplify the relation between y and x, we can set $c = \ln k$, $k > 0$. This gives

$$\arctan \frac{y}{x} = \ln|x| + \ln k$$

$$= \ln k|x|.$$

Then $y/x = \tan(\ln k|x|)$, so

$$y = x \tan(\ln k|x|).$$

Two remarks are in order. First, it is usually impossible in integrating a separated equation to find a nice formula that expresses y in terms of x, or vice versa, as we did in Example 1; we may have to make do with an implicit formula of the form $F(x, y, c) = 0$. Second, our solution to $x^2 y' = x^2 + xy + y^2$ quite naturally fails to exist when $x = 0$. Depending on the value of k, there are other values of x for which $\tan(\ln k|x|)$ fails to exist.

Equations of the type discussed in Example 2 are sometimes called **homogeneous,** although the terminology conflicts with another use we have for that word. Note that an equation of the form

$$\frac{dy}{dx} = f\left(\frac{ax + by}{cx + dy}\right)$$

is homogeneous in this sense, because it can be written

$$\frac{dy}{dx} = f\left(\frac{a + by/x}{c + dy/x}\right).$$

The equation

$$\frac{dy}{dx} = f\left(\frac{ax + by + m}{cx + dy + n}\right)$$

fails to be homogeneous unless $m = n = 0$. However, a preliminary substitution of the form $x = z + h, y = w + k$ will produce a homogeneous equation in z and w if $ad - bc \neq 0$.

---

e x a m p l e 3

To make the differential equation

$$\frac{dy}{dx} = \frac{x + y + 1}{x - y}$$

amenable to a substitution of the form $u = y/x$, we try to find constants h and k such that letting $z = x + h$ and $w = y + k$ will make the ratio on the right homogeneous in z and w. We have

$$\frac{x + y + 1}{x - y} = \frac{z + h + w + k + 1}{z + h - w - k},$$

so we want $h + k + 1 = 0$ and $h - k = 0$. The unique solution of this pair of linear equations for h and k is $h = k = -\frac{1}{2}$. Since $dz/dx = 1$, the chain rule written in Leibnitz notation gives

$$\frac{dy}{dx} = \frac{dy}{dz}\frac{dz}{dx} = \frac{dw}{dz}.$$

To solve the given differential equation, we just have to solve the transformed equation

$$\frac{dw}{dz} = \frac{z + w}{z - w}.$$

This can be done by rewriting it as

$$\frac{dw}{dz} = \frac{1 + w/z}{1 - w/z}$$

and substituting $u = w/z$, $dw/dz = u + z\, du/dz$. Then substitute back by $z = x + \frac{1}{2}, w = y + \frac{1}{2}$. This part of the problem is left as an exercise.

EXERCISES

Each of the following differential equations can be written in one or the other of two forms:

$$\frac{dy}{dx} = f(ax + by + c) \qquad \text{or} \qquad \frac{dy}{dx} = g\left(\frac{y}{x}\right).$$

Decide which form applies in each case and use a transformation of the type

$$u = ax + by + c \qquad \text{or} \qquad u = \frac{y}{x}$$

to put the equation in separated form. Then solve the equation and eliminate u from the solution.

1. $dy/dx = 2x + y + 3$.

2. $dy/dt = (t - y)^2$.

3. $x^2\, dy/dx = x^2 + y^2$.

4. $t\, dx = (t + x)\, dt$.

5. $dy/dx = (y - x)/(y + x)$.

6. $dz/dv = (2z + v + 1)^2$.

7. None of the following equations can be written in the homogeneous form

$$\frac{dy}{dx} = g\!\left(\frac{y}{x}\right).$$

However, a preliminary transformation of the form $x = z + h$, $y = w + k$, where h and k are constant, gives an equation of that form in z and w. Solve the equations by first making the transformation with general h and k and then finding h and k. Then solve the equation in z, w and transform back to x, y for the final result.

(a) $dy/dx = (x + y + 1)/(x + 2y + 3)$.

(b) $(x + y)\, dy = (x - y + 1)\, dx$.

(c) $dy/dx = (x + y + 1)/(x - y)$

(d) $dy/dx = (y - x)/(x + y + 1)$.

(e) $(4x + 3y - 7)dx + (3x - 7y + 4)\, dy = 0$.

(f) $(2x - 3y + 4)dx + (3x - 2y + 1)\, dy = 0$.

8. (a) Show that for the equation $(x + 2y + 1)\, dx + (2x + 4y - 1)\, dy = 0$ it's impossible to use a transformation of the form $x = z + h$, $y = w + k$ to make the equation homogeneous.

(b) Solve the equation in part (a) by first making a substitution of the form $z = ax + by + c$.

9. Find all values of x, depending on $k > 0$, for which the solution $y = x \tan (\ln k|x|)$ to $x^2\, dy/dx = x^2 + xy + y^2$ fails to exist.

10. (a) Show that if $ad - bc \neq 0$, there is a unique choice for the constants h and k in the substitution $x = z + h$, $y = w + k$ that transforms the quotient $(ax + by + m)/(cx + dy + n)$ into the homogeneous quotient $(az + bw)/(cz + dw)$.

(b) Show that if $ad - bc = 0$, then either the numerator in part (a) is already a constant multiple of the denominator or else the desired h and k fail to exist.

11. Show that the differential equation

$$P(x, y)\, dx + Q(x, y)\, dy = 0$$

can be written in the form $dy/dx = g(y/x)$ if both

$$P(tx, ty) = t^n P(x, y) \qquad \text{and} \qquad Q(tx, ty) = t^n Q(x, y)$$

for some number n. [*Hint:* Let $t = 1/x$.]

12. **Bugs in mutual pursuit.** Four identical bugs are on a flat table, each moving at the same constant speed $a > 0$. Use xy coordinates on the table, and locate bugs 1 through 4 initially in respective quadrants 1 through 4, each at one of the points $(\pm 1, \pm 1)$. Bug 1 always heads directly toward bug 2, bug 2 toward bug 3, bug 3 toward bug 4, and bug 4 toward bug 1, so their paths are mutually congruent. The next five parts ask you to examine the paths.

(a) Use the symmetry of the paths to show that at any moment the bugs are at the corners of a square, and in particular, if bug 1 is at (x, y), then bugs 2, 3, and 4 are, respectively, at $(-y, x)$, $(-x, -y)$, and $(y, -x)$.

(b) Using the notation of part (a), show that the paths of bugs 1 and 3 satisfy the differential equation

$$\frac{dy}{dx} = \frac{y - x}{x + y},$$

unless $x + y = 0$. What is the analogous differential equation for bugs 2 and 4?

(c) Show that the coordinates of the path of bug 1 satisfy an equation of the form

$$\ln \sqrt{x^2 + y^2} + \arctan \frac{y}{x} = k.$$

(d) Show that in polar coordinates $x = r \cos \theta$, $y = r \sin \theta$ the equation can be put in the form

$$\ln r + \theta = k \qquad \text{or} \qquad r = Ce^{-\theta}$$

of a **logarithmic spiral.** Then sketch the first bug's path starting at $(1, 1)$.

(e) Use the formula

$$s = \int_{\theta_0}^{\theta_1} \sqrt{r^2 + \left(\frac{dr}{d\theta}\right)^2} \, d\theta$$

for arc length in polar coordinates to show that a logarithmic spiral starting at a finite point and winding in toward the origin has finite length. In particular, show that if the bugs were reduced to points, they would all reach the origin from $(\pm 1, \pm 1)$ in time $2/a$, where a is the constant speed of each one of them.

*13. An **airplane** flies horizontally with constant airspeed v miles per hour, starting at distance a east of a fixed beacon and always pointing directly at the beacon. There is a crosswind from the south with constant speed $w < v$.

(a) Locate the positive x axis along the east-west line through the beacon and with the origin at the beacon. Show that the graph $y = y(x)$ of the airplane's path satisfies

$$\frac{dy}{dx} = \frac{y}{x} - \frac{w}{v} \sqrt{1 + \left(\frac{y}{x}\right)^2}.$$

(b) Show that the airplane's path traces the graph of

$$y = \frac{x}{2} \left[\left(\frac{a}{x}\right)^{w/v} - \left(\frac{x}{a}\right)^{w/v} \right].$$

[*Hint:* Let $y/x = u$ as in Section 3C.]

Analysis of Systems. One of the most important uses for equations of the form $dy/dx = G(x, y)/F(x, y)$ is their relation to systems of two equations such as

$$\frac{dx}{dt} = F(x, y), \qquad \frac{dy}{dt} = G(x, y).$$

The connection stems from the chain rule $dy/dt = (dy/dx)(dx/dt)$ that, slightly rewritten, leads to

$$\frac{dy}{dx} = \frac{dy/dt}{dx/dt} = \frac{G(x, y)}{F(x, y)}.$$

The single equation in x and y contains less information than the pair of equations in x, y, and t. However, the single equation is sometimes easier to solve explicitly, and its solution provides some information about the system. The next five exercises explore these ideas in some examples.

14. Show that each solution $x = x(t)$, $y = y(t)$ to the system

$$\frac{dx}{dt} = -y, \qquad \frac{dy}{dt} = x$$

traces a circle $x^2 + y^2 = r^2$ for some appropriate constant r.

15. Show that each solution $x = x(t)$, $y = y(t)$ to the system

$$\frac{dx}{dt} = y, \qquad \frac{dy}{dt} = x$$

traces a hyperbola $x^2 - y^2 = C$ for some constant C.

16. Show that each solution $x = x(t)$, $y = y(t)$ to the system

$$\frac{dx}{dt} = x, \qquad \frac{dy}{dt} = y$$

traces a path lying on a line.

17. Show that each solution $x = x(t)$, $y = y(t)$ to the system

$$\frac{dx}{dt} = x + y, \qquad \frac{dy}{dt} = x^2 - y^2$$

traces a path lying on a graph of the form $y = x - 1 + Ce^{-x}$ for some constant C.

Generalized Solutions. A continuous function $y(x)$ that satisfies a first-order differential equation except at isolated points is sometimes called a **generalized solution.** Typically the failure of $y(x)$ to satisfy the equation at a point x_0 is due to the failure of $y'(x_0)$ to exist.

18. The direction field defined by the differential equation

$$y' = F(x, y) = \begin{cases} 0, & \text{if } x < 0, \\ 1, & \text{if } 0 \le x. \end{cases}$$

is discontinuous, though $\partial F/\partial y = 0$, so $\partial F/\partial y$ turns out to be continuous.

(a) Sketch the direction field for the equation, and attempt to sketch a continuous solution graph for the differential equation on the interval $-1 \le x \le 1$ and satisfying the initial condition $y(-1) = 0$.

(b) Explain why your graph can't have a derivative on an interval containing $x = 0$.

(c) Verify that, for each constant C,

$$y(x) = \begin{cases} C, & x < 0, \\ x + C, & x \geq 0, \end{cases}$$

is (i) continuous and (ii) satisfies the differential equation for every real number $x \neq 0$ and so is a generalized solution. What choice C satisfies $y(-1) = 0$?

19. Find a generalized solution to the equation

$$y' = \begin{cases} 1, & x < 1, \\ 2x, & x \geq 1, \end{cases}$$

that also satisfies initial condition $y(0) = 1$. Sketch the graph of your solution for $0 \leq x \leq 2$.

4   EXACT EQUATIONS (OPTIONAL)

The ideas in this section provide a technique for solving some first-order differential equations. Beyond that we'll investigate the representation of solutions as level curves of a function of two variables and also see how a direction field can be modified to be the gradient field $\nabla F(x, y)$ of a function $F(x, y)$.

4A   Solutions by Integration

The solutions of some differential equations are quite naturally presented in the implicit form $F(x, y) = C$ as opposed to the explicit form $y = y(x)$. For example, solutions of $y' = x/y$ are neatly represented by the family of hyperbolas $x^2 - y^2 = C$. The explicit solution $y = \sqrt{1 - x^2}$, with $-1 < x < 1$, is one of the implicitly defined solutions to $y' = x/y$ determined by the quadratic equation; another is $y = \sqrt{1 - x^2}$ on the same interval. The implicit form is useful because it contains both solutions.

Suppose now that C is constant, and that we start with a relation $F(x, y) = C$ that implicitly defines $y = y(x)$ as a differentiable function of x, or perhaps $x = x(y)$ as a function of y. If we differentiate $F(x, y) = C$ with respect to x using the chain rule, we get

$$F_x(x, y) + F_y(x, y)\frac{dy}{dx} = 0,$$

where the subscripts denote partial differentiation with respect to the indicated variable while the other variable is held fixed. The resulting first-order differential equation is, of course, automatically satisfied by the function $y = y(x)$. Written in differential form, the differential equation looks like

$$P(x, y)\, dx + Q(x, y)\, dy = 0,$$

where $P = F_x$ and $Q = F_y$. Not every equation of the general form displayed above arises as indicated above from what is called the **exact differential** of a function $F(x, y)$, namely,

$$dF(x, y) = F_x(x, y)\, dx + F_y(x, y)\, dy,$$

but, for those that do, the differential equation itself is called **exact.** Once an equation has been identified as exact, we can then try to display its solution graphs as level curves of some function $F(x, y)$.

---

e x a m p l e 1

Sometimes it's possible to make an educated guess at the exactness of a differential and then solve the associated equation at a glance. An equation

$$g(x)\, dx + h(y)\, dy = 0$$

with variables separated is always exact if $g(x)$ and $h(y)$ are continuous functions on respective domain intervals $a \leq x \leq b$, $c \leq y \leq d$. The reason is that there is always a solution of the special implicit form

$$G(x) + H(y) = C,$$

where $\qquad G(x) = \displaystyle\int_a^x g(u)\, du \qquad$ and $\qquad H(y) = \displaystyle\int_c^y h(u)\, du.$

Of course the choice of the lower limit of integration need not be the left endpoint of each interval. For instance, $x\, dx + y\, dy = 0$ has the implicit solution $\frac{1}{2}x^2 + \frac{1}{2}y^2 = C$, where $0 < C = $ constant. The graphs of these solutions are circles centered at the origin, and these circles are the level curves of $F(x, y) = \frac{1}{2}(x^2 + y^2)$. A circle of radius $r > 0$ corresponds to the points at level $z = r^2$ on the graph in xyz space of the equation $z = x^2 + y^2$. These are shown in Figure 1.10.

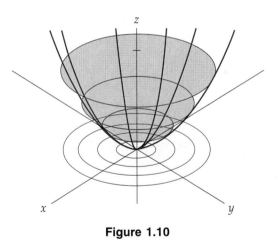

Figure 1.10

---

e x a m p l e 2

Recall that much of the technique of indefinite integration is based on our ability to recognize a function when we are given its derivative. In a slight extension of this technique, what we want to do here is recognize a function when we are given two partial derivatives. For example, since $\partial/\partial x(x \sin y) = \sin y$ and $\partial/\partial y(x \sin y) = x \cos y$, the equation

$$\sin y\, dx + x \cos y\, dy = 0$$

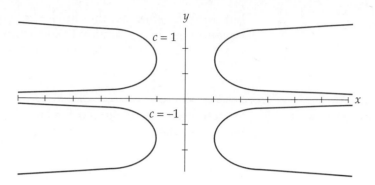

Figure 1.11 $\sin y \, dx + x \cos y \, dy = 0$; $y = \arcsin (c/x)$, $c = \pm 1$.

is easily seen to be equivalent to

$$d(x \sin y) = 0.$$

Thus the implicit form of the general solution to the equation

$$\sin y \, dx + x \cos y \, dy = 0$$

is $x \sin y = c$. An explicit solution that we can derive by solving for y is $y = \arcsin (c/x)$. Another representation is $x = c \csc y$. Figure 1.11 shows some graphs; the broken curves represent points on alternative branches of the arcsine function.

If you think a differential equation may be exact but you can't guess the solution, there is an integration routine for trying to find $F(x, y)$. The idea is to hypothesize the existence of F and then see what conclusions can be drawn about F. (Of course, one possible conclusion is that the equation you're looking at *isn't* exact and won't yield up its solution to this method.)

e x a m p l e 3

The differential equation $y' = (x + y)/(y - x)$ is equivalent, if $y \neq x$, to

$$(x + y) \, dx + (x - y) \, dy = 0.$$

If the left side is the exact differential of some function F, this means precisely that $F_x(x, y) = x + y$ and $F_y(x, y) = x - y$. In principle we can work with either of these two equations; since they seem to be of roughly equal complexity, we may as well pick the first one. We integrate both sides with respect to x and find

$$F(x, y) = \int F_x(x, y) \, dx$$

$$= \int (x + y) \, dx$$

$$= \tfrac{1}{2}x^2 + yx + \phi(y).$$

Note the "constant" of integration $\phi(y)$; since y is held constant during the integration, we can allow for different values of the integration constant, depending on y (but *not* on x), and in fact this freedom is crucial for making the remaining steps effective. Having found that

$$F(x, y) = \tfrac{1}{2}x^2 + yx + \phi(y),$$

it now makes sense to differentiate with respect to y so that we can compare the result with our second equation $F_y = x - y$. We find

$$F_y(x, y) = \frac{\partial}{\partial y} \, [\tfrac{1}{2}x^2 + yx + \phi(y)]$$

$$= x + \phi'(y) = x - y.$$

From this last equality we conclude that $\phi'(y) = -y$. Integration with respect to y shows that $\phi(y)$ must be of the form $-\tfrac{1}{2}y^2 + c$. It follows that

$$F(x, y) = \tfrac{1}{2}x^2 + yx - \tfrac{1}{2}y^2 + c,$$

and solutions of the original differential equation must satisfy the implicit relation

$$\tfrac{1}{2}x^2 + yx - \tfrac{1}{2}y^2 + c = 0.$$

Note that this is still not quite the same as having a formula for an explicit solution; we would have to solve the quadratic equation for y, or else for x, to get that. But completing the square in the first two terms shows that the graphs of all solutions must lie on a hyperbola of the form $(x + y)^2 - 2y^2 = C$. This may be the most informative way to look at the solutions.

4B Exactness Test

In Example 3 it was nice that in solving for $\phi'(y)$ we were able to come up with an expression independent of x. Otherwise the resulting inconsistency would have spoiled the rest of the routine. Fortunately there is a simple test that we can apply before going to the trouble of doing any formal integration. If the equation we're trying to solve fails the test, there's no point in continuing the solution method as outlined above. The test is easy to state and to apply.

■ 4.1 THEOREM

Assume that $P(x, y)$ and $Q(x, y)$ have continuous second-order partial derivatives in a rectangular region R. Then

$$P(x, y) \, dx + Q(x, y) \, dy$$

is an exact differential if and only if

$$\frac{\partial Q(x, y)}{\partial x} = \frac{\partial P(x, y)}{\partial y}$$

throughout the rectangle R.

Note. If the test fails, then the differential can't be exact, because exactness means

$$P(x, y) = \frac{\partial F(x, y)}{\partial x} \quad \text{and} \quad Q(x, y) = \frac{\partial F(x, y)}{\partial y} \quad \text{for some } F(x, y).$$

But equality of mixed partials would then imply that

$$\frac{\partial Q(x, y)}{\partial x} = \frac{\partial^2 F(x, y)}{\partial y\, \partial x} = \frac{\partial^2 F(x, y)}{\partial x\, \partial y} = \frac{\partial P(x, y)}{\partial y}.$$

This proves one of the implications of Theorem 4.1. The rest of the proof is deferred in favor of an example.

---

e x a m p l e 4

The differential equation

$$(x - y)\, dx + (x + y)\, dy = 0$$

isn't exact because

$$\frac{\partial(x + y)}{\partial x} = 1 \neq -1 = \frac{\partial(x - y)}{\partial y}.$$

This does *not* mean that the differential equation has no solutions, but just that we can't solve it by identifying it as an exact equation. (The equation can be solved by the method of Example 4 of Section 3B.)

To verify that the exactness test $P_y(x, y) = Q_x(x, y)$ is sufficient to guarantee exactness under our hypotheses, all we have to do is walk through the routine of Section 4A in full generality. It's worth doing this to fix the ideas in mind, but as a practical matter it's not a good idea to memorize the formulas, because the initial choice of which variable to work with can lead to quite different looking results. Assume the equation

$$P(x, y)\, dx + Q(x, y)\, dy = 0$$

satisfies the consistency condition

$$\frac{\partial Q(x, y)}{\partial x} = \frac{\partial P(x, y)}{\partial y}$$

in some rectangular region of the xy plane. We try to find $F(x, y)$ such that

$$F_x(x, y) = P(x, y), \qquad F_y(x, y) = Q(x, y).$$

Suppose we choose to integrate the second equation with respect to y, while holding x fixed. We get

$$F(x, y) = \int Q(x, y)\, dy + \phi(x).$$

To find $\phi(x)$, differentiate this equation with respect to x and use $F_x(x, y) = P(x, y)$ to get

$$P(x, y) = \frac{\partial}{\partial x} \int Q(x, y)\, dy + \phi'(x).$$

Now $\phi'(x)$ can be expressed entirely in terms of P and Q, so we can in principle integrate again, this time with respect to x, to find a function $\phi(x)$ to put into the equation above for $F(x, y)$. It's crucial, of course, that our expression

$$P(x, y) - \frac{\partial}{\partial x} \int Q(x, y)\, dy$$

for $\phi'(x)$ really turns out to be independent of y. To see that it is, we'll check that its derivative with respect to y is zero. Differentiating this formula with respect to y cancels the y integration and yields

$$\frac{\partial P(x, y)}{\partial y} - \frac{\partial Q(x, y)}{\partial x}.$$

This expression is identically zero if the consistency condition $Q_x = P_y$ holds, so $\phi'(x)$ will indeed be independent of y under that assumption. If we had started with integrating $F_x = P$ instead of $F_y = Q$, everything would be essentially the same, the only alterations being the interchange of x and y and the interchange of P and Q. This completes the proof of Theorem 4.1.

e x a m p l e 5

The differential equation

$$(e^x \cos y + 2x) \, dx - e^x \sin y \, dy = 0$$

passes the exactness test, because

$$\frac{\partial}{\partial x}(-e^x \sin y) = -e^x \sin y = \frac{\partial}{\partial y}(e^x \cos y + 2x).$$

We can expect to find a solution in the implicit form $F(x, y) = C$ by solving the equations

$$F_x(x, y) = e^x \cos y + 2x \qquad \text{and} \qquad F_y(x, y) = -e^x \sin y.$$

We start with whichever equation seems simpler; let's take the second equation and integrate with respect to y. We find

$$F(x, y) = \int F_y(x, y) \, dy$$

$$= \int -e^x \sin y \, dy$$

$$= e^x \cos y + \phi(x).$$

Here the constant of integration may turn out to depend explicitly on x. According to the equation for F_x, we'll have

$$F_x(x, y) = \frac{\partial}{\partial x}[e^x \cos y + \phi(x)]$$

$$= e^x \cos y + \phi'(x).$$

Thus $e^x \cos y + \phi'(x) = e^x \cos y + 2x$, so $\phi'(x) = 2x$. Since $\phi(x)$ must then be of the form $x^2 + c$, we find $F(x, y) = e^x \cos y + x^2 + c$. The solution can be written in implicit form as

$$e^x \cos y + x^2 = C.$$

We can then extract from this formula the explicit solution $y = \arccos [e^{-x}(C - x^2)]$.

The frequently used formulas

■■■ **4.2**   **(a)** $d(uv) = v\,du + u\,dv.$ **(b)** $d\left(\dfrac{u}{v}\right) = \dfrac{v\,du - u\,dv}{v^2}.$

can sometimes be used to make an informed guess about the solutions to some equations once they are in differential form.

---

e x a m p l e 6

The equation $dy/dx = -y/(x + y^2)$ can be written in differential form as

$$y\,dx + (x + y^2)\,dy = 0 \qquad \text{or} \qquad (y\,dx + x\,dy) + y^2\,dy = 0.$$

This last regrouping of terms is suggested by Equation 4.2a. Fortunately the second term is just $d(y^3/3)$, so the entire equation can be written

$$d(xy) + d\left(\frac{y^3}{3}\right) = 0, \qquad \text{with solution} \qquad xy + \frac{y^3}{3} = C.$$

---

e x a m p l e 7

The equation $y \cos x\,dx + \sin x\,dy = 0$ has the form $d(uv)$, where $u = \sin x$ and $v = y$. Solutions are given by $y \sin x = C$, and it's easy to check that $d(y \sin x) = y \cos x\,dx + \sin x\,dy$.

---

e x a m p l e 8

The equation $y\,dx - x\,dy = 0$ fails the exactness test, but recalling Equation 4.2b suggests multiplying by y^{-2} to get

$$\frac{y\,dx - x\,dy}{y^2} = 0 \qquad \text{or} \qquad d\left(\frac{x}{y}\right) = 0.$$

Thus $x/y = c$ or $x = cy$ for the solution. The factor y^{-2} is an example of an *integrating factor* discussed below.

4C   *Integrating Factors*

A first-order differential equation, whether written as $y' = f(x, y)$ or in differential form $P(x, y)\,dx + Q(x, y)\,dy = 0$, typically has associated with it a direction field, or slope field, as described in Section 2. (If $P(x, y)$ and $Q(x, y)$ are both zero at the same point (x_0, y_0), the equation doesn't say anything at (x_0, y_0), so we assume that any such points have been excluded in what follows.) The pair of functions $(P(x, y), Q(x, y))$ is related to the slope function by $f(x, y) = -P(x, y)/Q(x, y)$. In addition, $(P(x, y), Q(x, y))$ generates a **vector field** that we can sketch by drawing arrows from points (x, y) to $(P(x, y), Q(x, y))$, as shown in Figure 1.12. An arrow from this vector field starting at (x, y) will have slope $Q(x, y)/P(x, y)$ and so will be perpendicular to the segment of slope $f(x, y) = -P(x, y)/Q(x, y)$ associated with (x, y). However, if $P(x, y)\,dx + Q(x, y)\,dy = 0$ is an exact equation, with implicit solution formula $F(x, y) = C$, then the vector field is the **gradient field** of F, that is, $(P, Q) = (\partial F/\partial x, \partial F/\partial y)$, often abbreviated $(P, Q) = \nabla F$, or $(P, Q) = \text{grad } F$. We've seen previously that

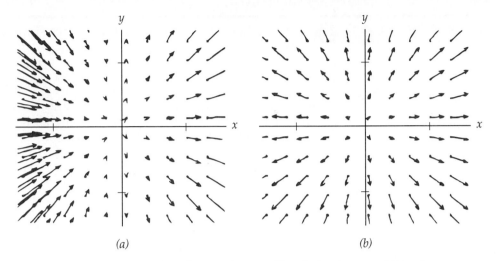

Figure 1.12 (a) $P(x, y) = x^2$, $Q(x, y) = xy$. (b) $x^{-1}P(x, y) = x$, $x^{-1}Q(x, y) = y$.

exactness of a first-order equation is an aid to finding an implicit solution formula, with solution graphs that are level curves of some function. We now ask whether we can adjust the *lengths* of the vectors in a vector field (P, Q), possibly also reversing some directions, so that the resulting field will be the gradient of some function; in other words, can we find a function $M(x, y)$ such that the vector field (MP, MQ) is a gradient field? If we can find such an M, then

$$M(x, y)P(x, y)\, dx + M(x, y)Q(x, y)\, dy = 0$$

will be an exact equation, solvable in principle by integration. For this reason, M is often called an **integrating factor,** abbreviated IF. It can be shown that if $P(x, y)$ and $Q(x, y)$ are sufficiently well-behaved in a region of the xy plane, then an integrating factor $M(x, y)$ always exists in a neighborhood of any given point (x_0, y_0) in the region. Unfortunately it's hard to find a formula for such an $M(x, y)$ except in very special circumstances; some of these are indicated in the following examples and exercises.

***example* 9**

The equation $x^2\, dx + xy\, dy = 0$ fails the exactness test, since letting $P(x, y) = x^2$ and $Q(x, y) = xy$ we find

$$\frac{\partial P(x, y)}{\partial y} = 0 \neq y = \frac{\partial Q(x, y)}{\partial x}$$

unless $y = 0$. However, multiplying by the integrating factor x^{-1} when $x \neq 0$ gives $x\, dx + y\, dy = 0$. This equation is clearly exact, because we can let $F(x, y) = (x^2 + y^2)/2$ and note that $\partial F(x, y)/\partial x = x$ and $\partial F(x, y)/\partial y = y$. Both the inexact and the exact vector fields are shown in Figure 1.12, scaled by the additional factor $\frac{1}{5}$. Note that the arrows at corresponding points of the two fields are parallel but have different lengths and sometimes even point in opposite directions.

e x a m p l e 10

The equation $y\,dx + 2x\,dy = 0$ is not exact, since with $P(x, y) = y$ and $Q(x, y) = 2x$,

$$\frac{\partial P(x, y)}{\partial y} = 1 \neq 2 = \frac{\partial Q(x, y)}{\partial x}.$$

However, the equation can be solved by separating variables: just divide by xy, so in effect $M(x, y) = (xy)^{-1}$ is an integrating factor. Now integrate $dx/x + dy/y = 0$ to get

$$\int \frac{dx}{x} + \int \frac{2\,dy}{y} = c \qquad \text{or} \qquad \ln|x| + \ln y^2 = c.$$

Combining logarithms we get $\ln|x|y^2 = c$ or $xy^2 = \pm e^c = C$. In this way we found the family of implicit solutions $F(x, y) = xy^2 = C$. But now we see that this family of solutions satisfies the exact differential equation

$$F_x(x, y)\,dx + F_y(x, y)\,dy = y^2\,dx + 2xy\,dy = 0.$$

Comparison of this equation with the original differential equation allows us to make the easy guess that $N(x, y) = y$ is an even simpler integrating factor for the original equation. This observation can now be used to solve some equations of the form

$$y\,dx + (2x + Y(y))\,dy = 0, \qquad \text{for example,} \qquad y\,dx + (2x + y^2)\,dy = 0.$$

The integrating factor $N(x, y) = y$ that we just found for the original equation now applies here to give us

$$y^2\,dx + (2xy + y^3)\,dy = 0.$$

Since $\partial y^2/\partial y = 2y = \partial(2xy + y^3)/\partial x$, we can proceed with some confidence to look for a $G(x, y)$ such that $G_x(x, y) = y^2$ and $G_y(x, y) = 2xy + y^3$. Integrating the first of these equations with respect to x gives $G(x, y) = xy^2 + g(y)$. Now differentiate this equation with respect to y and compare the result, $G_y(x, y) = 2xy + g'(y)$ with the previous expression $G_y(x, y) = 2xy + y^3$ for G_y. The conclusion is that $g'(y) = y^3$, so $g(y) = \frac{1}{4}y^4 + C$. The implicit-form solution we're looking for is then $xy^2 + \frac{1}{4}y^4 = C$.

EXERCISES

1. Most of the following differential equations are exact if written in the form $P\,dx + Q\,dy = 0$. Check the consistency condition $P_y = Q_x$, and, if possible, find a solution in the form $F(x, y) = C$. Then find the particular solution that satisfies the given initial condition.

   (a) $(y + 2xy^2)\,dx + (x + 2x^2y)\,dy = 0$; $y = 1$ when $x = 1$.

   (b) $dy/dx = (2x + 1)/(2y + 1)$; $y = 0$ when $x = 0$.

   (c) $(y + e^y)\,dx + x(1 + e^y)\,dy = 0$; $y = 0$ when $x = 1$.

   (d) $(\cos y - y \sin x)\,dx + (\cos x - x \sin y)\,dy = 0$; $y = \pi/2$ when $x = \pi/2$.

   (e) $(y + 1)\,dy - xy\,dx = 0$; $y = 0$ when $x = 2$.

   (f) $(y - x)e^x\,dx + (1 + e^x)\,dy$; $y = 1$ when $x = 1$.

2. Determine for which values of the constant k each of the following equations satisfies the consistency condition $P_y = Q_x$. Then solve the equation for those values of k.

(a) $y\,dx + (kx + y^2 + 1)\,dy = 0$.

(b) $\cos y\,dx + (\cos y + k(x + y)\sin y)\,dy = 0$.

(c) $((x + ky + 1)/(x + ky))\,dx + (k/(x + ky))\,dy = 0$.

3. Show that an equation for which the variables are separable can be made formally exact upon multiplication by a suitable integrating factor.

4. An integrating factor $M(x, y)$ for the equation $P(x, y)\,dx + Q(x, y)\,dy = 0$ is a function such that

$$M(x, y)P(x, y)\,dx + M(x, y)Q(x, y)\,dy = 0$$

is exact in some region.

(a) Show that a linear equation of the form

$$\frac{dy}{dx} + f(x)y = 0,$$

with $f(x)$ continuous on an interval $a < x < b$, can be written

$$f(x)y\,dx + dy = 0.$$

(b) Show that multiplying by $M(x) = e^{\int f(x)\,dx}$ produces an equation that satisfies the consistency condition for exactness.

(c) Use the result of part (b) to derive the general solution to the given first-order equation

$$y = Ce^{-\int f(x)\,dx}.$$

(d) Show that starting with the prior assumption that the solution is either always zero or else never zero, the result of part (c) can be obtained by first separating variables.

5. There is no universally applicable method for finding an integrating factor (IF), though it can be proved that one always exists if the given equation satisfies certain rather mild conditions. Here are some examples for which such a factor is provided. Solve the equation using the given factor.

(a) Show that $M(y) = y^{-2}$ is an IF for the equation $(y + y^2)\,dx - x\,dy = 0$.

(b) Show that $M(y) = e^{-y}$ is an IF for the equation

$$(1 + x)e^x\,dx + (ye^y - xe^x)\,dy = 0.$$

(c) Show that $M(x) = 2x$ is an IF for the equation

$$(2x^2 + 3xy - y^2)\,dx + (x^2 - xy)\,dy = 0.$$

(d) Show that $M(x, y) = 1/(x^2 + y^2)$ is an IF for the equation

$$y\,dx - (x^2 + x + y^2)\,dy = 0.$$

6. The differential equation $g(y)\,dy/dx = h(x)$ is of the variables-separable type, with implicit solution $G(y) = H(x) + C$, where $G'(y) = g(y)$ and $H'(x) = h(x)$.

(a) Show that this differential equation is also exact if written in differential form.

(b) Use the method for solution of exact equations by integration to solve the equation in part (a) in terms of $G(y)$ and $H(x)$. Compare the result with the solution by the variables-separable method.

Exact Equations Using Differential Formulas. Use Equations 4.2, or simple modifications, to identify solutions to the following. In some cases you can find solutions by separating variables, though that may be more work.

7. $y\,dx + (x + y)\,dy = 0$.

8. $y\,dx - (x + y^2)\,dy = 0$.

9. $y\sin x\,dx - \cos x\,dy = 0$.

10. $y^2\,dx + 2xy\,dy = 0$.

11. $y^2\,dx - 2xy\,dy = 0$.

12. $y^2\cos x\,dx + 2y\sin x\,dy = 0$.

13. $2xy^3\,dx + 3x^2y^2\,dy = 0$.

14. $y^2\,dx + 2(xy - y)\,dy = 0$.

Finding One-Variable Integrating Factors. It's possible to compute integrating factors for $P(x, y)\,dx + Q(x, y)\,dy = 0$ in the following two special cases.

(i) Assume $(P_y(x, y) - Q_x(x, y))/Q(x, y) = R(x)$ is independent of y. Then
$$M(x) = e^{\int R(x)\,dx} \text{ is an integrating factor.}$$

(ii) Assume $(Q_x(x, y) - P_y(x, y))/P(x, y) = S(y)$ is independent of x. Then
$$M(y) = e^{\int S(y)\,dy} \text{ is an integrating factor.}$$

Note that if the consistency condition $Q_x = P_y$ holds, then M will simply be a positive constant. Find an integrating factor for each of the following, and use it to solve the equation. Then check that your solution does satisfy the equation.

15. $y\,dx - x\,dy = 0$. [*Hint:* Both $R(x) = -2/x$ and $S(y) = -2/y$ will work.]

16. $(y^2 + 3x)\,dx + xy\,dy = 0$. [*Hint:* $R(x) = 1/x$.]

17. $2xy\,dx + (y^2 - x^2)\,dy = 0$. [*Hint:* $S(y) = -2/y$.]

18. $(y^2 + 4ye^x)\,dx + 2(y + e^x)\,dy = 0$. [*Hint:* $R(x) = 1$.]

19. $\sin y\,dx + x\cos y\,dy = 0$.

20. The one-variable integrating factors described in the preamble to Exercises 15 through 19 will typically produce an exact equation.

(a) Show this for case (i) if $Q(x, y) \neq 0$.

(b) Show this for case (ii) if $P(x, y) \neq 0$.

21. Solve the nonexact equation $(x - y)\,dx + (x + y)\,dy = 0$ by using the transformation $u = y/x$ and rewriting the equation as one to be solved for $u = u(x)$. (See text Example 4 of Section 4B.)

*22. A basic result of multivariable calculus says that the gradient vector $\nabla F(x, y) = (F_x(x, y), F_y(x, y))$ is perpendicular to the level curve of F that contains (x, y). Assume continuous second-order partials for F in what follows.

(a) Show that an implicit equation of a curve perpendicular to the level curves of F, called an **orthogonal trajectory** of the level curves, satisfies $F_y(x, y)\,dx - F_x(x, y)\,dy = 0$.

(b) Show that this last equation passes the exactness test if and only if F is a **harmonic function,** which means that $F_{xx} + F_{yy} = 0$ for all (x, y).

(c) Show that an implicit equation for a level curve of F always satisfies the exact equation $F_x(x, y)\,dx + F_y(x, y)\,dy = 0$.

*23. The assumption that R is rectangular in the statement of the exactness test is used only for showing the sufficiency of the condition. It is used there just to guarantee that an integration with respect to either x or y can proceed unimpeded as we go from point to point in the region R in which the condition $Q_x = P_y$ holds.

(a) Give an example of a region R that is not rectangular but such that the proof in the text will still work.

(b) Use Green's theorem to prove the sufficiency of the exactness test under the assumption that the region R is simply connected.

5 LINEAR EQUATIONS

The first-order differential equations we have looked at so far can nearly all be written in the form

$$\frac{dy}{dx} = F(x, y),$$

where F is some fairly simple function of x and y. If we make the requirement that F have the form $F(x, y) = -g(x)y + f(x)$, then we get what is called a **linear equation** in **normal form** with dy/dx having coefficient 1. It will be helpful below to write it as

▬ **5.1** $$\frac{dy}{dx} + g(x)y = f(x).$$

Linear equations are general enough that they have many interesting applications; all the same, they are special enough that we can say significant things about their solutions without knowing much in the way of details about $g(x)$ and $f(x)$. The name *linear* is used because the equation is of first degree in its dependence on dy/dx and y.

---

e x a m p l e 1

If f happens to be identically zero, we can find solutions $y = y(x)$ to Equation 5.1 by assuming $y \neq 0$ and rewriting the equation as

$$\frac{y'}{y} = -g(x).$$

Integrating with respect to x, we get

$$\ln|y| = -G(x) + c,$$

where G is an indefinite integral of g and c is a constant. Taking the exponential of both sides gives

$$|y| = e^c e^{-G(x)}.$$

Then removing the absolute value allows us to replace the positive constant e^c by an arbitrary nonzero constant K:

$$y = Ke^{-G(x)}$$
$$= Ke^{-\int g(x)dx}.$$

5A   *Exponential Integrating Factor*

The method of solution used in Example 1 fails if the function $f(x)$ in Equation 5.1 is not zero; it also has the technical defect that it forces us to assume $y \neq 0$. (Conceivably there are solutions that take on the value zero; in fact $y(x) = 0$ is one such.) Both objections can be avoided at once if we use the following method, suggested by the form of the solution found in Example 1. For the differential equation

$$y' + g(x)y = f(x),$$

written in normal form, we define an **exponential integrating factor** or **exponential multiplier** to be

$$M(x) = e^{\int g(x)dx},$$

where $\int g(x)\, dx$ is some indefinite integral of g. The trick is to multiply the differential equation by M to get

$$e^{\int g(x)dx}y' + g(x)e^{\int g(x)dx}y = f(x)e^{\int g(x)dx}.$$

The whole point is that the left side can now be written as the derivative of $e^{\int g(x)dx}y$, because, by the product rule, applied to the factors $e^{\int g(x)dx}$ and y,

$$\frac{d}{dx}[e^{\int g(x)dx}y] = e^{\int g(x)dx}y' + g(x)e^{\int g(x)dx}y.$$

Thus we have rewritten the normal-form linear differential equation in the form

$$\frac{d}{dx}[e^{\int g(x)dx}y] = e^{\int g(x)dx}f(x);$$

it remains only to integrate both sides with respect to x and then to solve for y. A more general class of integrating factor was taken up in Section 4, but these are not needed for the present discussion.

---

e x a m p l e 2

To find all solutions of the linear differential equation

$$y' = xy + x,$$

we first rewrite the equation in the normal form

$$y' - xy = x.$$

The exponential multiplier is then found by identifying the coefficient function $g(x) = -x$ and

$$\begin{aligned} M(x) &= e^{\int g(x)dx} \\ &= e^{-\int xdx} \\ &= e^{-(1/2)x^2}. \end{aligned}$$

Multiplying the differential equation by M gives

$$e^{-(1/2)x^2}y' - xe^{-(1/2)x^2}y = xe^{-(1/2)x^2}.$$

But we know from the preceding discussion, or we could verify directly, that this last equation is the same as

$$\frac{d}{dx}[e^{-(1/2)x^2}y] = xe^{-(1/2)x^2}.$$

Integrating both sides with respect to x gives

$$e^{-(1/2)x^2}y = \int xe^{-(1/2)x^2}\,dx + C$$

$$= -e^{-(1/2)x^2} + C.$$

Then multiplying by $e^{+(1/2)x^2}$ gives

$$y = -1 + Ce^{(1/2)x^2}$$

for the solution.

Three points should be emphasized about the exponential multiplier method:

1. The linear differential equation should be in the normal form

$$y' + g(x)y = f(x)$$

before identification of the coefficient function g for the purpose of computing $M(x) = e^{\int g(x)dx}$.

2. The differential equation

$$y' + g(x)y = f(x)$$

and its multiplied form

$$M(x)y' + M(x)g(x)y = M(x)f(x)$$

are completely equivalent to one another in the sense that any solution of one equation is also a solution of the other. The reason is that the multiplier M, being an exponential function, is never equal to zero. Hence we can multiply and divide by it as we please without altering the validity of an equation.

3. It is an easy matter to check that the method provides a unique solution to the initial-value problem $y' + g(x)y = f(x)$ with $y(x_0) = y_0$ on an interval on which $f(x)$ and $g(x)$ are both continuous:

$$y(x) = \frac{M(x_0)}{M(x)}\,y(x_0) + \frac{1}{M(x)}\int_{x_0}^{x} M(u)f(u)du.$$

Note that the choice of a constant of integration in the exponent of $M(x)$ doesn't spoil the uniqueness, because

$$M(x) = e^{\int g(x)dx + C} = e^{C}e^{\int g(x)dx}.$$

The result is that the nonzero constant factor e^{C} will always cancel out, since only ratios of the function M occur in the expression for $y(x)$.

EXERCISES

1. Find an exponential multiplier $M(x)$ for each expression such that $M(x)$ times that expression can be written in the form $d(M(x)y)/dx$.

(a) $y' + 2y$.

(b) $dy/dx + y$.

(c) $dy/dx + (2/x)y$.

(d) $y' + e^x y$.

(e) $e^x y' + e^{-x}y$.

(f) y'.

2. Find the general solution of each of the following first-order equations. Then find the particular solution that satisfies the given initial condition.

   (a) $ds/dt + ts = t$, $s(0) = 0$.

   (b) $y' = y + 1$, $y(0) = 1$.

   (c) $2dy/dx = xy$, $y(1) = 0$.

   (d) $t\,dP/dt + P = t^3$, $P(1) = 0$.

   (e) $dx/dy + x = 1$, $x(1) = 1$.

   (f) $dx/dt = x + t$, $x(0) = 0$.

   (g) $y' + y = e^x$, $y(0) = 0$.

   (h) $y' + \cos x = 0$, $y(0) = 2$.

3. Find the general solution of the differential equation

   $$y' + f(x)y = x,$$

   where

   $$f(x) = \begin{cases} 1, & 0 \le x < 1, \\ 1/x, & 1 \le x. \end{cases}$$

   Try

   $$M(x) = \begin{cases} e^x, & 0 \le x < 1, \\ x, & 1 \le x. \end{cases}$$

4. The exponential integrating factor $M(x)$ is designed to make the replacement

   $$M(x)y' + M(x)g(x)y = \frac{d}{dx}(M(x)y)$$

   valid. Expand the expression on the right, and solve the resulting differential equation for M by separating variables to derive the formula for the integrating factor.

Special Second-Order Equations. The following differential equations of order two can be solved by regarding them as first-order linear equations to be solved for $z = y'$ and then integrating z to find y. Find the most general solution to each one by first making the appropriate substitution for y' and y'' and then integrating once to find $y(x)$.

   5. $y'' + y' = 1$.

   6. $y'' + (1/x)y' = 1$.

   7. $xy'' = 1$.

   8. $y'' + xy' = x$.

Special Third-Order Equations. The following differential equations of order 3 can be solved by regarding them as first-order linear equations to be solved for $w = y''$ and then integrating w twice to find y. Find the most general solution to each one by first making the appropriate substitution for y'' and y''' and then integrating twice to find $y(x)$.

   9. $y''' + y'' = 0$.

   10. $y''' + (1/x)y'' = x$.

   11. $y''' = 0$.

   12. $2y''' + (3/x)y'' = 0$, $x \ne 0$.

13. Show that the linear differential equation $dy/dx + g(x)y = f(x)$ with $f(x)$ and $g(x)$ continuous on an interval satisfies the hypotheses of the general existence and uniqueness theorem of Section 2C for $dy/dx = F(x, y)$, namely, that both $F(x, y)$ and its partial derivative $\partial F(x, y)/\partial y$ are continuous in a rectangle $R = \{(x, y): a < x < b, c < y < d\}$.

5B   Linear Models

The linear equations are rather special among the first-order equations for a number of reasons. One is that there are solution formulas for them that display the properties of the solutions in a way that makes them fairly easy to interpret. Another is that there are many processes that can be modeled very well by linear equations. In addition, what happens with a linear application can often, though not always, be used as a good indication of what will happen under somewhat more complicated assumptions. Our first illustration of this point has to do with what is called a **mixing problem:** a container holding fluid with an amount S of some substance (salt will do as well as anything for our purposes) in solution undergoes enrichment or dilution by addition and/or subtraction from S at some known rates. Thus the analysis reduces to the simple relation

$$\frac{dS}{dt} = \{\text{inflow rate for } S\} - \{\text{outflow rate for } S\}.$$

Note that for consistency the rates on the right side must be stated in terms of quantities of dissolved material rather than of volumes of solution. Figure 1.13 is a schematic diagram showing three possible ways for the value of S to be influenced. To make our model linear we usually need to assume that the solution is kept thoroughly mixed at all times so that the concentration C of dissolved material per unit of volume V is kept homogeneous throughout the container. Under that assumption it's often convenient to deal with concentrations of the substance in question. For example, in a full container of capacity V, the concentration will be $C = S/V$. It follows that if the outward flow rate of solution is R_O, then

$$\{\text{Outflow rate for } S\} = R_O C = R_O \frac{S}{V}.$$

Note that the total volume V of solution in a compartment may increase or decrease depending on relative rates of inflow, outflow, and evaporation. Evaporation, by changing V, affects the outflow rate for S.

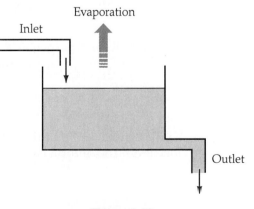

Figure 1.13

e x a m p l e 1

Suppose that a 100-gallon vat contains 10 pounds of salt dissolved in water and that a solution of the same salt is being run into the vat at a rate of 3 gallons per minute. The solution being run into the vat has a concentration in pounds per gallon that increases slowly with time according to the formula

$$C(t) = 1 - e^{-t/100}.$$

The solution is kept thoroughly mixed and the excess drawn off, also at a rate of 3 gallons per minute. Let $S(t)$ stand for the pounds of salt in the tank at time $t \geq 0$. We have

$$\frac{dS}{dt}(t) = 3C(t) - \frac{3S(t)}{100}$$

$$= 3(1 - e^{-t/100}) - \frac{3}{100}S(t).$$

The resulting first-order equation is linear, and in normal form it is

$$\frac{dS}{dt} + \frac{3}{100}S = 3(1 - e^{-t/100}).$$

An exponential multiplier is given by

$$M(t) = e^{\int(3/100)dt} = e^{3t/100}.$$

Multiplying the equation by M puts it in the form

$$\frac{d}{dt}(e^{3t/100}S) = 3(e^{3t/100} - e^{2t/100}),$$

and integration with respect to t gives

$$e^{3t/100}S(t) = \int 3(e^{3t/100} - e^{2t/100})\, dt + K$$

$$= 100e^{3t/100} - 150e^{2t/100} + K.$$

Then multiplication by $e^{-3t/100}$ gives

$$S(t) = 100 - 150e^{-t/100} + Ke^{-3t/100}.$$

To determine the constant K, we recall that the vat initially contains 10 pounds of salt, so $S(0) = 10$. Then setting $t = 0$ in the formula for $S(t)$ gives

$$10 = -50 + K \qquad \text{or} \qquad K = 60.$$

Thus the desired particular solution is

$$S(t) = 100 - 150e^{-t/100} + 60e^{-3t/100}.$$

Notice that

$$\lim_{t \to \infty} C(t) = 1,$$

so the concentration of the solution being added approaches 1 pound per gallon. From this information we could conclude on physical grounds that the total amount of salt in the 100-gallon tank should approach 100 pounds; indeed, the formula for $S(t)$ shows that

$$\lim_{t \to \infty} S(t) = 100.$$

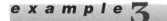

e x a m p l e 2

Consider the following modification of Example 1: Everything remains the same as before except that now the mixed solution is being drawn off at the rate of 4 gallons per minute and the salt concentration of the added solution is held constant at 1 pound per gallon. Since fresh solution is still being added at the rate of 3 gallons per minute, the tank is being emptied at 1 gallon per minute, with the result that t minutes after the process starts, the tank will contain $100 - t$ gallons of solution. In particular, after 100 minutes the tank will be empty, and the process will have effectively ended. Therefore, we'll consider the following differential equation on the time interval $0 \le t \le 100$:

$$\frac{dS}{dt}(t) = 3 - 4\frac{S(t)}{100 - t}.$$

The ratio $S(t)/(100 - t)$ is the concentration in pounds per gallon of the salt solution at time t; this varies both because $S(t)$ is changing and because the total amount of solution, $100 - t$, is decreasing. The differential equation is solved by rewriting it in the form from which we can calculate an integrating factor:

$$\frac{dS}{dt} + \frac{4}{100 - t}S = 3, \qquad M(t) = e^{\int 4/(100-t)\,dt} = e^{-4\ln(100-t)} = (100 - t)^{-4}.$$

Thus we only have to integrate both sides of $d((100 - t)^{-4}S)/dt = 3(100 - t)^{-4}$ to get

$$(100 - t)^{-4}S = (100 - t)^{-3} + K, \qquad \text{or} \qquad S = (100 - t) + K(100 - t)^4.$$

The constant K can be determined from the initial value $S(0) = 10$ as in Example 1; it now turns out that $K = -9 \cdot 10^{-7}$. From the general formula for $S(t)$ we can calculate, for example, the time at which the tank contains the maximum amount of salt, or focus on the behavior of the salt concentration $C(t) = S(t)/(100 - t) = 1 + K(100 - t)^3$ in pounds per gallon.

The next example also has to do with varying fluid concentrations, but in this case we're dealing with an empirically determined law for arriving at the relevant differential equation for **membrane transport,** that is, for passage of a chemical in solution through a porous membrane. Another feature of this example is that the controlled chemical concentration on one side of the membrane undergoes an abrupt change of form; this requires keeping careful track of the integration variable.

e x a m p l e 3

Let $f(t)$ be the concentration of a chemical solution on one side of a porous membrane, and let $u(t)$ be the concentration on the other side. Suppose that diffusion takes place through the membrane in such a way that

$$\frac{du}{dt} = 2(f(t) - u),$$

that is, such that the rate of change of u is proportional to the difference in concentrations. If $u(0) = 3$, and $f(t)$ is maintained so that

$$f(t) = \begin{cases} 4, & 0 \le t \le 10, \\ 1, & 10 < t, \end{cases}$$

then we can most easily solve the equation by writing it as

$$\frac{du}{dt} + 2u = 2f(t).$$

An exponential multiplier M is given by

$$M(t) = e^{\int 2 dt} = e^{2t}.$$

Hence the differential equation can be written

$$\frac{d}{dt}(e^{2t}u) = 2e^{2t}f(t).$$

Integration of both sides from $t = 0$ to $t = s$ gives

$$\int_0^s \frac{d}{dt}[e^{2t}u(t)]\, dt = \int_0^s 2e^{2t}f(t)\, dt$$

or $e^{2s}u(s) - u(0) = \displaystyle\int_0^s 2e^{2t}f(t)\, dt.$ Then

$$u(s) = u(0)e^{-2s} + 2e^{-2s}\int_0^s e^{2t}f(t)\, dt.$$

Using the integral with limits is convenient here because we can write, according to the definition of f,

$$\int_0^s e^{2t}f(t)\, dt = \begin{cases} \int_0^s 4e^{2t}\, dt, & 0 \le s \le 10, \\ \int_0^{10} 4e^{2t}\, dt + \int_{10}^s e^{2t}\, dt, & 10 \le s, \end{cases}$$

$$= \begin{cases} 2(e^{2s} - 1), & 0 \le s \le 10, \\ 2(e^{20} - 1) + \frac{1}{2}(e^{2s} - e^{20}), & 10 \le s. \end{cases}$$

$$= \begin{cases} 2(e^s - 1), & 0 \le s \le 10, \\ \frac{3}{2}e^{20} - 2 + \frac{1}{2}e^{2s}, & 10 \le s. \end{cases}$$

Then

$$u(s) = 3e^{-2s} + \begin{cases} 4 - 4e^{-2s} & 0 \le s \le 10, \\ (3e^{20} - 4)e^{-2s} + 1, & 10 \le s. \end{cases}$$

$$= \begin{cases} 4 - e^{-2s}, & 0 \le s \le 10, \\ (3e^{20} - 1)e^{-2s} + 1, & 10 \le s. \end{cases}$$

We can now return to the original independent variable t. Sketching the graph of the two-piece solution $u(t)$ is left as an exercise. Note that because of the abrupt change in concentration at $t = 10$, the solution $u(t)$ fails to be differentiable at that point. Since $u(t)$ is nevertheless continuous at $t = 10$, we have some confidence that our solution is a good representation of the physical process.

Experience tells us that hot or cold objects cool down or warm up to the temperature of their surroundings. A useful way to quantify these observations is called **Newton's law of cooling,** which asserts that the temperature $u(t)$

of an object has a rate of change proportional to the difference between $u(t)$ and the **ambient temperature** $a(t)$ of the surroundings:

$$\frac{du}{dt} = -k(u - a), \qquad k = \text{const}, k > 0.$$

Note that if $u > a$, the right-hand side of the equation is negative, so u will decrease, and actual cooling takes place. If $u < a$, then u will increase, and what we have is really a *law of heating*. (The constant k varies with the length of the time units employed, so, for example, in changing from hours to minutes, k would be divided by 60.)

---

e x a m p l e 4

Suppose a beaker of water initially at 50°C is placed in a bath of water whose temperature $a(t)$ at time t hours gradually warms from 40 to 50° for $t \geq 0$ according to $a(t) = 50 - 10e^{-t}$. With $k = \frac{1}{2}$, the temperature $u(t)$ of the beaker satisfies the equation

$$\frac{du}{dt} = -\tfrac{1}{2}[u - (50 - 10e^{-t})] \qquad \text{or} \qquad \frac{du}{dt} + \tfrac{1}{2}u = 25 - 5e^{-t}.$$

Multiplication by the exponential factor $M(t) = e^{t/2}$ gives

$$\frac{d}{dt}(e^{t/2}u) = 25e^{t/2} - 5e^{-t/2}.$$

Integration with respect to t gives

$$e^{t/2}u = 50e^{t/2} + 10e^{-t/2} + C \qquad \text{or} \qquad u = 50 + 10e^{-t} + Ce^{-t/2}.$$

At $t = 0$ we have $u(0) = 50$, so we find $C = -10$ and $u(t) = 50 + 10e^{-t} - 10e^{-t/2}$. Using this formula you can estimate the time it would take for the beaker to cool to its minimum temperature.

EXERCISES

1. A 100-gallon tank full of salt solution contains initially 10 pounds of salt in solution. At $t = 0$, pure water is added at a rate of 2 gallons per minute, with a resulting overflow of 2 gallons per minute of salt solution. Assuming that the solution is kept thoroughly mixed at all times, find the amount of salt in the tank at time $t > 0$.

2. Salt solution enters a 100-gallon tank of initially pure water from two different sources. One source provides water containing 1 pound of salt per gallon at a rate of 2 gallons per minute. A second source provides 3 gallons of salt solution per minute at a varying concentration $C(t) = 2e^{-2t}$, measured in pounds of salt per gallon. Assume that the contents of the tank are kept thoroughly mixed at all times and that the solution is drawn off at a rate of 5 gallons per minute. Find the amount of salt in the tank at time $t > 0$.

3. The current $I(t)$ in an electric circuit satisfies the differential equation

$$L\frac{dI}{dt} + RI = E(t),$$

where L and R are positive constants representing inductance and resistance, respectively, and $E(t)$ is a variable applied voltage. Show that

$$I(t) = \frac{1}{L} e^{-Rt/L} \int_0^t E(u) e^{Ru/L} \, du + I(0) e^{-Rt/L}.$$

4. A pellet of mass m falling under the influence of gravity through a resisting medium has velocity $v(t)$ at time t, satisfying

$$m \frac{dv}{dt} = mg - kv.$$

Here g is the constant acceleration of gravity and k is a positive constant that measures the resistance of the medium. Show that

$$v(t) = \left(v(0) - \frac{mg}{k} \right) e^{-kt/m} + \frac{mg}{k}.$$

5. Sketch the graph of the function $u(t)$ found at the conclusion of Example 4 in the text. What is the minimum value of $u(t)$, and when does it occur? What is $\lim_{t \to \infty} u(t)$?

6. A full 100 cubic foot tank initially has 10 pounds of salt in solution. At a certain time more salt solution begins to enter the tank at a rate of 1 cubic foot per hour, while thoroughly mixed solution runs out a drain at the same rate. However, the amount of salt in the added solution decreases at a constant rate from 1 pound per cubic foot initially all the way down to zero pound per cubic foot at the end of 1 hour.

   (a) Find the amount of salt in the tank at a given time during the first hour. In particular, about how much salt will be in the tank at the end of 1 hour?

   (b) Estimate the time at which the amount of salt in the tank is at a maximum during the first hour.

   (c) If pure water continues to run into the tank after the first hour at the rate of 1 cubic foot per hour, how much more time will it take for the total amount of salt in the tank to reach 5 pounds?

7. Two 100-gallon mixing tanks are initially full of pure water. A solution containing 1 pound of salt per gallon of water pours into the first tank at the rate of 1 gallon per minute. Thoroughly mixed solution runs from the first tank to the second at the rate of 1 gallon per minute, where it too is thoroughly mixed in before draining away at 1 gallon per minute. Obviously there will always be at least as much salt in the first tank as in the second; find the maximum amount of this excess.

8. Two 100-gallon tanks X and Y are initially full of pure water. Salt solution is added to X at 1 gallon per minute (gpm) from an external source, each gallon containing 1 pound of salt. Mixed solution is pumped from X to Y at 2 gpm and runs from Y into a waste drain at 4 gpm. Let $x = x(t)$ and $y = y(t)$ be the respective amounts of salt in X and Y at time $t \geq 0$.

   (a) At what times t_X and t_Y will the respective tanks first become empty?

   (b) Find a pair of differential equations satisfied by $x(t)$ and $y(t)$ for $0 \leq t \leq t_X$ and $0 \leq t \leq t_Y$, respectively.

(c) Find a formula for $x(t)$ alone.

(d) Use the result of part (c) to find $y(t)$.

9. A 100-gallon tank is initially full of pure water. Salt solution is added for 10 minutes at the rate of 1 gallon per minute with salt content of the added solution increasing linearly over the 10 minutes from 1 pound per gallon to 2 pounds per gallon. Thoroughly mixed salt solution is drawn off at the rate of 1 gallon per minute. Estimate the amount of salt in the tank at the end of the 10 minutes.

10. A 100-gallon mixing vat is initially half-full of pure water. Two gallons of salt solution per minute at a concentration of 1 pound of salt per gallon begin to flow in, while 1 gallon per minute of mixed solution flows out. Estimate the amount of salt in the vat at the moment it begins to overflow.

11. A 100-gallon tank initially contains 50 gallons of water with a total of 10 pounds of salt dissolved in it. A drain is opened in the bottom that is regulated so as to let out 1 gallon of solution per minute. Simultaneously, salt solution begins to be added at 2 gallons per minute with a concentration of 2 pounds per gallon.

    (a) How much salt is in the tank when it first becomes full and starts to overflow?

    (b) If the process is allowed to continue with overflow at an additional outflow of 1 gallon per minute, what is the upper limit for the total amount of salt in the tank? Estimate the additional time after the start of overflow for the amount of salt in the tank to reach 175 pounds.

12. A 100-gallon mixing vat is initially full of pure water. One gallon of water begins to flow into the tank per minute, with a pound of salt in each gallon. Simultaneously 2 gallons of mixed solution per minute are pumped out. Find the maximum amount of salt in the vat, and show that the maximum occurs when the vat is half-full of solution.

13. A 100-gallon mixing vat is initially full of pure water. One pound per minute of salt dissolved in 1 gallon of water flows into the tank, 1 gallon per minute of mixed solution flows out, and 1 gallon of water per minute evaporates from the vat. Estimate the amount of salt in the vat when it is half-empty.

14. A chemical in solution diffuses across a membrane with diffusion coefficient k from a compartment with known concentration $f(t)$ to a compartment with resulting concentration $u(t)$. Solve the diffusion equation

$$\frac{du}{dt} = k(f(t) - u)$$

under each of the following sets of assumptions. In each case find the smallest upper limit b not exceeded by the concentration $u(t)$ for $t \geq 0$.

    (a) $u(0) = 0$, $f(t) = 10$, $k = 2$.

    (b) $u(0) = 0$, $f(t) = 10e^{-t}$, $k = 2$.

    (c) $u(0) = 10$, $f(t) = 10$ if $0 \leq t \leq 5$, $f(t) = 20$ if $5 < t$, $k = 2$.

(d) $u(0) = 0$, $f(t) = 10(1 - e^{-t})$, $k = 2$.

(e) $u(0) = 5$, $f(t) = 10$, $k(t) = (1 + t)^{-1}$.

15. Find the general form of a solution $u(t)$ of Newton's cooling equation $du/dt = -k(u - a)$ under each of the following sets of assumptions. Also, use an appropriate scale in each case to sketch the graphs of $u = u(t)$ and $u = a(t)$ relative to the same pair of axes.

    (a) $u(0) = 100$, $a(t) = 50$, $k = 1$.

    (b) $u(0) = 100$, $a(t) = 50e^{-t}$, $k = 1$.

    (c) $u(0) = 100$, $a(t) = 50e^{-t}$, $k = 2$.

    (d) $u(0) = 0$, $a(t) = 50 \sin t$, $k = 1$.

    (e) $u(0) = 100$, $a(t) = 200 + 50e^{-2t}$, $k = 1$.

16. In Example 3 of the text the solution $u(t) = 50 + 10e^{-t} + 40e^{-t/2}$ is derived for a cooling process from 100° down to 50°. Estimate the value t for which $u(t) = 75$. [*Hint:* The relevant equation can be regarded as a quadratic equation to be solved for $e^{-t/2}$.]

17. A container of ice cream mix at 70°F is placed in a mixture of ice and brine at 30°F. Assume the validity of Newton's law described prior to Example 4 and that the ice cream has reached 40° after 15 minutes.

    (a) Find an approximate value for the constant k in Newton's law.

    (b) When will the ice cream mix reach 35°?

18. Suppose that an iron bar initially at 300°F is immersed in a water bath at 100°F for 30 minutes and then is transferred to another water bath at 50°F. Assume the validity of Newton's law described before Example 4 of the text.

    (a) What will the temperature of the bar be after an additional 30 minutes, assuming the cooling coefficient for the iron in water is $k = 0.1$?

    (b) Suppose that initially the iron is cooled for 30 minutes in air at 100°, for which the cooling coefficient is only $k = 0.07$, and is then immersed in water for 30 minutes. What will the temperature of the bar be at the end of the hour?

19. If a beaker containing s units of liquid at initial temperature u_0 has an additional unit at temperature $u_1 < u_0$ added and mixed in right away, the temperature of the mixture averages out to $(su_0 + u_1)/(s + 1)$. Assume the surrounding temperature is held at a constant $a < u_1$ and that Newton's law is valid as described before text Example 4.

    (a) Find the resulting temperature $u(t_1)$ of the mixture if the initial s units at temperature u_0 are first allowed to cool over time period t_1 upon which a single unit at temperature $u_1 < u_0$ is mixed in.

    (b) Find the temperature $v(t_1)$ of the mixture if initially one unit at temperature u_1 is mixed in and then the mixture is allowed to cool over time period t_1.

    (c) Which of the routines described in parts (a) and (b) produces the cooler mixture? Does your intuition, informed by Newton's law, lead you to the same answer?

20. A cup of water is initially at 180° in a 70° room. If the water cools to 160° in 10 minutes, estimate the constant k in Newton's law of cooling, and find the approximate temperature of the water after another 10 minutes.

21. A rubber band initially of length l has one end attached to a wall while the band is stretched uniformly over its entire length away from the wall by pulling the other end away from the wall at constant rate r. Note that particles of the band nearer to the fixed end move proportionately more slowly; in particular, the fixed end doesn't move at all and the other end moves at rate r.

(a) Find a differential equation satisfied by $x = x(t)$, the distance from the wall of a given particle of the band.

(b) After t time units of stretching, what is the distance $x(t)$ from the wall of a particle of rubber that is initially at distance x_0 from the wall?

22. (a) One end of a 1-foot rubber band is attached to a wall while the other end is stretched away from the wall at 1 foot per minute. Starting at the same time as the stretching, a small bug starts to crawl from the wall along the band at the constant rate of b feet per minute relative to the band. Will the bug ever reach the other end of the band, and if so, how will its arrival time depend on b? Assume that the stretching is uniform throughout the length of the band.

(b) Solve the same problem under the assumptions that the band is initially l feet long, is being stretched at r feet per minute, and the bug starts out already x_0 feet from the wall.

5C   Linearization (Optional)

Given that there is a formula for the general solution of a first-order linear equation, and that the solutions have in general fairly simple properties, it is important to consider the question whether some nonlinear equations can for some purposes be effectively replaced by linear equations. This replacement is called *linearization,* and the resulting linear equation is called a *linearized* version of the nonlinear equation. To be reasonably representative of the nonlinear equation $y' = f(x, y)$, a linearization needs to be restricted in its application to some interval of y values containing a value y_0 of particular interest. Thus, for each x such that $f(x, y_0)$ is differentiable as a function of y, make the first-degree **Taylor approximation**

$$f(x, y) \approx f(x, y_0) + f_y(x, y_0)(y - y_0),$$

where f_y is the derivative of f with respect to y, holding x fixed; this approximation is geometrically equivalent to replacing the graph of f, regarded as a function of y, by its tangent line at $y = y_0$. We define the **linearization** of $y' = f(x, y)$ near y_0 to be the differential equation

▬ **5.2** $y' = f(x, y_0) + f_y(x, y_0)(y - y_0).$

Even if a nonlinear equation can be "solved" by an elementary formula, it may be impossible to derive a usable expression for the solution in terms of the independent variable. In that case linearization can often provide an approximate solution formula that is accurate enough for some purposes.

example 1

In the nonlinear equation $y' = xy^2$ the function f of Equation 5.2 is $f(x, y) = xy^2$, so $f_y(x, y) = 2xy$. At $y_0 = 3$ we have $f(x, 3) = x \cdot 3^2 = 9x$ and $f_y(x, 3) = 2x \cdot 3 = 6x$. Near $y_0 = 3$ the nonlinear equation gets replaced by the linear equation $y' = 9x + 6x(y - 3)$ or $y' = 6xy - 9x$. The general solution is $y = \frac{3}{2} + Ce^{3x^2}$.

example 2

For the nonlinear equation $y' = f(x, y) = \tan y$ we have $f(x, 0) = 0$ and $f_y(x, 0) = \sec^2 0 = 1$. Thus the linearized Equation 5.2 at $y = 0$ becomes simply $y' = y$, with general solution $y(x) = Ce^x$. Near $y = 0$, we can impose an initial condition, say $y(0) = 0.1$, to get the particular solution $y(x) = 0.1e^x$. The given nonlinear equation can be solved by separation of variables and has the family of solutions $\ln |\sin y| = x + c$. The particular solution of this family satisfying the initial condition $y(0) = 0.1$ can be written $y(x) = \arcsin (e^x \sin (0.1))$. This last formula is "elementary" and is after all the exact solution; on the other hand, it is a bit cumbersome as compared with $y(x) = 0.1e^x$. Figure 1.14 shows that the two graphs are hard to tell apart up to $x = 1$. (The difference in values remains less than 0.003 up to that point.) By the time we get to $x = 2$, the discrepancy has increased to a little less than $\frac{1}{10}$. By comparing the differential equations, you can deduce which graph has the larger values; simply note that $\tan y > y$ for $0 < y < \pi/2$.

Bernoulli Equations. The limited range of applicability of a linearized equation sometimes makes it desirable to solve the original nonlinear equation directly if at all possible. A nonlinear equation that can be transformed directly into a linear equation is the **Bernoulli equation**

$$y' + g(x)y = g(x)y^n.$$

The parameter n can be any real number, and if $n = 0$ or $n = 1$, the equation is already linear, so we can ignore those values in applying the following method. The trick is first to multiply the equation by $B(y) = (1 - n)y^{-n}$. It then turns out that letting $u = y^{1-n}$ yields the linear equation

▬ 5.3
$$u' + (1 - n)g(x)u = (1 - n)f(x).$$

If you can solve this equation explicitly using the usual exponential multiplier $M(x)$, all that remains to be done is to set $y = u^{1/(1-n)}$ to find the solution to the Bernoulli equation.

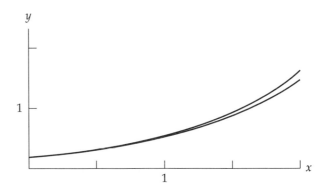

Figure 1.14 Solutions of $y' = \tan y$ and $y' = y$ with $y(0) = 0.1$.

**e x a m p l e 3**   The Bernoulli equation

$$\frac{dy}{dx} + \frac{1}{x}y = \frac{1}{y}$$

fits the general form displayed above with $n = -1$. Then the factor $(1 - n)$ equals 2, so the linear equation to be solved is

$$\frac{du}{dx} + \frac{2}{x}y = 2.$$

The appropriate exponential multiplier is $M(x) = x^2$. The general solution then turns out to be $u = \frac{2}{3}x + c/x^2$. Since u and y were related by $u = y^{1-n} = y^2$ (to make the transition to a linear equation), we find $y^2 = \frac{2}{3}x + c/x^2$. The right side must be positive if y is to be real-valued, so we can write the explicit formula $y = \pm\sqrt{\frac{2}{3}x + c/x^2}$.

EXERCISES

Find the linearization of each of the following differential equations near the indicated point.

1. $y' = y^3 + y$; $y_0 = 0$.

2. $y' = x \arctan y$; $y_0 = 0$.

3. $y' = y^2 + 1$; $y_0 = 1$.

4. $y' = x \tan y$; $y_0 = \pi/4$.

5. $y' = -\sin y$; $y_0 = 0$.

6. $y' = -\sin y$; $y_0 = \pi/4$.

7. $y' = x^2 y^2$; $y_0 = -2$.

8. $y' = x/y$; $y_0 = 1$.

9. The linearization of $y' = xy^2$ near $y = 3$ is shown in Example 2 of the text to be $y' = 6xy - 9x$.

   (a) Find the solution y_1 of the above nonlinear equation that satisfies $y_1(0) = 3$.

   (b) Find the solution y_2 of the corresponding equation linearized near $y = 0$ that satisfies $y_2(0) = 3$.

   (c) Sketch the graphs of the solutions y_1 and y_2 found in parts (a) and (b); how do they compare at $x = \frac{1}{2}$?

10. The velocity $v = v(t)$ of a certain falling body subject to nonlinear velocity-dependent air resistance satisfies $dv/dt = 32 - 0.01v^2$.

   (a) Find the linearization of this differential equation near velocity $v_0 = 100$.

   (b) Solve the linearized equation found in part (a) subject to the initial condition $v(0) = 100$.

11. The equation $m\, dv/dt = mg - kv^\alpha$, $\alpha > 0$, for a falling body with viscous damping is nonlinear for $\alpha \neq 1$ and is seldom solvable for $v = v(t)$ in elementary terms. (The cases $\alpha = 1$ and $\alpha = 2$ are exceptional.) Without solving the differential equation, show that the terminal velocity is $v_\infty = \sqrt[\alpha]{mg/k}$. [Hint: What happens to dv/dt as $t \to \infty$?]

12. The falling-body equation $m\, dv/dt = mg - kv^2$ has solutions for which the terminal velocity is $v_\infty = \sqrt{gm/k}$.

   (a) Find the linearized version of the differential equation near $v = v_\infty$.

(b) Solve the linear equation found in part (a) with initial condition $v(0) = \sqrt{gm/k}$.

13. The falling-body equation $m\, dv/dt = g - kv^\alpha$, $0 < m =$ constant, has solutions for which the terminal velocity is $v_\infty = \sqrt[\alpha]{gm/k}$.

   (a) Find the linearized version of the differential equation near $v = v_\infty$.

   (b) Solve the linear equation found in part (a) under the assumption that $v(0) = v_\infty$.

14. **(a)** Find the linearization of the falling-body equation $m\, dv/dt = g - kv^2$ near $v = v_0$.

   (b) Solve the linear equation found in part (a) under the assumption that $v(0) = v_0$.

15. **(a)** Show that the linearization of the falling-body equation $m\, dv/dt = g - kv^\alpha$ near $v = v_0$ is $m\, dv/dt = gm + k(\alpha - 1)v_0^\alpha - k\alpha v_0^{\alpha-1}v$.

   (b) Show that the solution of the linear equation of part (a) under the assumption that $v(0) = v_0$ is

$$v(t) = \frac{1}{\alpha}\left(v_0 - \frac{gm}{kv_0^{\alpha-1}}\right)e^{-\alpha\,(kv_0^{\alpha-1}/m)t} + \frac{1}{\alpha}\left(\frac{gm}{kv_0^{\alpha-1}} + (\alpha - 1)v_0\right).$$

16. **Stefan's law** for the Kelvin scale temperature $u(t)$ of a body in an environment with ambient temperature a states that $du/dt = C(a^4 - u^4)$.

   (a) Show that **Newton's law of cooling,** $du/dt = k(a - u)$, can be regarded as the linearization of Stefan's law near $u = a$, with $k = 4a^3C$.

   (b) Show that the solution to Newton's equation is

$$u = a + (u_0 - a)e^{-4a^3Ct},$$

   where $u_0 = u(0)$.

   (c) Show that the nonlinear Stefan equation has solutions that satisfy

$$\ln\left|\frac{u - a}{u + a}\right| - 2\arctan\frac{u}{a} = -4a^3Ct + c,$$

   where c is an integration constant.

   (d) Discuss the problem of expressing u as a function of t using the relation in part (c), and use a direction field sketch for Stefan's equation to sketch some typical solution graphs.

17. A nonlinear model for membrane transport of a chemical solution to a compartment with concentration $u = u(t)$ is $du/dt = c(k - u)^\alpha$, where c, k, and α are positive constants.

   (a) Show that the linearization at $u = u_0$ of the nonlinear equation is $du/dt = C(K - u)$, with $C = c\alpha(k - u_0)^{\alpha-1}$ and $K = (k - u_0 + \alpha u_0)/\alpha$.

   (b) Make a graphical comparison of the solutions $u = u(t)$ to the nonlinear equation and its linearization for the case $\alpha = 2$, $c = k = 1$, and $u_0 = 0$. In particular, describe the relative behaviors of these solutions as t gets large.

18. The nonlinear differential equation

$$y' + g(x)y = f(x)y^n$$

is a **Bernoulli equation.** The purpose of this exercise is to go through the details of how a Bernoulli equation can be reduced to solving a linear equation. (When $n = 0$ or $n = 1$, the Bernoulli equation is already linear, so in what follows n is not equal to 0 or 1.)

(a) Show that if $n \neq 1$, multiplying by $M(y) = (1 - n)y^{-n}$ and letting $u(y) = y^{1-n}$ yields the linear equation $u' + (1 - n)g(x)u = (1 - n)f(x)$. (This last equation is linear and can in principle always be solved for $u(y)$. Then replace u by y^{1-n} in the solution of the linear equation.)

(b) What problem could arise with defining u by $u = y^{1-n}$ if $y \leq 0$?

19. Solve the Bernoulli equation $y' = y - y^4$ given $y(0) = \frac{1}{2}$. (Could you use separation of variables here?)

20. Solve the Bernoulli equation $x^2 y' - xy = y^2$ given $y(1) = 1$. Also find the solution for which $y(1) = 0$, and explain why it doesn't come out of the general method.

21. Solve the Bernoulli equation

$$\frac{dv}{dy} = \frac{1}{v} - \frac{v}{y + y_0},$$

given that y_0 is a positive constant and that $v(0) = 0$.

22. Assume $n \neq 0$ or 1 in the Bernoulli equation $y' + g(x)y = f(x)y^n$, and show that the linearization at $y = y_0$ is $y' + G(x)y = F(x)$, where $G(x) = (g(x) - ny_0^{n-1} f(x))$ and $F(x) = (1 - ny_0^n)f(x)$. What value for y_0 must be avoided if $n = -1$?

CHAPTER REVIEW AND SUPPLEMENT

Find all solutions that satisfy:

1. $x\, dy/dx + y - x = 0$.

2. $dy/dx = 1/(y(1 - x)^2)$.

3. $dx/dt = tx + e^t$.

4. $(1 + x)y' + y = \cos x$.

5. $y^3 y' = (y^4 + 1)e^x$.

6. $dy/dx = 4x^3 y - y$, $y(1) = 1$.

7. $xy' + (2x - 3)y = x^4$.

8. $y' = xy + y$.

9. $t\, dx/dt = -2x + t^3$, $x(2) = 1$.

10. $t\, dx/dt = 1$.

11. $dx/dt = -3x^2$.

12. $dy/dt + ty = 1$.

13. $dx/dt = (x + t)^2$. [*Hint:* Let $x + t = y$.]

14. $dy/dt = \cos^2 y$.

15. Consider the differential equation $dy/dx = e^{x-y}$.

(a) In what region of the xy plane are all solutions strictly increasing?

(b) In what region of the xy plane are all solutions concave up?

(c) Is the line $y = x$ a solution graph?

(d) Is the line $y = x$ an isocline?

(e) Solve the differential equation by separation of variables. Can you get the information asked for above directly from your solution formula? Which approach seems simpler?

16. Consider the differential equation $dy/dx = e^{x-y}$.

   (a) What conclusions can you draw from Theorem 2.1 (Section 2C) on existence and uniqueness about solutions of this equation?

   (b) Can a solution graph passing through the point $(x, y) = (0, 1)$ cross the line $y = x$? Explain your reasoning.

17. Consider the family of linear equations $y' + ay = c$, with a, c constant, $a \neq 0$.

   (a) Show that the isoclines of the direction field of this equation are horizontal lines and that every such line is an isocline.

   (b) Sketch the direction field associated with the differential equation $y' + 2y = 1$.

18. Early experiments with objects dropped from rest above the earth led to the conjecture that after an object had fallen distance s, its velocity would be proportional to s. (Under the ordinary assumption that the acceleration of gravity is constant, the velocity is proportional to $\sqrt{s}$.)

   (a) Is the early conjecture consistent with initial velocity zero? Explain your reasoning.

   (b) Is the early conjecture consistent with positive initial velocity? How would acceleration be related to s under this assumption?

19. A 100-gallon mixing vat is initially full of pure water, whereupon 2 gallons of salt solution per minute is added, each gallon containing 1 pound of salt. Water evaporates from the tank at the rate of 1 gallon per minute, and the excess solution overflows into a drain. Find the amount of salt in the tank at time t under the given assumptions and also under the altered assumption that the tank initially contains 50 pounds of salt in solution.

20. Coffee cooling. We are presented two choices for cooling one cup of coffee over a period of 10 minutes: (i) let the coffee cool by itself for 10 minutes and then add cream or (ii) add the same amount of cream right away and then allow the mixture to cool for 10 minutes. Assume that mixing quantity p of liquid at temperature T_0 and quantity q at temperature T_1 instantly results in quantity $p + q$ with average temperature $(pT_0 + qT_1)/(p + q)$.

   (a) Give an explanation for what your intuition tells you about whether method (i) or (ii) will result in cooler coffee at the end of 10 minutes.

   (b) Show that the choice of adding the cream at the end of the waiting time will result in cooler coffee provided that the initial temperature c of the cream is strictly less than the ambient temperature a of the room.

   (c) Show that the *difference* between the final temperatures attained by the two choices is independent of the initial temperature of the coffee.

21. Newton versus Stefan. Newton's law of cooling is $du/dt = n(u - a)$, where the constant n is negative. Here $u = u(t)$ is the temperature of an object in a surrounding medium at temperature a. The purpose of this exercise is to compare Newton's law with **Stefan's law,** which says that $du/dt = n(u^4 - a^4)$. (Clearly both can't be precisely correct; the facts are that there is a sound empirical and theoretical basis for the validity of Stefan's law over the observable absolute temperature range and that Newton's law provides a useful linearized version of Stefan's law under certain circumstances.)

(a) Use the factorization $u^4 - a^4 = (u^3 + au^2 + a^2u + a^3)(u - a)$ to show that

$$u^4 - a^4 = 4a^3 \frac{(u/a)^3 + (u/a)^2 + (u/a) + 1}{4} (u - a).$$

(b) Show that the fractional factor in part (a) will be as close as you like to 1 if the ratio $(u - a)/a$ is small enough.

(c) Show that Newton's law in the form $du/dt = 4na^3(u - a)$ is a good approximation to Stefan's law if $u - a$ is small enough relative to a.

(d) Derive the implicit solution $\ln |(u - a)/(u + a)| - 2 \arctan (u/a) = 4na^3t + c$ to Stefan's equation. Then explain why the corresponding solution $\ln (u - a) = 4na^3t + c$ to the linearized equation might be preferred for computational purposes.

22. **Snowplow problem.** At h_0 hours before midnight it starts snowing very heavily on bare roads, but always at a constant rate. A snowplow starts clearing previously unplowed roads at midnight, its rate of progress dx/dt at a given time being always inversely proportional to the depth of snow that has fallen up to that time, hence also inversely proportional to the total time elapsed since it began snowing. The plow has cleared 2 miles of road by 2 a.m., but the snowfall is so heavy that only one additional mile is cleared by 4 a.m.

(a) Find h_0.

(b) Continuing under the same assumptions, what is the total amount cleared by 6 a.m.?

(c) What is the constant of proportionality that expresses dx/dt in inverse proportion to elapsed time?

23. The nonlinear equation $dy/dx = |y|$ is not covered by the existence and uniqueness theorem of Section 2C, because $|y|$ fails to be differentiable as a function of y at $y = 0$. Still, the equation has a unique solution graph passing through each point (x, y) of the plane.

(a) Show that if a solution $y(x)$ of $y' = |y|$ is nonzero at x_0 (i.e., $y(x_0) = y_0 \neq 0$), then that solution is uniquely determined and is nonzero for all x.

(b) Explain how to conclude from (a) that a solution such that $f(x_0) = 0$ must in fact be identically zero.

2 CHAPTER

Applied Dynamics and Equilibrium

The applications discussed in Chapter 1 have been kept fairly simple to provide immediate applications of the solution techniques. The present chapter is organized by area of application rather than by technique for solution. The examples show that interpreting a solution can be at least as complex as deriving the solution and that interpretation has a vital role to play in the study of dynamics. In each section a segment of the real number line is regarded as a mathematical **state space** $\mathscr{S}$ that can be identified with the physical state space of some dynamical system.

Each section is independent of the others, so you can take them up in any order, covering those that seem more interesting. However, you may want at certain points to jump ahead to Section 6 on numerical methods for further applications.

1 POPULATION DYNAMICS

In Example 1 of Chapter 1, Section 3B, we let $P(t)$ be the size of a population at time t and considered the consequences of the linear growth model

$$\frac{dP}{dt} = kP,$$

where k is a positive **growth constant.** The main conclusion is that the population will grow exponentially, more specifically that

$$P(t) = P_0 e^{kt},$$

where P_0 is the population size at time $t = 0$. Over short time intervals, these results fit certain populations, mainly bacteria, quite well. However, it is clear that $P(t)$ will tend to infinity as t tends to infinity, so without some restriction on t the model cannot be correct. A more realistic model for some populations is given by the **logistic differential equation,** sometimes called the **Verhulst-Pearl equation:**

$$\frac{dP}{dt} = kP\left(1 - \frac{P}{P_\infty}\right),$$

where P_∞ is a limiting size for the population, beyond which the growth rate is necessarily zero. As $P(t)$ approaches P_∞, the factor $1 - P/P_\infty$ approaches zero and so has an inhibiting effect on the growth of P. The logistic equation can be solved by separation of variables:

$$\frac{dP}{P - P^2/P_\infty} = k \, dt, \qquad P \neq P_\infty \neq 0.$$

The partial-fraction identity

$$\frac{1}{P} + \frac{1/P_\infty}{1 - P/P_\infty} = \frac{1}{P - P^2/P_\infty}$$

allows us to integrate the left side, and we get

$$\ln |P| - \ln \left|1 - \frac{P}{P_\infty}\right| = kt + c.$$

Then

$$\ln \left|\frac{P}{1 - P/P_\infty}\right| = kt + c$$

or, taking exponentials on both sides,

$$\frac{P}{1 - P/P_\infty} = Ke^{kt}, \qquad K = \pm e^c.$$

Solving for P gives

$$P(t) = \frac{P_\infty Ke^{kt}}{P_\infty + Ke^{kt}}.$$

We can solve the resulting equation for K to get $K = P_0 P_\infty/(P_\infty - P_0)$, where $P_0 = P(0)$. Substituting this value for K into our expression for $P(t)$ and simplifying, we find

$$P(t) = \frac{P_0 P_\infty}{(P_\infty - P_0)e^{-kt} + P_0}.$$

The whole point of the preceding analysis is to arrive at a general formula into which we can put some known information and from which we can then derive something we didn't previously know.

example 1

If we know k, P_0, and P_∞ in the formula derived previously, then we can let t tend to infinity, and we see right away that the first term in the denominator tends to 0. Hence

$$\lim_{t \to \infty} P(t) = P_\infty.$$

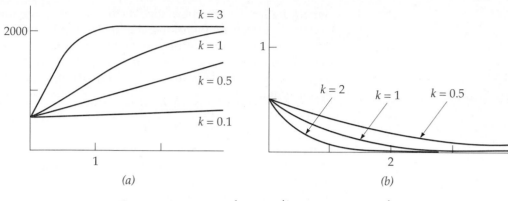

Figure 2.1 (*a*) $P = 10^6/(1500e^{-kt} + 500)$. (*b*) $y = e^{-kt}$.

This result is consistent with our original interpretation of P_∞ as a population size beyond which there is no further increase in size. We now see that, in theory the value P_∞ is never actually attained in this model, but that the population size $P(t)$ approaches $P_\infty = 2000$ asymptotically. Figure 2.1*a* shows the graphs of some typical relations for $P = P(t)$.

2   RADIOACTIVE DECAY

A radioactive substance such as carbon 14 decays in such a way that if $y(t)$ is the amount present in a sample at time t, then $y(t)$ satisfies a relation of the form **decay rate $r(t)$ is proportional to amount present at time t,** or

$$\frac{dy}{dt} = -ky, \qquad k > 0, t \geq 0.$$

The number k is called the **decay constant.** (Note that the actual **decay rate,** $r(t) = -dy/dt = ky(t)$, is minus the slope of the graph of y at time t and is not constant.) The differential equation is both separable and linear, and we find

$$y(t) = y_0 e^{-kt}$$

for the general solution, where y_0 is the amount present at time $t = 0$. Figure 2.1*b* shows some typical decreasing exponential graphs.

Let $t_1 < t_2$ be two arbitrary times and consider the equation $y(t_2) = y(t_1)/2$, or

$$y_0 e^{-kt_2} = \tfrac{1}{2} y_0 e^{-kt_1};$$

we are asking for the time t_2 at which we have half as much as we did at time t_1. Canceling y_0 and rearranging, we find

$$e^{-k(t_2 - t_1)} = \tfrac{1}{2}.$$

Taking logarithms of both sides gives

$$-k(t_2 - t_1) = -\ln 2,$$

so
$$t_2 - t_1 = \frac{\ln 2}{k}.$$

Thus the time interval required after t_1 for decay by 50% is the same regardless of what t_1 is. Hence we can speak of this time interval of length $H = t_2 - t_1$ as the **half-life** of the substance. By the previous equation, the half-life is

$$H = \frac{\ln 2}{k},$$

where k is the decay constant. The half-life of a sample is relatively easy to measure in a laboratory, and we can then compute the decay constant from H. For example, carbon 14 has a half-life of about $H \approx 5568$ years, so its decay constant is

$$k = \frac{\ln 2}{H} \approx 1.245 \cdot 10^{-4}.$$

The number of atoms of a substance in a typical sample is very large, on the order of 10^{23} per gram of the substance. However, for some purposes it is more convenient to think of $y(t)$ as a continuous approximation to a number of atoms rather than to a mass; this is so in the next example, in which a decay rate is expressed as a number of atoms per gram per minute.

---

e x a m p l e 1

In living wood the decay rate of carbon 14 is partly balanced by absorption of additional carbon 14 from the environment; the resulting decay rate for carbon 14 atoms in living wood has been measured empirically to be about 6.68 per minute per gram of wood. When a tree dies, exponential decay of carbon 14 is no longer balanced in the wood, because natural absorption through growth has stopped. If a tree dies at some time that we designate as $t = 0$, we'd like to estimate the present time t_1 on that scale. Let $y(t)$ be the amount of carbon 14 in a sample at time t, and let $r = -dy/dt$ be its decay rate in atoms per gram per minute. Thus $r(t) = ky(t)$, where $k = 1.245 \cdot 10^{-4}$ is the radioactive decay constant of carbon 14. Since we know that $ky(0) = 6.68$ and $y(t) = y(0)e^{-kt}$, we must have

$$r(t) = ky(t)$$
$$= ky(0)e^{-kt} = 6.68e^{-kt}.$$

Solving $r(t) = 6.68e^{-kt}$ for t gives

$$t = -\frac{1}{k} \ln \frac{r(t)}{6.68}.$$

Suppose we are able to measure the decay rate $r(t_1) = 6.3$ in a laboratory. Then the observed rate of decay of 6.3 atoms per gram per minute would give an age, measured from tree cutting time, of

$$t_1 = -\frac{1}{1.245 \cdot 10^{-4}} \ln \frac{6.3}{6.68}$$

$$\approx 470 \text{ years.}$$

Note that we have changed the time unit to years in the last equation, since the half-life was measured in years. The decay rates 6.3 and 6.68 can both be measured in any other common time unit since we used only their ratio.

example 2

An atom of a radioactive substance typically decays into an atom of some other radioactive substance with its own decay constant. For example, uranium 238, with a half-life of $4.51 \cdot 10^9$ years and decay constant $m = 1.537 \cdot 10^{-10}$, decays to thorium 234, with a half-life of 0.066 year and decay constant $n = 10.498$; after a sequence of 17 decay steps, the result is the element lead 206. Lead 206 is radioactively stable, and no further decay takes place in this sequence. Some of the longer-lived elements in the uranium series are useful for dating oil paintings that contain lead compounds, and it is important to understand the interplay between two or more elements in the series. We'll consider the first two named here: uranium 238 and thorium 234.

Let $U(t)$ be the amount of uranium 238 in a sample at time $t \geq 0$, and let $T(t)$ be the amount of thorium 234 in the same sample. Then

$$\frac{dU}{dt} = -mU.$$

Since the amount of thorium present, and hence its rate of change, is enhanced by the rate $mU > 0$ at which uranium decays to thorium, we have

$$\frac{dT}{dt} = -nT + mU.$$

The preceding pair of differential equations is an example of a *system* of first-order equations. In this example, the system is easy to solve, because the first equation does not contain T; we solve it and get

$$U(t) = U(0)e^{-mt}.$$

The second equation is then

$$\frac{dT}{dt} = -nT + mU(0)e^{-mt}.$$

This is a first-order linear equation, and in normal form it looks like

$$\frac{dT}{dt} + nT = mU(0)e^{-mt}.$$

The integrating factor is e^{nt}, and the solution is

$$T(t) = \frac{mU(0)}{n-m}e^{-mt} + \left(T(0) - \frac{mU(0)}{n-m}\right)e^{-nt}.$$

In this example, $m = 1.537 \cdot 10^{-8}$ and is so small that even over the interval $0 < t < 2000$ years e^{-mt} decreases only to about 0.99997. For this reason, if n is relatively large, the last term in the expression for $T(t)$ tends to zero fairly rapidly as t gets large. Because e^{-mt} varies so slowly, the first term appears to be practically constant. Thus, instead of tending to zero as we would expect, there seems to be a kind of positive *equilibrium* value E for $T(t)$ as t increases moderately from zero: E is about $mU(0)/(n-m)$ for 2000 years. Over a long enough time period, $T(t)$ does tend to zero of course.

EXERCISES

Population Dynamics

1. How should the constants k and P_0 be chosen in the solution $P(t) = P_0 e^{kt}$ to the linear growth equation $dP/dt = kP$ if $P(t)$ is to satisfy $P(10) = 100$ and $P(20) = 150$? With this choice for P_0 and k, what is $P(-10)$?

2. Show that if $P(t) = P_0 e^{kt}$ for positive constants P_0 and k, then $P(t)$ always doubles in a fixed time period $D = t_2 - t_1$, where $D = \ln 2/k$. The constant D is called the **doubling time** of this exponential growth law.

3. If a solution $P(t)$ for

$$\frac{dP}{dt} = kP, \qquad k > 0,$$

doubles during a time period of length D, how long does $P(t)$ take to triple in size?

4. Assume $P(t)$ differentiable for all real t and also that $P'(t) = kP(t)$ for some constant k. Show that the acceleration $P''(t)$ is $k^2 P(t)$ and, more generally, that $P^{(n)} = k^n P(t)$.

5. If the population of the United States was 4 million in 1790, 17 million in 1840, and 63 million in 1890, show that the size of the population from 1790 to 1890 cannot be approximated with reasonable accuracy by a formula of the form $P(t) = P_0 e^{kt}$.

6. A natural generalization of the logistic differential equation is

$$\frac{dP}{dt} = kP\left(1 - \frac{P^q}{P_\infty}\right), \qquad q > 0.$$

Note: In doing this problem, the formula

$$\int \frac{dP}{P(1 - P^q)} = \frac{1}{q} \ln\left(\frac{P^q}{1 - P^q}\right) + C$$

is useful; it can be derived using partial fractions or substitution of $P^q = x$.

(a) Show that the solution curves for this equation satisfy

$$\lim_{t \to \infty} P(t) = (P_\infty)^{1/q}, \qquad \text{if } P(t) \neq 0.$$

(b) Find $P(t)$ if $P(0) = P_0$, noting the special case $P_0^q = P_\infty$.

7. Show that the logistic differential equation can be written in the form

$$\frac{dP}{dt} = aP - bP^2,$$

and show how the parameters a and b are related to the rate constant k and the limit population P_∞.

8. The **general logistic equation** $dP/dt = aP^\alpha(1 - P/P_\infty)^\beta$, $a > 0$, $\alpha > 0$, $\beta > 0$, has elementary solution formulas only for special values of the exponents α and β. Nevertheless, we can get qualitative information about solutions without finding formulas.

(a) Show that solutions with an initial value satisfying $0 < P(0) < P_\infty$ are strictly increasing as long as they remain less than P_∞.

(b) Show that a solution graph has an inflection point at $P_I = (\alpha/(\alpha + \beta))P_\infty$ if $0 < P(0) < P_I$.

9. Writing the logistic equation in the form

$$\frac{1}{P}\frac{dP}{dt} = k\left(1 - \frac{P}{P_\infty}\right)$$

shows that for this model the growth rate of the population per individual member decreases with rate $-k/P_\infty$ as P increases. For some insect populations, it turns out that the equation

$$\frac{1}{P}\frac{dP}{dt} = k\left(\frac{P_\infty - P}{P_\infty + cP}\right), \qquad c = \text{const} > 0$$

is more realistic.

(a) Show that if $c = 0$ we get the logistic equation.

(b) Find the general solution of this differential equation in implicit form.

(c) What form does the solution take if we determine the constant of integration by $P(0) = P_0$?

10. The **Gompertz population equation** is

$$\frac{dP}{dt} = aP(L - \ln P),$$

where a and L are positive constants. Show that solutions are

$$P(t) = P_\infty e^{-Ke^{-at}}.$$

where $P_\infty = e^L$ and $K = \ln(P_\infty/P(0))$.

Radioactive Decay

11. Thallium 206 has a half-life of 4.19 minutes and decays atom for atom to lead 206. If we start with 1 gram of thallium 206, how much is left after 1 minute? How much is left after 1 hour? How much lead 206 will we have after 2 minutes?

12. Suppose that charcoal from an ancient campsite exhibits 0.9 disintegration per minute per gram of carbon 14 as compared with 6.68 disintegrations for live wood. Estimate the age of the charcoal.

13. Suppose that a sample of charcoal is known, for reasons having nothing to do with carbon 14, to be 1200 years old. What rate of carbon 14 decay would the sample be expected to exhibit?

14. If a sample of some radioactive element is reduced from 1.0 gram to 0.91 gram after 37 days, what is the half-life of the element?

15. One of the elements in the uranium series is uranium 234 with a half-life of $H_1 = 2.48 \cdot 10^5$ years. Uranium 234 decays to thorium 230, which has a half-life of $H_2 = 80,000$ years. Thorium 230 in turn decays to radium 226, which has a half-life of $H_3 = 1622$ years.

(a) If we have a sample of pure uranium 234, what fraction will remain after 10,000 years?

(b) If we start with 1 gram of pure uranium 234, how much thorium 230 will be present after 10,000 years?

(c) If a single sample of radioactive material consists of 1 gram of uranium 234 and 1 gram of thorium 230, how much of each will be present after 10,000 years?

(d) If a single sample of radioactive material consists of 1 gram of uranium 234, how much radium 226 will be present after 10,000 years?

16. Suppose we have an initial quantity U_0 of a radioactive element, say uranium, with decay constant m. Suppose the uranium decays to another radioactive element, say thorium, with initial quantity T_0 and decay constant n, where $n > m$. Show that if

$$nT_0 < mU_0$$

then, after time $t = 0$, the amount of thorium increases to an absolute maximum and then decays toward zero.

17. Branching sometimes occurs in radioactive decay. For example, bismuth 214 has a half-life of 19.7 minutes, but 0.9996 of the atoms decay to polonium 214, with a half-life of $1.64 \cdot 10^{-4}$ second, while the remaining 0.0004 decay to thallium 210, with a half-life 1.32 minutes. The decay of polonium and thallium each results in the same element, lead 210. Let k_1, k_2, k_3 represent the decay constants for bismuth, polonium, and thallium, and let $p = 0.9996$. Derive a formula for the amount of lead 210 present at time t, in terms of k_1, k_2, k_3, and p, if we were to start at $t = 0$ with 1 gram of pure bismuth 214.

3 VELOCITY IN A RESISTING MEDIUM

One of Newton's fundamental laws of motion asserts that the sum F of all forces acting at time t on a moving body of fixed mass is equal to its mass m times the acceleration of the body at that time. Since acceleration is by definition the time derivative $dv(t)/dt$ of velocity $v(t)$, the law can be written in the form $m\,dv/dt = F$. In this section we will consider only examples for which the motion is restricted to a single linear path with coordinate x. As a result of this simplifying assumption, force and velocity are also represented by real-valued functions. In general F may depend not only on time t but on position x and velocity v. However, there are several interesting applications in which F is independent of position; in that case we can often find $v(t)$ by a single integration and then can find the position $x(t)$ by another integration.

example 1 A car can be propelled along a straight horizontal track by magnetic interaction with the track. There is a retarding force due to air resistance and possible friction with the track, the sum of which is proportional to the velocity v of the car. (Thus the faster the car goes, the greater is the resisting force.) We'll assume that the propelling force is turned off at some time that we may as well call $t = 0$, leaving the car with velocity $v(0) = v_0$. In many applications the retarding force has the form $r(v) = -kv$, where k is a positive constant. In that case the Newton law can be written as the explicit differential equation

$$m\frac{dv}{dt} = -kv.$$

This equation can be solved by using the exponential multiplier $e^{kt/m}$ or else by separating variables. Both methods give

$$v(t) = v_0 e^{-kt/m}.$$

From this equation we learn that the velocity drops exponentially to zero because of the retarding forces. Note also that a larger value of k (more resistance) leads to an even more rapid decrease in velocity, as you can see from the position of k in the exponential. It is also interesting to note that the mass m plays a role: a larger mass corresponds to a *less rapid* decrease in velocity. (The effect of a larger mass is to increase **momentum** mv, thus tending to maintain the velocity at a higher rate.)

Since the car's velocity $v(t)$ at time t is the derivative of its position $x(t)$, it follows that, to within a constant of integration, position is the integral of velocity:

$$x(t) = \int v(t)\, dt + C$$

$$= \int v_0 e^{-kt/m}\, dt + C$$

$$= -v_0\frac{m}{k}e^{-kt/m} + C.$$

The constant of integration can be determined by observing $x(t)$ at some specific time t_1. Then

$$x(t_1) = -v_0\frac{m}{k}e^{-kt_1/m} + C.$$

Thus $C = x(t_1) + (v_0 m/k)e^{-kt_1/m}$, so

$$x(t) = v_0\frac{m}{k}(e^{-kt_1/m} - e^{-kt/m}) + x(t_1).$$

---

example 2

There is an apparent anomaly in Example 1: the velocity never actually reaches zero; in other words, the car never rolls to a stop, since the exponential is never zero. This discrepancy between normal experience and theory is due to the empirical nature of formulas for retarding forces. There is in general no simple formula to describe them perfectly under all circumstances. For example, the retarding force may depend very much on the surface characteristics of the moving object. However, we can compute the distance $x = x(t)$ from the starting point using the formula at the end of Example 1 or else by computing a definite integral:

$$x(t) = \int_0^t v(u)\,du$$

$$= \int_0^t v_0 e^{-ku/m}\,du$$

$$= v_0\left[-\frac{m}{k}e^{-ku/m}\right]_0^t$$

$$= \frac{mv_0}{k}(1 - e^{-kt/m}).$$

As t tends to infinity, the exponential term tends to zero, so the upper limit for the values of $x(t)$ is easily seen to be mv_0/k; given our assumptions, the range of motion is limited even though in theory the velocity never reaches zero.

e x a m p l e 3

A projectile of mass m fired upward from the surface of the earth encounters not only air resistance but also the downward force mg of gravity. The differential equation satisfied by velocity $v(t)$ has the form

$$m\frac{dv}{dt} - -r(v) - mg$$

$$= -kv - mg,$$

provided we assume air resistance proportional to the first power of velocity: $r(v) = -kv$. Again we have a first-order linear equation that can be solved by using the exponential multiplier $e^{kt/m}$ (or by separating variables):

$$\frac{d}{dt}(e^{kt/m}v) = -ge^{kt/m}.$$

The solution is

$$v(t) = -\frac{mg}{k} + Ce^{-kt/m}$$

$$= -\frac{mg}{k} + \left(\frac{mg}{k} + v_0\right)e^{-kt/m},$$

where $v_0 = v(0)$ is the initial velocity. As t gets large the second term tends to zero, leaving the limit $v_\infty = -mg/k$; this number is the **terminal velocity** of the projectile when it is viewed simply as a falling body after having reached its maximum height. Figure 2.2 shows graphs of velocity $v(t)$ with $v_0 > 0$ and $v_0 < 0$; the graphs on the right are of the corresponding height functions, showing their tendencies toward straight lines with slopes $-mg/k$.

To find out when the projectile reaches its maximum height, we set $v(t) = 0$ and solve for time t:

$$-\frac{mg}{k} + \left(\frac{mg}{k} + v_0\right)e^{-kt/m} = 0.$$

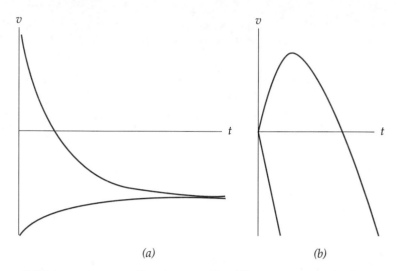

Figure 2.2 (a) $v = \dot{y} = -mg/k + (v_0 + mg/k)e^{-kt/m}$. (b) $y = y(t)$; slope approaches $-mg/k$.

Solving for the exponential factor and taking logarithms gives the solution

$$t_{\max} = \frac{m}{k} \ln\left(1 + \frac{kv_0}{mg}\right).$$

The height $h(t)$ attained at time t is just the integral of v from 0 to t. We leave it as an exercise to compute $h(t)$ and to show that the maximum height attained is

$$h(t_{\max}) = \frac{mv_0}{k} - \frac{m^2g}{k^2} \ln\left(1 + \frac{kv_0}{mg}\right).$$

Mass and Weight. The **mass** m of an object is a measure of the amount of matter in it, while its **weight** w is the magnitude of the gravitational force exerted on it by a standard physical object, usually the earth. Thus $w = ma$, where a is the acceleration due to gravity. If weight is measured at the surface of the earth, then $w = mg$, where $g \approx 32.2$ feet per second per second, or about 9.8 meters per second per second given that 1 meter is about 3.28 feet. (Variation in g is due to local variations in the structure of the earth.) The international standard unit of measurement for mass is the **kilogram,** and a 1-kilogram object weighs about 2.2 pounds near the surface of the earth. It is common practice to allow the term "kilogram" to denote also the *weight* of a 1-kilogram mass near the earth's surface. Thus an object of mass 1 kilogram weighs about 1 "kilogram" near the earth's surface.

---

e x a m p l e 4

An object of weight about $w = 2200$ pounds near the surface of the earth will have mass about $m = 2200/2.2 = 1000$ kilograms. If the velocity v of the object satisfies a differential equation of the form

$$m\frac{dv}{dt} = -kv - mg, \qquad k > 0,\ g = \text{acceleration of gravity,}$$

then the value of the constant k will depend on the units of distance, mass, and time that we use. If we choose to deal with weight $w = mg$ instead of mass, the equation becomes

$$\frac{w}{g}\frac{dv}{dt} = -kv - \frac{w}{g}g, \quad \text{or} \quad w\frac{dv}{dt} = -kgv - wg.$$

Thus our equation has the same general form that it had when we dealt with mass instead of weight:

$$w\frac{dv}{dt} = -Kv - wg, \quad K = kg.$$

This equation remains valid as long as w is measured in the same way at all times. It's convenient that the differential equation takes the same form whether we use mass or standardized weight, but we need to take care that the constant k or $K = kg$ is correctly matched to the units being used; since its value is usually determined empirically, the appropriate units will be built into the experiment. In particular, suppose K were to be determined in an experiment to be $K = 0.1$, using pounds w measured at the surface of the earth. If the units for distance and time were feet and seconds, then the mass m of the object would be about $2200/32.2 \approx 68.3$. The constant k would correspondingly be about $0.1/32.2 \approx 0.0031$. Switching instead to kilograms, we would have $m \approx 1000$ and $k \approx 0.0031/2.2 \approx 0.0014$.

EXERCISES

1. A 6440-pound car is guided along a track with velocity $v = v(t)$ subject to retarding force $-0.5v$ and with initial speed $v(0) = 100$ feet per second.

   (a) What is the approximate velocity at $t = 2$ seconds?

   (b) What is the upper limit of the car's progress along the track?

2. A car of mass 100 kilograms is guided along a track with velocity $v = v(t)$ subject only to a retarding force of the form $-kv$ and that $v(0) = 10$ feet per second.

   (a) The car's velocity has fallen to 7 feet per second at time $t = 120$ seconds. Estimate the value of k.

   (b) Assume that the car appears to have come to rest just after traveling 3600 feet. Estimate the value of k using that observation.

3. Suppose the differential equation $m\,dv/dt = -kv$ is satisfied by a moving object of mass m and that the constant k has the value 0.36 when the velocity v is measured in feet per second.

   (a) What value should k have if velocity is measured in miles per second?

   (b) What value should k have if velocity is measured in miles per hour?

   (c) Would you get the same answers in parts (a) and (b) if the questions referred to the differential equation $m\,dv/dt = -kv^2$?

4. A body of mass m falling under the influence of gravity and with resistance force proportional to velocity v satisfies the equation

$$m\frac{dv}{dt} = gm - kv, \qquad k > 0.$$

(a) Find the most general solution of the differential equation.

(b) Express $v(t)$ in terms of g, k, m, and the initial velocity v_0.

(c) Sketch typical graphs of the solutions in part (b) for each of the three distinct cases $0 \le v_0 < mg/k$, $v_0 = mg/k$, and $mg/k < v_0$.

5. A quadratic retarding force of the form $-kv^2$ has more effect at velocities $v > 1$ than does a linear retarding force $-kv$. Suppose that $k = 0.25$ for a car of mass 100 kilograms guided on a horizontal track with only the quadratic retarding force acting and with an initial velocity of 40 feet per second.

(a) Show that $v(t) = 400/(t + 10)$.

(b) The assumption of a v^2 retarding force is usually unrealistic for low velocities; using the result of part (a), show that in theory the car will never come to rest.

6. A body of mass m having velocity $v = v(t)$ moves subject only to frictional forces F proportional to the square of the velocity: $F = -kv^2$.

(a) Find the differential equation satisfied by v.

(b) Solve the differential equation in part (a) in terms of the parameters m, k, and $v_0 = v(0)$.

(c) Find the distance traversed in time t, and show that it tends to infinity as t does if $v_0 \ne 0$.

(d) Show that, for arbitrary $v_0 > 0$ and $v_1 > 0$, velocity satisfies $v(t) < v_1$ if $t > m/(kv_1)$.

7. Suppose that a 100-kilogram boat glides through the water with velocity $v = v(t)$ subject only to a retarding force of the form $-0.1v^{3/2}$.

(a) Find the velocity in terms of initial velocity $v_0 = v(0)$.

(b) Find the distance traveled in time t in terms of v_0.

(c) Show that the boat will theoretically never stop moving but that there is an upper limit to its forward progress.

8. The equation $m\, dv/dt = -kv^r$ can be rewritten

$$\frac{dv}{dt} = -Kv^r.$$

(a) Assuming K and r constant, $r > 1$, find $v(t)$ in terms of K, r, and initial velocity $v(0) = v_0 > 0$.

(b) For what values of $r > 1$ is $v(t)$ bounded above?

(c) Discuss the case $0 < r < 1$, and show that the velocity falls to zero in finite time $v_0^{1-r}(K(1 - r))^{-1}$.

*9. A baseball weighing 5 ounces is popped up vertically with an initial velocity of 161 feet per second. Assume that $g = 32.2$ feet per second per second and that the retarding force of air resistance is $-kv$, where $k = 0.5$. The units determining k are ounces and seconds. Part (c) requires you to estimate the solution of a transcendental equation.

(a) Estimate the time it takes for the ball to reach its maximum height.

(b) Estimate the maximum height of the ball.

(c) Express the height of the ball as a function of the time after maximum height, and use the result to estimate the additional time before hitting the ground and the speed of the ball when it hits the ground.

(d) Explain why the speed on returning to the ground is less than the given initial velocity.

(e) Answer parts (a) through (c) under the assumption that air resistance can be neglected (i.e., $k = 0$) and explain why the differences in corresponding results are reasonable.

10. Suppose that a projectile of mass m is fired upward with initial velocity v_0 from the surface of the earth under the influence of gravitational force $-mg$ and air-resistance force $-kv$, as in Example 3 of the text.

(a) Compute the height $h(t)$ attained after time t.

(b) Show that the maximum height attained is

$$h_{max} = \frac{mv_0}{k} - \frac{m^2 g}{k^2} \ln\left(1 + \frac{kv_0}{mg}\right).$$

(c) The dependence of the maximum height attained (see part (b)) on k and m reduces to dependence on just the ratio of these two quantities. Explain how this fact can be deduced directly by looking at the differential equation satisfied by $v(t)$.

11. In Example 3 of the text it was shown that for a projectile fired up from the surface of the earth with initial velocity v_0 and air resistance $-kv$, the time of maximum height is

$$t_{max} = \frac{m}{k} \ln\left(1 + \frac{kv_0}{mg}\right).$$

(a) Show that $\lim_{k \to 0} t_{max} = v_0/g$.

(b) Solve the projectile velocity problem under the assumption that $k = 0$, and compare the corresponding value for t_{max} with the result of part (a).

12. A ball of mass m is tossed up from the edge of a vertical cliff so that on the way back down it just misses the edge of the cliff. The ball is subject to acceleration of gravity g and to air resistance of magnitude $k|v|$, $k > 0$.

(a) Show that if initial velocity satisfies $|v_0| > mg/k$, then the initial speed will never again be attained.

(b) Show that if initial velocity satisfies $|v_0| < mg/k$, then the initial speed $|v_0|$ will be attained again at time

$$t_1 = \frac{m}{k} \ln \frac{mg + k|v_0|}{mg - k|v_0|}.$$

13. An object of mass m is propelled downward with initial velocity v_0, subject to constant gravity g and linear retarding force $-kv$.

(a) Show that if $v_0 = mg/k$, then the velocity remains constant.

(b) Show that the velocity increases if $v_0 < mg/k$ and decreases if $v_0 > mg/k$.

14. A projectile of mass m is fired up from the surface of the earth against the force due to gravity and against the force of air resistance in the form $-kv^2$.

(a) Show that the velocity v satisfies the equation $dv/dt = -(k/m)v^2 - g$, where g = acceleration of gravity.

(b) Solve the equation in part (a), and express the time of maximum altitude in terms of initial velocity v_0.

(c) Show that the time of maximum altitude increases as v_0 is increased, but that it has upper limit $\pi\sqrt{m/(4kg)}$.

(d) Show that the maximum altitude attained is $h_{\max} = (m/2k)\ln(1 + (kv_0^2/mg))$.

15. An object of mass m falls to earth under the force due to gravity and against an air resistance of the force of magnitude kv^2.

(a) Show that velocity v satisfies the equation $dv/dt = g - (k/m)v^2$.

(b) Rewrite the differential equation in the form

$$\frac{dv}{(gm/k) - v^2} = \frac{k}{m}\,dt,$$

and solve for v using the partial-fraction identity

$$\frac{1}{a^2 - v^2} = \frac{1}{2a}\left(\frac{1}{a - v} + \frac{1}{a + v}\right),$$

to perform the integration in v.

(c) Find the terminal velocity $v_\infty = \lim_{t\to\infty} v(t)$.

4　VARIABLE MASS AND ROCKET PROPULSION

We'll consider here just the motion of a body of varying mass moving along a straight path. Such a path might, for example, lie along a line passing through the center of gravity of the earth and extending out into space. Let $F = F(t)$ be the total external force acting on the body at time t, let $m = m(t)$ be the mass of the body, and let $v = v(t)$ be the body's velocity. For a body of *constant* mass m, we have the familiar equation

$$F = ma, \qquad \text{where } a = \frac{dv}{dt} \text{ is } \textbf{acceleration.}$$

However, in the case of an object such as a rocket, fuel is initially counted as a part of the rocket's mass and is then consumed at such a high rate that the assumption of constant mass is not applicable. Newton may not have had rockets in mind, but he formulated the relationship between F, m, and v in enough generality that it takes care of the possibility of nonconstant mass. The general **Newton second force law** is

$$F = \frac{d(mv)}{dt}.$$

Here $p = mv$ is the **momentum** of a body, defined so that the momentum of a body equals the sum of the momenta of its parts. It follows from this equation and the product rule for differentiation that if m is constant, then

$$F = m\frac{dv}{dt} + v\frac{dm}{dt} = ma + 0 = ma.$$

Thus the general force law $F = d(mv)/dt$ contains the law for constant mass, $F = ma$, as a special case. In general, however, $m = m(t)$ may vary with time.

---

e x a m p l e 1

The principle behind rocket propulsion can be illustrated by the following simple example. Consider a projectile of mass m_0 moving on a linear path with constant velocity $v_0 > 0$ and acted on by no external forces; thus the total momentum of all parts of the object remains equal to the initial constant value $m_0 v_0$. At time $t = 0$ a piece of the projectile having mass $m < m_0$ is ejected straight out the rear of the projectile with velocity w. Note that $w < 0$, because the particle is moving in the direction opposite to that of the projectile. After the ejection of mass m, the projectile has reduced its mass to $m_1 = m_0 - m$; to find its velocity v_1 after ejection, we express the constant total momentum of the system, consisting of projectile plus particle, as the sum of the momenta of the two parts: $m_0 v_0 = mw + (m_0 - m)v_1$. Solving for v_1 gives

$$v_1 = \frac{m_0 v_0 + m(-w)}{m_0 - m} = \frac{m_0}{m_0 - m}v_0 + \frac{m}{m_0 - m}(-w).$$

It's easy to check that the projectile of reduced mass $m_0 - m$ has had its velocity increased by the difference $v_1 - v_0 = m(v_0 - w)/(m_0 - m)$. Note that increasing the mass m of the ejected particle to a higher value, still below the obvious upper limit m_0, makes for a larger increase in velocity. In a somewhat different way, increasing the speed $-w$ with which the particle is ejected will also increase v_1. Continued ejection of particles will increase the velocity still more. This is the essence of how a rocket works.

To appreciate how an analysis of momentum plays a role in describing rocket motion, we need to consider not only the velocity $v = v(t)$ of the rocket but also the velocity $w = w(t)$ of the expelled exhaust gas, which for a simple nonsteerable rocket will have sign opposite to that of $v(t)$. In other words, we need to consider the forces acting on all parts of the system consisting of both rocket and exhaust. We then delete the part associated with the exhaust to arrive at an equation for the rocket itself. The calculation given at the end of this section corrects the fundamental equation $F = d(mv)/dt$ by a term $-w\,dm/dt$ that accounts for the expelled exhaust gas. The analysis there shows that the correct formula expressing the forces acting on the rocket alone has the smaller magnitude

$$F = \frac{d(mv)}{dt} - w\frac{dm}{dt}.$$

Note that v and w will typically have opposite signs, because the exhaust will be moving in the direction opposite to that of the rocket. Also dm/dt is negative, since the mass m of the rocket is decreasing, at least while the rocket is actually firing. Thus typically $w\, dm/dt > 0$. The correction term $-w\, dm/dt$ can be thought of as resulting from the ejection of fuel mass from the rocket.

To apply the previous equation, expand the first term on the right by the product rule and recombine terms to get

$$F = m\frac{dv}{dt} + (v - w)\frac{dm}{dt}$$

$$= m\frac{dv}{dt} + v_e\frac{dm}{dt},$$

where $v_e = v - w > 0$ is the velocity of the exhaust gas *measured relative to the rocket*. In practice v_e can often be assumed to be constant, and we'll make that assumption unless the contrary is explicitly stated. The force F represents the sum of external forces such as the gravitational attraction of the earth or frictional air resistance acting on the rocket. The resulting differential equation can be regarded either as a first-order equation to be solved for $m = m(t)$ or, more usually, a first-order equation to be solved for $v = v(t)$. In the latter formulation we rewrite the equation as

$$m\frac{dv}{dt} = -v_e\frac{dm}{dt} + F,$$

often called the **fundamental rocket equation.** Since the exhaust velocity v_e will be positive while the rocket motor is firing, and $dm/dt < 0$, the term $-v_e\, dm/dt$ will in fact be positive. In case the motor is turned off, both v_e and dm/dt will be zero, and the rocket equation reduces to $m\, dv/dt = F$. The term $-v_e\, dm/dt$ in the fundamental rocket equation represents a force and is called the **thrust** of the rocket motor. Note that a significant increase in velocity from the action of the motor is due not only to exhaust velocity v_e, which is often constant, but also to a rapid rate dm/dt of fuel consumption.

---

e x a m p l e **2**
A rocket so remote from the effects of external forces F that they can be neglected can be assumed to satisfy an equation of the form

$$m\frac{dv}{dt} = -v_e\frac{dm}{dt},$$

or

$$\frac{dv}{dt} = -\frac{v_e}{m}\frac{dm}{dt}.$$

Assuming v_e to be constant, we can easily integrate both sides with respect to t to get

$$v(t) = -v_e \ln m(t) + C.$$

If we know the mass m_0 and velocity v_0 of the rocket at some initial time t_0, we can determine the constant of integration to be

$$C = v_0 + v_e \ln m_0,$$

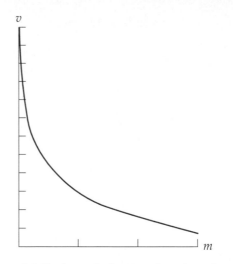

Figure 2.3 Rocket velocity as a function of mass.

where m_0 is the initial mass of rocket plus fuel. Hence

$$v = v_0 + v_e \ln m_0 - v_e \ln m$$

$$= v_0 + v_e \ln \frac{m_0}{m}.$$

The general form of the relation between v and m is shown in Figure 2.3. (The graph is just a vertical shift of the graph of $v = -v_e \ln m$.)

What this picture shows is that velocity increases quite rapidly as we decrease the mass of the rocket by burning fuel, and a more rapid rate of fuel consumption, attained, for example, by igniting more motors, will increase the velocity more rapidly. This fact can be deduced directly from the original differential equation. More precisely, we're considering an assertion about the rate of change of velocity, that is, the acceleration $a = dv/dt$ of the rocket. The differential equation is

$$\frac{dv}{dt} = -v_e \frac{1}{m} \frac{dm}{dt}.$$

Since m is decreasing, $dm/dt < 0$, so acceleration is always positive. The formula for a shows that acceleration can be made large by making m small, but that approach has the disadvantage of reducing the payload and the potential fuel supply. The alternative is to make dm/dt large in absolute value, that is, to burn the fuel very rapidly.

***e x a m p l e* 3**

Continuing from Example 2, we can ask for the velocity of a rocket as a function of the mass m_1 of the fuel burned up to time t_1 during the rocket's flight. Let m_R be the residual mass of the rocket together with its unburned fuel. (Thus the initial mass of the rocket is $m_0 = m_R + m_1$.) Using the formula for $v(t)$ from

Example 2, we find

$$v(t_1) = v_0 + v_e \ln \frac{m_R + m_1}{m_R}$$

$$= v_0 + v_e \ln \left(1 + \frac{m_1}{m_R}\right).$$

Thus, for given initial velocity v_0 and exhaust velocity v_e, velocity at time t_1 depends just on the ratio of burned fuel to the residual mass of the rocket. If a constant rate α is used for fuel consumption, then the total mass $m(t)$ of the rocket plus unburned fuel at time t has the form

$$m(t) = m_r + m_f - \alpha t, \qquad 0 \le t \le m_f/\alpha,$$

where m_r is the mass of the rocket without fuel and m_f is the initial mass of fuel in the rocket. When the fuel is all used up at $t = m_f/\alpha$, then $m = m_r$ from that time on. The velocity of the rocket is

$$v(t) = v_0 + v_e \ln \frac{m_r + m_f}{m_r + m_f - \alpha t}$$

$$= v_0 + v_e \ln \left(1 + \frac{\alpha t}{m_r + m_f - \alpha t}\right), \qquad 0 \le t \le m_f/\alpha.$$

At burnout time $t_b = m_f/\alpha$, when all the fuel has been burned, the velocity is

$$v(t_b) = v_0 + v_e \ln \left(1 + \frac{m_f}{m_r}\right),$$

and this is the maximum velocity attainable by firing the motor. Note that $v(t_b)$ is independent of the rate, whether constant or variable, at which the fuel is consumed.

e x a m p l e 4

Recall that the fundamental rocket equation has the form

$$m \frac{dv}{dt} = -v_e \frac{dm}{dt} + F.$$

The only external force that we'll consider in this example is that of gravity: $F = -mg$, where g is the acceleration of gravity near the earth's surface; in particular, we're ignoring air resistance. The equation to look at is thus

$$\frac{dv}{dt} = -\frac{v_e}{m} \frac{dm}{dt} - g.$$

Integration with respect to t gives

$$v(t) = -v_e \ln m(t) - gt + c.$$

The constant c can be determined by specifying the initial velocity $v_0 = v(0)$. Suppose, for example, that the rocket is launched from the surface of the earth with initial velocity $v(0) = 0$. In that case $c = v_e \ln m(0)$, and then

$$v(t) = v_e \ln \frac{m(0)}{m(t)} - gt.$$

Assuming a constant rate α of fuel consumption, empty rocket mass m_r, and initial fuel mass m_f, we find $m(t) = m_r + m_f - \alpha t$ whenever $0 \le t \le m_f/\alpha$. For these values of t we then have

$$v(t) = v_e \ln \frac{m_r + m_f}{m_r + m_f - \alpha t} - gt.$$

The parameters m_r, m_f, and v_e might well be chosen so that $v(t) \ge 0$. In particular, when the fuel is all used up at $t = m_f/\alpha$, the requirement $v(m_f/\alpha) \ge 0$ is easily verified to be equivalent to

$$v_e \ge \frac{g m_f}{\alpha \ln (1 + m_f/m_r)}.$$

Note: For the calculations in Examples 1 to 3, the mass m of the rocket plus fuel could just as well have been replaced by weight $W = mg$, where g is acceleration of gravity. The reason is that the external force F is zero in those differential equations, and they are all of degree one in m and dm/dt; thus multiplying by the constant g gives us the original equation with mass m replaced by weight $W = mg$. The simple replacement of mass by weight is not always possible when $F \ne 0$, as in Example 4 and in the next example.

 **e x a m p l e 5**

A rocket subject to linear, velocity-dependent frictional forces and constant gravity has velocity $v(t)$ obeying

$$m \frac{dv}{dt} = -v_e \frac{dm}{dt} - kv - gm, \qquad k > 0.$$

If $m = m(t)$ is known, the differential equation for v can be written in first-order linear form as

$$\frac{dv}{dt} + \frac{k}{m} v = -\frac{v_e}{m} \frac{dm}{dt} - g.$$

Now suppose there is a constant rate α of fuel consumption so that $m(t) = m_0 - \alpha t$, where m_0 is the initial mass of rocket plus fuel. The equation then becomes

$$\frac{dv}{dt} + \frac{k}{m_0 - \alpha t} v = \frac{\alpha v_e}{m_0 - \alpha t} - g.$$

The appropriate integrating factor is $(m_0 - \alpha t)^{-k/\alpha}$. With initial condition $v(0) = 0$ the solution works out to be

$$v(t) = \frac{\alpha v_e}{k} + \frac{g}{\alpha - k}(m_0 - \alpha t) - \left(\frac{\alpha v_e}{k} + \frac{g m_0}{\alpha - k} \right) \left(\frac{m_0 - \alpha t}{m_0} \right)^{k/\alpha}.$$

The way in which this solution depends on the various parameters is not at all obvious, and the investigation is left to the exercises.

Derivation of the Rocket Equation. Let $p = mv$ denote rocket momentum, exclusive of exhaust gas, at time t, and let Δp, Δm, and Δv denote the changes in p, m, and v in the time interval of length Δt between t and $t + \Delta t$. With a view to finding dp/dt we consider the momentum at time $t + \Delta t$:

$$p + \Delta p = (m + \Delta m)(v + \Delta v) + (-\Delta m)(w + \Delta w),$$

where w is the velocity of the expelled exhaust gas. The second term on the right is the change in rocket momentum due to the escape of exhaust gas. (Note that for a rocket that is burning fuel, $\Delta m < 0$, because the decrement Δm represents a decrease in mass. Also the gas velocity $w + \Delta w$ will typically be directed opposite to $v + \Delta v$, the velocity of the rocket itself. However, the assumptions are not used in the computation.) Now subtract $p = mv$ from both sides, simplify, and divide by Δt to get

$$\frac{\Delta p}{\Delta t} = m\frac{\Delta v}{\Delta t} + (v - w)\frac{\Delta m}{\Delta t} + \Delta m\frac{\Delta v}{\Delta t} - \Delta m\frac{\Delta w}{\Delta t}.$$

As Δt tends to zero, so does Δm. It follows that the last two terms tend to zero as Δt tends to zero, and we get for a limit the equation

$$\frac{dp}{dt} = m\frac{dv}{dt} + (v - w)\frac{dm}{dt}.$$

Write $v_e = v - w$ for the exhaust velocity relative to the rocket and $F = dp/dt$ for the external force on the rocket. We can then write the fundamental rocket equation as

$$F = m\frac{dv}{dt} + v_e\frac{dm}{dt}.$$

All forces in this last equation are to be thought of as applied at the center of mass of the rocket.

It is tempting to try to derive the rocket equation just by applying the product rule to the equation $F = d(mv)/dt$:

$$F = m\frac{dv}{dt} + v\frac{dm}{dt}.$$

This equation is correct if F, m, and v are interpreted as pertaining to the entire system consisting of rocket plus fuel plus exhaust gas. The trouble is that we really want an equation in which v can be interpreted as the velocity of just the rocket together with its unburned fuel. This is precisely what the previous derivation gives us.

The precise derivation given above is quite valid regardless of whether m is increasing or decreasing and whether w is positive or negative. Thus the equations

$$m\frac{dv}{dt} + (v - w)\frac{dm}{dt} = F \qquad \text{or} \qquad \frac{d(mv)}{dt} - w\frac{dm}{dt} = F$$

can be applied when mass is not dissipating but increasing, as in Exercises 20 to 25.

---

e x a m p l e 6

Consider a car of mass m_0 rolling with constant velocity v_0 on a straight frictionless track and with no external forces acting on the car. At time $t = 0$ the car begins to pick up sand at a constant rate r from a motionless overhead hopper. The mass of the car as a function of time is $m(t) = m_0 + rt$. The external force is

$F = 0$, and the initial velocity of the accreting mass is $w = 0$, so we have

$$m\frac{dv}{dt} + (v - 0)\frac{dm}{dt} = 0,$$

or $d(mv)/dt = 0$ by the product rule for derivatives. Thus the momentum mv is constant. In particular, $m(t)v(t) = m_0 v_0$, so for $t > 0$

$$v(t) = \frac{m_0 v_0}{m_0 + rt} = \frac{v_0}{1 + rt/m_0} < v_0.$$

The velocity decreases as t increases, and indeed the acceleration is negative:

$$\frac{dv(t)}{dt} = \frac{-rm_0 v_0}{(m_0 + rt)^2} < 0.$$

EXERCISES

1. A rocket subject to a retarding force proportional to velocity has velocity $v = v(t)$ satisfying

   $$m\frac{dv}{dt} = -v_e\frac{dm}{dt} - kv, \qquad k = \text{const} > 0.$$

   (a) Note that the differential equation is a first-order linear equation for v. Then find the general solution under the assumption that $m(t) = m_0 - \alpha t$, where α is a constant rate of fuel consumption.

   (b) Assuming initial velocity zero, show that $v(t) \leq \alpha v_e/k$ for $t \geq 0$.

2. Referring to Example 1 of the text, make the following comparisons.

   (a) Show that increasing the mass of the ejected particle by a factor $\rho > 1$ causes a greater increase $v_1 - v_0$ in velocity than would increasing the relative velocity $v_0 - w$ of ejection by the same factor ρ.

   (b) Show that increasing the mass of the ejected particle by a factor $\rho > 1$ causes a greater increase $v_1 - v_0$ in velocity than would multiplying the velocity w of the ejected particle by the same factor ρ.

   (c) Compare the effects on $v_1 - v_0$ of increasing velocity w of ejection and increasing *relative* velocity $v_0 - w$ by the same factor $\rho > 1$.

3. (a) The discussion in Example 2 of the text and Figure 2.3 suggests that in principle a rocket might attain arbitrarily large velocity by virtue of having consumed a large enough proportion of its mass as burning fuel. What practical considerations might prevent this from happening for a rocket based on earth?

   (b) Starting with initial velocity $v_0 = 0$, maintaining constant exhaust velocity v_e, and neglecting external forces, what ratio of residual rocket mass to initial mass would be required for a rocket to attain velocity c, the speed of light?

   (c) The extreme upper limit for exhaust velocity from a fuel-burning rocket is about $v_e = 2.18$ miles per second. Using this value for v_e and $c = 186{,}000$ miles per second, estimate the size of the ratio found in part (b).

4. A car propelled along a smooth track by ejecting sand from the rear of the car at a fixed rate of 20 pounds per second and with velocity relative to the car of 10 feet per second behaves somewhat like a rocket. Assume that the car together with its occupants weighs 1000 pounds and that it starts with an additional 500 pounds of sand.

(a) If the car starts from a standstill and frictional forces are negligible, what would the car's velocity be when the sand is all used up?

(b) If the car starts from a standstill and frictional forces retard the car by an amount $F(t) = -v(t)$, what would the car's velocity be at the instant when the sand is all used up?

(c) If the car described in part (b) continues to be subject to the same frictional forces, what would its velocity be 60 seconds after the instant at which the sand is all used up? How far would it have traveled by that time?

5. (a) Solve the fundamental rocket equation with constant exhaust velocity $v_e > 0$ and external force $F = 0$. Assume $v(0) = v_0 \geq 0$, but make no assumption about the form of $m(t)$.

(b) With the same assumptions as in part (a), assume also that there is a constant rate α of fuel consumption so that $m(t) = m_r + m_f - \alpha t$, where m_f is the initial fuel mass and m_r the mass of the rocket itself. Show that at burnout, when $t = m_f/\alpha$, the acceleration is $\alpha v_e/m_r$.

6. To launch a rocket from a standing position on the surface of the earth, a positive initial acceleration has to be applied to overcome g, the acceleration of gravity. Neglecting air resistance, show that a successful launch requires $\alpha v_e > g m_0$, where α is a constant rate of fuel consumption, v_e is the exhaust velocity, and m_0 is the initial mass of rocket plus fuel. (Note that the product αv_e is the **thrust** of the rocket motor and has the dimensions of force. The thrust must exceed the force of gravity gm.)

7. Example 3 of the text shows that starting with velocity $v_0 = 0$ the fundamental rocket equation has solution

$$v(t) = v_e \ln \frac{m_0}{m_0 - \alpha t}, \qquad 0 \leq t \leq m_f/\alpha$$

if $m_0 = m_r + m_f$, there are no external forces, and fuel is burned at constant rate α.

(a) Assume $y(0) = 0$ and integrate velocity $v(t)$ to get distance

$$y(t) = v_e t - \frac{v_e}{\alpha}(m_0 - \alpha t) \ln \frac{m_0}{m_0 - \alpha t}.$$

[*Hint:* $\int \ln x \, dx = x \ln x - x$.]

(b) Show that at burnout time $t_b = m_f/\alpha$, the distance covered is

$$y_b = \frac{v_e}{\alpha}\left[m_f - m_r \ln\left(1 + \frac{m_f}{m_r}\right) \right].$$

What does this equation say about the size of α if we want y_b to be large? See also the previous exercise for restrictions on the size of α.

8. (a) Solve the fundamental rocket equation for constant velocity $v_e > 0$ under the assumption that $F = -gm$, where $m(t) = m_r + m_f - at$ and $v(0) = v_0 \geq 0$.

(b) Integrate the result found in part (a) to find the height of a radially guided rocket starting from the earth's surface with initial velocity $v_0 = 0$.

9. A fully fueled rocket weighs 32 pounds, half of which is fuel, has exhaust velocity $v_e = 800$ feet per second, and consumes all its fuel at a constant rate within 10 seconds. Suppose the rocket is launched vertically from a standing position on the surface of the earth. Take $g = 32$ feet per second per second, and neglect air resistance. Describe the entire history of the rocket from launch at $t_0 = 0$ to return to earth at t_r. In particular:

(a) Find the velocity and acceleration of the rocket for $0 \leq t \leq 10$.

(b) Find the height of the rocket at $t = 10$ and its position and velocity for $10 \leq t \leq t_r$, during which time the rocket is under the influence of gravity only. In particular find the maximum height of the rocket above the earth and the speed of the rocket when it hits the earth.

*10. Use the data of the previous problem under the additional assumption that the rocket encounters a retarding force of air resistance proportional to its velocity, with proportionality factor $k = 0.01$.

(a) Show that if distance is measured up from the earth's surface, the external force function has the form $F = -0.01v - 32m$.

(b) Find the velocity and acceleration of the rocket for $0 \leq t \leq 10$.

(c) Find the height of the rocket at $t = 10$ and its position and velocity for $10 \leq t \leq t_r$, during which time the rocket is under the influence of gravity and air resistance only. In particular find the maximum height of the rocket above the earth and the speed of the rocket when it hits the earth.

11. (a) Show that the mass $m(t)$ of a rocket plus its unburned fuel can be computed from its velocity $v(t)$ by

$$m(t) = m(0)e^{(v(0)-v(t))/v_e}$$

under the assumption $F = 0$ for external forces.

(b) What is the corresponding result if $F = -gm(t)$, where g is a positive constant?

12. (a) Show that the fundamental rocket equation can be written

$$m \frac{dv}{dm} = -v_e$$

if there are no external forces. Solve the equation for v in terms of m and the initial mass $m_0 = m(0)$ of rocket plus fuel if $v(0) = 0$.

(b) If velocity v is significantly large relative to the speed of light c, the relativistic version of the equation in part (a) must be used.

$$m \frac{dv}{dm} = -v_e\left(1 - \frac{v^2}{c^2}\right).$$

Assume initial velocity $v(0) = 0$, and use separation of variables to solve this differential equation for v in terms of m and the initial mass $m_0 = m(0)$ of rocket plus fuel. Sketch the shape of the graph of v as a function of m on $0 \leq m \leq m_0$ for the special case $v_e = c/2$. (You can use c as the unit of measurement on the v axis and m_0 as the unit on the m axis.)

(c) Show that as c tends to infinity the solution to the problem in part (b) tends to the solution to the problem in part (a).

13. Solve the relativistic rocket equation

$$m \frac{dv}{dm} = -v_e \left(1 - \frac{v^2}{c^2}\right)$$

for v as a function of m with initial total mass m_0 and initial velocity v_0.

14. A multistage rocket can be thought of as a succession of single-stage rockets in which the initial velocity for each stage is the same as the final velocity for the previous stage and the sum of the total masses of the later stages are counted as part of the unexpendable rocket mass of the previous stage. A body stage is jettisoned as soon as its fuel is gone.

(a) Consider a two-stage rocket with initial velocity v_0, rocket mass M_r, fuel mass M_f, and exhaust velocity V_e for the first stage, and corresponding quantities m_r, m_f, v_e for the second stage. Assuming no external forces, find a formula for the velocity of the second stage at burnout when all the fuel is gone. [*Hint:* Use the first part of Example 3 of the text.]

(b) Suppose in part (a) that $M_r = m_r$, $M_f = m_f$, and $V_e = v_e$. Show that the final velocity of the second stage is greater than the final velocity attained by a single-stage rocket of mass $2m_R$ and initial fuel mass $2m_f$, and compute the difference in velocities.

*15. For a rocket operating fairly near the surface of the earth, the estimate mg for the magnitude of the gravitational attraction of the earth is adequate. At greater distances Newton's **inverse-square law** should be used. Let r be the earth's radius and x the distance of the rocket from the surface of the earth. Then the inverse-square law gives

$$F_g = \frac{GMm}{(r + x)^2}$$

for the magnitude of the force of gravity at height x above the earth. Here M is the mass of the earth and G is the **gravitational constant,** which of course depends on the units of measurement being used. (Thus G can be estimated from $GMr^{-2} = g$ if we know M and r.) Using this notation, show that $x = x(t)$ satisfies this *second-order* differential equation when the rocket is above the atmosphere:

$$\frac{d^2x}{dt^2} + \frac{GM}{(r + x)^2} = -\frac{v_e}{m} \frac{dm}{dt}.$$

16. Consider a rocket with initial velocity zero and no external forces acting on it. Let m_f be the initial fuel mass and let m_r be the mass of the rocket with no fuel in it. Show that if the velocity $v(t)$ of the rocket is ever to exceed the exhaust velocity v_e it is necessary to have $m_f > (e - 1)m_r \approx 1.72m_r$. [*Hint:* Integrate $dv/dt = -v_e(dm/dt)/m$.]

17. As suggested at the end of the derivation of the fundamental rocket equation in the text, it is tempting simply to apply the product rule for differentiation to the equation $d(mv)/dt = F$ to arrive at an equation for rocket motion. (This simplified derivation is incorrect because the $m(t)$ and $v(t)$ in the equation thus derived would necessarily refer to the system consisting not only of the rocket plus fuel unburned up to time t but also to the expelled exhaust.) This exercise examines the consequences of trying to predict rocket motion with no external forces by using the uncorrected equation

$$m \frac{dv}{dt} = -v \frac{dm}{dt}.$$

   (a) Show that we would conclude that $v(t) = v(0)m(0)/m(t)$. What would you then conclude about motion starting with initial velocity $v(0) = 0$?

   (b) Let $a_c(t)$ and $a_u(t)$ be the predicted rocket accelerations according to the corrected and uncorrected rocket equations, respectively. Show that given the same rate of fuel consumption $\dot{m}(t)$, then $a_c(t) > a_u(t)$ as long as $v_e m(t) > v(0)m(0)$.

18. Consider the rocket equation

$$m \frac{dv}{dt} = -v_e \frac{dm}{dt} + F(t)$$

   with constant exhaust velocity $v_e > 0$ and external force F depending only on time.

   (a) Assuming $v(t)$ known from observation and $F(t)$ identically zero, derive the formula

$$m(t) = m(0)e^{(v(0)-v(t))/v_e}, \qquad t \geq 0.$$

   (b) Derive a formula analogous to the one in part (a) for the case in which $F(t)$ is continuous for $t \geq 0$.

*19. A rocket with no forces but a rocket motor with exhaust velocity v_e acting on it has mass m_0 when fully fueled. If fuel mass m_1 is consumed to accelerate the rocket from velocity v_0 to v_1, how much fuel is required to fire the motor in the opposite direction to bring the velocity down to v_2?

Motion Subject to Increasing Mass

20. A car of mass m_0 is sliding with constant velocity v_0 over a straight track with negligible friction and with no external forces acting on the car. (See Example 6 of the text.) At time $t_0 = 0$ the car begins to gain cargo having initial velocity zero at a constant rate $r > 0$.

   (a) Find the distance $x(t)$ covered by the car t time units after the mass gain starts.

   (b) Show that as r tends to zero the distance found in part (a) tends to the distance that would be attained if no additional mass were added.

21. A container of mass m_0 starts with velocity $v_0 > 0$ and moves in a straight linear path. As the container moves it absorbs mass that is streaming in the opposite direction with constant speed s.

(a) Find a differential equation relating increasing mass $m(t)$ and its velocity $v(t)$, assuming no effects on the motion other than the absorption of additional mass.

(b) Solve the equation found in part (a), and show that the velocity reaches the value zero when the mass has increased by $v_0 m_0/s$.

(c) Show that if m as allowed to become arbitrarily large, then the speed of the container tends to s.

22. Imagine a body of mass m_0 moving through space on a linear path with constant velocity v_0 and with no external forces acting on it. At time $t = t_0$ the body begins to absorb hitherto motionless mass, to increase its mass to $m = m(t)$.

    (a) Show that after time t_0 the velocity of the body is, for some time at least, given by $v(t) = m_0 v_0/m(t)$.

    (b) Assume that $m(t) = m_0 + rt$, $r = \text{const} > 0$, for $t \geq t_0$, and show that the body would attain velocity $v_1 < v_0$ at time $t_1 = (m_0/r)(v_0/v_1 - 1)$.

23. Imagine a body of mass m_0 moving through space on a linear path with constant velocity v_0 and with no external forces acting on it. At time $t = t_0$ the body begins to increase in mass, through uniform accretion of motionless mass according to some formula $m = m(t)$. Note that at all times, both before and after time t_0, the velocity satisfies

$$m \frac{dv}{dt} + v \frac{dm}{dt} = 0.$$

    (a) Show that after time t_0 the velocity of the body is $v(t) = m_0 v_0/m(t)$.

    (b) Find a formula for the distance traveled after time $t_0 = 0$ if the mass at time t is $m(t) = m_0 + at^2$, where a is some positive constant. Show in particular that the distance tends to a finite value s_0 as t tends to infinity, and find s_0.

24. A hailstone with zero initial velocity falling from a considerable height above the earth has an initial mass of 1 gram and is gaining mass by accreting motionless vapor according to the formula $m(t) = 1 + t/10$, where t is measured in seconds. Assume that distance is measured down toward the earth and that $g = 32$ feet per second per second.

    (a) Assuming free fall with no air resistance, find the hailstone's velocity $v(t)$ at general time $t > 0$.

    (b) Assuming the air-resistance term $-0.01v$, find the hailstone's velocity $v(t)$ at general time $t > 0$.

    (c) Estimate the two values for $v(10)$ obtained from parts (a) and (b).

25. A hailstone falling with zero initial velocity from a considerable height above the earth has an initial mass of 1 gram and is losing mass because of evaporation according to the formula $m(t) = 1 - t/10$, where t is measured in seconds. Assume that the vapor particles have velocity zero after they leave the hailstone. Assume also that distance is measured down toward the earth and that $g = 32$ feet per second per second.

    (a) Assuming free fall with no air resistance, find the hailstone's velocity $v(t)$ at time $0 \leq t < 10$.

(b) Assuming the air-resistance term $-0.01v$, find the hailstone's velocity $v(t)$ at time $0 \le t < 10$.

(c) Compare the two values for $v(7)$ obtained from parts (a) and (b). Can you reconcile with reality the behavior of the two formulas for $v(t)$ as t tends to 10?

5   EQUILIBRIUM SOLUTIONS (OPTIONAL)

Given a differential equation $dy/dt = f(t, y)$, we may want to know what happens to the solutions $y(t)$ for very large values of t. If we can find a formula for $y(t)$, we may be willing to accept $\lim_{t \to \infty} y(t)$ as an estimate. Lacking such a formula, we can try a numerical computation for large t and get some satisfaction that way, the drawbacks being that accuracy may deteriorate significantly over a long interval and no sure conclusions can be drawn about limiting behavior. The following example illustrates a third alternative that avoids finding $y(t)$ either by formula or by approximation.

e x a m p l e 1

If an object of mass m falls under constant gravity g against resistance of the form $-kv^{\alpha}$, $\alpha > 0$, then the velocity $v = v(t)$ satisfies

$$m \frac{dv}{dt} = mg - kv^{\alpha}.$$

Solution formulas are sometimes available via separation of variables, but not for general $\alpha > 0$. For example, if $\alpha = 1$, then

$$v = \frac{mg}{k} + Ce^{-kt/m}.$$

The constant C can be determined by imposing an initial condition. Note that no matter what C is, the limit of $v(t)$ as t tends to infinity is mg/k. Not only that, with $C = 0$, the constant solution $y_0(t) = mg/k$ is precisely the limit constant that we arrived at by letting t tend to infinity.

The correspondence found above between the constant solutions mg/k and the limit at infinity of all solutions extends to the case for which α is any positive number. Note that for low velocities, in particular when $mg - kv^{\alpha} > 0$, or $v < \sqrt[\alpha]{mg/k}$, the velocity is increasing, since the differential equation requires that $dv/dt > 0$. Similarly, $dv/dt < 0$, so the velocity decreases when $v > \sqrt[\alpha]{mg/k}$. Figure 2.4 shows the direction field for a particular choice of constants, along with a few solutions. The picture makes it plausible that the solutions actually tend to $\sqrt[\alpha]{mg/k}$; that this is true follows from Theorem 5.1 below. Finally, note that all constant solutions $v = v_0$ can be found without finding any nonconstant solutions; since $dv_0/dt = 0$, the right side of the differential equation becomes $mg - kv_0^{\alpha} = 0$, so solving for $v_0 = \sqrt[\alpha]{mg/k}$

Example 1 shows that a constant solution of a differential equation can have a very special significance because of the possibility that other solutions may

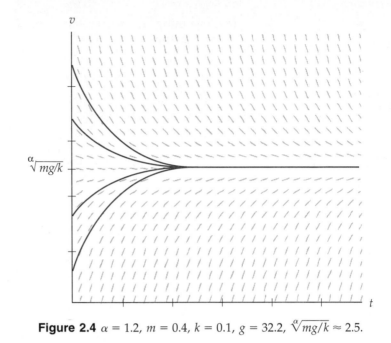

Figure 2.4 $\alpha = 1.2$, $m = 0.4$, $k = 0.1$, $g = 32.2$, $\sqrt[\alpha]{mg/k} \approx 2.5$.

approach it in the long run. Because of the physical context in which constant solutions often appear, the term **equilibrium solution** is often used instead of *constant solution*. For instance, the equilibrium solution $v = \sqrt[\alpha]{mg/k}$ in Example 1 has the property that if you start out at that velocity you maintain it, and if you start out with any other velocity you eventually get arbitrarily close to the equilibrium value. Such an equilibrium is called **stable**. Finding equilibrium solutions for the first-order equation $dy/dt = f(t, y)$ is in principle a simple matter. Since y is to be constant, $dy/dt = 0$, so we just have to find constant solutions y that satisfy the equation $f(t, y) = 0$ identically.

---

example 2

To find all equilibrium solutions to $dy/dt = y^2 - y$, set $dy/dt = 0$ and solve $y^2 - y = y(y - 1) = 0$. The solutions are $y = 0$ and $y = 1$, so the two equilibrium solutions are $y_1(t) = 0$ and $y_2(t) = 1$. For values of y between 0 and 1, all the slopes of the direction field are negative, so solution graphs with initial values in that interval decrease, apparently toward the value 0 of the equilibrium solution $y_1(t) = 0$. See Figure 2.5a.

---

example 3

The differential equation $dy/dt = (y - 1)/(t^2 + 1)$ has an equilibrium solution corresponding to each fixed value of y for which the right side is identically zero. The unique equilibrium solution is $y_0(t) = 1$. Note that the direction field has positive slopes if $y > 1$ and negative slopes if $y < 1$. Thus solutions with initial values more than 1 increase in value, and solutions with initial values less than 1 decrease in value. Hence all solutions but $y_0(t) = 1$ diverge away from 1 as t increases, as in Figure 2.5b. The solution $y_0(t) = 1$ is an example of an **unstable equilibrium,** defined generally in Chapter 6, Section 8.

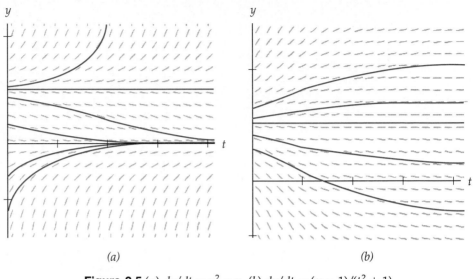

Figure 2.5 (a) $dy/dt = y^2 - y$. (b) $dy/dt = (y - 1)/(t^2 + 1)$.

The differential equation $dy/dt = e^y$ has no equilibrium solutions at all. The reason is that the right-hand side is never zero. Hence dy/dt can't be zero, so there can't be a constant solution.

If initial-value problems for a first-order differential equation don't always have unique solutions, the behavior of the direction field may not be an accurate indication of the relation of equilibrium solutions to the other solutions. The next example illustrates this point.

example 5

The differential equation $dy/dt = y^{2/3}$ has a single equilibrium solution: $y_0(t) = 0$. Some other solutions can be found by separation of variables: $y_c(t) = (t + c)^3/27$. Still more solutions can be pasted together from pieces of these:

$$y = \begin{cases} y_c(t), & t < c, \\ 0, & c \le t. \end{cases} \qquad y = \begin{cases} 0, & t < c, \\ y_c(t), & c \le t. \end{cases}$$

Some of these solutions are sketched in Figure 2.6 along with the direction field for $0 \le t \le 6$. The slopes of the direction field tend to zero as y tends to 0, so the slopes of the solution graphs flatten out as y approaches zero. However, some solutions actually cross the path of the equilibrium solution, while others become identical with it from some point on.

Equations of the special form $dy/dt = f(y)$ arise often in practice and are called **autonomous**, becuase f has no explicit dependence on t. The following theorem allows us to draw conclusions about long-term behavior of their solutions relative to equilibrium values. The result seems quite intuitive except that some condition on $f(y)$ is needed to avoid a difficulty posed by Example 5.

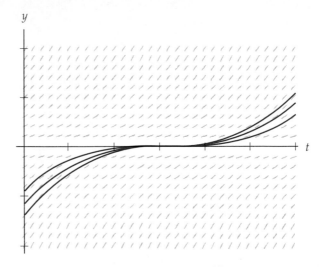

Figure 2.6 $dy/dt = y^{2/3}$.

Theorem 8.7 in Chapter 6, Section 8B, complements the theorem in the context of higher-dimensional equations.

■ 5.1 THEOREM

Let $y(t)$ be a solution of $dy/dt = f(y)$, and suppose that $f(y)$ has a continuous derivative in some interval containing an equilibrium point y_0.

(a) If $f(y) > 0$ for $y(t_0) \le y < y_0$ or $f(y) < 0$ for $y_0 < y \le y(t_0)$, then $\lim_{t \to \infty} y(t) = y_0$.

(b) If $f(y) < 0$ for $y(t_0) \le y < y_0$ or $f(y) > 0$ for $y_0 < y \le y(t_0)$, then $\lim_{t \to -\infty} y(t) = y_0$.

Proof. We'll treat only the first assumption in case (a), since the others are quite similar. Because $dy/dt = f(y) > 0$ for some value t_0, $y(t)$ continues to increase for $t > t_0$ as long as $y(t) < y_0$. But $y(t) = y_0$ is impossible, because the uniqueness theorem (Theorem 2.1 of Chapter 1, Section 2C) won't allow the graphs of $y(t)$ and the constant solution y_0 to have points in common. Furthermore, since $y(t)$ is continuous, its graph can't skip over the value y_0. Hence, $y(t)$ remains strictly bounded above by y_0, and as an increasing function it has a finite limit $\bar{y} \le y_0$ at infinity. To show $\bar{y} = y_0$, suppose otherwise. Since $f(y)$ is continuous,

$$\lim_{t \to \infty} y'(t) = \lim_{t \to \infty} f(y(t)) = f(\lim_{t \to \infty} y(t)) = f(\bar{y}) > 0.$$

We assumed $\bar{y} < y_0$, so $f(\bar{y}) > 0$ by (a). By the mean-value theorem, there is some number τ with $t < \tau < t + 1$ such that $y(t + 1) - y(t) = y'(\tau)$. But if $\lim_{t \to \infty} y'(t) = f(\bar{y}) > 0$, then $y(t)$ would have to be unbounded, with slope tending to $f(\bar{y})$, a contradiction.   ■

e x a m p l e **6**

The logistic population growth equation $dy/dt = ky(c - y), k, c > 0$ has equilibrium solutions $y_0(t) = 0$ and $y_c(t) = c$. Since $f(y) = ky(c - y) > 0$ in the interval $0 < y < c$, case (a) of Theorem 5.1 applies and implies that a solution $y(t)$ that starts with initial value in the interval $0 < y < c$ will tend to c as t tends to $+\infty$.

For an initial value bigger than c, we're in the interval $c < y < \infty$, where $f(y) = ky(c - y) < 0$. Hence the second assumption in part (a) of the theorem applies, so we still have limit c, approached from above this time. The solution $y_c(t) = c$ is an example of an **asymptotically stable equilibrium** in that solutions starting close enough to it approach it as t tends to infinity.

---

example 7

The logistic equation of Example 6 can be modified to include a fixed negative term $-h$ on the right side: $dy/dt = ky(c - y) - h$. The constant h is usually interpreted as a *harvest rate* that might be applied to manage a population of some species. To find equilibrium solutions we solve for y in the quadratic equation $ky(c - y) - h = 0$ or $-ky^2 + kcy - h = 0$. It turns out that there are distinct real solutions only when $h < kc^2/4$, in which case we have the two positive equilibrium solutions

$$y_m = \frac{c}{2} - \sqrt{\left(\frac{c}{2}\right)^2 - \frac{h}{k}}, \qquad y_M = \frac{c}{2} + \sqrt{\left(\frac{c}{2}\right)^2 - \frac{h}{k}}.$$

Thus $f(y)$ factors to $f(y) = -k(y - y_m)(y - y_M)$ and is positive if $y_m < y < y_M$. Part (a) of Theorem 5.1 shows that given an initial value between the two equilibrium values $y_m < y_M$ a solution will tend to the larger value y_M as t tends to infinity. Other possibilities are considered in the exercises.

Here is a convenient way to display the direction field of a differential equation $y' = f(y)$ together with the graph of the function $f(y)$ so that you can see how the information in one relates to the other. Doing this requires us to align the y axes in both so that they are parallel to each other. Our choice in Figure 2.7 is to rotate the graph of f through 90° clockwise from its normal position with the y axis horizontal, while leaving the sketch of the direction field alone. Putting the $f(y)$ axis on the same level as the t axis for the vector field results in a horizontal alignment of direction field slopes y' with the corresponding points on the graph of $f(y)$. In particular, horizontal slopes occur at the level at which the graph of f crosses its y axis, and positive slopes correspond to points at which $f(y) > 0$.

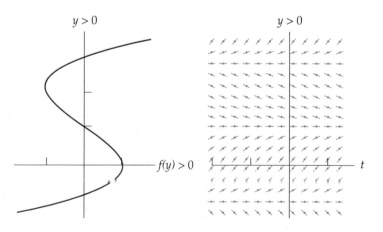

Figure 2.7 Graph of $f(y)$ and slopes $f(y)$.

EXERCISES

Determine which of the following differential equations has one or more equilibrium solutions, and for those that do, find all equilibrium solutions. Then determine the long-term behavior of the nonconstant solutions by applying Theorem 5.1 or by finding explicit solution formulas.

1. $dy/dt = -y$.

2. $dy/dt = y^2$.

3. $dy/dt = ty$.

4. $dy/dt = \sin y$.

5. $dy/dt = 1 - y^2$.

6. $dy/dt = y - t$.

7. $du/dt = u(u - 1)$.

8. $dv/dt = 1 + 3v + v^2$.

9. $dy/dt = y(1 - e^y)$.

10. A car of mass m is propelled along a track with constant acceleration a against a frictional force of magnitude kv^3.

   (a) Find the differential equation satisfied by the velocity $v = v(t)$.

   (b) Find all equilibrium velocities.

   (c) Use Theorem 5.1 to describe the behavior of solutions to the differential equation in part (a) having positive initial velocities.

11. A body of mass m falls under the influence of constant gravitational acceleration g and with air resistance of the form $k(v^2 + 2v)$, $v \geq 0$.

   (a) Find all equilibrium velocities.

   (b) Are all your answers to part (a) physically meaningful? Explain.

12. A body of mass m falls under the influence of constant gravitational acceleration g and with air resistance of the form kv^α. If the initial velocity is equal to the equilibrium velocity, how long does it take for the body to fall distance h?

13. A body of mass m falls under the influence of constant acceleration g and with air resistance of the form $k \ln (1 + v)$. Find the equilibrium velocity.

14. A body of mass m falls under the influence of constant gravitational acceleration g and with air resistance of the form $k \tan (\alpha v)$, $\alpha > 0$.

   (a) Find the equilibrium velocity.

   (b) Show that the equilibrium velocity can be at most $\pi/(2\alpha)$, regardless of the choices for k, g, and m, if $v(0) < \pi/2$.

15. Define a function of velocity v by

$$F(v) = \begin{cases} v, & 0 \leq v \leq 1, \\ (v^2 + 1)/2, & 1 < v. \end{cases}$$

Suppose an object of mass m falls under the influence of constant gravity g and retarding force $kF(v)$, $k > 0$.

   (a) Sketch the graph of $F(v)$ for $0 \leq v \leq 3$.

   (b) Find the equilibrium velocities.

16. (a) Newton's law of cooling is expressed by the differential equation $du/dt = -k(u - a)$, where k and a are positive constants. The equation has a single equilibrium solution; what is it?

(b) Stefan's law is expressed by the differential equation $du/dt = -k(u^4 - a^4)$, where k and a are positive constants. The equation has a single equilibrium solution; what is it?

17. Consider Newton's law of cooling with variable ambient temperature $a(t) = e^{-t}$: $du/dt = -k(u - e^{-t})$, $k > 0$.

   (a) Show that the differential equation has no equilibrium solutions.

   (b) Show that every solution of the differential equation tends to zero as t tends to infinity.

18. Consider solutions of the logistic equation $dy/dt = y(1 - y) - h$ modified by the constant harvest rate h.

   (a) For what positive values of h are there two equilibrium solutions?

   (b) What are the equilibrium solutions corresponding to the values of h you found in part (a)?

   (c) Describe the behavior of a solution with initial value between the equilibrium values found in part (a).

   (d) Describe the behavior of a solution with initial value below the lower equilibrium value found in part (b).

   (e) Describe the behavior of a solution with initial value above the upper equilibrium value found in part (b).

   (f) What is the special significance of the value $h = \frac{1}{4}$? In particular, would you expect this precise value to be significant in an attempt to use the differential equation to model the history of a real population?

19. Consider soutions of the logistic equation $dy/dt = ky(c - y) - h$ modified by the constant harvest rate h.

   (a) For what positive values of h are there two distinct equilibrium solutions?

   (b) What are the equilibrium solutions corresponding to the values of h you found in part (a)?

   (c) Describe the behavior of a solution with initial value between the equilibrium values found in part (b).

   (d) Describe the behavior of a solution with initial value below the lower equilibrium value found in part (b).

   (e) Describe the behavior of a solution with initial value above the upper equilibrium value found in part (b).

   (f) What is the special significance of the value $h = kc^2/4$? In particular, would you expect this precise value to be significant in an attempt to use the differential equation to model the history of a real population?

20. Consider solutions of the general logistic equation $dy/dt = y^\alpha(1 - y)^\beta$, α, β positive constants, $0 \le y \le 1$.

   (a) What are the equilibrium solutions?

   (b) Describe the behavior of a solution with initial value between the equilibrium values found in part (b).

21. Consider solutions of the general logistic equation $dy/dt = ky^\alpha(c - y)^\beta - h$, $\alpha, \beta > 0$, modified by the constant harvest rate h, assuming $c > 0$.

    (a) Show that $f(y) = ky^\alpha(c - y)^\beta - h$ has a unique maximum value on the interval $0 \le y \le c$ at $y = \alpha c/(\alpha + \beta)$.

    (b) For what positive values of h are there two distinct equilibrium solutions?

    (c) Describe the behavior of a solution with initial value between the equilibrium values identified in part (b).

22. Example 5 preceding Theorem 5.1 shows that requiring only continuity of $f(y)$ in the statement of the theorem would be insufficient. Explain in detail how the example allows us to reach this conclusion.

23. **Bifurcation.** A value λ_0 for a real parameter λ in a family of differential equations is a **bifurcation point** if solutions change character in a fundamental way as λ passes through λ_0. Show the $\lambda = 0$ is a bifurcation point for $dy/dt = \lambda y - y^3$ as follows.

    (a) Show that for $\lambda \le 0$ all solutions tend to the equilibrium value $y = 0$ as t tends to infinity.

    (b) Show that for $\lambda > 0$, if a solution has a value $y(t_0)$ in the interval $-\sqrt{\lambda} < y < \sqrt{\lambda}$, then that solution tends away from the equilibrium value $y = 0$ as t tends to infinity.

6 NUMERICAL METHODS

A solution

$$y = f(x)$$

for a first-order differential equation

$$\frac{dy}{dx} = F(x, y)$$

can sometimes be expressed by a formula such as in the simple example

$$y' = -2y; \qquad y = ce^{-2x},$$

where c is a parameter that depends on an initial condition. The importance of such formulas is that they often show quite easily how the values of the solution depend not only on the independent variable x but also on the parameters; in our example, it is quite clear that y tends to 0 as x tends to ∞, regardless of what value c has. However, when solution formulas are not readily available, we rely on numerical approximation, and if our purpose in finding a solution formula is simply to generate numerical values, it may be preferable to bypass the formula and work directly with the differential equation, as we do in this section.

6A Euler Method

Suppose we are given a first-order differential equation, together with an initial condition, that we can't solve in an elementary way, for example,

$$y' = x^2 + y^2, \qquad y(-2) = -1.$$

We want to find approximate values for a solution $y = y(x)$. Since we can make only finitely many estimates, we restrict ourselves to some interval of values, say $x_0 \leq x \leq b$. We then try to approximate $y(x)$ at the points

$$x_1 = x_0 + h,\, x_2 = x_0 + 2h,\, \ldots,\, x_m = x_0 + mh,$$

where m is some chosen number of subdivisions of the interval from x_0 to b, each of the same length

$$h = \frac{b - x_0}{m}.$$

To make our estimates, we approximate the graph of $y = y(x)$ by a sequence of straight-line segments chosen from the direction field of our differential equation

$$y' = F(x, y),$$

starting with the initial condition $y(x_0) = y_0$. The segment of the direction field through (x_0, y_0) has slope $F(x_0, y_0)$, and we extend that segment until it intersects the vertical line $x = x_1$ at a point (x_1, y_1). We then simply repeat the process with (x_1, y_1), replacing (x_0, y_0) to get (x_2, y_2), and so on, to (x_m, y_m). Figure 2.8 shows such a sequence of approximating segments along with the solution curve that they approximate. The entire sequence is called an **Euler polygon.** Each segment is determined by its initial point (x_k, y_k) and the slope $F(x_k, y_k)$; the equation of the line containing the segment is thus

$$y - y_k = F(x_k, y_k)(x - x_k)$$

or

$$y = y_k + F(x_k, y_k)(x - x_k).$$

On this line, the y value corresponding to $x = x_{k+1}$ is then

$$y_{k+1} = y_k + F(x_k, y_k)(x_{k+1} - x_k),$$

or, since $h = x_{k+1} - x_k$,

$$y_{k+1} = y_k + hF(x_k, y_k).$$

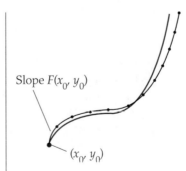

Slope $F(x_0, y_0)$

(x_0, y_0)

Figure 2.8 Solution and Euler polygon.

The preceding formula defines the **Euler approximation.** Starting with x_0 and y_0, we get a table of approximations to the solution $y = y(x)$.

k	x_k	$y_{k+1} = y_k + hF(x_k, y_k)$
0	x_0	y_0
1	x_1	$y_1 = y_0 + hF(x_0, y_0)$
2	x_2	$y_2 = y_1 + hF(x_1, y_1)$
3	x_3	$y_3 = y_2 + hF(x_2, y_2)$
$\vdots$	$\vdots$	$\vdots$

e x a m p l e 1

The differential equation

$$y' = xy$$

has solutions $y = ce^{x^2/2}$, which we will compare with the results of our numerical approximation method. We have

$$y_{k+1} = y_k + hF(x_k, y_k)$$
$$= y_k + hx_k y_k.$$

With $h = 0.1$, the next table shows the comparison, with the values of the exact solution rounded off. What we get is a fair approximation for some purposes, for example, for drawing a graph. Notice that the accuracy deteriorates as we go down the columns, because the errors compound each other. We have started with initial values $(x_0, y_0) = (0, 1)$.

x_k	y_k	$e^{(1/2)x_k^2}$
0.0	1	1
0.1	1	1.005
0.2	1.01	1.020
0.3	1.03	1.046
0.4	1.06	1.083
0.5	1.10	1.133
0.6	1.16	1.197
0.7	1.23	1.277
0.8	1.31	1.377
0.9	1.42	1.499
1.0	1.55	1.647

A computing routine to print approximate values of x and y for x between 0 and 1, with step size $h = 0.01$, might look like this:

```
DEFINE F(X, Y) = X * Y

SET X = 0

SET Y = 1

SET H = 0.01

DO WHILE X < 1

SET Y = Y + H * F(X, Y)

SET X = X + H

PRINT X, Y

LOOP
```

To improve accuracy, we could increase m, the number of subdivisions of the interval, from x_0 to b, thus making h smaller. Of course, increasing the number of subdivisions increases the likelihood of significant round-off error in the arithmetic. Rather than recklessly decreasing h, we prefer to use a simple modification of the method, described next, aimed at giving a smaller error at each step without decreasing the step size.

6B   Improved Euler Method

To improve the accuracy of the Euler method we use a process often called **prediction-correction;** this assigns a slope to each approximating segment that is a corrected average of the Euler slope and of what the next Euler slope would be if we were to take a single Euler step as prediction.

We will now use y_{k+1} to denote our improved approximate value at the $(k + 1)$th step, and we will use p_{k+1} for the corresponding simple Euler approximation, based on a previously computed value y_k. We follow these steps to approximate the solution to $y' = F(x, y)$, $y(x_0) = y_0$:

1. Compute the slope $F(x_k, y_k)$.
2. Determine a predictor estimate p_{k+1} by $p_{k+1} = y_k + hF(x_k, y_k)$.
3. Compute the average of the two slopes $F(x_k, y_k)$ and a modifier, or corrector, slope $F(x_{k+1}, p_{k+1})$ and use it to determine y_{k+1} by

$$y_{k+1} = y_k + h \frac{F(x_k, y_k) + F(x_{k+1}, p_{k+1})}{2}.$$

The formula for y_{k+1} comes from writing the equation for the line through (x_k, y_k) with the average slope and then setting $x = x_{k+1}$ to get the corresponding value $y = y_{k+1}$. The general idea is illustrated in Figure 2.9. A computing routine to implement the method for the initial-value problem $y' = xy$, $y(0) = 1$

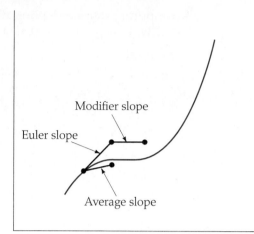

Figure 2.9 Improved Euler slope is an average.

might look like this, with step size $h = 0.01$, printing values for x between 0 and 1:

```
DEFINE F(X, Y) = X * Y

SET X = 0

SET Y = 1

SET H = 0.01

DO WHILE X < 1

SET P = X + H * F(X, Y)

SET Y = Y + (H/2) * (F(X, Y) + F(X + H, P))

SET X = X + H

PRINT X, Y

LOOP
```

At each stage the value p_k is the *prediction* and y_k is the *correction*. If $h < 0$, the approximations move from larger to smaller x values.

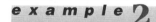 **2**

Here is the improved Euler approximation for the same problem we looked at in Example 1: $y' = xy$ with $(x_0, y_0) = (0, 1)$, and step size $h = 0.1$. We have $F(x, y) = xy$, so

$$p_{k+1} = y_k + hF(x_k, y_k)$$
$$= y_k + 0.1x_k y_k$$

and

$$y_{k+1} = y_k + \frac{h}{2} \left(F(x_k, y_k) + F(x_{k+1}, p_{k+1}) \right)$$
$$= y_k + 0.05(x_k y_k + x_{k+1} p_{k+1}).$$

x_k	p_k	y_k	$e^{(1/2)x_k^2}$
0	1	1	1
0.1	1	1.005	1.005
0.2	1.015	1.020	1.020
0.3	1.041	1.046	1.046
0.4	1.077	1.083	1.083
0.5	1.127	1.133	1.133
0.6	1.190	1.197	1.197
0.7	1.269	1.277	1.277
0.8	1.367	1.377	1.377
0.9	1.487	1.499	1.499
1.0	1.634	1.648	1.649

The values of p_k, y_k, and the exact solution are rounded back to three decimal places. Notice that there is a difference between the last two columns only at $x = 1$. The arithmetic required in the preceding examples can be done quite easily with a hand calculator. However, if you want many millions of steps, a programmed digital computer is probably essential. It is important to remember that reducing step size h cannot indefinitely improve the accuracy of the approximations. The increased amount of arithmetic required will eventually cause enough round-off error such that further reduction of h is self-defeating. It can be shown that the error made at each step by using an approximate formula, called the **local formula error,** is of order h^2 for the Euler method and h^3 for the improved Euler method. Thus, if taking $h = 0.01$ can be expected to produce about 0.0001 amount of error in one Euler step, it can be expected to make an error of about 0.000001 in one improved Euler step. The advantage of the modification is obvious, because to get the latter accuracy from the simple Euler method would require 1000 times as many steps as the improved method over a given interval, thereby increasing by a factor of 1000 the **round-off error** inherent in the arithmetic, in particular in the number of significant digits in the calculations. As a practical matter we often try a reasonable-looking step size and then accept the results if cutting the step size in half produces no significant change in the final outcome. Such checks are important, because it can happen that under adverse assumptions the accumulated error over an x interval of length L can grow exponentially as a function of L.

EXERCISES

At the Web site http://math.dartmouth.edu/~rewn/ the Applets EULER, 1STORD, 1ORDPLOT, and RUNKUT implement, in order, the Euler method with numeric output, the improved Euler method with numeric output, the improved Euler method with graphic output, and the 4th-order Runge-Kutta method with numeric output.

1. Each of the following differential equations has a unique solution, expressible in terms of elementary functions, that satisfies the prescribed condition. First find the solution formula; then compare the values given by this formula with approximate values that you compute using the Euler method. Use at least five equally spaced steps on the given interval if you compute by hand and at least 20 steps if you use a computer.

   (a) $y' = y$ for $0 \leq x \leq 0.5$ with $y = 1$ when $x = 0$.

   (b) $y' = y + x$ for $0 \leq x \leq 0.5$ with $y = 0$ when $x = 0$.

2. Substitute the improved Euler method for the Euler method in each part of Exercise 1.

3. Consider the differential equation
   $$y' = 1 + y^2 \qquad \text{for} \qquad -1 \leq x \leq 1.$$

   (a) Sketch the direction field of the differential equation.

   (b) Sketch a solution curve passing through $(x, y) = (0, 0)$.

   (c) Apply the Euler method to approximate the solution in part (b) at points $x_k = k/5$ for $k = 1$ through $k = 5$.

   (d) Use the improved Euler method to print a table of values for the solution satisfying $y(0) = 0$ for $-1 \leq x \leq 1$.

   (e) Compare the results of (d) with the solution $y = \tan x$.

4. The initial-value problem
   $$y' = f(x), \qquad y(x_0) = y_0$$
   has the solution
   $$y = y_0 + \int_{x_0}^{x} f(t)\, dt.$$

   (a) Show that applying the Euler method to the differential equation is equivalent to approximating the solution integral by Riemann sums (i.e., by using the rectangle rule).

   (b) Show that the improved Euler method for the differential equation is equivalent to using the trapezoid rule for approximating the solution integral.

Euler Polygon. This term is used to denote the connected sequence of line segments joining the successive points (x_k, y_k) of an Euler approximate solution of $y' = f(x, y)$. (It can be proved that for a carefully chosen sequence of step sizes tending to 0, the resulting Euler polygons will converge on some interval to the graph of a solution if $f(x, y)$ is continuous.) Plot the Euler polygon for the following initial-value problems on the indicated intervals and with the specified step size h. If possible, use a solution by formula for comparison; otherwise use a sketch of the direction field.

5. $y' = y$, $y(0) = 1$, $0 \leq x \leq 2$; $h = \frac{1}{4}$.

6. $y' = x + y$, $y(0) = 0$, $0 \leq x \leq 2$; $h = \frac{1}{4}$.

7. $y' = x - y$, $y(0) = 0$, $0 \le x \le 4$; $h = \frac{1}{2}$.

8. $y' = xy$, $y(0) = 1$, $0 \le x \le 1$; $h = \frac{1}{4}$.

9. Use computer graphics to plot a sketch of the direction field of the first-order differential equation $y' = F(x, y)$ listed below and then plot approximate solutions to initial-value problems determined by conditions of the form $y(x_0) = y_0$. Apply the routine to each of the following differential equations, plotting the direction field for points (x, y) in the rectangle $|x| \le 4$, $|y| \le 4$, and plotting the solution satisfying $y(-1) = -1$.

    (a) $y' = -y + x$.   **(d)** $y' = \sin xy$.

    (b) $y' = xy$.   **(e)** $y' = \sqrt{1 + y^2}$.

    (c) $y' = \sin y$.   **(f)** $y' = \sqrt{1 + x^2}$.

10. The 4th-order **Runge-Kutta** approximations to the solution of $y' = F(x, y)$, $y(x_0) = y_0$, are computed as follows:

$$y_{k+1} = y_k + \frac{h}{6}(m_1 + 2m_2 + 2m_3 + m_4),$$

where
$$m_1 = F(x_k, y_k)$$

$$m_2 = F\left(x_k + \frac{h}{2}, y_k + \frac{h}{2}m_1\right)$$

$$m_3 = F\left(x_k + \frac{h}{2}, y_k + \frac{h}{2}m_2\right)$$

$$m_4 = F(x_k + h, y_k + hm_3).$$

(The local formula error is of order h^5.) Apply a computer algorithm to implement the routine. Then compare the results obtained from it for the problem

$$y' = xy, \qquad y(0) = 1,$$

with the results obtained from the improved Euler method, as well as with the solution formula $y = e^{x^2/2}$. Try step sizes $h = 0.1$, 0.01, and 0.001.

11. Show that the successive application of the Taylor approximations

$$y_{n+1} \approx y_n + hy'_n + \tfrac{1}{2}h^2 y''_n \qquad \text{and} \qquad y'_{n+1} \approx y'_n + hy''_n$$

lead to the improved Euler formula $y_{n+1} \approx y_n + \tfrac{1}{2}h(y'_n + y'_{n+1})$.

12. The initial-value problem

$$y' = \sqrt{y}, \qquad y(0) = 0,$$

has two distinct solutions for $x \ge 0$:

$$y_1(x) = 0 \qquad \text{and} \qquad y_2(x) = \frac{x^2}{4}.$$

    (a) Show that direct application of the Euler method produces y_1 exactly at the points of approximation.

    (b) Show that applying the Euler method to

$$y' = \sqrt{y}, \qquad y(x_0) = \frac{r_0^2}{4},$$

for small positive x_0 produces an approximation to y_2.

13. Consider the generalized **logistic equation**

$$\frac{dP}{dt} = kP^\alpha \left(1 - \frac{P^\beta}{m}\right).$$

(a) Let $k = 1$, $m = 5$, and $P(0) = 1$. Find numerical approximations to the solution in the range $0 \le t \le 10$ for the parameter pairs $(\alpha, \beta) = (0.5, 1)$, $(0.5, 2)$, $(1.5, 1)$, $(1.5, 2)$, $(2, 2)$.

(b) Estimate a parameter pair (r, q) that yields approximately the values $P(0) = 1$, $P(2) = 2.4$, $P(4) = 2.9$.

14. A 100-gallon mixing tank is full at the time $t = 0$. Salt solution at a concentration of 1 pound per gallon is thereafter added at a decreasing rate of $e^{-0.5t}$ gallons per minute, and solution is drawn from the tank so as to maintain the 100-gallon level.

(a) Find the first-order equation satisfied by the amount $S(t)$ of salt in the tank at time t minutes.

(b) Find a numerical approximation to $S(t)$ in the range $0 \le t \le 20$, assuming that $S(0) = 0$. Sketch the graph of the solution.

15. The rocket equation

$$m\frac{dv}{dt} = -v_e \frac{dm}{dt} + F$$

becomes a first-order equation for $v(t)$ if we specify time-dependent mass $m(t) = m_0 + m_1 - at$ for $0 \le t \le m_0/a$, and $m(t) = m_1$ for $m_0/a < t$. Let the external force F be given by

$$F(t) = -gm(t) - kv^\alpha,$$

where g is the acceleration of gravity and k and α are positive constants.

(a) Find numerical approximations to $v(t)$ in the range $0 \le t \le m_0/a$, given that $g = 32.2$, $v(0) = 0$, $m_0 = 25$, $m_1 = 75$, $a = 1$, $v_e = 3500$, and $\alpha = 1.5$. Use successively $k = 2, 1, 0.5, 0.1$, and sketch the corresponding graphs for $v = v(t)$ using an appropriate scale.

(b) Estimate the values of k in part (a) that would produce approximately the values $v(5) = 17.5$, $v(10) = 40$, $v(15) = 63$, $v(20) = 87$, if $v(0) = 0$.

(c) Repeat part (a) for the range $0 \le t \le 50$, noting that $m(t)$ cannot be described by a single elementary function.

16. Following is a generalization of the **logistic equation:**

$$\frac{dP}{dt} = kP\left(1 - \frac{P}{L(t)}\right).$$

Here $L(t)$ acts to limit population size $P(t)$. Let $k = 1$ and $P(0) = \frac{1}{2}$, and use numerical analysis to sketch the graph of $P(t)$ for $0 \le t \le 25$. In each case compare the graph of $P(t)$ with that of $L(t)$.

(a) $L(t) = 3 + \sin t$.

(b) $L(t) = \ln(2 + t)$.

(c) $L(t) = t + 1$.

17. A hailstone falling with zero initial velocity from a considerable height above the earth has an initial mass of 1 gram and is gaining mass through freezing according to the formula $m(t) = (1 + t/50)^3$, where t is measured in seconds. Assume that distance is measured down toward the earth and that $g = 32$ feet per second per second.

    (a) Assuming free fall with no air resistance, find a formula for the hailstone's velocity $v(t)$ for time $t > 0$.

    (b) Assuming the air-resistance term $-0.01v$, generate a computer plot of the hailstone's velocity $v(t)$ for $0 \le t \le 40$, and compare it with a computer plot of the graph of the solution found in part (a).

18. A hailstone falling with zero initial velocity from a considerable height above the earth has an initial mass of 1 gram and is losing mass because of melting according to the formula $m(t) = (1 - t/50)^3$, where t is measured in seconds. Assume that distance is measured down toward the earth and that $g = 32$ feet per second per second.

    (a) Assuming free fall with no air resistance, find a formula for the hailstone's velocity $v(t)$ for time $t > 0$.

    (b) Assuming the air-resistance term $-0.01v^{0.5}$, generate a computer plot of the hailstone's velocity $v(t)$ for $0 \le t \le 40$, and compare it with a computer plot of the graph of the solution found in part (a).

19. A car can be propelled along a smooth track by ejecting sand from the rear of the car at a fixed rate of 20 pounds per second and with velocity relative to the car of 10 feet per second. Assume that the car together with its occupants weighs 1000 pounds and that it carries initially an additional 500 pounds of sand.

    (a) If the car starts with initial velocity $v(0) = 100$ feet per second and frictional forces retard the car by an amount $F(t) = -0.01v(t)^2$, write down the differential equation satisfied by $v(t)$.

    (b) Find the approximate velocity of the car 25 seconds later, when the sand is all used up.

    (c) Find the approximate velocity of the car 25 seconds after the instant at which the sand is all used up. Is it reasonable for this velocity to be less than 100 feet per second?

20. A generalized logistic equation for growth of a population $P = P(t)$ is

$$\frac{dP}{dt} = KP^{\alpha}(c - P)^{\beta}, \qquad \alpha, \beta > 0.$$

With $\alpha = \beta = 2$, $K = 3$, and $c = 4$, estimate the long-term value of $P(t)$ if

    (a) $P(0) = 1$.

    (b) $P(0) = 5$.

    (c) Explain the results of parts (a) and (b) on the basis of the behavior of the direction field of the differential equation.

21. Consider a 100-gallon tank from which mixed solution is being drawn off at the rate of 4 gallons per minute. Salt solution is being added at the rate of 3 gallons per minute, and the salt concentration of the added solution is gradually increasing from time $t = 0$ according to the formula $(1 - e^{-t})$ pounds per gallon.

(a) When will the tank be effectively empty?

(b) Find a differential equation for the amount $S(t)$ of salt in the tank over the relevant time interval.

(c) Use a numerical solution method to approximate the solution satisfying the initial condition $S(0) = 10$. Then estimate the maximum total amount of salt in the tank during the process and the time at which the maximum occurs.

(d) Explain what goes wrong when you try to find an elementary solution formula for the differential equation.

22. An object weighing 1600 pounds is dropped under the influence of constant gravitational acceleration 32 feet per second per second and with air-resistance force of magnitude $0.1 \ln (1 + v)$. Use a numerical method to estimate the time it takes for the velocity v to reach 500 feet per second. Recall that weight equals mass times gravitational acceleration.

23. An object weighing 640 pounds is dropped under the influence of constant gravitational acceleration 32 feet per second per second and with air-resistance force of magnitude $0.1v^{1.2}$. Use a numerical method to estimate the time it takes for the velocity v to reach 500 feet per second.

6C Newton's Method

Finding approximate solutions to an equation $f(x) = 0$ is one of the most frequently occurring problems in pure and applied mathematics. Newton's method for approximating solutions to this equation can be looked at as a special case of the numerical solution of a differential equation as follows. To locate a point x^* such that $f(x^*) = 0$, try to find a function $x = x(t)$ such that $f(x(t))$ converges rapidly to 0 as t tends to infinity. An effective way to do this is to ask that $f((x(t)) = Ce^{-t}$ for some constant C and for t tending to infinity. Differentiating this equation with respect to t gives

$$f'(x)\dot{x} = -Ce^{-t} = -f(x), \qquad \text{or} \qquad \dot{x} = -\frac{f(x)}{f'(x)}.$$

We started with a family of implicit solutions to these differential equations, namely, $f(x) = Ce^{-t}$. However, since what we want is $x = x(t)$, and we supposedly can't even solve $f(x) = 0$ directly, we try something else. The idea here is apply Euler's method, an example of a **discrete dynamical system,** that is, a system with time measured in discrete jumps. We apply the method to the second form of the differential equation, with step size h:

$$x_{n+1} = x_n - h\frac{f(x_n)}{f'(x_n)}.$$

The value x_n is only an approximation to $x(nh)$, but since $f(x(nh))$ approaches zero exponentially, we can expect similarly rapid convergence of $f(x_n)$ to zero.

Two choices remain to be made, the starting value x_0, which should be somewhere near a number x^* such that $f(x^*) = 0$, and a value for h. Customary usage has it that Newton's method uses $h = 1$, but other choices are also interesting. Generally speaking, making $0 < h < 1$ requires more steps to reach an accurate result, while making h significantly larger than 1 increases the risk of an erratic response from the method. However, we'll see that larger values of h are sometimes particularly desirable.

example 1

Table 1 lists the values of x_n when $f(x) = e^x - 2$ for three choices of h. We've used $x_0 = 2$ for the starting value each time. The exact solution to $e^x - 2 = 0$ is $\ln 2 \approx 0.693147$. Note that we arrive at this value after just five steps when $h = 1$. When $h = 2$, the numbers x_n oscillate about the true value for many thousands of steps before producing even two-place accuracy.

Table 1: Newton-Euler Approximations to $\ln 2 \approx 0.693147$

$h = \frac{1}{2}$	$h = 1$	$h = 2$
2	2	2
1.63534	1.27067	.541341
1.33022	.831957	.86921
1.09464	.702351	.546341
.929301	.693189	.862601
.82413	.693147	.550853
.762747	.693147	.856685
.72913	.693147	.554954
.711459	.693147	.851349
.702386	.693147	.558704
.697788	.693147	.846503
.695473	.693147	.562152
.694311	.693147	.842077
.69373	.693147	.565336
.693438	.693147	.838013
.693293	.693147	.56829
.69322	.693147	.834263
.693184	.693147	.57104
.693165	.693147	.83079
.693156	.693147	.57361
.693152	.693147	.827561

example 2

The bias in favor of step size $h = 1$ is strongly supported by the geometric significance of the method in that case. This is the usual introduction to the method, and it goes as follows. To estimate a solution x^* of $f(x) = 0$, make a

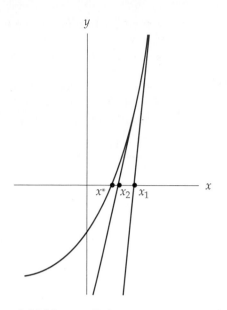

Figure 2.10 Newton-Euler approximations; $h = 1$.

good initial guess x_0 and draw the tangent line to the graph of $y = f(x)$ at $(x_0, f(x_0))$. This line is the graph of

$$y = f(x_0) + f'(x_0)(x - x_0);$$

the graph of this line crosses the x axis when $y = 0$, that is, when $f(x_0) + f'(x_0)(x - x_0) = 0$. Solving the last equation for $x = x_1$ gives the classical Newton formula for the next estimate

$$x_1 = x_0 - \frac{f(x_0)}{f'(x_0)} \quad \text{and in general} \quad x_{n+1} = x_n - \frac{f(x_n)}{f'(x_n)},$$

replacing x_0 by x_1, then x_1 by x_2, and so on. This is our Euler approximation formula for the differential equation $\dot{x} = -f(x)/f'(x)$ with step size $h = 1$. Figure 2.10 shows the graph of $y = e^x - 2$ together with the intersections of its tangent lines with the x axis at $x_1 \approx 1.27067$ and $x_2 \approx 0.831957$.

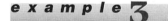 **e x a m p l e 3**

Transcendental equations such as $320t + 2200(1 - e^{-0.1t}) = 10{,}000$ come up often in applied mathematics. To apply Newton's method, let

$$f(t) = 320t + 2200(1 - e^{-0.1t}) - 10{,}000,$$

and estimate the positive values of t where $f(t) = 0$. Since $f'(t) = 320 + 220e^{-0.1t} > 0$, $f(0) = -10{,}000$, and $f(t) \to \infty$ as $t \to \infty$, there must be a unique positive t^* such that $f(t^*) = 0$. The Newton iteration is

$$t_{n+1} = t_n - \frac{320t + 2200(1 - e^{-0.1t_n}) - 10{,}000}{320 + 220e^{-0.1t_n}}.$$

Starting with $t_0 = 1$, we arrive after six steps at the estimate $t_6 \approx 24.9426$.

example 4

Formal application of Newton's method without regard to the special characteristics of the function $f(x)$ can lead to anomalous results. For example, exponential functions such as $f(x) = e^{x^2/4}$ are never zero, but we can try to use Newton's method on this function anyway; this leads us to the iteration formula $x_{n+1} = x_n - 2/x_n$. Putting $x_0 = 1$ we get $x_n = (-1)^n$; this sequence oscillates between 1 and -1 and so doesn't converge to anything. We might be suspicious if we did get convergence, since there is no solution to converge to. With other starting values the behavior is no better, and will usually be worse.

The exercises show other ways that Newton's method may fail to produce a sequence tending to a solution of $f(x) = 0$, though when it does what it's supposed to, it typically converges quite rapidly. Furthermore, the scope of Newton's method can be extended in a surprisingly large number of ways, for example, to finding the complex solutions of an equation, which is essentially a 2-dimensional problem, and also to still higher-dimensional problems. Newton's method must be used with care, as the previous example shows, but in spite of that warning it remains one of the most widely used recipes of contemporary numerical analysis. Finally, it has been known for many years that even when the method fails, it may do so in intriguing ways that illustrate some of the basic features of the chaotic dynamics of a **discrete dynamical system** $x_{n+1} = N(x_n)$. The "dynamics" of this system consists of the possible outcomes of starting with particular values $x = x_0$ and forming the sequences $x_0, x_1 = N(x_0), x_2 = N(x_1), x_3 = N(x_2), \ldots, x_n = N(x_{n-1}), \ldots$. Another way to look at this is to denote the n-fold composition of N with itself by N^n; thus $N^1(x) = N(x)$, $N^2(x) = N(N(x))$, $N^3(x) = N(N(N(x)))$, and so on. In this notation $x_n = N^n(x_0)$ when $n \geq 1$, and the particular choice for x_0 can be regarded as an initial condition.

To be specific about the relation of Newton's method to discrete dynamics, the recurrence relation $x_{n+1} = x_n - f(x_n)/f'(x_n)$ can be abbreviated $x_{n+1} = N(x_n)$, where $N(x) = x - f(x)/f'(x)$. Then the sequence of approximations to a solution of $f(x) = 0$ is $N^n(x_0)$, where x_0 is the initial estimate of the solution. Here follows an example of how the method can lead to chaotic dynamics.

example 5

The function $f(x) = x^2 + 1$ is never zero for real values of x, but we can still apply Newton's method to see what happens; there are, of course, purely imaginary roots at $\pm i$. In this example the function to construct is

$$N(x) = x - \frac{f(x)}{f'(x)}$$

$$= x - \frac{x^2 + 1}{2x} = \frac{1}{2}\left(x - \frac{1}{x}\right).$$

Since $N(x)$ has only real values if x is real, we can't expect to approximate the imaginary roots by applying Newton's method with a real starting value x_0.

What does happen when x_0 is real is this: For some starting values x_0, $N''(x_0)$ runs repeatedly through a finite set of real values. For all other values of x_0, $N(x_0)$ wanders in an apparently aimless way through the real numbers. (This latter behavior could be interpreted as a frantic and futile attempt to get at the imaginary roots while confined to the real number line. It can be shown that starting with a nonreal complex number will take us to one of the imaginary roots.) In addition, every real value results from some starting value x_0. Some of the details of this behavior are treated in the exercises.

In numerical root-finding the error differences $\varepsilon_n = x_n - x^*$ between a root x^* and its successive approximations x_n are useful for describing rapidity of convergence to the root. It turns out that if $f'(x^*) \neq 0$, then a convergent Newton sequence x_n **converges quadratically** to a root of $f(x)$, which means that there is a constant K such that

$$\varepsilon_{n+1} \sim K\varepsilon_n^2.$$

The symbol "$\sim$" is to be interpreted here to mean simply that the Taylor expansion in powers of c_n of the expression on the left begins with the term on the right. If x^* is a pth-order root of $f(x)$, that is, $f(x) = (x - x^*)^p g(x)$ for some $g(x)$ not zero at x^*, we can still get quadratic convergence if we use Euler step size $h = p$. Quadratic convergence is usually desirable once we have reached a small error ε_n, since the next error will be to some extent determined by the square of ε_n, which is even smaller.

■ 6.1 THEOREM

Suppose $f^{(p)}(x^*)$ is the first nonzero derivative of $f(x)$ at x^* and that higher-order derivatives are continuous. If the Newton-Euler recursion

$$x_{n+1} = x_n - p\frac{f(x_n)}{f'(x_n)}$$

produces a sequence convergent to x^* for some starting value x_0, then

$$x_{n+1} - x^* \sim \frac{(x_n - x^*)^2}{p(p+1)}\frac{f^{(p+1)}(x^*)}{f^{(p)}(x^*)}.$$

Proof. Subtracting x^* from both sides of the Newton-Euler equation gives $\varepsilon_{n+1} = \varepsilon_n - pf(x_n)/f'(x_n)$. We now expand $f(x)$ and $f'(x)$ about x^*, in other words, in powers of $\varepsilon_n = x_n - x^*$. Writing $f^{(k)}$ for $f^{(k)}(x^*)$, we get

$$f(x_n) = \frac{\varepsilon_n^p f^{(p)}}{p!} + \frac{\varepsilon_n^{(p+1)} f^{(p+1)}}{(p+1)!} + \frac{\varepsilon_n^{(p+2)} f^{(p+2)}}{(p+2)!} + \cdots,$$

$$f'(x_n) = \frac{\varepsilon_n^{(p-1)} f^{(p)}}{(p-1)!} + \frac{\varepsilon_n^p f^{(p+1)}}{p!} + \frac{\varepsilon_n^{(p+1)} f^{(p+2)}}{(p+1)!} + \cdots.$$

Substitute these expansions into the recursion formula and write the result as a single fraction. Because of the choice of Euler step size $h = p$, the terms of degree p in the numerator cancel out. Then cancel the common factors $(p-1)!$

and ε_n^{p-1} in the top and bottom. We're left with

$$\varepsilon_{n+1} = \frac{\varepsilon_n^2/(p(p+1))f^{(p+1)} + \varepsilon_n^3/(p(p+1)(p+2))f^{(p+2)} + \cdots}{f^{(p)} + (1/p)\varepsilon_n f^{(p+1)} + \cdots}$$

$$= \frac{\varepsilon_n^2 f^{(p+1)}}{p(p+1)f^{(p)}} + \cdots.$$

Replacing ε_n and ε_{n+1} by their definitions gives the desired result. ∎

EXERCISES

At Web site http://math.dartmouth.edu/~rewn/ the Applet NEWTON can be used to implement Newton's method.

Each of the following equations has either a single solution or else no solution in the given interval. Find those solutions that do exist to five-place accuracy.

1. $x^3 - 24x^2 + 12x - 8 = 0$, $0 \le x < \infty$.

2. $x^3 + 24x^2 + 12x + 8 = 0$, $0 \le x < \infty$.

3. $e^{-x} = x$, $-\infty < x < \infty$.

4. $e^x = x$, $-\infty < x < \infty$.

5. $e^{x^2} = e^x$, $0 < x < \infty$.

6. $x + 1 = \tan x$, $-\pi/2 < x < \pi/2$.

Locate all the real solutions of the following equations to five-place accuracy. Use a preliminary graphical analysis to identify an appropriate starting point for applying Newton's method to each solution.

7. $x^4 - x - 1$, $-\infty < x < \infty$.

8. $x^3 - 4x^2 + 5x - 2$, $-\infty < x < \infty$.

9. $e^{-x} - \ln x$, $0 < x < \infty$.

10. $\sin x - x/10$, $-\infty < x < \infty$.

11. $x^2 - \cos(x + 1)$, $-\infty < x < \infty$.

12. $5 \sin 16x - \tan x$, $-\pi/2 < x < \pi/2$.

13. The function $f(x) = \arctan x$ has just one zero value, located at $x^* = 0$. To see the importance of having a close enough starting point x_0 in applying Newton's method, experiment with the following choices on this function.

   (a) $x_0 = 1$, $h = 1$.   **(b)** $x_0 = 2$, $h = 1$.   **(c)** $x_0 = 2$, $h = \frac{1}{2}$.

14. A 10-pound weight is fired straight up from the surface of the earth with an initial velocity of 100 feet per second. There is an air-resistance force of magnitude equal to 0.1 times velocity. Use Newton's method as the last step in the process of finding out how long it takes for the weight to fall back to earth. Assume $g = 32$.

15. Since the function $f(x) = x^2 + 1$ is positive for all real x, applying Newton's method with a real number for a starting point can't find the imaginary roots at $\pm i$. Investigate what happens starting with **(a)** $x_0 = 1/\sqrt{3}$, **(b)** $x_0 = 1$, **(c)** $x_0 = \frac{1}{2}$.

16. Under the right circumstances, choosing a step size h bigger than 1 can lead to convergence.

    (a) With $f(x) = x^2$, take $h = 1$ and explain what happens to Newton's method, starting at an arbitrary point x_0.

    (b) With $f(x) = x^2$, take $h = 2$ and explain what happens to Newton's method, starting at an arbitrary point x_0.

    (c) Let n be a positive integer. With $f(x) = x^n$, take $h = n$ and explain what happens to Newton's method, starting at an arbitrary point x_0.

17. Show that Newton's method with step size $h > 1$ replaces the tangent line approximation through $(x_n, f(x_n))$ with a line that is not as steep as the tangent. Explain the analogous conclusion for the choice $0 < h < 1$.

18. Suppose a sequence of numbers $x_1, x_2, x_3, \ldots$ is generated by successive application of the Newton formula $x_{n+1} = x_n - f(x_n)/f'(x_n)$ and that the sequence converges to a number x^*. Show that if $f'(x)$ is continuous in an open interval containing x^*, then x^* is necessarily a solution of $f(x) = 0$. [*Hint:* Let n tend to infinity in the Newton formula.]

19. A **fixed point** of a function $F(x)$ is a domain point x^* such that $F(x^*) = x^*$.

    (a) Show that if F is a real-valued function of a real variable, then F has a fixed point at just those domain points, if any, for which the graph of F crosses the line $y = x$.

    (b) Assume that $f(x)$ is a differentiable real function such that $f'(x) \neq 0$, and let $N(x) = x - f(x)/f'(x)$. Show that $f(x^*) = 0$ if and only if x^* is a fixed point for $N(x)$.

    (c) Assume that $f(x)$ is a continuously differentiable real function, and let $N(x) = x - f(x)/f'(x)$. Show that if the sequence $N^n(x_0)$ converges as $n \to \infty$ to a value x^*, then x^* must be a fixed point of N.

    (d) Use the result of part (c) to show that if $f(x) = x^2 + 1$ and $N(x) = x - f(x)/f'(x)$, then there is no real value x_0 such that $N^n(x_0)$ can converge as $n \to \infty$.

20. (a) Derive the Newton iteration formula $x_{n+1} = N(x_n)$ for $f(x) = e^{x^2/4}$.

    (b) Show that the formula derived in part (a) can't ever produce a convergent sequence x_n, no matter what the starting value x_0 is. [*Hint:* Suppose x_n converged to x^*, and let $n \to \infty$ in the iteration formula. Can you find x^*?]

21. Consider the iteration formula $x_{n+1} = N(x_n)$, where $N(x)$ is some continuous function of x. We can try to express the right side of this formula as the result of applying Newton's method to the approximate solution of $f(x) = 0$ for some function $f(x)$. In other words, given $N(x)$, find a differentiable $f(x)$ such that

$$N(x) = x - \frac{f(x)}{f'(x)}.$$

    (a) Show that $y = f(x)$ described above must satisfy the first-order linear differential equation

$$y' + \frac{y}{N(x) - x} = 0.$$

(b) Solve the differential equation in part (a) to show that if $N(x) = x^2$, then $f(x) = x/(x - 1)$ is a possible choice.

***(c)** Repeated application of $x_{n+1} = N(x_n)$ starting with some choice for x_0 produces different results, depending on x_0. Explain these differences for $N(x) = x^2$ on the basis of the graph of the function $f(x)$ of part (b).

22. The equation $dx/dt = hx(m - x)$ is the **logistic differential equation,** and it has the well-behaved solutions described in Section 1. A popular computer recreation is to iterate the related discrete equation $x_n = hx_n(1 - x_n)$ with some fixed $h \geq 1$, starting with an x_0 strictly between 0 and 1. You get a sequence of numbers

$$x_0, \ x_1 = hx_0(1 - x_0), \ x_2 = hx_1(1 - x_1), \ldots, \ x_n = hx_{n-1}(1 - x_{n-1}).$$

It turns out that if $0 \leq h \leq 3$, the sequence x_n converges to $1 - 1/h$. For larger values of h the sequence exhibits *period doubling*. Thus, for h somewhat larger than 3, alternate elements of x_n converge to two limit points. For still larger choices of h, subsequences of x_n consisting of every fourth, then eighth, and in general every 2^nth value, converge to different values, with apparently *random* or *chaotic* behavior for values of h closer to $h = 4$. Here is a way to study and understand this behavior by relating it to an application of the Euler method to a family of logistic equations.

(a) Make a computer graphics simulation of the long-term behavior of $x_n = hx_n(1 - x_n)$ by starting with some number x_0 between 0 and 1, for example, $x_0 = \frac{1}{2}$. Let $h = 1$, and plot x_{80} through x_{100} vertically over $h = 1$ on a horizontal h axis. Then step h ahead by successive increments of size 0.01 to get $h = 1.01, 1.02, 1.03, \ldots$, repeating the plot of x_{80} through x_{100} until you reach $h = 3.99$. You should get a picture something like the one below.

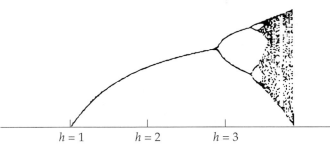

$h = 1 \qquad h = 2 \qquad h = 3$

(b) Show that if the Euler method with step size h is applied to the particular case of the **logistic equation** $dx/dt = x((1 - 1/h) - x)$, you get the recurrence equation $x_{n+1} = hx_n(1 - x_n)$ discussed above. Note that as $n \to \infty$, the sequence x_n estimates the behavior of a solution $x(t)$ to the differential equation for large values $t = nh$. (For $h > 3$ the step size is so large that the validity of the estimate is hopelessly compromised.)

(c) Each parameter value h for which branching or period doubling occurs in the discrete system is an example of a **bifurcation point,** that is, a point of fundamental change in system behavior. Estimate the first three such values of h from the picture, or, better still, by running a numerical simulation.

23. (a) Show that using Newton's method to find a root of

$$f_\lambda(x) = \left(\frac{x - (1 - 1/\lambda)}{x} \right)^{1/(\lambda - 1)}$$

for $\lambda > 1$ leads to the "logistic" iteration formula $x_{n+1} = N(x_n) = \lambda x_n(1 - x_n)$.

(b) Explain why the iteration formula $x_{n+1} = N(x_n)$ derived in part (a) can be expected to produce a sequence converging to $1 - 1/\lambda$ for at least some values of $\lambda > 1$.

(c) In part (a) of Exercise 22 you are asked to demonstrate that increasing the value of λ beyond the number 3 leads to bifurcated and eventually "chaotic" behavior in the sequences generated by the relation $x_{n+1} = N(x_n)$ and using the same x_0 for each value of λ. Demonstrate that if the starting value x_0 is adjusted appropriately for each λ with root-finding in mind, the erratic behavior disappears. [*Hint:* $f_\lambda(x)$ is typically ill-defined as a real function for $x < (1 - 1/\lambda)$.]

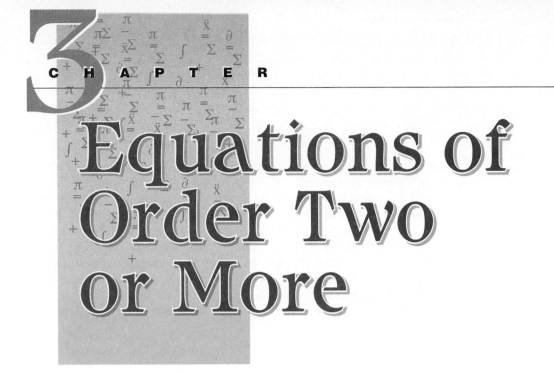

3

Equations of Order Two or More

1 INTRODUCTION

Recall that the **order** of a differential equation equals that of the highest-order derivative that the equation depends on. Thus equations of the form

$$y'' = f(x, y, y') \qquad \text{and} \qquad y''' = g(x, y, y', y'')$$

have order two and three, respectively. Experience with differentiation strongly suggests that it is derivatives of order one or two that appear most prominently in applications. The reason is that first and second derivatives have well-understood interpretations: slope and velocity for order one, or concavity and acceleration for order two. More broadly speaking, a derivative of order *more* than two usually has less intuitive appeal. Interpretation of derivatives carries over to interpretation of differential equations, and second-order equations as a class turn out to be the most important ones in pure and applied mathematics, since they are in principle capable of incorporating many basic geometric and physical ideas in a single equation. Nevertheless, equations of still higher order do arise in applications and are useful also in developing techniques for studying second-order equations.

A large class of equations with significant applications exists for which there is a generally applicable method for finding solution formulas; these are the linear constant-coefficient equations, examples of which are

$$y'' + y = 0,$$
$$y'' - 3y' + 2y = e^x,$$
$$y'' - y = x,$$
$$y^{(4)} - y = 1.$$

We will at first concentrate on the second-order case, partly because those are the ones that most often come up in applied mathematics, and partly because the higher-order extension does not involve fundamentally different ideas. The standard example of the kind of application considered later in this chapter is

$$m\frac{d^2x}{dt^2} + k\frac{dx}{dt} + hx = F(t),$$

where m is the mass of an oscillating weighted spring, k is a friction coefficient, h is a spring-stiffness coefficient, and F represents an externally applied force. The methods of the next few sections are aimed at giving a complete analysis of the initial-value problem for this and many related differential equations. In particular, we can predict what will happen to the mechanism in the future if we know the initial position $x(t_0)$ and initial velocity $x'(t_0)$ at some time t_0. For many of the examples in this chapter, including the linear ones listed above, we'll be able to prove the existence and uniqueness of solutions to the appropriate initial-value problems by careful derivation of elementary solution formulas; for the remainder, such as the nonlinear pendulum equation $y'' = -\sin y$, we can refer to the following theorem, proved in more theoretical treatments.

■ EXISTENCE AND UNIQUENESS THEOREM

The initial-value problem

$$y'' = f(x, y, y'), \qquad y(x_0) = y_0, \qquad y'(x_0) = z_0$$

has a unique solution on an interval if f and its partial derivatives $\partial f/\partial y$ and $\partial f/\partial z$ are bounded and continuous on some 3-dimensional rectangle containing (x_0, y_0, z_0).

1A *Exponential Solutions*

The differential equation

$$y' - ry = 0,$$

with r constant, was treated in Sections 3 and 5 of Chapter 1. Using the exponential multiplier method of Section 5, Chapter 1, it is easy to show that every solution is of the form

$$y = ce^{rx}, \qquad c \text{ const.}$$

The reason is that multiplication of the equation by e^{-rx} gives the equivalent equation

$$e^{-rx}y' - re^{-rx}y = 0$$

or
$$(e^{-rx}y)' = 0.$$

It follows that $e^{-rx}y = c$, so $y = ce^{rx}$. It turns out that, with a suitable extension of the definition of the exponential function, every solution of an nth-order constant-coefficient equation of the form

$$a_ny^{(n)} + a_{n-1}y^{n-1} + \cdots + a_1y' + a_0y = 0, \qquad a_n \neq 0,$$

can be written as a sum of polynomial multiples of exponential functions. Note that the left side is a sum of constant multiples of y and its derivatives. We will often refer to a sum of multiples of functions as a *linear combination* of those functions. The relevance of the idea is developed in the rest of the chapter; here we will consider some examples.

---

example 1

We look for exponential solutions to
$$y'' - 5y' + 6y = 0,$$
having the form $y = e^{rx}$, r constant. We have $y' = re^{rx}$, $y'' = r^2 e^{rx}$, so to satisfy the equation we need
$$r^2 e^{rx} - 5re^{rx} + 6e^{rx} = 0$$
or
$$(r^2 - 5r + 6)e^{rx} = 0.$$

The exponential factor e^{rx} is never zero, so the last equation is satisfied if and only if
$$r^2 - 5r + 6 = (r - 2)(r - 3) = 0.$$
Letting $r = 2$ and $r = 3$, we get the respective solutions
$$y = e^{2x}, \qquad y = e^{3x}.$$
It is easy to verify directly that these two functions are solutions of
$$y'' - 5y' + 6y = 0.$$
It is also straightforward to verify that, for any constants c_1 and c_2, the function
$$y = c_1 e^{2x} + c_2 e^{3x}$$
is a solution. To find the particular solution satisfying
$$y(0) = 1, \qquad y'(0) = -1,$$
we compute $y' = 2c_1 e^{2x} + 3c_2 e^{3x}$ and substitute $x = 0$ in the expressions for y and y'. We find
$$c_1 + c_2 = 1,$$
$$2c_1 + 3c_2 = -1.$$
Solving this pair of equations for c_1 and c_2 gives $c_1 = 4$, $c_2 = -3$. Hence
$$y = 4e^{2x} - 3e^{3x}$$
is the desired solution.

Generalizing from Example 1 suggests that to find exponential solutions to
$$Ay'' + By' + Cy = 0,$$
the thing to do is solve the associated purely algebraic equation
$$Ar^2 + Br + C = 0,$$
called the **characteristic equation** of the differential equation. If the roots $r = r_1$ and $r = r_2$ are distinct real numbers, then the general solution will turn out to be
$$y = c_1 e^{r_1 x} + c_2 e^{r_2 x}.$$
Particular solutions will be determined by imposing two conditions on y to fix the values of c_1 and c_2. The roots r_1, r_2 are called **characteristic roots.**

e x a m p l e 2

The differential equation

$$y'' - 2y' - y = 0$$

has the characteristic equation

$$r^2 - 2r - 1 = 0,$$

with roots

$$r = \frac{2 \pm \sqrt{8}}{2} = 1 + \sqrt{2}, \quad 1 - \sqrt{2}.$$

The corresponding solutions are

$$y_1(x) = e^{(1+\sqrt{2})x}, \quad y_2(x) = e^{(1-\sqrt{2})x}.$$

The general solution is

$$y = c_1 e^{r_1 x} + c_2 e^{r_2 x}, \quad r_1 = 1 + \sqrt{2}, \quad r_2 = 1 - \sqrt{2}.$$

We'll see shortly that every solution of the differential equation has the form

$$y(x) = c_1 e^{r_1 x} + c_2 e^{r_2 x},$$

where $r_1 = 1 + \sqrt{2}$, $r_2 = 1 - \sqrt{2}$, and c_1, c_2 is a pair of constants.

Some features of the solution formula at the end of Example 2 are important enough that there is special terminology for referring to them. If $y_1(x)$ and $y_2(x)$ are functions defined on a common domain, then

$$y(x) = c_1 y_1(x) + c_2 y_2(x)$$

is called a **linear combination** of $y_1(x)$ and $y_2(x)$ with **coefficients** c_1, c_2. It is significant in the example that neither $y_1(x) = e^{r_1 x}$ nor $y_2(x) = e^{r_2 x}$ is a constant multiple of the other one. Otherwise, the linear combination could simply be reduced to a constant multiple of either function. This property is usually stated in a formal definition using a single equation: $y_1(x)$ and $y_2(x)$ are **linearly independent** on an interval if the identity $c_1 y_1(x) + c_2 y_2(x) = 0$ holds there only with the obvious choice $c_1 = c_2 = 0$ for the constants. It is left as an exercise to show that **two functions are linearly independent if and only if neither one is a constant multiple of the other.** If we think of the constants c_1 and c_2 as both being allowed to vary, then the corresponding linear combinations $y(x) = c_1 y_1(x) + c_2 y_2(x)$ are an example of what is called a **2-parameter family** with **parameters** c_1 and c_2.

1B Factoring Operators

Differential operator notation will be useful here as well as later for solving systems of differential equations, so we start with some of their basic properties. We let $D = d/dx$; that is, D will stand for differentiation with respect to some specified variable, call it x here. We interpret $D - 2$, $D^2 - 1$, and similar polynomials in D as operations on sufficiently differentiable functions $y = y(x)$, producing new functions. Here are four examples:

(a) $$(D + 1)e^{2x} = De^{2x} + e^{2x} = 3e^{2x}$$

(b)
$$(D^2 + 2) \sin x = D(D \sin x) + 2 \sin x$$
$$= D \cos x + 2 \sin x = \sin x.$$

(c)
$$(D - 2)y = Dy - 2y$$
$$= y' - 2y.$$

(d)
$$(D^2 - 1)y = D^2y - y$$
$$= D(Dy) - y = y'' - y.$$

One of the most important facts about ordinary polynomials is that they can in principle be factored if their coefficients are constants; the analog is true for polynomial operators, a fact which will lead us to a complete understanding of constant-coefficient linear equations.

■ 1.1 THEOREM

If r and s are constants, then
$$(D + r)(D + s) = (D + s)(D + r)$$
$$= D^2 + (r + s)D + rs.$$

It follows that if r_1 and r_2 are the characteristic roots of the differential equation $y'' + ay' + by = 0$, then
$$y'' + ay' + by = (D - r_1)(D - r_2)y$$
$$= (D - r_2)(D - r_1)y.$$

Proof. To prove that two operators are equal, prove that they produce the same result when applied to a given sufficiently differentiable function $y(x)$. If r and s are constants,
$$(D + r)(D + s)y = (D + r)(y' + sy)$$
$$= D(y' + sy) + r(y' + sy)$$
$$= y'' + (r + s)y' + (rs)y$$
$$= [D^2 + (r + s)D + rs]y.$$

Interchanging r and s really changes nothing in this last expression, so we can conclude also that $(D + s)(D + r) = (D + r)(D + s)$ in the sense that these two operators have the same effect on twice-differentiable functions $y(x)$. In particular, it follows that if $r = r_1$ and $s = r_2$ are the roots of the characteristic polynomial $r^2 + ar + b = (r - r_1)(r - r_2)$, then $y'' + ay' + by = (D^2 + aD + b)y$ can be written as either $(D - r_1)(D - r_2)y$ or $(D - r_2)(D - r_1)y$. ■

---

e x a m p l e 3

The differential equation
$$y'' + 2y' + y = 0$$
can be written
$$(D^2 + 2D + 1)y = 0$$
$$(D + 1)^2 y = 0$$
Note that the characteristic equation is
$$r^2 + 2r + 1 = 0 \quad \text{or} \quad (r + 1)^2 = 0,$$

so there is a double root $r = -1$ with associated solution $y = e^{-x}$. To find another solution, suppose that y is some solution and let $z = (D + 1)y$. Then

$$(D + 1)z = (D + 1)^2 y = 0$$

presents us with a first-order equation satisfied by z, namely,

$$z' + z = 0.$$

The solutions of this equation are all of the form

$$z = c_1 e^{-x}.$$

It then follows from $(D + 1)y = z$ that y must satisfy the additional first-order equation

$$y' + y = c_1 e^{-x}$$

for some choice of constant c_1. But this first-order linear equation can be solved by multiplying by the integrating factor $M(x) = e^x$ to get $e^x y' + e^x y = c_1$, or

$$(e^x y)' = c_1.$$

Hence $e^x y = c_1 x + c_2$ and

$$y = c_1 x e^{-x} + c_2 e^{-x}.$$

This is the general solution of the differential equation.

We can prove the following theorem just by applying the **order-reduction** method, that is, by solving a succession of first-order equations as in Example 3.

■ 1.2 THEOREM

Suppose the characteristic equation $r^2 + ar + b = 0$ of

$$y'' + ay' + by = 0$$

has real roots r_1 and r_2. There are two distinct cases.
(a) If $r_1 \neq r_2$, the general solution is $y = c_1 e^{r_1 x} + c_2 e^{r_2 x}$.
(b) If $r_1 = r_2$, the general solution is $y = c_1 x e^{r_1 x} + c_2 e^{r_1 x}$.

The constants c_1 and c_2 are uniquely determined by prescribing values $y(x_0) = y_0$ and $y'(x_0) = z_0$ of the solution and its derivative at a single point x_0.

Proof. In operator form, the differential equation is

$$(D - r_1)(D - r_2)y = 0.$$

We assume $y(x)$ is a solution and show that it has the form claimed above. Set $z(x) = (D - r_2)y(x)$ and substitute z for $(D - r_2)y$ in the previous equation. Now solve the resulting equation $(D - r_1)z = 0$ to get

$$z = c_1 e^{r_1 x}.$$

Note that c_1 is determined by $c_1 = z(x_0)e^{-r_1 x_0} = (y'(x_0) - r_2 y(x_0))e^{-r_1 x_0}$. Given the relation between y and z, the solution y then satisfies

$$(D - r_2)y = c_1 e^{r_1 x},$$

and multiplication by $e^{-r_2 x}$ gives

$$D(e^{-r_2 x}y) = c_1 e^{(r_1 - r_2)x}.$$

For case (a), assuming $r_1 \neq r_2$, we integrate to get

$$e^{-r_2 x} y = \frac{c_1}{r_1 - r_2} e^{(r_1 - r_2)x} + c_2.$$

Now multiply by $e^{r_2 x}$ to get

$$y = \frac{c_1}{r_1 - r_2} e^{r_1 x} + c_2 e^{r_2 x}.$$

For neatness, we can rename the constant $c_1/(r_1 - r_2)$ and call it c_1 to get

$$y(x) = c_1 e^{r_1 x} + c_2 e^{r_2 x}. \tag{$*$}$$

For case (b), assuming $r_1 = r_2$, we have $D(e^{-r_2 x} y) = c_1$. Integrating both sides gives

$$e^{-r_2 x} y = c_1 x + c_2.$$

In that case, $\qquad\qquad y(x) = c_1 x e^{r_2 x} + c_2 e^{r_2 x}. \tag{$**$}$

Finally, note that once c_1 is determined from $y(x_0)$ and $y'(x_0)$ as noted above, the constant c_2 can in any case be determined from the value $y(x_0)$ alone; just solve the appropriate equation ($*$) or ($**$) for c_2 with $x = x_0$. ∎

A condition imposed on the values of a solution and its successive derivatives at a single domain point x_0 is called an **initial condition.** The problem of finding a solution to a differential equation that also satisfies given initial conditions is called an **initial-value problem.** Geometrically, the initial conditions described in Theorem 1.2 require the graph of a solution to go through a given point (x_0, y_0) with slope $y'(x_0) = z_0$. It is possible to extract from the proof of Theorem 1.2 some formulas for determining coefficients c_1 and c_2 from initial conditions in a linear combination of two solutions. Since the formulas aren't particularly memorable, it's usually just as efficient to work directly, as in the next example.

e x a m p l e 4

Suppose we want a solution graph of $y'' - 4y = 0$ that passes through $(0, 1)$ with slope 2. In other words, we want $y(0) = 1$ and $y'(0) = 2$. Since the characteristic equation of the differential equation is $r^2 - 4 = 0$ and has roots $r = \pm 2$, all solutions have the form

$$y(x) = c_1 e^{2x} + c_2 e^{-2x}.$$

The corresponding derivative formula is

$$y'(x) = 2c_1 e^{2x} - 2c_2 e^{-2x}.$$

Setting $x = 0$, we get the two equations

$$y(0) = c_1 + c_2 = 1 \qquad \text{and} \qquad y'(0) = 2c_1 - 2c_2 = 2.$$

Solving these for c_1 and c_2 gives the unique solution $c_1 = 1$, $c_2 = 0$, so $y(x) = e^{2x}$.

e x a m p l e 5

Earlier in Example 3 we saw that the equation $y'' + 2y' + y = 0$ had characteristic equation $r^2 + 2r + 1 = 0$ with repeated characteristic roots $r_1 = r_2 = -1$. Theorem 1.2 says that the general solution is $y(x) = c_1 e^{-x} + c_2 x e^{-x}$. Initial conditions $y(0) = 1$, $y'(0) = 0$ require first that $y(0) = c_1 = 1$. Since $y'(x) = -c_1e^{-x} + c_2e^{-x} - c_2xe^{-x}$, the second condition requires $y'(0) = -c_1 + c_2 = 0$. Hence, $c_1 = c_2 = 1$, and the initial-value problem has solution $y = e^{-x} + xe^{-x}$.

Suppose that in Example 5 we didn't want to satisfy initial conditions at a single point but wanted instead a solution graph passing through two given points in the xy plane. Such conditions applied to a single solution at more than one point are called **boundary conditions.** The problem of finding a solution of a differential equation that also satisfies boundary conditions is called a **boundary-value problem.** Boundary-value problems are theoretically more complicated than initial-value problems, and the relevant theorems are deferred to Chapter 9. Nevertheless some boundary-value problems may be quite simple computationally. The next example is of this kind.

---

e x a m p l e 6   Our object is to find a solution to $y'' - 4y = 0$ with boundary conditions $y(0) = 1$ and $y(1) = 2$. Thus in this example we need work only with the expression

$$y(x) = c_1e^{2x} + c_2e^{-2x}$$

for the general solution itself, since values of $y'(x)$ aren't involved. The resulting equations for the coefficients are

$$y(0) = c_1 + c_2 = 1 \qquad \text{and} \qquad y(1) = e^2c_1 + e^{-2}c_2 = 2.$$

Though it is not guaranteed by Theorem 1.2, these equations turn out here to have a unique solution, namely, $c_1 = (2 - e^{-2})/(e^2 - e^{-2}) \approx 0.26$, $c_2 = (e^2 - 2)/(e^2 - e^{-2}) \approx 0.74$. The desired solution is

$$y(x) = \frac{2 - e^{-2}}{e^2 - e^{-2}}\, e^{2x} + \frac{e^2 - 2}{e^2 - e^{-2}}\, e^{-2x}.$$

Examples 5 and 6 might suggest that the distinction between initial conditions and boundary conditions is just a minor computational one. A more insightful way to think about the difference is as follows. An initial-value problem asks us to start at a given point with a given slope and then go where the differential equation tells us to go. Under the rather broad technical restrictions of the existence and uniqueness theorem, it will always be possible to go *somewhere.* A boundary-value problem asks whether a differential equation can get us from one prescribed point to another prescribed point. Exercise 10 in Section 2B shows that this point-to-point transition may not always be possible. A companion exercise at the end of the present section shows that for second-order equations with real characteristic roots, boundary problems are always solvable. We'll deal with roots r_1, r_2 of the characteristic equation that are complex numbers in Section 2. For now we stay with examples that lead to real roots. In any case it will be important for us that the general solution is a linear combination of simple solutions. In Theorem 1.2, a formula

$$c_1e^{r_1x} + c_2e^{r_2x}, \qquad \text{or} \qquad c_1xe^{r_1x} + c_2e^{r_1x},$$

expresses every possible solution as a linear combination of two solutions: e^{r_1x} and e^{r_2x} in the first case, e^{r_1x} and xe^{r_1x} in the second. (For example, just take $c_1 = 1$ and $c_2 = 0$ to get the solution e^{r_1x}.)

EXERCISES

1. With $D = d/dx$ compute:

(a) $(D + 1)e^{-2x}$.

(b) $(D^2 + 1)e^x$.

(c) $D^3 e^{3x}$.

(d) $(D^2 + D - 1)\sin x$.

(e) $(D^2 + 1)x\cos x$.

2. Find the characteristic equation of each of the following differential equations. Then solve the characteristic equation and use the roots to write down the general solution of the differential equations.

(a) $y'' + y' - 6y = 0$.

(b) $2y'' - y = 0$.

(c) $y'' + 2y' + y = 0$.

(d) $y'' + 3y' + y = 0$.

(e) $y'' - y' = 0$.

(f) $y'' - 3y' - y = 0$.

(g) $2y'' - 3y' + y = 0$.

(h) $3y'' + 3y' = 0$.

3. For each differential equation in Exercise 2, find the particular solution satisfying the corresponding initial or boundary condition given below. Then sketch the graph of that particular solution.

(a) $y(0) = 2$, $y'(0) = 2$.

(b) $y(0) = 1$, $y'(0) = 0$.

(c) $y(0) = 1$, $y'(0) = 2$.

(d) $y(1) = 1$, $y'(1) = 1$.

(e) $y(0) = 1$, $y(1) = 0$.

(f) $y(0) = 0$, $y(1) = 0$.

(g) $y(0) = 0$, $y'(0) = 0$.

(h) $y(1) = 1$, $y(2) = 2$.

4. Sketch the graph of each function of x given. Then find a differential equation of the form

$$y'' + ay' + by = 0$$

of which each is a solution; write the general solution of the differential equation, and verify that the given function is a special case of your general solution.

(a) xe^{-x}.

(b) $e^x + e^{-x}$.

(c) $1 + x$.

(d) $2e^{2x} - 3e^{3x}$.

(e) $xe^{-x} - e^{-x}$.

(f) $e^{-3x} + e^{5x}$.

[*Hint:* What characteristic roots go with each solution?]

5. Put each of these equations in operator form: $(aD^2 + bD + c)y = 0$. Then factor the operator [e.g., $D^2 - 1 = (D - 1)(D + 1)$].

(a) $y'' + 2y' + y = 0$.

(b) $y'' - 2y = 0$.

(c) $2y'' - y = 0$.

(d) $y'' + 3y' = 0$.

(e) $y'' = 0$.

(f) $y'' - y' = 0$.

6. Each of the following equations can be written in factored operator form:

$$(D - r_1)(D - r_2)y = f(x).$$

In each case let $(D - r_2)y = z$ and solve

$$(D - r_1)z = f(x)$$

for the most general possible z. Having found z, solve $(D - r_2)y = z$.

(a) $D(D - 3)y = 0$.

(b) $D^2 y = 1$.

(c) $y'' - y = 1$.

(d) $y'' + 2y' + y = x$.

7. The differential equation $y'' + (1/x)y' - (1/x^2)y = 0$, $x > 0$, can be written in operator form as $(D^2 + (1/x)D - 1/x^2)y = 0$.

(a) Show that the equation can also be written $D(D + 1/x)y = 0$.

(b) Solve the equation in part (a) by letting $z = (D + 1/x)y$ and solving a succession of first-order equations.

(c) Show that $D(D + 1/x) \neq (D + 1/x)D$.

(d) Solve $(D + 1/x)Dy = 0$.

8. The hyperbolic cosine and hyperbolic sine are defined by
$$\cosh x = \tfrac{1}{2}(e^x + e^{-x}), \qquad \sinh x = \tfrac{1}{2}(e^x - e^{-x}).$$

(a) Show that if constants d_1 and d_2 are suitably chosen in terms of c_1 and c_2, then
$$c_1 e^{rx} + c_2 e^{-rx} = d_1 \cosh rx + d_2 \sinh rx.$$

(b) Express the general solution of
$$y'' - k^2 y = 0$$
in terms of hyperbolic functions.

9. (a) Show that the characteristic equation of
$$Ay'' + By' + Cy = 0,$$
with A, B, C constant, $A \neq 0$, has real roots if and only if $B^2 \geq 4AC$.

(b) Show that when $B^2 > 4AC$, the general solution of the differential equation in part (a) can be written
$$y = e^{\alpha x}(d_1 \cosh \beta x + d_2 \sinh \beta x),$$
where $\alpha = -B/2A$, $\beta = \sqrt{B^2 - 4AC}/2A$.

10. Assume $|A| < \tfrac{1}{4}$ in the equation
$$Ay'' + y' + y = 0$$
and show that, as A tends to zero, and with proper choice of arbitrary constants, there are solutions of this equation tending, for each fixed x, to solutions of $y' + y = 0$.

11. The differential equation $y'' - 2y' + y = 0$ has infinitely many solutions $y(x)$ with graphs passing through the point $(0, 1)$. Find the three that have slopes -1, 0, and 1 at that point and sketch their graphs.

12. Initial conditions $y(x_0) = a_0$ and $y'(x_0) = a_1$, imposed at a single point x_0, will always be satisfied by some solution of $y'' - 3y' + 2y = 0$; show that the analogous conditions $y(\ln 2) = a_0$ and $y'(0) = a_1$, at two different points 0 and $\ln 2$, can be satisfied only if $a_0 = 2a_1$.

13. A chain of length l and mass density δ per unit of length lies unattached and in a straight line on the deck of a ship in such a way that it will slide with negligible friction.

(a) If the chain runs out over the side with no force acting on it but gravity at constant acceleration g, show that the amount y hanging over the side satisfies $d^2y/dt^2 = (g/l)y$ as long as $0 \leq y \leq l$. (Assume the deck is more than height l above the water.)

(b) How fast is the chain accelerating as the last link goes over the side?

(c) Find $y(t)$ if $y(0) = y_0$ and $dy/dt(0) = v_0$.

(d) Show that if the chain starts from rest with length $y_0 > 0$ hanging over the side, then the last link goes over the side at time $t_1 = \sqrt{l/g}\,\ln\left((l + \sqrt{l^2 - y_0^2})/y_0\right)$.

14. **Unstable behavior.** The behavior of solutions of a differential equation can sometimes be so sensitive to changes in the initial conditions that their slightest alteration leads to drastically different behavior of the solutions; as a result, small errors in physical input data can produce very bad long-term predictions. Here is an example.

   (a) Solve the differential equation $y'' - y = 0$ subject to the initial conditions $y(0) = 1$, $y'(0) = \alpha$, where α is a constant.

   (b) Show that the solution $y(x)$ determined in part (a) by the choice $\alpha = -1$ satisfies $\lim_{x\to\infty} y(x) = 0$ but that any other choice for the real number α leads to $\lim_{x\to\infty} y(x) = \pm\infty$.

   (c) Sketch the graphs of the solutions in part (a) for $\alpha = -1.1$, $\alpha = -1$, and $\alpha = -0.9$ in such a way as to display their differing behaviors.

15. Let $r(x)$ and $s(x)$ be differentiable functions defined on an interval $a < x < b$. Show that $(D - r(x))(D - s(x)) = (D - s(x))(D - r(x))$ precisely when the functions $r(x)$ and $s(x)$ differ by a constant on the interval. [*Hint:* Show that two second-order operators are equal by showing that they give the same result when applied to an arbitrary twice-differentiable function.]

16. Here is an alternative way to arrive at the modified exponential solution xe^{mx} when the characteristic equation of $y'' + ay' + by = 0$ has m as a double root. First write the equation in operator form as $(D - m)^2 y = 0$, or as

$$y'' - 2my' + m^2 y = 0.$$

   Now try to find a solution of the form $y = e^{mx}u(x)$ by substitution into the displayed equation. (This technique is used in an essential way in Chapter 9 for dealing with homogeneous linear equations with nonconstant coefficients.) [*Hint:* Show that $u''(x) = 0$.]

17. This exercise gives a clue as to why the factor x occurs in the "equal-root" case.

   (a) Show that $(D - r)(D - (r + h))y = 0$, or $(D^2 - (2r + h)D + r(r + h))y = 0$, can also be written $y'' - (2r + h)y' + r(r + h)y = 0$.

   (b) Show that if $h \neq 0$, the general solution is $y_h = c_1 e^{(r+h)x} + c_2 e^{rx}$.

   (c) Let $c = 1/h$, $c_2 = -1/h$, and show that with these choices $\lim_{h\to 0} y_h(x) = xe^{rx}$ for all x.

   (d) Show that the limit in part (c) is, by definition, the derivative of the solution e^{rx} with respect to r.

*18. Assume that the characteristic roots of the constant-coefficient differential equation $y'' + ay' + by = 0$ are real numbers r_1, r_2. Show that if $x_1 \neq x_2$, the **boundary-value problem**

$$y'' + ay' + by = 0, \qquad y(x_1) = y_1, \qquad y(x_2) = y_2$$

always has a unique solution for given numbers y_1 and y_2. [*Hint:* Consider the cases $r_1 \neq r_2$ and $r_1 = r_2$ separately, and show that you can always solve for the desired constants c_1 and c_2 in the general solution.]

19. Two functions $y_1(x)$ and $y_2(x)$ are **linearly independent** on an x interval if the identity $c_1y_1(x) + c_2y_2(x) = 0$ holds only for coefficients $c_1 = c_2 = 0$. Show that y_1 and y_2 are linearly independent on an interval if and only if neither one is a constant multiple of the other.

2 COMPLEX SOLUTIONS

2A *Complex Exponentials*

The formal computations carried out in Section 1 can all be carried out under the more general assumption that the roots of the characteristic equation

$$r^2 + ar + b = 0$$

associated with

$$y'' + ay' + by = 0$$

are complex numbers. (Appendix A reviews complex arithmetic.) To make sense of these computations, we need to have a **complex exponential** function defined for real numbers x by using the **De Moivre formula**

$$e^{ix} = \cos x + i \sin x.$$

Plotting the complex number e^{ix} in the complex plane by its real and imaginary parts $\cos x$ and $\sin x$ shows, in Figure 3.1, that x can be interpreted as an angle measured in radians. Our motivation for this definition of e^{ix} comes from its properties, established next as Equations 2.1, 2.2, and 2.3 (see also Exercise 14, Section 2B).

The numbered formulas listed below are the basic properties of e^{ix} that are routinely used. We begin with the **absolute value,** recalling that $|\alpha + i\beta| = \sqrt{\alpha^2 + \beta^2}$. We have

 2.1
$$\begin{aligned}
|e^{ix}| &= |\cos x + i \sin x| \\
&= \sqrt{\cos^2 x + \sin^2 x} = 1.
\end{aligned}$$

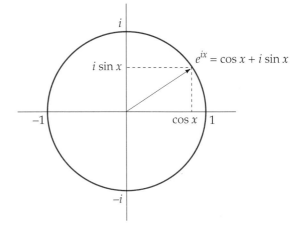

Figure 3.1 Relation of e^{ix} to $\cos x$ and $\sin x$.

Using the addition formulas for sine and cosine shows that

$$e^{ix}e^{ix'} = (\cos x + i \sin x)(\cos x' + i \sin x')$$
$$= (\cos x \cos x' - \sin x \sin x') + i(\cos x \sin x' + \sin x \cos x')$$
$$= \cos (x + x') + i \sin (x + x') = e^{i(x+x')}.$$

It follows that

■ **2.2** $$e^{ix}e^{ix'} = e^{i(x+x')}.$$

In particular, when $x' = -x$, we get $e^{ix}e^{-ix} = e^0 = 1$ so that

■ **2.3** $\dfrac{1}{e^{ix}} = e^{-ix};$ that is, $\dfrac{1}{\cos x + i \sin x} = \cos x - i \sin x.$

Equations 2.2 and 2.3, aside from their usefulness, serve to justify the use of the exponential notation; the function e^{ix} behaves something like the real-valued exponential e^x, for which $e^x e^{x'} = e^{x+x'}$ and $(e^x)^{-1} = e^{-x}$.

---

example 1

The function e^{ix} is, in general, complex-valued, and Figure 3.1 shows that 1 and -1 are the only real values it assumes. In particular, if k is an integer,

$$e^{ik\pi} = \cos k\pi + i \sin k\pi$$
$$= (-1)^k.$$

However, the real-valued functions $\cos x$ and $\sin x$ can be represented in terms of e^{ix} as follows:

$$\frac{1}{2}(e^{ix} + e^{-ix}) = \frac{1}{2}(\cos x + i \sin x + \cos x - i \sin x) = \cos x,$$

$$\frac{1}{2i}(e^{ix} - e^{-ix}) = \frac{1}{2i}(\cos x + i \sin x - \cos x + i \sin x) = \sin x.$$

Using complex exponentials may seem like an awkward way to do trigonometry, but it turns out that Formula 2.2 is often a very natural and convenient substitute for the addition formulas for cosine and sine. The complex exponential notation is used not only because of the simple formulas described previously but because it works naturally with differentiation and integration. To differentiate or integrate a complex-valued function $u(x) + iv(x)$, we have only to differentiate or integrate the real and imaginary parts. Here are the precise definitions:

$$\frac{d}{dx}(u(x) + iv(x)) = \frac{du}{dx}(x) + i\frac{dv}{dx}(x),$$

$$\int (u(x) + iv(x))\, dx = \int u(x)\, dx + i \int v(x)\, dx.$$

Accordingly, we have

$$\frac{d}{dx} e^{ix} = \frac{d}{dx}(\cos x + i \sin x)$$
$$= -\sin x + i \cos x$$
$$= i(\cos x + i \sin x) = ie^{ix}.$$

The end result is

$$\frac{d}{dx} e^{ix} = ie^{ix},$$

just as for a real constant i. Similarly,

$$\int e^{ix}\,dx = \frac{1}{i} e^{ix} + C,$$

where C is an arbitrary real or complex constant. More generally, we define

$$e^{(\alpha+i\beta)x} = e^{\alpha x}e^{i\beta x}$$

and compute

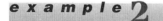 **2.4**
$$\frac{d}{dx} e^{(\alpha+i\beta)x} = (\alpha + i\beta)e^{(\alpha+i\beta)x}$$

and

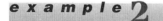 **2.5**
$$\int e^{(\alpha+i\beta)x}\,dx = \frac{1}{\alpha + i\beta} e^{(\alpha+i\beta)x} + C, \qquad C \text{ real or complex,}$$

assuming α and β are not both zero in Equation 2.5. These computations are left as exercises; they are simple but nevertheless important to understand.

2B   Complex Characteristic Roots

Given the facts about complex exponentials just described, the derivation of the general solution of

$$y'' + ay' + by = 0$$

given in Section 1 proceeds word for word the same in case the roots of the characteristic equation are complex. The only thing that remains to be done then is to interpret the solution when a and b are real numbers. Rather than go through the identical general computation again, we will illustrate it with an example and then state the general result.

---

e x a m p l e 2

The differential equation $y'' + y = 0$ in operator form is $(D^2 + 1)y = 0$, and in factored form this is

$$(D - i)(D + i)y = 0.$$

If we let, for some solution y,

$$(D + i)y = z,$$

we then need to solve

$$(D - i)z = 0.$$

As in the real case previously treated, we multiply by an exponential factor designed to make the left side the derivative of a product. The correct factor is

$$e^{\int -i\,dx} = e^{-ix},$$

and we observe that $e^{-ix}(D - i)z = D(e^{-ix}z)$. To solve

$$D(e^{-ix}z) = 0,$$

all we have to do is integrate with respect to x and then multiply by e^{ix}. Since $e^{ix}e^{-ix} = 1$, we get

$$z = c_1 e^{ix}.$$

Substitution of this result into the equation for y gives

$$(D + i)y = c_1 e^{ix}.$$

This time the multiplier is e^{ix}, and we find

$$D(e^{ix}y) = c_1 e^{2ix}.$$

Since the integral of e^{2ix} is $(2i)^{-1}e^{2ix}$, we have

$$e^{ix}y = \frac{1}{2i} c_1 e^{2ix} + c_2.$$

Now multiply by e^{-ix} to get

$$y = c_1 e^{ix} + c_2 e^{-ix},$$

where we have absorbed the factor $1/(2i)$ into the constant c_1. The roots of the characteristic equation $r^2 + 1 = 0$ are $\pm i$, so the same rule for writing down the solution works here as it does in the case of real roots. It remains only to extract the real-valued solutions from this general complex-valued solution. This is done simply by using the definition of $e^{\pm ix}$ in terms of $\cos x$ and $\sin x$:

$$
\begin{aligned}
y &= c_1 e^{ix} + c_2 e^{-ix} \\
&= c_1(\cos x + i \sin x) + c_2(\cos x - i \sin x) \\
&= (c_1 + c_2) \cos x + i(c_1 - c_2) \sin x.
\end{aligned}
$$

To simplify the formula, we can set $d_1 = c_1 + c_2$, and $d_2 = i(c_1 - c_2)$. This does not involve any change in generality in the constants, because we can always solve for the c's in terms of the d's. In fact,

$$c_1 = \tfrac{1}{2}(d_1 - id_2) \qquad \text{and} \qquad c_2 = \tfrac{1}{2}(d_1 + id_2).$$

Note that if the d's are to be real numbers, which is the important case in most applications, then the c's will in general be complex. The solution formula

$$y = d_1 \cos x + d_2 \sin x$$

lets us, for example, find a formula for the solution satisfying initial conditions like $y(0) = 1$, $y'(0) = 2$. Since

$$y' = -d_1 \sin x + d_2 \cos x,$$

we see right away that $d_1 = 1$ and $d_2 = 2$. Thus

$$y = \cos x + 2 \sin x$$

is the required solution.

Whenever the equation $y'' + ay' + by = 0$ has coefficients that are real numbers, even though the characteristic equation

$$r^2 + ar + b = 0$$

may have complex roots as in Example 2, these complex roots will be **conjugate** to each other, that is, of the form $r_1 = \alpha + i\beta$ and $r_2 = \alpha - i\beta$, where α and β are real numbers. This follows from the quadratic formula

$$r = \frac{-a \pm \sqrt{a^2 - 4b}}{2},$$

and the assumption that $a^2 - 4b < 0$. Thus

$$r_1 = -\frac{a}{2} + \frac{i}{2}\sqrt{4b - a^2}, \qquad r_2 = -\frac{a}{2} - \frac{i}{2}\sqrt{4b - a^2}.$$

It follows that the complex solutions

$$y = c_1 e^{(\alpha + i\beta)x} + c_2 e^{(\alpha - i\beta)x}, \qquad \alpha, \beta \text{ real},$$

can always be written

$$
\begin{aligned}
y &= e^{\alpha x}(c_1 e^{i\beta x} + c_2 e^{-i\beta x}) \\
&= e^{\alpha x}[c_1(\cos \beta x + i \sin \beta x) + c_2(\cos \beta x - i \sin \beta x)] \\
&= e^{\alpha x}[(c_1 + c_2) \cos \beta x + i(c_1 - c_2) \sin \beta x] \\
&= d_1 e^{\alpha x} \cos \beta x + d_2 e^{\alpha x} \sin \beta x.
\end{aligned}
$$

This is a form of the solution that is often used in practice, so we include it in the general statement. The proof is formally the same as that of Theorem 1.2 in Section 1B, so we omit it. The only difference here is that we can interpret the solutions as being complex-valued, though of course the case of real roots is automatically included also.

■ 2.6 THEOREM

The differential equation

$$y'' + ay' + by = 0, \qquad a, b \text{ constant},$$

has for its general solution

$$y = c_1 e^{r_1 x} + c_2 e^{r_2 x}, \qquad r_1 \neq r_2$$
$$y = c_1 x e^{r_1 x} + c_2 e^{r_1 x}, \qquad r_1 = r_2,$$

where r_1, r_2 are the roots of $r^2 + ar + b = 0$. Assuming a, b are real with $a^2 - 4b < 0$, we can write $r_1 = \alpha + i\beta$, $r_2 = \alpha - i\beta$. The general solution can then be written

$$y = c_1 e^{\alpha x} \cos \beta x + c_2 e^{\alpha x} \sin \beta x.$$

Initial conditions $y(x_0) = y_0$, $y'(x_0) = z_0$ can always be satisfied by a unique choice of c_1 and c_2.

---

e x a m p l e **3**

The solutions of

$$Ay'' + By' + Cy = 0$$

can be classified according to the sign and relative size of the constants A, B, and C. The characteristic equation is

$$Ar^2 + Br + C = 0$$

with roots

$$r_1, r_2 = \frac{-B \pm \sqrt{B^2 - 4AC}}{2A}.$$

If $B^2 - 4AC > 0$, the roots are real and unequal, so the solutions are conveniently written

$$y = c_1 e^{r_1 x} + c_2 e^{r_2 x}.$$

If $B^2 - 4AC = 0$, the roots are equal and real, so solutions are all of the form

$$y = c_1 x e^{r_1 x} + c_2 e^{r_1 x}.$$

Finally, with $B^2 - 4AC < 0$, the solutions have the form

$$y = c_1 e^{\alpha x} \cos \beta x + c_2 e^{\alpha x} \sin \beta x$$

where $\alpha = -B/2A$ and $\beta = \sqrt{4AC - B^2}/2A$. Clearly, the case $B^2 - 4AC = 0$ is critical in that it represents the division between **oscillatory solutions** $(B^2 - 4AC < 0)$ and **nonoscillatory solutions** $(B^2 - 4AC \geq 0)$. The physical significance of this classification is explained in Chapter 4.

EXERCISES

1. Show that each of the following complex numbers has absolute value 1. Then find a real number x such that the complex number can be written in the form $e^{ix} = \cos x + i \sin x$; for example,

$$\frac{\sqrt{3} + i}{2} = \cos \frac{\pi}{6} + i \sin \frac{\pi}{6} = e^{i\pi/6}.$$

   (a) i.

   (b) $(1 + i)/\sqrt{2}$.

   (c) $(1 - i)/\sqrt{2}$.

   (d) $(\sqrt{3} - i)/2$.

2. Given a pair d_1, d_2 of real numbers, find complex numbers c_1, c_2 such that

$$c_1 e^{(\alpha + i\beta)x} + c_2 e^{(\alpha - i\beta)x} = e^{\alpha x}(d_1 \cos \beta x + d_2 \sin \beta x).$$

   (a) $d_1 = 1$, $d_2 = 0$.

   (b) $d_1 = 4$, $d_2 = -2$.

   (c) $d_1 = 0$, $d_2 = \pi$.

   (d) $d_1 = 1$, $d_2 = 1$.

3. Recall that a function is periodic, with period p, if $f(x + p) = f(x)$ for all x in the domain of f.

   (a) Show that e^{ix} has period 2π.

   (b) Show that $e^{i\beta x}$ is periodic for β real, and find the smallest positive period if $\beta \neq 0$.

4. Show that $e^{-i\beta x} = \cos x - i \sin \beta x$. What properties of cos and sin are used here?

5. **(a)** Verify Equation 2.4.

   (b) Verify Equation 2.5.

6. Solve each of the following differential equations by factoring the differential operator associated with it and then successively solving a pair of first-order linear equations.

   (a) $y'' + y = 1$.

   (b) $y'' + 2y' + 2y = 0$.

   (c) $y'' + 2y = 0$.

   (d) $y'' + y' = x$.

7. Find all real or complex solutions of $y'' + iy' = 0$.

8. Find the roots of the characteristic equation of each of the following differential equations. Then write the general solution of the differential equation, replacing complex exponentials by $e^{\alpha x} \cos \beta x$ and $e^{\alpha x} \sin \beta x$ where it's appropriate.

(a) $y'' + 2y = 0.$

(b) $2y'' + 3y' = 0.$

(c) $y'' - 2y' + 2y = 0.$

(d) $y'' - y' + y = 0.$

(e) $2y'' + y' - y = 0.$

(f) $y'' + y' = 2y.$

(g) $2y'' + y' + y = 0.$

(h) $3y'' - y' + y = 0.$

9. For each of the differential equations in Exercise 8, find the solution satisfying the following corresponding initial-value or boundary-value conditions; do this by solving a pair of linear equations for the unknown constants.

(a) $y(0) = 0,\ y'(0) = 1.$

(b) $y(0) = 1,\ y'(0) = 0.$

(c) $y(\pi) = 0,\ y'(\pi) = 0.$

(d) $y(0) = 2,\ y'(0) = -1.$

(e) $y(0) = 0,\ y'(0) = 2.$

(f) $y(0) = 0,\ y(1) = 1.$

(g) $y(0) = 0,\ y'(0) = 0.$

(h) $y(0) = 0,\ y(1) = 0.$

10. The differential equation $y'' + y = 0$ may fail to have a unique solution if we require the solution to satisfy certain critically chosen boundary conditions. (Exercise 18 in Section 1 shows that if the constant-coefficient equation $y'' + ay' + by = 0$ has real characteristic roots, then the associated boundary-value problem $y(x_1) = y_1,\ y(x_2) = y_2$ always has a unique solution.)

(a) Show that the boundary-value problem $y'' + y = 0,\ y(0) = 1,\ y(\pi) = 1$ has no solution. [*Hint:* You know all solutions of $y'' + y = 0.$]

(b) Show that the boundary-value problem $y'' + y = 0,\ y(0) = 1,\ y(2\pi) = 1$ has infinitely many solutions.

11. (a) Show that $c_1 \cos \beta x + c_2 \sin \beta x$ can be written in the form $A \cos (\beta x - \phi)$, where $A = \sqrt{c_1^2 + c_2^2}$ if ϕ satisfies $\cos \phi = c_1/A$ and $\sin \phi = c_2/A.$

(b) The result of part (a) is useful because it shows that

$$c_1 \cos \beta x + c_2 \sin \beta x$$

has a graph that is the same as that of $\cos \beta x$ shifted by a **phase angle** ϕ and multiplied by an **amplitude** A. Sketch the graph of

$$\cos 2x + \sqrt{3} \sin 2x$$

by first finding ϕ and A.

12. (a) Show that $y(x) = c_1 \cos \beta x + c_2 \sin \beta x$ can be written in the form $A \sin (\beta x + \theta)$, where $A = \sqrt{c_1^2 + c_2^2}$ is the **amplitude** of $y(x)$ if θ satisfies $\sin \theta = c_1/A$ and $\cos \theta = c_2/A.$ (This is an alternative to the more usual transformation discussed in the previous exercise and in Section 2C that uses cosine instead of sine as the basic reference function.)

(b) The result of part (a) says that

$$c_1 \cos \beta x + c_2 \sin \beta x = A \sin (\beta x + \theta)$$

for appropriately chosen real numbers A and θ. Use this result to sketch the graph of $y(x) = \cos 2x + \sqrt{3} \sin 2x$, by first finding A and θ.

(c) Show that
$$A \sin (\beta x + \theta) = A \cos (\beta x - \phi),$$
where $\phi = \pi/2 - \theta$. The number ϕ (or sometimes $-\phi$) is called a **phase shift,** and A is the **amplitude** of the trigonometric function $A \sin (\beta x + \theta)$.

13. Separate the real and imaginary terms in the infinite series
$$\sum_{k=0}^{\infty} \frac{(ix)^k}{k!}$$
into two power series. Then use the result as a partial justification of the definition
$$e^{ix} = \cos x + i \sin x.$$

14. Complex-valued differentiable functions $f(x) = u(x) + iv(x)$ and $g(x) = s(x) + it(x)$ obey the same basic rules relative to differentiation that real-valued functions do. Use the corresponding relations for real-valued functions to show that the following formulas hold on an interval $a < x < b$ on which both f and g are differentiable complex-valued functions.
    (a) $(f + g)' = f' + g'$.
    (b) $(cf)' = cf'$, c const.
    (c) $(fg)' = fg' + f'g$.
    (d) $(f/g)' = (f'g - fg')/g^2$, $g \neq 0$.

15. (a) Show that if $y = y(x)$ is a solution to the constant-coefficient equation $y'' + ay' + by = 0$ and c is a constant, then the function y_c defined by $y_c(x) = y(x + c)$ is also a solution. [*Hint:* It's not necessary to know the form of $y(x)$ in terms of elementary functions.]
    (b) Generalize the result of part (a) to a solution $y(x)$ of the nth-order constant-coefficient equation
    $$y^{(n)} + a_{n-1}y^{(n-1)} + \cdots + a_1 y' + a_0 y = 0.$$
    (c) Generalize the result of part (a) to a solution $y(x)$, $a < x < b$, of a second-order equation, linear or nonlinear, of the form $y'' = F(y, y')$. [*Hint:* $y_c(x)$ will in general be defined on an interval different from $a < x < b$.]

16. Let $y(x)$ be a solution of a second-order equation $y'' + ay' + by = 0$. Could constants a and b have been chosen so that $y(x) = x \cos x$? What about $\sin x + \cos 2x$? Justify your answers.

17. Find second-order differential equations $y'' + ay' + by = 0$ that have the following as solutions:
    (a) $\sin 2x$.
    (b) $\cos 2x$.
    (c) $e^x \cos 2x$.
    (d) $e^{-x} \cos (x/3)$.
    (e) $\sin (x/2)$.
    (f) xe^{2x}.
    (g) $\sin 3x - \cos 3x$.
    (h) $x - 7$.
    [*Hint:* What are the characteristic roots associated with each solution?]

18. For fixed α and fixed $\beta \neq 0$, show directly that $e^{\alpha x} \cos \beta x$ and $e^{\alpha x} \sin \beta x$ are linearly independent. What if $\beta = 0$?

2C Order More Than Two

The general nth-order constant-coefficient homogeneous linear equation can be written $L(y) = 0$, where the differential operator L looks like

$$L = D^n + a_{n-1}D^{n-1} + \cdots + a_1 D + a_0.$$

In applications the coefficients a_k are almost always real numbers, though that's not essential to the theory.

e x a m p l e 1

If $L = D^3 + 2D^2 + 1$, the associated homogeneous differential equation is $y''' + 2y'' + y = 0$. If $L = D^4 + 3D^2$, the differential equation is $y'''' + 3y'' = 0$, sometimes written $y^{(4)} + 3y'' = 0$.

The method used in the second-order case, with result explained in the statement of Theorem 2.6, can be successively applied to produce the complete solution to the nth-order equation, provided that we can find the roots of the corresponding characteristic equation

$$r^n + a_{n-1}r^{n-1} + \cdots + a_1 r + a_0 = 0.$$

In complete generality, finding these roots can be a rather daunting technical problem whose practical solution requires numerical approximation techniques. Nevertheless, we can deal with a number of significant examples fairly simply.

Knowing about the solutions to the nth-order equation turns out to be important for understanding some special types of second-order nonhomogeneous equations $y'' + ay' + by = f(x)$. Otherwise it's mainly 4th-order equations that arise directly in applications. The method of repeated integration used in the second-order case can be used to prove the following theorem. The inductive argument involves no additional ideas beyond those used in proving Theorem 2.6, so we omit the proof.

■ 2.7 THEOREM

The nth-order constant-coefficient equation $L(y) = 0$ has for its general solution a sum of constant multiples

$$c_1 y_1(x) + \cdots + c_n y_n(x)$$

of solutions $y_k(x)$, where the c_k are arbitrary constants; these constants are uniquely prescribed by initial conditions: $y(x_0) = z_0$, $y'(x_0) = z_1, \ldots,$ $y^{n-1}(x_0) = z_{n-1}$. If $r_1, \ldots, r_n$ are the roots of the characteristic equation $r^n + a_{n-1}r^{n-1} + \cdots + a_1 r + a_0 = 0$, the terms $y_k(x)$ in the solution can each be written in the form $x^l e^{r_k x}$, $l = 0, 1, \ldots, m - 1$, where m is the multiplicity of the root r_k. If roots $\alpha + i\beta$ and $\alpha - i\beta$ occur in complex conjugate pairs, then the corresponding pair of exponential solutions can be replaced by

$$x^l e^{\alpha x} \cos \beta x, \qquad x^l e^{\alpha x} \sin \beta x.$$

The constants of integration in the solution formulas we've just been dealing with appear so often we extend our terminology to cover them as follows. An expression of the form

$$c_1 y_1 + c_2 y_2 + \cdots + c_n y_n$$

is called a **linear combination** of $y_1, y_2, \ldots, y_n$ with **coefficients** $c_1, c_2, \ldots, c_n$. The numbers c_k are regarded as parameters, and the linear combination displayed above is an example of an n-**parameter family** of functions.

---

e x a m p l e 2

The differential equation $y'''' - y = 0$ has characteristic equation $r^4 - 1 = 0$. To find the roots, we note that since $r^4 = 1$, then either $r^2 = 1$ or $r^2 = -1$. Hence the roots are $r_1 = 1, r_2 = -1, r_3 = i$, and $r_4 = -i$. The first two roots provide the solutions e^x and e^{-x}. The second pair provides e^{ix} and e^{-ix} or the alternative form $\cos x$ and $\sin x$. The complete solution is then

$$y(x) = c_1 e^x + c_2 e^{-x} + c_3 \cos x + c_4 \sin x.$$

Initial conditions $y(0) = 0, y'(0) = 1, y''(0) = 2, y'''(0) = 1$ impose conditions on the constants c_k. For example,

$$y'(x) = c_1 e^x - c_2 e^{-x} - c_3 \sin x + c_4 \cos x,$$

so $y'(0) = c_1 - c_2 + c_4 = 1$. The complete set of conditions reduces to

$$c_1 + c_2 + c_3 = 0$$
$$c_1 - c_2 + c_4 = 1$$
$$c_1 + c_2 - c_3 = 2$$
$$c_1 - c_2 - c_4 = 1.$$

Straightforward elimination shows that $c_1 = 1, c_2 = 0, c_3 = -1$, and $c_4 = 0$. Thus the particular solution that satisfies the initial conditions is $y(x) = e^x - \cos x$.

---

e x a m p l e 3

The 3rd-order differential equation $(D - 1)^3 y = 0$, which looks like $y''' + 3y'' + 3y' - y = 0$ when written without operator notation, has the single characteristic root $r = 1$ with multiplicity 3. The general solution is

$$y = c_1 e^x + c_2 x e^x + c_3 x^2 e^x.$$

---

e x a m p l e 4

The equation $y'''' + \lambda y'' = 0$ is satisfied under certain conditions by a function $y(x)$ that describes the lateral deflection, measured x units from one end, of a uniform column under a vertical compressive force. (The constant $\lambda = P/\rho$ depends on the structure of the column and on the vertical load P applied to it.) The characteristic equation is $r^4 + \lambda r^2 = 0$, or $r^2(r^2 + \lambda) = 0$. With $\lambda > 0$, the roots are $r_1 = r_2 = 0$ and $r_3 = \sqrt{\lambda}i, r_4 = -\sqrt{\lambda}i$. So the general solution is

$$y(x) = c_1 + c_2 x + c_3 \cos \sqrt{\lambda} x + c_4 \sin \sqrt{\lambda} x.$$

Initial conditions at a single point x_0 are physically uninteresting in this problem. What is usually done is to impose *boundary conditions* on $y(x)$ and $y'(x)$, or else on $y(x)$ and $y''(x)$ at points corresponding to the two ends of the column, say at $x = 0$ and $x = L$. The existence of a unique solution depends critically on the value of λ. These matters are taken up in the following exercises and also in Chapter 10, Section 3.

Finding the roots of an nth-degree characteristic equation is in general a difficult problem. The analogs of the *quadratic formula* for solutions of quadratic equations exist for equations of degree 3 and 4 but are somewhat complicated. The following examples illustrate some fairly simple special cases.

e x a m p l e 5

The cubic equation $r^3 + 2r^2 + 5r = 0$ can be written $r(r^2 + 2r + 5) = 0$. Apart from the obvious root $r_1 = 0$, there are the roots $r_2 = -1 + 2i$ and $r_3 = -1 - 2i$ of the quadratic equation $r^2 + 2r + 5 = 0$. The corresponding solutions of the differential equation

$$y''' + 2y'' + 5y = 0$$

are
$$y = c_1 + c_2 e^{-x} \cos 2x + c_3 e^{-x} \sin 2x.$$

e x a m p l e 6

The 4th-degree, or *quartic*, equation $r^4 - 13r^2 + 36 = 0$ can be looked at as a quadratic equation in r^2, with solutions $r^2 = (13 \pm 5)/2$. Since $r^2 = 4$ or $r^2 = 9$, the four distinct roots are $r = \pm 2, \pm 3$. Hence the solutions to

$$y^{(4)} - 13y'' + 36y = 0$$

consist of all members of the 4-parameter family

$$y = c_1 e^{2x} + c_2 e^{-2x} + c_3 e^{3x} + c_4 e^{-3x}.$$

It will be useful for us to apply the idea behind Theorem 1.2 (Section 1B) in reverse order, starting with solutions and arriving at a differential equation that has those solutions.

e x a m p l e 7

To find a constant-coefficient equation $L(y) = 0$ of least possible order having e^x, e^{2x}, and e^{-2x} for solutions, we write the linear operator equation

$$(D - 1)(D - 2)(D + 2)y = 0.$$

Clearly e^{-2x} is a solution, because $(D + 2)e^{-2x} = 0$. Since the operator factors can be written in any order, and $(D - 2)e^{2x} = 0$ and $(D - 1)e^x = 0$, the three functions e^x, e^{2x}, and e^{-2x} are solutions of the differential equation

$$(D - 1)(D - 2)(D + 2)y = 0 \quad \text{or} \quad y''' - y'' - 4y' + 4y = 0.$$

e x a m p l e 8

Let us find a constant-coefficient equation $L(y) = 0$ of least possible order having $\cos x$ and $\sin 2x$ for solutions. These solutions would have arisen from characteristic roots $\pm i$ and $\pm 2i$, respectively. We write the linear operator equation

$$(D^2 + 1)(D^2 + 4)y = 0.$$

Clearly $\sin 2x$ is a solution, because $(D^2 + 4) \sin 2x = 0$. Since the operator factors can be written in any order, and $(D^2 + 1) \cos x = 0$, both $\sin 2x$ and $\cos x$ are solutions of the differential equation. Linear combinations $y = c_1 \cos x + c_2 \sin x + c_3 \cos 2x + c_4 \sin 4x$ constitute the general solution.

Independence of Basic Solutions. In Section 1 we defined linear independence for two functions. For n functions the extended definition is this: $y_1(x)$, $y_2(x)$, ..., $y_n(x)$ are **linearly independent** on an interval if the identity $c_1 y_1(x) + c_2 y_2(x) + \cdots + y_n(x) = 0$ holds there only with the obvious choice $c_1 = c_2 = \cdots = c_n = 0$ for the constants. It is left as an exercise to show that a set of two or more functions is linearly independent if and only if no one of them is expressible as a linear combination of the others. Thus independence ensures that there is no redundancy in the expression of a given function as a linear combination of functions from an independent subset. An independent subset is called a **basis** for the linear combinations of its elements. That the solutions $x^l e^{\alpha x} \cos \beta x$, $x^l e^{\alpha x} \sin \beta x$ listed at the conclusion of Theorem 2.7 are indeed linearly independent is a simple consequence of the theorem, as follows.

■ 2.8 COROLLARY

The n solutions $x^l e^{r_k x}$ listed in Theorem 2.7, with real form $x^l e^{\alpha x} \cos \beta x$, $x^l e^{\alpha x} \sin \beta x$, are linearly independent.

Proof. Suppose a linear combination $y(x)$ of these n solutions is identically zero:

$$c_1 y_1(x) + \cdots + c_n y_n(x) = 0.$$

To prove independence we need to show that all $c_k = 0$. The linear combination is a solution of a linear homogeneous differential equation $L(y) = 0$, and since this solution is identically zero, it obviously satisfies initial conditions $y(x_0) = y'(x_0) = \cdots = y^{(n-1)}(x_0) = 0$ at a given point x_0. Since choosing all the $c_k = 0$ suffices to produce this solution, and since Theorem 2.7 guarantees uniqueness of the numbers c_k, all c_k must be zero.   ■

EXERCISES

1. Find the most general solution to each of the following differential equations.

(a) $y''' - y = 0$.

(b) $y^{(4)} - 2y'' + y = 0$.

(c) $y''' - 2y' = 0$.

(d) $y^{(4)} - y'' = 0$.

(e) $y''' - 8y = 0$.

(f) $y^{(4)} - 4y'' + 4y = 0$.

(g) $(D^2 - 4)(D^2 - 1)y = 0$.

(h) $D(D - 1)^3 y = 0$.

(i) $y''' - 2y'' = 0$.

(j) $y^{(4)} - y = 0$.

2. Find constant-coefficient linear differential equations with real coefficients and of the smallest possible order that have the following functions $y(x)$ as solutions.

(a) $y(x) = \cos 4x$.

(b) $y(x) = x \cos 4x$.

(c) $y(x) = x^2 \cos 4x$.

(d) $y(x) = x \sin 4x$.

(e) $y(x) = xe^x \sin x$.

(f) $y(x) = x^3 \cos 4x$.

(g) $y(x) = x^5$.

(h) $y(x) = \cos 4x + \sin 3x$.

(i) $y(x) = e^{-x} \cos x$.

(j) $y(x) = x \cos x + \cos 2x$.

[*Hint:* Find the characteristic roots associated with each equation.]

3. Listed below are families of solutions to some linear constant-coefficient differential equations. In each case find a differential equation of least possible order satisfied by the family.

(a) $c_1 + c_2 x + c_3 e^x$.

(b) $c_1 \cos x + c_2 \sin x + c_3 + c_4 x$.

(c) $c_1 \cos x + c_2 \sin x + c_3 e^x$.

(d) $c_1 \cos 2x + c_2 \sin 2x + c_3 e^{-x}$.

(e) $c_1 + c_2 x + c_3 x^2$.

(f) $c_1 \cos 3x + c_2 \sin 3x + c_3$.

(g) $c_1 e^x + c_2 e^{-x} + c_3 \cos x + c_4 \sin x$.

(h) $c_1 e^x \cos 2x + c_2 e^x \sin 2x + c_3$.

(i) $c_1 \cos x + c_2 \sin x$.

(j) $c_1 \cos x + c_2 \sin x + c_3 \cos 3x + c_4 \sin 3x$.

4. Each of the solution families listed in the previous exercise can be made to satisfy the corresponding set of conditions listed below; find the appropriate values for the constants c_k.

(a) $y(0) = 1$, $y'(0) = 2$, $y''(0) = 1$.

(b) $y(0) = y'(0) = 2$, $y''(0) = y'''(0) = -1$

(c) $y(0) = 2$, $y'(0) = y''(0) = 0$.

(d) $y(0) = 2$, $y'(0) = y''(0) = -3$.

(e) $y(0) = y'(0) = 1$, $y''(0) = 2$.

(f) $y(0) = 1$, $y'(0) = 3$, $y''(0) = -9$.

(g) $y(0) = y'(0) = 1$, $y''(0) = -1$, $y'''(0) = 3$.

(h) $y(0) = 2$, $y'(0) = 3$, $y''(0) = 1$.

5. Find the general solution to each of the following equations.

(a) $y^{(6)} - y''' = 0$.

(b) $y'''' - 2y''' = 0$.

(c) $y^{(5)} - y' = 0$.

(d) $y^{(4)} + y = 0$.

(e) $(D^2 - 4)(D^2 - 16)y = 0$.

(f) $D^2(D - 1)^3 y = 0$.

[*Hint:* In (d), if $r^4 + 1 = 0$, note that $r^2 = \pm i = \pm e^{i(\pi/2)}$, k an integer. Then $r = \pm i e^{i(\pi/4)}$.]

6. Explain why $y(x) = \cos x + \sin 2x$ can't be the solution to a constant-coefficient equation of the form $y'' + ay' + by = 0$. Find an equation of higher order that $y(x)$ does satisfy.

7. The general solution $y(x) = c_1 + c_2 x + c_3 \cos \sqrt{\lambda} x + c_4 \sin \sqrt{\lambda} x$ to $y'''' + \lambda y'' = 0$ is derived in Example 4 of the text; use it to do the following.

(a) If $\lambda = 4\pi^2$, find the infinitely many solutions that satisfy the boundary conditions $y(0) = y(1) = 0$, $y'(0) = y'(1) = 0$.

(b) Under the assumption that $y(x)$ represents the horizontal deflection of a column under a vertical compressing force, we can interpret the boundary conditions in part (a) to mean that the ends of the column are rigidly embedded in floor and ceiling. Sketch some typical solutions, assuming small displacements.

(c) Show that if $\lambda = \pi^2$, then the only solution satisfying the boundary conditions in part (a) is the identically zero solution.

8. The general solution $y(x) = c_1 + c_2 x + c_3 \cos \sqrt{\lambda} x + c_4 \sin \sqrt{\lambda} x$ to $y'''' + \lambda y'' = 0$ is derived in Example 4 of the text; use it to do the following.

(a) If $\lambda = \pi^2$, find the infinitely many solutions that satisfy the boundary conditions $y(0) = y(1) = 0$, $y''(0) = y''(1) = 0$.

(b) Under the assumption that $y(x)$ represents the horizontal deflection of a column under a vertical compressing force, we can interpret the boundary conditions in part (a) to mean that the ends of the column are hinged at both floor and ceiling. Sketch some typical solutions, assuming small displacements.

(c) Show that if $\lambda = \pi^2/4$, then the only solution satisfying the boundary conditions in part (a) is the identically zero solution.

9. A differential equation for the lateral displacement $y = y(x)$ at distance x from one end of a uniform rotating shaft is
$$y'''' - \lambda y = 0,$$
where the constant $\lambda > 0$ is proportional to the speed of rotation.

(a) Show that the most general solution to the differential equation can be written
$$y = c_1 \cosh \sqrt[4]{\lambda} x + c_2 \sinh \sqrt[4]{\lambda} x + c_3 \cos \sqrt[4]{\lambda} x + c_4 \sin \sqrt[4]{\lambda} x,$$
where $\cosh u = (e^u + e^{-u})/2$ and $\sinh u = (e^u - e^{-u})/2$.

(b) Let $\lambda = n^4 \pi^4$, where n is a positive integer, and find solutions of the differential equation subject to boundary conditions $y(0) = y''(0) = 0$, $y(1) = y''(1) = 0$.

(c) Sketch the graphs of the solutions found in part (b) for $n = 1, 2, 3$.

(d) Show that if λ is not of the form prescribed in part (b), then the only solution to the problem posed there is the identically zero solution.

10. Show that a set of two or more functions is linearly independent if and only if no one of them is expressible as a linear combination of the others.

11. (a) Show that if $y = y(x)$ is a solution to $y''' + ay'' + by' + cy = 0$, where $a, b,$ and c are constant, then, for any fixed constant c, the function $y = y(x + c)$ is also a solution of the same equation. This property of the set of solutions is called **translation invariance**. [*Note:* You can do this without using explicit solution formulas.]

(b) Is the set of solutions of an nth-order constant-coefficient linear equation $L(y) = 0$ translation-invariant? Explain.

(c) What can you say about translation invariance of the set of solutions of equations such as $y^{(n)} = f(y, y', \dots, y^{(n-1)})$ if f is not itself explicitly dependent on x?

(d) Explain why translation invariance of the set of solutions fails to hold for equations with nonconstant coefficients, even, for example, for a first-order equation like $y' + g(x)y = 0$. [*Hint:* Consider the direction field of a first-order linear equation.]

12. A uniform horizontal beam sags by an amount $y = y(x)$ at a distance x measured from one end. For a fairly rigid beam with uniform loading, $y(x)$ typically satisfies a 4th-order differential equation $y'''' = R$, where R is a constant depending on the load being carried and on the characteristics of the beam itself. If the ends of the beam are supported at $x = 0$ and at $x = L$, then $y(0) = y(L) = 0$. The extended beam also behaves as if its profile had an inflection point at each support so that $y''(0) = y''(L) = 0$.

(a) Use the multiple characteristic root of the associated homogeneous equation to find the general solution of the homogeneous equation.

(b) Show that the vertical deflection at point x is
$$\tfrac{1}{24}R(x^4 - 2Lx^3 + L^3x), \qquad 0 \le x \le L.$$

2D   *Sinusoidal Oscillations (Optional)*

Two-parameter families of the form
$$y(x) = c_1 \cos \beta x + c_2 \sin \beta x$$
occur often in the description of **oscillation**, a term that means the repeated switching back and forth from one side to another of some relatively stable reference. The particular oscillation displayed above is called **harmonic** because of the central role it plays in the analysis of musical tones. More generally, a function $f(x)$ is called **periodic** with period p if there is a number $p \neq 0$ such that $f(x + p) = f(x)$ for all x. Thus the number $2\pi/\beta$ is a **period** of the oscillation $y(x)$, since for its two terms we have

$$\cos \beta \left(x + \frac{2\pi}{\beta} \right) = \cos (\beta x + 2\pi) = \cos \beta x,$$

$$\sin \beta \left(x + \frac{2\pi}{\beta} \right) = \sin (\beta x + 2\pi) = \sin \beta x, \qquad \text{for all } x.$$

The **frequency,** or number of oscillations, per x unit of a periodic function $f(x)$ is just the reciprocal of the period, and for our harmonic oscillation $y(x)$ the frequency will be $\nu = \beta/(2\pi)$. (The number β itself is called the **circular frequency** of the oscillation and is equal to the number of oscillations per 2π units.)

---

example 1   The harmonic oscillation
$$y(x) = \cos 2x - 3 \sin 2x$$
has period π and frequency $1/\pi$, since both $\cos 2x$ and $\sin 2x$ have period π. In addition, $y(x)$ has other periods $n\pi$ for integer n, but the period π, the smallest

positive period, is usually singled out as a fundamental period. The function

$$f(x) = \cos 2x - 3 \sin 4x$$

is periodic with period π but is not a harmonic oscillation, because the fundamental periods of the two terms are unequal.

Since the parameters c_1 and c_2 in our general harmonic oscillation don't usually have clear meanings, it is sometimes preferable to write $y(x)$ in terms of other parameters that have useful interpretations. We choose an angle ϕ, shown in Figure 3.2a, called a **phase shift** or **phase angle,** such that

$$\cos \phi = \frac{c_1}{\sqrt{c_1^2 + c_2^2}}, \qquad \sin \phi = \frac{c_2}{\sqrt{c_1^2 + c_2^2}}.$$

(Such a ϕ can always be found. The reason is that the sum of the squares of the expressions for $\cos \phi$ and $\sin \phi$ is 1, so these two numbers represent the coordinates of a point that determines, with origin as vertex, an angle ϕ with the positive c_1 axis. See Figure 3.2a.) Now set $A = \sqrt{c_1^2 + c_2^2}$ so that $c_1 = A \cos \phi$ and $c_2 = A \sin \phi$. Using an addition formula for the cosine we can then write

$$c_1 \cos \beta x + c_2 \sin \beta x = A(\cos \phi \cos \beta x + \sin \phi \sin \beta x)$$

$$= A \cos (\beta x - \phi).$$

Thus our original 2-parameter family has been rewritten in terms of two new parameters A and ϕ. Furthermore, it's clear that the graph of $y(x)$ is really just a horizontal shift of the simple oscillation $A \cos \beta x$. (See Figure 3.2b for some sample graphs; you can match the graphs with the formulas by checking the value at $x = 0$.) The number A is the **amplitude** of the oscillation about the value $y = 0$, and the actual width of the oscillation from side to side is consequently $2A$. As for the phase shift ϕ, it satisfies $\tan \phi = c_2/c_1$, so in general $\phi = \arctan(c_2/c_1)$. Note that if c_1 and c_2 are both positive, then the point $(\cos \phi, \sin \phi)$ lies in the first quadrant, in which case we can pick a unique ϕ so that $0 < \phi < \pi/2$. Specifically, we can determine ϕ by $\phi = \arctan(c_2/c_1)$, where

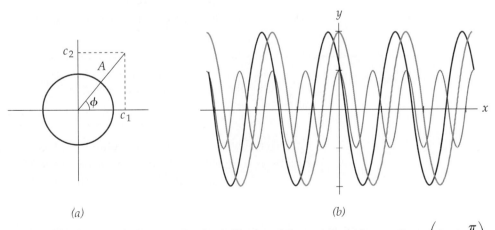

(a)   (b)

Figure 3.2 (a) Phase shift ϕ; amplitude A. (b) $y = \cos 8x$, $y = 2 \cos 4x$, $y = 2 \cos \left(4x + \dfrac{\pi}{3}\right)$.

the principal branch of the arctangent function is assumed. In other cases it's better to check directly that ϕ is in the correct quadrant; for example, if c_1 and c_2 are both negative, then we can choose to have $\pi < \phi < 3\pi/2$. To summarize,

■ **2.9** $A = \sqrt{c_1^2 + c_2^2}$, $\phi = \arctan \dfrac{c_2}{c_1}$; $c_1 = A \cos \phi$, $c_2 = A \sin \phi$.

To determine ϕ when $c_1 = 0$, we set $\phi = \pi/2$ if $c_2 > 0$ and $\phi = -\pi/2$ if $c_2 < 0$.

---

e x a m p l e 2

The trigonometric function $y(x) = 3 \cos 4x + 3 \sin 4x$ has period $2\pi/4 = \pi/2$ and hence a frequency of $2/\pi \approx 0.64$ complete oscillations per x unit. The amplitude is $A = \sqrt{3^2 + 3^2} = 3\sqrt{2}$. The phase shift is $\phi = \arctan \frac{3}{3} = \pi/4$. Hence the given function can be written as

$$y(x) = A \cos (\beta x - \phi) = 3\sqrt{2} \cos \left(4x - \frac{\pi}{4} \right).$$

Note that the definition of phase shift is stated entirely in terms of cosine, and if $\phi > 0$, it indicates that the graph of $y(x)$ is a shift to the right of the graph of $\cos \beta x$ by the amount ϕ/β. Because the cosine function has period 2π, the phase shift ϕ can always be altered by adding an integer multiple of 2π.

---

e x a m p l e 3

To analyze $\cos 4x - \sqrt{3} \sin 4x$, note that a phase shift is $\phi = \arctan (-\sqrt{3}) = -\pi/3$ and that the amplitude is $A = \sqrt{1^2 + (-\sqrt{3})^2} = 2$. Then express the oscillation in the form $A \cos (\beta x - \phi) = 2 \cos (4x + \pi/3)$. Thus the given combination of sine and cosine has the same graph as $2 \cos 4 (x + \pi/12)$, which itself is a shift to the *left* of the graph of $2 \cos 4x$ by $\pi/12$.

Two cosine oscillations $A \cos (\beta x - \phi)$ and $B \cos (\gamma x - \psi)$ having a common period P are said to be **in phase** if $\phi - \psi$ is an integer multiple of 2π; otherwise they are **out of phase** by $|\phi - \psi|/P$.

---

e x a m p l e 4

Figure 3.2b shows the graphs of $\cos 8x$, $2 \cos 4x$, and $2 \cos (4x + \pi/3)$. The first two are in phase, with common period $\pi/2$. The third is out of phase with the other two by $(\pi/3)/(\pi/2) = \frac{2}{3}$.

The functions that arise as solutions of second-order constant-coefficient equations $y'' + ay' + by = 0$ are purely harmonic oscillations
$$y(x) = c_1 \cos \beta x + c_2 \sin \beta x = A \cos (\beta x - \phi)$$
only in the special case $a = 0$, $b > 0$ that corresponds to the case of purely imaginary conjugate roots for the characteristic equation. In all other cases the solutions don't oscillate at all, or else oscillate but without a constant amplitude. In the latter case, nonperiodic oscillation occurs when $a \neq 0$ and

$a^2 - 4b < 0$, which corresponds to complex conjugate roots with nonzero real part: $\alpha \pm i\beta$. The solutions then look like

$$y(x) = e^{\alpha x}(c_1 \cos \beta x + c_2 \sin \beta x)$$
$$= Ae^{\alpha x} \cos(\beta x - \phi).$$

Thus the contrast between harmonic and nonharmonic oscillation of these solutions boils down to the behavior of the exponential factor $e^{\alpha x}$. Because $-1 \leq \cos(\beta x - \phi) \leq 1$, the graphs of these oscillations always lie between the graphs of $-Ae^{\alpha x}$ and $Ae^{\alpha x}$.

---

example 5

Here are two examples of nonharmonic oscillation:

$$y_1(x) = e^{-2x/5} \cos 6x, \qquad y_2(x) = \tfrac{3}{2}e^{x/5} \cos(5x + \tfrac{2}{5}).$$

Note that the harmonic factors in each example are periodic with respective periods $\pi/6$ and $\pi/5$, but that the function as a whole is not periodic. Since $\cos x$ takes the value zero whenever $6x = (k + \tfrac{1}{2})\pi$, k an integer, it follows that $y_1(x)$ is zero when $x = (k/6 + \tfrac{1}{12})\pi$. Similarly $y_2 = 0$ when $x = (k/5 + \tfrac{1}{10})\pi - \tfrac{2}{25}$. However, the precise locations of the successive maxima and minima depend on the exponential factor and the shift. See Figure 3.3.

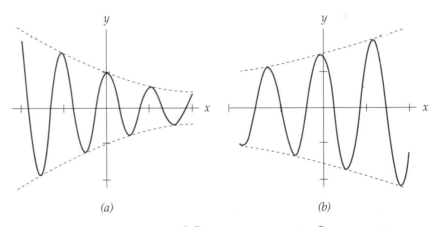

(a)   (b)

Figure 3.3 (a) $y_1(x) = e^{-2x/5} \cos 6x$. (b) $y_2(x) = \tfrac{3}{2}e^{x/5} \cos(5x + \tfrac{2}{5})$.

EXERCISES

Determine (i) the smallest positive period p, (ii) the frequency ν, and (iii) the amplitude A of each of the following periodic functions.

1. $\sin 3x$.

2. $\sin 3x + 2 \cos 3x$.

3. $\sin 3x - 2 \cos 3x$.

4. $\sin \sqrt{2}x$.

5. $\sin 3(x + 1) + \cos 3(x + 1)$.

6. $\sin 4x - 4 \cos 4x$.

Sketch the graphs of the following oscillatory functions $y(x)$. Also state whether it is a solution of a constant-coefficient differential equation of the form $y'' + ay' + by = 0$ and, if so, what the equation is.

7. $y(x) = 2\cos(x - \pi/2)$.

8. $y(x) = 3\cos(x + \pi/2) + 1$.

9. $y(x) = \cos(2x - \pi/3)$.

10. $y(x) = 3\cos(x/2 + \pi/4) - 1$.

11. $y(x) = e^{-x}\cos(x - \pi)$.

12. $y(x) = -e^x\cos(x + \pi/3)$.

13. $y(x) = 2e^{-x/2}\cos(x - \pi/3) + 2$.

14. $y(x) = 3\sin x + \cos 2x$.

Phase Angles. Each of the trigonometric functions $y(x)$ listed below can be written in the form $A\cos(\beta x - \phi)$. In each case find the amplitude A and a phase shift ϕ. Then sketch the graph.

15. $y(x) = \cos x + \sqrt{3}\sin x$.

16. $y(x) = \sqrt{2}(\cos x + \sin x)$.

17. $y(x) = 3\sin 2x$.

18. $y(x) = 2\sqrt{2}(\cos 5x - \sin 5x)$.

19. $y(x) = -\cos 4x$.

20. $y(x) = 2\sqrt{3}\cos x + 2\sin x$.

21. $y(x) = -3\cos 2x + 3\sqrt{3}\sin 2x$.

22. $y(x) = \sin x$.

Initial Values. For each of the functions listed below find the initial-value problem that the function satisfies; this means find (i) a constant-coefficient differential equation of the form $y'' + ay' = by = 0$ and (ii) $y(0)$ and $y'(0)$.

23. $y(x) = 3\cos(2x - \pi/2)$.

24. $y(x) = -\cos(3x + \pi/2)$.

25. $y(x) = 4\cos(x + \pi/3)$.

26. $y(x) = \cos(\sqrt{2}x + \pi/4)$.

27. $y(x) = e^{-x}\cos(2x + \pi)$.

28. $y(x) = -e^{-2x}\cos(3x + \pi/3)$.

29. $y(x) = e^x\cos(x - \pi/2)$.

30. $y(x) = e^{2x}\cos(2x - \pi/4)$.

31. Show that the function $y(x) = e^{\alpha x}\cos\beta x$ is never periodic unless $\alpha = 0$.

32. Show that the function $y(x) = e^{\alpha x}\cos\beta x$ has its maxima and minima at points with x coordinate $(1/\beta)\arctan(\alpha/\beta) + k\pi/\beta$ for integer k.

33. Let $0 < \beta < \gamma$ and define $f(x) = A\sin\beta x + B\sin\gamma x$, with A and B nonzero constants.

  (a) Show that if γ/β is a rational number, then $f(x)$ has period $2n\pi/\beta$ for some integer n.

  (b) Show that if $f(x)$ is periodic with period $\alpha > 0$, then γ/β must be a rational number. [*Hint:* Show that $f(\alpha) = 0$ and $f''(\alpha) = 0$. Then conclude that $\sin\beta\alpha = 0$ and $\sin\gamma\alpha = 0$.]

*34. Let $f(x)$ be differentiable for all real x. Show that if f has a sequence of periods α_n tending to zero, then f must have derivative zero everywhere and so must be constant. Conclude that if f is nonconstant and periodic, then it has a least-positive period.

*35. Let $f(x)$ have a least-positive period α. Show that if f also has period $\beta > \alpha$, then β is an integer multiple of α. [*Hint:* If not, then $n\alpha < \beta < (n + 1)\alpha$ for some integer n. Show that $(n + 1)\alpha - \beta < \alpha$ is also a period.]

*36. Assume that the differential equation $y'' + ay' + by = 0$ has conjugate complex roots $r_1 = \alpha + i\beta$ and $r_2 = \alpha - i\beta$.

  (a) Show that if $x_1 < x_2$, the boundary-value problem

$$y'' + ay' + by = 0, \qquad y(x_1) = y_1, \qquad y(x_2) = y_2$$

  always has a unique solution for given y_1 and y_2, provided that $\beta(x_2 - x_1)$ is not an integer multiple of π. [*Hint:* Use explicit solution formulas, and show that you can solve for the desired constants c_1 and c_2 in the general solution.]

(b) Show that the boundary-value problem $y'' + y = 0$, $y(0) = y(\pi) = 1$, has no solution.

(c) Show that the boundary-value problem $y'' + y = 0$, $y(0) = y(2\pi) = 1$, has infinitely many solutions.

37. The differential equation $y'' + \beta^2 y = 0$ has solutions of the form $A \cos (\beta x - \phi)$, $\beta > 0$. Conditions $y(0) = y_0 \neq 0$ and $y'(0) = z_0$ determine a unique solution as follows.

(a) Solve the equations $A \cos \phi = y_0$, $\beta A \sin \phi = z_0$ to get $A = \sqrt{y_0^2 + z_0^2/\beta^2}$ and $\phi = \arctan [(1/\beta)z_0/y_0]$ if $y_0 \neq 0$.

(b) Show that if $y_0 = 0$, the formulas in part (a) can reasonably be interpreted to mean $A = \beta$ and $\phi = \pi/2$ when $z_0 > 0$, $\phi = -\pi/2$ when $z_0 < 0$.

3 NONHOMOGENEOUS LINEAR EQUATIONS

3A General Solution

Differential operators of the form

$$L = D^2 + aD + b$$

are **linear,** which means that

$$L(c_1 y_1 + c_2 y_2) = c_1 L(y_1) + c_2 L(y_2)$$

for every pair of constants c_1, c_2 and every pair of twice-differentiable functions y_1, y_2. Verification of the linearity of this second-order operator L is a direct consequence of the linearity of differentiation and of the arithmetic operations involved and is left as an exercise. The linearity property has allowed us to conclude that $c_1 y_1 + c_2 y_2$ is a solution of the **homogeneous equation** $L(y) = 0$, or

$$y'' + ay' + by = 0,$$

whenever y_1 and y_2 are solutions. The linearity of L also plays a part in describing the general solution of the **nonhomogeneous equation** $L(y) = f$, or

$$y'' + ay' + b = f(x),$$

where $f(x)$ is given on some interval. We'll see the solution has two parts, only one of which depends on f. The function f is often called a **forcing function** because of its interpretation in mechanical and electrical problems.

■ 3.1 THEOREM

Let y_p be some particular solution of the nonhomogeneous equation

$$L(y) = f,$$

where L is linear. Then the general solution of $L(y) = f$ is a sum

$$y = y_p + y_h,$$

where y_h is the general solution of the homogeneous equation $L(y) = 0$.

Proof. Assume that $L(y_p) = f$ and that y is some solution of $L(y) = f$. Then

$$L(y - y_p) = L(y) - L(y_p)$$
$$= f - f = 0.$$

Thus $y - y_p$ is equal to some solution y_h of $L(y) = 0$. It follows that $y = y_p + y_h$. ∎

Note that Theorem 3.1 applies to any linear operator at all; it will have several applications later, but for now we will use it to conclude that

$$y'' + ay' + by = f(x)$$

has the general solution

$$y = y_p(x) + c_1 y_1(x) + c_2 y_2(x),$$

where y_p is some particular solution and y_1 and y_2 are basic exponential solutions of the associated homogeneous equation.

***e x a m p l e* 1**   The equation

$$y'' - y = x$$

has $y'' - y = 0$ for its associated homogeneous equation, with the most general solution

$$y_h = c_1 e^x + c_2 e^{-x}.$$

Also it is easy to see by inspection that $y_p = -x$ is a solution of the nonhomogeneous equation, since $y_p' = -1$ and $y_p'' = 0$. (We will see shortly how to discover such a solution by integration.) It follows from Theorem 3.1 that every solution has the form

$$y = y_p + y_h$$
$$= -x + c_1 e^x + c_2 e^{-x}.$$

To find the particular solution satisfying conditions of the form

$$y(x_0) = a_0, \qquad y'(x_0) = a_1,$$

just compute $y'(x) = -1 + c_1 e^x - c_2 e^{-x}$ and solve

$$-x_0 + c_1 e^{x_0} = a_0$$
$$-1 + c_1 e^{x_0} - c_2 e^{-x_0} = a_1$$

for the correct constants c_1 and c_2.

3B   Undetermined-Coefficient Method

Example 1 has the advantage of simplicity, but it is unsatisfying because it used a solution $y_p(x) = -x$ that was just a guess, or, in more elegant parlance, was "found by inspection." The present subsection puts the inspection method on a more formal and routine footing for a special class of equations. This routine is the most efficient way to solve many equations, but for second-order equations its applicability is not as wide as that of the Green's function method of Section 3C.

The undetermined-coefficient method amounts to making educated guesses to find trial solutions containing certain constants to be determined by substitution into a constant-coefficient equation such as

$$y'' + ay' + by = f(x).$$

The method works only when the forcing function $f(x)$ is a linear combination, with constant coefficients, of functions that can be written in either of the two forms

$$x^n e^{\alpha x} \cos \beta x \quad \text{or} \quad x^n e^{\alpha x} \sin \beta x,$$

where n is a nonnegative integer. A typical example is

$$f(x) = 2x e^{-2x} + 3 \sin 2x.$$

For the first term we've chosen $n = 1$, $\alpha = -2$, $\beta = 0$, and for the second one $n = 0$, $\alpha = 0$, $\beta = 2$. (It's no accident that the special form of these terms is the same as that of solutions to homogeneous constant-coefficient equations; we'll see that that's precisely why the method works.)

Since we're dealing with linear differential operators like $L = D^2 + \alpha D + b$, a differential equation of the form

$$L(y) = a_1 f_1(x) + \cdots + a_m f_m(x), \qquad a_k \text{ const,}$$

can be solved by first finding a solution $y_k(x)$ to each of the m related equations

$$L(y) = f_1(x), \ldots, L(y) = f_m(x).$$

The linearity of L then implies that the function

$$y(x) = a_1 y_1(x) + \cdots + a_m y_m(x)$$

is a solution to the given equation. This application of linearity is sometimes called the **superposition principle.**

---

e x a m p l e 2

To solve $y'' - y = 4e^{2x} + 5e^{-3x}$, we can use linearity to split the problem into the solution of the two equations

$$y'' - y = e^{2x} \quad \text{and} \quad y'' - y = e^{-3x}.$$

The associated homogeneous equation $y'' - y = 0$ is of course the same for both equations, with common solution $y_h(x) = c_1 e^x + c_2 e^{-x}$. All that remains to be done is to find single solutions $y_1(x)$ and $y_2(x)$ to each of the two simpler equations and then write the general solution in the form $y = y_h + 4y_1 + 5y_2$.

For the first equation, $y'' - y = e^{2x}$, a reasonable guess is that there might be a solution of the form $y_1(x) = Ae^{2x}$ for some constant A. Indeed, we compute $y_1' = 2Ae^{2x}$, $y_1'' = 4Ae^{2x}$, and substitute:

$$y_1'' - y_1 = 4Ae^{2x} - Ae^{2x}$$
$$= 3Ae^{2x} = e^{2x}.$$

To satisfy the differential equation, we need $3A = 1$, so $A = \frac{1}{3}$ gives us $y_1 = \frac{1}{3}e^{2x}$. To solve $y'' - y = e^{-3x}$, a trial solution of a similar general form, $y_2 = Ae^{-3x}$, leads to $y_2' = -3Ae^{-3x}$, $y_2'' = 9Ae^{-3x}$, and thus to $y_2'' - y_2 = 8Ae^{-3x} = e^{-3x}$. Hence $A = \frac{1}{8}$ and $y_2 = \frac{1}{8}e^{-3x}$. Adding $4y_1 + 5y_2$ together with y_h gives the general solution

$$y = c_1 e^x + c_2 e^{-x} + \tfrac{4}{3}e^{2x} + \tfrac{5}{8}e^{-3x}.$$

Why Did Our Guesses Work? That our guesses were correct for the form of the two particular solutions that we needed in Example 2 follows from the fact that we did find solutions of that form. Now let's look at a procedure that will lead to correct guesses automatically and will also work as well in more complicated examples. The general setup was

$$(D - r_1)(D - r_2)y = e^{ax},$$

with $r_1 = 1$, $r_2 = -1$, and either $a = 2$ or $a = -3$. Recall that we already know all about solving constant-coefficient homogeneous equations if we know the roots of the characteristic equation. Consider what happens if we apply the operator $(D - a)$ to both sides of the previous equation:

$$(D - a)(D - r_1)(D - r_2)y = (D - a)e^{ax}$$
$$= ae^{ax} - ae^{ax} = 0.$$

The operator $(D - a)$ is called an **annihilator** of e^{ax}, since $(D - a)e^{ax} = 0$. We now have two relevant facts at our disposal:

1. All solutions of the above 3rd-order homogeneous equation are of the form

$$y = c_1 e^{r_1 x} + c_2 e^{r_2 x} + c_3 e^{ax},$$

   provided that a is different from the two unequal numbers r_1 and r_2.
2. A sufficiently differentiable solution of the second-order *nonhomogeneous* equation must also be a solution of the 3rd-order equation, since the latter was obtained by applying $(D - a)$ to the former.

Since the functions in the 2-parameter family $c_1 e^{r_1 x} + c_2 e^{r_2 x}$ are solutions of the associated *homogeneous* linear second-order equation, they can't by themselves provide a solution of the *nonhomogeneous* equation. Hence a solution to the nonhomogeneous equation must be among the functions of the form $c_3 e^{ax}$. Here is an outline of the routine for finding the terms in a linear combination for a trial solution y_p to $L(y) = f(x)$, where f has the special form

$$x^n e^{\alpha x} \cos \beta x \qquad \text{or} \qquad x^n e^{\alpha x} \sin \beta x.$$

1. Include in y_p the function f itself and all terms in its successive derivatives, discarding any repeated terms.
2. If a term included in step 1 happens to be a solution of the homogeneous equation, multiply that term and all terms surviving from step 1 by the single lowest power x^k such that the resulting terms are no longer homogeneous solutions.
3. Form a linear combination with undetermined constant coefficients of the terms from step 2 and determine the values of the coefficients by substitution into $L(y) = f$.

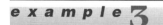

e x a m p l e 3

Here are some examples of functions $f(x)$ and corresponding trial solutions $y_p(x)$, assuming no term in $y_p(x)$ satisfies the homogeneous equation.

$$f(x) = ce^{rx}; \qquad y_p(x) = Ae^{rx}.$$
$$f(x) = cx^2; \qquad y_p(x) = Ax^2 + Bx + C.$$
$$f(x) = cx^2 e^{rx}; \qquad y_p(x) = (Ax^2 + Bx + C)e^{rx}.$$
$$f(x) = c \cos \beta x; \qquad y_p(x) = A \cos \beta x + B \sin \beta x.$$
$$f(x) = cx \sin \beta x; \qquad y_p(x) = (Ax + B) \sin \beta x + (Cx + D) \cos \beta x.$$
$$f(x) = ce^{\alpha x} \cos \beta x; \qquad y_p(x) = (A \cos \beta x + B \sin \beta x)e^{\alpha x}.$$

example 4

To solve $y'' + 4y = 3 \cos 2x$ note that the equation has the associated homogeneous solution $y_h = c_1 \cos 2x + c_2 \sin 2x$. The choice $y_p = A \cos 2x + B \sin 2x$ for $y_p(x)$ would thus be inadequate, because substitution into the left side produces zero, not a multiple of $\cos 2x$. (This is why it is important to be aware of the solutions of the homogeneous equation before finding a trial solution.) In this example, we should take

$$y_p(x) = Ax \cos 2x + Bx \sin 2x.$$

Routine computation shows that

$$y_p'(x) = A \cos 2x + B \sin 2x + 2Bx \cos 2x - 2Ax \sin 2x,$$
$$y_p''(x) = -4A \sin 2x + 4B \cos 2x - 4Ax \cos 2x - 4Bx \sin 2x.$$

We find $y_p'' + 4y_p = -4A \sin 2x + 4B \cos 2x$. Since we want $y_p'' + 4y_p = 3 \cos 2x$, equating coefficients in the two formulas on the right gives $A = 0$ and $B = \frac{3}{4}$. The general solution is $y = y_h + y_p = c_1 \cos 2x + c_2 \sin 2x + \frac{3}{4}x \cos 2x$.

example 5

The equation $y'' + 2y' + y = 3e^{-x} + 2x$ has $y_h(x) = c_1 e^{-x} + c_2 x e^{-x}$ for the homogeneous solution. Solving first the equation $y'' + 2y' + y = 2x$, we try $y_1(x) = Ax + B$ and find $y_1'(x) = A$, $y_1''(x) = 0$. Substitution into the equation with $2x$ on the right gives

$$0 + 2A + Ax + B = 2x.$$

Equating coefficients of x gives $A = 2$. The constants satisfy $2A + B = 0$, so $B = -2A = -4$. Thus $y_1(x) = 2x - 4$. To solve $y'' + 2y' + y = 3e^{-x}$, we try

$$y_2(x) = Ax^2 e^{-x}.$$

Then

$$y_2'(x) = A(2x - x^2)e^{-x}$$
$$y_2''(x) = A(2 - 4x + x^2)e^{-x}.$$

Substitution into the equation with $3e^{-x}$ on the right gives

$$A(2 - 4x + x^2)e^{-x} + 2A(2x - x^2)e^{-x} + Ax^2 e^{-x} = 3e^{-x}.$$

The terms with x and x^2 as factors all cancel out, and we are left with $2A = 3$. Thus $A = \frac{3}{2}$, so $y_2(x) = \frac{3}{2}x^2 e^{-x}$. The general solution of the original equation is then

$$y = y_1 + y_2 + y_h$$
$$= 2x - 4 + \tfrac{3}{2}x^2 e^{-x} + c_1 e^{-x} + c_2 x e^{-x}.$$

We could have found y_p by using a single trial solution

$$y = Ax + B + Cx^2 e^{-x},$$

which would probably be more efficient. However, splitting the problem into two parts does have the effect of reducing the number of undetermined coefficients you need to work with at one time.

The undetermined coefficient method depends on the observation that if we want to solve

$$L(y) = f(x),$$

where $f(x)$ is itself a solution of a homogeneous equation $N(y) = 0$, then

$$N(L(y)) = N(f(x)) = 0.$$

Thus the desired solution $y(x)$ must be among the solutions of

$$N(L(y)) = 0.$$

If N and L are linear constant-coefficient operators, then the solutions of the preceding equation are sums of constant multiples of specific functions of the form $x^k e^{rx}$, where r may be real or complex. Thus the only additional problem is to determine the "undetermined coefficients" of combination that will actually give a solution of the original equation $L(y) = f(x)$. Theorem 3.2 of Section 3C guarantees that the desired coefficients do exist.

e x a m p l e **6**

The differential equation

$$y'' - y = e^x$$

can be written

$$(D^2 - 1)y = e^x.$$

Since $(D - 1)e^x = 0$, we have for any solution $y(x)$,

$$(D - 1)(D^2 - 1)y = (D - 1)^2(D + 1)y = 0.$$

Hence any particular solution y must have the form

$$y(x) = c_1 e^{-x} + c_2 e^x + c_3 x e^x.$$

Since the first two terms are solutions of the associated homogeneous equation $y'' - y = 0$, we can concentrate on the remaining term $c_3 x e^x$. To find c_3, we substitute into the nonhomogeneous differential equation. The resulting algebraic equation determines c_3.

Outline of Steps. To solve $L(y) = f(x)$, where L is a constant-coefficient linear operator and $f(x)$ is a solution of some constant-coefficient homogeneous equation:

1. Solve the associated homogeneous equation $L(y) = 0$ to get y_h.
2. Treat each term in $f(x)$ separately. Thus we can assume in the following that $f(x)$ consists of a single term with coefficient 1.
3. If $f(x) = x^k e^{rx}$, then this term corresponds to characteristic root r of multiplicity $k + 1$, so take for the annihilator the operator $N = (D - r)^{k+1}$.
4. If $f(x) = x^k e^{\alpha x} \cos \beta x$ or $f(x) = x^k e^{\alpha x} \sin \beta x$, then this term corresponds to characteristic root $\alpha + i\beta$ of multiplicity $k + 1$, so take $N = ((D^2 - 2\alpha D) + (\alpha^2 + \beta^2))^{k+1}$.
5. Solve the homogeneous equation $NL(y) = 0$. Discard all terms that occur in the solution of $L(y) = 0$. Substitute what remains into the nonhomogeneous equation $L(y) = f(x)$ to determine the coefficients.
6. Combine the homogeneous solution with all terms of the particular solution to get $y = y_h + y_p = y_h + a_1 y_1(x) + \cdots + a_m y_m(x)$, where $f(x) = a_1 f_1(x) + \cdots + a_m f_m(x)$.

EXERCISES

1. Show that each of the following operators L is linear by verifying that
$$L(c_1 y_1 + c_2 y_2) = c_1 L(y_1) + c_2 L(y_2)$$
for arbitrary constants c_1, c_2 and sufficiently differentiable functions y_1, y_2.
   (a) $L(y) = D^2 y$. (c) $L(y) = 2Dy + xy$.
   (b) $L(y) = D^2 y + Dy$. (d) $L(y) = 2D^2 y + xDy$.

2. Which of the following operators L are linear?
   (a) $L(y) = xDy$. (c) $L(y) = x^2 D^2 y + y^2$.
   (b) $L(y) = xD^2 y + xy^2$. (d) $L(y) = y^2$.

3. Each of the following linear differential equations has the given function y_p as a solution. Verify this and find the most general solution.
   (a) $y'' + y = x$; $y_p = x$. (c) $2y'' - y = e^x$; $y_p = e^x$.
   (b) $y'' - 2y' + y = 1$; $y_p = 1$. (d) $xy' + y = 1$; $y_p = 1$.

4. Find homogeneous differential equations of least possible order for which the following functions are solutions. In other words, find an annihilator of least possible order for these functions.
   (a) $e^x + 2e^{2x}$. (e) $x \sin 3x$.
   (b) $e^x \cos x - e^x \sin x$. (f) $x^2 \cos 4x$.
   (c) $x + 1$. (g) $xe^x \sin x$.
   (d) $xe^x - 2e^x$. (h) $x^3 e^{-x} \cos 2x$.

5. Find the appropriate form for a trial solution for each of the following equations. For example, you would use $y_p = A \cos 2x + B \sin 2x$ for $y'' - y = \sin 2x$.
   (a) $y'' - y = \cos x$. (d) $y'' - y = xe^x$.
   (b) $y'' + y = \cos x$. (e) $y'' - 2y' + y = xe^x$.
   (c) $y'' - y = e^x$. (f) $y'' = x^5$.

6. Find the general solution of the following equations by first finding the general solution of the associated homogeneous equation and then adding to it a particular solution found by the undetermined-coefficient method.
   (a) $y'' - y = e^{2x}$. (h) $y'' = \cos x + \sin x$.
   (b) $y'' - y = 3e^x$. (i) $y'' + y = x \cos x$.
   (c) $y'' + 2y' + y = e^x$. (j) $y''' - y = xe^x$.
   (d) $y'' - 3y = x$. (k) $y'' + 4y' + 4y = 3x$.
   (e) $y'' - y = e^x + x$. (l) $y'' - y' - 12y = 2e^{4x}$.
   (f) $y'' - 2y = \cos 2x$. (m) $y'' + 2y' + 2y = e^x$.
   (g) $y'' + y = \cos x$. (n) $y''' - y' = x$.

7. For each of the differential equations in Exercise 5, find the general form for all solutions but leave the coefficients undetermined.

8. For each of the differential equations in Exercise 5, find the particular solution $y(x)$ satisfying the initial conditions $y(0) = 0$, $y'(0) = 1$. Then sketch the graph of that solution.

3C   Green's Functions

The undetermined-coefficient method is often the most efficient way to solve a nonhomogeneous equation whose forcing function $f(x)$ has a special form involving trigonometric and exponential functions. For $f(x)$ of a more general type, in particular for certain discontinuous functions, we can use a formula for particular solutions of second-order constant-coefficient equations that may in principle be applied to any integrable $f(x)$. This allows us to conveniently generalize the concept of solution of a second-order equation to certain functions having just one continuous derivative. The method depends on solving $(D - r_1)(D - r_2)y = f(x)$ by successive integrations and then finding a formula for the solution in terms of r_1, r_2, and f. (The alternative **Laplace transform** method in Chapter 8 is less widely applicable. The **variation-of-parameters** alternative in Chapter 9 is used for equations with nonconstant coefficients.) The proof of Theorem 3.2 is deferred to the end of this section.

■ 3.2 THEOREM

Assume $f(x)$ is continuous on an interval containing x_0, and let $y_p(x)$ be the solution of the **normalized equation** with coefficient 1 for y''

$$y'' + ay' + by = f(x)$$

satisfying $y_p(x_0) = y_p'(x_0) = 0$. If r_1, r_2 are the roots of the characteristic equation $r^2 + ar + b = 0$, then:

(a) If r_1, r_2 are real and unequal,

$$y_p(x) = \frac{1}{r_1 - r_2} \int_{x_0}^{x} (e^{r_1(x-t)} - e^{r_2(x-t)})f(t)\, dt,$$

(b) If $r_1 = r_2$,

$$y_p(x) = \int_{x_0}^{x} (x - t)e^{r_1(x-t)}f(t)\, dt$$

(c) If r_1, $r_2 = \alpha \pm i\beta$, $\beta \neq 0$,

$$y_p(x) = \frac{1}{\beta} \int_{x_0}^{x} e^{\alpha(x-t)} \sin(\beta(x - t))f(t)\, dt.$$

In each formula, the factor in the integrand exclusive of $f(t)$ is a function $G(x, t) = g(x - t)$ called the **Green's function** of the operator $D^2 + aD + b$. In general, then,

$$y_p(x) = \int_{x_0}^{x} g(x - t)f(t)\, dt$$

where the function g has one of the three forms:

(a) $g(u) = \dfrac{1}{r_1 - r_2}(e^{r_1 u} - e^{r_2 u}), \qquad r_1 \neq r_2.$

(b) $g(u) = ue^{r_1 u}, \qquad r_1 = r_2.$

(c) $g(u) = \dfrac{1}{\beta}e^{\alpha u}\sin \beta u, \qquad r_1 = \alpha + i\beta, \qquad r_2 = \alpha - i\beta.$

An integral of the form displayed above is called a **convolution** integral, and some care is needed not to mix up the variables. Specifically, the variable x that occurs in the upper limit and in the difference $x - t$ must be treated as a constant with respect to integration in the other variable. Before proving Theorem 3.2, we will give some examples.

---

example 1

Here is an equation in which the right side is itself a solution of the associated homogeneous equation:

$$y'' + 2y' + y = e^{-x}.$$

The homogeneous solution is $y_h = c_1 xe^{-x} + c_2 e^{-x}$, corresponding to a double root $r = -1$ of the equation $r^2 + 2r + 1 = 0$. The Green's function is $G(x, t) = (x - t)e^{-(x-t)}$. The particular solution satisfying $y(0) = y'(0) = 0$ can be computed explicitly from

$$y_p(x) = \int_0^x (x - t)e^{(x-t)}e^{t}\,dt.$$

Note that integration is with respect to t, so x is temporarily fixed. We find

$$
\begin{aligned}
y_p(x) &= \int_0^x xe^{-(x-t)}e^{-t}\,dt - \int_0^x te^{-(x-t)}e^{-t}\,dt \\
&= xe^{-x}\int_0^x dt - e^{-x}\int_0^x t\,dt \\
&= xe^{-x}[t]_0^x - e^{-x}[\tfrac{1}{2}t^2]_0^x \\
&= xe^{-x}(x - 0) - e^{-x}(\tfrac{1}{2}x^2 - 0) \\
&= x^2 e^{-x} - \tfrac{1}{2}x^2 e^{-x} = \tfrac{1}{2}x^2 e^{-x}.
\end{aligned}
$$

This is the particular solution with $y_p(0) = y_p'(0) = 0$. Incidentally, the general solution is

$$y(x) = y_p(x) + y_h(x) = \tfrac{1}{2}x^2 e^{-x} + c_1 xe^{-x} + c_2 e^{-x}.$$

To satisfy initial conditions of the form $y(0) = y_0$, $y'(0) = z_0$, we'd just need to determine appropriate values for c_1 and c_2.

---

example 2

Here is an example similar to the previous one but involving trigonometric functions:

$$y'' + 9y = \sin 2x.$$

The characteristic roots of the associated homogeneous equation are $r_1 = 3i$, $r_2 = -3i$. The Green's function then is $g(x - t) = \tfrac{1}{3}\sin 3(x - t)$. The integrand in the Green's function formula is then $\frac{1}{3}\sin 3(x - t) \sin 3t$. Integration with respect to t can be carried out using a table of integrals, or more directly using the identity

$$\sin A \sin B = \tfrac{1}{2}\cos(A - B) - \tfrac{1}{2}\cos(A + B).$$

Thus the particular solution satisfying $y(0) = y'(0) = 0$ is

$$y_p = \frac{1}{3}\int_0^x \sin 3(x - t) \sin 3t\, dt$$

$$= \frac{1}{6}\int_0^x (\cos(3x - 6t) - \cos 3x)\, dt$$

$$= \tfrac{1}{6}[-\tfrac{1}{6}\sin(3x - 6t) - t\cos 3x]_0^x$$

$$= \tfrac{1}{18}\sin 3x - \tfrac{1}{6}x\cos 3x.$$

The next example cannot be done by undetermined coefficients except by breaking the problem into three separate parts and adjusting the initial conditions for each part. The key to taking care of this difficulty using a Green's function is breaking the integration interval at points b where the form of the integrand changes and treating the integration limits this way:

$$\int_a^x g(x - t)f(t)\, dt = \int_a^b g(x - t)f(t)\, dt + \int_b^x g(x - t)f(t)\, dt.$$

e x a m p l e 3   The equation $y'' - y = f(x)$ has the associated homogeneous solution $y_h = c_1 e^x + c_2 e^{-x}$, so the Green's function is

$$g(x - t) = \tfrac{1}{2}[e^{(x-t)} - e^{-(x-t)}].$$

Suppose $f(x)$ is given by

$$f(x) = \begin{cases} 0, & x < 0, \\ 1, & 0 \le x \le 1, \\ 0, & 1 < x. \end{cases}$$

The solution satisfying $y(0) = y'(0) = 0$ is

$$y(x) = \frac{1}{2}\int_0^x [e^{(x-t)} - e^{-(x-t)}]f(t)\, dt.$$

Since $f(t) = 0$ for $t < 0$, we clearly have $y(x) = 0$ for $x < 0$. When $0 \le x \le 1$, we find

$$y(x) = \frac{1}{2}\int_0^x [e^{(x-t)} - e^{-(x-t)}]\, dt$$

$$= \frac{1}{2}e^x \int_0^x e^{-t}\, dt - \frac{1}{2}e^{-x}\int_0^x e^t\, dt$$

$$= \tfrac{1}{2}e^x[-e^{-t}]_0^x - \tfrac{1}{2}e^{-x}[e^t]_0^x$$

$$= \tfrac{1}{2}e^x - \tfrac{1}{2} - \tfrac{1}{2} + \tfrac{1}{2}e^{-x} = \cosh x - 1.$$

When $1 < x$, we note that $f(t) = 0$ for $1 < t$, so the integral from 1 to x is zero. All that remains is the integral with respect to t of $\frac{1}{2}(e^{x-t} - e^{-(x-t)})$:

$$y(x) = \frac{1}{2} e^x \int_0^1 e^{-t}\, dt - \frac{1}{2} e^{-x} \int_0^1 e^t\, dt$$

$$= \tfrac{1}{2} e^x [-e^{-t}]_0^1 - \tfrac{1}{2} e^{-x} [e^t]_0^1$$

$$= \tfrac{1}{2} e^x - \tfrac{1}{2} e^{x-1} - \tfrac{1}{2} e^{-(x-1)} + \tfrac{1}{2} e^{-x} = \cosh x - \cosh(x - 1).$$

Just as $f(x)$ is defined by different formulas on different intervals, so is y:

$$y(x) = \begin{cases} 0, & x < 0, \\ \cosh x - 1, & 0 \le x \le 1, \\ \cosh x - \cosh(x - 1), & 1 < x. \end{cases}$$

Since the forcing function $f(t)$ is discontinuous in Example 3, we can't expect a solution $y(x)$ to be twice continuously differentiable. However, the three pieces of $y(x)$ do fit together at the joints in such a way that it has one continuous derivative. Examples like this prompt the broadening of the idea of "solution" to "generalized solution," discussed in the exercises.

Proof of Theorem 3.2. The proof is an application of the method we used earlier to derive the most general solution to $y'' + ay' + by = 0$. We factor the operator and write

$$(D - r_1)(D - r_2)y = f(x), \qquad (D - r_2)y = z.$$

Then solve $(D - r_1)z = f(x)$ by first multiplying by $e^{-r_1 x}$. The result is

$$e^{-r_1 x} z = \int e^{-r_1 x} f(x)\, dx + c.$$

We choose the constant c so that $z(x_0) = 0$. This can be done by using a definite integral. We get

$$z(x) = e^{r_1 x} \int_{x_0}^x e^{r_1 u} f(u)\, du.$$

Now solve $(D - r_2)y = z$ by the same method:

$$y(x) = e^{r_2 x} \int_{x_0}^x e^{-r_2 t} z(t)\, dt$$

$$= e^{r_2 x} \int_{x_0}^x e^{(r_1 - r_2)t} \left[\int_{x_0}^t e^{-r_1 u} f(u)\, du \right] dt.$$

Here we have replaced x by t in the integral for z. Setting $x = x_0$ shows that $y(x_0) = 0$ and $z(x_0) = 0$. Since $z(x_0) = y'(x_0) - r_2 y(x_0)$, it follows that $y'(x_0) = 0$ also.

Now integrate by parts in t using $u = \int_{x_0}^t e^{-r_1 u} f(u)\, du$, $dv = e^{(r_1 - r_2)t}\, dt$. Assuming $r_1 \ne r_2$, we get

(a)
$$y(x) = \frac{1}{r_1 - r_2} e^{r_1 x} \int_0^x e^{-r_1 u} f(u)\, du - \frac{1}{r_1 - r_2} e^{r_2 x} \int_0^x e^{-r_2 t} f(t)\, dt$$

$$= \frac{1}{r_1 - r_2} \int_{x_0}^x [e^{r_1(x-t)} - e^{r_2(x-t)}] f(t)\, dt.$$

A similar integration by parts when $r_1 - r_2 = 0$ (see Exercise 11) gives

(b)
$$y(x) = \int_{x_0}^{x} (x - t)e^{r_1(x-t)}f(t)\,dt.$$

If $r_1 = \alpha + i\beta$ and $r_2 = \alpha - i\beta$, then $r_1 - r_2 = 2i\beta$, and

$$\frac{1}{r_1 - r_2}[e^{r_1(x-t)} - e^{r_2(x-t)}] = \frac{1}{\beta}e^{\alpha(x-t)}\sin\beta(x-t).$$

Substituting this into (a) gives part (c) of Theorem 3.2.   ■

EXERCISES

1. For each of the differential equations in Exercise 5 of Section 3B, find the Green's function $G(x, t) = g(x - t)$ associated with the equation.

2. For each of the following differential equations, first find the general solution of the associated homogeneous equation; then find the associated Green's function $G(x, t) = g(x - t)$ and use it to find the particular solution satisfying the given initial condition of the special form $y(x_0) = y'(x_0) = 0$.

   (a) $y'' - y = x$; $x_0 = 0$.
   (d) $y'' + 2y' + y = 0$; $x_0 = 1$.

   (b) $y'' + y = x$; $x_0 = 0$.
   (e) $y'' - 4y = e^x$; $x_0 = -1$.

   (c) $y'' + 2y' + y = e^{2x}$; $x_0 = 0$.
   (f) $y'' - 3y' + 2y = e^{-x}$; $x_0 = 2$.

3. Find the particular solution to $y'' - 4y = f(x)$ satisfying $y(0) = y'(0) = 0$, where

   (a) $f(x) = \begin{cases} 1, & 0 \le x, \\ 0, & x < 0. \end{cases}$
   (c) $f(x) = \begin{cases} 1, & 0 \le x, \\ -1, & x < 0. \end{cases}$

   (b) $f(x) = \begin{cases} 0, & 0 \le x, \\ 1, & x < 0. \end{cases}$
   (d) $f(x) = \begin{cases} x, & 0 \le x, \\ 0, & x < 0. \end{cases}$

   [*Hint:* The forcing function f in part (c) is a linear combination of those in (a) and (b).]

   (e) $f(x) = \begin{cases} x - 1, & 1 \le x, \\ 0, & x < 1. \end{cases}$
   (f) $f(x) = x^2$.

4. **(a)** The independent variable in a differential equation is often denoted by t, for time. For example, in

   $$\frac{d^2x}{dt^2} + a\frac{dx}{dt} + bx = f(t),$$

   t usually stands for time. Show that the Green's function solution can be written

   $$x_p(t) = \int_{t_0}^{t} g(t - u)f(u)\,du.$$

   (b) Solve

   $$\frac{d^2x}{dt^2} + x = \sin t,$$

   if $x(0) = 0$ and $dx/dt(0) = 1$.

5. Since the right side of the equation $y'' + 3y' + 2y = (1 + e^x)^{-1}$ is not it-self a solution of a homogeneous constant-coefficient linear equation, the undetermined-coefficient method won't work. Find the particular solution satisfying $y(0) = y'(0) = 0$. [*Hint:* $e^{2t}/(1 + e^t) = e^t(1 - 1/(1 + e^t))$.]

6. Since the right side of the equation $y'' - 2y' + y = e^x/x$ is not itself a solution of a homogeneous constant-coefficient linear equation, the undetermined-coefficient method won't work. Note also that the equation has no solution on an interval containing $x = 0$, since the right side is undefined there.

(a) Find the particular solution satisfying $y(1) = y'(1) = 0$ for $0 < x$.

(b) Find the particular solution satisfying $y(-1) = y'(-1) = 0$ for $x < 0$.

7. Find the general solution of each of the following differential equations by writing the equation in the form

$$(D - r_1)(D - r_2)y = f(x),$$

letting $z = (D - r_2)y$ and then solving in succession the two first-order equations $(D - r_1)z = f(x)$ and $(D - r_2)y = z$.

(a) $y'' - y' - 2y = e^x$.

(b) $y'' - 4y = e^{3x}$.

8. Find the values of c_1 and c_2 that make the general solution at the end of Example 6, Section 3B, of the text satisfy $y(0) = y_0$, $y'(0) = y_1$.

9. Show that if $(D - r)y(x) = f(x)$ then the conditions $y(x_0) = 0$, $f(x_0) = 0$ hold if, and only if, $y(x_0) = 0$ and $y'(x_0) = 0$.

10. (a) Show that if $r_1 = \alpha + i\beta$, $r_2 = \alpha - i\beta$, then

$$\frac{1}{r_1 - r_2}(e^{r_1 u} - e^{r_2 u}) = \frac{1}{\beta} e^{\alpha u} \sin \beta u.$$

(b) Show that

$$\lim_{\beta \to 0} \frac{1}{\beta} e^{\alpha u} \sin \beta u = u e^{\alpha u}.$$

(c) Let r_1 and r_2 be real-valued. Show that

$$\lim_{r_1 \to r_2} \frac{1}{r_1 - r_2}(e^{r_1 u} - e^{r_2 u}) = u e^{r_2 u}.$$

These limit relations make plausible the Green's function for the case of equal roots.

11. To complete the derivation of the Green's function for the case of equal roots, use integration by parts to show that

$$e^{rx} \int_{x_0}^{x} \left[\int_{x_0}^{t} e^{-ru} f(u) \, du \right] dt = \int_{x_0}^{x} (x - t)e^{r(x-t)} f(t) \, dt.$$

[*Hint:* Let $u = \int_{x_0}^{t} e^{-ru} f(u) \, du$, $dv = dt$.]

*12. There are Green's function formulas for equations of order higher than two. Here it is for normalized 3rd-order constant-coefficient equations $L(y) =$

$f(x)$ with initial conditions $y(x_0) = y'(x_0) = y''(x_0) = 0$ and distinct characteristic roots r_1, r_2, r_3:

$$\int_{x_0}^{x} \left(\frac{e^{r_1(x-t)}}{(r_1 - r_2)(r_1 - r_3)} + \frac{e^{r_2(x-t)}}{(r_2 - r_1)(r_2 - r_3)} + \frac{e^{r_3(x-t)}}{(r_3 - r_1)(r_3 - r_2)} \right) f(t) \, dt.$$

The derivation follows the same lines as that given in the text for second-order equations.

(a) Express the 3rd-order formula in terms of r_1, α, and β, in case r_1 is real, $r_2 = \alpha + i\beta$, and $r_3 = \alpha - i\beta$, $\beta > 0$.

(b) What do you think the analogous formula is in case $r_1 \neq r_2 = r_3$?

Generalized Solutions. A differentiable function that fails to have a second derivative at some points of definition of a differential equation but which otherwise satisfies the equation is an example of a **generalized solution.** (This idea is further extended in Chapter 8 with the introduction of *generalized functions*. The terminology should not be confused with *general solution* that we often use in connection with linear equations.)

13. Verify that the solution obtained in Example 3 of the text has a continuous first derivative at all points but that its second derivative fails to exist at $x = 0$ and at $x = 1$.

14. Consider the differential equation

$$y'' = \begin{cases} -1, & x < 0, \\ 1, & 0 \leq x, \end{cases} \quad \text{for all real } x.$$

Verify that

$$y = \begin{cases} -x^2/2, & x < 0, \\ x^2/2, & 0 \leq x, \end{cases}$$

is a generalized solution if: (i) $y(x)$ satisfies the differential equation everywhere but at $x = 0$; (ii) $y'(x)$ is continuous for all x; (iii) $y''(0)$ fails to exist.

*15. The purpose of this exercise is to show that if $f(x)$ is continuous on an interval containing a, then the Green's function formula

$$y(x) = \int_{a}^{x} g(x - t) f(t) \, dt$$

represents a continuously differentiable function on the interval.

(a) Consider here the case $g(x - t) = (e^{r_1(x-t)} - e^{r_2(x-t)})/(r_1 - r_2)$, rewriting the formula for $y(x)$ so that the variable x doesn't occur under an integral sign.

(b) Treat as in part (a) the case $g(x - t) = (x - t)e^{r(x-t)}$.

4   FORMAL INTEGRATION METHODS (OPTIONAL)

The general differential equation of the form $\ddot{y} = f(t, y, \dot{y})$ has as special cases many second-order equations other than the constant-coefficient linear ones discussed earlier in the chapter. In particular, the equation may be nonlinear or may be linear but with nonconstant coefficients. Even though the computer-driven numerical methods treated in the next chapter are quite adequate for some purposes, there are often good reasons for pursuing general solution formulas. One is that the dependence of a solution on parameters is clarified when it is displayed in a formula. Another is that solution formulas allow us to settle questions about the behavior of solutions as the independent variable tends to infinity. We use t for the independent variable because it will represent time in most of our applications. Then $y = y(t)$ will be the sought-after solution, and $\dot{y}$ and $\ddot{y}$ will denote its time derivatives.

Apart from numerical approximations, there aren't any universally applicable methods for solving $\ddot{y} = f(t, y, \dot{y})$ explicitly. Section 4A deals with the approximation of nonlinear equations by linear ones. Additional special methods use successive integration, already used in Sections 1 and 2. In Section 4B the equation is assumed independent of the position variable y. In Section 4C the equation is assumed **autonomous,** which means that it is independent of the time variable t. In these subsections the standard integral tables and symbolic computer routines can sometimes be used to good advantage.

4A   Linearization

Most natural phenomena are nonlinear in their inherent structure and in their response to influences from their surroundings. Replacement of a nonlinear model by a linear one can often be justified empirically, with the understanding that sufficient accuracy of the solution may be restricted to a fairly limited range of values. The advantage of the replacement is that it may become possible to solve the linear equation directly in terms of elementary functions, together with any variable parameters that may enter the problem.

example 1

In Chapter 4 the undamped pendulum equation $\ddot{y} = -(g/l) \sin y$ is replaced for small amplitudes $y = y(t)$ by the linear equation $\ddot{y} = -(g/l)y$ with easily computed solutions $y = A \cos (\sqrt{g/l}\, t - \phi)$. The mathematics behind this replacement is simply that the tangent line to the graph of $\sin y$ is the graph of y. Adding a linear damping term $-k\dot{y}$ leaves us with the second-order constant-coefficient equation

$$\ddot{y} = -\frac{g}{l} y - k\dot{y}.$$

The solutions of this equation are described in detail in Chapter 4, Section 2. In the undamped case for which $k = 0$, all solutions to the linear equation are periodic; this happens with the nonlinear equation only for certain initial conditions.

The replacement of the nonlinear equation by a linear one in Example 1 is an instance of **linearization** for a second-order equation $\ddot{y} = f(t, y, \dot{y})$ near a point (y_0, z_0), using

■ **4.1** $f(t, y, z) \approx f(t, y_0, z_0) + f_y(t, y_0, z_0)(y - y_0) + f_z(t, y_0, z_0)(z - z_0).$

This is simply the extension to the two variables of the tangent approximation, the subscripts y and z referring to partial differentiation with respect to those variables. As a practical matter most of the examples will be **autonomous** equations, that is, equations $\ddot{y} = f(y, \dot{y})$ that don't exhibit an explicit dependence on the independent variable t. (More general equations are considered in Section 4B and in Chapter 9, Section 2.)

---

example 2

In the pendulum equation of Example 1, $f(y, \dot{y}) = -(g/l) \sin y$ is a function of y alone, and the linearization is localized near $y_0 = 0$. Since

$$\sin y \approx \sin y_0 + \cos y_0(y - y_0)$$
$$\approx 0 + 1(y - 0) = y,$$

the linearization of $\ddot{y} = -(g/l) \sin y$ at $y_0 = 0$ is $\ddot{y} = -(g/l)y$ as written in Example 1.

---

example 3

The falling-body equation $\ddot{y} = -g - (k/m)\dot{y}^2$ has $f(y, z) = -g - (k/m)z^2$. This function is independent of y. Linearizing at $z = z_0$, Equation 4.1 becomes

$$-g - \frac{k}{m}z^2 \approx -g - \frac{k}{m}z_0^2 - 2\frac{k}{m}(z - z_0)$$

$$\approx -g - \frac{k}{m}z_0^2 + 2\frac{k}{m}z_0 - 2\frac{k}{m}z_0 z.$$

For example, the equation linearized near $z_0 = 0$ becomes simply $\ddot{y} = -g$. Initial conditions $y(0) = 100$, $\dot{y}(0) = 0.01$ yield the particular solution $y(t) = -\frac{1}{2}gt^2 + 0.01t + 100$. This solution could be expected to remain approximately correct only for small velocities, that is, velocities near $z_0 = 0$.

---

example 4

The nonlinear equation $\ddot{y} = \ln y$ linearizes near $y_0 > 0$ to the first-degree equation $\ddot{y} = \ln y_0 + y_0^{-1}(y - y_0)$. If $y_0 = 1$, we get $\ddot{y} = y - 1$. The general solution of the latter equation is $y(t) = 1 + c_1 e^t + c_2 e^{-t}$. With initial conditions $y(0) = y_0 = 1$ and $\dot{y}(0) = z_0$, we find $y(t) = 1 + z_0 \sinh t$. The details are left as an exercise.

EXERCISES

1. For each of the following nonlinear equations, find the linearization near the given values $y = y_0$ or $\dot{y} = z_0$, or both, as appropriate.

   (a) $\ddot{y} = y - \sin y$, $y_0 = \pi/2$.

   (b) $\ddot{y} = \frac{1}{2}\dot{y}^2 + \frac{1}{2}$, $z_0 = 1$.

   (c) $\ddot{y} = \tan(y\dot{y})$, $y_0 = 0$, $z_0 = 1$.

   (d) $\ddot{y} = e^y$, $y_0 = 0$.

   (e) $y^2\ddot{y} = 512$, $y_0 = 16$.

   (f) $\ddot{y} + \dot{y} + ky^2 = 0$, $y_0 = z_0 = 1$, k const.

2. For parts (a) through (e) of the previous exercise, and under the respective initial conditions listed below, solve the linearized equation.

   (a) $y(0) = \pi/2$, $\dot{y}(0) = 0$.   **(d)** $y(0) = 0$, $\dot{y}(0) = 0$.

   (b) $y(0) = 0$, $\dot{y}(0) = 1$.   **(e)** $y(0) = 16$, $\dot{y}(0) = 0$.

   (c) $y(0) = 0$, $\dot{y}(0) = 1$.

3. Show that the solution of the linearized initial-value problem $\ddot{y} = y - 1$, $y(0) = 0$, $\dot{y}(0) = z_0$ of Example 4 of the text can be written $y(t) = 1 + z_0 \sinh t$.

4. **(a)** Show that for $y_0 > 0$ the differential equation $\ddot{y} = \ln y$ linearizes at $y = y_0$ to $\ddot{y} = (\ln y_0 - 1) + y/y_0$.

   (b) Show that the solution of the linearized equation in part (a) with initial values $y(0) = y_0$, $\dot{y}(0) = 0$ can be written as $y(t) = y_0 + y_0 \ln y_0 (\cosh (t/\sqrt{y_0}) - 1)$.

5. The **Duffing oscillator** $\ddot{y} = y - y^3$ has *stable equilibrium points* at $y_s = \pm 1$ and an *unstable equilibrium* at $y_u = 0$. (These ideas are developed in Chapter 4, Section 5A.)

   (a) Show that, when linearized at a stable point, the resulting equation has only oscillatory solutions for its constant solutions.

   (b) Show that, when linearized at the unstable point, a solution $y(t)$ of the linear equation satisfying initial conditions $y(0) = 0$, $\dot{y}(0) = z_0 > 0$ increases as t increases from zero.

   (c) Show that, when linearized at the unstable point, a solution $y(t)$ of the linear equation satisfying initial conditions $y(0) = y_0 > 0$, $\dot{y}(0) = 0$ increases from y_0 as t increases from zero.

6. The **Van der Pol oscillator** is $\ddot{y} + k(y^2 - 1)\dot{y} + y = 0$, where k is a positive constant.

   (a) Show that the linearization of this equation near $(y, \dot{y}) = (y_0, z_0)$ is
   $$\ddot{y} + k(y_0^2 - 1)\dot{y} + (2ky_0z_0 + 1)y = 2ky_0^2z_0.$$

   (b) Show that the linearized equation has harmonically oscillating solutions precisely either when $y_0 = 1$ and $z_0 > -(2k)^{-1}$ or when $y_0 = -1$ and $z_0 < (2k)^{-1}$.

   (c) Assume that $z_0 = 0$. Show that the linearized equation has oscillatory solutions precisely when $1 - 2/k < y_0^2 < 1 + 2/k$.

7. The gravitational acceleration acting at distance y from the center of a planet of mass M can be expressed by the second-order equation $\ddot{y} = -GM/y^2$.

   (a) Show that the linearization of this equation near $y = y_0$ is $\ddot{y} = -g + (2g/y_0)(y - y_0)$, where $g = GM/y_0^2$ is the acceleration of gravity at distance y_0.

   (b) If y_0 is very large relative to g, the second term in the linearization of part (a) is often omitted to get $\ddot{y} = -g$. Show that to keep the additional error in acceleration due to using this estimate less than 0.01, we should use the estimate only when $|y - y_0| < y_0/(200g)$.

    (c) Show that the solution to the linearized equation of part (a) with initial conditions $y(0) = y_0$ and $\dot{y}(0) = 0$ is $y(t) = \frac{1}{2}y_0(3 - \cosh \sqrt{2g/y_0}\,t)$.

    (d) Show that applying the second-degree Taylor approximation $\cosh x \approx 1 + \frac{1}{2}x^2$ to the result of part (c) yields the solution $y(t) = y_0 - \frac{1}{2}gt^2$ to the equation $\ddot{y} = -g$ with initial conditions $y(0) = y_0$, $\dot{y}(0) = 0$.

8. Coulomb's law for the repulsion of electric charges of the same sign can be expressed by $\ddot{y} = k/y^2$, where k is a positive constant.

    (a) Show that the Coulomb equation linearized at $y = y_0$ is $\ddot{y} = 3ky_0^{-2} - 2ky_0^{-3}y$.

    (b) Show that the linearized equation with initial conditions $y(0) = y_0 > 0$ and $\dot{y}(0) = 0$ has solution $y(t) = \frac{1}{2}y_0(3 - \cos \sqrt{2k/y_0^3}\,t)$.

    (c) The solution displayed in part (b) is periodic; explain what this implies about its validity for large t as an approximation to a solution of $\ddot{y} = k/y^2$, $k > 0$.

    (d) Show that applying the second-degree Taylor approximation $\cos x \approx 1 - \frac{1}{2}x^2$ to the result of part (b) yields the solution $y(t) = y_0 + \frac{1}{2}(k/y_0^2)t^2$ to the equation $\ddot{y} = k/y_0^2$ with initial conditions $y(0) = y_0$, $\dot{y}(0) = 0$.

4B   Independence of Position: $\ddot{y} = f(t, \dot{y})$

Let's assume we are given an equation that doesn't involve the unknown function y directly but contains only $\dot{y}$ and $\ddot{y}$. The trick is to introduce the new function name $z = z(t)$ by setting $\dot{y} = z$. Then $\ddot{y} = \dot{z}$, and we replace $\ddot{y}$ and $\dot{y}$ in the original equation to get the pair of equations

$$\dot{y} = z,$$
$$\dot{z} = f(t, z).$$

This is a first-order system of equations but of a very special kind. Most significantly, the second equation, $\dot{z} = f(t, z)$, doesn't contain y. So we try to solve this equation for z by treating it as a linear equation if it is one, or by any other method, for example, separation of variables, that seems appropriate. Once $z(t)$ is known, then we can find y by integrating both sides of $\dot{y} = z(t)$. Note that $\dot{y}$ and $\ddot{y}$ are unaltered by adding a constant to y. Since $\ddot{y} = f(t, \dot{y})$ is independent of y, adding a constant to a solution always gives another solution, and solution formulas in the following examples will consequently show an arbitrary additive constant. Geometrically this means that shifting a solution graph up or down along the y axis always produces another solution graph.

---

e x a m p l e **1**   In the equation

$$t\ddot{y} + \dot{y} = 0$$

we set $\dot{y} = z$ so that $\ddot{y} = \dot{z}$ to get

$$t\dot{z} + z = 0.$$

We can then treat this in the standard way as a first-order linear equation. Or we may recognize directly from the product rule that

$$t\dot{z} + z = \frac{d}{dt}(tz),$$

so integration gives

$$tz = c_1.$$

Then using our substitution $\dot{y} = z$ in the other direction gives

$$t\dot{y} = c_1 \qquad \text{or} \qquad \dot{y} = c_1 t^{-1}.$$

Another integration gives the solution

$$y = c_1 \ln|t| + c_2.$$

example 2

Consider the nonlinear equation

$$\ddot{y} - 2t(\dot{y})^2 = 0.$$

Since y is absent, we set $\dot{y} = z$ to get $\ddot{y} = \dot{z}$ and

$$\dot{z} = 2tz^2.$$

The variables separate giving

$$z^{-2}\dot{z} = 2t.$$

One integration gives

$$-z^{-1} = t^2 + c_1 \qquad \text{or} \qquad z = \frac{-1}{t^2 + c_1}.$$

Since $\dot{y} = z$, we integrate again to get

$$y = -\int \frac{dt}{t^2 + c_1} + c_2.$$

Consulting a table of indefinite integrals shows that there are three separate cases. Some typical graphs appear in Figure 3.4, where we have assumed $c_2 = 0$.

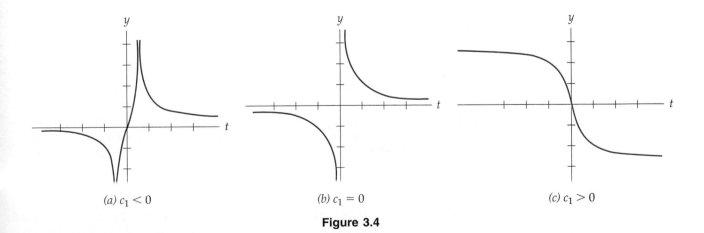

(a) $c_1 < 0$ (b) $c_1 = 0$ (c) $c_1 > 0$

Figure 3.4

$$c_1 > 0: \qquad y = \frac{-1}{\sqrt{c_1}} \arctan \frac{t}{\sqrt{c_1}} + c_2.$$

$$c_1 = 0: \qquad y = \frac{1}{t} + c_2.$$

$$c_1 < 0: \qquad y = \frac{1}{2\sqrt{|c_1|}} \ln \left| \frac{\sqrt{|c_1|} + t}{\sqrt{|c_1|} - t} \right| + c_2.$$

If $y(0) = 0$, the formula with $c_1 > 0$ provides solutions that pass through points in the second and fourth quadrants, and the one with $c_1 < 0$ provides the rest. The appearance of the first of these two formulas could be improved by renaming the positive constant $\sqrt{c_1}$ simply c_1, and the other one by renaming the positive constant $\sqrt{|c_1|}$ simply $-c_1$. However, that would complicate the simple relationship between c_1 and the initial values for $\dot{y}$. For example, as things stand, $\dot{y}(0) = -1/c_1$ as long as $c_1 \neq 0$.

4C   Independence of Time: $\ddot{y} = f(y, \dot{y})$

Equations of this kind are called **autonomous**. A solution to such an equation is still a function of the independent variable t and is defined on one or more intervals. Because the equation itself is independent of t, shifting a solution graph right or left along the t axis will always produce another solution graph. The idea here is again to think of the "velocity" $\dot{y}$ as a quantity of primary interest and to give it a name, say $\dot{y} = z$, or $\dot{y} = v$ if $\dot{y}$ really can be interpreted as a velocity. Then $\ddot{y} = \dot{z}$, so the equation $\ddot{y} = f(y, \dot{y})$ can be written $\dot{z} = f(y, z)$. While this is a first-order equation, it is awkward, because it contains not only y and z but also a derivative with respect to a third variable t. The way around this is to use the chain rule to eliminate explicit reference to t:

$$\dot{z} = \frac{dz}{dt}$$

$$= \frac{dy}{dt} \frac{dz}{dy} = z \frac{dz}{dy}.$$

Replacing $\dot{z}$ by $z\, dz/dy$ allows us to eliminate explicit use of t in the velocity equation $\dot{z} = f(y, z)$:

$$z \frac{dz}{dy} = f(y, z).$$

If we can solve this equation with an implicit relation $G(y, z, C) = 0$, then we can use the definition of z as $z = dy/dt$ and deal once again with a first-order equation of the form $G(y, \dot{y}, C) = 0$. The equation $G(y, z, C) = 0$ is called a **first integral** of $\ddot{y} = f(y, \dot{y})$, because in principle it takes just one more integration to solve the original equation $\ddot{y} = f(y, \dot{y})$. Thus if a first integral can be rewritten explicitly as $dy/dt = h(y, C)$, then the variables separate easily, and we are in fact faced with a single integration problem:

$$\int \frac{dy}{h(y, C)} = \int dt.$$

Actually carrying out the integration still leaves us only with t as a function of y, and we may well want to invert that relation to find $y = y(t)$.

Summary. To solve $\ddot{y} = f(y, \dot{y})$:

1. Set $dy/dt = z$, $d^2y/dt^2 = z\,dz/dy$, and solve the first-order differential equation $z\,dz/dy = f(y, z)$ to get a relation $G(y, z, C) = 0$ between y and z.
2. Set $z = dy/dt$, and solve the first-order equation $G(y, \dot{y}, C) = 0$.

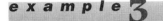

To solve $d^2y/dt^2 = y\,dy/dt$ with initial conditions $y(0) = 0$, $\dot{y}(0) = \frac{1}{2}$, let $dy/dt = z$ and $d^2y/dt^2 = z\,dz/dy$ to get

$$z\frac{dz}{dy} = yz.$$

Note that $z(0) = \dot{y}(0) = \frac{1}{2} \neq 0$, so we can assume $z(t) \neq 0$ for t near 0. We divide by z and are left with $dz/dy = y$. Integrating both sides with respect to y gives the first integral $z = \frac{1}{2}y^2 + C_1$. The initial conditions show that $z = \frac{1}{2}$ and $y = 0$ at $t = 0$, so $\frac{1}{2} = 0 + C_1$ and $C_1 = \frac{1}{2}$. Thus $z = dy/dt = \frac{1}{2}(y^2 + 1)$, so the integration is

$$\int \frac{dy}{y^2 + 1} = \int \tfrac{1}{2}\,dt.$$

This second integration gives $\arctan y = t/2 + C_2$. Since $y = 0$ when $t = 0$, we find that $C_2 = 0$, so

$$y = \tan\frac{t}{2}$$

is our solution to the initial-value problem.

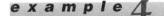

If instead of the initial values for $\ddot{y} = y\dot{y}$ given in Example 3, we had $y(0) = 0$, $\dot{y}(0) = -\frac{1}{2}$, the value for the first integration constant would have been $C_1 = -\frac{1}{2}$. In that case, we would have had a different integration problem:

$$\int \frac{dy}{y^2 - 1} = \int \tfrac{1}{2}\,dt, \quad \text{with result } \tfrac{1}{2}\ln\left|\frac{y - 1}{y + 1}\right| = \tfrac{1}{2}t + C_2.$$

Solving for y gives

$$y = \frac{1 + Ke^t}{1 - Ke^t}, \quad \text{where } K = \pm e^{2C_2}.$$

To satisfy the remaining initial condition, $y(0) = 0$, we make $K = -1$, in which case $y = (1 - e^t)(1 + e^t)^{-1}$. Figure 3.5 shows not only the two solutions we've just computed but also some others with $y(0) = 0$ and the slope $\dot{y}(0)$ at the origin taking on various values. Note that if $dy/dt(0) = 0$, we get the constant solution $y(t) = 0$, this one could be regarded as having come from dropping the condition $z \neq 0$ that we made at first.

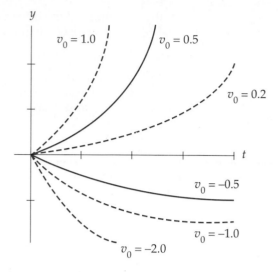

Figure 3.5 Solutions of $\ddot{y} = y\dot{y}$, $y(0) = 0$, $\dot{y}(0) = z_0$.

One thing setting the family of solutions to the previous nonlinear example apart from those of linear constant-coefficient equations $\ddot{y} = -k\dot{y} - hy$ is that simply by changing the initial value $\dot{y}(0)$ we can pass from a solution that tends to infinity in finite time to one that tends to a limit as t tends to infinity. No such behavior occurs with $\ddot{y} = -k\dot{y} - hy$.

example 5

The undamped pendulum equation $\ddot{y} = -(g/l)\sin y$ can be written using $\dot{y} = z$ as $z\,dz/dy = (g/l)\sin y$. A first integral is $\frac{1}{2}z^2 = -(g/l)\cos y + C$ or $\frac{1}{2}\dot{y}^2 = (g/l)\cos y + C$. The constant C could be determined by a variety of conditions, one of which is that the velocity $\dot{y} = 0$ when y reaches its maximum value $y_{\max} = \eta > 0$. Thus $0 = (g/l)\cos\eta + C$, so $C = -(g/l)\cos\eta$ and $\frac{1}{2}\dot{y}^2 = (g/l)(\cos y - \cos\eta)$. We are faced with the problem of integrating the separated equation

$$\int \frac{dy}{\sqrt{\cos y - \cos\eta}} = \pm\sqrt{\frac{2g}{l}}\int dt.$$

The integral on the left cannot be evaluated in terms of elementary functions but is an **elliptic integral** that can be rewritten in a standard form with tabulated values. For many purposes there is no advantage to using information about the elliptic integral over using the numerical methods of the next chapter. However, determining the time required for a complete cycle from position η to $-\eta$ and back to η again, i.e., the period P of oscillation, is an exception. By symmetry we see that

$$P = 4\int_0^{P/4} dt = 4\sqrt{\frac{l}{g}}\int_0^{\eta} \frac{dy}{\sqrt{2(\cos y - \cos\eta)}}, \qquad 0 < \eta < \pi.$$

At this point it's instructive to compare the above formula for P with the one derived from the linearized equation $\ddot{y} = -(g/l)\sin y$. The complete solution to the linearized version of the problem posed above is $y = \eta\cos(\sqrt{g/l}\,t)$.

Under that assumption, the period is $P = 2\pi\sqrt{l/g}$. Two things stand out right away when we compare the two expressions for P: (a) The dependence of P on length l and gravity g is proportional to $\sqrt{l/g}$ in both versions. (b) The linearized version predicts a period that is independent of the maximum amplitude η, while the nonlinear version predicts a period that depends on η. The nonlinear version is the correct one, and the nature of P's dependence on η is investigated in the next example.

example 6

The period of an undamped pendulum with angular amplitude η was determined in Example 5 to be $P = 4\sqrt{l/g}I(\eta)$, where

$$I(\eta) = \int_0^\eta \frac{dy}{\sqrt{2\,(\cos y - \cos \eta)}}.$$

Numerical evaluation of this integral for $0 \le \eta < \pi$ allows us to sketch the graph of $I(\eta)$ shown in Figure 3.6. It turns out that $I(0+) = \pi/2$ and that $I(\pi-) = +\infty$. The picture shows that for small values of the maximum amplitude η the number $I(\eta)$ is only slightly larger than $\pi/2$; this makes the period only slightly larger than $P = 4\sqrt{l/g}(\pi/2) = 2\pi\sqrt{l/g}$, which is the period predicted by the solution $y = \eta \cos(\sqrt{g/l}t)$ of the linearized equation $\ddot{y} = -(g/l)\sin y$. But with $\eta = 1$ radian, about $57°$, the linear estimate $\pi/2$ is about 6% too small when compared with $I(1)$, and the deterioration is obviously much more severe thereafter, becoming unbounded as η approaches π.

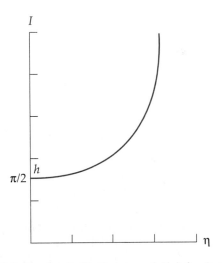

Figure 3.6 Graph of elliptic integral $I(\eta)$ for $0 < \eta < \pi$.

example 7

Finding approximate values of $I(\eta)$, introduced in Example 6, by using Simpson's rule is awkward because the integrand is unbounded as y approaches η. This difficulty can be avoided by changing the variable of integration in such a way as to write $I(\eta)$ in terms of the **complete elliptic integral**

$$K(k) = \int_0^{\pi/2} \frac{d\phi}{\sqrt{1 - k^2 \sin^2 \phi}}, \qquad k^2 < 1.$$

It's left as an exercise to show that a change of variable gives

$$I(\eta) = \int_0^\eta \frac{dy}{\sqrt{2 (\cos y - \cos \eta)}} = K(k), \qquad \text{where } k = \sin \frac{\eta}{2}.$$

Note that the interval of integration $0 \le y \le \eta$ changes to the fixed interval $0 \le \phi \le \pi/2$. Tables of this integral are usually printed for values of a parameter α between 0 and 90° such that $k = \sin \alpha°$.

For a more detailed, but still brief, treatment of elliptic integrals, see M. L. Boas, *Mathematical Methods in the Physical Sciences*, 2nd ed., Wiley, 1983.

EXERCISES

1. Find solutions for the following by the method of Section 4A.

   (a) $t\ddot{y} - \dot{y} = 0$.   **(c)** $\ddot{y} + \dot{y}^2 = 0$.   **(e)** $t\ddot{y} + \dot{y} = t^3$.

   (b) $t^2\ddot{y} + \dot{y}^2 = 0$.   **(d)** $t\ddot{y} + \dot{y} = 0$.   **(f)** $\ddot{y} + \cos t = 0$.

2. Find solutions for the following by the method of Section 4B.

   (a) $y\ddot{y} - \dot{y}^2 = 0$.   **(c)** $\ddot{y} = \dot{y}^3$.   **(e)** $\ddot{y} + \dot{y}^2 = 1$.

   (b) $y^2\ddot{y} + \dot{y}^3 = 0$.   **(d)** $\dot{y} + \dot{y}^2 = 0$.   **(f)** $\ddot{y} = y$.

3. Find solutions that satisfy the given conditions.

   (a) $\ddot{y} - 2t = 0$; $y(0) = 0$, $\dot{y}(0) = 1$.

   (b) $t\ddot{y} + 2\dot{y} = 0$; $y(1) = 1$, $\dot{y}(1) = 2$.

   (c) $\ddot{y} = y\dot{y}$; $y(0) = 0$, $\dot{y}(0) = 1$.

   (d) $\ddot{y} = 2(y^3 - y)$; $y(0) = 0$, $\dot{y}(0) = 1$.

4. Show that the equation in Exercise 1b has more than one solution satisfying $y(0) = 0$ and $\dot{y}(0) = 0$.

5. The **catenary equation** $d^2y/dx^2 = k\sqrt{1 + (dy/dx)^2}$, $0 < k = $ constant, is satisfied by the shape $y(x)$ of a chain suspended between two supports. The constant k depends on the location of the fixed ends and the length of the chain.

   (a) Solve the equation with the side conditions $y(0) = 1/k$, $y'(0) = 0$, and sketch the graph of this solution.

   (b) Use the differential equation directly to show that the curvature $\kappa = y''/(1 + (y')^2)^{3/2}$ of all solutions is at most k.

   (c) Show that the catenary equation can also be written $d^2y/dx^2 = k\, ds/dx$, where s is the arc length measured along the chain, and hence that if the catenary equation satisfies the boundary conditions $y(a) = \alpha$, $y(b) = \beta$, $a < b$, then $ks = y'(b) - y'(a)$, where s is the length of the chain. [*Hint:* Don't solve the differential equation.]

6. The formula

$$y = \frac{1}{2\sqrt{|c_1|}} \ln \left| \frac{\sqrt{|c_1|} + t}{\sqrt{|c_1|} - t} \right|$$

appears in Section 4B, Example 2 of the text as a solution formula for solutions to the nonlinear differential equation $\ddot{y} = 2t(\dot{y})^2$, valid for solutions whose graphs contain points in the first and third quadrants of the ty plane.

Show that for $|t| < \sqrt{|c_1|}$ the formula can be expressed by the **inverse hyperbolic tangent** function as

$$y = \frac{1}{\sqrt{|c_1|}} \tanh^{-1} \frac{t}{\sqrt{|c_1|}},$$

and that for $|t| > \sqrt{|c_1|}$ it can be expressed by the **inverse hyperbolic cotangent** function as

$$y = \frac{1}{\sqrt{|c_1|}} \coth^{-1} \frac{t}{\sqrt{|c_1|}}.$$

Here $\tanh x = (e^x - e^{-x})/(e^x + e^{-x})$ and $\coth x = 1/\tanh x$. The point is to express their inverse functions in terms of the function $\ln$.

7. This exercise shows that applying the method of Section 4C to an equation can produce an extraneous "solution" that fails to satisfy the equation.

   (a) Show that the solution $y = y(t)$ of $\ddot{y} = -y$, $y(0) = 1$, $\dot{y}(0) = 0$ also satisfies an equation of the form $(\dot{y})^2 = -y^2 + C$, and that with $C = 1$ this last equation has the constant solution $y(t) = 1$.

   (b) Verify that $y(t) = 1$ fails to satisfy $\ddot{y} = -y$ but that the first- and second-order equations in part (a) have in common the solution $y(t) = \cos t$ satisfying the conditions $y(0) = 1$, $\dot{y}(0) = 0$.

8. The integration method for equations of the form $\ddot{y} = f(y)$ can be phrased as follows.

   (a) To integrate $\ddot{y} = f(y)$, multiply both sides by $\dot{y}$ to get $\dot{y}\ddot{y} = f(y)\dot{y}$. Show that if there is a function $F(y)$ such that $F'(y) = f(y)$, then

   $$\tfrac{1}{2}\dot{y}^2 = F(y) + C_1.$$

   (b) Apply the technique of part (a) to show that the solution of the initial-value problem $\ddot{y} = -y$, $y(0) = 1$, $\dot{y}(0) = 0$ is $y = \sin(t + \pi/2) = \cos t$.

   (c) Apply the technique of part (a) to show that a solution $y = y(t)$ of the undamped pendulum equation $\ddot{y} = -\sin y$ satisfies $\dot{y}^2 = 2\cos y + C$, where C is constant.

   (d) Find the constant C in part (c) if $\ddot{y} = -\sin y$ is subject to initial conditions $y(0) = y_0$ and $\dot{y}(0) = z_0$.

   (e) Letting $z = \dot{y}$, use the results of parts (c) and (d) to sketch the curves $z^2 = 2\cos y + C$ in the yz plane for $C = -1$, 0, 1, and 2. What is the physical significance of the value $C = -2$? What can you say about the case $C = -3$?

*9. (a) Find all solutions of the differential equation $\ddot{y} = 2y\dot{y}$. (This problem is of a different type from the one in Section 4B, Example 2 of the text. Nevertheless, the indefinite integrals needed for that example can be used here also.)

(b) Sketch the graph of a typical solution of each kind from part (a).

10. The equation $\ddot{y} = -1/y^2$ is a normalized version of the inverse-square law of gravitational attraction $\ddot{y} = -GM/y^2$ for the distance $y(t)$ from an attracting body of mass M.

(a) Find the linearized equation near the value $y_0 = 1$, and find the solution of the linearized equation subject to the initial conditions $y(0) = 1$, $\dot{y}(0) = 0$.

(b) Show that under the same initial conditions as in part (a), the solution $y(t)$ to the nonlinear equation satisfies $\dot{y} = -\sqrt{(2 - 2y)/y}$.

(c) Show that $y(t)$ satisfies

$$\arccos \sqrt{y} + \sqrt{y(1 - y)} = \sqrt{2}t.$$

Estimate the value of t corresponding to $y = 0$.

(d) Sketch the graph of the equation in part (c) to show the relation between y and t.

11. Follow the steps of parts (b) and (c) of the previous exercise for the more general equation $\ddot{y} = -GM/y^2$, with initial conditions $y(0) = y_0$, $\dot{y}(0) = 0$, to show that

$$y_0 \arccos \sqrt{\frac{y}{y_0}} + \sqrt{y(y_0 - y)} = \sqrt{\frac{2GM}{y_0}}\, t.$$

Use this equation to show that the time it takes to fall from y_0 to 0 is $\pi y_0^{3/2}/(2\sqrt{2GM})$.

12. A spherical raindrop falls through the air under the influence of the earth's gravitational acceleration g and subject to negligible air resistance. The raindrop increases in mass m through the accumulation of additional moisture in such a way that its rate of change with respect to distance y fallen is proportional to the square of its radius. ($dm/dy = kr^2$.) Assume that the raindrop starts with radius zero and velocity zero at $t = 0$.

(a) Show that radius r is proportional to distance y.

(b) Combining Newton's second law (Chapter 2, Section 4) with the constant gravitational acceleration g gives $d/dt(mv) = gm$. Show that this equation is equivalent to

$$v\frac{d}{dy}(y^3 v) = gy^3.$$

(c) Solve the differential equation in part (b) with initial conditions $y = v = 0$. [*Hint:* Multiply both sides by y^3.]

(d) Find how far the raindrop falls in time t, and show that its acceleration is a constant $g/7$.

13. The second-order linear equation
$$y'' + b(x)y' + c(x)y = 0$$
with coefficients $b(x)$ and $c(x)$ that are not necessarily constant can some-times be solved by making the substitution $y' = -zy$.

(a) Show that this substitution leads to the **associated Riccati equation**
$$\frac{dz}{dx} + z^2 + b(x)z + c(x) = 0.$$

(b) Show that if $z = z(x)$ is a solution of the Riccati equation, then a solution of the related second-order linear equation is given by
$$y(x) = Ce^{Z(x)}$$
where $Z(x)$ is some indefinite integral of $z(x)$.

14. Show that if y is a solution of the linear equation
$$a(x)y'' + (a(x)b(x) - a'(x))y' + a^2(x)c(x)y = 0$$
then $z = y'/(a(x)y)$ is a solution of the **general Riccati equation**
$$\frac{dz}{dx} = a(x)z^2 + b(x)z + c(x).$$

15. (a) Use $\cos y = 1 - 2\sin^2(y/2)$ to show that the pendulum period factor $I(\eta)$ is
$$I(\eta) = \int_0^\eta \frac{dy}{\sqrt{2(\cos y - \cos \eta)}} = \int_0^\eta \frac{\frac{1}{2}dy}{\sqrt{\sin^2(\eta/2) - \sin^2(y/2)}}.$$

(b) Use part (a) and the change-of-variable relation $\sin(y/2) = k\sin\phi$ with $k = \sin(\eta/2)$ to show that
$$I(\eta) = \int_0^{\pi/2} \frac{d\phi}{\sqrt{1 - k^2\sin^2\phi}}, \qquad 0 < k < 1.$$

16. Show the following about the complete elliptic integral
$$K(k) = \int_0^{\pi/2} \frac{d\phi}{\sqrt{1 - k^2\sin^2\phi}}.$$

(a) $K(0) = \pi/2$.

(b) $K(k)$ is strictly increasing for $0 \leq k < 1$.

(c) $K(1) = +\infty$.

17. Do a computer plot of the results of a Simpson's rule computation of the elliptic integral $K(k)$, as defined in the text, for $0 \leq k \leq 0.9$. Use at least 90 equally spaced values of k and at least 20 Simpson rule intervals in the interval $[0, \pi/2]$. Note that $K(0) = \pi/2$.

*18. The complete elliptic integral
$$K(k) = \int_0^{\pi/2} \frac{d\phi}{\sqrt{1 - k^2\sin^2\phi}}, \qquad 0 \leq k < 1,$$
tends to plus infinity as k tends to one from the left. Show this in the following steps.

(a) Show that

$$K(k) \geq \frac{1}{\sqrt{2}} \int_0^{\pi/2} \frac{d\phi}{\sqrt{1 - k \sin \phi}}.$$

(b) Make the substitution $x = \sin \phi$ in the integral in part (a) to conclude that

$$K(k) \geq \frac{1}{2} \int_0^1 \frac{dx}{\sqrt{1 - x}\sqrt{1 - kx}} \geq \frac{1}{2} \int_0^1 \frac{dx}{1 - kx}.$$

(c) Evaluate the last integral in part (b), and show that it tends to infinity as $k \to 1$ from the left.

19. The elliptic integral $K(k)$ defined in the text is more specifically called the **complete elliptic integral of the first kind**. The **complete elliptic integral of the second kind** is defined by

$$E(k) = \int_0^{\pi/2} \sqrt{1 - k^2 \sin^2 \phi} \, d\phi, \qquad 0 \leq k \leq 1.$$

The term "elliptic integral" comes from the application of the integral of the second kind to computing the arc length of an ellipse.

(a) Show that the length l of the ellipse parametrized by $x = a \cos t$, $y = b \sin t$, with $a > 0$, $b > 0$, and $0 \leq t \leq 2\pi$, can be expressed by

$$l = \int_0^{2\pi} \sqrt{a^2 \sin^2 t + b^2 \cos^2 t} \, dt.$$

(b) Assume $a \leq b$ in part (a), and show that $l = 4bE(k)$, where $k = \sqrt{1 - (a^2/b^2)}$.

20. Do a computer plot of the results of a Simpson's rule computation of the elliptic integral $E(k)$, as defined in the previous exercise, for $0 \leq k \leq 1$. Use at least 10 equally spaced values of k and at least 10 Simpson rule intervals in the interval $[0, \pi/2]$. Note that $E(0) = \pi/2$ and $E(1) = 1$.

CHAPTER REVIEW AND SUPPLEMENT

Find all solutions that satisfy:

1. $y'' + 2y' + y = e^{-x} + 3e^x$

2. $y'' + y = x \sin x$

3. $y'' - y = \sin x$

4. $y'' - y' - y = 1$, $y(0) = y'(0) = 1$

5. $y'' + 2y' + 3y = 1$.

6. $(D - 1)^2 y = x^3 - x$.

7. $y'' + 9y = \sin 3x$, $y(0) = 1$, $y'(0) = 0$.

8. $y''' = x$.

9. $(D^2 + 4)y = \cos 3x$, using Section 3C.

10. $y'''' = 81y$.

11. $y'' + y = 0$, $y(0) = -1$, $y(\pi) = 1$.

12. $y'' + y = 0$, $y(0) = 0$, $y(\pi/2) = 2$.

Find the general form for a trial solution y_p for each of the following. (For example, for $y'' - y = e^x$, choose $y_p = Axe^x$.) You need not determine any coefficient values.

13. $y'' - 4y = xe^{2x} + e^{2x}$.

14. $y'' + y = x^2 \cos x$.

15. $y'' - 5y' + 6y = xe^{2x} + e^{3x}$.

16. $y'' + 4y = x^2 \cos 2x - 2\sin 2x$.

17. $y'' - 4y = e^{2x} + 5\cos x$.

18. $y'' + y = 3x \sin(x - 3)$.

19. $y'' - y' = x^2 + 2e^x$.

20. $y''' - y = e^{x/2} \sin \sqrt{3}x$.

21. $y''' = 1 + x + x^3$.

22. $y'' + y = x^{99} \cos x$.

23. The simplest real forms for basic solution pairs to the constant-coefficient equation $y'' + ay' + by = 0$ are $\{e^{r_1 x}, e^{r_2 x}\}$, $\{e^{r_1 x}, xe^{r_1 x}\}$ and $\{e^{\alpha x} \cos \beta x, e^{\alpha x} \sin \beta x\}$.

(a) Make a corresponding complete list of triples for the equation
$$y''' + ay'' + by' + cy = 0.$$

(b) Make a corresponding complete list of quadruples for
$$y'''' + ay''' + by'' + cy' + dy = 0.$$

24. Derive from scratch the fundamental sinusoidal solutions to the harmonic oscillator problem $\ddot{y} + y = 0$, $y(0) = 0$, $\dot{y}(0) = 1$. Do this in the following steps (our earlier derivations were made using complex exponentials):

(a) Multiply the equation by $\dot{y}$. Integrate with respect to t to get $\frac{1}{2}\dot{y}^2 + \frac{1}{2}y^2 = C$.

(b) Find C, solve for $\dot{y}$, and solve the resulting first-order equation to get $y(t) = \sin(t + c)$.

25. Suppose that $y_1(x)$ and $y_2(x)$ are real-valued functions defined for all real x and you know that $y_1(x)$ is not a constant multiple of $y_2(x)$. Are the two functions necessarily linearly independent? Explain your answer, using an example if necessary.

26. Let the functions y_1, y_2 by defined for all real x by $y_1(x) = e^{rx}$ and $y_2(x) = e^{sx}$, where r and s are unequal complex numbers. Show that y_1 and y_2 are linearly independent even if complex constants c_1, c_2 are allowed in $c_1 y_1(x) + c_2 y_2(x)$.

27. For what values of the constant b do the non–identically zero solutions of $y'' + y' + by = 0$ oscillate as functions of x?

28. The current $I(t)$ flowing through a certain electric circuit at time t satisfies
$$I'' + RI' + I = \sin t,$$
where $R > 0$ is a constant resistance. The equation has a solution of the form $I(t) = A \sin(t - \alpha)$ for certain constants $A > 0$ and α. Find A and α.

29. Show that the functions e^x and e^{-x} are linearly independent on an interval $a < x < b$ by showing directly that neither function can be a constant multiple of the other on such an interval.

30. (a) Show that the functions e^x and e^{-x} are linearly independent on an interval $a < x < b$ by showing directly that the equation

$$c_1 e^x + c_2 e^{-x} = 0$$

can't be satisfied for all real x unless the constants c_1 and c_2 are both zero.

(b) Show that the equation $2e^x - 3e^{-x} = 0$ is satisfied for exactly one real x and that $2e^x + 3e^{-x} = 0$ is satisfied for no real x.

(c) For what *complex* values of x is $2e^x + 3e^{-x} = 0$ satisfied?

31. Suppose that

$$\frac{d^2 y}{dt^2} = -y.$$

(a) Find the general solution $y(t)$ of this equation.

(b) Let

$$z(t) = \frac{dy}{dt}(t).$$

Show that the parametrized curve $(y(t), z(t))$ either traces a clockwise circular path or reduces to a single point.

32. Theorem 2.7 (Section 2C) implies that there is a one-to-one correspondence between the set of all n-tuples of initial values $(z_0, z_1, \ldots, z_{n-1})$ and all solutions to the nth-order homogeneous equation $L(y) = 0$. Explain how this conclusion follows.

33. Theorem 2.7 implies that there is a one-to-one correspondence between the set of all n-tuples of initial values $(z_0, z_1, \ldots, z_{n-1})$ and all n-tuples $(c_1, c_2, \ldots, c_n)$ of coefficients of linear combination. Explain how this conclusion follows.

34. Assume only that L is a linear operator such that $L(y_1) = w_1$, $L(y_2) = w_2$, and $L(y_3) = w_3$. Using just this information, find linear combinations z of the y_1, y_2, and y_3 so that:

(a) $L(z) = 2w_1 - 3w_2$.   **(c)** $L(z) = L(2z) + w_1 + w_2$.

(b) $L(z) = w_1 + 2w_2 - 4w_3$.   **(d)** $L(z) = 0$.

35. (a) Show that the family of differential equations

$$y'' - (2r + h)y' + r(r + h)y = 0,$$

depending on the parameter h, can also be written

$$(D - r)(D - (r + h))y = 0.$$

(b) Show that the equations of part (a) have solutions $y = c_1 e^{(r+h)x} + c_2 e^{rx}$ if $h \neq 0$.

(c) Let $c_1 = 1/h$, $c_2 = -1/h$, and show that for each fixed x, the resulting solution $y_h(x)$ tends as $h \to 0$ to

$$y = \frac{d}{dr} e^{rx} = x e^{rx}.$$

4

Applied Dynamics and Phase Space

This chapter is about applications of differential equations for which the solution $y = y(t)$ is a function of time t determined by its relation to acceleration d^2y/dt^2 and velocity dy/dt, hence the term "dynamic," implying motion. Fundamental to much of our discussion will be the simplest form of Newton's **second law of motion,** according to which force F is equal to constant mass m times acceleration a. Thus, since $a = d^2y/dt^2$, we have

$$F = m \frac{d^2y}{dt^2}, \quad \text{a special case of} \quad F = \frac{d}{dt}\left(m \frac{dy}{dt}\right),$$

or **force equals the rate of change of momentum,** considered in Chapter 2, Section 4. It is often convenient to follow Newton's convention of writing time derivatives using overdots; thus $\dot{y} = dy/dt$, and $\ddot{y} = d^2y/dt^2$. Thus many of the differential equations we'll be looking at take the form

$$m\ddot{y} = F(t, y, \dot{y}).$$

The mathematical **state space** in each section is 2-dimensional, described by coordinates such as $(y, \dot{y})$. Thus it is fundamental to our study that **a state of a system governed by a second-order equation can be thought of as determined by both position y and velocity $\dot{y}$.**

You may want to skip some applications at a first reading, and you may want to skip ahead to Section 7 on numerical methods to supplement the study of some applications. Note also that Sections 5 and 6 on phase space are relevant to all the preceding Sections 1 through 4. First we'll consider some examples of nonoscillatory phenomena.

1 GRAVITATIONAL FORCES

A freely falling object near the surface of the earth, and at a distance $y = y(t)$ above it, can be assumed for some purposes to have a constant acceleration of magnitude g. The acceleration g is about 32.2 feet per second per second, or 9.8 meters per second per second. (This value varies within narrow limits from place to place on earth, and it also decreases as the height y above earth increases, but it stays within those same narrow limits up to about 4 miles above the earth.) If distance $y(t)$ is measured down from some point above the earth's surface, the force F can be written in two equal ways:

$$m\ddot{y} = mg.$$

Both sides have the same sign because gravity acts to increase the velocity $\dot{y}$ and so makes its derivative satisfy $\ddot{y} > 0$. (If we measure y up from the earth, we have $m\ddot{y} = -mg$ instead.) Canceling m and integrating twice, we get

$$\dot{y} = gt + c_1, \qquad y = \tfrac{1}{2}gt^2 + c_1 t + c_2.$$

(Alternatively, we could work from the double characteristic root $r_1 = r_2 = 0$ of the associated homogeneous equation $\ddot{y} = 0$.)

---

e x a m p l e 1

If a freely falling body has height $y_0 = 200$ feet below some reference point and initial velocity $\dot{y}_0 = 100$ feet per second at time $t = 0$, the height at time $t > 0$ satisfies

$$\dot{y}_0 = g \cdot 0 + c_1 \qquad\qquad \text{so} \qquad c_1 = \dot{y}_0 = 100$$

and

$$y_0 = \tfrac{1}{2}g \cdot 0 + 100.0 + c_2 \qquad \text{so} \qquad c_2 = y_0 = 200.$$

Hence

$$y(t) = \tfrac{1}{2}gt^2 + 100t + 200.$$

The assumption that a body is "freely falling" is sometimes admissible, but in practice air resistance often has to be taken into account. One model of somewhat restricted validity (see also Exercise 5) prescribes a **drag force** proportional to the velocity $\dot{y}$:

$$F_d = -k\dot{y}, \qquad k \text{ a positive constant.}$$

The equation of forces is now

$$m\ddot{y} = mg - k\dot{y};$$

the minus sign is needed because the drag decreases the velocity and so tends to push $\ddot{y}$ toward negative values. The solutions of our equation

$$\ddot{y} + \frac{k}{m}\dot{y} = g$$

can be found in two steps. The characteristic equation is $r^2 + (k/m)r = 0$, with roots $r_1 = 0$, $r_2 = -k/m$. The homogeneous solution is

$$y_h = c_1 + c_2 e^{-kt/m}.$$

The trial solution $y_p = At$ shows that $A = mg/k$. The total solution is $y_p + y_h$, or

$$y(t) = \frac{mg}{k} t + c_1 + c_2 e^{-kt/m}.$$

example 2

The **terminal velocity** of a falling body subject to drag force F_d is defined by

$$v_\infty = \lim_{t \to \infty} \dot{y}(t).$$

From the solution $y(t)$ found previously, we compute

$$\dot{y}(t) = \frac{mg}{k} - \frac{k}{m} c_2 e^{-kt/m}.$$

Then $v_\infty = mg/k$. Note that a large drag constant k produces an expectedly small terminal velocity. Note also that the analysis just given does not apply if $k = 0$. But the solution of the freely falling body problem gives

$$v_\infty = \lim_{t \to \infty} y(t)$$

$$= \lim_{t \to \infty} gt + c_1 = \infty.$$

The product mg is the **weight** W of the body (measured in an unaccelerated state relative to the earth), so the formula for terminal velocity can be written $v_\infty = W/k$.

Examples 1 and 2 deal with a gravitational force that is constant, because the mass of the moving body is constant and the distances involved are so small that there is no significant variation in the force due to the attracting body. In the following two examples, the force due to gravity varies because the effective mass is varying.

example 3

Consider a chain of length l and total mass m lying on a smooth horizontal plane surface, for example, a ship's deck. Suppose one end of the chain drops through a hole in the surface, hanging down under the force of gravity, while the rest of the chain is stretched out straight on the surface. The position of the chain at time t can be described by letting $y = y(t)$ be the length of chain hanging below the surface. We'll assume the frictional force created by the weight of the chain on the surface to be negligible. The force acting to move the chain is $F = g(m/l)y$, where g is the acceleration of gravity and m/l is the mass density of the chain per unit of length. On the other hand, the total force acting is also given by $F = m\ddot{y}$, since the entire chain moves as a single body as one end falls through the hole. Thus y satisfies the differential equation $m\ddot{y} = g(m/l)y$, or $\ddot{y} = (g/l)y$. The general solution of this equation is

$$y = c_1 e^{\sqrt{g/l}\, t} + c_2 e^{-\sqrt{g/l}\, t},$$

and the constants c_1, c_2 can be determined by initial conditions of the form $y(0) = y_0$, $\dot{y}(0) = z_0$. For example, with $g = 32$, $l = 8$, $y(0) = 1$, and $\dot{y}(0) = 0$ we find $y(t) = \frac{1}{2}e^{2t} + \frac{1}{2}e^{-2t} = \cosh 2t$. Note that the entire chain becomes simply a falling body as soon as it has all passed through the hole; from that time on the operative differential equation becomes just $\ddot{y} = g$. Thus in our numerical example, the explicit solution formula holds as long as $y(t) \leq 8$. Starting with $t = 0$, the time interval in question ends at the time t_1 when $\cosh 2t = 8$, or $e^{2t} + e^{-2t} = 16$, or $(e^{2t})^2 - 16e^{2t} + 1 = 0$. This is a quadratic equation that can be solved for e^{2t}, from which it follows that $t_1 = \frac{1}{2} \ln (8 + \sqrt{63}) \approx 1.38$ seconds. (See Exercise 19 for more details.)

To take account of friction of the chain sliding over the surface in Example 3, we use the observation that **sliding frictional force** is often proportional to the force pressing the sliding object onto the bearing surface.

---

e x a m p l e 4

In the case of the chain of Example 3 the sliding force when length y of chain has passed through the hole has magnitude $F_f = kg(m/l)(l - y)$, $k = \text{const} > 0$; this is so since the mass of the chain lying on the surface is the mass m/l per unit of length times the length $l - y$ of chain on the surface. The force F_f can now be incorporated into the differential equation of Example 3 as a retarding force:

$$m\ddot{y} = g\frac{m}{l}y - kg\frac{m}{l}(l - y),$$

where k is a proportionality constant. The factor m can be canceled and the equation rewritten in the form

$$\ddot{y} - (1 + k)\frac{g}{l}y = -kg.$$

If $k = 0$, we get the frictionless equation; if k is very small, we can expect a solution satisfying given initial conditions to be close to the corresponding solution of the frictionless equation. Indeed the solutions of the more general nonhomogeneous equation are

$$y(t) = c_1 e^{\sqrt{(1+k)g/l}\,t} + c_2 e^{-\sqrt{(1+k)g/l}\,t} + \frac{kl}{1 + k}.$$

When $k = 0$, we get the solutions to the corresponding frictionless problem. For positive values of k, the differential equation shows that to have increasing velocity, that is, $\ddot{y} > 0$, we need

$$\ddot{y} = (1 + k)\frac{g}{l}y - kg > 0, \quad \text{in particular} \quad y(0) > \frac{kl}{1 + k};$$

if this condition isn't met, it can happen that, even with positive initial velocity $\dot{y}(0)$, the chain may slow up and never reach the free-fall state where $\ddot{y} = g$.

EXERCISES

1. A ball of mass m, and hence weight $w = mg$, is dropped from a point 200 feet above the surface of the earth so that its initial velocity is 0. Assume that the drag force is insignificant.

(a) When does the ball hit the ground?

(b) What is the impact velocity at ground level?

(c) From what height should the ball be dropped to get an impact velocity of 100 feet per second?

(d) Is the mass of the ball relevant to any of the preceding questions?

2. Suppose in Exercise 1 that the weight of the ball is 32.2 pounds, so we can take $m = 1$, and that the drag force is significant, with drag constant $k = 0.1$.

(a) Show that the ball hits the ground after time t_1, where t_1 is the positive solution of the equation

$$\frac{t}{10} + e^{-t/10} = \frac{171}{161}, \qquad \text{with } g = 32.2.$$

(b) Show that t_1 in part (a) is about 3.74 seconds.

(c) Compare the result obtained in part (b) with the answer obtained by using the free-fall formula $y(t) = gt^2/2$.

(d) Find the terminal velocity of the ball.

(e) About how long does it take for the ball to reach a velocity of 75 feet per second?

3. Suppose a falling body is subject to a drag constant k and has mass m.

(a) If the initial velocity is 0, show that the distance covered in time t is

$$y(t) = \frac{mg}{k}t - \frac{m^2g}{k^2}(1 - e^{-kt/m}).$$

(b) Show that the formula for $y(t)$ in part (a) satisfies $y(t) \le \frac{1}{2}gt^2$ for $t \ge 0$. [*Hint:* $\ddot{y} = g - (k/m)\dot{y} \le g$. Now integrate from 0 to t.]

(c) Find the analog of the formula given in part (a) for the case of initial velocity $\dot{y}(0) = v_0$.

4. (a) For a falling body of mass m subject to linear drag with constant k, show that initial velocity v_0, leads to velocity at time t given by

$$\dot{y}(t) = \frac{mg}{k} + \left(v_0 - \frac{mg}{k}\right)e^{-kt/m}.$$

(b) Find the limit as k tends to zero of the formula for $\dot{y}(t)$ in part (a). Does this agree with the free-fall formula $v_0 + gt$?

5. The drag constant k can be estimated by using an observed value of the terminal velocity $v_\infty = mg/k$.

(a) Find k if a body of weight $w = 100$ pounds achieves terminal velocity $v_\infty = 180$ feet per second.

(b) How far must the body in part (a) fall to obtain the velocity of 150 feet per second?

*6. The nonlinear equation

$$m\ddot{y} = mg - k\dot{y}^2$$

is sometimes used to model fall in a resisting medium. Here the retarding action of the drag term is much greater at large velocities than for the linear model. To solve the equation, let $v = \dot{y}$, $\dot{v} = \ddot{y}$ and solve

$$m\dot{v} = mg - kv^2$$

by separation of variables for $v(t)$. Then solve $\dot{y} = v(t)$ by a single integration.

(a) Solve the equation $m\dot{v} = mg - kv^2$ under the assumption that $v(0) = v_0$, and show that the solution can be written

$$v(t) = \sqrt{\frac{mg}{k}} \left(\frac{e^{at} - c_0 e^{-at}}{e^{at} + c_0 e^{-at}} \right),$$

where $c_0 = (\sqrt{mg} - \sqrt{k}v_0)/(\sqrt{mg} + \sqrt{k}v_0)$ and $a = \sqrt{kg/m}$.

(b) Find the terminal velocity v_∞ in this model and compare it with one obtained from the linear model $m\dot{v} = mg - kv$.

(c) Assuming $y(0) = 0$, show that

$$y(t) = \frac{v_\infty^2}{g} \ln \left(\cosh \frac{gt}{v_\infty} + \frac{v_0}{v_\infty} \sinh \frac{gt}{v_\infty} \right).$$

7. A 32-pound weight is projected straight up from the earth with an initial velocity of 100 feet per second.

(a) If air resistance is negligible, about how many seconds would it take for the weight to reach its highest point? How high is the highest point? How many more seconds to strike the earth again?

(b) If the weight encounters an air-resistance force equal to one-tenth times its velocity, about how many seconds would it take for the weight to reach its highest point? About how high is the highest point?

(c) Show by careful computation that the weight retarded by air resistance strikes the ground first. (This isn't altogether obvious, because while the retarded weight doesn't go as high, it falls more slowly because of the air resistance.)

8. A projectile of mass m fired straight up from the ground under the influence of gravitational acceleration $-mg$ and air resistance $-kv$ is at distance y at time t from its starting point, where $m\ddot{y} = -k\dot{y} - gm$.

(a) Find $y(t)$ and $\dot{y}(t)$.

(b) At what time after firing does the projectile reach its highest point, and how high is that?

9. A projectile subject just to the influence of gravitational deceleration $-g$ is fired straight up with initial velocity v_0. Show that the maximum height attained is $v_0^2/(2g)$.

10. An object of mass m is dropped from rest subject to constant gravity g and linear retarding force $-kv$.

(a) Show that the only attainable velocities v_1 satisfy $0 \leq v_1 < mg/k$.

(b) Show that velocity v_1 is attained at time

$$t_1 = \frac{m}{k} \ln \frac{mg}{mg - kv_1}.$$

11. A 100-pound weight is to be dropped from a great height. We want to design a parachute for the weight so that it will hit the earth at no more than 20 feet per second. If the retarding force is proportional to velocity, how large should the constant of proportionality be to achieve the desired landing speed?

12. A 64-pound weight, including parachute, is dropped from 5000 feet above the earth. The retarding force due to air resistance with the parachute closed has magnitude $0.1v$.

    (a) Assume that with the parachute closed the velocity reaches 160 feet per second. How long does it take to reach this speed, and at what height does it occur?

    (b) Assume the parachute opens completely at the instant the velocity reaches 160 feet per second and that thereafter the retarding force is such that the terminal velocity is 16 feet per second. Estimate the additional time elapsed before the weight hits the ground.

13. A 100-pound weight experiences a retarding force equal to its velocity in feet per second when falling through the earth's atmosphere. Do a numerical computation to estimate the time it takes to hit the earth after being dropped from a height of 500 feet. How does this compare with the time to hit the earth in the absence of retarding force?

14. An object of mass m is dropped from rest under constant gravity g and against retarding force $-kv$, $k > 0$. The object has velocity $v_1(t) = (mg/k)(1 - e^{-kt/m})$ at time t after the initial drop. The corresponding formula when $k = 0$ is $v_2(t) = gt$.

    (a) Show that $v_1(t) \leq v_2(t)$ for $t \geq 0$. [*Hint:* What differential equations do v_1 and v_2 satisfy?]

    (b) Show that the limit as k tends to zero of $v_1(t)$ is $v_2(t)$.

15. A car of mass m is propelled along a straight horizontal track with initial velocity v_0 and subject thereafter to a retarding force $-kv$ proportional to velocity.

    (a) Show that the car will stop within distance mv_0/k of the starting point.

    (b) Show that the velocity will reach velocity v_1, where $v_0 > v_1 > 0$, at time $t = (m/k) \ln (v_0/v_1)$.

16. A 400-pound car is to be propelled along a straight horizontal track with initial velocity 500 feet per second and subject thereafter to a retarding force $-kv$ proportional to velocity. A braking mechanism must be designed to stop the car within 2000 feet of the starting point; what value should k have to achieve this?

17. A car of mass m is propelled along a straight horizontal track with initial velocity v_0 and subject thereafter to a retarding force $-kv^2$ proportional to velocity squared: $m\dot{v} = -kv^2$.

    (a) Show that $v(t) = v_0/(1 + kv_0t/m)$ for $t \geq 0$.

    (b) Show that distance traveled in time t is $y(t) = (m/k) \ln (1 + kv_0t/m)$, which tends to infinity as t tends to infinity.

(c) Recall (Chapter 1, Section 3, Example 2) that with linear retarding force $-kv$, the distance traveled in time t is $y(t) = (mv_0/k)(1 - e^{-kt/m})$, which remains bounded above by m/k as t tends to infinity. Give an intuitive reason for the discrepancy between the results of parts (b) and (c).

18. The differential equation $m\, dv/dt = gm - kv^\alpha$, $\alpha > 0$, has solutions $v = v(t)$ that tend to a (constant) terminal velocity v_∞ as t tends to infinity. Without trying to solve the differential equation, find v_∞. [*Hint:* As $v(t)$ approaches a constant, what happens to dv/dt?]

19. In Example 3 of the text we needed to solve $\cosh 2t = 8$ or $e^{2t} + e^{-2t} = 16$ for a positive value of t. This can be done as follows: Multiply by e^{2t} to get $(e^{2t})^2 - 16e^{2t} + 1 = 0$. The solutions of this quadratic equation for e^{2t} are $e^{2t} = (16 \pm \sqrt{16^2 - 4})/2 = 8 \pm \sqrt{63}$. Applying the natural logarithm to both sides gives $t = \frac{1}{2}\ln(8 \pm \sqrt{63})$. However, $\ln x > 0$ only if $x > 1$, so we must choose the plus sign to get a positive value for t. Recalling that $\sinh u = \frac{1}{2}e^u - \frac{1}{2}e^{-u}$ and $\cosh u = \frac{1}{2}e^u + \frac{1}{2}e^{-u}$, find the real solution values t for the following equations.

(a) $e^t + 2e^{-t} = 3$.

(b) $\sinh t = 2\cosh t - 2$.

(c) $6e^{-t} = e^t + 1$.

20. A 50-foot chain lies stretched out on the deck of a ship with 10 feet hanging over the side. The chain is released from rest so that it slides with negligible friction over the deck and down toward the water 40 feet below the deck. Assume that the acceleration of gravity is 32 feet per second per second.

(a) Find the differential equation satisfied by the length of chain hanging over the side.

(b) Express the length of chain hanging over the side as a function of time.

(c) Express the velocity of the chain as a function of time.

(d) About how long does it take for the chain to first hit the water? [*Hint:* To solve for t as a function of y, first solve a quadratic equation for $e^{4t/5}$.]

(e) What is the chain's velocity when it first hits the water? [*Hint:* Use the expression for $e^{4t/5}$ found in the previous part.]

(f) What is the chain's acceleration when it first hits the water?

Newtonian Gravity. In studying motion over a wide range of distances from the earth, the equation $m\ddot{y} = mg$ should be replaced by the more accurate **inverse-square law** $m\ddot{y} = -GMm/y^2$, or $\ddot{y} = -GM/y^2$, where M is the mass of the earth, y is distance measured from the earth's center of mass, and G is the **gravitational constant.** The product GM can be estimated by observing that $GM/R^2 \approx g$, where R is the radius if the earth. Thus $GM \approx R^2 g$. It follows that the **weight** of an object of mass m at distance y from the center of the earth is about $W_R = mgR^2/y^2$.

21. A body of mass m influenced only by the earth's attraction and moving radially at distance y above the earth's surface satisfies the nonlinear equation $\ddot{y} = -R^2 g y^{-2}$.

(a) Solve this equation by multiplying through by $\dot{y}$ and then integrating both sides with respect to t to get $\int \dot{y}\ddot{y}\,dt = R^2 g \int y^{-2}\dot{y}\,dt$ and $\frac{1}{2}\dot{y}^2 = R^2 g y^{-1} + C$.

(b) Initial conditions $y(0) = y_0 \geq R$, $\dot{y}(0) = v_0$ require that $\frac{1}{2}\dot{y}^2 = R^2 g y^{-1} - R^2 g y_0^{-1} + \frac{1}{2}v_0^2$; explain why.

(c) Show that if $y_0 = R$ and $0 < v_0 < \sqrt{2Rg}$, then $\dot{y} = 0$ at the maximum height $y_{\max} = R/(1 - v_0^2/(2Rg))$ upon which the body falls back to earth.

(d) Show that for the gravity law $m\ddot{y} = gm$, the maximum height attained with $y_0 = R$ and $v_0 > 0$ is $R + v_0^2/(2g)$.

(e) Show that the difference between the results of parts (c) and (d) is equal to $v_0^2/(4Rg^2 - 2gv_0^2)$.

(f) The critical velocity $v_E = \sqrt{2Rg}$ is called the **escape velocity** from the earth. Explain why this terminology is appropriate, and show that v_E is about 7 miles per second at the surface of the earth, where R is about 4000 miles.

22. A change in altitude is related to the different values of g as follows.

(a) Let $g_0 = GM/R^2$ and $g_1 = GM/(R + h)^2$ be measurements of g at altitudes R and $R + h$, respectively. Show that $h = R(\sqrt{g_0/g_1} - 1)$.

(b) If $g_0 = 32$ at a point where $R = 3960$ miles and $g_1 = 28$, estimate h.

2 PENDULUM

The motion of a freely swinging or **undamped pendulum** with no outside force but gravity acting on it is easy to describe in terms of the angle $\theta = \theta(t)$ that the pendulum makes with a vertical line. Figure 4.1 shows the setup. We'll con-

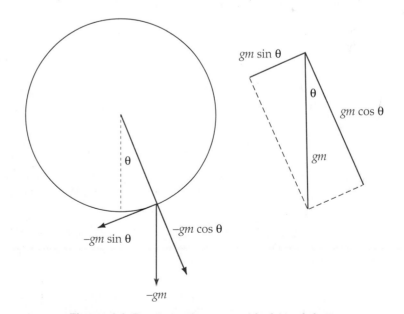

Figure 4.1 Gravity acting on an ideal pendulum.

sider an **ideal pendulum** with a wire of negligible mass attaching a weight of mass m to the pivot and with all mass concentrated as the weight's center of gravity at distance l from the pivot. See Exercise 23 for something more general.

Figure 4.1 gives the analysis of the downward-directed gravitational force of magnitude mg into components parallel and perpendicular to the pendulum rod. At any given position, the gravitational force component in the direction of motion is perpendicular to the component directed along the length of the pendulum; it follows by the Pythagorean relation that the sum of the squares of the component magnitudes must be $g^2 m^2$. The coordinates in these directions are, respectively, $-gm \sin \theta$ and $-gm \cos \theta$, as can be seen in Figure 4.1. The force along the length of the pendulum must be exactly balanced by an opposite force at the pivot and plays no other role in the motion. Since distance along the circular path of motion is $l\theta$ if θ is measured in radians, the force coordinate in the direction of actual motion can also be expressed as $m \, d^2(l\theta)/dt^2$. Equating our two expressions for force coordinates in the direction of motion gives the differential equation satisfied by $\theta = \theta(t)$:

$$m \frac{d^2(l\theta)}{dt^2} = -gm \sin \theta.$$

The minus sign indicates that the velocity $d(l\theta)/dt$ is decreasing if $0 < \theta < \pi$ and increasing if $-\pi < \theta < 0$. Division by the constant lm gives the alternative form

$$\ddot{\theta} = -\frac{g}{l} \sin \theta.$$

(For an alternative derivation see Exercise 16 of Chapter 5, Section 1C.) This differential equation is nonlinear, and we can find approximations to its solutions by the methods of Section 7. However, if θ remains small, say $|\theta| < 0.1$, we may find acceptable approximate solutions by using the tangent-line estimate to the graph of $\sin \theta$ near $\theta = 0$, namely, $\sin \theta \approx \theta$. This replacement leads to the linear equation

$$\ddot{\theta} = -\frac{g}{l} \theta.$$

Whether the process of consistently replacing nonlinear terms in an equation by linear estimates, called **linearization** of the equation, is acceptable or not in practice depends very much on the demands for accuracy in the specific problem at hand. In the case of relatively small pendulum oscillations, it turns out that we can always get a good *qualitative* picture of pendulum behavior from the linearized equation. We'll take advantage of this fact in the next two examples.

---

e x a m p l e **1**   The solutions of the linearized pendulum equation, derived above, depend on the characteristic equation $r^2 + g/l = 0$, with purely imaginary roots $r_1, r_2 = \pm i \sqrt{g/l}$. These roots lead us to the 2-parameter family of oscillatory solutions

$$\theta(t) = c_1 \cos \sqrt{\frac{g}{l}} \, t + c_2 \sin \sqrt{\frac{g}{l}} \, t.$$

If we arrange it so that $\theta = 0$ when $t = 0$, we can write

$$\theta(t) = c_2 \sin \sqrt{\frac{g}{l}}\, t.$$

With only slight loss of generality we'll assume $c_2 > 0$, so c_2 can be interpreted as the maximum angular displacement of the pendulum. The displacement of the center of mass of the pendulum along its circular path is

$$y(t) = l\theta(t) = A \sin \sqrt{\frac{g}{l}}\, t,$$

where $A = lc_2$ is the spatial amplitude of the pendulum's swing. The velocity of the center of mass is

$$\dot{y}(t) = l\dot{\theta}(t) = A \sqrt{\frac{g}{l}} \cos \sqrt{\frac{g}{l}}\, t.$$

This shows that the maximum velocity, $A\sqrt{g/l}$, occurs when the pendulum is in the downward vertical position. Also, the velocity is 0 when $\sqrt{g/l}\,t = k\pi/2$, k an odd integer, that is, when $t = k\pi\sqrt{l/g}/2$; naturally this is just when the absolute value of the displacement reaches its maximum. It's easy to verify that the period of $y(t)$ is $2\pi\sqrt{l/g}$.

The statements about pendulum behavior in Example 1 are sufficiently accurate for many purposes if the amplitude of oscillation remains small. (For example, accurate surveys of the local variation in the acceleration of gravity have been based on the linearized model.) For larger oscillations the linear model begins to show significant discrepancies with what happens in practice. In particular, the chaotic behavior that an externally driven pendulum may exhibit does not occur in the linear model. Our main resources in investigating chaotic behavior are physical experiment and numerical methods of the kind described in Section 7. Solution formulas for the unforced nonlinear equation can be derived using *elliptic functions*; quite independently of these formulas it is elementary to show that solutions of $\ddot{y} = -(g/l)\sin y$ are periodic as functions of time t, just as for the linear model. (See Section 5 for the details.)

We often have to account for frictional damping of a pendulum. This can be allowed for fairly accurately by including a term of the form $-k(l\dot{\theta})$ on the right side of the undamped equation. Dividing by mass m gives

$$\ddot{\theta} = -\frac{g}{l} \sin \theta - \frac{k}{m}\, \dot{\theta}.$$

Numerical solution of this equation is taken up in Section 7; for now we'll look at the **linearized damped pendulum equation** that we get when we replace $\sin \theta$ by θ:

$$\ddot{\theta} + \frac{k}{m}\, \dot{\theta} + \frac{g}{l}\, \theta = 0.$$

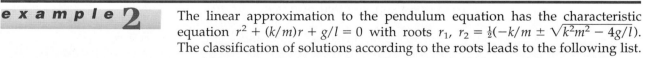

example 2

The linear approximation to the pendulum equation has the characteristic equation $r^2 + (k/m)r + g/l = 0$ with roots $r_1, r_2 = \frac{1}{2}(-k/m \pm \sqrt{k^2 m^2 - 4g/l})$. The classification of solutions according to the roots leads to the following list.

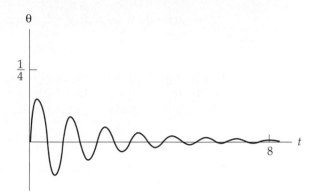

Figure 4.2 Damped oscillation for a linearized pendulum.

(a) $\theta(t) = c_1 e^{\alpha_1 t} + c_2 e^{\alpha_2 t}$ if $k^2/m^2 > 4g/l$, where the real numbers α_1, $\alpha_2 = \frac{1}{2}(-k/m \pm \sqrt{k^2/m^2 - 4g/l})$ are both negative.
(b) $\theta(t) = c_1 t e^{-kt/2m} + c_2 e^{-kt/2m}$ if $k^2/m^2 = 4g/l$.
(c) $\theta(t) = e^{-kt/2m}(c_1 \cos \beta t + c_2 \sin \beta t)$ if $k^2/m^2 < 4g/l$, where the real number $\beta = \frac{1}{2}\sqrt{4g/l - k^2/m^2}$.

If k should happen to be 0, then only case (c) can occur, and we get the periodic solution we found in Example 1: $\theta(t) = c_1 \cos \sqrt{g/l}t + c_2 \sin \sqrt{g/l}t$. For a real pendulum it is usually only case (c) that is important, so we will illustrate that case. Take $g = 32$, $K = 1.0$, $m = 1$, and $l = 1$. Then $\beta = \frac{1}{2}\sqrt{4 \cdot 32 - 1} \approx 5.6$. The solution

$$\theta(t) = e^{-0.5t}(c_1 \cos \beta t + c_2 \sin \beta t)$$

is still subject to initial conditions, for example, $\theta(0) = 0$, $\dot{\theta}(0) = 1$. These are satisfied if $c_1 = 0$ and $c_2 \beta = 1$, so we get

$$\theta(t) = \frac{1}{\beta} e^{-0.5t} \sin \beta t$$

$$\approx 0.18 e^{-0.5t} \sin 5.6t$$

A part of the graph appears in Figure 4.2. The damping constant $K = 1.0$ evidently leads to quite significant damping of the oscillations. The motion is not periodic as in the undamped case, but the points of zero displacement from downward vertical are still equally spaced along the time axis, the space between them being about $2\pi/5.65 \approx 1.11$ radians.

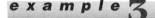 **e x a m p l e 3**

Assuming small enough oscillations, the displacement angle θ of an undamped pendulum nearly satisfies the linearized equation $\ddot{\theta} = -(g/l)\theta$, with general solution expressible as

$$\theta(t) = A \cos\left(\sqrt{\frac{g}{l}}\, t - \phi\right).$$

The period of this oscillation is $P = 2\sqrt{l/g}\,\pi$. To see the effect of various initial conditions, suppose that $\theta(0) = \theta_0$ and $\dot{\theta}(0) = \psi_0$. Then

$$A \cos \phi = \theta_0 \quad \text{and} \quad \sqrt{\frac{g}{l}}\, A \sin \phi = \psi_0.$$

Solving these equations for A and ϕ we get

$$A = \sqrt{\theta_0^2 + \frac{l\psi_0^2}{g}} \quad \text{and} \quad \phi = \arctan\left(\sqrt{\frac{l}{g}}\,\frac{\psi_0}{\theta_0}\right).$$

For example, if $\theta_0 = 0.2$ is the initial displacement angle and $\psi_0 = -0.3$ is the initial angular velocity for a 4-foot pendulum, then $A = \sqrt{0.2^2 + 0.3^2/8} \approx 0.23$ and $\phi \approx \arctan(-0.53) \approx -0.49$ radian. Thus the angular displacement as a function of time is approximately $0.23 \cos(\sqrt{8}t + 0.49)$.

For the *nonlinear* pendulum equation $\ddot{y} = -(g/l)\sin y - (k/m)\dot{y}$ we address the matter of existence and uniqueness of solutions with a theorem for second-order initial-value problems proved in more theoretical treatments of the subject.

■ EXISTENCE AND UNIQUENESS THEOREM

The initial-value problem

$$\ddot{y} = f(t, y, \dot{y}), \qquad y(t_0) = y_0, \qquad \dot{y}(t_0) = z_0,$$

has a unique solution on the interval $t_0 \leq t \leq t_1$ if the partial derivatives $\partial f(t, y, z)/\partial y$ and $\partial f(t, y, z)/\partial z$ are bounded and continuous for all (t, y, z) with t in the interval.

In the pendulum equation, we note that $f(t, y, z) = -(g/l)\sin y - (k/m)z$, that $\partial f(t, y, z)/\partial y = -(g/l)\cos y$, and that $\partial f(t, y, z)/\partial z = -k/m$. Since these partial derivatives are uniformly bounded and continuous, the initial-value problem for the damped pendulum equation always has a unique solution.

EXERCISES

Unless something to the contrary is explicitly stated, the assumption in each of the following exercises is that the motion in question is governed by a *linear* equation and that acceleration of gravity is $g = 32$ feet per second per second.

1. Determine the amplitude A of the displacement $y(t)$ in terms of $\theta(0)$ and $\dot{\theta}(0)$ of an undamped pendulum satisfying the linear equation $\ddot{\theta} + (g/l)\theta = 0$. What is its period?

2. How long should an undamped pendulum be if its displacement angle satisfies $\ddot{\theta} + (g/l)\theta = 0$ and it is to have a period of 2 seconds?

3. **(a)** Given the relation $y(t) = l\theta(t)$, express the equation $\ddot{\theta} + (k/m)\dot{\theta} + (g/l)\theta = 0$ in terms of y, k, l, and m.

   (b) What condition must k satisfy for the pendulum described by the differential equation in part (a) to oscillate?

   (c) Show that oscillations described by the differential equation in part (a) undergo an increase in the time between successive vertical positions if l increases slightly. What is the effect of an increase in m?

4. Consider the equation

$$\ddot{\theta} + \frac{k}{m}\,\dot{\theta} + \frac{g}{l}\,\theta = 0.$$

where m, k, g, and l are positive constants.

(a) Show that if the roots of the characteristic equation are real, then these roots are both negative, and that if they are complex, then they both have negative real parts.

(b) What can you conclude about the oscillation of a pendulum from part (a)?

(c) If $k^2/m^2 \geq 4g/l$, explain why $\theta(t)$ is nonoscillatory.

5. Show that the general solution of

$$\ddot{\theta} + \frac{k}{m}\,\dot{\theta} + \frac{g}{l}\,\theta = 0,$$

in the case $k^2/m^2 < 4g/l$, can be written in the form

$$\theta(t) = Ae^{-kt/2m}\cos\left(\sqrt{\frac{B}{l}}\,t + \alpha\right).$$

[*Hint:* Use the addition formula for the cosine.]

6. Solve the equation

$$\ddot{y} + \dot{y} + y = a_0\cos\omega t, \qquad a_0 \neq 0,\ \omega \neq 0,\ \sqrt{3}/2,$$

for the oscillation of a pendulum with externally applied force $a_0\cos\omega t$. First find a particular solution $y_p(t)$ by the method of undetermined coefficients using the trial solution $y_p(t) = A\cos\omega t + B\sin\omega t$.

7. Let $y = y(t)$ be a periodic solution of

$$y'' = -\frac{g}{l}\sin y,$$

and let y_0 be the maximum value of $y(t)$ such that $0 < y_0 < \pi/2$.

(a) Show that if $y(t)$ satisfies

$$y' = \sqrt{\frac{2g}{l}\,(\cos y - \cos y_0)}$$

then it satisfies the second-order equation.

(b) Show that the time T required for $y(t)$ to go from $-y_0$ to y_0 is

$$T = \sqrt{\frac{2l}{g}}\int_0^{y_0}\frac{dy}{\sqrt{\cos y - \cos y_0}}.$$

Thus the period of the solution is $p = 2T$.

8. A cylindrical object floating in liquid with its central axis vertical satisfies a differential equation for its vertical displacement $y(t)$ from equilibrium:

$$\ddot{y} + \frac{k}{m}\,\dot{y} + A\frac{w}{m}\,y = 0,$$

where m is the mass of the object, A is its circular cross-sectional area, w is the density of the liquid, and k is a positive friction constant.

(a) Show that the differential equation has oscillatory solutions if and only if $k^2 < 4Awm$.

(b) Discuss the effect on $y(t)$ of increasing m and of increasing w, assuming that $k^2 < 4Awm$. Show that for increasing m, the frequency alters critically when $k^2 = 2Awm$.

9. A 2-foot undamped pendulum is released from angle $\theta = 0.1$ radian (about 5.7°) from the vertical.

(a) Express θ as a function of time t after release.

(b) About how long does it take for the pendulum to reach the vertical position $\theta = 0$ for the first time?

(c) Find the angular speed $|\dot{\theta}|$ when $\theta = 0$.

10. A 2-foot undamped pendulum has maximum linear velocity 0.5 foot per second at its tip. What is the maximum angular displacement?

11. A pendulum 6 inches long is pushed down from an angular displacement of $\frac{1}{10}$ radian with angular speed $\frac{1}{2}$ radian per second. Find the period, the maximum amplitude, and the maximum angular velocity.

12. Leave everything in the previous problem the same except that the pendulum is initially pushed up instead of down. Do any of the answers change? Explain why.

13. An undamped pendulum 2 feet long swings with small amplitude; estimate its period of oscillation.

14. An undamped pendulum 2 feet long swings with amplitude 5°. Estimate its maximum linear velocity at the tip.

15. About how long should an undamped pendulum with small oscillations be if it is to have a period of 1 second?

16. Express the solution to the initial-value problem $\ddot{\theta} = -(g/l)\theta$, $\theta(0) = \theta_0$, $\dot{\theta}(0) = \psi_0$ in the form $\theta(t) = c_1 \cos \omega t + c_2 \sin \omega t$, that is, find c_1, c_2, and ω.

17. Suppose an undamped pendulum with small oscillations has length l_0 and period 1 second at a location where the acceleration of gravity is known to be 32 feet per second per second. The same pendulum at a point far above the surface of the earth has period 1.1 seconds; what is the acceleration of gravity there?

18. A pendulum of length l and mass m has its oscillations damped by a force equal to a constant k times its linear velocity $l\dot{\theta}$. Show that the pendulum will actually exhibit small oscillations only if $2m\sqrt{g/l} > k$, where g is the acceleration of gravity.

19. A pendulum of length l and mass m has its oscillations damped by a force equal to a constant k times its linear velocity $l\dot{\theta}$.

(a) Show that the time difference between vertical positions is about

$$T = \frac{2\pi m}{\sqrt{4gm^2/l - k^2}}.$$

(b) What happens to T in part (a) as k tends to zero?

20. A pendulum of length l and mass m has its oscillations damped by a force equal to a constant k times its angular velocity $\dot{\theta}$. Show that the width of the pendulum's small oscillations decreases less rapidly over time if the pendulum is made longer or heavier.

21. **(a)** Carry out the solution of the equations $A \cos \phi = \theta_0$, $\sqrt{g/l}A \sin \phi = \psi_0$ to get the formulas in Example 3 of the text: $A = \sqrt{\theta_0^2 + l\psi_0^2/g}$, $\phi = \arctan(\sqrt{l/g}\psi_0/\theta_0)$. Assume $\theta_0 \neq 0$.

   (b) Show that if $\theta_0 = 0$, the formulas in part (a) should mean $A = \sqrt{l/g}\psi_0$ and $\phi = \pi/2$ when $\psi_0 > 0$, $\phi = -\pi/2$ when $\psi_0 < 0$.

22. According to Newton's **inverse-square law**, the acceleration of gravity generated by the attraction of a spherically homogeneous body of mass M is $g = GM/r^2$, where r is the distance from the center of the body and G, which depends on the units of measurement, is the universal gravitational constant. Since the period $P = 2\pi\sqrt{l/g}$ of a pendulum can be used to estimate g, the period can be used to estimate significantly high altitudes h above the earth as follows.

   (a) Let $g_0 = GM/r^2$ and $g_1 = GM/(r + h)^2$ be measurements of g at altitudes r and $r + h$, respectively. Show that $h = r(\sqrt{g_0/g_1} - 1)$. If $g_0 = 32$ and $g_1 = 26$, estimate h under the assumption that g_0 is measured where $r = 3957$ miles.

   (b) Show that $r_1/r_0 = P_1/P_0$, where P_0 and P_1 are period measurements of the same pendulum at altitudes r_0 and r_1, respectively.

23. Our derivation of the pendulum equation depended on the assumption that the weight at the end was attached to the pivot by a thin wire of negligible mass, as in a Foucault pendulum. In case the mass of the connecting rod is too big to be neglected, it can be shown that the pendulum equation differs only in the constant on the right side:

$$\ddot{\theta} = -\frac{mgl}{I}\sin\theta,$$

where I is the moment of inertia of the entire pendulum about the pivot. Under our original assumption, $I = ml^2$, which yields the original equation. For a component of the pendulum that is effectively 1-dimensional of length l,

$$I = \int_0^l \delta(x)x^2 \, dx,$$

where $\delta(x)$ is the mass density of the pendulum at distance x from the pivot. Also, there is an *additivity property:* the moment of inertia of the whole is equal to the sum of the moments of inertia of the component parts.

   (a) Show that the pendulum equation for a uniform 1-dimensional rod of mass m and length l is $\ddot{\theta} = -3(g/l)\sin\theta$.

   (b) Use additivity for moments of inertia to show that if our original ideal pendulum of mass m and weightless wire connector is augmented by a uniform connector of mass μ and length l, the differential equation becomes

$$\ddot{\theta} = -\frac{(\mu + m)g}{(\mu/3 + m)l}\sin\theta.$$

3   MECHANICAL OSCILLATORS

A second-order constant-coefficient linear differential equation

$$a\frac{d^2x}{dt^2} + b\frac{dx}{dt} + cx = f(t)$$

has solutions that can be neatly classified into distinct types, depending on the relations between the constants $a, b, c,$ and the function $f(t)$. The most important class consists of the **oscillations,** namely, those solutions whose values persist in switching back and forth from one side to the other of some fixed reference. Equations of this kind are important not only because of their direct physical applicability but also because of the insight they yield about related nonlinear phenomena. A typical mechanism that can be analyzed using a constant-coefficient equation appears in cross section in Figure 4.3. The working parts consist of a piston that oscillates in a cylinder that generates frictional forces, and a spring that extends and compresses.

An ideal spring exerts a force with magnitude proportional to its extension or compression from its equilibrium position, which we'll denote by 0 on a fixed scale. Thus, if x is the amount of displacement from 0, then the force f_S exerted by the spring is representable, according to **Hooke's law,** by

$$f_S = -hx, \qquad h > 0.$$

For an ideal spring it is assumed that h is constant. The minus sign in front of the factor h comes about because a positive extension of the spring, $x > 0$, induces a restoring force that *decreases* the velocity dx/dt, making $d^2x/dt^2 < 0$; likewise compression, $x < 0$, induces an increase in velocity, making $d^2x/dt^2 > 0$. The schematic diagrams in Figure 4.4 track a spring through two stages each of extension and compression. Note carefully the sign changes in $x, dx/dt,$ and d^2x/dt^2 from one stage to the next.

Retarding frictional force f_F in the mechanism may be due to the viscosity of some surrounding fluid or to friction in mechanical parts of the mechanism; we assume that the sum of these forces is proportional to the velocity:

$$f_F = -k\frac{dx}{dt}, \qquad k > 0.$$

A time-dependent external force $f_E = f(t)$ may act independently of f_S and f_F, so the total force acting parallel to the scale is

$$f_S + f_F + f_E = -hx - k\frac{dx}{dt} + f(t).$$

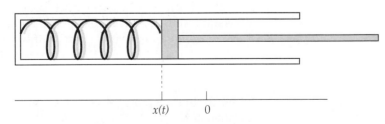

$$x(t) \qquad 0$$

Figure 4.3 Spring-loaded plunger.

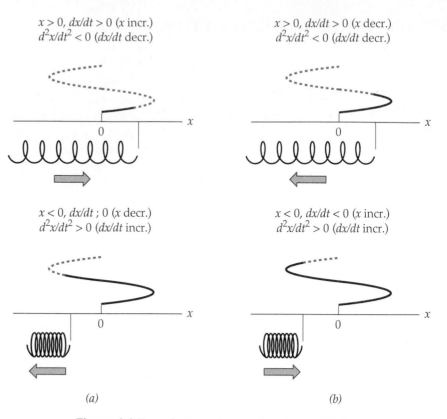

$x > 0$, $dx/dt > 0$ (x incr.)
$d^2x/dt^2 < 0$ (dx/dt decr.)

$x > 0$, $dx/dt > 0$ (x decr.)
$d^2x/dt^2 < 0$ (dx/dt decr.)

$x < 0$, dx/dt ; 0 (x decr.)
$d^2x/dt^2 > 0$ (dx/dt incr.)

$x < 0$, $dx/dt < 0$ (x incr.)
$d^2x/dt^2 > 0$ (dx/dt incr.)

(a)   (b)

Figure 4.4 Four distinct phases of spring oscillation.

On the other hand, general physical principles assert that this force must also be equal to the mass m of the moving parts times the acceleration $a = d^2x/dt^2$: $F = ma$. Thus

■ **3.1**
$$m\frac{d^2x}{dt^2} + k\frac{dx}{dt} + hx = f(t).$$

The external forcing function $f(t)$ can be thought of as an "input" to the mechanism (or to the differential equation) and a solution $x(t)$ can be thought of as a "response" that the mechanism (or the equation) makes to the input.

The assumption that there is a constant $h > 0$ such that hx describes the force required to extend or compress a spring by an amount x is called **Hooke's law.** If Hooke's law holds for a spring, then an extension x requires force hx and an extension $x + \Delta x$ requires force $h(x + \Delta x)$. It follows that the incremental force required to extend from position x to $x + \Delta x$ is $h\,\Delta x$. An additional consequence is that, if Hooke's law applies, the coordinate x can be chosen to have its zero value located at an arbitrary position of the extension or compression. Of course there are always practical limits to the applicability of Hooke's law; in extreme cases, overextension could even break a spring, and overcompression could buckle it. Thus it's customary to make various limiting assumptions about the nonnegative constants m, k, and h and also about the externally applied force $f(t)$.

Harmonic Oscillation. This mode of oscillation is called *harmonic* because of its association with musical tones, as described in Chapter 10. We assume that $f(t) = 0$ and that $k = 0$ in Equation 3.1. The assumption $k = 0$ represents an ideal (frictionless) state that can only be approximately realized in the mechanism of Figure 4.3. Nevertheless, undamped oscillation has fundamental importance as the limiting case for small k. Our assumptions $f(t) = 0$ and $k = 0$ will allow us to conclude that every nontrivial solution is **periodic** with a **period** p such that $x(t + p) = x(t)$ for all t. The reason is that if $f(t) = 0$ and $k = 0$, Equation 3.1 becomes

$$\frac{d^2 x}{dt^2} + \frac{h}{m} x = 0,$$

and all solutions consist of the periodic trigonometric functions

$$x(t) = c_1 \cos\left(\sqrt{\frac{h}{m}}\, t\right) + c_2 \sin\left(\sqrt{\frac{h}{m}}\, t\right)$$

$$= c_1 \cos \omega t + c_2 \sin \omega t, \qquad \text{where } \omega = \sqrt{h/m}.$$

Expressions of this last form occur often in physical settings, and a process that can be represented in this way is called a **harmonic oscillator.** The analysis described in Chapter 3, Section 2B, and in expanded form in Section 2D, is helpful in interpreting oscillations, because the arbitrary constants in the general solution formulas will then have some direct physical significance. Briefly, we choose an angle ϕ such that

$$\cos \phi = \frac{c_1}{\sqrt{c_1^2 + c_2^2}}, \qquad \sin \phi = \frac{c_2}{\sqrt{c_1^2 + c_2^2}}.$$

Then set $A = \sqrt{c_1^2 + c_2^2}$ so that $c_1 = A \cos \phi$ and $c_2 = A \sin \phi$. Using an addition formula for the cosine we can then write

$$x(t) = A \,(\cos \phi \cos \omega t + \sin \phi \sin \omega t)$$

$$= A \cos (\omega t - \phi).$$

Since the cosine factor is between -1 and 1, the number $A > 0$ is the maximum deviation of the oscillation from the value $x = 0$, and A is called the **amplitude** of the oscillation; the total width of the oscillation is then $2A$. The number $2\pi/\omega$ is a **period** of the oscillation, since $x(t + 2\pi/\omega) = x(t)$ for all t. The **frequency,** or number of oscillations per time unit t, is just the reciprocal of the period and is denoted by $\nu = \omega/2\pi$. The number ω itself is called the **circular frequency** of the oscillation and is equal to the number of oscillations per 2π time units. The number ϕ in the expression $\omega t - \phi$ is the **phase angle** of the oscillation (see Figure 4.5a), and if $\phi > 0$, it indicates that the graph of $x(t)$ is a shift to the right of the graph of $\cos \omega t$ by the amount ϕ/ω. The two trigonometric oscillations $A \cos (\omega t - \phi)$ and $B \cos (\omega t - \psi)$, $A, B > 0$, having a common period $2\pi/\omega$, are said to be **in phase** if $\phi - \psi$ is an integer multiple of 2π. Otherwise these oscillations are **out of phase** with a **phase difference** of time $|\phi - \psi|/\omega$; in particular their respective maxima and minima never occur at the same times.

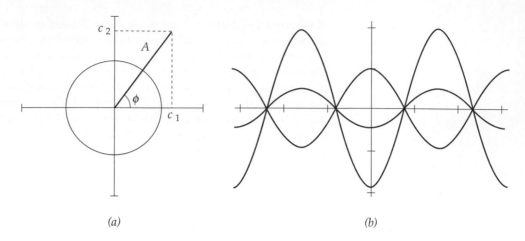

(a)   (b)

Figure 4.5 (*a*) Phase angle ϕ; amplitude A. (*b*) Oscillations in and out of phase.

---

e x a m p l e 1   Figure 4.5*b* shows the graphs of $\cos 2t$, $2\cos(2t - \pi)$, and $\frac{1}{2}\cos(2t - \pi)$. The last two of these are in phase with each other, but both of them are out of phase with the first one, $\cos 2t$, with common phase difference $\pi/2$. Note that the two oscillations in phase attain their maxima and minima at the same time.

---

e x a m p l e 2   Note that the definition of phase angle is stated entirely in terms of cosine. To find, for example, the phase angle for $\cos 4t - \sqrt{3}\sin 4t$, first express the oscillation in the form $A\cos(\omega t - \phi) = 2\cos(4t + \pi/3)$. Thus the phase angle is $\phi = -\pi/3$, so the given combination of sine and cosine has a graph that is a shift to the left of the graph of $2\cos 4t$ by $|\phi/\omega|$, which is $\pi/12$ for this example.

Returning to the differential equation $\ddot{x} + (h/m)x = 0$ and its solutions, we see that they exhibit the **natural circular frequency** $\omega = \sqrt{h/m}$ and frequency $\nu = \sqrt{h/m}/2\pi$ that depend only on h and m. On the other hand, the amplitude A and phase angle ϕ depend on the undetermined constants c_1 and c_2, which in turn depend on initial conditions as well as on h and m.

---

e x a m p l e 3   When a buoyant object such as a marine navigation buoy floats in water, it has a tendency to oscillate vertically with a steady rhythm. To understand how this happens we use the **Archimedes principle,** which states that an object immersed partly or entirely in a fluid is buoyed up by a force equal to the weight of fluid displaced. In particular, an object of mass m floating in equilibrium must be buoyed (up) by a force equal to its own weight $w = mg$ (acting down); the buoyant force has magnitude equal to the weight of displaced fluid. If the object is depressed below its equilibrium level, the net upward force equals the weight of the additional displaced fluid. On the other hand, in an elevated position the force of gravity (acting down) exceeds the buoyant force by the excess of the object's weight over the reduced weight of the displaced fluid.

We'll discuss a simple case in which the surface of the water remains flat and a buoy floats as a vertical cylinder. Choose a vertical coordinate line with origin $y = 0$ at the equilibrium position of the bottom of the buoy and positive y values directed up. In a depressed position ($y < 0$) the additional displaced water has weight $-\rho S y$, where ρ is the weight density of water and S is the cross-sectional area of the cylindrical buoy. The resulting upward directed force can be equated to mass times acceleration:

$$m \frac{dy^2}{dt^2} = -\rho S y.$$

When $y > 0$, the term on the right is negative, consistent with the expected excess downward force of gravity, so the differential equation is valid in all cases. The solutions of this equation all have the form

$$y(t) = c_1 \cos \sqrt{\frac{\rho S}{m}} \, t + c_2 \sin \sqrt{\frac{\rho S}{m}} \, t$$

$$= A \cos \left(\sqrt{\frac{\rho S}{m}} \, t - \phi \right).$$

The amplitude $A = \sqrt{c_1^2 + c_2^2}$ of the oscillations is completely determined by initial conditions. The period of oscillation will be $p = 2\pi \sqrt{m/(\rho S)}$ time units. Rapid oscillation goes with a small period, which is what we get with a fluid of high density or a buoy of large cross section S. On the other hand, a relatively massive buoy will oscillate fairly slowly.

Note that since the buoy mass m and the mass density ρ of water appear on both sides of the differential equation, we can just as well use units of weight, as measured in air, for both of these. Since water weighs about 62.5 pounds per cubic foot, a 2000-pound buoy with a 2-foot radius will have a natural oscillation period of about $2\pi \sqrt{2000/(62.5 \cdot 4\pi)}\phi \approx 10$ seconds.

Damped Oscillation. The piston shown in Figure 4.3 exerts a damping force that can be varied by changing the viscosity of the medium in which the piston moves. If we continue to assume that $f = 0$ in Equation 3.1, then we have to deal with the differential equation

$$\frac{d^2 x}{dt^2} + \frac{k}{m} \frac{dx}{dt} + \frac{h}{m} x = 0.$$

The characteristic equation is

$$r^2 + \frac{k}{m} r + \frac{h}{m} = 0,$$

which has roots

$$r_1 = \frac{1}{2m} (-k + \sqrt{k^2 - 4mh}), \qquad r_2 = \frac{1}{2m} (-k - \sqrt{k^2 - 4mh}).$$

We can distinguish three distinct cases depending on the discriminant $k^2 - 4mh$.

Overdamping, $k^2 - 4mh > 0$. Because we have assumed that k and h are both positive, this case occurs when $k > 2\sqrt{mh}$. Physically, this inequality means that the friction constant k exceeds the constant $\sqrt{mh}$, depending on the spring stiffness h and the mass m, by a factor more than 2. The effect of the assumption $k > 2\sqrt{mh}$ is to make the roots of the characteristic equation satisfy $r_2 < r_1 < 0$. As a result, the general solution is

$$x(t) = c_1 e^{r_1 t} + c_2 e^{r_2 t},$$

where the exponentials decrease as t increases. It follows that in case of overdamping the solution really fails to oscillate at all in the sense described at the beginning of the section.

e x a m p l e 4

In the previous discussion we assumed $m = 2$, $h = 2$, and $k = 5$ so that $k > 2\sqrt{mh}$. The characteristic equation is $2r^2 + 5r + 2 = 0$, with roots $r_1 = -\frac{1}{2}$, $r_2 = -2$, so that the displacement from equilibrium at time t is

$$x(t) = c_1 e^{-t/2} + c_2 e^{-2t}.$$

A typical graph is shown in Figure 4.6a. The maximum displacement occurs at just one point, after which the displacement tends steadily to zero as t increases.

Underdamping, $k^2 - 4mh < 0$. This case occurs when $k < 2\sqrt{mh}$ so that, relative to $\sqrt{mh}$, the friction constant k is small. A typical solution is illustrated in Figure 4.6b. The characteristic roots are now complex conjugates of one another:

$$r_1 = \frac{1}{2m}(-k + i\sqrt{4mh - k^2}), \qquad r_2 = \frac{1}{2m}(-k - i\sqrt{4mh - k^2}).$$

The general form of the displacement function is then

$$x(t) = e^{-kt/2m}\left(c_1 \cos\frac{\sqrt{4mh - k^2}}{2m}t + c_2 \sin\frac{\sqrt{4mh - k^2}}{2m}t\right)$$

$$= Ae^{-kt/2m}\cos\left(\frac{\sqrt{4mh - k^2}}{2m}t - \phi\right),$$

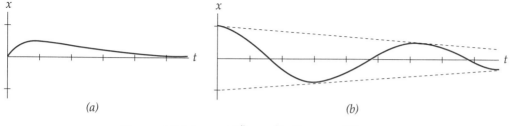

(a)   (b)

Figure 4.6 (a) $x = e^{-t/2} - e^{-2t}$. (b) $x = e^{-t/10}\cos t$.

e x a m p l e 5

Take $h = 2$, $m = 1$, and $k = 2$. Then $k < 2\sqrt{mh}$, and the displacement at time t is

$$x(t) = Ae^{-t}\cos(t - \phi).$$

Figure 4.6 shows the graph of such a function with $A = 1$ and $\phi = 0$. It is easy to see that this choice for the constants A and ϕ gives a solution satisfying the initial conditions

$$x(0) = 1, \qquad \frac{dx}{dt}(0) = -1.$$

Critical Damping, $k = 2\sqrt{mh}$. This case lies between overdamping and underdamping, and it is unstable in the sense that an arbitrarily small change in one of the parameters k, m, or h will disturb the equality $k = 2\sqrt{mh}$ and produce one of the other two cases. Numerically, the case of critical damping is distinguished by the equality of the characteristic roots: $r_1 = r_2 = -k/2m$. It follows that the displacement function is given by

$$x(t) = c_1 t e^{-kt/2m} + c_2 e^{-kt/2m}.$$

example 6

Take $m = h = 1$ and $k = 2$. Then

$$x(t) = (c_1 t + c_2)e^{-t}.$$

If $x(0) = x_0$ and $dx/dt(0) = v_0$, then

$$c_1 = (v_0 + x_0), \qquad c_2 = x_0$$

so that

$$x(t) = [(v_0 + x_0)t + x_0]e^{-t}.$$

Figure 4.7 shows four possibilities, depending on the size of v_0, the initial velocity.

A critically damped displacement is like an overdamped one in that there is no oscillation from one side to the other of the equilibrium position. But with fixed mass m and spring constant h, the critical viscosity value $k = 2\sqrt{mh}$ generally produces a relatively efficient return toward equilibrium: the exponen-

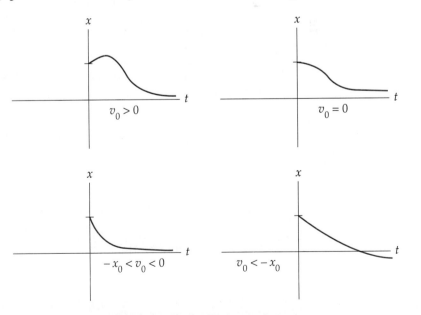

Figure 4.7 Critically damped states.

tial decay rate for critical damping and underdamping is $\frac{1}{2}k/m$, while for over-damping it is $\frac{1}{2}(k + \sqrt{k^2 - 4mh})/m$. In physical terms a higher viscosity than the critical value $k = 2\sqrt{mh}$ produces a more sluggish return from a nonequilibrium value, and lower viscosity allows oscillation.

Forced Oscillation. In the specific instances considered so far, the differential equation

$$m\frac{d^2x}{dt^2} + k\frac{dx}{dt} + hx = f(t)$$

has been subject to initial conditions $x(0) = x_0$, $dx/dt(0) = v_0$, but the external force function f has been assumed to be identically zero. The resulting free oscillation is described by a solution of a homogeneous differential equation. (Note that "free" does not mean "undamped.") When f is not identically zero, we speak of **forced oscillation**. From a purely mathematical point of view, there is no reason why the function f on the right side of the preceding differential equation cannot be chosen to be, say, an arbitrary continuous function for $t \geq 0$. However, a force function f that assumes large values could easily drive the oscillations quite outside the range in which we can maintain the original assumptions used to derive the differential equation. (For example, stretching a spring too far might change its characteristics to the point of destroying its elasticity altogether.) For this reason, the function f is chosen in the examples and exercises to have a rather restricted range of values. In every example, of course, we can use the decomposition of a solution $x(t)$ into homogeneous and particular parts,

$$x(t) = x_h(t) + x_p(t).$$

Fortunately, $x_h(t)$ has already been discussed earlier in this section for various choices of m, k, and h in the homogeneous differential equation. What remains to be done, then, is to discuss the effect of adding a solution of the nonhomogeneous equation. If $k > 0$, the analysis given in the earlier examples shows that every homogeneous solution tends to zero like an exponential of the form $e^{-kt/(2m)}$. Thus, for values of t that make $kt/(2m)$ moderately large, the addition of the homogeneous solution has a negligible effect. Such an effect is called **transient,** and the particular solution is therefore called the **steady-state** solution.

e x a m p l e 7   If $f(t) = a_0 \cos \omega t$, then the differential equation

$$m\frac{d^2x}{dt^2} + k\frac{dx}{dt} + hx = a_0 \cos \omega t, \qquad k > 0,$$

has a particular solution of the form

$$x_p(t) = A \cos \omega t + B \sin \omega t$$

that can be found by the method of undetermined coefficients. Substitution of x_p into the equation yields

$$(h - \omega^2 m)A + \omega k B = a_0,$$
$$\omega k A - (h - \omega^2 m)B = 0.$$

Solving this pair of equations for A and B gives

$$A = \frac{(h - \omega^2 m)a_0}{(h - \omega^2 m)^2 + \omega^2 k^2}, \qquad B = \frac{\omega k a_0}{(h - \omega^2 m)^2 + \omega^2 k^2},$$

so the particular solution we get is

$$x_p(t) = \frac{a_0}{(h - \omega^2 m)^2 + \omega^2 k^2}((h - \omega^2 m)\cos \omega t + \omega k \sin \omega t)$$

$$= \frac{a_0}{\sqrt{(h - \omega^2 m)^2 + \omega^2 k^2}}\cos(\omega t - \phi),$$

where $\phi = \arctan[\omega k/(h - \omega^2 m)]$. What we have found is just one of many solutions, each satisfying a different set of initial conditions. However, our particular solution is the steady-state solution. Notice that $x_p(t)$ is very much like the external force $f(t) = a_0 \cos \omega t$. In fact the two amplitudes differ only by the factor $\rho(\omega) = ((h - \omega^2 m)^2 + k^2 \omega^2)^{-1/2}$.

There are two problems that often occur in the design of a damped oscillator. One problem is making the amplitude of the steady-state response $x_p(t)$, called the **response amplitude,** large relative to the amplitude of the external forcing amplitude a_0. For example, in the design of a sensitive seismograph the mechanism needs to be able to "amplify" a small vibration, i.e., to increase its amplitude to the point where it can be detected mechanically or electrically. On the other hand, we may want to design a detector that can receive an input $f(t)$ of large amplitude and then reduce, or "attenuate," the amplitude of the output $x_p(t)$ to manageable size while maintaining the other characteristics of $f(t)$. In either case a key formula is the factor $\rho(\omega)$ derived in Example 7. To amplify we make ρ large, and to attenuate we make ρ small.

Note that the frequency ω of the steady-state response $x_p(t)$ is the same as the frequency of the external forcing function $a_0 \cos \omega t$ so that in the long run all frequencies are to a large extent transmitted by the mechanism. The phenomenon called **resonance** can be defined to be a relationship between a mechanism and an external oscillatory force $f(t)$ such that the steady-state response $x_p(t)$ of the mechanism oscillates with a significantly large amplitude. The test for what constitutes "significantly large" depends on the physical characteristics of the mechanism.

---

e x a m p l e **8** The expression

$$\rho(\omega) = \frac{1}{\sqrt{(h - \omega^2 m)^2 + k^2 \omega^2}}$$

in Example 7 represents the ratio of the amplitudes of $x_p(t)$ and $f(t)$. It's reasonable to say that resonance occurs for the mechanism if the ratio $\rho(\omega)$ is larger than 1, in which case we have **amplification,** that is, an increase in amplitude. More particularly, if we choose $\omega = \sqrt{h/m}$, then the first term inside the square root becomes zero, and we are left with

$$\rho(\omega) = \frac{1}{\omega k} = \frac{1}{k} \sqrt{\frac{m}{h}}.$$

If $\rho(\omega) > 1$, that is, if $\sqrt{m} > k\sqrt{h}$, we would say that ω is a **resonance** frequency for the mechanism.

Some people prefer to use the term "resonance" only in cases for which the response amplitude is maximized under various conditions, but the usual usage is our less restrictive one in which the output amplitude is bigger than the input amplitude. Nevertheless, the problem of maximizing response amplitude by maximizing the factor $\rho(\omega)$ is a significant one and is taken up in the exercises. Figure 4.8 shows the graphs of the amplification factor $\rho(\omega) = 1/\sqrt{(h - \omega^2 m)^2 + k^2\omega^2}$ with the choice $h = m = 1$ for four different values of k. Finally, if $\omega = \sqrt{h/m}$ and $\rho < 1$, we get reduced response amplitude as compared with input amplitude, called **attenuation,** rather than amplification.

Beats (Optional). The term **beats** refers to a phenomenon in which a relatively high frequency oscillation has an amplitude that varies in such a way that the successive local maximum and minimum values of the oscillation partially delineate another oscillation of lower frequency. Figure 4.9 shows the graph of such an oscillation, on which we have superimposed smooth dotted curves that outline the paths of the extreme values. If the underlying oscillation repre-

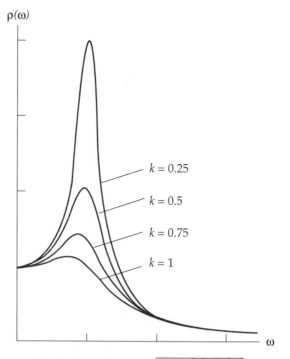

$\rho(\omega)$

$k = 0.25$

$k = 0.5$

$k = 0.75$

$k = 1$

ω

Figure 4.8 $\rho(\omega) = 1/\sqrt{(1 - \omega^2)^2 + k^2\omega^2}$.

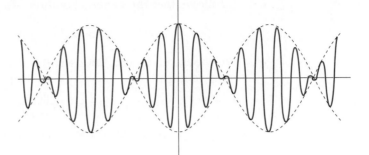

Figure 4.9 Graph of $\frac{1}{2}\cos 22t + \frac{1}{2}\cos 18t = \cos 20t \cos 2t$.

sents a sonic vibration, these outlining or "enveloping" oscillations can sometimes be heard as a distinct lower-frequency sound, in which case a listener is said to be *hearing beats*. In electric circuits there is an analog to beats that often goes by the name *amplitude modulation*.

Beats can be generated by adding or subtracting two harmonic oscillations with relatively high but distinct frequencies. For example, the identity

$$\sin \omega_0 t - \sin \omega t = 2 \sin \tfrac{1}{2}(\omega_0 - \omega)t \cos \tfrac{1}{2}(\omega_0 + \omega)t$$

exhibits the difference of two sine oscillations, with circular frequencies ω_0 and ω, as a single cosine oscillation with circular frequency $\frac{1}{2}(\omega_0 + \omega)$, and with an amplitude $A(t) = 2 \sin \frac{1}{2}(\omega_0 - \omega)t$ that varies with time. (A suggestion for deriving this and similar identities is given in the exercises.) The frequency of the beat oscillation is just $\frac{1}{2}(\omega_0 - \omega)$. Thus a low frequency (i.e., a low note in the case of an audible sound) will be produced by making $\omega_0 - \omega$ small. It could happen that the high-frequency oscillation, with frequency $\frac{1}{2}(\omega_0 + \omega)$, produces a sound outside anyone's hearing range but that the beat tone is audible. The next example describes an ideal mechanism for producing beats.

e x a m p l e **9**   Consider an undamped oscillator driven by an external force $f(t)$:

$$m \frac{d^2x}{dt^2} + hx = f(t).$$

The solutions of the associated homogeneous equation all have the form

$$x_h = c_1 \cos \sqrt{\frac{h}{m}}t + \sin \sqrt{\frac{h}{m}}t$$

$$= c_1 \cos \omega_0 t + c_2 \sin \omega_0 t, \qquad \text{where } \omega_0 = \sqrt{\frac{h}{m}}.$$

We'll assume $\omega \neq \omega_0$ and let $f(t) = a_0 \cos \omega t$. It's easy to verify that, since $h = m\omega_0^2$, a particular solution of the differential equation is

$$x_p = \frac{a_0}{m(\omega_0^2 - \omega^2)} \cos \omega t.$$

It follows that the general solution can be written in the form $x = x_h + x_p$, or

$$x(t) = c_1 \cos \omega_0 t + c_2 \sin \omega_0 t + \frac{a_0}{m(\omega_0^2 - \omega^2)} \cos \omega t.$$

We choose initial conditions $x(0) = 0$ and $\dot{x}(0) = 0$ and check that these conditions require

$$c_1 = \frac{-a_0}{m(\omega_0^2 - \omega^2)} \qquad \text{and} \qquad c_2 = 0.$$

The unique solution to the initial-value problem is then

$$x(t) = \frac{a_0}{m(\omega^2 - \omega_0^2)} (\cos \omega_0 t - \cos \omega t).$$

This formula expresses the response to the driving force as the difference of two periodic functions with the same amplitude but different circular frequencies ω_0 and ω. The appropriate trigonometric identity for understanding this oscillation is treated in Exercise 21:

$$\cos \omega_0 t - \cos \omega t = 2 \sin \tfrac{1}{2}(\omega - \omega_0)t \sin \tfrac{1}{2}(\omega + \omega_0)t;$$

using it we find

$$x(t) = \frac{2a_0}{m(\omega^2 - \omega_0^2)} \sin \tfrac{1}{2}(\omega - \omega_0)t \sin \tfrac{1}{2}(\omega + \omega_0)t.$$

If ω is close to ω_0, we get a low-frequency *beat tone* of the form

$$b(t) = \frac{2a_0}{m(\omega^2 - \omega_0^2)} \sin \tfrac{1}{2}(\omega - \omega_0)t,$$

that modifies, or *modulates*, the higher-frequency oscillation $\sin \tfrac{1}{2}(\omega_0 + \omega)t$.

EXERCISES

1. Each of the following differential equations represents a free oscillation. Classify them according to type: harmonic, overdamped, underdamped, or critically damped. (*Note*: $\dot{x} = dx/dt$, $\ddot{x} = d^2x/dt^2$.)

   (a) $d^2x/dt^2 + 2dx/dt + x = 0$.

   (b) $d^2x/dt^2 + 2dx/dt + 2x = 0$.

   (c) $\ddot{x} + 9x = 0$.

   (d) $\ddot{x} + 3\dot{x} + x = 0$.

   (e) $\ddot{x} + a\dot{x} + a^2x = 0$, $a > 0$.

   (f) $d^2x/dt^2 + \tfrac{1}{4}dx/dt + \tfrac{1}{8}x = 0$.

2. Find the general solution for each differential equation in Exercise 1.

3. For each of the following general solutions of a second-order constant-coefficient equation, find the choice of the arbitrary constants that satisfies the corresponding initial conditions. Sketch the solution that you find. Also find a differential equation that the solution satisfies.

   (a) $x(t) = c_1 \cos 2t + c_2 \sin 2t$; $x(0) = 0$, $dx/dt(0) = 1$.

   (b) $x(t) = A \cos (3t - \phi)$; $x(0) = 1$, $dx/dt(0) = -1$.

   (c) $x(t) = c_1 t e^{-2t} + c_2 e^{-2t}$; $x(0) = 0$, $dx/dt(0) = -1$.

   (d) $x(t) = A e^{-t} \cos (2t - \phi)$; $x(0) = 2$, $dx/dt(0) = 2$.

   (e) $x(t) = c_1 e^{-2t} + c_2 e^{-4t}$; $x(0) = -1$, $dx/dt(0) = -1$.

4. Find the steady-state solution to each of the following differential equations. Also estimate the earliest time beyond which the transient solution remains less than 0.01, assuming initial conditions $x(0) = 1$, $\dot{x}(0) = 0$.

   (a) $d^2x/dt^2 + 2dx/dt + 2x = 2 \cos 3t$. [*Hint:* Show $|x(t)| \leq \sqrt{2}e^{-t}$.]

   (b) $d^2x/dt^2 + 3dx/dt + 2x = \cos t$.

5. There is a well-known analogy between the behavior of a damped mass-spring system and of an *RLC* electric circuit. Here L is the inductance (analog of mass) of a coil, R is the resistance (analog of friction constant) in the circuit, and C is the capacitance (analog of reciprocal of spring stiffness) or ability of a capacitor to store a charge. The differential equation satisfied by the charge $Q(t)$ on the capacitor at time t is

$$L\frac{d^2Q}{dt^2} + R\frac{dQ}{dt} + \frac{1}{C}Q = E(t),$$

   where $E(t)$ is the voltage impressed on the circuit from an external source. (See Figure 4.10c.) The charge $Q(t)$ is related to the current flow $I(t)$ by $I = dQ/dt$.

   (a) Derive the relations that must hold between $R, L,$ and C in order that the response of $Q(t)$ should be, respectively, underdamped, critically damped, and overdamped.

   (b) Show that if $C = \infty$ (capacitor is absent), the equation for $I(t)$ is $L\,dI/dt + RI = E(t)$.

   (c) Solve the equation in part (b) when $E(t) = E \sin \omega t$ and $I(0) = 0$, and show that, if t is large enough, the current response differs negligibly from

$$\frac{E}{Z} \sin (\omega t - \theta),$$

   where $Z = \sqrt{R^2 + \omega^2 L^2}$ and $\cos \theta = R/Z$. The function $Z(\omega)$ is called the **impedance** of the circuit in response to the sinusoidal input of frequency ω.

   (d) Show that, if $0 < C < \infty$, a long-term current response to input voltage $E(t) = E \sin \omega t$ is $E(t - \alpha)/Z(\omega)$, where $Z(\omega) = \sqrt{R^2 + [\omega L - 1/(\omega C)]^2}$.

6. The range of validity of Hooke's law for a given spring, and the corresponding value of a Hooke constant h, can be determined as follows. Hang the spring with known weights $W_j = m_j g, j = 1, \ldots, n$, attached to the free end. If the additional weight is always proportional to the additional extension, then Hooke's law is valid for this range of extensions and h is the constant of proportionality. A similar procedure applies to compression.

   (a) Assume distance units are in feet. A spring with a 5-pound weight appended has length 6 inches but with an 8-pound weight appended has length 1 foot. If the spring satisfies Hooke's law with constant h, find h.

   (b) What if distance is measured in meters and force in kilograms in part (a)? (There are about 3.28 feet in a meter and 2.2 pounds in a kilogram.)

   (c) Suppose we know Hooke's constant to be $h = 120$ for a certain spring. We observe that between hanging a 20-pound weight from it and then a larger weight we get an additional extension of 6 inches. How big is the larger weight?

(d) A spring is compressed to length 20 centimeters by a force of 5 kilograms, to length 10 centimeters by a force of 6 kilograms, and to 5 centimeters by a force of 7 kilograms. Discuss the possible validity of Hooke's law given this information.

7. The equation $\ddot{x} + hx = \cos \omega t$ has solutions $x_p(t) = (h - \omega^2)^{-1} \cos \omega t$ if $\omega^2 \neq h$. (This follows from Example 7 if we extend it to the case $k = 0$.)

(a) If $h = 2$, what is a value of ω that will produce a response of amplitude 4 in x_p?

(b) If $\omega = 2$, what is a value of h that will produce a response of amplitude 5 in x_p?

(c) If $h = 2$, what is the unique positive value of ω for which solutions to $\ddot{x} + hx = \cos \omega t$ become unbounded as $t \to \infty$?

(d) If $\omega = 10$, what is the range of h values that will produce response amplitudes for x_p above 4?

8. Answer the following questions about solutions $x(t)$ to the damped and unforced equation $m\ddot{x} + k\dot{x} + hx = 0$, where m, k, and h are positive constants.

(a) If $m = 2$, how should h and k be related so that the nontrivial solutions will be oscillatory?

(b) If the friction and Hooke constants satisfy $h = k = 1$, how should the mass m be chosen so that all nontrivial solutions will oscillate?

(c) If $m = h = 1$, how should k be chosen so that $x(t)$ will have a factor with circular frequency $\frac{1}{2}$?

(d) If $m = h = 1$ how should k be chosen so that $x(t)$ is oscillatory?

9. Answer the following questions about solutions $x(t)$ of the damped and forced equation $m\ddot{x} + k\dot{x} + hx = \cos \omega t$, where $h > 0$ and $k > 0$.

(a) If $m = k = h = 1$, how should ω be chosen so that the amplitude of the steady-state solution will be 1?

(b) If $k = \omega = 1$, what relation must hold between m and h so that the steady-state amplitude will be 1?

(c) If $m = h = \omega = 1$, how should k be chosen so that the amplitude of the steady-state solution will be 2?

(d) What polynomial relation must hold among m, h, k, and ω if the frequency of the transient solution is to be the same as the steady-state frequency? As a special case, show that if the homogeneous solutions have circular frequency ω, and also $h = m\omega^2$, then $k = 0$, so there is no damping and no nontrivial transient solution.

10. Find the amplitude A, the frequency $\omega/2\pi$, and a phase angle ϕ for each of the following periodic functions.

(a) $2 \cos t + 3 \sin t$.   **(c)** $\sin \pi t + 2 \cos \pi t$.

(b) $-2 \cos 2t + 3 \sin 3t$.

11. By how much are each of the following pairs of oscillations out of phase? You can decide this by expressing each pair in the general form $A \cos(\omega t - \phi)$, $B \cos(\omega t - \psi)$.

(a) $(\sqrt{3}/2) \cos t + \frac{1}{2} \sin t$, $\cos t$.

(b) $\frac{1}{2} \cos t + (\sqrt{3}/2) \sin t$, $\cos t$.

(c) $(1/\sqrt{2}) \cos t + (1/\sqrt{2}) \sin t$, $\sin t$.

12. A weight of mass $m = 1$ is attached by springs with Hooke constants h_1, h_2 to two fixed vertical supports. The weight oscillates along a horizontal line with negligible friction. By analyzing the force due to each spring, show that the displacement $x = x(t)$ of the weight from equilibrium satisfies $\ddot{x} = -(h_1 + h_2)x$.

13. A weight of mass $m = 1$ is attached by springs with Hooke constants h_1, h_2 to two fixed vertical supports. The weight oscillates along a horizontal line with negligible friction. Let the respective unstressed lengths of the two springs be l_1, l_2, and let b denote the distance between the supports.

(a) By analyzing the force due to each spring, show that the displacement $x = x(t)$ of the weight from the support attached to the first spring satisfies

$$\ddot{x} = -(h_1 + h_2)x + h_1 l_1 + h_2(b - l_2).$$

(b) Use the result of part (a) to show that the constant solution for $x(t)$ is

$$x_e = \frac{h_1 l_1 + h_2(b - l_2)}{h_1 + h_2}.$$

(c) Show that the constant value x_e of part (b) is the value of the solution to the differential equation that satisfies the two initial conditions $x(0) = x_e$, $\dot{x}(0) = 0$.

*14. A clock pendulum is often suspended from its fixed support by a short segment of flat spring. Let h be the Hooke constant applicable to the distance of deflection of such a pendulum along the path of motion of its center of gravity.

(a) Show that the circular frequency of oscillation is approximately $\sqrt{h/m + g/l}$, where m and l are the mass and length of the pendulum.

(b) Let the motion of such a pendulum be subject to linear damping with damping constant $k > 0$. Show that if oscillation is to take place, it would be desirable to have $4(hm + (g/l)m^2) > k^2$.

15. The recoil mechanism of an artillery piece is designed containing a linearly damped spring mechanism. The spring stiffness h and the damping factor k should be chosen so that after firing the gun barrel will tend to its original position before firing without additional oscillation. We'll assume a given initial velocity V_0 and mass m for the gun barrel during recoil and also a fixed maximum recoil distance E always attained by the barrel.

(a) How should h and k be chosen so that under these conditions the gun barrel undergoes critical damping after firing?

(b) Show that the differential equation for displacement $x(t)$ can be written so as to display dependence only on the parameters V_0 and E.

16. During construction of a suspension bridge, two towers have been erected, and a 10-ton weight is suspended between the towers by a cable anchored to both towers. Because of the elastic properties of the cable and the towers, it takes a $\frac{1}{2}$-ton force to move the weight sideways by 0.1 foot. An earthquake moves the base of each tower sideways with identical displacements of the form $0.25 \cos 6t$ feet in t seconds. Assume a linear model for the lateral force on the weight and that damping is negligible.

   (a) Find Hooke's constant h and the natural unforced frequency of oscillation for the weight.

   (b) Find the amplitude of the quake motion as a function of t, and compare it with the amplitude of the forced oscillation of the weight.

17. The differential equation

   $$m\ddot{x} + k\dot{x} + hx = a_0 \sin \omega t$$

   determines the displacement $x(t)$ of a damped spring with external forcing $f(t) = a_0 \sin \omega t$ as a function of time t. A frequency ω that maximizes the amplitude of $x(t)$ is called a **maximum resonance frequency**. This exercise asks you to investigate the maximum resonance frequency for fixed ω, and under various assumptions about the mechanism.

   (a) Consider the mechanically ideal case where $k = 0$. Show that choosing ω to equal the natural circular frequency $\omega_0 = \sqrt{h/m}$ produces a response $x(t)$ that contains the factor t and hence has deviations from the equilibrium position $x = 0$ that become arbitrarily large as t increases. (Thus there is no theoretical maximum resonance frequency in this case, though in practice the maximum response will be limited by the structural capacity of the mechanism to accommodate wide deviations from equilibrium.)

   (b) Assume that k is a fixed positive number. Show that the steady-state displacement $x_p(t)$ has maximum amplitude when h and m are chosen so that $\sqrt{h/m} = \omega$ and that the maximum amplitude is $a_0(\omega k)^{-1}$. (Making such choices for h and m can be thought of as **tuning** the mechanism for maximum response.)

   (c) Continuing with the ideas of part (b), show that a small response amplitude to a given forcing frequency ω can be achieved by making $|h - \omega^2 m|$ large.

18. In the previous exercise we assumed the input frequency to be fixed and considered the effect of varying h and m in the differential equation. Suppose now that h, m, and k are fixed positive numbers and that we want to choose ω so as to maximize the amplitude of the response $x_p(t)$.

   (a) Show that the amplitude ratio

   $$\rho(\omega) = ((h - \omega^2 m)^2 + k^2 \omega^2)^{-1/2}$$

   of $x_p(t)$ over $f(t)$ is maximized when the function $F(\omega) = (h - \omega^2 m)^2 + k^2 \omega^2$ is minimized.

(b) Show that if $k^2 \le 2hm$, then $F'(\omega) = 0$ when $\omega^2 = (2hm - k^2)/(2m^2)$. Conclude from this that in this case the maximum response amplitude

$$\frac{2m|a_0|}{k\sqrt{4hm - k^2}} \qquad \text{occurs for} \qquad \omega = \omega_0 \sqrt{1 - \frac{k^2}{2hm}},$$

where $\omega_0 = \sqrt{h/m}$ is the natural circular frequency of the undamped ($k = 0$) mechanism.

(c) Use the result of part (b) to show that maximum value of the response amplitude is greater than input amplitude (i.e., there is amplification) if $2m > k\sqrt{4mh - k^2}$. (Here we assume that $4mh - k^2 > 0$, which is equivalent to assuming that the transient response $x_h(t)$ is oscillatory. Why are these two assumptions equivalent?)

(d) Show that if $k^2 \ge 2hm$, then $F'(\omega) = 0$ only when $\omega = 0$ and that $F(\omega)$ is strictly increasing for $\omega \ge 0$. Hence conclude that in this case maximum response is zero and occurs only for the constant forcing function $f(t) = 0$.

19. The purpose of this exercise is to show that if m, k, and h are positive constants, then each particular solution of

$$m\ddot{x} + k\dot{x} + hx = b_0 \sin \omega t$$

remains bounded as $t \to \infty$.

(a) Show that the solutions of the associated homogeneous equation all tend to zero as $t \to \infty$. (These are the transient solutions.)

(b) Show that

$$x_p(t) = \frac{b_0}{k^2\omega^2 + (h - m\omega^2)^2} (-k\omega \cos \omega t + (h - m\omega^2) \sin \omega t)$$

is a particular solution and that it satisfies the inequality $|x_p(t)| \le 1/\sqrt{k^2\omega^2 + (h - m\omega^2)^2}$. [*Hint:* See Example 7 of the text; it's easy.]

(c) Show how to conclude from the results of (a) and (b) that all solutions are bounded.

20. The purpose of this exercise is to observe the effect on the individual solutions of the initial-value problem

$$\ddot{x} + hx = \sin \omega t, \qquad x(0) = \dot{x}(0) = 0$$

of letting the parameter ω approach the positive constant $\sqrt{h}$. (The differential equation represents a highly idealized situation from a physical point of view, because there is no damping term.)

(a) Show that the unique solution to the initial-value problem with $\omega \ne \sqrt{h}$ is

$$x(t) = \frac{1}{h - \omega^2} \left(\sin \omega t - \frac{\omega}{\sqrt{h}} \sin \sqrt{h}t \right),$$

and that the solution satisfies

$$|x(t)| \le \frac{1 + \omega/\sqrt{h}}{|h - \omega^2|} \qquad \text{for all values of } t.$$

(b) Show that as ω approaches $\sqrt{h}$, the solution values found in part (a) approach

$$\frac{1}{2h}(-\sqrt{h}t\cos\sqrt{h}t + \sin\sqrt{h}t).$$

Show also that in contrast to the inequality in part (a), this function oscillates with arbitrarily large amplitude as t tends to infinity.

(c) Find an initial-value problem that has the function obtained in part (b) as a solution. [*Hint:* What happens to the original differential equation as $\omega \to \sqrt{h}$?]

21. Suppose that an undamped, but forced, oscillator has the form

$$\ddot{x} + 2x = \sum_{k=0}^{n} a_k \cos kt.$$

(a) Use the linearity of the differential equation to show that it has the particular solution

$$x_p(t) = \sum_{k=0}^{n} \frac{a_k}{2 - k^2} \cos kt.$$

(The trigonometric sum on the right in the differential equation is an example of a *Fourier series*, discussed in general in Chapter 10. An extension of such a sum to an infinite series can be used to represent a very general class of functions.)

(b) How does the solution in part (a) change if the left side of the differential equation is replaced by $\ddot{x} + 4x$ and the right side remains the same?

22. A 64-pound cylindrical buoy with a 2-foot radius floats in water with its axis vertical. Assume the buoy is depressed $\frac{1}{2}$ foot from equilibrium and then released.

(a) Find the differential equation for the displacement from equilibrium during undamped motion.

(b) Find the amplitude and period of the motion.

23. A cylindrical buoy with a 1-foot radius floats in water with its axis vertical. Assume the buoy is depressed from equilibrium, released, and allowed to oscillate undamped. If the period of oscillation is 4 seconds, how much does the buoy weigh?

24. (a) Prove the identity $\cos\alpha - \cos\beta = -2\sin\frac{1}{2}(\alpha + \beta)\sin\frac{1}{2}(\alpha - \beta)$, used in the analysis of beats. [*Hint:* Apply the addition formula for the cosine function to $\cos((\alpha + \beta)/2 + (\alpha - \beta)/2) = \cos\alpha$ and similarly to $\cos((\alpha + \beta)/2 - (\alpha - \beta)/2) = \cos\beta$.]

(b) Prove the identity $\sin\alpha - \sin\beta = 2\sin\frac{1}{2}(\alpha - \beta)\cos\frac{1}{2}(\alpha + \beta)$, used in the analysis of beats. [*Hint:* Apply the addition formula for the cosine function to $\cos((\alpha + \beta)/2 + (\alpha - \beta)/2) = \cos\alpha$ and similarly to $\cos((\alpha + \beta)/2 - (\alpha - \beta)/2) = \cos\beta$.]

25. Consider the differential equation

$$\ddot{x} + 25x = 16\cos 3t.$$

(a) Show that the equation has general solution
$$x(t) = c_1 \cos 5t + c_2 \sin 5t + \cos 3t.$$

(b) Show that the particular solution satisfying $x(0) = 0$, $\dot{x}(0) = 0$ is
$x_p(t) = \cos 3t - \cos 5t$.

(c) Show that $\cos 3t - \cos 5t = 2 \sin 4t \sin t$. [*Hint:* Use an identity from the previous problem.]

(d) Use the result of part (c) to sketch the graph of the particular solution found in part (b) for $0 \le t \le 2\pi$.

26. (a) The phase difference between $\cos(\omega t - \alpha)$ and $\cos(\omega t - \beta)$ is $\alpha - \beta$. What is the *time* shift required to put the two oscillations in phase?

(b) What is the phase difference between $\cos(\omega t - \alpha)$ and $\sin(\omega t - \beta)$? [*Hint:* Express the second one in terms of cosine.]

27. Let $f(t) = \sin \alpha t + \sin \beta t$, where α and β are positive numbers.

(a) Show that if $\beta = r\alpha$ for some rational number r, then $f(t)$ is periodic for some period $p > 0$; that is, $f(t + p) = f(t)$ for all t. Show also that p can be expressed as (possibly different) integer multiples of both π/α and π/β.

*(b) Prove that if an $f(t)$ of the form given above is periodic with period $p > 0$, then $\beta = r\alpha$ for some rational number r. Thus, for example, $\sin t + \sin \sqrt{2}t$ can't be periodic. [*Hint:* Check that $f(p) = 0$ and $f''(p) = 0$. Then conclude that $(\alpha^2 - \beta^2) \sin \alpha p = (\alpha^2 - \beta^2) \sin \beta p = 0$ so that either $\alpha = \pm\beta$ or αp and βp are integer multiples of π.]

28. Suppose a physical process is accurately modeled by a differential equation of the form
$$m \frac{d^2x}{dt^2} + k \frac{dx}{dt} + hx = 0,$$

with m, k, and h positive constants. It may be possible by observation to draw conclusions about the parameters in the underlying process. Given each of the following sets of information about the constants and a solution, find the implications for the other constants.

(a) $m = 1$, $x(t) = e^{-3t} \cos 6t$.

(b) $h = 1$, $x(t) = e^{-t} \sin 5t$.

(c) $k = 1$, $x(t) = e^{-t/2} \cos(t/2)$.

(d) $k = 3$, $h = 2$, $x(t) = e^{-4t} \sin 4t$.

29. Suppose we want to construct a damped harmonic oscillator with Hooke constant $h = 2$ and damping constant $k = 3$. What is the lower limit m_0 for the mass m such that oscillatory solutions are possible? Does oscillation occur for $m = m_0$?

30. A weight of mass m is attached to an elastic cord of length l that obeys Hooke's law with constant h. The other end of the cord is attached to the top of a tower from which the weight is then dropped. Assume negligible air resistance. (In practice this is sometimes done with a long **bungee cord.**)

(a) What is the velocity of the weight when the cord first begins to take the strain of the fall?

(b) Show that from the time the cord begins to stretch until the weight bounces back up again the cord's extension beyond its relaxed state is given by

$$y(t) = \frac{gm}{h} \left(1 - \cos \sqrt{\frac{h}{m}} t \right) + \sqrt{\frac{2glm}{h}} \sin \sqrt{\frac{h}{m}} t.$$

(c) Show that the maximum extension of the cord from its relaxed state is

$$y_{\max} = \frac{gm + \sqrt{g^2m^2 + 2glmh}}{h}.$$

[*Hint:* Express the result of part (b) in terms of phase and amplitude.]

(d) Show that the maximum possible extent of the oscillation is $2\sqrt{g^2m^2 + 2glmh}/h$.

(e) For some types of elastic cord material, Hooke's constant h is inversely proportional to the length l of the cord. Show that, for cords made with a given type of such material, the maximum extension resulting from the experiment described above is proportional to l. Show also that the maximum stretching force experienced with cords of different unstressed lengths l is independent of l. The latter fact has made it possible for people to make a stunt out of the experiment by jumping off a tower while tied to a very long cord, but with limited physical stress to the jumper. (There are no known limits to the mental stress involved.)

4   ELECTRIC CIRCUITS

The study of electric circuits, and of the more general networks taken up in Chapter 6, Section 3, has to do with time-dependent flow of **current** $I(t)$ in conductors. Electric circuits discussed in this section can all be viewed as consisting of a single wire loop containing a (possibly time-dependent) source of voltage, or electric force, with **voltage** $E(t)$, a resistor with **resistance** R, a capacitor capable of storing electric charge $Q(t)$ and having **capacitance** C, and an inductor coil with **inductance** L that induces additional voltage in response to a decreasing current, as well as inducing a decrease in voltage in response to an increasing current. (Thus an inductor has the effect of smoothing out the current and making it more nearly constant.) Direction of current flow will be described relative to a fixed loop direction; in our single-loop diagrams, such as the ones in Figure 4.10, this direction has been consistently chosen to be

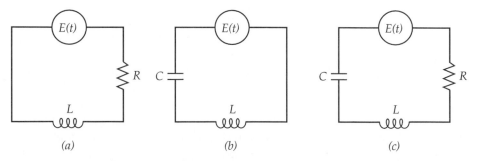

(a)   *(b)*   *(c)*

Figure 4.10 *(a)* LR circuit; *(b)* LC circuit; *(c)* LRC circuit.

clockwise. With this convention, current flowing in the clockwise direction will be designated **positive,** and the reverse flow will be **negative.** A circuit containing resistance, inductance, and capacitance so that L, R, and C are all positive is sometimes called an LRC **circuit.**

The relation between circuit elements that allows us to predict the values $I(t)$ of current is given by the following law.

■ KIRCHHOFF'S LOOP LAW

The sum of voltage differences across the elements of a closed loop is equal to the voltage source in the loop.

_____

The **voltage differences** referred to in the loop law are due to the presence of resistance, inductance, or capacitance in the loop. The differences are related to current I in the following ways:

Voltage difference $E_R = IR$ is due to resistance R.

Voltage difference $E_L = L\, dI/dt$ is due to inductance L.

Voltage difference $E_C = Q/C$ is due to capacitance C, where Q is the charge on the capacitor, and charge is related to current by

■ 4.1
$$I = \frac{dQ}{dt}.$$

Remarks

1. Equation 4.1 expresses the definition of **current** $I(t)$ in a loop as the rate of motion of positive charge $Q(t)$ through a cross section of wire in the loop. (According to the standard convention, electrons, which have negative charge, move opposite to current.)
2. The constants L, R, C are never negative; making one of them zero is equivalent to asserting that the corresponding type of circuit element is not included in the loop.
3. Note that for a steady, i.e., constant, current, we have $dI/dt = 0$ so that there is no voltage difference across an inductor in that case. For this reason it makes sense to include an inductor only in circuits in which the current varies over time, the particular case of most interest being a circuit driven by an alternating voltage source, for example, $E(t) = E_0 \sin \omega t$.
4. Many people denote current by i rather than I and then write $\sqrt{-1} = j$ instead of $\sqrt{-1} = i$. We've kept the latter convention, and so use I for current.

_____

e x a m p l e 1

The loop in Figure 4.10a contains, in addition to the voltage source, only resistance and inductance, so the loop law dictates that

$$L\frac{dI}{dt} + RI = F.$$

This first-order equation has an elementary solution if L, R, and E are constant:

$$I = \frac{E}{R} + c_1 e^{-Rt/L}.$$

The constant c_1 can be determined by an initial condition of the form $I(0) =$ const. Given that R and L are positive constants, the solution formula predicts that the current will approach the value E/R exponentially as t tends to infinity. Note that if $L = 0$, the differential equation reduces to a purely algebraic relation from which we deduce that $I = E/R$ for all values of t. This last equation is the statement of **Ohm's law.**

e x a m p l e **2**

Computing $I(t)$ was particularly simple in Example 1 because we assumed that E was constant. However, if we have a variable voltage input, say from a microphone, of the form $E(t) = E_0 \cos \omega t$, then we resort to finding an exponential multiplier to solve the resulting first-order linear equation in the standard form

$$\dot{I} + \frac{R}{L} I = \frac{E_0}{L} \cos \omega t.$$

We find $M(t) = e^{Rt/L}$, with the result that

$$\frac{d}{dt} (e^{Rt/L})I = \frac{E_0}{L} e^{Rt/L} \cos \omega t.$$

The integral of the function of t on the right is computed using two integrations by parts and is available from tables or symbolic computing programs:

$$e^{Rt/L}I = \frac{(E_0/L)e^{Rt/L}}{R^2/L^2 + \omega^2} \left(\frac{R}{L} \cos \omega t + \omega \sin \omega t \right) + c_1.$$

Hence,
$$I(t) = \frac{E_0}{R^2 + L^2\omega^2} (R \cos \omega t + L\omega \sin \omega t) + c_1 e^{-Rt/L}.$$

What we learn from this formula is that, just as with a constant input voltage, there is a **transient** term, i.e., one that dies off to zero as t tends to infinity. However, the other term is periodic with the same period ω as the input voltage. If we write the formula for current in terms of a phase angle α, we get

$$I(t) = \frac{E_0}{\sqrt{R^2 + L^2\omega^2}} \cos (\omega t - \alpha) + c_1 e^{-Rt/L}, \qquad \text{where } \alpha = \arctan \frac{L\omega}{R}.$$

From this formula we learn how the phase angle α depends on the design of the circuit: to increase α, increase L, or reduce R. Similar interpretations are taken up in the exercises.

e x a m p l e **3**

The loop in Figure 4.10b contains all the basic circuit elements except resistance. In classical circuits such a state of affairs is impossible; there is always some resistance. However, we can think of such a circuit as being "superconducting" in the sense that the resistance is negligible for our purposes. Furthermore, the simple model we get helps to elucidate the more general one discussed in the following example.

The loop law asserts that

$$L\dot{I} + \frac{1}{C} Q = E(t).$$

To solve this equation for I, or for Q, we need some other relation between those two quantities. That relation is provided by Equation 4.1, $I = \dot{Q}$, referred to above, and we can use it to eliminate I from the loop-law equation, getting

$$L\frac{d^2Q}{dt^2} + \frac{Q}{C} = E.$$

We assume L and C are positive constants, so the characteristic equation $Lr^2 + C^{-1} = 0$ has roots $r_1, r_2 = \pm i/\sqrt{LC}$. The resulting solutions of the homogeneous equation all have the form

$$Q_h = c_1 \cos\frac{t}{\sqrt{LC}} + c_2 \sin\frac{t}{\sqrt{LC}}$$

$$= c_1 \cos\omega_0 t + c_2 \sin\omega_0 t.$$

In the previous line and hereafter we use the abbreviation $\omega_0 = 1/\sqrt{LC}$, because this **natural circular frequency** ω_0 appears in a number of fundamental formulas. The voltage source often supplies an alternating current that can be characterized by voltage $E(t) = E_0 \cos\omega t$. If ω happened to equal the natural circular frequency ω_0, we would expect the nonhomogeneous equation to have unbounded oscillatory solutions containing the factor t. Since this phenomenon can't be realized for very long without exceeding the capacity of a real physical mechanism, we'll assume $\omega \neq \omega_0$; in that case it's easy to check that a particular solution is

$$Q_p = \frac{E_0}{L(\omega_0^2 - \omega^2)}\cos\omega t.$$

It follows that the general solution can be written in the form $Q = Q_h + Q_p$, or

$$Q(t) = c_1 \cos\omega_0 t + c_2 \sin\omega_0 t + \frac{E_0}{L(\omega_0^2 - \omega^2)}\cos\omega t.$$

Initial conditions for $Q(t)$ that are of particular interest are $Q(0) = 0$ (this says that there is no initial charge on the condenser) and $\dot{Q}(0) = 0$ (since $I(0) = \dot{Q}(0) = 0$, this says that the initial current is zero). It is easy to check that these initial conditions require

$$c_1 = \frac{-E_0}{L(\omega_0^2 - \omega^2)}, \qquad c_2 = 0.$$

The unique solution is then

$$Q(t) = \frac{E_0}{L(\omega_0^2 - \omega^2)}(\cos\omega t - \cos\omega_0 t).$$

The previous formula describes the response to the input voltage as the difference of two periodic functions with the same amplitude but different frequencies ω_0 and ω. The current, which is the time derivative of the charge, is then

$$I(t) = \frac{E_0}{L(\omega_0^2 - \omega^2)}(\omega_0 \sin\omega_0 t - \omega \sin\omega t).$$

An analysis of this last formula in Exercise 3 leads to a description of the phenomenon of *amplitude modulation*.

The loop in Figure 4.10c contains all the basic circuit elements. According to the loop law, the sum of the three voltage difference E_R, E_L, and E_C equals the voltage source $E(t)$, so we have

■■ 4.2
$$L\frac{dI}{dt} + RI + \frac{Q}{C} = E(t).$$

As in Example 3, we eliminate I in favor of Q using Equation 4.1: $I = dQ/dt$. We get

■■ 4.3
$$L\frac{d^2Q}{dt^2} + R\frac{dQ}{dt} + \frac{Q}{C} = E(t).$$

If we then manage to find $Q(t)$, we can later find $I(t)$ by using Equation 4.1 again; just differentiate $Q(t)$ to find $I(t)$. But if what we really want is just $I(t)$, and we know that L, R, and C are constant, we can differentiate both sides of Equation 4.2 and then use Equation 4.1 to get a differential equation satisfied by $I(t)$:

■■ 4.4
$$L\frac{d^2I}{dt^2} + R\frac{dI}{dt} + \frac{1}{C}I = \dot{E}(t).$$

This last equation has the disadvantage that satisfying initial conditions requires not only a knowledge of $I(0)$, which can be measured, but also of $\dot{I}(0)$, which is hard to determine.

---

e x a m p l e 4

When L, R, and C are constant in Equation 4.4, the roots of the characteristic equation

$$Lr^2 + Rr + \frac{1}{C} = 0$$

are important in determining the precise behavior of the circuit under an applied voltage $E(t)$. If $E(t)$ is also constant, then $\dot{E}(t) = 0$ and Equation 4.4 is homogeneous. The roots of the characteristic equation are

$$r_1, r_2 = -\frac{1}{2}\frac{R}{L} \pm \frac{1}{2}\sqrt{\left(\frac{R}{L}\right)^2 - \frac{4}{LC}}.$$

The corresponding solutions will be oscillatory whenever the discriminant $(R/L)^2 - 4/(LC)$ is negative. This is the most interesting case for applications, and we'll assume it holds for the purposes of this example. We assume also that the constants R, L, C are all positive. Thus the homogeneous equation, with $\dot{E}(t) = 0$, has solutions of the form

$$I_h = e^{-Rt/2L}(c_1 \cos \omega_0 t + c_2 \sin \omega_0 t), \qquad \text{where } \omega_0 = \frac{1}{2}\sqrt{\frac{4}{LC} - \left(\frac{R}{L}\right)^2}.$$

These solutions are *transient*, because the exponential factor makes them tend rapidly to zero as time goes on. It may be that they die out so rapidly that we can safely ignore them in computing the response of the circuit to an input voltage. On the other hand, it may be the initial response of the circuit that we are particularly interested in. And of course the solutions of the homogeneous equation have a role to play in determining the effect of initial conditions on the general solution.

It is important to remember that the solution $I_h(t)$ we have just been discussing represents current as a function of time. If we want then to find the charge $Q(t)$ on the capacitor, we solve the first-order equation $\dot{Q} = I(t)$. To do this explicitly we of course need an initial value $Q(0)$, which may be difficult to measure in practice.

---

e x a m p l e 5

If the transient solutions found in Example 4 are considered negligible, we may want to make specific calculations only for a solution of the related nonhomogeneous equation. Let's assume an alternating voltage source $E(t) = E_0 \sin \omega t$. Hence $\dot{E}(t) = \omega E_0 \cos \omega t$, and to find $I(t)$ we solve

$$L\ddot{I} + R\dot{I} + \frac{1}{C}I = \omega E_0 \cos \omega t.$$

We use a trial solution of the form $I_p = A \cos \omega t + B \sin \omega t$. (There is no chance that this could turn out to satisfy the homogeneous equation, because all homogeneous solutions contain the exponential factor $e^{-Rt/2L}$ found in Example 4.) Next we compute $\dot{I}$ and $\ddot{I}$, substitute into the differential equation, and equate coefficients of $\cos \omega t$ and $\sin \omega t$. The result is a pair of equations to be solved for A and B:

$$(C^{-1} - \omega^2 L)A + R\omega B = \omega E_0$$
$$-R\omega A + (C^{-1} - \omega^2 L)B = 0.$$

Solving for A and B we find

$$I_p(t) = \frac{(C^{-1} - \omega^2 L)\omega E_0}{(C^{-1} - \omega^2 L)^2 + R^2\omega^2} \cos \omega t + \frac{R\omega^2 E_0}{(C^{-1} - \omega^2 L)^2 + R^2\omega^2} \sin \omega t.$$

Combining the two terms into one cosine term to introduce a phase angle α gives

$$I_p(t) = \frac{E_0}{\sqrt{(L\omega - (C\omega)^{-1})^2 + R^2}} \cos (\omega t - \alpha),$$

where $\alpha = \arctan (R\omega/(C^{-1} - L\omega^2))$, $0 \le \alpha \le \pi$. This solution is called the **steady-state** solution, because it differs from the general solution only by the transient solution, which itself tends exponentially to zero as t tends to infinity.

What we can immediately conclude from all this is that if the input voltage is represented by a pure sine of frequency ω and amplitude E_0, then the current also tends to vary with frequency ω. The factor

$$Z(\omega) = \sqrt{(L\omega - (C\omega)^{-1})^2 + R^2}$$

in the denominator of the expression for I_p is called the **impedance** of the circuit; it is a function not only of the parameters R, L, C but also of the input frequency ω. Note that a large impedance produces an attenuated output amplitude. On the other hand, minimizing $Z(\omega)$ produces a maximum value for the output amplitude. The value ω_0 for which this maximum is attained is called the **maximum resonance frequency** of the circuit.

The discussion in Example 5 is reminiscent of the analysis of vibrations of a spring mechanism in Section 3. The analogy is in fact complete and is helpful in

Table 1

Spring		Circuit	
$m\ddot{x} + k\dot{x} + hx = F(t)$		$L\ddot{Q} + R\dot{Q} + \dfrac{1}{C}Q = E(t)$	
Position	x	Charge	Q
Velocity	$\dot{x}$	Current	$\dot{Q} = I$
Mass	m	Inductance	L
Damping	k	Resistance	R
Elasticity	$1/h$	Capacitance	C
Applied force	$F(t)$	Applied voltage	$E(t)$

thinking about circuits if you have some intuition about what happens in a spring mechanism. Table 1 shows the correspondences in parallel columns.

Notice particularly that the relationship between stiffness h and capacitance C is reciprocal. A very strong spring has a large h and hence small elasticity, which in the correspondence table is analogous to small capacitance. Put the other way, large capacitance is analogous to a very elastic spring, that is, one with a small h. You may find it illuminating to review the preceding circuit examples with the mechanical analogy in mind.

Before proceeding further, standard units of measurement should be agreed on for the quantities we have to deal with. Voltage is in **volts,** current is in **amperes,** and charge is in **coulombs.** Because of Equation 4.1 relating current and charge, there is a relation between amperes and coulombs; it turns out that one ampere equals one coulomb per second. Resistance is measured in **ohms,** inductance in **henrys,** and capacitance in **farads.** Just as with a mechanical system, it is the balance between the sizes of these quantities that determines the inherent qualities of an electric circuit. In practice, capacitance C is usually rather small, often on the order of 10^{-6} farad. The resistance R is usually measured in tens, hundreds, or thousands of ohms. If inductance L is in the range 0.1 to 1.0 henry, the natural circular frequency $\omega_0 = 1/\sqrt{LC}$ will be on the order of 3000 to 1000 cycles per second.

example **6**

Analog computers can be designed to mimic the operation of mechanical systems by using the analogy described above. The parameters L, R, C are analogs to mass m, damping k, and elasticity $1/h$. Suppose we want to model the displacement $x(t)$ of a spring mechanism with mass 1000 pounds, damping factor $k = 24$, and stiffness $h = 200$. If there is an externally applied force $F(t) = 220 \cos 95t$, with t measured in seconds, an appropriate LRC circuit model might have inductance 1000 henrys, resistance 24 ohms, and capacitance $\frac{1}{200} = 0.005$ farad. The external alternating-current source would be characterized by voltage $E(t) = 220 \cos 95t$, having maximum voltage 220 volts, period $2\pi/95$, and hence frequency $95/2\pi \approx 15.1$ cycles per second. Remember that for any

LRC circuit, it is the charge $Q(t)$ on the condenser that models the physical displacement. To determine a particular solution satisfying given initial conditions on $x(0)$, $\dot{x}(0)$, just start the circuit with $Q(0) = x(0)$, $I(0) = \dot{x}(0)$.

Alternatively, we could scale the constants down by the factor 100, since this would just amount to dividing the associated differential equation by 100. The resulting circuit would have $L = 10$ henrys, $R = 0.24$ ohm, $C = 0.00005$ farad (or 50 microfarads), and maximum input voltage 2.2 volts, still with frequency ≈ 15.1 cycles per second. This low-voltage circuit has the advantage that it is simpler to operate.

e x a m p l e 7

Let's turn Example 6 around. Suppose we observe a mechanical response x from some system that looks like $x(t) = 2.2e^{-31t} \sin(3t - 2)$. Can we design a circuit that gives precisely this response to some voltage input and initial conditions? In other words, can we find a differential equation having $x(t)$ as a solution? The answer is "yes" of course, since a solution of that form is associated with a homogeneous equation with characteristic roots $\alpha \pm i\beta = -31 \pm 3i$. The associated differential operator is

$$(D + 31 - 3i)(D + 31 + 3i) = D^2 + 62D + 970.$$

Hence the choice $L = 1$, $R = 62$, $C = \frac{1}{970} \approx 0.001$ will do for the circuit elements. The circuit should be started out with $Q(0) = x(0) = 0$ and $I(0) = \dot{x}(0) \approx 59.27$ amperes to produce something like the given oscillations. The voltage across the resistor required to produce that much initial current is $62 \cdot 59.27 \approx 3675$ volts. This voltage is rather high; to reduce it we can divide the circuit parameters L, R, C by some suitably large number, say 100.

e x a m p l e 8

Figure 4.11a is a diagram of an *RC* circuit. This of course is really just an *LRC* circuit with negligible inductance present. (The mechanical analog would be a massless spring with damping, something that would be impossible using a real spring.) The differential equation for the charge Q on the capacitor is a special case, with $L = 0$, of the one in Equation 4.3, and we'll assume $E(t) = E_0 \sin \omega t$ there. Thus we want to solve

$$R\dot{Q} + \frac{1}{C}Q = E_0 \sin \omega t.$$

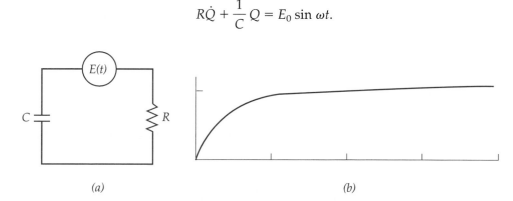

(a)   (b)

Figure 4.11 (*a*) *RC* circuit. (*b*) $V_0/E_0 = RC\omega/\sqrt{1 + R^2C^2\omega^2}$.

We can solve this equation from scratch, or rely on the result of Example 5 as follows. We introduce the defining relation $I = \dot{Q}$ into Example 5 and also set $L = 0$. Integrating the steady-state solution $I_p(t) = \dot{Q}(t)$ of Example 5 with respect to t yields

$$Q_p(t) = \frac{E_0}{\sqrt{C^{-2} + R^2\omega^2}} \sin(\omega t - \alpha) = \frac{CE_0}{\sqrt{1 + R^2C^2\omega^2}} \sin(\omega t - \alpha),$$

where $\alpha = \arctan RC\omega$. Transient solutions have the form $Q_h(t) = Ke^{-t/RC}$.

It's interesting to compare the amplitude E_0 of the input voltage with the amplitude of the output voltage $V(t)$ as measured across the resistance R in the circuit. According to Ohm's law, $V(t) = RI(t) = R\dot{Q}(t)$. Computing $\dot{Q}$ from the above solution formula, we find

$$V(t) = \frac{RC\omega E_0}{\sqrt{1 + R^2C^2\omega^2}} \cos(\omega t - \alpha).$$

Thus the voltage amplitude as a function of ω is $V_0 = RC\omega E_0/\sqrt{1 + R^2C^2\omega}$, and the ratio of output amplitude to input amplitude is

$$\frac{V_0}{E_0} = \frac{RC\omega}{\sqrt{1 + R^2C^2\omega^2}}.$$

What we see here is that $V_0/E_0 = F(RC\omega)$, where $F(x) = x/\sqrt{1 + x^2}$. The function F is increasing as x increases, it tends to 1 as x tends to infinity, and $F(0) = 0$. With R and C fixed, the same can be said about $F(RC\omega)$ as a function of ω. Figure 4.11b shows the graph of V_0/E_0 as a function of ω for values of R and C such that $RC = 2$.

When the amplitude ratio is near 1, in other words, when ω is large, the output amplitude is almost as large as the input amplitude. On the other hand, small values of ω produce a response of small amplitude as compared with the amplitude of the input voltage. For this reason, an RC circuit can be used as a *high-pass filter* that tends to *filter out* low-frequency input and to *pass through* high-frequency input.

EXERCISES

1. Consider a circuit with inductance $L = \frac{1}{5}$ henry, resistance $R = 2$ ohms, and capacitance $C = 1/1285$ farad. See Example 6 of the text.

   (a) Find the general formula for the capacitor charge $Q(t)$ if $E(t) = 0$.

   (b) Use the result of part (a) to find $Q(t)$ if $Q(0) = 1$ and $I(0) = 0$.

   (c) Find $Q(t)$ under the initial conditions of part (b) if $E(t) = e^{-t}$.

   (d) Find $Q(t)$ under the initial conditions of part (b) if $E(t) = \sin 20t$.

2. Referring to Equation 4.4 of the text, show that if the voltage source is constant and $CR^2 > 4L$, then all solutions $I(t)$ are transient and tend to zero exponentially without oscillation as t tends to infinity.

3. In Example 3 of the text we derived the formula

$$Q(t) = \frac{E_0}{L(\omega_0^2 - \omega^2)}(\cos \omega t - \cos \omega_0 t).$$

for the response of a superconducting LC circuit to input voltage $E(t) = E_0 \cos \omega t$ in which $\omega_0 = 1/\sqrt{LC}$ is the natural circular frequency of the circuit.

(a) Show that

$$\cos \omega t - \cos \omega_0 t = 2 \sin \tfrac{1}{2}(\omega_0 - \omega)t \sin \tfrac{1}{2}(\omega_0 + \omega)t.$$

[*Hint:* Let $\alpha = \tfrac{1}{2}(\omega - \omega_0)t$, $\beta = \tfrac{1}{2}(\omega + \omega_0)t$, and use $\cos(\alpha \pm \beta) = \cos \alpha \cos \beta \mp \sin \alpha \sin \beta$.]

(b) Assume that ω_0 is fairly large and that ω is close to ω_0 with $\omega \neq \omega_0$. Show that $Q(t)$ can then be interpreted as a low-frequency oscillation (with frequency $|\omega - \omega_0|/(4\pi)$) and rapidly varying amplitude. (See Figure 4.9.) (In practice the low-frequency oscillation transmits information and the high-frequency factor is thought of as a **carrier wave** subject to **amplitude modulation** by the low-frequency oscillation.)

4. **(a)** Work out the details for finding the solution

$$Q(t) = \frac{E_0}{L(\omega_0^2 - \omega^2)}(\cos \omega t - \cos \omega_0 t).$$

(given in Example 3 of the text) to the problem $L\ddot{Q} + Q/C = E_0 \cos \omega t$, with $Q(0) = \dot{Q}(0) = 0$. (The natural frequency ω_0 is equal to $1/\sqrt{LC}$.)

(b) Show that for fixed t, as ω approaches ω_0 the solution of part (a) approaches

$$Q_0(t) = \frac{E_0}{2L\omega_0} t \sin \omega_0 t.$$

5. In Example 8 of the text we got the steady-state solution of the RC circuit equation simply by specializing from the solution of the general LRC equation. If we want to find the general solution for the current in an RC circuit, we need to proceed differently.

(a) Show that in an RC circuit the current satisfies

$$R\dot{I} + \frac{1}{C}I = E(t).$$

(b) Find the general solution of the differential equation in part (a) for the case $E(t) = E_0 \cos \omega t$. (You already know the steady-state solution from Example 8; can you use that to save some work?)

(c) Use the result of part (b) to find $I(t)$ for general R, C, and $E(t) = E_0 \cos \omega t$ if I satisfies $I(0) = 0$.

6. In Example 8 of the text the phase angle is $\alpha = \arctan RC\omega$. Show that as ω tends to zero the oscillations of the input and steady-state output tend to be more nearly in phase. Show, on the other hand, that as ω tends to infinity the corresponding oscillations become more nearly out of phase by $\pi/2$. [*Hint:* Output current is the time derivative of condenser charge Q.]

7. Show that the resonant frequency that maximizes the amplitude of the output response of an LRC circuit to a sinusoidal input $E_0 \sin \omega t$ is $\omega_M = 1/\sqrt{LC}$ and that the resulting impedance is $Z(\omega_M) = R$. See Example 5 of the text.

8. Show that if the voltage source in a circuit loop is turned off at some time t_0, then the current $I(t)$ and charge $q(t)$ will both satisfy the same second-order homogeneous differential equation from that time on.

9. An LRC circuit loop with no voltage source ($E(t) = 0$) is closed at time $t = 0$ when the charge on the capacitor is $q_0 > 0$. Show that if $CR^2 = 4L$, then $q(t) = q_0(1 + Rt/(2L))e^{-Rt/(2L)}$ and that this quantity tends decreasingly to zero. Assume $I(0) = 0$.

5   PHASE SPACE AND PERIODICITY

5A   *Phase Space*

The differential equation for the displacement $y = y(t)$ of a linearly damped piston (see Section 3) is

$$m\ddot{y} + k\dot{y} + hy = 0.$$

Previously, our pictorial view of a solution $y = y(t)$ of such an equation came simply from looking at the graph of $y(t)$ in the ty plane. What we'll do here is consider simultaneously with $y(t)$ the new dependent variable $z(t) = \dot{y}(t)$ and plot the curve in the yz plane represented parametrically by $(y, z) = (y(t), z(t))$. The yz plane in this context is called the **phase space** or **state space** of the differential equation we're looking at, and an image curve represented by $(y, z) = (y(t), z(t))$ is called a **phase curve** or **phase path.** This way of regarding solutions does two things for us. One is that it allows us to focus attention on the relation between two quantities that determine future states of the dynamical system: position y and velocity $z = \dot{y}$. The other is that it permits graphical study of long-term behavior of solutions without having to plot points for extremely large values of t extended along a time axis. For example, we'll be interested in studying solutions as their values approach, for large t, those of a constant solution $y(t) = y_0$ with velocity $\dot{y}(t)$ identically zero. Constant solutions are also called **equilibrium solutions.** Behavior near equilibrium solutions is taken up in detail in Section 6 and in Chapter 6, Section 8.

We could in principle study phase curves of nonlinear second-order equations of the form $\ddot{y} = f(t, y, \dot{y})$. However, phase-space analysis yields the most useful results for the **autonomous** equation $\ddot{y} = f(y, \dot{y})$, from which the time variable t is explicitly absent. (See Exercise 37 to understand why.) The first example is both simple and familiar.

e x a m p l e **1**   The undamped linear oscillator equation $\ddot{y} + \omega^2 y = 0$, $\omega \neq 0$, is discussed in detail in Section 3. There we found that the solutions $y(t) = c_1 \cos \omega t + c_2 \sin \omega t$ could all be written in the form $y(t) = A \cos(\omega t - \alpha)$, and from this it follows that the derivative $\dot{y}(t) = z(t)$ is $z(t) = -A\omega \sin(\omega t - \alpha)$. The parametric representation

$$(y(t), z(t)) = (A \cos(\omega t - \alpha), -A\omega \sin(\omega t - \alpha))$$

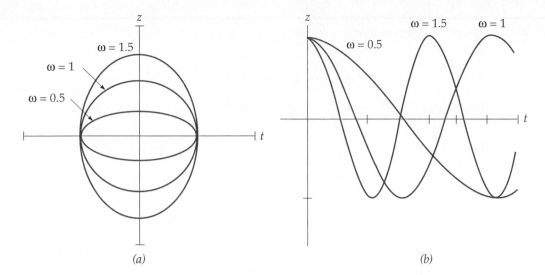

Figure 4.12 $\ddot{y} + \omega^2 y = 0$, $y(0) = 1$, $\dot{y}(0) = 0$; $\omega = 0.5, 1, 1.5$. (*a*) Phase curves. (*b*) Graphs.

describes an ellipse with semiaxes A and $|\omega|A$ centered at the origin of the yz plane. To see this, eliminate t between the equations for y and z to get

$$y^2 + \frac{z^2}{\omega^2} = A^2.$$

Phase curves for three different values of ω are shown in Figure 4.12*a*. Note that the velocity z attains both its maximum and its minimum when the position y is zero. Similarly, the position y is at its maximum and minimum when the velocity z is zero. Furthermore, it's clear from this finite picture that the motion is periodic, not just in the sense that the position y repeatedly traces over the same interval from $-A$ to A but also that a given position y always goes with the same velocity z. Note, however, that some information about the solution is lost by going to this picture. Specifically, Figure 4.12*a* contains little information about how y and z are each related to a specific time t, the notable exception being that as t increases the ellipses are traced *clockwise* regardless of whether $\omega > 0$ or $\omega < 0$. (For example, at a point in the first quadrant, y must be increasing, because $\dot{y} = z$ is positive there. Similar remarks apply to the other three quadrants.) Figure 4.12*b* shows the corresponding graphs of some solutions; note that a steeper graph goes with increased vertical elongation of a phase curve.

The phase curves of an autonomous second-order equation $\ddot{y} = f(y, \dot{y})$ are determined by the first-order equation

$$z\frac{dz}{dy} = f(y, z).$$

To see this, use the chain rule, and the relations $\dot{y} = z$, $\dot{z} = f(y, z)$, to write

$$\frac{dz}{dy}\frac{dy}{dt} = \frac{dz}{dt}, \quad \text{or} \quad z\frac{dz}{dy} = f(y, z).$$

e x a m p l e **2**

The oscillator equation $\ddot{y} + \omega^2 y = 0$ leads, as described above, to the first-order equation

$$z \frac{dz}{dy} + \omega^2 y = 0, \qquad \text{or in separated form,} \qquad z\, dz = -\omega^2 y\, dy.$$

The solutions are found by integration to be

$$\tfrac{1}{2}z^2 = -\tfrac{1}{2}\omega^2 y^2 + C, \qquad \text{or} \qquad z^2 + \omega^2 y^2 = 2C.$$

These are just the equations of the ellipses found in Example 1 by starting with explicit solutions in terms of t.

Finding the constant, or equilibrium, solutions of a second-order equation $\ddot{y} = f(y, \dot{y})$ is straightforward. A constant solution y must satisfy the two equations $\dot{y} = 0$ and $\ddot{y} = 0$. In terms of phase variables y and z and the function f, these two requirements say that $z = 0$ and $f(y, z) = 0$. Thus we have the criterion: An equilibrium solution of $\ddot{y} = f(y, \dot{y})$ is represented in (y, z) phase space as an **equilibrium point** on the y axis with y determined by $f(y, 0) = 0$.

e x a m p l e **3**

To locate in phase space the equilibrium solutions of the equation $\ddot{y} + \omega^2 y = 0$ in Example 2, we can proceed directly from the relations

$$\dot{y} = z \qquad \text{and} \qquad \dot{z} = -\omega^2 y.$$

We want both of these expressions to be identically zero if y is to be constant. Clearly the only solution is $(y, z) = (0, 0)$. (We could also have drawn this conclusion from the explicit solution formula obtained in Example 1, but in the most important examples such formulas won't be available, so dealing directly with the differential equation is preferable in general.)

e x a m p l e **4**

The damped linear oscillator equation $\ddot{y} + k\dot{y} + y = 0, 0 < k < 2$, has solutions of the form

$$y(t) = e^{-kt/2}(c_1 \cos \omega t + c_2 \sin \omega t), \qquad \omega = \sqrt{1 - \left(\frac{k}{2}\right)^2}.$$

These solutions can be rewritten as $y(t) = A e^{-kt/2} \cos(\omega t - a)$. Hence, we can compute the derivative $z(t) = \dot{y}(t)$ to be

$$z(t) = A e^{-kt/2}\left(-\frac{k}{2} \cos(\omega t - \alpha) - \omega \sin(\omega t - \alpha)\right).$$

This in turn can be rewritten as $z(t) = -B e^{-kt/2} \sin(\omega t - \beta)$. Since $y(t)$ and $z(t)$ have a common exponential factor, the phase curves may be represented parametrically by

$$(y(t), z(t)) = e^{-kt/2}(A \cos(\omega t - a), -B \sin(\omega t - \beta)).$$

If k were zero, with the exponential factor absent, we would typically get elliptic phase curves as in Figure 4.12a. But since the exponential tends to zero as t tends to infinity, each curve spirals in toward the origin as t increases, as in Figure 4.13a.

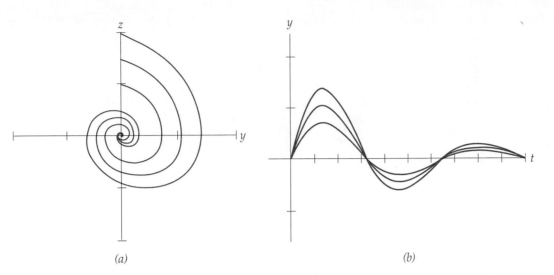

Figure 4.13 $\ddot{y} = -0.5\dot{y} - y$, $y(0) = 0$, $\dot{y}(0) = 1$, 1.5, 2. (*a*) Phase curves. (*b*) Graphs.

Some typical results are shown in Figure 4.13*a* for various values of $z(0) = \dot{y}(0)$. Graphs of corresponding solutions are shown in Figure 4.13*b*. These pictures are not altogether analogous to the ones in Figure 4.12. The reason is that there the phase curves and graphs illustrate responses to changes in the differential equation itself, that is, changes in ω, rather than to changes in initial conditions as in Figure 4.13. The result of plotting several phase curves, appropriately spaced, of a *single* differential equation is sometimes called a **phase portrait**. Note that Figures 4.12*a* and 4.13*a* are phase portraits, while Figures 4.12*b* and 4.13*b* are not.

example 5

The equilibrium solutions of the equation $\ddot{y} = -k\dot{y} - y$ in Example 4 are the solutions in the yz plane of the pair of equations

$$z = 0 \quad \text{and} \quad -kz - y = 0.$$

The only solution is evidently $(y, z) = (0, 0)$.

The next example is about a nonlinear equation and is particularly important because it illustrates one way that a physical system and its corresponding mathematical representation can exhibit drastically different solution responses to minute changes in certain initial conditions. (A system that behaves this way near *all* initial conditions is often called *chaotic*.)

example 6

It can be shown (see Section 2) that a pendulum of length l has angular displacement from the vertical $y(t)$ at time t satisfying

$$\ddot{y} = -\frac{g}{l} \sin y, \quad g = \text{gravitational acceleration,}$$

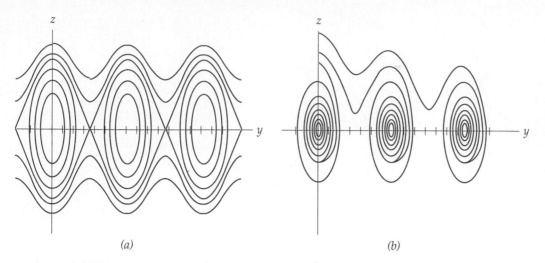

(a)   (b)

Figure 4.14 (a) Undamped pendulum. $\ddot{y} = -\sin y$, $z^2 = 2\cos y + 2C$; $C = -\frac{1}{2}, 0, \frac{1}{2}, 1, \frac{6}{5}, 2$.
(b) Damped pendulum. $\ddot{y} = -0.1\dot{y} - \sin y$, $y(0) = 0$; $\dot{y}(0) = 2, 2.3, 2.8$.

assuming there are no external frictional forces. We can't find an elementary formula for $y(t)$, so we proceed as follows to find an equation satisfied by y and $z = \dot{y}$. Multiply both sides of the differential equation by $\dot{y}$ and then integrate with respect to t:

$$\int \dot{y}\ddot{y}\, dt = \int -\frac{g}{l}\dot{y}\sin y\, dt.$$

The result can be written as

$$\tfrac{1}{2}\dot{y}^2 = \frac{g}{l}\cos y + C.$$

(Check by differentiating both sides of this last equation with respect to t.) Replacing $\dot{y}$ by z gives an equation in phase space:

$$\tfrac{1}{2}z^2 = \frac{g}{l}\cos y + C, \qquad \text{or} \qquad z = \pm\sqrt{2\frac{g}{l}\cos y + 2C}.$$

In Figure 4.14a, where $l = g$, the values of $2C$ for which $|2C| < 2$ correspond to disjoint loops that represent ordinary swings of a pendulum, somewhat like the oscillations shown in Figure 4.12. If $2C > 2$, the velocity is never zero, so the pendulum goes "over the top" repeatedly. Hence there are two distinct classes of phase path that are periodic as functions of y, one above and one below the y axis. The upper paths are linked to increasing y and clockwise rotation of the pendulum, the lower ones to counterclockwise rotation.

Sketching the graphs of the phase paths for various values of the integration constant C is fairly easy using the fact that they are tangent to the direction field of $dz/dy = \dot{z}/\dot{y} = -\sin y/z$. (See Exercise 31.) Another alternative is to use computer graphics, perhaps together with some numerical solution techniques described in Section 7.

Finally, the equilibrium solutions are just the points on the y axis satisfying $-\sin y = 0$, in other words, the points $(y, z) = (k\pi, 0), k = 0, \pm1, \pm2, \dots$. These points lie at the centers of the closed loops in Figure 4.14a.

Approaching the equilibrium points we get curves that *appear* in Figure 4.14a to cross each other on the *x* axis. A curve of this kind is called a **separatrix** because it separates phase curves representing two distinct behaviors: periodic, sometimes called **libration,** and nonperiodic, called **rotation.** In fact these separating curves themselves don't actually cross on the horizontal axis. (What motions do they represent?) Each separatrix approaches an equilibrium point, but such a point itself represents a constant solution quite distinct from the separatrix.

The transition from back-and-forth pendulum swings to the over-the-top behavior described in Example 6 can sometimes be achieved by making very small changes in initial conditions. For example, a pendulum may swing so close to the top that a minute boost in velocity will send it over the top. The corresponding change in phase curve is from a closed loop to a wavelike curve as you can see from Figure 4.14a. This behavior is an example of *instability*, some aspects of which are taken up in detail in Section 6B and in Chapter 6, Section 8.

The undamped, nonlinear pendulum equation is fairly exceptional in that we can find elementary formulas for its phase paths and can thus sketch a reasonably accurate portrait by hand. For the next two examples it's extremely helpful to use an approximate numerical method for drawing a phase portrait. The Applet PHASEPLOT at Web site http://math.dartmouth.edu/~rewn/ is designed to do the job. The basic structure of the method is similar to that of Chapter 2, Section 6B. Section 7A of this chapter describes the necessary simple modification.

example 7

The nonlinear pendulum equation with linear damping is

$$\ddot{y} = -k\dot{y} - \frac{g}{l}\sin y.$$

We haven't any way to find formulas for either the solutions or the phase curves of this equation. In both instances we need the numerical methods described in Section 7 to get usable results. Figure 4.14b shows a sample. This picture, particularly the curve starting at $(y, z) = (0, 2)$, should be compared with the one in Figure 4.13a. Note that the damping factor k for Figure 4.13a is five times as large as it is for Figure 4.14b, so the curves in the latter picture spiral in less quickly toward their equilibrium positions.

example 8

One form of the **Van der Pol equation** is
$$\ddot{y} = -k(y^2 - 1)\dot{y} - y, \qquad k > 0,$$
and it can be regarded as a modification of the harmonic oscillator equation $y = -y$ by a term $-k(y^2 - 1)\dot{y}$ that has a damping effect when $|y| > 1$ and an amplifying effect when $|y| < 1$. Realized in an electric circuit, this term has the effect of driving an output signal toward periodic oscillation, an oscillation that will be displayed as a closed loop in a phase portrait. A larger value of k will

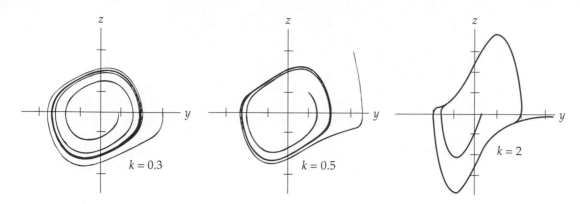

Figure 4.15 Phase curves of the Van der Pol equation $\ddot{y} = -k(y^2 - 1)\dot{y} - y$, $k > 0$.

force a more rapid transition toward periodicity, as you can see from Figure 4.15. These pictures were made using a numerical method described in Section 7. Each one contains just two phase curves, one that spirals out toward a closed loop from the inside, and one that spirals in toward the same loop. The closed loops toward which the phase curves tend are called **limit cycles** and are themselves phase curves of solutions. A limit-cycle trajectory is an example of an **equilibrium set** to be contrasted with a single equilibrium point; the latter is necessarily static in character, while a limit cycle is a dynamically active trajectory.

5B *Existence of Periodic Solutions*

Periodic phenomena are important for the very basic reason that they represent a kind of regularity. The fact that the pendulum equation $\ddot{y} = -\sin y$ has periodic solutions is not obvious and depends not at all on the periodicity of the sine function. (Indeed the differential equation has lots of nonperiodic solutions of the rotational or over-the-top variety, and the harmonic oscillator equation $\ddot{y} = -y$ has all its solutions periodic with period 2π even though $-y$ is not periodic.) What is important for the existence of at least *some* periodic solutions to $\ddot{y} = -\sin y$, or more generally to $\ddot{y} + f(y) = 0$, is that the function $f(y)$ should be positive in some interval $0 < y < a$ and negative in some interval $-b < y < 0$. (In particular, this will be the case if $f(y)$ is positive for positive y and $f(y)$ is also an **odd** function, that is, $f(-y) = -f(y)$.) Thus, using phase-space ideas, the second-order equations $\ddot{y} + y^3 = 0$, $\ddot{y} + \sin^5 y = 0$, and $\ddot{y} + ye^y = 0$, with $f(y)$ shown in Figure 4.16, will all be seen to have at least some periodic solutions, just as the pendulum equation $\ddot{y} + \sin y = 0$ does. To see why this is so, we'll turn our attention from the second-order equation $\ddot{y} + f(y) = 0$ to the equivalent first-order system

$$\frac{dy}{dt} = z$$

$$\frac{dz}{dt} = -f(y).$$

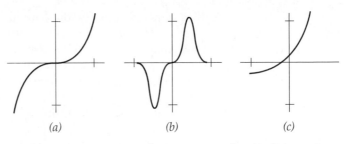

Figure 4.16 (a) $f(y) = y^3$. (b) $f(y) = \sin^5 y$. (c) $f(y) = ye^y$.

We can use the chain rule relation $(dz/dt)/(dy/dt) = dz/dy$ to draw some conclusions about periodicity from the related first-order equation that we get by dividing dz/dt by dy/dt:

$$\frac{dz}{dy} = \frac{-f(y)}{z}.$$

Geometrically the idea is very simple. The assumptions will allow us to conclude that the path of a solution $y = y(t)$, when looked at in the yz phase space with $(y(t), z(t)) = (y(t), \dot{y}(t))$, traces a closed curve. Thus certain paths loop back to an initial point (y_0, z_0) and start over again with the same initial values on the same path. Since the function $f(y)$ in $\ddot{y} + f(y) = 0$ is not explicitly dependent on t, the solution is periodic, and the period is just the time it takes to complete one loop of the closed curve. We can summarize with the following qualitative statement: **periodic solutions correspond to closed loops in phase space.**

■ 5.1 PERIODICITY THEOREM

Suppose that $f(y)$ is continuous and is positive for $c < y \le c + a$ but negative for $c - a \le y < c$. Let $y = y(t)$ be the solution to a second-order initial-value problem $\ddot{y} + f(y) = 0$ with $y(0) = y_0$, $\dot{y}(0) = z_0$. For initial points (y_0, z_0) sufficiently close to $(c, 0)$ in the yz phase space, the phase-space paths are nonintersecting closed loops around $(c, 0)$; hence the corresponding solutions $y(t)$ are periodic functions of t.

Proof. Replacing y by $y + c$ throughout allows us to assume that $c = 0$. The equation $\ddot{y} + f(y) = 0$ is equivalent to the system $\dot{y} = z$, $\dot{z} = -f(y)$. The phase curves satisfy the separated first-order equation $z\,dz + f(y)dy = 0$. When this is integrated we get

$$\tfrac{1}{2}z^2 + F(y) = C, \qquad \text{where } F(y) = \int_0^y f(u)\,du.$$

Note that $F(y)$ is strictly increasing on the interval $0 \le y \le a$, because the integrand $f(u)$ is positive there. Thus each solution path in phase space must lie on the graph of an equation $z^2/2 + F(y) = C$, with distinct nonintersecting paths corresponding to different values of the constant C. Pick a curve through $(y, z) = (0, z_0)$ with $z_0 > 0$ so that $\tfrac{1}{2}z_0^2 < F(a)$. Then $C - \tfrac{1}{2}z_0^2$, so $F(y)$ increases to $\tfrac{1}{2}z_0^2$, and z decreases to 0 as y increases toward a. The reason y is strictly increasing along this part of the path is that $dy/dt = z > 0$. (Note that the path has a vertical tangent where it crosses the positive y axis, because $dy/dt = 0$ and

$dz/dt < 0$ there; similarly, the path has a horizontal tangent where it crosses the positive z axis, because $dz/dt = -f(0) = 0$ and $dy/dt > 0$ there.)

The equation $\ddot{y} = -f(y)$ is also equivalent to the "reversed" system $\dot{y} = -z$, $\dot{z} = f(y)$, in which the phase-space paths are traced in the opposite direction. If we now allow y to *decrease* from $y = 0$, we get a piece of a phase path of the reversed system for which z decreases as y decreases. The reason is that $f(y)$ is negative for $-a < y < 0$, so its integral,

$$F(y) = \int_0^y f(u)\,du = \int_y^0 -f(u)\,du,$$

increases as y decreases. By choosing z_0, perhaps smaller than before, so that $\frac{1}{2}z_0^2 < F(-a)$, we guarantee that z decreases to 0 as y decreases. (If we needed a smaller z_0 for the left side, we agree to use that one on the right also.) So far, we have a phase path that lies above the y axis and joins a point $(y, z) = (-B, 0)$ to a point $(y, z) = (A, 0)$. The lower half of the path, starting with $y = A$, $z = 0$, traces the graph of $\frac{1}{2}z^2 + F(y) = \frac{1}{2}z_0^2$ right to left, since here $\dot{y} = z < 0$, and so completes a closed loop.   ∎

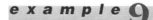

Let y stand for the extension $(y > 0)$ or compression $(y < 0)$ from the unstressed position of a spring for which the linear relation $F = -hy$ of Hooke's law is replaced by $F = -\gamma y^3$, $\gamma > 0$. In the absence of retarding forces, the differential equation of motion is $\ddot{y} = -\gamma y^3$. The motion in phase space is governed by the system

$$\dot{y} = z, \qquad \dot{z} = -\gamma y^3.$$

The phase curves satisfy

$$\frac{dz}{dy} = \frac{-\gamma y^3}{z},$$

with solutions $\frac{1}{2}z^2 + (\gamma/4)y^4 = C$, $C = \text{const.}$ In this example, all the phase paths are bounded loops, corresponding to periodic oscillations of the spring. Figure 4.17 shows phase plots and corresponding solution graphs for the choice $\gamma = 1$. The graphs on the left were made using the numerical methods described in Chapter 2, Section 6; they could also be made by hand using the direction field. The pictures of solutions $y = y(t)$ were made using the numerical methods of Section 7 of this chapter. The successive initial values $(y_0, z_0) = (0, 0.5), (0, 1), (0, 1.5), (0, 2)$ correspond to larger and larger phase curves along with steeper and steeper initial slopes to the solution graphs. All the graphs in Figure 4.17b are periodic according to the periodicity theorem, but the period of oscillation varies, with a longer period corresponding to a smaller initial velocity; these contrast with the solutions of the linear harmonic oscillator equation $\ddot{y} = -\gamma y$ for which all solutions have $\sqrt{\gamma}$ for a common period.

Computing the period of a periodic solution of the type guaranteed by the periodicity theorem is most often done using the numerical techniques of Section 7. A few examples where hand computation suffices are described in the exercises and in the following example.

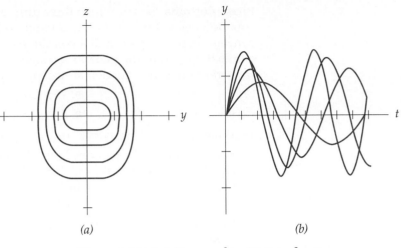

Figure 4.17 (a) $dz/dy = -y^3/z$. (b) $\ddot{y} + y^3 = 0$.

example 10

The harmonic oscillator equation $\ddot{y} = -\alpha^2 y$ has nothing but periodic solutions. In particular, the initial conditions $y(0) = 0$, $\dot{y}(0) = z_0$ specify the unique solution $y(t) = z_0 \sin \alpha t$ with period $2\pi/\alpha$.

A related example is a differential equation with a "two-piece" right side:

$$\ddot{y} = \begin{cases} -\alpha^2 y, & \text{if } 0 \le y, \\ -\beta^2 y, & \text{if } y < 0. \end{cases}$$

This can be thought of as representing a spring that behaves linearly under both extension and compression but with different characteristics for the two modes. It is left as an exercise to show that one complete oscillation of the solution satisfying initial conditions $y(0) = 0$, $\dot{y}(0) = z_0$ is

$$y(t) = \begin{cases} \dfrac{z_0}{\alpha} \sin \alpha t, & 0 \le t \le \dfrac{\pi}{\alpha}, \\[2mm] -\dfrac{z_0}{\beta} \sin \beta \left(t - \dfrac{\pi}{\alpha} \right), & \dfrac{\pi}{\alpha} \le t \le \dfrac{\pi}{\alpha} + \dfrac{\pi}{\beta}. \end{cases}$$

Thus the period of the oscillation is $\pi/\alpha + \pi/\beta$. The shape of the phase curves is taken up in the exercises.

EXERCISES

Phase Space versus Space-Time. Make a sketch in the yz phase space of the solution to each of the following initial-value problems. (You can do this directly either by using expressions for $y(t)$ and $z(t)$ or by eliminating t to get a single equation in y and z.) Also make a separate sketch relative to ty axes of the solution graph itself.

1. $\ddot{y} + y = 1$, $y(0) = 2$, $\dot{y}(0) = 0$.

2. $\ddot{y} - y = 0$, $y(0) = 1$, $\dot{y}(0) = 0$.

3. $\ddot{y} = 1$, $y(0) = \dot{y}(0) = 0$.

4. $\ddot{y} + y = 0$, $y(0) = 2$, $\dot{y}(0) = 1$.

5. $\ddot{y} + \dot{y} = 0$, $y(0) = 2$, $\dot{y}(0) = 0$.

6. $\ddot{y} - \dot{y} = 0$, $y(0) = 1$, $\dot{y}(0) = 0$.

Phase Portraits. Sketch at least three different phase curves for each differential equation below by first finding a family of equations relating y and $z = \dot{y}$ to a constant depending on initial conditions. You can find a suitable equation by eliminating t from the general formulas for $y(t)$ and $z(t) = \dot{y}(t)$ and, if necessary, rewriting the equation so that it displays only one arbitrary constant.

7. $\ddot{y} + y = 1$.   **9.** $\ddot{y} = 1$.   **11.** $\ddot{y} + \dot{y} = 0$.

8. $\ddot{y} - y = 0$.   **10.** $\ddot{y} + y = 0$.   **12.** $\ddot{y} - \dot{y} = 0$.

13. What are the phase curves of the equations $\ddot{y} = \dot{y}$ and $\ddot{y} = -\dot{y}$?

14. Some differential equations have constant solutions, which we have called **equilibrium solutions,** and some don't. Find all equilibrium solutions of the differential equations in Exercises 1 through 6.

15. The differential equation $\ddot{y} = -y + y^2$ has constant equilibrium solutions at two distinct points. What are they? Does the equation have periodic solutions?

16. The differential equation $\ddot{y} + \dot{y} - y + y^3 = 0$ has three distinct constant solutions. What are they?

17. The differential equation $\ddot{y} = -y + \delta y^3$ is a harmonic oscillator equation if $\delta = 0$, and it is said to model a **hard spring** if $\delta < 0$ and a **soft spring** if $\delta > 0$.

(a) Show that in all cases the equation has periodic solutions with initial conditions near enough to $(y_0, z_0) = (0, 0)$. Do you expect to find periodic solutions near the other two equilibria when $\delta > 0$? Explain.

(b) Explain why the terms "hard" and "soft" are appropriate.

*18. The differential equation $\ddot{y} = -y - k \operatorname{sgn}(\dot{y})$ is a harmonic oscillator equation modified by a friction term depending only on the direction of motion and not on the magnitude of its velocity, since the discontinuous real function sgn is defined by

$$\operatorname{sgn}(z) = \begin{cases} 1, & x > 0, \\ 0, & z = 0, \\ -1, & x < 0. \end{cases}$$

This assumption about friction is sometimes appropriate for motion at low speed with high resistance and is called **Coulomb friction.**

(a) Show that the phase paths are patched together from two families of semicircles, one family centered at $y = -k$ for $\dot{y} \geq 0$ and the other family centered at $y = k$ for $\dot{y} < 0$.

(b) Let $k = 1$ and plot the phase curves starting at (i) $(y, \dot{y}) = (-6, 0)$, (ii) $(y, \dot{y}) = (-4, 0)$, (iii) $(y, \dot{y}) = (-2, 0)$.

(c) Show by using properties of the phase plane that the motion stops completely in finite time regardless of the initial conditions.

Behavior near Equilibrium. A first-order autonomous equation $\dot{y} = f(y)$ has in its phase portrait just a single curve, one that is particularly easy to identify: It is the graph in the yz plane of $z = f(y)$. The equilibrium points are the points where this graph crosses the horizontal axis. Sketch the phase curve of each of the following equations, and mark the intervals that have an equilibrium point at one or both ends. Then for an arbitrary initial point y_0 in each such interval, decide how the corresponding solution will behave relative to each equilibrium endpoint; will it move toward such a point or away from it?

19. $\dot{y} = -y + 1$.   **21.** $\dot{y} = y^2$.   **23.** $\dot{y} = 1 - y^2$.

20. $\dot{y} = -y + y^3$.   **22.** $\dot{y} = y - y^3$.   **24.** $\dot{y} = \sin y$.

25. (a) Show that the general solution $y(t) = Ae^{-kt/2} \cos(\omega t - \alpha)$, $A > 0$, of a damped linear oscillator equation has derivative $\dot{y}(t)$ that can be written in the form $z(t) = -Be^{-kt/2} \sin(\omega t - \beta)$, $B > 0$. In particular, find the constant B in terms of the constants A, k, and ω.

*(b) Show that $(y(t), z(t)) = (A \cos(\omega t - a), -B \sin(\omega t - \beta))$, A, B nonzero constants, is a parametric representation of an ellipse unless $\alpha - \beta$ is an odd multiple of $\pi/2$, in which case it degenerates to a line segment.

26. Show that each phase curve in the yz plane of a solution of $\ddot{y} = \omega^2 y$ satisfies an equation $\omega^2 y^2 - z^2 = C$ for some appropriate constant C. Sketch these curves for $\omega = \frac{1}{2}$, 1, and $\frac{3}{2}$ if the curves are to pass through $(y, z) = (1, 0)$. This should produce a picture analogous to Figure 4.12a but containing hyperbolas instead of ellipses.

27. The phase curves $z^2 = 2(g/l) \cos y + 2C$ with $C = g/l$ (see Example 6 of the text) relate to two distinct types of pendulum behavior.

(a) A single point of the form $(y, z) = (k\pi, 0)$, k an integer, is called an **equilibrium point** and corresponds to a constant solution of $\ddot{y} = -(g/l) \sin y$. What physical states do these solutions describe? In particular, what is the physical distinction between odd and even values of k?

(b) If $z^2 = 2(g/l)(\cos y + 1)$ and $z > 0$, the pendulum is in a very special kind of motion. Describe that motion in words, and do the same for the case $z < 0$.

*28. The pendulum equation $\ddot{y} = -\sin y$ has been integrated once in Example 6 of the text to produce the relation $\frac{1}{2}\dot{y}^2 = \cos y + C$ between y and $\dot{y}$.

(a) Requiring $y(0) = 0$, $\dot{y}(0) = \pm 2$, show that the specific relation satisfied under those initial conditions is $\dot{y}^2 = 2(1 + \cos y)$.

(b) The equation found in part (a) has a graph in phase space called a **separatrix** because it separates the librational phase curves from the rotational ones. Taking square roots, separating variables, and using $\cos(y/2) = \pm\sqrt{(1 + \cos y)/2}$, show that a separatrix satisfies a first-order equation $\sec(y/2)\,dy = \pm 2dt$.

(c) Integrate the differential equation in part (b) to show that a separatrix for the equation $\ddot{y} = -\sin y$ satisfies $1 + \sin(y/2) = e^{\pm t} \cos(y/2)$. (There is no such elementary formula relating time and position for the rotational and librational solutions.)

29. By studying Figure 4.14b carefully you can sketch the phase curves for the damped pendulum equation $\ddot{y} = 0.1\dot{y} - \sin y$ passing through the distinct points

$$(y, z) = (0, -2), (0, -2.3), (0, -2.8).$$

Make your three sketches relative to a single set of yz axes, and explain in words how the corresponding pendulum motions differ from the ones represented in Figure 4.14b. [*Hint:* You'll need to use more of the negative y axis than Figure 4.14b does.]

30. (a) Periodicity Theorem 5.1 guarantees the existence of periodic solutions to the pendulum initial-value problem $\ddot{y} = -(g/l)\sin y$, $y(0) = y_0$, $\dot{y}(0) = z_0$ if (y_0, z_0) is near phase points of the form $(2k\pi, 0)$ but not necessarily if $2k\pi$ is replaced by an odd multiple of π. Explain this difference on the basis of the statement of the periodicity theorem.

(b) Explain the difference described in part (a) on physical grounds.

*31. A differential equation of the form $\ddot{y} = \phi(y)$ expresses the acceleration of a point on a line as a function of position y on the line.

(a) Show that the equation is equivalent to the pair of equations $\dot{y} = z$, $\dot{z} = \phi(y)$.

(b) Show that the phase curves of $\ddot{y} = \phi(y)$ satisfy $dz/dy = \phi(y)/z$ for $z \neq 0$.

(c) Show that the differential equation of part (b) has solutions $\frac{1}{2}z^2 = \Phi(y) + \text{const}$, where $\Phi'(y) = \phi(y)$.

(d) Assume $\phi(y) = -y^{-2}$, and sketch the phase curve through $(y, z) = (2, -1)$.

(e) Find the solution of $\ddot{y} = -y^{-2}$ under the assumption that $(y, z) = (2, -1)$ when $t = 0$. Sketch the graph of this solution for $0 \leq t \leq \frac{4}{3}$.

32. (a) Find all equilibrium points for the damped pendulum equation $\ddot{y} = -k\dot{y} - (g/l)\sin y$.

(b) Based on what you know about a physical pendulum, separate the equilibrium points found in part (a) into two classes: (i) those for which a solution starting near the equilibrium point tends toward it; (ii) those for which a solution starting near the equilibrium point tends away from it.

33. Show that the **Duffing oscillator** $\ddot{y} = -y^3 + y$ has constant equilibrium solutions $y(t) = 1$ and $y(t) = -1$ and that a solution with initial values $y(0) = y_0$, $\dot{y}(0) = z_0$ close enough to $(1, 0)$ or $(-1, 0)$ is periodic, with phase curve looping around $(1, 0)$ or $(-1, 0)$.

34. The equation $\ddot{y} + k(y^2 + \dot{y}^2 - 1)\dot{y} + y = 0$, $k > 0$, is a modification of the Van der Pol equation $\ddot{y} + k(y^2 - 1)\dot{y} + y = 0$. The modified equation exhibits damping when $y^2 + \dot{y}^2 > 1$ and amplification when $y^2 + \dot{y}^2 < 1$.

(a) Show that the circle of radius 1 centered at the origin is a phase curve for the equation.

(b) Find every solution to this modified equation that has a phase plot that lies on the circle of part (a).

(c) What are the equilibrium solutions of the modified equation?

35. The **Morse model** for the displacement y from equilibrium of the distance between the two atoms of a diatomic molecule is $\ddot{y} = K(e^{-2ay} - e^{-ay})$, where K and a are positive constants. Show that the equation has periodic solutions.

36. In all our examples, phase curves of $\ddot{y} = f(y, \dot{y})$ that wind repeatedly about an equilibrium point, for example, the origin in the yz plane, wind in the *clockwise* direction. Explain why this winding is always clockwise. [*Hint:* Show first that $z = \dot{y}$ must change sign as the path winds around the point.]

37. The reason for restricting phase-space analysis for second-order equations to the autonomous case $\ddot{y} = f(y, \dot{y})$ is that for a nonautonomous equation $\ddot{y} = f(t, y, \dot{y})$ there will usually be many phase curves passing through a given point of the yz plane, with the result that a phase portrait becomes too cluttered to be useful.

  (a) Show that the equation $\ddot{y} = 6t$ has phase curves parametrized by equations of the form $y(t) = t^3 + c_1 t + c_2$, $z(t) = 3t^2 + c_1$, where c_1 and c_2 are arbitrary constants.

  (b) Show that distinct curves of the form found in part (a) satisfy the initial conditions $y(t_0) = 0$, $z(t_0) = 0$ for arbitrary choices of t_0.

  (c) Sketch three different phase curves near $(y, z) = (0, 0)$ and passing through that point.

38. The differential equation in Example 10 of the text is $\ddot{y} + f(y) = 0$, where

$$f(y) = \begin{cases} \alpha^2 y, & \text{if } 0 \le y, \\ \beta^2 y, & \text{if } y < 0. \end{cases}$$

Show that a phase curve passing through $(y, z) = (0, z_0)$, with $z_0 > 0$, consists of a closed loop composed of right and left halves of two ellipses, one passing through $(y, z) = (z_0/\alpha, 0)$ and the other passing through $(y, z) = (-z_0/\beta, 0)$. This can be done by considering the first-order equation $dz/dy = -f(y)/z$ in two separate pieces and then coordinating the results.

39. Verify directly that the solution given in the second part of Example 10 of the text is in fact a solution to the oscillator equation with a two-piece right-hand side and that the solution can be extended periodically. Note that this is not simply a matter of plugging some formulas into the differential equation; it's also necessary to check that the slopes of the solution graphs match up properly at $t = 0$, π/α, and $\pi/\alpha + \pi/\beta$.

Plotting Phase Curves Using Direction Fields and Isoclines. Letting $\dot{y} = z$, we've seen that phase curves of the autonomous second-order equation $\ddot{y} = f(y, \dot{y})$ satisfy the equation $dz/dy = f(y, z)/z$. These curves can sometimes be sketched using the direction field of the first-order equation. (See Chapter 1, Section 2B, where it is explained how to do this using isoclines.) Use this method, with or without isoclines, to sketch phase portraits for the equations in Exercises 40 through 48.

40. $\ddot{y} = -y$.

41. $\ddot{y} = y + 1$.

42. $\ddot{y} = y^2$.

43. $\ddot{y} = y$.

44. $\ddot{y} = -y^2$.

45. $\ddot{y} = y + \dot{y}$.

46. $\ddot{y} = -\sin y$.

47. $\ddot{y} = y$.

48. $\ddot{y} = -|y|y$. [*Hint:* Show that phase curves for $\ddot{y} = -|y|y$ are symmetric in the z axis.]

49. The inverse-square law equation $\ddot{y} = -y^{-2}$, $y > 0$, has paths in phase space that are symmetric about the y axis and become unbounded as y tends to zero.

   (a) Sketch the ones passing through the points $(y, z) = (1, 1)$, $(1, 2)$, $(2, 2)$. Also indicate the directions that $(y(t), \dot{y}(t))$ traces these curves as t tends to plus infinity.

   (b) Show that a phase curve of $\ddot{y} = -y^{-2}$ crosses the y axis if and only if it contains a point (y_0, z_0) for which $y_0 z_0^2 < 2$.

50. If the function $f(y)$ in Theorem 5.1 is an odd function, that is, $f(-y) = -f(y)$, then the phase paths of the equation $\ddot{y} = -f(y)$ will be not only symmetric about the horizontal y axis but also about the z axis. Explain why this is so.

51. In Example 6 of the text it is shown that a solution $y = y(t)$ of the undamped pendulum equation $\ddot{y} = -(g/l) \sin y$ satisfies $\frac{1}{2}\dot{y}^2 = (g/l) \cos y + C$, $C =$ const.

   (a) If $y(0) = y_0$, and $\dot{y}(0) = z_0$, find the value of C for the resulting solution.

   (b) Given that the initial angle is $y(0) = y_0$, find out how large the initial speed $|z_0|$ must be so that the pendulum goes over the top. [*Hint:* You must have $\dot{y}^2(t) > 0$ for all t.]

52. Calvin has asked Hobbes for a push on a swing with ropes 9 feet long. If Hobbes pushes the swing when it's stationary at its lowest position, what angular velocity does he need to give it to make it go over the top? Neglect friction.

53. A certain second-order differential equation has, among others, the phase curves shown in the accompanying figure. Answer the following questions about the specified positions $y(t)$ and their corresponding velocities $\dot{y}(t)$. The scales on the two axes are slightly different and are indicated at $y = \dot{y} = 3$.

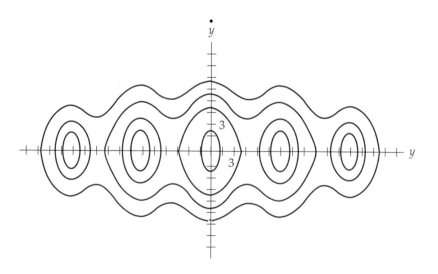

   (a) For a solution for which the velocity is 4.5 when the position is 21, estimate the velocity when the position is 10.

(b) Does the solution referred to in part (a) appear to be periodic?

(c) For a solution for which the position is 7 when the velocity is 2.5, estimate the maximum and minimum velocities.

(d) Estimate the maximum and minimum positions attained by the solution referred to in part (c).

(e) Would you expect a solution satisfying $y(0) = \dot{y}(0) = 1$ to be periodic?

(f) The differential equation has constant solutions; try to identify the ones that appear in the phase portrait.

(g) If a phase curve were parametrized by $(y(t), \dot{y}(t))$, would it be traced clockwise or counterclockwise?

54. A certain second-order differential equation has, among others, the phase curves shown in the accompanying figure. Answer the following questions about the specified positions $y(t)$ and their corresponding velocities $\dot{y}(t)$.

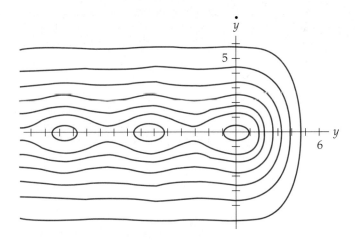

(a) For a solution for which the velocity is 3.4 when the position is 0, estimate the velocity when the position is 3.

(b) Does the solution referred to in part (a) appear to be periodic?

(c) For a solution for which the position is 1 when the velocity is 1, estimate the maximum and minimum velocities.

(d) Estimate the maximum position attained by the solution referred to in part (c); what can you say about the minimum position?

(e) Would you expect a solution satisfying $y(0) = \dot{y}(0) = -4$ to be periodic?

(f) The differential equation has constant solutions; try to identify the ones that appear in the phase portrait.

55. A nonconstant periodic solution $y(t)$, along with its derivative $\dot{y}(t)$, to the nonautonomous (i.e., time-dependent) second-order equation $\ddot{y} = g(t, y, \dot{y})$ always generates a closed loop in $y\dot{y}$ phase space.

(a) Verify the assertion above.

(b) Verify that the equation $\ddot{y} - \dot{y} + e^{2t}y = 0$ has solutions of the form
$$y(t) = c_1 \cos e^t + c_2 \sin e^t.$$
Show that these solutions are not periodic functions of t unless $c_1 = c_2 = 0$, and that they do not trace closed loops in $y\dot{y}$ phase space.

(c) Assume that the autonomous differential equation $\ddot{y} = g(y, \dot{y})$ has a unique solution satisfying each initial condition of the form $y(t_0) = y_0$, $\dot{y}(t_0) = z_0$. Show that every solution that traces a closed loop in $y\dot{y}$ phase space is periodic, with period τ equal to the time it takes to traverse the loop once.

6 ENERGY AND STABILITY

6A *Potential Energy*

The equation $m\ddot{y} = -f(y)$, with m a positive constant, has the interpretation in mechanics that force (mass m times acceleration $\ddot{y}$) depends just on position y. A simple example is the harmonic oscillator equation $m\ddot{y} = -hy$, with h a positive constant. This equation can be solved completely using the formula $y(t) = A \cos (\sqrt{h/m}t - \alpha)$. Unfortunately such explicit elementary solutions to $m\ddot{y} = -f(y)$ are available only for a few additional cases. However, the concept of *potential energy* defined below allows us to draw conclusions about solutions even in the absence of explicit solution formulas. The basic idea is to introduce velocity $\dot{y}$, even though it's not explicit in $m\ddot{y} = -f(y)$, and regard the second-order equation as a *system* of two equations. In the present context it's natural to think of $\dot{y}$ as velocity and so denote it by v. Thus if $\dot{y} = v$, then $\ddot{y} = \dot{v}$ and $\dot{v} = -f(y)$. By the chain rule
$$\frac{d^2y}{dt^2} = \frac{dv}{dt} = \frac{dv}{dy}\frac{dy}{dt} = v\frac{dv}{dy}.$$
Using this result allows us to express $m\ddot{y} = -f(y)$ as
$$mv\frac{dv}{dy} + f(y) = 0, \qquad \text{so} \qquad m\int v \, dv + \int f(y) \, dy = 0.$$
Carrying out the integration gives $\frac{1}{2}mv^2 + U(y) = C$, where $U(y)$ is some function such that $U'(y) = f(y)$. This function of velocity $T = \frac{1}{2}mv^2$ is called the **kinetic energy** of the resulting motion, and the function of position $U(y)$ is called the **potential energy,** or **potential,** of the force field $f(y)$. So *kinetic energy plus potential energy is constant,* and a physical system governed by the equation $\ddot{y} = -f(y)$ is consequently called **conservative,** because the **total energy** $E(y, v) = \frac{1}{2}mv^2 + U(y)$ is constant, that is, *conserved,* during a specific motion. For example, when potential energy decreases, the kinetic energy must increase by the same amount. Because velocity $v = \dot{y}$ and position y satisfy $E(y, \dot{y}) = C$, a parametrized curve $(y(t), v(t))$ lies on a level curve of the function $E(y, v)$ of two variables. Note that, because it is defined to be an indefinite integral of the function $f(y)$, the potential $U(y)$ is defined only to within an additive constant. Thus potential energy may be regarded as a relatively abstract concept, though ultimately a very useful one. The first two examples are

atypical in that an explicit formula $y(t)$ for the motion is possible for both of them.

---

example 1

The elementary falling-body problem can be described by the second-order equation $m\ddot{y} = -mg$. Here $y = y(t)$ represents the 1-dimensional position, measured up from some initial point $y(0) = y_0$, of a body of mass m under the influence of a constant gravitational acceleration of magnitude g acting in the downward direction. A potential is $U(y) = \int_0^y mg\,dy = mgy$. Conservation of energy is expressed by

$$E(y, v) = \tfrac{1}{2}mv^2 + mgy = C.$$

This equation allows us to express either y or v as a function of the other. The elementary solution $y(t) = \tfrac{1}{2}gt^2 + v_0 t + y_0$ is easily seen to satisfy the conservation equation with $C = \tfrac{1}{2}mv_0^2 + mgy_0$. Choosing a different potential by adding a constant c to it would add mgc to C. In any case, total energy is constant during the motion, and the relation between position and velocity is governed by an equation $E(y, v) = C$.

---

example 2

According to Newton's inverse-square law of gravitational attraction, the distance $y(t)$ between the centers of mass of two bodies moving directly toward or away from each other, for example, the earth and a rocket, satisfies a differential equation of the form

$$\frac{d^2y}{dt^2} = -\frac{k}{y^2}, \qquad k \text{ a positive constant.}$$

Some background for this equation is discussed in Chapter 6, Section 5, and here we'll just consider its solution. We can interpret this equation as the equation of motion of a body of mass 1 under the influence of the 1-dimensional force $-f(y) = -ky^{-2}$. Thus the kinetic energy is $T = \tfrac{1}{2}v^2$, and the potential energy is $U(y) = -ky^{-1}$. The potential function increases through negative values toward zero as y increases, so the kinetic energy would correspondingly decrease.

It's natural to assume that the separation distance y is always positive, and since $-ky^{-2} < 0$, the second derivative d^2y/dt^2 is always negative. Consequently the graph of a solution $y = y(t)$ is always concave down, regardless of whatever initial conditions are imposed. Denoting $\dot{y}$ by v, we make our generic substitution $\ddot{y} = v\,dv/dy$, getting

$$v\frac{dv}{dy} = -\frac{k}{y^2}, \qquad \text{which integrated with respect to } y \text{ is} \qquad \tfrac{1}{2}v^2 = \frac{k}{y} + C_1.$$

Initial conditions $y(0) = y_0$, $\dot{y}(0) = v(0) = v_0$ require that $C_1 = \tfrac{1}{2}v_0^2 - k/y_0$, so

$$v^2 = \frac{2k}{y} - \frac{2k}{y_0} + v_0^2.$$

Since the left side of this equation is a square, the right side must be nonnegative, placing a restriction on the relationship between y and v. We'll see that the special case in which $v_0^2 = 2k/y_0$, so $v^2 = 2k/y$, is critical. Thus to find $y(t)$ we

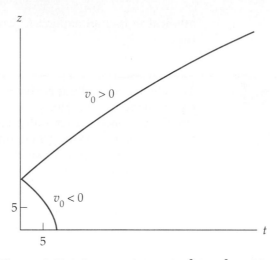

Figure 4.18 Solutions of $\ddot{y} = -ky^{-2}$ for $v_0^2 = 2k/y_0$.

need to solve $(dy/dt)^2 = 2ky^{-1}$, or $dy/dt = \pm\sqrt{2k}y^{-1/2}$. The two cases, distinguished by the plus-or-minus sign, correspond to significantly different physical situations, depending on whether the two bodies in question are always moving apart ($dy/dt > 0$) or together ($dy/dt < 0$). Separating variables gives

$$y^{1/2}\,dy = \pm\sqrt{2k}\,dt, \qquad \text{which integrated is} \qquad \tfrac{2}{3}y^{3/2} = \pm\sqrt{2k}t + C_2.$$

The constant C_2 can be determined by setting $y(0) = y_0$ so that $C_2 = \tfrac{2}{3}y_0^{3/2}$. Then solving for y gives

$$y = \left(y_0^{3/2} \pm 3\sqrt{\frac{k}{2}}\,t\right)^{2/3},$$

where we use the plus sign if $v_0 > 0$ and the minus sign if $v_0 < 0$. In the latter case, the value $y = 0$ is reached at time $t = \tfrac{1}{3}\sqrt{2/k}y_0^{3/2}$. Alternatively, if $v_0 > 0$, the distance $y(t)$ increases indefinitely. Indeed, in that case the critical initial velocity that we chose relative to y_0, namely, $v_0 = \sqrt{2k/y_0}$, is called the **escape velocity** of one body with respect to the other. This and other matters having to do with $d^2y/dt^2 = -k/y^2$ are taken up in the exercises. The two cases are illustrated in Figure 4.18.

It's particularly helpful in some examples to consider curves parametrized by $(y, v) = (y(t), \dot{y}(t))$ in what is called the 2-dimensional **phase space,** as in the next example. This concept is introduced in Section 5A, but the present discussion is independent of that.

e x a m p l e 3

The harmonic oscillator equation $\ddot{y} = -y$ has $U(y) = \tfrac{1}{2}y^2$ for a potential, with corresponding total energy $E(y, v) = \tfrac{1}{2}v^2 + \tfrac{1}{2}y^2$. The typical phase curve is a circle $\tfrac{1}{2}v^2 + \tfrac{1}{2}y^2 = E > 0$ of radius $\sqrt{2E}$ centered at $(y, v) = (0, 0)$. Figure 4.19a shows some parabolic curves $y = \text{const}$ on the graph of $E = E(y, v)$ in 3-dimensional space, along with two level curves in the yv plane and the corresponding

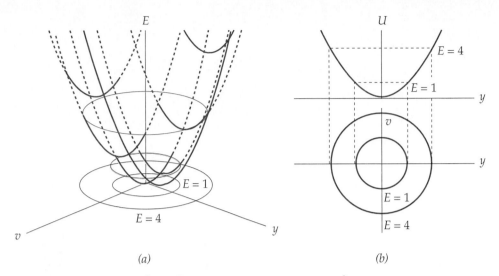

Figure 4.19 (a) $E = \frac{1}{2}v^2 + \frac{1}{2}y^2$ with level curves. (b) $U = \frac{1}{2}y^2$ with level curves of E.

circles on the graph of $E(y, v)$. Because a term $\frac{1}{2}mv^2$ always occurs in the function $E(y, v)$, the parabolas $y =$ const always occur in the graph of E. Therefore in the interest of avoiding clutter, and reducing the picture to its essentials, we can suppress these parabolas and display only the graph of $U(y)$, properly aligned with some plane level curves of E as in Figure 4.19b.

The upward-opening curve in Figure 4.19b is called a **potential well** for the following reason. Since kinetic energy T is nonnegative, $E = T + U \geq U(y)$. For a given energy level E, the position y is then restricted to those values for which $U(y) \leq E$. Thus a particle with energy E may be "trapped in the well" at level E. (One of the many reasons potential energy is a useful concept is that it fosters this kind of heuristic thinking.)

e x a m p l e 4

Continuing from Example 3, the restriction $E \geq U(y) = \frac{1}{2}y^2$ is equivalent to $\sqrt{2E} \geq |y|$. For the more general harmonic oscillator $\ddot{y} = -\omega^2 y$, a potential is $U_\omega(y) = \frac{1}{2}\omega^2 y^2$. At energy level E we have $E \geq \frac{1}{2}\omega^2 y^2$, which translates into $\sqrt{2E/\omega^2} \geq |y|$. Thus to attain a given amplitude of oscillation, the total energy E must increase like the square of the circular frequency ω.

e x a m p l e 5

Figure 4.20a shows a fairly weak steel spring leaf that is planted vertically in a firm base and is allowed to vibrate back and forth in a vertical plane. Fixed symmetrically above the path of the spring's tip are two magnets of equal strength that act on the steel spring. Experiments show that positive constants γ and δ, dependent on the characteristics of the spring and magnets, can be chosen so that solutions of the differential equation $\ddot{y} = -\gamma y^3 + \delta y$ provide a good description for the displacement $y = y(t)$ of the spring tip. Typical ex-

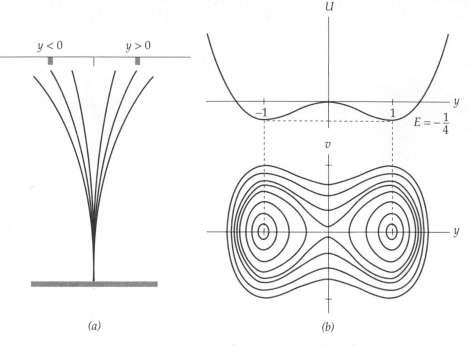

Figure 4.20 (*a*) **Duffing oscillator:** $\ddot{y} = -y^3 + y$. (*b*) $U = \frac{1}{4}y^4 - \frac{1}{2}y^2$ with level curves of E.

pected behavior would be for the spring to wiggle near one magnet or the other or else to oscillate from a neighborhood of one magnet to the other. The general behavior is well enough illustrated by taking $\gamma = \delta = 1$, in which case $\ddot{y} = -y^3 + y$. A potential is $U(y) = \frac{1}{4}y^4 - \frac{1}{2}y^2$, so the phase curves satisfy equations of the form $\frac{1}{2}v^2 + \frac{1}{4}y^4 - \frac{1}{2}y^2 = C$. Some of these are sketched in Figure 4.20*b* along with the graph of $U(y)$. The closed convex loops around the equilibrium points $(\pm 1, 0)$ correspond to periodic solutions that oscillate about the constant equilibrium values $y = \pm 1$ and remain, respectively, either positive or negative. The closed nonconvex loops correspond to periodic solutions that oscillate about the constant equilibrium value $y = 0$. Given our choice of integration constant in the potential function, the small loops go with energies E such that $-\frac{1}{4} < E < 0$, and the large loops go with $E > 0$. Energies $E = -\frac{1}{4}$ and $E = 0$ go with the equilibrium solutions themselves. The differential equation itself is called a **Duffing oscillator.** The damped and externally driven version is considered in the exercises for Section 7 using numerical methods.

6B   Stability

Suppose $E(y, v) = \frac{1}{2}mv^2 + U(y)$ is the total energy function associated with the conservative equation $\ddot{y} = -U'(y)$. For each fixed $y = y_1$, the graph of $e = E(y_1, v)$ is an upward-pointing parabola symmetric about $v = 0$. It follows that any local or global minimum points on the graph of $E(y, v)$ can occur only where $v = 0$ and so must correspond to the respective local or global minimum values for the function $E(y, 0) = U(y)$. The test described below shows that these minimum values of potential energy imply *stability* of an equilibrium point for the equation. (This is one of the reasons that the concept of potential energy is important.) In Section 5 an **equilibrium solution** for a differential equation is defined to be a constant function $y(t) = y_0$ that satisfies the equation, and the corresponding **equilibrium point** in phase space is a point of the form $(y, v) = (y_0, 0)$.

■ DEFINITION

An equilibrium point $(y_0, 0)$ of the conservative equation $\ddot{y} = -f(y)$ is called **stable** if a phase path $(y, v) = (y(t), \dot{y}(t))$ can be made to stay arbitrarily close to the point $(y_0, 0)$ for all t simply by making the initial point $(y(t_0), \dot{y}(t_0))$ close enough to $(y_0, 0)$. An **unstable** equilibrium is, of course, one that is *not* stable in that there are phase points more than some fixed distance δ away from the equilibrium but on paths that start arbitrarily close to it. In applications, stability of an equilibrium means that a small enough displacement from equilibrium, either in position or velocity or both, will from some time on result in a phase path for which position and velocity lie as near as you please to the equilibrium position and to zero velocity, respectively. It turns out that an equilibrium y_0 of $\ddot{y} = -f(y)$ will be stable if the potential has a *strict* minimum value at $y = y_0$; **strict minimum** just means that $U(y) > U(y_0)$ throughout some neighborhood of y_0.

***example* 6**

The Duffing oscillator $\ddot{y} = -y^3 + y$ has potential $U(y) = \frac{1}{4}y^4 - \frac{1}{2}y^2$ with strict local minimum values at $y = \pm 1$. The small closed loops around $(\pm 1, 0)$ are characteristic for stable equilibrium at these points. The large loops that pinch in arbitrarily closely to the equilibrium at $(0, 0)$ display instability, because they fail to remain close to $(0, 0)$ no matter how closely they pass by that point; in particular, they loop out well beyond $(\pm 1, 0)$.

For conservative equations $\ddot{y} = -f(y)$ we have the following criterion.

■ 6.1 LAGRANGE STABILITY TEST

Suppose that $f(y)$ is continuous near $y = y_0$ and that $\ddot{y} = -f(y)$ has $y(t) = y_0$ for an equilibrium solution. If a potential energy function $U(y)$ for the equation (i.e., $U'(y) = f(y)$) has a strict local or global minimum at $y = y_0$, then $(y, \dot{y}) = (y_0, 0)$ is a *stable* equilibrium point for the differential equation.

Proof. By a shift of coordinates we can assume that $y_0 = 0$ and that $U(0) = 0$. Suppose a is chosen so that $U(y)$ is continuous and positive for $0 < |y| \leq a$. Let δ be the minimum of $E(y, v) = \frac{1}{2}v^2 + U(y)$ on the circle $y^2 + v^2 = a^2$. Now suppose a phase path with $y^2(t_0) + v^2(t_0) < a^2$ satisfies $0 < E(y, v) = C < \delta$ for some t_0. For this path to leave the region R_a in which $E(y, v) < \delta$, the corresponding value of $E(y, v)$ would have to become at least as large as the minimum value δ on the circle. But this is impossible since $E(y, v) = C < \delta$ on the path. Finally, note that R_a shrinks to $(0, 0)$ as $a \to 0$. ■

Note that the usual calculus techniques for identifying strict local minima (i.e., $U'(y) = f(y) = 0$, $U''(y) = f'(y) > 0$) may be applicable in applying the Lagrange stability test.

example 7

The pendulum equation $\ddot{y} = -\sin y$ is a nice example to which we can apply Lagrange's test. A potential is $U(y) = -\cos y$, or $1 - \cos y$ if we want it to be nonnegative. The constant-energy curves in phase space satisfy equations of the form $\frac{1}{2}v^2 - \cos y = C$. Note that the minimum values for $U(y)$ occur at even multiples of π, $y = 2n\pi$. These angles correspond to a pendulum in its downward, and therefore stable, vertical position, since the angle y is measured from that position. Figure 4.21 shows the correspondence between potential energy and constant-energy curves. The odd multiples of π, namely, $y = (2n + 1)\pi$, correspond to the unstable vertical positions of a pendulum.

In determining which points on a potential energy graph correspond to stable or unstable equilibria, it's helpful to think of a little ball rolling along the graph of $U(y)$ under the influence of constant gravity and with negligible friction. The equilibrium points are the critical points, where $U'(y) = 0$, since those points correspond to the constant solutions of $\ddot{y} = -U'(y)$. Among the critical points are those stable positions identified by the Lagrange stability test, where the graph of $U(y)$ has a strict minimum. Figure 4.22 shows a graph with four critical points, one of which is a stable minimum; of the other three, two are inflection points and one is a strict maximum. These three points, which are clearly unstable for the motion of the ball, turn out to correspond to unstable equilibrium solutions for the differential equation $\ddot{y} = -U'(y)$ according to the following criterion.

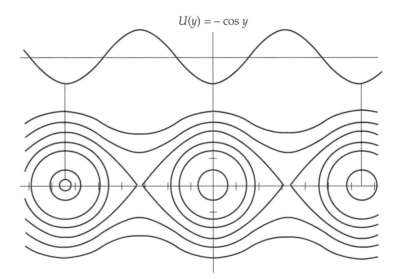

$U(y) = -\cos y$

Figure 4.21 Phase curves and potential energy graph for $\ddot{y} = -\sin y$.

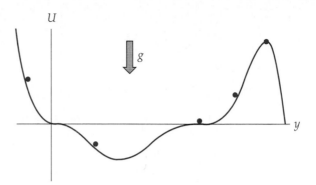

Figure 4.22 Stable and unstable equilibria for a ball rolling on a curve.

■ 6.2 INSTABILITY TEST

Let y_0 be an equilibrium solution of $\ddot{y} = -f(y)$. If a potential $U(y) = \int f(y)\, dy$ is strictly decreasing in an interval $[y_0, b]$ to the right of $y_0 (f(y) < 0)$, or strictly increasing in an interval $[a, y_0]$ to the left of $y_0 (f(y) > 0)$, then y_0 is an *unstable* equilibrium solution.

Proof. The assumptions imply that either $U'(y) = f(y)$ is strictly negative for y in an interval $[y_0, b]$ or else strictly positive in an interval $[a, y_0]$. If $y(t)$ satisfies $y_0 < y(t_0) < b$ in the first case, then the corresponding acceleration $\ddot{y}(t) = -f(y(t))$ is positive. It follows that if $\dot{y}(t_0) > 0$, then $y(t)$ increases from $y(t)$ to b, no matter how close $(y(t_0), \dot{y}(t_0))$ is to $(y_0, 0)$. Thus y_0 is unstable. A similar argument applies to the interval $[a, y_0]$. ■

---

example 8

The conservative system $\ddot{y} = -2y + \frac{3}{5}y^2$ has $U(y) = y^2 - \frac{1}{5}y^3$ for a potential. This equation can be regarded as a small perturbation of the harmonic oscillator equation $\ddot{y} = -2y$ when y is very small, because the additional term $\frac{3}{5}y^2$ is even smaller. For large values of y the behavior can be expected to be quite different. The graph of $U(y)$ shown in Figure 4.23 has a strict minimum at $(y, v) = (0, 0)$, so $y_0 = 0$ is a stable equilibrium solution. Also there is an unstable equilibrium at $y_1 = \frac{10}{3}$. At energy levels strictly between $U(0) = 0$ and $U(y_1) = \frac{100}{27} \approx 3.7$, the phase curves are closed loops that represent periodic solutions of the differential equation. At energy levels above $\frac{100}{27}$ the phase curves loop around the origin from the lower right and head off to infinity on the upper right. At negative energy levels the phase curves are unbounded but don't loop around the origin.

---

example 9

The concept of total energy $E = T + U$ is useful for studying nonconservative systems as well as conservative ones; equations of the form $\ddot{y} = -f(y) - g(\dot{y})$ are good examples. A more specific example is the damped harmonic oscillator equation $m\ddot{y} = -ky - h\dot{y}$. Define a potential to be what it would be if $g(\dot{y}) = 0$, namely, $U(y) = \int f(y)\, dy$. The total energy is still defined as $E(y, v) =$

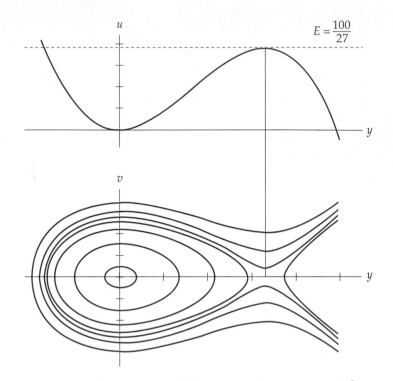

Figure 4.23 Potential and phase curves for $\ddot{y} = -2y + \frac{3}{5}y^2$.

$\frac{1}{2}mv^2 + U(y)$. If the differential equation were conservative, that is, if $g(\dot{y}) = 0$, the total energy would be constant and we'd have $dE/dt = 0$. More generally, since $v = \dot{y}$ and $m\ddot{y} + U'(y) = g(\dot{y})$, we have

$$\frac{dE}{dt} = \frac{d}{dt}\left(\tfrac{1}{2}m\dot{y}^2 + U(y)\right) = m\ddot{y}\dot{y} + U'(y)\dot{y}$$

$$= (m\ddot{y} + U'(y))\dot{y} = g(\dot{y})\dot{y}.$$

For example, the damped harmonic oscillator has $g(\dot{y}) = -k\dot{y}$, so $dE/dt = -k\dot{y}^2$. We conclude from this that the energy E is decreasing as a function of time. A system for which E is nonincreasing but not constant is called **dissipative** for this reason. Stability for dissipative equations is covered by the discussion of Chapter 6, Section 8.

EXERCISES

1. Verify directly that the general solution $y(t) = -\tfrac{1}{2}gt^2 + v_0t + y_0$ to $m\ddot{y} = -mg$ satisfies a conservation equation of the form $\tfrac{1}{2}mv^2 + mgy = C$, and find the constant C in terms of y_0 and v_0.

2. The falling-body equation $m\ddot{y} = -mg$ has phase curves that are unbounded in both the y and v directions. Assume units chosen so that $g = 1$.

   (a) Sketch the curves of constant total energy for energies $E = -1, 0, 1, 2$ assuming potential $U(0) = 0$.

(b) What is the relation between the constant energies E of part (a) and the maximum height of the falling body?

3. The falling-body equation $m\ddot{y} = -mg$ has total energy expressible as $E(y, v) = \frac{1}{2}mv^2 + mgy$. Use this relation to express maximum possible height y in terms of energy level E.

4. Sketch the surface that is the graph relative to 3-dimensional yvE axes of the total energy function $E(y, v) = \frac{1}{2}v^2 + gy$ associated with $\ddot{y} = -g$. Assume the units are chosen so that $g = 1$.

5. In Example 1 of the text we found the total energy function for the frictionless falling-body equation $m\ddot{y} = -mg$ to be $E(y, v) = \frac{1}{2}mv^2 + mgy$. In what follows assume $m = 1$, $g = 32$ feet per second per second.

   (a) Use appropriate scales on ty axes to sketch the graph of the solution of the differential equation that satisfies the initial conditions $y(0) = 0$, $\dot{y}(0) = 16$.

   (b) Use appropriate scales on yv axes to sketch the graph of the phase curve of the differential equation that satisfies the initial conditions $y(0) = 0$, $\dot{y}(0) = 16$. Mark the points on this graph that correspond to $t = 0, 0.5, 1$.

   (c) Make a sketch relative to 3-dimensional axes of the total energy function $E = E(y, v)$.

6. The harmonic oscillator equation $\ddot{y} = -2y$ has bounded phase curves.

   (a) Sketch the curves of constant total energy for energies $E = 1, 2, 3$ assuming potential $U(0) = 0$.

   (b) What is the relation between the constant energies E of part (a) and the maximum displacement y? Answer the same question with displacement replaced by velocity.

   (c) What becomes of the phase curves of part (a) for constant energy $E = 0$? For $E < 0$?

7. Sketch the surface that is the graph relative to 3-dimensional yvE axes of the total energy function $E(y, v) = \frac{1}{2}v^2 + y^2$ associated with the harmonic oscillator equation $\ddot{y} = -2y$. On the graph, and below it in the plane $E = 0$, sketch the graphs corresponding to energy level $E = 1$.

8. **(a)** Show that the second-order equation $\ddot{y} = -f(y)$, with $f(y)$ continuous, is equivalent to the first-order system $\dot{y} = v$, $\dot{v} = -f(y)$.

   (b) Recall that a potential energy function for the system in part (a) is a function U such that $U'(y) = f(y)$, and the total energy is then $E(y, v) = \frac{1}{2}v^2 + U(y)$. Verify directly, by using the chain rule for functions of two variables, that if $y = y(t)$, $v = v(t)$ is a solution of the system, then

   $$\frac{d}{dt} E(y(t), v(t)) = 0,$$

   with the consequence that energy $E(y, v)$ is constant during the motion of the system.

9. The differential equation $\ddot{y} = -y + \delta y^3$ is a harmonic oscillator equation if $\delta = 0$, and is said to model a **hard spring** if $\delta < 0$ and a **soft spring** if $\delta > 0$.

(a) Show that there are three equilibrium solutions if $\delta > 0$ and only one if $\delta \leq 0$. Discuss the stability of these solutions.

(b) Explain why the terms "hard" and "soft" are appropriate.

10. An alternative to the use of the chain rule formula $d^2y/dt^2 = v\,dv/dy$ in integrating the equation $m\ddot{y} = -f(y)$ is simply to multiply the equation by $\dot{y}$: $m\dot{y}\ddot{y} = -f(y)\dot{y}$. Show that integrating both sides of this equation with respect to t yields the conservation of energy equation $\frac{1}{2}mv^2 + U(y) = C$, where $v = \dot{y}$ and $U'(y) = f(y)$.

11. In Example 2 of the text it is shown that the escape velocity at distance y_0 of such a body subject to the inverse-square law is $v_E = \sqrt{2k/y_0}$.

(a) Without using integration, use the equation $v^2 = 2k/y - 2k/y_0 + v_0^2$, derived in Example 2, to show that if $v_0 > v_E$ at distance y_0, then v exceeds $\sqrt{v_0^2 - v_E^2}$ thereafter. Conclude that y tends to infinity.

(b) Use integration to show that if $v_0 = v_E$ at distance y_0, then y still tends to infinity.

12. The inverse-square law equation $\ddot{y} = -ky^{-2}, y > 0, 0 < k = \text{const}$, has phase paths in yv phase space that tend to infinity in the v direction as y tends to zero.

(a) Assuming $k = 1$, sketch a single phase curve passing through each of the points $(y, v) = (1, 1)$, $(1, 2)$, $(2, 2)$.

(b) Show that a phase curve of $\ddot{y} = -ky^{-2}$ contains a point for which $v = 0$ if and only if it contains a point (y_0, v_0) for which $v_0^2 < 2k/y_0$.

(c) Explain the connection between the statement in part (b) and escape velocity.

13. The inverse-square law equation $\ddot{y} = -ky^{-2}, k > 0$, has phase curve $v^2 - 2ky^{-1} = 2E$, or $y = 2k/(v^2 - 2E)$, corresponding to energy level E. Since y represents the distance between the centers of mass of two bodies, we assume $y > 0$.

(a) Let $k = 1$, and sketch typical phase curves corresponding to positive values of E. Describe the general features of the motion in this case. (Each curve has two separate pieces.)

(b) Let $k = 1$, and sketch typical phase curves corresponding to negative values of E. Describe the general features of the motion in this case.

(c) The phase curve corresponding to $E = 0$ is a separating curve, or separatrix, for the two families of curves described in parts (a) and (b). Sketch the separatrix.

14. The differential equation $\ddot{y} = ky^{-2}, k > 0$ expresses the **Coulomb force law** for mutually repelling bodies, for example, two bodies bearing electric charges of the same sign.

(a) Show that the phase curves satisfy the equation $y = 2k/(2E - v^2)$.

(b) Sketch some typical phase curves, following the convention that only points (y, v) for which $y > 0$ are physically relevant.

(c) Show that the velocity attainable on a phase path with total energy E is bounded above by $\sqrt{2E}$.

(d) Conclude from the result of part (c) that the velocity attainable on a phase path with initial values $y(0) = y_0$ and $\dot{y}(0) = v_0$ is bounded above by $\sqrt{v_0^2 + 2k/y_0}$.

15. A car of mass m is propelled along a straight horizontal tract with initial velocity v_0 and subject thereafter to a retarding force $-kv^2$, proportional to velocity squared: $m\ddot{y} = -k\dot{y}^2$.

(a) Show that distance traveled in time t is $y(t) = (m/k) \ln (1 + kv_0 t/m)$, which tends to infinity as t tends to infinity.

(b) Show that with linear retarding force $-kv$, the distance traveled in time t is $y(t) = (mv_0/k)(1 - e^{-kt/m})$, which remains bounded above by m/k as t tends to infinity.

(c) Give an intuitive account of the discrepancy in behavior for large t between the results of parts (a) and (b).

16. An equation of the form $\frac{1}{2}\dot{y}^2 + U(y) = C$ that we derived from $\ddot{y} = -f(y)$ at the beginning of the section is sometimes called a **first integral** of the second-order differential equation, because in principle one more integration will solve the differential equation. Just solve for $\dot{y} = dy/dt$ to get $dy/dt = \pm\sqrt{C - 2U(y)}$. Use separation of variables to solve for y as a function of t in the case of the harmonic oscillator equation $\ddot{y} = -y$, for which $U(y) = \frac{1}{2}y^2$.

17. The Duffing oscillator $\ddot{y} = -y^3 + y$ has potential $U(y) - \frac{1}{4}y^4 - \frac{1}{2}y^2$. Show that the point $(y, v) = (0, 0)$ in phase space is an unstable equilibrium point for the equation by showing that a phase path through $(0, v_0)$ will, for arbitrarily small $|v_0| \neq 0$, contain points at distance more than $\sqrt{2}$ from $(y, v) = (0, 0)$.

18. The **Morse model** for the displacement $y(t)$ from equilibrium of the distance between the two atoms of a diatomic molecule has potential function $U(y) = k(1 - e^{-ay})^2$, k, a positive.

(a) Sketch the graph of $U(y)$ and show that $(y, \dot{y}) = (0, 0)$ is a stable equilibrium.

(b) What second-order equation does $y(t)$ satisfy?

19. Find all equilibrium solutions for the following equations, sketch the graph of the potential $U(y)$, $U(0) = 0$, for each of the following equations, and determine whether each of these equilibrium solutions is stable or unstable.

(a) $\ddot{y} = y^5$.   **(d)** $\ddot{y} = y^2 - y - 1$.   **(g)** $\ddot{y} = y^4 - y^2$.

(b) $\ddot{y} = y^4$.   **(e)** $\ddot{y} = \cos y$.   **(h)** $\ddot{y} = e^y - 1$.

(c) $\ddot{y} = y - y^2$.   **(f)** $\ddot{y} = \tan y$.   **(i)** $\ddot{y} = 1 - y^2$.

20. Make a careful plot of the graph of each potential $U(y)$ associated with the conservative system $\ddot{y} = -U'(y)$, paying particular attention to the location of maxima and minima. Use this information to determine for which energy levels $E = \frac{1}{2}\dot{y}^2 + U(y)$, if any, the corresponding level curves of $\frac{1}{2}\dot{y}^2 + U(y)$ are closed loops, and, hence, which values of C correspond to periodic solutions.

(a) $U(y) = y^2 - y^3$.   **(b)** $U(y) = y - y^3$.

(c) $U(y) = 1.$ **(e)** $U(y) = \cosh y.$

(d) $U(y) = y^2/(1 + y^2).$ **(f)** $U(y) = \sinh y.$

21. Spherical pendulum. Consider a pendulum of length l swinging freely in space from a fixed pivot. (The term "spherical" pendulum is used because the moving end of the pendulum traces a curve lying on a sphere of radius l centered at the pivot.) It can be shown that, for an ideal pendulum of length l whose motion is not confined to a plane, the latitude angle $\theta(t)$ from the vertical satisfies $\ddot{\theta} = -(g/l) \sin \theta + M^2 \cos \theta/\sin^3 \theta$, where $M \neq 0$ is a constant rotation velocity. (See Exercise 18, Chapter 5, Section 1C.)

(a) Show that there is precisely one equilibrium point θ^* in the interval $0 < \theta < \pi$ and that in fact $0 < \theta^* < \pi/2$. [*Hint:* Sketch the graphs of $\cos \theta$ and $\sin^4 \theta$ using the same axes.]

(b) Assume $g = l$ and $M = 1$. Make a numerical estimate of θ^* using a computer plot or, better still, Newton's method.

(c) Use Test 6.1 to show that θ^* is stable.

*22. For integer $n \geq 2$, the identity $y - y^n = y(1 - y^{n-1})$ shows that a solution of the equation $y - y^n = 0$ must either be 0 or have absolute value 1.

(a) Show that for each odd integer $n \geq 3$ the equation $\ddot{y} = y^n - y$ has exactly three equilibrium points, and show that only one of these three is stable. Also sketch a typical potential for this case.

(b) Show that for each even integer $n \geq 2$ the equation $\ddot{y} = y^n - y$ has exactly two equilibrium points, and show that only one of these two is stable. Also sketch a typical potential for this case.

(c) Sketch a typical phase-plane portrait for each of the two cases.

23. Show by example that the strictness requirement for the minimum in the Lagrange stability test is really necessary if the test is to give a correct result. [*Hint:* What if $U(y)$ is constant on an interval?]

*24. Bead on a wire. (a) Suppose a bead of mass m slides without friction on a wire bent into the shape of the graph of $y = f(x)$, the only force being constant gravity g acting in the negative y direction. Show that if f is twice differentiable, the x coordinate of the bead satisfies the equation

$$\ddot{x} = \frac{-(g + f''(x)\dot{x}^2)f'(x)}{1 + f'(x)^2}.$$

[*Hint:* $(m/2)(1 + f'(x)^2)\dot{x}^2 + mgf(x)$ is constant; apply d/dt.]

(b) Solve the equation of part (a) if $f(x) = ax + b$, a, b constant.

(c) Identify the equilibrium points of the motion, and discuss their stability.

Bifurcation Points

25. A value λ_0 for a real parameter λ in a family of differential equations is a **bifurcation point** if system trajectories change character in a fundamental way as λ passes through λ_0. You are asked to show that $\lambda_0 = 1$ is a bifurcation point for $\dot{y} = \sin y \,(\cos y - \lambda)$, because the equilibrium at $y = 0$ changes from stable to unstable as λ increases past 1. Note that additional equilibrium points exist for $|\lambda| \leq 1$.

(a) Use Test 6.1 to show that if $\lambda > 1$ the equilibrium point $y = 0$ is stable.

(b) Use Test 6.2 to show that if $\lambda < 1$ the equilibrium point at $y = 0$ is unstable.

(c) Show that if $|\lambda| < 1$ the equilibrium point at $y_\lambda = \arccos \lambda$ is stable.

(d) Investigate equilibrium stability when $\lambda = 1$.

Stability for First-Order Equations

A first-order autonomous equation $\dot{y} = f(y)$ has in its phase portrait just a single curve: the graph in the yz plane of $z = f(y)$. The equilibrium points are the points where this graph crosses the horizontal y axis. Sketch the phase curve of each of the following equations, and mark the intervals that have an equilibrium point at one or both ends. Then state whether each endpoint is a stable or unstable equilibrium.

26. $\dot{y} = -y + 1$.   **28.** $\dot{y} = y^2$.   **30.** $\dot{y} = 1 - y^2$.

27. $\dot{y} = -y + y^3$.   **29.** $\dot{y} = y - y^3$.   **31.** $\dot{y} = \sin y$.

7 NUMERICAL METHODS

Second-order differential equations are either linear, of the form

$$\frac{d^2y}{dt^2} + a(t)\frac{dy}{dt} + b(t)y = f(t),$$

or nonlinear, for example, the damped pendulum equation

$$\frac{d^2y}{dt^2} + k\frac{dy}{dt} + \frac{g}{l}\sin y = 0.$$

Equations of both types can be written in the very general form

$$\frac{d^2y}{dt^2} = F\left(t, y, \frac{dy}{dt}\right).$$

It is this form that we will treat here along with **initial conditions** of the form

$$y(t_0) = y_0, \qquad \frac{dy}{dt}(t_0) = z_0,$$

where t_0, y_0, and z_0 are given constants. General theory guarantees a unique solution to this problem if $F(t, y, z)$ and its partial derivatives with respect to y and z are continuous on some 3-dimensional box containing the point (t_0, y_0, z_0). It is important to realize that the only type of problem for which we so far have a universally effective method of actually displaying such solutions is the linear equation with $a(t)$ and $b(t)$ both constant. Even then, if $f(t)$ is not a function that can be integrated formally, we may have trouble finding a formula for a particular solution. What we may then settle for is a numerical approximation y_k to the value $y(t_k)$ of the true solution at a discrete set of points t_k. Such approximations were treated in Chapter 2 for first-order equations, and the methods used here for second-order equations are simple modifications of the first-order methods.

If the purpose in solving a differential equation is just to obtain numerical values for some particular solution, then a purely numerical approach may be more efficient than first finding a solution formula and then digging the desired numbers, or graph, out of the formula. On the other hand, if what you want is to display the nature of a solution's dependence on certain parameters in the differential equation, or on initial conditions, then solution by formula is preferable if at all possible. Beyond that, detailed properties of a solution, such as whether it is periodic or only approximately so, can be hard to get from a numerical approximation and easy to get from a formula like $y(t) = 2 \cos 3t$.

7A   *Euler Methods*

Numerical methods for first-order equations can be motivated by using the interpretation of the first derivative as a slope. Rather than trying to make something of the interpretation of the second derivative in a second-order equation, what we will do is find a pair of first-order equations equivalent to a given second-order equation and then apply first-order methods to the simultaneous solution of the pair of equations. The principle is easiest to understand in the general case

$$y'' = F(t, y, y'), \qquad y(t_0) = y_0, \qquad y'(t_0) = z_0.$$

There are many ways to find an equivalent pair of first-order equations, but the most natural is usually to introduce the first derivative y' as a new unknown function z. We write $z = y'$, $z' = y''$, so we can replace $y'' = F(t, y, y')$ by $z' = F(t, y, z)$. The pair to be solved numerically is then

$$y' = z, \qquad\qquad y(t_0) = y_0,$$
$$z' = F(t, y, z), \qquad z(t_0) = z_0.$$

Since $y'(t_0) = z(t_0) = z_0$, there is an initial condition that goes naturally with each equation. To find an approximate solution, we can do what we would do with a single first-order equation, except that at each step we find new approximate values for both unknown functions y and z and then use these values to compute new approximations in the next step. The iterative formulas are as follows for the simple **Euler method,** with step size h:

$$y_{k+1} = y_k + hz_k,$$
$$z_{k+1} = z_k + hF(t_k, y_k, z_k),$$

where $t_k = t_0 + kh$. The starting values y_0, z_0 come from the initial conditions.

---

e x a m p l e 1   The initial-value problem

$$\ddot{y} = -\sin y, \qquad y(0) = \dot{y}(0) = 1, \qquad \text{for } 0 \le t < 20$$

describes the motion of a pendulum with fairly large amplitude, so large that the linear approximation $\ddot{y} = -y$ would be inadequate. We go ahead to solve the system

$$\dot{y} = z, \qquad y(0) = 1,$$
$$\dot{z} = -\sin y, \qquad z(0) = 1,$$

The essence of the computation can be expressed as follows.

```
DEFINE F(T, Y, Z) = -SIN(Y)
SET T = 0
SET Y = 1
SET Z = 1
SET H = 0.01
DO WHILE T < 20
  SET S = Y
  SET Y = Y + H * Z
  SET Z = Z + H * F(T, S, Z)
  SET T = T + H
  PRINT T, Y
LOOP
```

Note the command SET S = Y, saving the current value of y for use two lines later; without this precaution, the advanced value $y + hz$ would be used, which is not correct. The printout results in 2000 values of t from 0.01 to 20 by steps of size 0.01 along with the corresponding y values. We could also get the z values, which are approximations to the values of the derivative $\dot{y}$ at the same t points. This is useful in making a phase-space plot of the solution, plotting approximations to the points $(y(t), \dot{y}(t))$. In the formal routine listed above, the line PRINT T, Y would be replaced by something like PLOT Y, Z for a phase plot.

Rather than displaying a table with 2000 entries, Figure 4.24 shows as a continuous curve the graph of y for the initial conditions $y(0) = y'(0) = 1$. The replacement of sin y by y in the differential equation is inappropriate in this instance, because the values of y that occur are too large to make the approximation a good one. The linearized initial-value problem $\ddot{y} = -y$, $y(0) = y'(0) = 1$ has solution cos t + sin t, whose graph is shown in Figure 4.24 as a dotted curve. It's clear that solution to the linearized problem differs drastically from the shape and period shown by the solution to the true nonlinear pendulum problem.

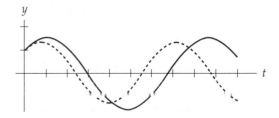

Figure 4.24 Solutions $y(t)$ to $\ddot{y} = -\sin y$ (—) and $\ddot{y} = -y$ ($\cdots$) with $y(0) = y'(0) = 1$.

The **improved Euler method** follows this routine:

$$p_{k+1} = y_k + hz_k$$

$$q_{k+1} = z_k + hF(t_k, y_k, z_k)$$

$$y_{k+1} = y_k + \frac{h}{2}(z_k + q_{k+1})$$

$$z_{k+1} = z_k + \frac{h}{2}[F(t_k, y_k, z_k) + F(t_{k+1}, p_{k+1}, q_{k+1})].$$

Here p_k and q_k provide the simple Euler estimates that are then used to compute the final estimates for y_k and $z_k = \dot{y}_k$; $t_k = t_0 + kh$ as before. (As with the simple Euler method, the value y_k has to be kept for use in computing z_{k+1} and cannot be replaced by y_{k+1} without significant error.)

The advantage of the modification is that the error in the final estimates is substantially reduced without adding much complexity to the computation on a digital computer.

---

e x a m p l e 2

The initial-value problem

$$y'' + y = 0, \qquad y(0) = 0, \qquad y'(0) = 1$$

has solution $y = \sin x$. Table 1 is a table of values that compares the Euler method, the improved Euler method, and the correct value rounded to six decimal places. The value of h used is 0.001, but only every hundredth value is given. The only discrepancies between the last two columns occur in the sixth and the last entries.

An algorithm to produce the first, third, and fourth columns in Table 1 might look like this. (The routine produces only every tenth row of the computed values.)

```
DEFINE F(T, Y, Z) = -Y
SET T = 0
SET Y = 0
SET Z = 1
SET H = .001
FOR J = 1 TO 30
FOR K = 1 TO 100
  SET P = Y + H * Z
  SET Q = Z + H * F(T, Y, Z)
  SET S = Y
  SET Y = Y + .5 * H * (Z + Q)
  SET Z = Z + .5 * H * (F(T, S, Z) + F(T + H, P, Q))
  SET T = T + H
  NEXT K
  PRINT T, Y, SIN(T)
NEXT J
```

Table 1

x	**Euler** y	**Imp-Euler** y	$y \approx \sin x$
0.1	0.099838	0.099834	0.099834
0.2	0.198689	0.198669	0.198669
0.3	0.295564	0.29552	0.29552
0.4	0.389496	0.389418	0.389418
0.5	0.479545	0.479426	0.479426
0.6	0.564812	0.564643	0.564642
0.7	0.644443	0.644218	0.644218
0.8	0.717643	0.717356	0.717356
0.9	0.783679	0.783327	0.783327
1.0	0.841892	0.841471	0.841471
1.1	0.891697	0.891208	0.891207
1.2	0.932598	0.932039	0.932039
1.3	0.964185	0.963558	0.963558
1.4	0.98614	0.98545	0.98545
1.5	0.998243	0.997495	0.997495
1.6	1.00037	0.999574	0.999574
1.7	0.992508	0.991665	0.991665
1.8	0.974725	0.973848	0.973848
1.9	0.9472	0.9463	0.9463
2.0	0.910297	0.909297	0.9093
2.1	0.864117	0.863209	0.863209
2.2	0.809387	0.808496	0.808496
2.3	0.746564	0.745705	0.745705
2.4	0.676275	0.675463	0.675463
2.5	0.599221	0.598472	0.598472
2.6	0.516173	0.515501	0.515501
2.7	0.427958	0.427379	0.42738
2.8	0.335458	0.334988	0.334988
2.9	0.239597	0.239249	0.239249
3.0	0.141333	0.141119	0.14112

Recall that in this algorithm we are dealing with a pair of equations of the form

$$\dot{y} = z \qquad \dot{z} = F(t, y, z).$$

Standard system-solving software uses routines of this kind. The Applet 2ORDPLOT at Web site http.//math.dartmouth.edu/~rewn/ uses this routine.

7B   Stiffness

We remarked in Section 6, Chapter 2, that the local formula errors due to using the Euler and improved Euler methods are, respectively, of order h and h^2. In addition to formula error and the round-off error that we have to deal with, there occurs for some second-order equations an additional difficulty due to what is called **stiffness** of a differential equation. The problem is that in a 2-parameter family of solutions a small error at some point may lead to approximation of a solution that is very different from the one we started on. Here is a simple example.

---

e x a m p l e 3

The differential equation $y'' - 100y = 0$ has the general solution
$$y(t) = c_1 e^{10t} + c_2 e^{-10t}.$$
The initial conditions $y(0) = 1, y'(0) = -10$ single out the solution $y_1(t) = e^{-10t}$. But the "nearby" initial conditions $y(0) = 1, y'(0) = -10 + 2\epsilon$, for small $\epsilon > 0$, single out the solution
$$y_2(t) = 0.1\epsilon e^{10t} + (1 - 0.1\epsilon)e^{-10t}.$$
The behavior of these two solutions is very different for only moderately large values of t. In particular, $y_1(t)$ tends to zero as t tends to infinity, but $y_2(t)$ gets large very soon unless the "error" ϵ is required to be small to an impractical degree. Figure 4.25 shows comparative behaviors, with $\epsilon = 0.00001$. Solutions of nonlinear equations may exhibit the same kind of behavior, and one way to deal with the problem is suggested in the exercises.

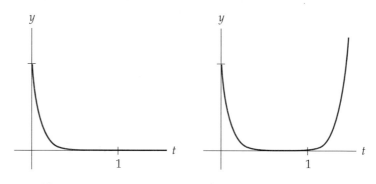

Figure 4.25 Initially close solutions that diverge later.

Accuracy in the Euler methods can be improved, within reasonable limits, by decreasing the step size h. This requires more steps to reach a given value of t and may produce more approximate solution values than is convenient. Thus we'd print results only after m steps of calculation. For example, $h = 0.001$ and $m = 10$ would produce approximate values with argument differences of 0.01.

EXERCISES

The Web site http://math.dartmouth.edu/~rewn/ has Applets called, respectively, 2NDORD and 2NDORDPLOT that can be used to do the numeric and graphic-numeric exercises for this section.

1. Consider the **Airy equation** $\ddot{y} + ty = 0$ for $0 \le t \le 12$.

    (a) Estimate the location of t values corresponding to the zero values of the solution satisfying $y(0) = 0$ and $\dot{y}(0) = 2$. What changes if you replace the condition $\dot{y}(0) = 0$ by $\dot{y}(0) = a$ for various choices of a? [*Hint:* Look for successive approximations to y with opposite sign.]

(b) What can you say about the questions in part (a) on an interval $n \leq t \leq 0$?

(c) Estimate the location of the maxima and minima of the solution satisfying $y(0) = 0$ and $\dot{y}(0) = 2$. [*Hint:* Look for successive approximations to $z = \dot{y}$ with opposite sign.]

2. The **Bessel equation** of order zero (see Chapter 9, Section 7B) is

$$x^2 y'' + xy' + x^2 y = 0.$$

For $1 \leq x \leq 40$, estimate the location of the zero values of a solution satisfying $y(1) = 1$, $y'(1) = 0$.

3. Make a numerical comparison of the solution $\ddot{y} = -\sin y$, $y(0) = 0$, $\dot{y}(0) = 1$, with the solution of $\ddot{y} = -y$ using the same initial conditions. In particular, estimate the discrepancies between the location of successive zero values for the two solutions, one of which is just $y = \sin t$.

4. The nonlinear equation

$$\frac{d^2 y}{dx^2} + k\frac{dy}{dx} + hy^2 = 0, \qquad h, k \text{ const,}$$

has solutions defined near $x = 0$. Compare the behavior of numerical solutions of the initial value problem $y(0) = 0$, $y'(0) - 1$ with the corresponding behavior when the nonlinear term hy^2 is replaced by a linear term hy. To do this you should investigate the result of choosing several different values of $k > 0$ and $h > 0$.

5. The linear equation

$$\frac{d^2 y}{dx^2} + a(x)\frac{dy}{dx} + b(x)y = 0$$

with continuous coefficients $a(x)$ and $b(x)$ occurs often with rather special choices for the coefficient functions. (Chapters 9 and 11 contain several examples.) For other choices of the coefficients, numerical methods are used. Study the behavior of numerical solutions of the initial-value problem $y(0) = 0$, $y'(0) = 1$ as follows.

(a) Let $a(x) = \sin x$ and $b(x) = \cos x$ for $0 \leq x \leq 2\pi$.

(b) Let $a(x) = e^{-x/2}$ and $b(x) = e^{-x/3}$ for $0 \leq x \leq 1$.

6. Nonuniqueness. The initial-value problem $\ddot{y} = 3\sqrt{y}$, $y(0) = 0$, $\dot{y}(0) = 0$ has the identically zero solution.

(a) Verify that $y(t) = \frac{1}{16}t^4$ is also a solution.

(b) Discuss application of Euler methods to the problem.

Stiff Equations

7. The initial-value problem

$$y'' = 25y, \qquad y(0) = 1, \qquad y'(0) = -5$$

has solution $y(t) = e^{-5t}$. Try to approximate this solution using the improved Euler method with step size $h = 0.001$ over the interval $0 \leq t \leq 10$. Interpret your results.

Note. The next two exercises deal with order reductions of somewhat different forms from those that occur in the previous exercises. Ready-made computer routines may have to be altered accordingly.

8. The second-order initial-value problem in the previous exercise can be converted to a pair of first-order equations by means of substitutions other than $y' = z$.

(a) Let $y' = -5y + u$ [i.e., $(D + 5)y = u$]. Then $u' = 5u$ [i.e., $(D - 5)u = 0$] gives $y'' = 25y$. Show that the initial conditions $y(0) = 1$, $y'(0) = -5$ are converted to $y(0) = 1$, $u(0) = 0$. Then approximate y and u by the improved Euler method over the interval $0 \le t \le 10$ and compare the results with those of Exercise 7.

(b) Let $y' = wy$, which is called a **Riccati substitution,** and show that $y'' = 25y$ becomes $w' = 25 - w^2$, provided that $y \ne 0$. Then show that $y(0) = 1$, $y'(0) = -5$ is equivalent to $y(0) = 1$, $w(0) = -5$, and approximate the solution to the system

$$y' = wy, \qquad y(0) = 1,$$
$$w' = 25 - w^2, \qquad w(0) = -5.$$

9. Show that the **Riccati substitution** $y' = wy$ reduces the general linear equation

$$\frac{d^2y}{dt^2} + a(t)\frac{dy}{dt} + b(t)y = 0$$

to the system of equations

$$y' = wy,$$
$$w' = -a(t)w - b(t) - w^2$$

provided that $y \ne 0$. Show also that the initial conditions $y(t_0) = y_0$, $y'(t_0) = z_0$ get converted by the same substitution into $y(t_0) = y_0$, $w'(t_0) = z_0/y_0$.

Falling Objects

10. Suppose that the displacement $y(t)$ of a falling object is subject to a nonlinear friction force

$$m\ddot{y} = -k\dot{y}^\alpha + mg.$$

(a) Find numerical approximations to $y(t)$ in the range $0 \le t \le 20$, with $g = 32.2$, $y(0) = 0$, $\dot{y}(0) = 0$, $m = 1$, and $\alpha = 1.5$. Use the values $k = 0$, 0.1, 0.5, 1, and sketch the graphs of $y = y(t)$ using an appropriate scale.

(b) Estimate the value of k that, along with $\dot{y}(0) = 0$ and the other parameter values in part (a), produces approximately the values $y(0) = 0$, $y(5) = 66$, $y(10) = 137$, and $y(15) = 208$.

11. A typical nonlinear model for an object of mass 1 dropped from rest with frictional drag is $\ddot{y} = -k\dot{y}^\alpha + g$, $y(0) = \dot{y}(0) = 0$; the model is linear if $\alpha = 1$. In numerical work assume $g = 32$ feet per second per second.

(a) If the terminal velocity for the linear model is 36 feet per second, what is k?

(b) Estimate the time it takes for the linear model to reach velocity 35.99 feet per second.

(c) If the terminal velocity for the nonlinear model with $\alpha = 1.1$ is 36 feet per second, estimate k.

(d) Estimate the time it takes for the nonlinear model of part (c) to reach velocity 35.99 feet per second.

12. **Bead on a wire.** Exercise 24 in Section 6 considers a bead sliding without friction under constant vertical gravity along a wire bent into the shape of the twice differentiable graph $y = f(x)$; the result is that x obeys

$$\ddot{x} = \frac{-(g + f''(x)\dot{x}^2)f'(x)}{1 + f'(x)^2}.$$

If $f(x) = -x^3 + 4x^2 - 3x$, $g = 32$, and $x(0) = 0$, estimate how large $\dot{x}(0) > 0$ should be for the bead to go over the hump in the wire.

Pendulum

13. (a) Use the Euler method for

$$y'' = F(t, y, y'), \qquad y(t_0) = y_0, \qquad y'(t_0) = z_0,$$

and apply it to the pendulum equation with $F(t, y, y') = -16 \sin y$, $y(0) = 0$, $y'(0) = 0.5$.

(b) Do part (a) using the improved Euler method.

14. For small oscillations of y, the approximation $\sin y \approx y$ is fairly good, leading to the replacement of the pendulum equation $y'' + (g/l) \sin y = 0$ by the linearized equation $y'' + (g/l)y = 0$, with solutions of the form

$$y_a(t) = c_1 \cos \sqrt{\frac{g}{l}} t + c_2 \sin \sqrt{\frac{g}{l}} t.$$

Assuming $g/l = 16$, compare $y_a(t)$ with the improved Euler approximation to the solution of the nonlinear equation under initial conditions.

(a) $y(0) = 0$, $y'(0) = 0.1$.   (b) $y(0) = 0$, $y'(0) = 4$.

15. Consider the damped pendulum equation $\ddot{\theta} = -(g/l) \sin \theta - (k/m) \dot{\theta}$, with $g = 32.2$, $l = 20$, $k = 0.03$, and $m = 5$. For the solution with $\theta(0) = 0$, $\dot{\theta}(0) = 0.2$.

(a) Estimate the maximum angles θ for $0 \le t \le 15$.

(b) Estimate the successive times between occurrences of the value $\theta = 0$ for $0 \le t \le 15$.

(c) Repeat part (b) but with initial conditions $\theta(0) = 0$, $\dot{\theta}(0) = 2.0$.

16. Consider the following modification of the pendulum equation $ml\ddot{\theta} = -gm \sin \theta$, written here in terms of forces. If the pendulum pivot is moved vertically from its usual fixed position at level 0 so that at time t it is at $f(t)$, with $f(0) = 0$, the additional vertical force component is $m\ddot{f}(t)$. Thus the vertical force due to gravity alone is replaced by $(-gm + m\ddot{f}(t)) \sin \theta$. It follows that the equation for displacement angle $\theta = \theta(t)$ becomes

$$ml\ddot{\theta} = -gm \sin \theta + m\ddot{f}(t) \sin \theta \qquad \text{or} \qquad \ddot{\theta} = \frac{1}{l}(-g + \ddot{f}(t)) \sin \theta.$$

(a) Show that if $f(t) = at$, with a constant, then there is no change in acceleration as compared with the fixed-pivot case and hence that the equation for the displacement angle θ remains the same: $\ddot{\theta} = -(g/l) \sin \theta$.

(b) Show that in the case of a general twice-differentiable $f(t)$, the position of the pendulum weight at time t is $\mathbf{x}(t) = (l \sin \theta(t), f(t) - l \cos \theta)$, where θ satisfies either of the differential equations displayed above.

(c) Let $g = 32$, $l = 5$, $m = 1$ and $f(t) = 2 \sin 4t$. Plot $(t, \theta(t))$ for t values 0.01 apart between 0 and 100, assuming $\theta(0) = 0$ and $\dot{\theta}(0) = 0.01, 0.001, 0.0001$. Note the long-term deviation in behavior as compared with the identically zero solution corresponding to $\dot{\theta}(0) = 0$.

(d) Plot the path of the pendulum weight under the assumptions in part (c).

Oscillators and Phase Space

17. An unforced oscillator displacement $x = x(t)$ satisfies $m\ddot{x} + k\dot{x} + hx = 0$, where k and h may depend on time t.

(a) Suppose that $k(t) = 0.2(1 - c^{-0.1t})$, $h = 5$, $m = 1$ and that $x(0) = 0$, $\dot{x}(0) = 5$. Compute a numerical approximation to $x(t)$ on the range of $0 \leq t \leq 20$. Then sketch the graph of $x = x(t)$.

(b) Do part (a) using instead $k = 0$ and $h(t) = 5(1 - e^{-0.2t})$.

18. The **spherical pendulum** equation is discussed in Exercise 21 for Section 6:

$$\ddot{y} = -\frac{g}{l} \sin y + M^2 \frac{\cos y}{\sin^3 y}.$$

(a) Rewrite the equation as a first-order system suitable for numerical solution and phase plots.

(b) Assume $l = g$ and compare solution graphs in the ty plane for the two cases $M^2 = 0.5$ and $M^2 = 0$. Discuss the effect of rotation on the pendulum's motion.

(c) Assume $l = g$ and make a phase portrait for the case $M^2 = 2$.

(d) Assume $l = g$ and make a phase portrait for the case $M^2 = 0.5$, showing clearly the behavior near the stable equilibrium points at $y = \pm \pi/3$.

19. Make phase plots of the **soft spring oscillator equation** $\ddot{y} = -\gamma y^3 + \delta y$ under each of the following assumptions.

(a) $\gamma = \delta = 1$.   **(b)** $\gamma = 1$, $\delta = 2$.   **(c)** $\gamma = 2$, $\delta = 1$.

20. Make solution graphs and $(y, \dot{y})$ phase plots for the periodically driven **hard spring oscillator equation** $\ddot{y} = -y - y^3 + k\dot{y} + \frac{3}{10} \cos t$, and use the results to detect long-term approach to periodic behavior for some $k < 0$.

21. **Time-dependent linear spring mechanism.** The equation $\ddot{y} + k(t)\dot{y} + h(t)y = \sin t$ represents an oscillator externally forced by $\sin t$, damped by $k(t)$, and with stiffness $h(t)$. Use initial conditions $y(0) = 0$, $\dot{y}(0) = 1$ to plot solutions with the following choices for $k(t)$ and $h(t)$.

(a) $k(t) = 0$, $h(t) = e^{t/10}$.

(b) $k(t) = 0$, $h(t) = e^{-t}$.

(c) $k(t) = \frac{1}{10}$, $h(t) = e^{t/10}$.

(d) $k(t) = (1 - e^{-t})$, $h(t) = 1$.

(e) $k(t) = 1$, $h(t) = 1/(1 + t^2)$.

(f) $k(t) = 0$, $h(t) = t/(1 + t^2)$.

22. Plot closed periodic phase paths for the **Morse model** of displacement y from equilibrium of the distance between the two atoms of a diatomic molecule:

$$\ddot{y} = K(e^{-2ay} - e^{-ay}), \qquad K = a = 1.$$

23. Consider the nonlinear oscillator equation $m\ddot{x} + k\dot{x}|\dot{x}|^{\beta} + hx = 0$, $0 \leq \beta =$ const. Let $m = 1$, $k = 0.2$, $h = 5$ and suppose that $x(0) = 0$, $\dot{x}(0) = 5$. Compute numerical approximations to $x(t)$ on the range $0 \leq t \leq 20$ for $\beta = 0$, 0.5, 1. Sketch the resulting graphs using computer graphics.

24. **Chaos.** The nonlinear **Duffing oscillator** considered in Section 6 generalizes to the periodically driven, damped initial-value problem

$$\ddot{y} + k\dot{y} - y + y^3 = A \cos \omega t, \qquad y(0) = \tfrac{1}{2}, \qquad \dot{y}(0) = 0.$$

(a) Make a computer plot of of the solution $y(t)$ for $0 \leq t \leq 500$ for the parameter choices $k = 0.2$, $A = 0.3$, $\omega = 1$ and initial conditions $y(0) = \dot{y}(0) = 0$. Compare your result graphically with Figure IS.4 in the Introductory Survey at the beginning of the book. You will probably find that while your picture has the same general sort of irregularity as the one in the book, it will not agree in detail beyond some t value. This behavior of the damped, periodically driven Duffing equation is often described as *chaotic*, which in practical terms means unpredictable. In particular, the specific output that you get will depend significantly not only on the parameter and initial values but on the choice of numerical method and step size, and even on the internal arithmetic of the machine used to generate the output. For this reason it seems impossible to describe accurately the global shape of the output from the damped, periodically driven Duffing oscillator. To find some order in the chaos we can resort to the method of time sections described in Section 7C.

(b) Change just the damping constant in part (a) to $k = 0$, and make both a plot of the solution. Comment on the qualitative changes that you see as compared with the output in part (a).

(c) Make several phase plane plots, starting at different points in the $y\dot{y}$ plane, of solutions of the Duffing equation. Use the parameter values $k = 0.2$, $A = 0.3$, and $\omega = 1$.

(d) Experiment with part (c) by trying your own choices for the three parameter values.

7C Time Sections

Some solution graphs and phase plots for nonautonomous equations $\ddot{y} = f(t, x, y)$ are so irregular that they display very little apparent order. This irregularity is sometimes a symptom of an underlying inability of numerical approximations to predict the actual time evolution of a solution that is theoretically completely determined. (See Exercise 6.) To uncover a more stable structure within a continuous-time system it may be helpful to sample a numerically plotted phase curve at equally spaced discrete times, with fixed time difference τ. Thus we look as at a discrete-time *subsystem*. A plot of the resulting sample is called a **Poincaré time section** and consists of a sequence $\mathbf{x}_n =$

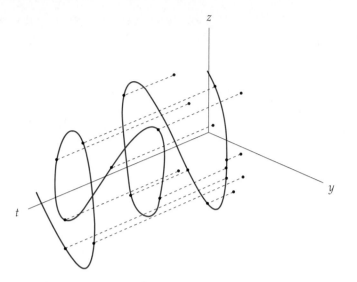

Figure 4.26 Projecting equally time-spaced points onto the yz plane.

$(y(n\tau), z(n\tau))$ of points in phase space. Thus we have an example of a **discrete dynamical system**, that is, a system observed at discrete times τ units apart.

　　We will typically be monitoring a time variable t that increases by a fixed amount h at each step in a numerical approximation. We want to check to see just which times make t close to an integer multiple of the sample step τ; to do this we can check to see when $t - m\tau < h$, where m is the largest integer with $m\tau < t$ and h is the step size of the numerical method in use. (This criterion may be written in software routines as mod $(t, \tau) < h$, where mod (t, τ) is the "remainder" on division of t by τ; that is, $t = m\tau +$ mod (t, τ).) We plot the corresponding phase point $\mathbf{x} = (y, z)$ when the criterion is met. Figure 4.26 shows a few points on a solution *graph* (*not* on a trajectory) chosen to be equally spaced in time and then projected parallel to the time axis onto the yz phase space. Here the differential equation isn't specified, but projecting 3200 times produced Figure 4.27a for the equation specified there. The term **stroboscopic section,** named for a visual sampling procedure of physical processes, is often used instead of *time section.*

　　If a second-order differential equation $\ddot{y} = f(t, x, y)$ is **time-periodic** with period $\tau > 0$, that is, if $f(t + \tau, x, y) = f(t, x, y)$ for all (t, x, y), then τ is a likely choice for the fixed time difference τ for a Poincaré time section.

---

e x a m p l e **1**　The periodically driven **Duffing oscillator** $\ddot{y} + k\dot{y} - y + y^3 = A \cos \omega t$ is treated in the previous set of exercises. See also Figure IS.4 in the Introductory Survey for a solution graph. (Theorem 8.9 of Chapter 6, Section 8B, explains that the long-term limit behavior for the unforced equation, with $A = 0$, is confined to a few fairly simple types.) In contrast, if $A \neq 0$ the problem of analyzing solution behavior gains an added dimension of complexity that we can study using time sections. We choose $k = \frac{1}{5}$, $A = \frac{3}{10}$, $\omega = 1$, and the initial

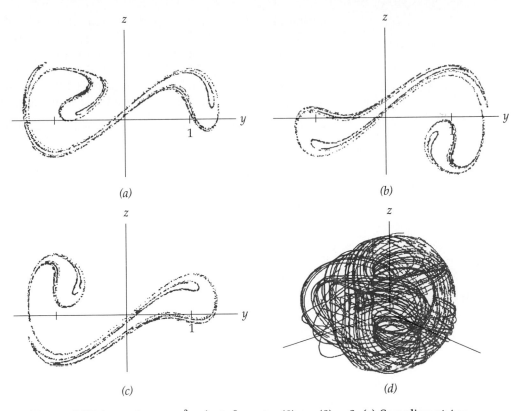

(a)   (b)

(c)   (d)

Figure 4.27 $\dot{y} = z$, $\dot{z} = y - y^3 - \frac{1}{5}z + \frac{3}{10}\cos t$, $y(0) = z(0) = 0$. (a) Sampling at $t = 2n\pi$. (b) Sampling at $t = 2n\pi + 2$. (c) Sampling at $t = 2n\pi + 5$. (d) Cylindrical image, $0 \le t \le 500$.

values $y(0) = \dot{y}(0) = 0$ and then convert to a first-order system by setting $\dot{y} = z$. Figure 4.27a shows a resulting phase-space sample of 3200 points starting at $t = 0$ and using sample step $\tau = 2\pi$ since $\cos t$ has period 2π. Parts (b) and (c) show the qualitatively similar pictures obtained by advancing the sampling times by 2 and 5, respectively. These samplings convey a sense of structure that is hard to quantify but nevertheless quite striking. The exercises contain suggestions for supplementing these pictures to show more structure. As the pictures evolve point by point you can see succeeding points hop around the general pattern, seemingly in a highly "random" manner.

Here is a way to think about time sections of solutions that is in some ways more satisfying than simply regarding them as projections parallel to a t axis. We interpret time t as a cylindrical coordinate angle, measuring this angle about the vertical z axis, and we plot $(y(t) \cos t, y(t) \sin t, z(t))$ in 3-dimensional space for an extended range of t values. In this way we may be able to transfer an extended image of a solution in rectangular tyz space to an image in a smaller region. Figure 4.27d is a perspective drawing of just such a transferred image, which we'll call a **cylindrical representation** of the solution. The preceding three pictures are cross sections of this 3-dimensional image taken in planes containing the vertical coordinate axis at angles 0, 2, and 5 radians to the left-hand horizontal axis in part (*d*). Extending the 3-dimensional trace in Figure 4.27*d* significantly beyond $t = 500$ results, at this scale, in a blob that conveys almost no information. In contrast, the time-section pictures span roughly the interval $0 \le t \le 3200 \cdot 2\pi \approx 20{,}106$. Taken altogether, the four pictures convey some feeling for what the solution is like. The analytical details are described in the exercises under cylindrical representation.

Caution. Unless you have an exceptionally fast computer, you need some patience in plotting a satisfactory time section. The reason is that typically an enormous amount of computed information is discarded between recorded samples. In this regard, note that if you're computing a periodic, or very nearly periodic, solution with least positive period τ, and you're recording only a phase-space time section with sample step τ, you can expect to get just repetition of a single dot in your picture. Time sections can be plotted point by point using the Applet TIMESECT available at the Web site http://math.dartmouth.edu/~rewn/

EXERCISES

1. Suppose you're making a phase plot of a solution having least positive period τ, but are recording only a time section with sample step τ/n, where n is a positive integer. How many dots would you expect your time section to display?

2. The unforced Duffing oscillator $\ddot{y} = y - y^3$ appears in text Example 5 of Section 6 as well as in the exercises for Section 5, where you're asked to show that solutions starting close enough to phase-space equilibrium are periodic.

   (a) Plot some typical graphs of the periodic solutions referred to above, and also make separate plots of the corresponding phase curves, as shown in Figure 4.20.

   (b) Plot some higher-energy solutions and corresponding phase plots that loop around both equilibrium points.

3. Time sections of the solution to $\ddot{y} = y - y^3 - 0.2\dot{y} + 0.3\cos t$, $y(0) = \dot{y}(0) = 0$ are considered in Example 1 of the text. The sampling is done at times $2n\pi$, $2n\pi + 2$, $2n\pi + 5$ for nonnegative integer n. Make additional time sections at (a) $2n\pi + 1$, (b) $2n\pi + 3$, (c) $2n\pi + 4$. Explain the relation of each picture you make to Figure 4.27*d*. Note that sampling times $m\tau + \alpha$ for integer m can be estimated by selecting times for which mod $(t - \alpha, \tau) < h$. For example, plot a point only when mod $(t - 4, 2\pi) < h$ in part (c).

4. Changing the forcing term in the initial-value problem in the previous exercise to $0.3\cos(t/2)$ results in a differential equation with time period 4π instead of 2π. Investigate graphically what this change does to the solution. In particular, what characteristics does a time section have after the change?

5. The time sequences used in making Figure 4.27*b* and *c* differ by 3. Noting that the resulting pictures are nearly symmetric to each other about the origin in the *yz* plane, one might guess that with a difference of exactly π

the corresponding pictures should be precisely symmetric to each other. Show that this guess is correct given that the initial conditions are $y(0) = z(0) = 0$, regardless of the choice of the constants k and A in the differential equation $\ddot{y} = y - y^3 - k\dot{y} + A\cos t$.

6. (a) Show that if $y - y^3$ is replaced by $-y$ in the periodically driven Duffing equation with $k > 0$, $A > 0$ to get $\ddot{y} = -y - k\dot{y} + A\cos t$, then all solutions of the resulting linear equation are oscillatory and tend to the periodic function $y_p(t) = (A/k)\sin t$ as t tends to infinity.

(b) Explain qualitatively what time sections for this equation can be expected to look like.

7. Figure IS.4 in the Introductory Survey at the beginning of the book shows a numerically simulated time evolution for the initial-value problem

$$\frac{d^2y}{dt^2} + \frac{1}{5}\frac{dy}{dt} - y + y^3 = \tfrac{3}{10}\cos t, \qquad y(0) = \dot{y}(0) = 0.$$

Such simulations are fairly satisfying in a qualitative way but tend to be quantitatively imprecise, because in the long run the sensitivity of the differential equation to small changes in input data magnifies the inevitable error that occurs at each step. Plot solutions $y(t)$ to this differential equation for $0 \leq t \leq 500$, replacing the initial values $(y(0), \dot{y}(0)) = (0, 0)$ successively by:

(a) $(0, 0.01)$, $(0, 0.001)$, $(0, 0.0001)$, etc.

(b) $(0.01, 0)$, $(0.001, 0)$, $(0.0001, 0)$, etc.

Record in general terms whatever long-term discrepancies you see among these pictures.

8. Make a time-section portrait for the Duffing initial-value problem

$$\frac{d^2y}{dt^2} + \frac{1}{5}\frac{dy}{dt} - y + y^3 = \tfrac{3}{10}\cos t, \qquad y(0) = -1, \qquad \dot{y}(0) = 1,$$

using time difference $\tau = 2\pi \approx 6.283185307$ and starting at $t = 0$. Compare the result with Figure 4.27a, noting the lack of sensitivity to even a large change in initial conditions. This kind of experimental evidence of "stability" is one reason for the interest in time sections.

9. The differential equation $\ddot{y} = -y^3 - \tfrac{1}{20}\dot{y} = 7\cos t$ can be thought of as a periodically driven, slightly damped spring that is very soft for small displacements and very hard for larger ones. Use a step size $h \leq 0.01$.

(a) Let $\dot{y} = z$ to convert the equation to a first-order system.

(b) Plot a time section at times $t \approx 2n\pi$, e.g., mod $(t, 2\pi) < h$.

(c) Plot a time section at times $t \approx 2n\pi + 1$, e.g., mod $(t - 1, 2\pi) < h$.

(d) Plot a time section at times $t \approx 2n\pi + 4$, e.g., mod $(t - 4, 2\pi) < h$.

10. The differential equation $\ddot{y} = y^3 - \tfrac{1}{5}\dot{y} = 7\cos t$ is a more strongly damped version of the one in the previous exercise.

(a) Plot a time section at times $t = 2n\pi$, and compare the result with the corresponding picture from the previous exercise.

(b) Is the comparison in part (a) consistent with the increased damping factor? Explain.

11. In contrast to the pictures in Figure 4.27, time sections for linear second-order equations tend to be more regulated.

(a) Show that the time-section plot with sample step $\tau = 2\pi$ for the linear initial-value problem

$$\ddot{y} + \omega^2 y = (\omega^2 - 1)\sin t, \qquad y(0) = 0, \qquad \dot{y}(0) = \omega$$

is confined to the phase-space ellipse $y^2 + (z-1)^2/\omega^2 = 1$ if $\omega \neq 0$.

(b) Plot this ellipse for $\omega = \frac{1}{3}$ and for $\omega = \frac{4}{3}$. On each ellipse plot the corresponding time sections with $\tau = 2\pi$: $(y(2\pi n), z(2\pi n))$.

(c) Plot the time section for $\omega = \sqrt{2}/2$. (The next exercise is particularly relevant to this part.)

*12. Since digital computing is carried out entirely with rational numbers, the distinction between rational and irrational numbers becomes definitely blurred in numerical work. An irrational number is represented by a rational approximation with as large a denominator as the software can accommodate. It follows from Exercise 33 in Chapter 3, Section 2D, that oscillatory solutions such as $y_\omega(t) = \sin \omega t + \sin t$ are periodic if and only if ω is rational in which case, if $\omega = p/q$, a period is $2\pi q$.

(a) Show that as an irrational ω is replaced by increasingly more accurate rational approximations r, the least positive period of the resulting function $y_r(t)$ tends to infinity.

(b) Show that a time section with sample step $\tau = 2\pi$ for y_ω consists of only finitely many points if ω is rational, and infinitely many different points if ω is irrational.

13. When choosing a sample step τ for sampling in making a time section it's important to take account of the natural time periodicity in $\ddot{y} = f(t, y, \dot{y})$. For example, in the forcing term $A \cos \omega t$ in the Duffing equation of text Example 1 the natural period is $2\pi/\omega$. Thus with $\omega = 1$ the natural period is 2π, so we used sample step $\tau = 2\pi$ in making Figure 4.27a, b, and c.

(a) Make a time-section plot with the specific choices $k = \frac{1}{5}$, $A = 0.3$, and $\omega = 1$ in the Duffing equation of Example 1 and using sample step $\tau = \pi$ starting at $t = 0$. The pattern of your plot should appear to contain the one in Figure 4.27a as a subset; explain why.

(b) Make a time-section plot with the specific choices $k = \frac{1}{5}$, $A = 0.3$, and $\omega = 1$ in the Duffing equation of Example 1 and using sample step $\tau = 4$ starting at $t = 0$. Note that the resulting pattern is relatively featureless as compared with Figure 4.27a, where the choice of τ is keyed to the period of the system. (Note that we have perversely assumed in our choice $\tau = 4$ that we're dealing with a period-4 system.)

Cylindrical Representation of Solution Graphs. The picture in Figure 4.27d was made by plotting $(y(t)\cos t, y(t)\sin t, z(t))$ in 3-dimensional perspective for $0 \leq t \leq 500$, where $(y(t), z(t))$ is a phase-plane point at time t. More generally, we can do a perspective plot of

$$\left(y(t) \cos \frac{2\pi t}{\tau}, y(t) \sin \frac{2\pi t}{\tau}, z(t) \right),$$

where τ is the time chosen for wrapping a single loop around the vertical axis in the cylindrical representation. As explained in the text, the Poincaré time sections are just the intersections of the cylindrical representation with planes containing its vertical axis. The Applet CYLREP at http://math.dartmouth.edu/~rewn/ can be used to do the following three exercises.

14. Reproduce Figure 4.27d using the above prescription with $\tau = 2\pi$.

15. Make a cylindrical representation of the solution of the initial-value problem $\ddot{y} = -y^3 - \frac{1}{20}\dot{y} = 7 \cos t$, $y(0) = \dot{y}(0) = 0$.

16. Make a cylindrical representation of the solution of the initial-value problem $\ddot{y} = -y^3 - \frac{1}{5}\dot{y} = 7 \cos t$, $y(0) = \dot{y}(0) = 0$. What change do you notice in comparison to the result of the previous exercise?

CHAPTER

Introduction
to Systems

The earlier chapters are about differential equations whose solutions are real-valued functions of a real variable. A useful generalization is to consider vector differential equations (or, equivalently, systems of real-valued differential equations) whose solutions are vector-valued functions of a real variable. There are two main reasons for making this generalization. One is that many phenomena in applied mathematics are most naturally presented in vector form. The other reason is that, even if the original problem is not in vector form, there may be a valuable technical trade-off in replacing a system with higher-order derivatives by a vector system with first-order derivatives. Both these statements will be illustrated in the present chapter. We will follow the convention of writing tuples of coordinates for vectors and abbreviating with boldface letters when the dimension doesn't need to be specified. Thus $\mathbf{x} = (x, y)$ means the same thing as the often-used notation $x\mathbf{i} + y\mathbf{j}$, and $\mathbf{x} = (x, y, z)$ means the same thing as $x\mathbf{i} + y\mathbf{j} + z\mathbf{k}$.

1 VECTOR EQUATIONS

1A Geometric Setting

Recall that a solution of a single first-order differential equation of the form

$$\frac{dx}{dt} = F(t, x)$$

is a real-valued function $x = x(t)$ defined on some interval $a < t < b$ and such that

$$\frac{dx}{dt}(t) = F(t, x(t)), \qquad a < t < b.$$

For example, $dx/dt = x + t$ has the general solution $x = Ce^t - t - 1$, defined for all real t. The pair of equations

$$\frac{dx}{dt} = F_1(t, x, y),$$

$$\frac{dy}{dt} = F_2(t, x, y)$$

is called a **system** of **dimension 2,** and a solution will have the form

$$x = x(t),$$
$$y = y(t), \qquad a < t < b.$$

As t increases from a to b, the point $(x(t), y(t))$ in the xy plane will trace out some path, perhaps like the one in Figure 5.1. Such a path, together with its direction of traversal, is called a **trajectory** of the system. Planetary orbits and particle paths in fluid flows have historically been prime examples of solution trajectories, so the terms **orbit** and **flow line** are often used instead of trajectory.

It the coordinates in a problem are chosen so that each equation in a system contains only one unknown function, then the system is called **uncoupled,** and we can try solving the equations independently of each other. Section 1 of Chapter 7 explains how we can sometimes change coordinates so as to find an uncoupled system that is **equivalent** in that it has the same solutions in different notation. In the next example we are presented with an uncoupled system, and each coordinate equation of the system is easy to solve. The variable t often represents time in applied problems, and we often use dots instead of primes to indicate time derivatives: $dx/dt = \dot{x}$, $d^2x/dt^2 = \ddot{x}$, and so on.

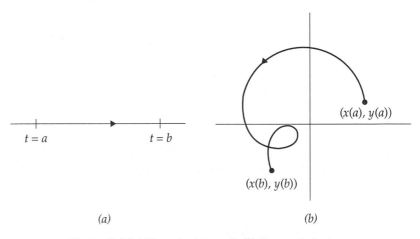

(a)   (b)

Figure 5.1 (*a*) Domain interval. (*b*) Image trajectory.

e x a m p l e 1

The uncoupled system

$$\dot{x} = -x,$$
$$\dot{y} = -2y$$

is particularly simple; each unknown function occurs in just one equation. Thus we can solve each equation by itself to get the general solution

$$x = c_1 e^{-t},$$
$$y = c_2 e^{-2t}.$$

At $t = 0$ we have $x(0) = c_1$, $y(0) = c_2$, so the initial conditions $x(0) = 3$, $y(0) = 1$ require $c_1 = 3$ and $c_2 = 1$. Hence

$$x = 3e^{-t},$$
$$y = e^{-2t}.$$

Squaring the first equation shows that the trajectory satisfies the equation of a parabola, $y = e^{-2t} = (x/3)^2$. Since the exponentials are both positive, the trajectory obeys the restrictions $x > 0$ and $y > 0$. Figure 5.2a shows the trajectory for $0 \le t < \infty$. The arrow point shows the direction of traversal for increasing t.

It is important to realize that a solution's trajectory by itself fails to give a complete geometric description of the solution; the correspondence between t values and points on the trajectory path is not made explicit by the picture. If we enhance our picture with a t axis perpendicular to the xy axes, we can sketch a graph of the solution for $0 \le t$ as shown in Figure 5.2b. This drawing shows in a single picture the correspondence between t and $(x(t), y(t))$ for the interval in question. The trajectory in the xy plane is also included here. In applied problems we may prefer to settle for the trajectory alone; for example, the trajectory may represent the path followed by some physical object, whereas the solution graph may have no particular physical significance.

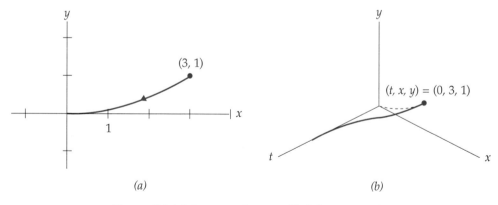

Figure 5.2 (a) Image trajectory. (b) Solution graph.

The system of differential equations in Example 1 can be written using vector notation by letting $\mathbf{x} = (x, y)$, $d\mathbf{x}/dt = (\dot{x}, \dot{y})$, and $\mathbf{F}(x, y) = (-x, -2y)$. Then equating corresponding coordinates gives

$$(\dot{x}, \dot{y}) = (-x, -2y) \qquad \text{or} \qquad \frac{d\mathbf{x}}{dt} = \mathbf{F}(\mathbf{x}).$$

e x a m p l e 2

The system

$$\frac{dx}{dt} = x + y + t,$$

$$\frac{dy}{dt} = x - y - t.$$

can be written $dx/dt = \mathbf{F}(t, \mathbf{x})$, where

$$\mathbf{F}(t, x, y) = (x + y + t, x - y - t).$$

While the systems encountered in applications are often low-dimensional, perhaps 2 or 3, systems of very high dimension do occur, so it's important to use notation that accommodates arbitrary dimension. For example, the motion of three bodies subject only to their own mutual gravitational attraction are described by the solution of an 18-dimensional first-order system, discussed in Chapter 6, Section 5. (For our own solar system we need dimension at least 60.) When we speak of a **first-order system of dimension** n in **normal form,** we mean a system of the form

$$\frac{dx_1}{dt} = F_1(t, x_1, x_2, \ldots, x_n),$$

$$\frac{dx_2}{dt} = F_2(t, x_1, x_2, \ldots, x_n),$$

$$\vdots$$

$$\frac{dx_n}{dt} = F_n(t, x_1, x_2, \ldots, x_n).$$

(Thus a 1-dimensional system is a single real equation.) Using the vector notations

$$\mathbf{x} = (x_1, x_2, \ldots, x_n), \qquad \dot{\mathbf{x}} = \left(\frac{dx_1}{dt}, \frac{dx_2}{dt}, \ldots, \frac{dx_n}{dt}\right)$$

we can write the system more concisely for some purposes as

$$\dot{\mathbf{x}} = \mathbf{F}(t, \mathbf{x}), \qquad \text{where } \mathbf{F}(t, \mathbf{x}) = (F_1(t, \mathbf{x}), \ldots, F_n(t, \mathbf{x})).$$

A solution is then a vector-valued function $\mathbf{x} = \mathbf{x}(t)$ defined on an interval $a < t < b$. Verifying that $\mathbf{x}(t)$ really is a solution amounts to checking that

$$\dot{\mathbf{x}}(t) = \mathbf{F}(t, \mathbf{x}(t)), \qquad a < t < b.$$

Trajectory curves for n-dimensional systems can be displayed only for $n = 2$ and $n = 3$, but the basic idea is still useful in higher dimensions. The space where trajectories of a *first-order* system are traced is the **state space** of the associated system, and under the assumptions of Theorem 1.1 in Section 1E, specifying a state will determine a unique trajectory starting at that state. A 3-dimensional trajectory is shown in Figure 5.3a.

One advantage of the vector interpretation of a system is that a solution $\mathbf{x}(t)$ can be interpreted as a position, at time t, of a point in a space of some dimension, and we can even draw a picture of this space in the case of dimension 2 or 3. Another advantage is that the vector derivative $\dot{\mathbf{x}}$ has a geometric meaning that the coordinate derivatives $\dot{x}_k$ do not have when considered separately: $\dot{\mathbf{x}}$

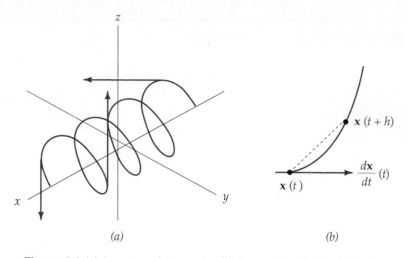

Figure 5.3 (*a*) Image and tangents. (*b*) Tangent as limit of chords.

can be interpreted as a tangent vector to the trajectory $\mathbf{x} = \mathbf{x}(t)$. Furthermore, if t is interpreted as a time variable, then $\dot{\mathbf{x}}$ is a **velocity vector** in the sense that its length is the speed of traversal of the trajectory, and $\ddot{\mathbf{x}}$ is by definition an **acceleration vector.** The following discussion shows how the formal definitions of tangent and velocity are motivated by the intuitive ideas behind them.

Figure 5.3*b* shows the points $\mathbf{x}(t)$ and $\mathbf{x}(t + h)$. The parallelogram law for addition of vectors shows that, if $\mathbf{z}$ is the vector (arrow) joining $\mathbf{x}(t)$ to $\mathbf{x}(t + h)$, then $\mathbf{z} + \mathbf{x}(t) = \mathbf{x}(t + h)$. It follows that $\mathbf{z} = \mathbf{x}(t + h) - \mathbf{x}(t)$. Now multiply this vector by the number $1/h$ to get a parallel vector that approaches the tangent direction to the trajectory at $\mathbf{x}(t)$ as $h \to 0+$. The result is easier to read if we write the entries of $\mathbf{x}(t)$ in a column:

$$\lim_{h \to 0} \frac{\mathbf{x}(t + h) - \mathbf{x}(t)}{h} = \begin{pmatrix} \lim_{h \to 0} \dfrac{x_1(t + h) - x_1(t)}{h} \\ \vdots \\ \lim_{h \to 0} \dfrac{x_n(t + h) - x_n(t)}{h} \end{pmatrix} = \begin{pmatrix} \dfrac{dx_1(t)}{dt} \\ \vdots \\ \dfrac{dx_n(t)}{dt} \end{pmatrix} = \dot{\mathbf{x}}(t).$$

We define the **tangent vector** or **velocity vector** to the trajectory at $\mathbf{x}(t)$ to be $\dot{\mathbf{x}}(t)$. Geometrically, $\dot{\mathbf{x}}$ appears as an arrow with its initial point at $\mathbf{x}(t)$, tangent to the trajectory and pointing in the direction of increasing t along the trajectory. The **speed** of $\mathbf{x}(t)$ at time t is the length of $\dot{\mathbf{x}}(t)$, defined by

$$|\dot{\mathbf{x}}| = \sqrt{\dot{x}_1^2 + \cdots + \dot{x}_n^2}.$$

This is just the length of the velocity vector, where **length** of a vector $\mathbf{y} = (y_1 \cdots y_n)$ is defined as usual by $|\mathbf{y}| = \sqrt{y_1^2 + \cdots + y_n^2}$.

---

e x a m p l e **3**   The pair of equations

$$\dot{x} = x - y$$
$$\dot{y} = x + y$$

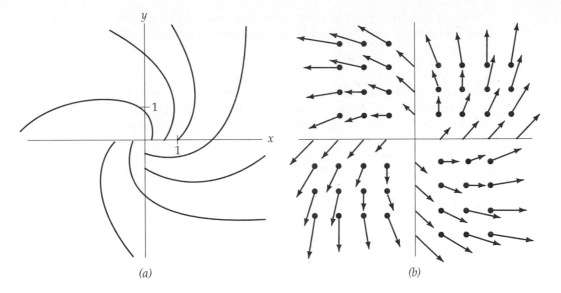

Figure 5.4 (*a*) Trajectories. (*b*) Vector field.

has for one of its solutions $x(t) = e^t \cos t$, $y(t) = e^t \sin t$. This is so because

$$\dot{x}(t) = e^t \cos t - e^t \sin t = x(t) - y(t),$$
$$\dot{y}(t) = e^t \cos t + e^t \sin t = x(t) + y(t).$$

The trajectory of this solution is a spiral curve satisfying the initial condition $(x(0), y(0)) = (1, 0)$. See Figure 5.4*a*, which shows some other trajectories also. The distance of the point $(x(t), y(t))$ from the origin increases exponentially, since

$$|(x(t), y(t))| = \sqrt{(e^t \cos t)^2 + (e^t \sin t)^2} = e^t$$

The speed at time t can be found by using the differential equations:

$$|(\dot{x}, \dot{y})| = \sqrt{(x - y)^2 + (x + y)^2}$$
$$= \sqrt{2x^2 + 2y^2} = \sqrt{2}e^t$$

Figure 5.4*b* shows an array of tangent vectors for the same system. From this picture you can get an idea of what the trajectories will look like in general, since an arrow in the sketch must be tangent to a trajectory that passes through the tail of the arrow. You can imagine superimposing Figure 5.4*a* on Figure 5.4*b* to see how tangent arrows fit the trajectories. The longer arrows farther from the origin show that the speed of traversal of a trajectory increases as the path moves away from the origin.

1B   Vector Fields

In the general first-order system

$$\frac{d\mathbf{x}}{dt} = \mathbf{F}(t, \mathbf{x}),$$

the vector-valued function $\mathbf{F}$ is understood to depend explicitly on t, so different tangent vectors may be assigned to a trajectory if it happens to pass through the same point **x** at two different times, as in Figure 5.1*b*. If **F**(*t*, **x**) is independent of *t*, the system is called **autonomous** and can be written

$$\frac{d\mathbf{x}}{dt} = F(\mathbf{x}).$$

Examples 1 and 3 are about autonomous systems, whereas Example 2 is time-dependent. For an autonomous system, the vector **F**(**x**), which is the tangent vector at **x**, is always the same regardless of what time it is when the trajectory passes through **x**. Such an assignment of vectors (i.e., arrows) **F**(**x**) to points **x** is called a **vector field.** Figure 5.4*b* shows a sketch of a vector field derived from Example 3. A vector **F**(**x**), rather than starting at the origin, is translated parallel to itself so that its tail is located at the point of tangency **x**. In other words, *the arrow is drawn from the point* **x** *to the point* **x** + **F**(**x**). Note that the arrows of the field **F** are tangent to solution trajectories of an autonomous system *d***x**/*dt* = **F**(**x**) in the *range space* of the solutions. (In contrast, the slope segments of a direction field are tangent to solution graphs in a space of which one coordinate is the independent variable *t*.)

---

e x a m p l e 4

The system

$$\frac{dx}{dt} = -y$$

$$\frac{dy}{dt} = x$$

can be written *d***x**/*dt* = **F**(**x**), where **x** = (*x*, *y*) and **F**(**x**) = (−*y*, *x*). The system can be looked at as saying that the tangent vector to a solution trajectory through the point (*x*, *y*) has the direction of the vector (−*y*, *x*). The tangent then has slope −*x*/*y*, while the line joining the origin to the point (*x*, *y*) has slope *y*/*x*. It follows that these two directions are perpendicular. Figure 5.5*c* shows a sketch of the vector field. The picture suggests that the flow lines may be circular in shape, which turns out to be true. (Indeed, their tangent vectors **F**(*x*, *y*) = (−*y*, *x*) at (*x*, *y*) are perpendicular to (*x*, *y*).) It's easy to verify that initial conditions *x*(0) = *u*, *y*(0) = *v* are satisfied by the family of solutions

$$x(t) = u \cos t - v \sin t, \qquad y(t) = u \sin t + v \cos t.$$

A routine for deriving such solutions is described in Section 2.

A sketch of a vector field plays somewhat the same role for a system that a direction field does for a single equation. In a direction field the segments all have the same length, all are tangent to the graph of a solution, and speed is indicated by the slopes of the segments. In a vector field, the arrows are tangent to the *trajectory* (i.e., image curve) of a solution, and speed is indicated by the lengths of the arrows.

For the general time-dependent system *d***x**/*dt* = **F**(*t*, **x**), the function **F**(*t*, **x**) specifies a tangent vector to a trajectory through **x** that may be different for different *t*. The stationary vector field of the kind pictured in Figure 5.4*b* is no longer appropriate, but it can be replaced by a sequence of "snapshots" taken at different times. Each snapshot will be a sketch of a single vector field but will show changes in the arrows as time t varies.

---

example 5

The system

$$\frac{dx}{dt} = (1 - t)x - ty,$$

$$\frac{dy}{dt} = tx + (1 - t)y$$

is determined by the time-dependent vector field

$$\mathbf{F}(t, x, y) = ((1 - t)x - ty, tx + (1 - t)y).$$

See Figure 5.5 for sketches of the vector field for three values of t.

The Applet VECFIELD at http://math.dartmouth.edu/~rewn/ will make sketches such as the ones in Figure 5.5.

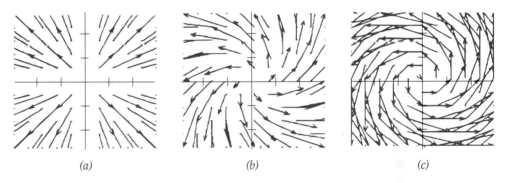

(a) (b) (c)

Figure 5.5 (a) $\mathbf{F}(0, x, y) = (x, y)$. (b) $\mathbf{F}(\frac{1}{2}, x, y) = (\frac{1}{2}(x - y), \frac{1}{2}(x - y))$. (c) $\mathbf{F}(1, x, y) = (-y, x))$.

1C   Order Reduction and Normal Form

It is often a useful trade-off to replace a differential equation of order higher than one by a first-order system at the expense of increasing the number of unknown functions. This is done to good effect, for example, in Chapter 3, Section 1, in Chapter 4, Sections 5 and 7, and at several places in the applications chapter following this one.

---

example 6

In the second-order equation

$$\frac{d^2y}{dt^2} = f\left(t, y, \frac{dy}{dt}\right)$$

we can let $dy/dt = z$. Then $d^2y/dt^2 = dz/dt$, so the two equations

$$\frac{dy}{dt} = z, \qquad \frac{dz}{dt} = f(t, y, z),$$

form a system equivalent to the original second-order equation. Notice that a solution of the system, being a pair of functions, produces not only a solution $y(t)$ of the given equation but also its derivative $dy/dt = z(t)$. (The basic idea is also treated in some detail in Section 5 of Chapter 4, outside the general context of systems.)

example 7

Here are some second-order equations and their companion systems.

(a) $$\ddot{y} + y = 0; \qquad \begin{cases} \dot{y} = z, \\ \dot{z} = -y. \end{cases}$$

(b) $$\ddot{y} = -\sin y; \qquad \begin{cases} \dot{y} = z, \\ \dot{z} = -\sin y. \end{cases}$$

(c) $$\ddot{y} + t\dot{y} + t^2 y = 0; \qquad \begin{cases} \dot{y} = z, \\ \dot{z} = -tz - t^2 y. \end{cases}$$

Initial conditions for these three second-order equations all have the form $y(t_0) = y_0$, $dy/dt(t_0) = z_0$. Since $dy/dt = z$, these translate directly into the conditions $y(t_0) = y_0$, $z(t_0) = z_0$ for the corresponding systems.

While the general form of the systems we came up with in Example 7 may require a little more writing than the second-order equations they came from, it often turns out to be conceptually and computationally simpler to deal with the first-order vector derivative $\dot{\mathbf{x}} = (\dot{y}, \dot{z})$ in the system than with the pair $(\dot{y}, \ddot{y})$ containing derivatives of different orders. A similar remark applies to the 4-dimensional vector derivative $\dot{\mathbf{x}} = (\dot{x}, \dot{y}, \dot{u}, \dot{v})$ introduced in the next example. The terms **state space** or **phase space** are used to refer to the higher-dimensional space that incorporates derivative coordinates in order to distinguish it from the basic space, called the **configuration space,** in which the position trajectories lie.

example 8

Second-order systems of the form

$$\ddot{x} = f(t, x, y, \dot{x}, \dot{y}),$$
$$\ddot{y} = g(t, x, y, \dot{x}, \dot{y})$$

occur often in applications, for example, in Chapter 6 in the study of vibrating mechanisms as developed in Section 4 and in the study of planetary motion as in Section 5. Dealing instead with an equivalent first-order system makes it easy to extend the numerical methods of Chapter 2, Section 6, to second-order systems and often provides some additional geometric insight into the problem under study. The idea in this case is to introduce two new dependent variables $u = \dot{x}$ and $v = \dot{y}$. It follows then that we have $\ddot{x} = d\dot{x}/dt = \dot{u}$ and $\ddot{y} = d\dot{y}/dt = \dot{v}$. The configuration space consists of points (x, y) and the state space of points (x, y, u, v). The original system can then be rewritten as the system of four first-order equations

$$\dot{x} = u,$$
$$\dot{y} = v,$$
$$\dot{u} = f(t, x, y, u, v),$$
$$\dot{y} = g(t, x, y, u, v).$$

Initial conditions for the second-order system specify positions and velocities for x and y, and might look, for example, something like $x(0) = 1$, $y(0) = 0$, $\dot{x}(0) = -1$, $\dot{y}(0) = 2$. In terms of x, y, u, v these conditions convert into $x(0) = 1$, $y(0) = 0$, $u(0) = -1$, $v(0) = 2$.

For doing computations as well as for geometric interpretation, it is important to be able to reduce a system to a standardized **normal form**

$$\frac{d\mathbf{x}}{dt} = \mathbf{F}(t, \mathbf{x})$$

containing only first-order derivatives isolated on the left, even if the original system contains higher-order derivatives.

example 9

The systems

$$\begin{cases} \dfrac{dx}{dt} = y + t, \\[2mm] \dfrac{dy}{dt} = x + t, \end{cases} \qquad \begin{cases} \dfrac{dx}{dt} = xy, \\[2mm] \dfrac{dy}{dt} = x^2 + t \end{cases}$$

are both in normal form.

example 10

To convert the second-order system

$$\frac{d^2x}{dt^2} + 2\frac{dy}{dt} = t,$$
$$\frac{d^2y}{dt^2} - \frac{dx}{dt} + y = 0,$$

to normal form, let $dx/dt = z$, $dy/dt = w$. Then $d^2x/dt^2 = dz/dt$ and similarly $d^2y/dt^2 = dw/dt$. Substitution gives us altogether

$$\frac{dx}{dt} = z, \qquad\qquad \frac{dy}{dt} = w,$$
$$\frac{dz}{dt} = -2w + t, \qquad \frac{dw}{dt} = z - y.$$

This reduction to normal form depended only on our ability to solve for the second derivatives.

e x a m p l e **11**   An example of a first-order system not in normal form is

$$\frac{dx}{dt} + \frac{dy}{dt} = 2x + 4y,$$

$$2\frac{dx}{dt} + 3\frac{dy}{dt} = 2x + 6y.$$

This system can be put in normal form by applying simple elimination to the derivative terms. Multiply the first equation by 2 and subtract from the second to get $dy/dt = -2x - 2y$. Now subtract this equation from the first one to get $dx/dt = 4x + 6y$. The result is a system in normal form:

$$\frac{dx}{dt} = 4x + 6y,$$

$$\frac{dy}{dt} = -2x - 2y.$$

EXERCISES

1. The following uncoupled systems can be solved by treating each equation separately. Find the general solution, and then find the particular solution that satisfies the given initial conditions.

   (a) $dx/dt = x + 1$, $x(0) = 1$,
   $dy/dt = y$, $y(0) = 2$.

   (b) $dx/dt = t$, $x(1) = 0$,
   $dy/dt = y$, $y(1) = 0$.

   (c) $dx/dt = x$, $x(0) = 0$,
   $dy/dt = \frac{1}{2}y$, $y(0) = 1$,
   $dz/dt = \frac{1}{3}z$, $z(0) = -1$.

   (d) $dx/dt = x + t$, $x(0) = 0$,
   $dy/dt = y - t$, $y(0) = 0$,
   $dz/dt = z$, $z(0) = 1$.

2. For each of the systems in Exercise 1, there is a vector-valued function $\mathbf{F}(t, \mathbf{x})$, with $\mathbf{x} = (x, y)$ or (x, y, z) such that the system can be written in the form $d\mathbf{x}/dt = \mathbf{F}(t, \mathbf{x})$.

   (i) Find $\mathbf{F}$ in each case.

   (ii) Find the speed of a trajectory through $\mathbf{x}$ at time t.

3. The solution trajectory shown in Figure 5.1 belongs to the system

$$\frac{dx}{dt} = -\beta y - (1 - \beta) \sin t, \qquad \frac{dy}{dt} = \beta x + (1 - \beta) \cos t$$

   for the parameter choice $\beta = 0.35$. Verify that the system has solutions of the form

$$x = \cos t + c \cos \beta t, \qquad y = \sin t + c \sin \beta t, \; c = \text{const}.$$

4. Sketch the following vector fields, associated with the systems of Exercise 1, by drawing a few arrows for $\mathbf{F}(\mathbf{x})$ or $\mathbf{F}(t, \mathbf{x})$ with their tails at selected points $\mathbf{x}$ of the form (x, y) or (x, y, z). In parts (b) and (d), make separate sketches for $t = -1$, $t = 0$, and $t = 1$.

   (a) $\mathbf{F}(x, y) = (x + 1, y)$.

   (b) $\mathbf{F}(t, x, y) = (t, y)$.

   (c) $\mathbf{F}(x, y, z) = (x, \frac{1}{2}y, \frac{1}{3}z)$.

   (d) $\mathbf{F}(t, x, y, z) = (x + t, y - t, z)$.

5. Sketch the vector field $\mathbf{F}(x, y) = (-y, x)$. Then sketch the trajectory curve tangent to arrows in the field sketch, starting at $(x, y) = (1, 0)$.

*6. Show that the system $\dot{x} = -ty$, $\dot{y} = tx$ has circular solution trajectories of radius $r > 0$, traced with increasing speed rt as time increases. [*Hint:* Show that $x\dot{x} + y\dot{y} = 0$.]

7. Computer software for plotting vector fields is widely available. Scaling the lengths of the arrows by a constant factor often makes a better picture.

   (a) Use computer graphics to sketch vector fields of the form $\mathbf{F}(x, y) = (F_1(x, y), F_2(x, y))$. In particular, do this for the field
   $$\mathbf{F}(x, y) = (y - x, y + x).$$

   (b) Do the same for the field $\mathbf{F}(x, y) = (xy, x + y)$.

   (c) Do the same for fields of the form $F(x, y, z) = (F_1(x, y, z), F_2(x, y, z), F_3(x, y, z))$. In this case, the problem is essentially no harder, but care is needed to keep the sketch from getting too crowded. Sketch the field $\mathbf{F}(x, y, z) = (-y, x, z)$.

8. By letting $\dot{y} = x$, express each of the following second-order differential equations as a first-order system of dimension 2. Also find the corresponding initial conditions for $x(0)$ and $y(0)$.

   (a) $\ddot{y} + \dot{y} + y = 0$, $y(0) = 1$, $\dot{y}(0) = 1$.

   (b) $\ddot{y} + t\dot{y} = t$, $y(0) = 0$, $\dot{y}(0) = 1$.

9. For each part of the previous exercise, solve either the given differential equation or the associated system, whichever seems easier.

10. Find a first-order system equivalent to the following differential equations:

   (a) $d^2x/dt^2 + (dx/dt)^2 + x^2 = e^t$. (Let $dx/dt = y$.)

   (b) $d^2x/dt^2 = x\, dx/dt$.

   (c) $d^3x/dt^3 = (d^2x/dt^2)^2 - x\, dx/dt - t$. (Let $dx/dt = y$, $dy/dt = z$.)

   (d) $d^3x/dt^3 = 12x\, dx/dt$.

11. Reduce each system to normal form, with each first derivative by itself on the left side.

   (a) $dx/dt + dy/dt = t$,
   $dx/dt - dy/dt = y$.

   (b) $dx/dt + dy/dt = y$,
   $dx/dt + 2dy/dt = x$.

   (c) $2dx/dt + dy/dt + x + 5y = t$,
   $dx/dt + dy/dt + 2x + 2y = 0$.

   (d) $dx/dt - dy/dt = e^{-t}$,
   $dx/dt + dy/dt = e^t$.

12. Consider the 2-dimensional coupled system
   $$\dot{x} = x + y, \qquad \dot{y} = 4x + y.$$

   (a) Change from coordinates (x, y) to (z, w) with the relations
   $$x = z + w, \qquad y = 2z - 2w,$$
   and show that this change results in the uncoupled system
   $$\dot{z} = 3z, \qquad \dot{w} = -w.$$

   (b) Solve the uncoupled system in part (a) for z and w, and use the coordinate change to solve for x and y. Then verify by substitution that your solution for x and y satisfies the given system.

13. Suppose the 2-dimensional system $dx/dt = F(t, x, y)$, $dy/dt = G(t, x, y)$ is such that the ratio $G(t, x, y)/F(t, x, y) = R(x, y)$ happens to be independent of t. This would occur in particular if neither F nor G depended explicitly on t. Since the chain rule allows us to write

$$\frac{dy}{dx} = \frac{dy/dt}{dx/dt}$$

under fairly general conditions, we can sometimes conclude that there are trajectory curves of the system satisfying the differential equation

$$\frac{dy}{dx} = R(x, y).$$

If this equation can be solved, we have a way to plot trajectories without finding solutions $x(t)$, $y(t)$. For example, $dx/dt = ty$, $dy/dt = -tx$ leads us to consider

$$\frac{dy}{dx} = -\frac{x}{y},$$

which has solutions $x^2 + y^2 = c$, representing circular trajectories. Using this method, sketch some trajectories for the following systems. Looking at the vector field will tell you how a trajectory is traced and let you find constant solutions, for which $\dot{x} = \dot{y} = 0$.

(a) $dx/dt = x - y$, $dy/dt = x^2 - y^2$.

(b) $dx/dt = e^{2y}$, $dy/dt = e^{x+y}$

(c) $dx/dt = e^t y$, $dy/dt = e^t x$.

(d) $dx/dt = xy + y^2$, $dy/dt = x + y$.

14. The 1-dimensional equation $\ddot{y} = -g$ is used to determine the motion of an object moving perpendicularly to a large attracting body and subject to no other forces. Suppose the range of motion is extended to a vertical plane with horizontal coordinate x. If there are still no forces acting horizontally, the single equation is replaced by the 2-dimensional uncoupled system

$$\ddot{x} = 0,$$
$$\ddot{y} = -g.$$

(a) Solve the 2-dimensional system, subject to the four initial conditions
$$x(0) = 0, \qquad y(0) = 0, \qquad \dot{x}(0) = z_0 > 0, \qquad \dot{y}(0) = w_0 > 0.$$

(b) Show that the trajectory of the solution found in part (a) follows a parabolic path.

(c) Show that the maximum height is attained when the horizontal displacement is $z_0 w_0/g$ and that the maximum height is $w_0^2/(2g)$.

(d) Show that the horizontal distance traversed before returning to height $y(0) = 0$ is $2z_0 w_0/g$. Show also that for a given initial speed $v_0 = \sqrt{z_0^2 + w_0^2}$, this horizontal distance is maximized by having $z_0 = w_0$.

15. A **projectile fired against air resistance** proportional to velocity satisfies the uncoupled system

$$\ddot{x} = -k\dot{x},$$
$$\ddot{y} = -k\dot{y} - g, \qquad k > 0.$$

(a) Solve the 2-dimensional system, subject to the four initial conditions
$$x(0) = 0, \qquad y(0) = 0, \qquad \dot{x}(0) = z_0 > 0, \qquad \dot{y}(0) = w_0 > 0.$$

(b) Show that the trajectory of the solution found in part (a) rises to a unique maximum at time $t_{max} = (1/k) \ln (1 + kw_0/g)$.

(c) Show that the position of maximum height has coordinates
$$x_{max} = \frac{z_0 w_0}{g + kw_0}, \qquad y_{max} = \frac{w_0}{k} - \frac{g}{k^2} \ln \left(1 + \frac{kw_0}{g}\right).$$

(d) Show that as k tends to zero the maximum height tends to $\frac{1}{2}w_0^2/g$.

16. Here is an outline of a derivation of the **ideal pendulum equation** $\ddot{\theta} = -(g/l) \sin \theta$ using a system of differential equations.

(a) Show that if x, y are rectangular coordinates and θ is the angle formed by the vector (x, y), measured counterclockwise from the downward vertical direction, then $x = l \sin \theta$ and $y = -l \cos \theta$. [*Hint:* These equations are a slight modification of the usual polar coordinate relations.]

(b) Show that $\ddot{x} = -l \sin \theta \dot{\theta}^2 + l \cos \theta \ddot{\theta}$ and $\ddot{y} = l \cos \theta \dot{\theta}^2 + l \sin \theta \ddot{\theta}$.

(c) Use the representation $\ddot{x} = 0$, $\ddot{y} = -g$ for the coordinates of the acceleration of gravity, together with the result of part (b), to derive the pendulum equation. (This derivation safely ignores the lengthwise, or radial, force on the pendulum, since that force is always perpendicular to the path of motion; the next exercise takes this force into account.) [*Hint:* Eliminate terms containing $\dot{\theta}^2$.]

*17. **Pumping on a swing.** A playground swing with someone "pumping" on it can be looked at as a pendulum of varying length, because pumping up and down has the effect of alternately raising and lowering the center of mass, thus changing the effective length of the pendulum. Here is an outline of the derivation of the differential equation satisfied by $\theta = \theta(t)$, the angle the swing makes with the vertical direction, where $l = l(t)$ is the effective length of the swing at time t.

(a) Let $x = l \sin \theta$ and $y = -l \cos \theta$ be the rectangular coordinates of the center of mass, with pivot at $x = 0$, $y = 0$. Use the chain rule to derive the two equations
$$\ddot{x} = l (\cos \theta)\ddot{\theta} - l (\sin \theta)\dot{\theta}^2 + 2\dot{\theta}\dot{l} \cos \theta + \ddot{l} \sin \theta,$$
$$\ddot{y} = l (\sin \theta)\ddot{\theta} + l (\cos \theta)\dot{\theta}^2 + 2\dot{\theta}\dot{l} \sin \theta - \ddot{l} \cos \theta.$$

(b) By successively eliminating first $\ddot{l}$ and then $\ddot{\theta}$ between the two relations $\ddot{x} = 0$ and $\ddot{y} = -g$, derive the system
$$\ddot{\theta} = -\frac{g}{l} \sin \theta - \frac{2\dot{\theta}\dot{l}}{l},$$
$$\ddot{l} = g \cos \theta + l\dot{\theta}^2.$$

(c) The effective length $l(t)$ can be controlled through the swinger's application of radial force to alternately shift the center of mass of this variable "pendulum." The length $l(t)$, controlled in this way, then determines a second-order equation for $\theta = \theta(t)$; find this nonlinear differential equation if $l(t) = 10 + \sin t$.

*18. A **spherical pendulum** is one that pivots freely about a fixed point in 3-dimensional space. (Note that the end of the pendulum moves on a sphere centered at the pivot.) Let l be the effective length of the pendulum, let θ be the angle the pendulum makes with the downward vertical direction, and let ϕ be the angle a vertical plane containing the pendulum makes with a positive horizontal axis through the pivot. The purpose of this exercise is to show that the angles θ and ϕ satisfy the following system of differential equations:

$$\ddot{\theta} = -\frac{g}{l} \sin \theta + \dot{\phi}^2 \sin \theta \cos \theta,$$

$$\ddot{\phi} = -2\dot{\theta}\dot{\phi} \cot \theta, \qquad \theta \neq k\pi, \ k \text{ integer.}$$

The restriction $\theta \neq k\pi$ has a physical significance: if the pendulum is ever in a vertical position ($\theta = k\pi$), then the motion is confined to a vertical plane and is governed by the single equation to which the system reduces if ϕ is constant, namely, $\ddot{\theta} = -(g/l) \sin \theta$.

(a) Show that if x, y, z are rectangular coordinates of the center of mass of the pendulum, then

$$x = l \sin \theta \cos \phi,$$
$$y = l \sin \theta \sin \phi,$$
$$z = -l \cos \theta.$$

(b) The gravitational acceleration vector has the coordinates $\ddot{x} = 0$, $\ddot{y} = 0$, $\ddot{z} = -g$. Show that

$$(\sin \phi)\ddot{x} - (\cos \phi)\ddot{y} = -l\ddot{\phi} \sin \theta - 2l\dot{\theta}\dot{\phi} \cos \theta = 0,$$

and use this to derive the second equation of the spherical pendulum system.

(c) Show that

$$(\cos \phi)\ddot{x} + (\sin \phi)\ddot{y} = l\ddot{\theta} \cos \theta - l\dot{\theta}^2 \sin \theta - l\dot{\phi}^2 \sin \theta = 0,$$
$$\ddot{z} = l\ddot{\theta} \sin \theta + l\dot{\theta}^2 \cos \theta = -g.$$

Then derive the first equation of the spherical pendulum system from these two equations.

19. Heat exchange. The temperatures $u(t) \geq v(t)$ of two bodies in thermal contact with each other may be governed for the warmer body by Newton's law of cooling and for the cooler body by the analogous heating law:

$$\frac{du}{dt} = -p(u - v), \qquad \frac{dv}{dt} = q(u - v),$$

where p and q are positive constants and the equations are subject to initial conditions $u(0) = u_0$, $v(0) = v_0$. (Note that p may be different from q if the two bodies have different capacities to absorb heat.) This is a coupled system, but because of its simple form it can easily be solved as follows.

(a) Show that $q \, du/dt + p \, dv/dt = 0$. Then integrate with respect to t to show that $qu(t) + pv(t) = c_0$, where $c_0 = qu_0 + pv_0$.

(b) Use the relation between u and v derived in part (a) together with the given system to derive a single differential equation satisfied by $u(t)$. Then solve this equation using the initial conditions.

(c) Find a formula for $v(t)$.

(d) Find out how long it takes for the initial temperature difference between the two bodies to be cut in half.

20. **Bugs in mutual pursuit.** Four identical bugs are on a flat table, each moving at the same constant speed v. Use xy coordinates on the table, and locate bugs 1 through 4 initially in respective quadrants 1 through 4, each at one of the points $(\pm1, \pm1)$. Bug 1 always heads directly toward bug 2, bug 2 toward bug 3, bug 3 toward bug 4, and bug 4 toward bug 1, so their paths are mutually congruent.

(a) Use the symmetry of the paths to show that at any moment the bugs are at the corners of a square, and in particular, if bug 1 is at (x, y), then bugs 2, 3, and 4 are, respectively, at $(-y, x)$, $(-x, -y)$ and $(y, -x)$.

(b) Show that $\dot{y}/\dot{x} = (y - x)/(x + y)$ and that $\dot{x}^2 + \dot{y}^2 = v^2$.

(c) Use part (b) to show that the path $(x, y) = (x(t), y(t))$ followed by bug 1 satisfies the system

$$\frac{dx}{dt} = \frac{-v}{\sqrt{2}} \frac{x + y}{\sqrt{x^2 + y^2}}, \qquad \frac{dy}{dt} = \frac{v}{\sqrt{2}} \frac{x - y}{\sqrt{x^2 + y^2}}.$$

Note: Solution formulas for bug 1, with $0 \le t \le 2/v$, are

$$x(t) = \sqrt{2}\left(1 - \frac{vt}{2}\right)\cos\left(\ln\left(1 - \frac{vt}{2}\right) + \frac{\pi}{4}\right),$$

$$y(t) = \sqrt{2}\left(1 - \frac{vt}{2}\right)\sin\left(\ln\left(1 - \frac{vt}{2}\right) + \frac{\pi}{4}\right).$$

Another description of the trajectories is obtained in Exercise 12 of Chapter 1, Section 3C.

21. An **airplane** flies horizontally with constant airspeed v, pointing always at a fixed nondirectional radio beacon at position $\mathbf{x} = 0$. Wind with constant velocity vector $\mathbf{w}$ is blowing horizontally.

(a) Show that if $\mathbf{x}(t)$ is the airplane's position at time t, then

$$\dot{\mathbf{x}} = -\frac{v}{|\mathbf{x}|}\mathbf{x} + \mathbf{w}.$$

(b) If $\mathbf{x} = (x, y)$ and $\mathbf{w} = (w_1, w_2)$ show that the system is written

$$\dot{x} = \frac{-vx}{\sqrt{x^2 + y^2}} + w_1, \qquad \dot{y} = \frac{-vy}{\sqrt{x^2 + y^2}} + w_2.$$

(c) See Exercise 7b in Chapter 6, Section 7, for numerical solutions.

1D   Equilibrium Solutions

There are special solutions of autonomous systems that seem rather unimportant at first sight but which turn out to be quite significant in practice; these are the constant solutions of the autonomous vector equation

$$\frac{d\mathbf{x}}{dt} = \mathbf{F}(\mathbf{x}).$$

A solution $\mathbf{x}(t) = \mathbf{x}_0$ that is constant for all t is called an **equilibrium solution** because it is invariant for all time. (The stability of that equilibrium may be very vulnerable to slight perturbations of the differential equation or its associated initial conditions. See Chapter 6, Section 8, and Chapter 7, Section 5.) The trajectory of a constant equilibrium solution is always just the single point in state space representing its constant value; once we have identified that point, the sole point on that trajectory has been completely determined. It is the behavior of other trajectories near the equilibrium point that then becomes the main focus of attention. The direct approach to finding the equilibrium solutions of the system displayed above is to note that a constant solution $\mathbf{x}(t) = \mathbf{x}_0$ necessarily satisfies $d\mathbf{x}/dt = 0$ for all t in the domain of $\mathbf{x}(t)$. It follows that the equilibrium solutions are just the constant solutions $\mathbf{x} = \mathbf{x}_0$ of the equation $\mathbf{F}(\mathbf{x}) = 0$. Note that this last equation is not a differential equation; in practice it may be a purely algebraic equation, though there is no guarantee that it will be easy to solve.

---

example 1

If the system

$$\dot{x} = 2x + y$$
$$\dot{y} = x + y + 1$$

is to have constant solutions $x(t)$, $y(t)$, then for such solutions we must have zero derivatives $\dot{x} = \dot{y} = 0$ identically in t. It follows that an equilibrium solution must satisfy the purely algebraic system

$$2x + y = 0$$
$$x + y + 1 = 0.$$

Subtract the first equation from the second to get $x = 1$, from which it follows that $y = -2$. Thus there is just one equilibrium solution: $(x, y) = (1, -2)$.

---

example 2

The second-order equation $\ddot{x} = -(g/l)\sin x$ governs the motion of an undamped pendulum of length l subject to gravitational acceleration of magnitude g. In this equation $x = x(t)$ is the angle the pendulum makes with a vertical line directed down from the pendulum's fixed pivot point. Letting $y = \dot{x}$ so that $\dot{y} = \ddot{x}$, we arrive at the equivalent first-order system

$$\dot{x} = y,$$

$$\dot{y} = -\frac{g}{l}\sin x.$$

This reduction to a first-order system is necessary, at least in principle, so that the role of $\dot{x}$ is clearly displayed. (Note that $\dot{x}$ doesn't occur explicitly in $\ddot{x} = -(g/l)\sin x$.) Setting $\dot{x} = \dot{y} = 0$, we conclude that the equilibrium solutions in the state space must satisfy $\sin x = 0$ and $y = 0$. Thus the infinitely many equilibrium points are located at $(x, y) = (k\pi, 0)$, where k is an integer, positive, negative, or zero. Even values of k correspond to motionless downward vertical positions of the pendulum. Odd values of k correspond to upward vertical positions that are very precariously balanced.

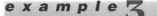

The system
$$\dot{x} = xy, \qquad \dot{y} = x - y - 1$$
has equilibrium solutions that must satisfy,
$$xy = 0, \qquad x - y - 1 = 0.$$
The first equation is satisfied by taking $x = 0$, so $y = -1$ follows from the second equation. Taking $y = 0$ in the first equation makes $x = 1$ in the second. Hence the equilibrium solutions are $(x, y) = (0, -1)$ and $(x, y) = (1, 0)$.

EXERCISES

Find all equilibrium solutions of the following systems.

1. $dx/dt = 2x - 3y + 1,$
$dy/dt = x + y.$

2. $dx/dt = x(y - 1),$
$dy/dt = (x - 1)y.$

3. $dx/dt = x - 2xy,$
$dy/dt = tx.$

4. $dx/dt = x \sin y,$
$dy/dt = y \sin x.$

5. $dx/dt = x - 2xy,$
$dy/dt = x - z,$
$dz/dt = x + y.$

6. $dx/dt = 1 - e^{(x-y)},$
$dy/dt = \sin (x - y),$
$dz/dt = xyz.$

7. The system
$$\dot{x} = 1 - x^2 - y^2, \qquad \dot{y} = x + cy$$
has two different equilibrium solutions for each value of the constant c; what are they, and on what curve in state space do they all lie?

8. Show that the general linear constant-coefficient system
$$\dot{x} = ax + by + h, \qquad \dot{y} = cx + dy + k$$
has either a unique equilibrium solution, or an entire line full of them, or else none at all.

9. Find the equilibrium solutions of the **Lotka-Volterra** system
$$\frac{dH}{dt} = (a - bP)H, \qquad \frac{dP}{dt} = (cH - d)P, \qquad a, b, c, d \text{ positive const.}$$
This system models the behavior of interacting host H and parasite P population sizes.

10. The Lotka-Volterra equations can be refined to take account of fixed limits $L > H(t)$ and $M > P(t)$ to the growth of the host and parasite populations as follows:
$$\frac{dH}{dt} = (a - bP)H(L - H), \qquad \frac{dP}{dt} = (cH - d)P(M - P),$$
where a, b, c, d are positive constants, with $L > d/c > 1$ and $M > a/b > 1$. Find all equilibrium solutions. The Lotka-Volterra system is discussed in detail in Chapter 6, Section 2.

11. Find all equilibrium solutions of the nonlinear system
$$\dot{x} = x(3 - y), \qquad \dot{y} = y(x + z - 3), \qquad \dot{z} = z(2 - y).$$

12. In the **Lorenz system** $\dot{x} = \sigma(y - x), \dot{y} = -xz + \rho x - y, \dot{z} = xy - \beta z$, and the constants β, ρ, σ are assumed to be positive. Show that equilibrium solutions are $(0, 0, 0)$ and, if $\rho > 1$, $(\pm\sqrt{\beta(\rho - 1)}, \pm\sqrt{\beta(\rho - 1)}, \rho - 1)$ and that there aren't any others.

13. The position $\mathbf{x} = (x, y)$ and velocity $\mathbf{v} = (z, w)$ vectors of a single planet orbiting a fixed sun are shown in Chapter 6, Section 5, to obey the 4-dimensional system

$$\dot{x} = z, \qquad \dot{y} = w, \qquad \dot{z} = \frac{-kx}{(x^2 + y^2)^{3/2}}, \qquad \dot{w} = \frac{-ky}{(x^2 + y^2)^{3/2}},$$

where k a positive constant. Does the system have equilibrium solutions (x_0, y_0, z_0, w_0)?

14. The nonlinear systems below are defined on curves or surfaces of which every point is an equilibrium solution. Identify all such solutions and sketch the relevant curves or surfaces.

   (a) $\dot{x} = (1 - x^2 - y^2)(x + y), \dot{y} = (1 - x^2 - y^2)(x^2 - y^2)$.

   (b) $\dot{x} = y(z - x - y), \dot{y} = y(x - y), \dot{z} = x^2y - y^3$.

15. The 1-dimensional linear system $\dot{x} = tx$ is **equivalent** to the 2-dimensional nonlinear autonomous system $\dot{x} = xy, \dot{y} = 1$ in the sense that knowing all solutions of one system allows us to find all solutions of the other.

   (a) Find all solutions of $\dot{x} = tx$. What are the constant solutions, if any?

   (b) Find all solutions of the nonlinear system $\dot{x} = xy, \dot{y} = 1$. Show that this system has no constant solutions. Thus the single equation has an equilibrium solution but the associated autonomous system does not. Note that the given equation is linear but the system isn't.

16. Letting $x_{n+1} = t$ and incorporating the single real equation $\dot{x}_{n+1} = 1$ into an n-dimensional nonautonomous vector system $\dot{\mathbf{x}} = \mathbf{F}(\mathbf{x}, t)$ produces an $(n + 1)$-dimensional autonomous system $\dot{\mathbf{y}} = \mathbf{F}(\mathbf{y})$, where $\mathbf{y} = (\mathbf{x}, x_{n+1}) = (x_1, x_2, \ldots, x_n, x_{n+1})$. Find such an autonomous system for

   (a) $\dot{y} = \sin(ty)$.

   (b) $\dot{x} = x + y + t, \dot{y} = x - y + t^2$.

   (c) $\dot{x} = ty, \dot{y} = tz, \dot{z} = tx$.

1E   *Existence, Uniqueness, and Flows (Optional)*

This section gives a precise description of what a dynamical system is in the mathematical sense. To begin, note that reduction of a system to first-order normal form isn't always possible, but when it is, and when the vector field is smooth, we can apply the following theorem that implies not only existence and uniqueness of solutions but also smoothness of the flow lines. For an illuminating discussion of the theorem see chapter 2 of V. I. Arnold, *Ordinary Differential Equations*, Springer-Verlag, 1991.

■ 1.1 EXISTENCE AND UNIQUENESS THEOREM

Suppose $\mathbf{F}(t, \mathbf{x})$ and its first-order partial derivatives with respect to the $\mathbf{x}$ variables are bounded and continuous on an open set B in $\mathcal{R}^{n+1}$ containing $(t_0, \mathbf{x}_0)$. Then the system $\dot{\mathbf{x}} = \mathbf{F}(t, \mathbf{x})$ has a unique solution satisfying $\mathbf{x}(t_0) = \mathbf{x}_0$. Furthermore, the value $\mathbf{x}(t)$ of this solution is a continuously differentiable function of not only t but also the initial point $\mathbf{x}_0$.

The uniqueness part of the theorem tells us that, for an autonomous system $\dot{\mathbf{x}} = \mathbf{F}(\mathbf{x})$ that satisfies the hypotheses of the theorem, two solution trajectories that have a point $\mathbf{x}_0$ in common can differ only in their extent; in other words, "distinct trajectories can't cross." These ideas can be stated more precisely as follows.

■ 1.2 COROLLARY

If $\dot{\mathbf{x}} = \mathbf{F}(\mathbf{x})$ satisfies the conditions of Theorem 1.1, and two solution trajectories of the system have a point $\mathbf{x}_0$ in common, then on either side of $\mathbf{x}_0$ one trajectory is contained in the other.

Proof. If two trajectories agree at $\mathbf{x}_0$, this common value taken as an initial value would dictate that the trajectories are the same from that time on until one of them terminates. Likewise, the reverse trajectory that satisfies $\dot{\mathbf{x}} = \mathbf{F}(\mathbf{x})$, and which coincides as a curve with a trajectory approaching $\mathbf{x}_0$ from the other side, would also be uniquely determined until termination of one of them.   ■

---

example 1

It's easy to verify that the system $\dot{x} = -y$, $\dot{y} = x$ has circular trajectories of radius $\sqrt{u^2 + v^2}$ traced by $x(t) = u \cos t - v \sin t$, $y(t) = u \sin t + v \cos t$ and satisfying $x(0) = u, y(0) = v$. These are shown in Figure 5.6a, and it's clear that distinct trajectories fail to intersect but are traced repeatedly if t is not restricted.

---

example 2

Figure 5.6b shows some computer plots of trajectories for the *nonautonomous* system $\dot{x} = (1 - t)x - ty$, $\dot{y} = tx + (1 - t)y$. Each one of the four trajectories is shown crossing one of the others.

One trajectory of a nonautonomous system may very well cross another, or even intersect itself at a nonzero angle, because in arriving at the same point in state space at different times, the direction of the vector field may turn out to be different. Snapshots of the vector field of this system are shown in Figure 5.5. Note, however, that the graphs of different solutions in txy space will have no points in common, because the first coordinate t will be different at each distinct point of the graph. This idea can be used to show how Theorem 1.1, which applies directly to autonomous vector fields, can be applied to time-dependent vector fields. See Exercise 12.

Flows. A trajectory of an autonomous system $\dot{\mathbf{x}} = \mathbf{F}(\mathbf{x})$ is called a **flow line** of the vector field $\mathbf{F}$, and we can think of their paths, as shown, for example, in Figure 5.6a, as the possible paths followed by fluid particles in a steady fluid

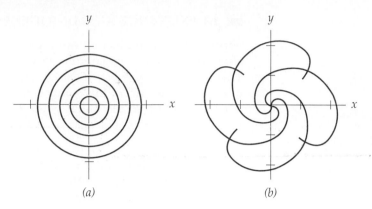

(a)   (b)

Figure 5.6 (*a*) Autonomous system trajectories. (*b*) Nonautonomous system trajectories.

flow with velocity vector $\mathbf{F}(\mathbf{x})$ at $\mathbf{x}$. In what follows we'll assume that the autonomous vector field $\mathbf{F}$ satisfies the conditions of Theorem 1.1 in some region B in $\mathcal{R}^n$, thus guaranteeing (1) that there is a unique flow line through each $\mathbf{x}$ in B and (2) that distinct flow lines have no points in common. We associate with each such vector field $\mathbf{F}$ a family of **flow transformations** T_t from B to B and a **flow** ϕ defined by

■ **1.3** $T_t(\mathbf{x}) = \phi(t, \mathbf{x}) = \mathbf{y}(t),$ where $\mathbf{y}(t)$ solves $\dot{\mathbf{x}} = \mathbf{F}(\mathbf{x})$ with $\mathbf{y}(0) = \mathbf{x}$.

In words, $T_t(\mathbf{x})$ is the point on the flow line of $\mathbf{F}$ starting at $\mathbf{x}$ that is reached after time t. Thus a family $\{T_t\}$ is a formal representation of a continuous-time **dynamical system** as described less formally at the beginning of the Introductory Survey of the entire book. The families $\{T_t\}$ are important in applications, because they, rather than systems of equations, are what we use to describe real-world phenomena.

---

e x a m p l e **3**

The system $\dot{x} = -y, \dot{y} = x$ has circular trajectories, as in Figure 5.6a. Thus a flow line of radius A for the vector field $\mathbf{F}(x, y) = (-y, x)$ can be parametrized by $x(t) = A \cos (t + \alpha), y(t) = A \sin (t + \alpha)$. To start one of these flow lines at a fixed point (u, v) when $t = 0$, we note that $A = \sqrt{u^2 + v^2}$ and write

$$T_t(u, v) = (\sqrt{u^2 + v^2} \cos (t + \alpha), \sqrt{u^2 + v^2} \sin (t + \alpha)),$$

where α is an angle that the radius from the origin to (u, v) makes with the positive x axis. (Thus $\alpha(u, v) = \arctan (v/u)$ if $u \neq 0$, and $\alpha(0, v)$ is the odd multiple of $\pi/2$ that makes $\alpha(u, v)$ continuous; $\alpha(0, 0)$ is not defined.)

The function $\phi: \mathcal{R}^3 \to \mathcal{R}^2$ defined by $\phi(t, u, v) = T_t(u, v)$ in Example 3 is obviously, for each fixed (u, v), not one-to-one as a function of t unless t is somehow restricted. However, for fixed $t = t_0$, T_{t_0} turns out not only to be one-to-one but to have a nice inverse, namely, T_{-t_0}. Indeed, it's a straightforward exercise to show that T_{t_0} is just a rotation about the origin through angle t_0. Hence the inverse of T_{t_0} is T_{-t_0}. This simple relationship between T_t and its inverse holds very generally, as described below.

The transformations T_t introduced above have the important **composition property:**

■ **1.4** $T_t T_s = T_{t+s}$,   in other words, $T_t(T_s(\mathbf{x})) = T_{t+s}(\mathbf{x})$,

whenever all three transformations are defined. Equation 1.4 holds because the system

$$\frac{d\mathbf{x}}{dt} = \mathbf{F}(\mathbf{x}),$$

of which $\mathbf{y}(t) = T_t(\mathbf{x})$ is a solution, has a unique solution starting at $T_t(\mathbf{x}_0)$ whose value at s time units later must coincide with the unique solution value achieved by starting at $\mathbf{x}_0$ and running for time $t + s$. Note that t and s can be negative in Equation 1.5, since they can decrease from zero. Another interpretation for negative time values is based on the *reversed system*

$$\frac{d\mathbf{x}}{dt} = -\mathbf{F}(\mathbf{x}),$$

which has solution trajectories traced in the direction $-\mathbf{F}(\mathbf{x})$ exactly opposite to that of the solutions of $d\mathbf{x}/dt = \mathbf{F}(\mathbf{x})$. We can use solutions of the reversed system to define T_t for $t < 0$ by $T_t(\mathbf{x}_0) = \mathbf{z}(t)$, where $\mathbf{z}(t)$ satisfies

$$\frac{d\mathbf{z}}{dt} = -\mathbf{F}(\mathbf{z}), \qquad \mathbf{z}(0) = \mathbf{x}_0.$$

It follows that each of T_{-t} and T_t is an **inverse operator** to the other, that is,

■ **1.5** $T_{-t}T_t = T_t T_{-t} = I,$

where I is an identity operator that leaves points fixed.

---

example 4

The circular flow ϕ defined by

$$T_t(u, v) = (\sqrt{u^2 + v^2}\cos(t + \alpha), \sqrt{u^2 + v^2}\sin(t + \alpha)), \qquad \alpha = \arctan\frac{v}{u},$$

is discussed in Example 3. While the parameter t is customarily interpreted as the time it takes for point $\mathbf{x}$ on a flow line to move to $T_t(\mathbf{x})$, in this special example t is also the angle subtended at the origin from $\mathbf{x}$ to $T_t(\mathbf{x})$. Two such angles are shown in Figure 5.7, t from (u, v) to $T_t(u, v)$ and s from (x, y) to

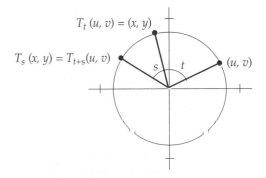

Figure 5.7 $T_{t+s}(u, v) = T_s(T_t(u, v))$.

$T_s(x, y)$. If $(x, y) = T_t(uv)$, as shown in the figure, then it's clear that Equation 1.4 should hold, since both times and angles should be added.

---

e x a m p l e 5

The uncoupled system $\dot{x} = x$, $\dot{y} = 2y$ is easily seen to have the solution $x(t) = ue^t$, $y = ve^{2t}$ with initial values $x(0) = u$, $y(0) = v$. The flow generated by the vector field $\mathbf{F}(x, y) = (x, 2y)$ is therefore

$$\phi(t, u, v) = T_t(u, v) = (ue^t, ve^{2t}).$$

The xy coordinates of the flow line of the flow starting at (u, v) satisfy $x = ue^t$, $y = ve^{2t}$, so $u^2y = vx^2$. Thus as $t > 0$ increases the flow lines trace parabolas heading away from the origin unless u or v is zero, in which case we get segments of the axes. See Figure 5.8a, where the initial points (u, v) shown all lie on a circle of radius $\frac{1}{2}$ centered at the origin. With decreasing negative t we would get portions of the parabolas heading toward the origin but never reaching it. The origin is an equilibrium solution, since $T_t(0, 0) = (0, 0)$ for all t, so the origin is itself a flow line, though a rather trivial one.

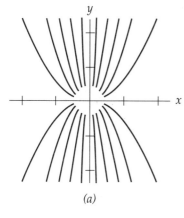
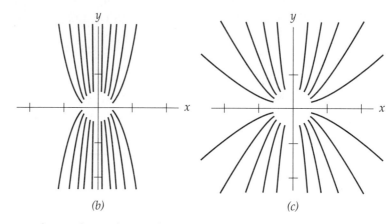

(a)   (b)   (c)

Figure 5.8 (a) $u^2y = vx^2$; (b) $u^3z = wx^3$; (c) $v^3z^2 = w^2y^3$.

---

e x a m p l e 6

The 3-dimensional system $\dot{x} = x$, $\dot{y} = 2y$, $\dot{z} = 3z$ has the solution $x(t) = ue^t$, $y = ve^{2t}$, $z(t) = we^{3t}$ with initial values $x(0) = u$, $y(0) = v$, $z(0) = w$. The flow χ generated by the vector field $\mathbf{G}(x, y, z) = (x, 2y, 3z)$ has a family $\{X_t\}$ of flow transformations such that

$$\chi(t, u, v, w) = X_t(u, v, w) = (ue^t, ve^{2t}, we^{3t}).$$

An individual flow line of the flow is either one of the positive or one of the negative axes or else has the shape of a curve called a *twisted cubic*. Plotting any one of these curves is fairly easy, but a representative sample of them in 3-dimensional perspective looks confusing, so as an alternative Figure 5.8 shows projections of such a sample into the three coordinate planes of $\mathcal{R}^3$. These projections are traced away from the origin starting at distance $\frac{1}{2}$ from there. In particular Figure 5.8a is the same picture used to illustrate the flow lines of the field $\mathbf{F}(x, y) = (x, 2y)$ in Example 5.

example 7

The 3-dimensional initial-value problem

$$\dot{x} = -z, \qquad \dot{y} = \tfrac{1}{2}, \qquad \dot{z} = x, \qquad x(0) = u, \qquad y(0) = v, \qquad z(0) = w$$

has the solution $x(t) = u \cos t - w \sin t, y = \tfrac{1}{2}t + v, z(t) = w \cos t + u \sin t$. The flow ψ generated by the vector field $\mathbf{G}(x, y, z) = (-z, \tfrac{1}{2}, x)$ has a family ψ_t of flow transformations such that

$$\psi(t, u, v, w) = X_t(u, v, w) = (u \cos t - w \sin t, \tfrac{1}{2}t + v, w \cos t + u \sin t).$$

The second coordinate, $y = \tfrac{1}{2}t + v$, makes the flow move in the general direction of the positive y axis. The x and z coordinates satisfy

$$x^2 + z^2 = (u \cos t - w \sin t)^2 + (w \cos t + u \sin t)^2 = u^2 + w^2.$$

Hence the projection on the xz plane of a flow line through (u, v, w) is a circle of radius $\sqrt{u^2 + v^2}$ centered at the origin. It follows that each flow line is a helix, a type of spiral curve, traced as t increases in the direction of the positive y axis. See Figure 5.9.

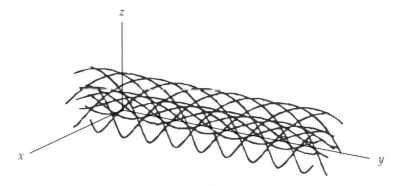

Figure 5.9 Helical flow lines.

EXERCISES

1. The domain of t values for which the solution of even an autonomous system is valid may be quite restricted. Illustrate this point by deriving the explicit solution to the 1-dimensional initial-value problem $\dot{x} = ax^2$, $x(0) = 1$, where a is a positive constant.

2. Theorem 1.1 on existence and uniqueness of solutions fails to apply everywhere to the 1-dimensional equation defined by $\dot{x} = \sqrt{x}, x \geq 0$ and $\dot{x} = 0$, $x < 0$.

    (a) Explain why the theorem doesn't apply if the initial condition is $x(0) = 0$.

    (b) Find two distinct solutions to the equation, both satisfying $x(0) = 0$.

3. The flow transformation of the vector field $\mathbf{F}(x, y) = (x, 2y)$ in Example 5 of the text is $T_t(u, v) = (ue^t, ve^{2t})$. Verify directly for this example that $T_{t+s}(u, v) = T_s(T_t(u, v))$ for all t and all (u, v).

4. Consider the system $\dot{x} = y$, $\dot{y} = x$.

(a) Verify that the system's flow is

$$T_t(u, v) = (\tfrac{1}{2}(u + v)e^t + \tfrac{1}{2}(u - v)e^{-t}, \tfrac{1}{2}(u + v)e^t - \tfrac{1}{2}(u - v)e^{-t}).$$

(b) Show that the flow line through (u, v) lies on a straight line if $|u| = |v|$ and otherwise traces one branch of a hyperbola as t varies.

(c) Verify directly that $T_{t+s}(u, v) = T_s(T_t(u, v))$ for all t and (u, v).

5. The flow transformation of the vector field $\mathbf{F}(x, y) = (-y, x)$ in Example 3 of the text is $T_t(u, v) = (u \cos t - v \sin t, u \sin t + v \cos t)$. Verify by direct substitution for this example that $T_{t+s}(u, v) = T_s(T_t(u, v))$ for all t and (u, v).

6. What are the flow "lines" of the identically zero vector field on $\mathscr{R}^2$?

7. A 2-dimensional **Hamiltonian system** has the form $\dot{x} = \partial H / \partial y$, $\dot{y} = -\partial H / \partial x$, where the real-valued **Hamiltonian function** $H(x, y)$ is assumed to be twice continuously differentiable.

(a) Show that the system $\dot{x} = y$, $\dot{y} = -x$ is Hamiltonian, and find the Hamiltonian function $H(x, y)$ for this system.

(b) Show that the flow lines of a 2-dimensional Hamiltonian system follow level curves of the associated Hamiltonian function.

8. Show that the second-order equation $\ddot{x} = -f(x)$ can be looked at as a first-order system by setting $y = \dot{x}$. Then show that the first-order system can be looked at as a Hamiltonian system, as defined in the previous exercise, with Hamiltonian

$$H(x, y) = \tfrac{1}{2}y^2 + U(x), \qquad \text{where } U'(x) = f(x).$$

The function $U(x)$ is the **potential energy** of the system and $H(x, y)$ is the **total energy.**

*9. Consider the flow transformation $T_t(u, v) = (x(t), y(t))$ generated by solutions of a system $\dot{x} = F(x, y)$, $\dot{y} = G(x, y)$ with initial conditions $x(0) = u$, $y(0) = v$. Assume that F and G have continuous first-order partials.

(a) For fixed t let $J_t(u, v)$ be the Jacobian determinant of $T_t(u, v)$ with respect to u and v. Show that

$$\frac{d}{dt} J_t(u, v) = F_u y_v - F_v y_u + G_v x_u - G_u x_v.$$

(b) Apply the chain rule, for example, $F_u = F_x x_u + F_y y_u$, to the partials of F and G in part (a) to show that $(d/dt)J_t = (F_x + G_y)J_t$, where F_x and G_y are evaluated at $(x(t), y(t))$.

(c) Noting that T_0 is an identity transformation so that $J_0 = 1$, solve the differential equation in part (b) to show that $J_t = e^{t(F_x + G_y)}$. The real-valued function div $(F, G) = F_x + G_y$ is called the **divergence** of the vector field with coordinate functions F and G.

(d) Use the result of part (c) together with the change-of-variable theorem for double integrals to show that a flow transformation acting in a region R sends a set of positive area into a set of smaller area, the same area, or larger area depending on whether div (F, G) is negative, zero, or positive throughout R.

10. A 2-dimensional **gradient system** has the form $\dot{x} = \partial U/\partial x$, $\dot{y} = \partial U/\partial y$, where the real-valued **potential function** $U(x, y)$ is assumed to be twice continuously differentiable. Show that the flow lines of a gradient system are perpendicular to the level curves of its potential function.

11. Consider the 2-dimensional uncoupled system $\dot{x} = x^3$, $\dot{y} = y^3$.

    (a) Sketch the vector field of the system near the origin.

    (b) For each fixed t, compute the Jacobian determinant J_t of the flow transformation T_t of the system.

    (c) Let B be a region of positive area in $\mathcal{R}^2$. Use the result of part (b) to show that the area of the image of B under T_t is bigger than $A(B)$ if $t > 0$ and less than $A(B)$ if $t < 0$.

    (d) Can you draw the same conclusion as in part (c) if the original system is replaced by $\dot{x} = x^2$, $\dot{y} = y^2$? Explain your reasoning.

12. Corollary 1.2 is relevant to nonautonomous systems because an n-dimensional nonautonomous system $\dot{\mathbf{x}} = \mathbf{F}(\mathbf{x}, t)$ can be regarded as an $(n + 1)$-dimensional autonomous system by setting $t = x_{n+1}$ in $\mathbf{F}(\mathbf{x}, t)$ and adding the equation $\dot{x}_{n+1} = 1$ to the system.

    (a) Show how the system $\dot{x}_1 = tx_2$, $\dot{x}_2 = tx_1$ would look after this conversion.

    (b) Show however that while the system given in part (a) has $(x_1, x_2) = (0, 0)$ for an equilibrium solution, the augmented 3-dimensional system has no equilibrium solutions.

2 LINEAR SYSTEMS

2A Definitions

A differential equation for a single real- or complex-valued function $x = x(t)$ is called a **linear differential equation** if it is a polynomial of degree one in x and one or more of its derivatives, with coefficients that may be functions of an independent variable t. Thus $\ddot{x} + t^2\dot{x} = \sin t$ and $e^t\ddot{x} - tx = t$ are linear, while $\ddot{x} + x^2 = 0$ and $x\ddot{x} + tx = 0$ are nonlinear. If we define a linear operator L by $L = a_n D^n + a_{n-1}D^{n-1} + \cdots + a_1 D + a_0$, a linear differential equation can be written $L(x) = f$ provided that some coefficient other than a_0 is different from zero. A **linear system** that determines solutions $x = x(t)$, $y = y(t)$, ... is a collection of equations in which each real-valued equation can be written in the form

$$L_1(x) + L_2(y) + \cdots = f(t).$$

Here each L_k is a linear operator. If $f = 0$ in each equation of such a system, the system is called **homogeneous.** In applying these definitions, the written order of the terms can always be changed. (For example, the equation $dx/dt = ty$ is equivalent to $dx/dt - ty = 0$.) Thus a homogeneous linear system is one for which both sides of each equation consists entirely of a sum of multiples of the unknown functions or their derivatives. The multiplier coefficients are functions, possibly constant, of the independent variable t alone. The system becomes nonhomogeneous upon the addition to at least one of the equations of a term $f(t)$ dependent on t at most; deletion of all such terms $f(t)$ produces the **associated homogeneous system.**

example 1

Here are some linear systems.

(a) $\dfrac{dx}{dt} + \dfrac{dy}{dt} = e^t,$

$\dfrac{dx}{dt} - \dfrac{dy}{dt} = e^{-t}.$

(b) $\dfrac{dx}{dt} - ty = \sin t,$

$\dfrac{d^2y}{dt^2} + t^2x = 0.$

(c) $\dfrac{d^2x}{dt^2} + \dfrac{dx}{dt} + \dfrac{dy}{dt} = 0,$

$x + \dfrac{dy}{dt} + y = 0.$

(d) $\dfrac{dx}{dt} = x + y + z,$

$\dfrac{dy}{dt} = x + y - z,$

$\dfrac{dz}{dt} = x - y + z.$

Only (c) and (d) are homogeneous systems. The associated homogeneous systems in (a) and (b) are obtained by replacing the exponential and sine terms by 0.

The following systems are nonlinear. (The property of homogeneity isn't really significant for nonlinear equations.)

(e) $\dfrac{dx}{dt} + \left(\dfrac{dy}{dt}\right)^2 = t,$

$\dfrac{dx}{dt} - \dfrac{dy}{dt} = 0.$

(f) $\dfrac{dx}{dt} - ty = 0,$

$\dfrac{d^2y}{dt^2} + \sin x = 0.$

2B Elimination Method

By adding nonzero multiples of one equation to another, we can sometimes rewrite a system so that it becomes easier to interpret and solve.

example 2

The sum and difference of the two equations in Example 1a are, respectively,

$$2\frac{dx}{dt} = e^t + e^{-t}, \qquad 2\frac{dy}{dt} = e^t - e^{-t}.$$

Integrating both equations with respect to t and dividing by 2 gives

$$x = \tfrac{1}{2}(e^t - e^{-t}) + c_1, \qquad y = \tfrac{1}{2}(e^t + e^{-t}) + c_2.$$

We can also do a part of the computation using the hyperbolic cosine and sine functions $\cosh t = (e^t + e^{-t})/2$, $\sinh t = (e^t - e^{-t})/2$. We then write the uncoupled system as

$$\frac{dx}{dt} = \cosh t, \qquad \frac{dy}{dt} = \sinh t.$$

The solutions can be written $x(t) = \sinh t + c_1$, $y(t) = \cosh t + c_2$; this follows, for example, from using two differentiation formulas for hyperbolic functions: derivative of $\sinh t$ is $\cosh t$ and vice versa.

Differential operators are particularly helpful if the coefficients in a linear system are constant, as the following example shows.

---

e x a m p l e 3

The vector equation

$$\frac{d}{dt}(x, y) = (2x + 4y + 2, x - y + 4)$$

represents the system

$$\frac{dx}{dt} = 2x + 4y + 2,$$

$$\frac{dy}{dt} = x - y + 4.$$

Using $D = d/dt$, we can write the system as

$$(D - 2)x - 4y = 2,$$
$$-x + (D + 1)y = 4.$$

Since $y(t)$ and $y(t)$ must be at least once differentiable, the first equation in the form $\dot{x} = 2x + 4y + 2$ shows that $x(t)$ must be twice differentiable. A similar remark applies to $y(t)$. If D were a number, we could add multiples of one equation to another so as to eliminate either x or y and then substitute back to get the other variable. The algebra of differential operators allows us to do something similar here. Operate on the second equation with $(D - 2)$:

$$(D - 2)x - 4y = 2,$$
$$-(D - 2)x + (D - 2)(D + 1)y = (D - 2)4.$$

Noting that $(D - 2)4 = -8$, we add the first equation to the second to get

$$(D - 2)(D + 1)y - 4y = -6.$$

But this equation is the same as

$$(D^2 - D - 6)y = -6.$$

The characteristic equation is $r^2 - r - 6 = (r + 2)(r - 3) = 0$. The roots are $r_1 = -2$, $r_2 = 3$; so the solution of the homogeneous equation for y is

$$y_h = c_1 e^{-2t} + c_2 e^{3t}.$$

By inspection we find a particular solution $y_p = 1$, so $y(t)$ must be of the form

$$y = y_h + y_p = c_1 e^{-2t} + c_2 e^{3t} + 1.$$

Now solve the second of the two equations for x:

$$x - (D + 1)y = 4$$

Substitution of the formula for $y(t)$ gives

$$x(t) = -c_1 e^{-2t} + 4c_2 e^{3t} - 3,$$

The vector solution is then

$$(x, y) = (-c_1 e^{-2t} + 4c_2 e^{3t} - 3, \; c_1 e^{-2t} + c_2 e^{3t} + 1).$$

Initial conditions $(x(0), y(0)) = (2, 1)$ require that

$$(-c_1 + 4c_2 - 3, \; c_1 + c_2 + 1) = (2, 1)$$

or $-c_1 + 4c_2 = 5$, $c_1 + c_2 = 0$. Thus $c_2 = 1$ and $c_1 = -1$ for the particular solution

$$(x_p, y_p) = (e^{-2t} + 4e^{3t} - 3, \; -e^{-2t} + e^{3t} + 1).$$

2C   *General Form of Solutions*

A step-by-step analysis of the computation in Example 3 would show that we have indeed found the most general solution and that it contains exactly the right number of arbitrary constants. It can happen that extraneous constants do turn up in a solution formula. This may happen if applying an operator increases the order of the system and so increases the number of constants. To find relations among the extra constants, substitute the solution back into the given system. To determine in advance what the right number of independent constants is, we appeal to Theorem 3.2 of Chapter 7, Section 3; this shows that the complete solution of a constant-coefficient linear system, *if it is in normal form*, contains a number of arbitrary constants equal to the number of equations in the system. We have the following.

Rule of Thumb. To determine the correct number of arbitrary constants for the solution of a constant-coefficient linear system, attempt to put the system in normal form. If your attempt succeeds, the correct number of arbitrary constants will equal the number of equations in your normal-form system.

Exercise 8 for this section shows, however, that not every linear system is equivalent to one in normal form. In particular, a system that is not in normal form may be inconsistent, having no solutions at all, or may have a nonstandard number of arbitrary constants in its general solution. In the latter case we still have recourse to substitution for finding relations among constants.

example 4

The second-order system

$$\frac{d^2x}{dt^2} = x + 2y + t,$$

$$\frac{d^2y}{dt^2} = 3x + 2y$$

can be written in operator form as

$$(D^2 - 1)x - 2y = t,$$
$$-3x + (D^2 - 2)y = 0.$$

If we multiply the second equation by 2 and operate on the first with $(D^2 - 2)$, then adding the resulting equations eliminates y:

$$(D^2 - 2)(D^2 - 1)x - 6x = -2t \qquad \text{or} \qquad (D^4 - 3D^2 - 4)x = -2t.$$

The characteristic equation is

$$r^4 - 3r^2 - 4 = (r^2 + 1)(r^2 - 4) = 0,$$

and has roots $r_1 = i, r_2 = -i, r_3 = 2, r_4 = -2$. The homogeneous solution is then

$$x_h(t) = c_1 \cos t + c_2 \sin t + c_3 e^{2t} + c_4 e^{-2t}.$$

By inspection, we find a particular solution $x_p = \frac{1}{4}t$. Thus

$$x = x_h + x_p = c_1 \cos t + c_2 \sin t + c_3 e^{2t} + c_4 e^{-2t} + \tfrac{1}{4}t.$$

To find y we could start over again, eliminating x between the two equations and finding an expression for $y_h + y_p$ containing four additional constants c_5, c_6, c_7, c_8. The original system written in first-order normal form contains just four equations, so there must be some relations among the constants that allow us to reduce to just four independent ones. This could be done by substitution into the original system. However, it's easier in this example not to start over with an equation for x alone but to proceed as follows. Solve the first of the given equations for y. A straightforward computation gives

$$y = \tfrac{1}{2}(D^2 - 1)x - \tfrac{1}{2}t,$$

or $\qquad y = -c_1 \cos t - c_2 \sin t + \tfrac{3}{2}c_3 e^{2t} + \tfrac{3}{2}c_4 e^{-2t} - \tfrac{3}{4}t.$

Putting this together with the formula we found above for x gives the complete solution. The four constants can be determined by imposing four initial conditions, for example, $x(0) = x_0$, $\dot{x}(0) = x_1$, $y(0) = y_0$, and $\dot{y}(0) = y_1$.

For both technical and conceptual reasons it's sometimes helpful to break a complicated problem into pieces, solve each piece separately, and then combine the results into a complete solution. As an example recall the single linear equation $\ddot{y} + 4y = 1$. This solution procedure can be separated into (1) finding the complete solution of the homogeneous equation $\ddot{y} + 4y = 0$, for which $y_h = c_1 \cos 2t + c_2 \sin 2t$, and (2) finding a single particular solution of the nonhomogeneous equation, for which we can in this instance simply guess the constant solution $y_p = \frac{1}{4}$. Because the equation is linear, the complete solution is then $y = y_h + y_p$. The decomposition of solutions $x(t) = x_h(t) + x_p(t)$ that holds for a single linear equation is valid for linear systems also. The essence of the proof is simply a vector reinterpretation of the proof of Theorem 3.1 in Chapter 3, Section 3A, for a single linear equation.

■ 2.1 THEOREM

If $\mathbf{x}(t) = (x(t), y(t), \ldots)$ is some solution of a nonhomogeneous system and $\mathbf{x}_p(t) = (x_p(t), y_p(t), \ldots)$ is an arbitrary particular solution of the same system, then

$$\mathbf{x}(t) = (x_h(t) + x_p(t), y_h(t) + y_p(t), \ldots)$$

for some appropriately chosen solution $\mathbf{x}_h(t) = (x_h(t), y_h(t), \ldots)$ of the associated homogeneous system. Thus the general solution of a linear system can be expressed as the sum of the general solution of the associated homogeneous system and some particular solution.

Proof. Suppose a nonhomogeneous system consists of m equations

$$\sum_{k=1}^{n} L_{jk}(x_k) = f_j(t), \qquad j = 1, \ldots, m$$

with solutions $\mathbf{x}(t) = (x_1(t), x_2(t), \ldots)$ and $\mathbf{x}_p(t) = (p_1(t), p_2(t), \ldots)$. Since each of the operators L_{jk} acts linearly on a real or complex function, it follows that $L_{jk}(x_k - p_k) = L_{jk}(x_k) - L_{jk}(p_k)$. Summing those equations over k gives

$$\sum_{k=1}^{n} L_{jk}(x_k - p_k) = \sum_{k=1}^{n} L_{jk}(x_k) - \sum_{k=1}^{n} L_{jk}(p_k)$$

$$= f_j(t) - f_j(t) = 0, \qquad j = 1, \ldots, m.$$

Thus $\mathbf{x}(t) - \mathbf{x}_p(t) = \mathbf{x}_h(t)$ for some solution $\mathbf{x}_h(t)$ of the associated homogeneous system. So $\mathbf{x}(t) = \mathbf{x}_h(t) + \mathbf{x}_p(t)$. ■

Theorem 2.1 allows us to break the solution of a linear system into homogeneous and particular parts that can be computed independently of each other. For now, finding the particular part will involve some informed guessing, though this feature can be avoided by using the algorithmic routines of Chapter 7.

Here is a nonhomogeneous system and the associated homogeneous system:

$$\begin{cases} \dot{x} = -y + \cos t \\ \dot{y} = x + \sin t \end{cases} \qquad \begin{cases} \dot{x} = -y \\ \dot{y} = x. \end{cases}$$

The homogeneous system is most easily solved by noting that $\ddot{x} = -\dot{y} = -x$, or $\ddot{x} + x = 0$; this has solutions

$$x_h = c_1 \cos t + c_2 \sin t.$$

The equation $\dot{x} = -y$ says $y = -\dot{x}$, from which we get

$$y_h = c_1 \sin t - c_2 \cos t.$$

These last two displayed equations give us the complete solution to the homogeneous system. To find a particular solution to the system, we use our experience with a single constant-coefficient equation as a guide. Multiples of $\cos t$ and $\sin t$ won't suffice, because they are already identified as solutions of the homogeneous system. So we try

$$x_p = At \cos t + Bt \sin t \qquad \text{and} \qquad y_p = Ct \cos t + Dt \sin t.$$

Substitute these expressions for x_p and y_p into the nonhomogeneous system to get

$$A \cos t - At \sin t + B \sin t + Bt \cos t = -Ct \cos t - Dt \sin t + \cos t,$$

$$C \cos t - Ct \sin t + D \sin t + Dt \cos t = At \cos t + Bt \sin t + \sin t.$$

Now equate coefficients of like terms in the first equation to get $A = 1$ and $B = 0$, so $y_p = t \cos t$. Similarly, from the second equation we get $C = 0$ and $D = 1$, so $y_p = t \sin t$. The complete solution is

$$x = x_h + x_p = c_1 \cos t + c_2 \sin t + t \cos t,$$

$$y = y_h + y_p = c_1 \sin t - c_2 \cos t + t \sin t.$$

EXERCISES

1. Classify each of these first- or second-order systems as linear or nonlinear.

   (a) $dx/dt = t + x^2 + y$,
   $dy/dt = t^2 + x + y$.

   (b) $dy/dt = t^2 + z$,
   $dz/dt = t^3 + y$.

   (c) $dx/dt = t^2x - y + e^t$,
   $dy/dt = 1$.

   (d) $d^2x/dt^2 = tx + y$,
   $dy/dt = x + ty$.

   (e) $dx/dt + dz/dt = 1$,
   $dx/dt - t\,dz/dt = x$.

   (f) $d^2x/dt^2 + dy/dt = 0$,
   $y + t\,d^2y/dt^2 = t$.

2. Use elimination by operator multiplication to get rid of one of the dependent variables. Solve the resulting equation for the remaining variable, and then determine the general solution $(x(t),\ y(t))$.

   (a) $dx/dt = 6x + 8y$,
   $dy/dt = -4x - 6y$.

   (b) $dx/dt = x + 2y$,
   $dy/dt = -2x + y$.

   (c) $dx/dt = x + 2y$,
   $dy/dt = x + y + t$.

   (d) $dx/dt = -y - t$,
   $dy/dt = x + t$.

3. It is in general not possible to find closed-form solutions for linear systems with nonconstant-coefficient functions. Here is one that can nevertheless be solved fairly easily by solving a second-order equation for y. Find the general solution.

   $$\dot{x} = (t^{-1} - t)x - t^2y, \qquad t > 0,$$
   $$\dot{y} = x + ty.$$

4. Reduce each system to the normal form with first derivatives on the left side:

   (a) $dx/dt + dy/dt = t$,
   $dx/dt - dy/dt = x$.

   (b) $dx/dt + dy/dt = y$,
   $dx/dt + 2dy/dt = x$.

   (c) $2dx/dt + dy/dt + x + 5y = t$,
   $dx/dt + dy/dt + 2x + 2y = 0$.

   (d) $dx/dt + dy/dt = \sin t$,
   $dx/dt - dy/dt = \cos t$.

5. After putting it in normal form, solve each system in Exercise 4. Then determine the constants in your solution so as to satisfy the corresponding initial conditions given here.

   (a) $x(0) = 1,\ y(0) = -1$.

   (b) $x(0) = 0,\ y(0) = 5$.

   (c) $x(0) = -1,\ y(0) = 0$.

   (d) $x(\pi) = 1,\ y(\pi) = 2$.

6. Use elimination by operator multiplication to get rid of one of the dependent variables. Solve the resulting equation for the remaining variable and then determine the general solution of the system. Substitution may be necessary to find relations among constants. Then determine the constants so that the initial conditions are satisfied.

   (a) $d^2x/dt^2 - x + dy/dt + y = 0$,
   $dx/dt - x + d^2y/dt^2 + y = 0,\ x(0) = y(0) = 0,\ \dot{x}(0) = 0,\ \dot{y}(0) = 1$.

   (b) $d^2x/dt^2 - dy/dt = 0$,
   $dx/dt + d^2y/dt^2 = 0,\ x(0) = 1,\ y(0) = 0,\ \dot{x}(0) = \dot{y}(0) = 0$.

(c) $d^2x/dt^2 - dy/dt = t$,
$dx/dt + dy/dt = x + y$, $x(0) = y(0) = 0$, $\dot{x}(0) = 1$.

(d) $d^2x/dt^2 - y = e^t$,
$d^2y/dt^2 + x = 0$, $x(0) = y(0) = \dot{x}(0) = \dot{y}(0) = 0$.

7. By introducing new dependent variables, $u = \dot{x}$, $v = \dot{y}$, attempt to reduce each system in Exercise 6 to first-order normal form.

8. None of these linear systems is equivalent to a first-order system in normal form. Discuss their solutions, or lack thereof.

(a) $dx/dt + dy/dt = 0$,
  $dx/dt + dy/dt = 1$.

(c) $dx/dt + dy/dt = t$,
  $dx/dt + dy/dt = x$.

(b) $dx/dt + dy/dt = 0$,
  $dx/dt + dy/dt = x$.

(d) $dx/dt + dy/dt = y$,
  $dx/dt + dy/dt = x$.

9. Suppose that you have used elimination on the system

$$\dot{x} + x - y = 0,$$
$$x + \dot{y} + 3y = e^t$$

to find a differential equation for y alone and that you have solved that equation for $y(t)$.

(a) Explain which equation it is then better to substitute $y(t)$ back into to finish solving the system.

(b) Explain why you generate an extraneous constant of integration if you don't make the best choice for your back substitution, and explain what you should do to eliminate the extra constant.

10. (a) Find a first-order system of dimension 4 equivalent to the second-order system

$$\ddot{y} - 3x - 2y = 0,$$
$$\ddot{x} - y + 2x = 0.$$

(b) Write the system found in part (a) in normal form.

(c) Solve the system in part (a).

11. (a) Find a 2-dimensional system of order 2 satisfied by the x and y coordinates of the solutions to

$$\dot{x} = z + w,$$
$$\dot{y} = z - w,$$
$$\dot{z} = x - y,$$
$$\dot{w} = x + y.$$

Then solve the second-order system and use its solution to solve the given system.

(b) Find a 2-dimensional system of order 2 satisfied by the z and w coordinates of the solutions to the system in part (a).

12. The motion of a particle located at point (x, y) on a rotating turntable with its own (x, y) coordinates fixed to the turntable and having constant angular velocity ω about $(x, y) = (0, 0)$ can be regarded as governed by a **centrifugal acceleration** $\omega^2(x, y)$ directed away from the center of rotation, and a **Corio-**

lis acceleration perpendicular to the instantaneous direction of the particle's motion and given by $2\omega(\dot{y}, -\dot{x})$.

(a) Show that x and y satisfy the system $\ddot{x} = \omega^2 x + 2\omega\dot{y}$, $\ddot{y} = \omega^2 y - 2\omega\dot{x}$. Operators and elimination can be used to solve such systems, but the method suggested below is neater for this special system.

(b) Show that if we let z denote the complex number $x + iy$, the system in part (a) can be rewritten as the single complex equation $\ddot{z} + 2i\omega\dot{z} - \omega^2 z = 0$.

(c) Show that the equation in part (b) has complex solutions $z(t) = (c_1 + c_2 t)e^{-i\omega t}$ and hence that the initial-value problem $x(0) = R$, $y(0) = 0$, $\dot{x}(0) = u$, $\dot{y}(0) = v$ for the system in part (a) has solutions in real form given by

$$x(t) = (R + ut)\cos \omega t + (v + \omega R)t \sin \omega t,$$

$$y(t) = -(R + ut)\sin \omega t + (v + \omega R)t \cos \omega t.$$

(d) Use either the complex or else the real form of the solution found in part (c) to show that the motion with respect to the turntable can be regarded as a linear motion having constant velocity vector **v**, but with a correction for the rotation of the turntable. What is the constant velocity vector **v** in terms of ω, R, u, and v?

CHAPTER REVIEW AND SUPPLEMENT

Solve the initial-value problems in Exercises 1 through 12.

1. $dx/dt = x^2 + 1$, $x(0) = 1$,
$dy/dt = y$, $y(0) = 2$.

2. $dx/dt = t$, $x(1) = 0$,
$dy/dt = x + y$, $y(1) = 0$.

3. $dx/dt = y + 1$, $x(0) = 1$,
$dy/dt = x$, $y(0) = 2$.

4. $dx/dt = -y$, $x(1) = 0$,
$dy/dt = -x$, $y(1) = 0$.

5. $\dot{x} = 3x - 4y$, $x(0) = 1$,
$\dot{y} = 4x - 7y$, $y(0) = 2$.

6. $\dot{x} = 3x - 5y$, $x(0) = 0$,
$\dot{y} = x - y$, $y(0) = 1$.

7. $\dot{x} = y + t$, $x(0) = 1$,
$\dot{y} = 4x - 1$, $y(0) = 2$.

8. $\dot{x} = 2$, $x(0) = 0$,
$\dot{y} = x + y$, $y(0) = 1$.

9. $\ddot{x} = -3x + y$, $x(0) = 3$, $\dot{x}(0) = 0$,
$\ddot{y} = 2x - 2y$, $y(0) = 3$, $\dot{y}(0) = 0$.

10. $\ddot{x} = y$, $x(0) = 0$, $\dot{x}(0) = 1$,
$\ddot{y} = x$, $y(0) = 1$, $\dot{y}(0) = 0$.

11. $dx/dt = -y$, $x(0) = 0$,
$dy/dt = x$, $y(0) = 1$,
$dz/dt = z$, $z(0) = -1$.

12. $dx/dt = t$, $x(0) = 0$,
$dy/dt = y$, $y(0) = 0$,
$dz/dt = y + z$, $z(0) = 1$.

13. Suppose that the autonomous differential equation $\dot{\mathbf{x}} = F(\mathbf{x})$ has solution $\mathbf{x} = \mathbf{x}(t)$ satisfying $\mathbf{x}(0) = \mathbf{x}_0$ and $\mathbf{x}(1) = \mathbf{x}_1$. What, if anything, can you say about a solution to $\dot{\mathbf{x}} = F(\mathbf{x})$?

14. Consider the system $(\dot{x}, \dot{y}, \dot{z}) = (-\omega y, \omega x, \sigma)$, where ω and σ are nonzero constants.

(a) Without solving the system, show that the acceleration vector of a non-zero solution trajectory (i) is always perpendicular to its velocity vector, (ii) is parallel to the xy plane, with length equal to ω^2 times the length of the corresponding position vector (x, y, z).

(b) Find the complete solution of the system. Then sketch the trajectory passing through $(1, 0, 0)$, assuming $\omega = \sigma = 1$.

15. Let $f(\mathbf{x}) = f(x, y))$ be a continuously differentiable function of two real variables. The **gradient vector field** $\nabla f(x, y) = (f_x(x, y), f_y(x, y))$ generates the autonomous vector differential equation $\dot{\mathbf{x}} = \nabla f(\mathbf{x})$, called a **gradient system.** (Subscripts denote partial derivatives.)

(a) Show that the solution trajectories of a gradient system are perpendicular to the level curves $f(x, y) = c$ of f. [*Hint:* Show $(d/dt)f(x(t), y(t)) = 0$.]

(b) Illustrate part (a) using the example $f(x, y) = x^2 + y^2$.

(c) Suppose that $\mathbf{x}(t)$ is a solution of a gradient system satisfying $\mathbf{x}(t_0) = \mathbf{x}_0$ and $\mathbf{x}(t_1) = \mathbf{x}_1$, where $\mathbf{x}_0$ and $\mathbf{x}_1$ lie on the same level curve of f; that is, $f(\mathbf{x}_0) = f(\mathbf{x}_1)$. Show that $\int_{t_0}^{t_1} |\dot{\mathbf{x}}(t)|^2 \, dt = 0$ and hence that the solution $\mathbf{x}(t)$ must reduce to a constant equilibrium solution.

16. Let $H(\mathbf{x}) = H(x, y)$ be a continuously differentiable function of two real variables. The vector field $\mathbf{H}(x, y) = (H_y(x, y), -H_x(x, y))$ is called a **Hamiltonian field,** and the autonomous vector differential equation $\dot{\mathbf{x}} = \mathbf{H}(\mathbf{x})$ is called a **Hamiltonian system.**

(a) Show that the solution trajectories of a Hamiltonian system are level curves of $H(x, y)$. [*Hint:* See part (a) of the previous exercise.]

(b) Illustrate part (a) using the example $H(x, y) = x^2 - y^2$.

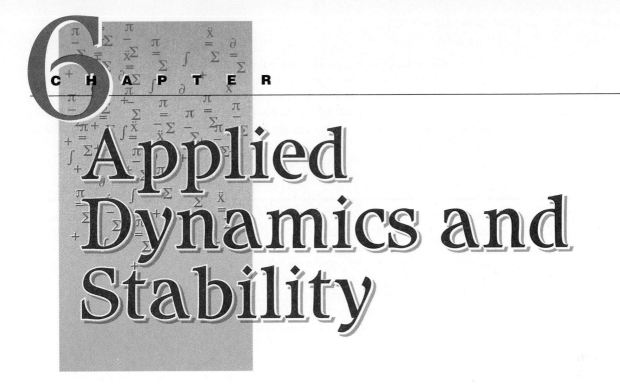

6 CHAPTER

Applied Dynamics and Stability

The examples in this chapter illustrate ideas and techniques that appear often in applied mathematics; they all involve varying quantities that have mutual effects on each other. First we will find a system of differential equations that the quantities satisfy. We will be able to give complete solutions for some systems in terms of elementary functions; for others we will be able to give qualitative descriptions of the solutions, then use numerical methods described in Section 7 for more accurate quantitative descriptions. The dimension of the state space will vary depending on the details of a particular application. The various subsections are independent of each other, so individual tastes can be met by skipping some in favor of those that seem more interesting. It may also be desirable in the case of some applications to jump ahead to Section 7 on numerical methods for additional examples.

1 MULTICOMPARTMENT MIXING

When fluid flows back and forth among several tanks or compartments of a physical system, the amounts $x(t), y(t), \ldots$ of a particular dissolved substance in tanks X, Y, . . . will typically vary as a function of time. Given enough information about the processes involved, we may be able to find a system of differential equations that has the functions $x(t), y(t), \ldots$ as the coordinates of its solution vector. Setting up the system will usually require for each tank that we express the rate of change of a typical amount x as the difference between rate of inflow and rate of outflow of the dissolved substance $x(t)$:

$$\frac{dx}{dt} = \{\text{inflow rate to tank X}\} - \{\text{outflow rate from tank X}\}.$$

329

There will be one such equation for each tank. Note that the rates on the right side must also be stated in terms of quantities of dissolved material rather than of volumes of solution. An underlying assumption is that the fluid in each tank is kept thoroughly mixed all the time. (Because it's so familiar, we'll usually think of salt dissolved in water, though any number of chemical solutions could be substituted.)

example 1

Figure 6.1 shows two 50-gallon tanks connected by flow pipes and with inlets and outlets all having the rates of flow as marked in gallons per minute (gal/min). The flow rates are arranged so that each tank is maintained at its capacity at all times. We suppose that each tank initially contains salt solution at a concentration in pounds per gallon that we leave unspecified for the moment, that the left-hand tank is receiving salt solution at a concentration of 1 pound per gallon, and that the right-hand tank is receiving pure water. The problem is to find out what happens to the amount of salt, in pounds, as time goes on. We assume that each tank is kept thoroughly mixed at all times so that the concentration of salt is the same throughout the whole tank at any time. In the left-hand tank, with salt content $x(t)$, the rate of change of the amount of salt is, by definition, dx/dt. On the other hand, because of the various flow rates, we can break this rate of change into three parts

$$\frac{dx}{dt} = -4\frac{x}{50} + 3\frac{y}{50} + 1,$$

where $x/50$ is the concentration of salt in the left tank and $y/50$ the concentration in the right tank, both in pounds per gallon. The term $-4(x/50)$ is the rate of outflow of salt, and the remaining terms represent the rate of inflow. Similarly,

$$\frac{dy}{dt} = 2\frac{x}{50} - 3\frac{y}{50}.$$

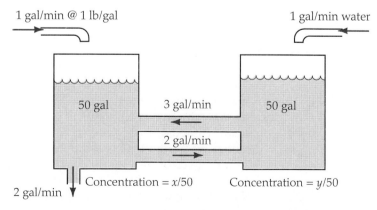

1 gal/min @ 1 lb/gal   1 gal/min water

50 gal   3 gal/min   50 gal

2 gal/min

2 gal/min   Concentration = $x/50$ Concentration = $y/50$

Figure 6.1

Thus we have a system of differential equations that we can write as

$$\frac{dx}{dt} = -\frac{4}{50}x + \frac{3}{50}y + 1,$$

$$\frac{dy}{dt} = \frac{2}{50}x - \frac{3}{50}y.$$

To solve it, we can use the elimination method, first writing the system in the form

$$(D + \tfrac{4}{50})x - \tfrac{3}{50}y = 1,$$

$$-\tfrac{2}{50}x + (D + \tfrac{3}{50})y = 0.$$

We multiply the first equation by $\frac{2}{50}$ and operate on the second by $D + \frac{4}{50}$. Addition of the two equations then gives

$$\left(D + \frac{4}{50}\right)\left(D + \frac{3}{50}\right)y - \frac{6}{(50)^2}y = \frac{2}{50},$$

or

$$\left(D^2 + \frac{7}{50}D + \frac{6}{(50)^2}\right)y = \frac{2}{50}.$$

The roots of the characteristic equation can be found in this case by the factorization

$$r^2 + \frac{7}{50}r + \frac{6}{(50)^2} = \left(r + \frac{1}{50}\right)\left(r + \frac{6}{50}\right)$$

to be $r_1 = -\frac{1}{50}$ and $r_2 = -\frac{6}{50}$. A particular solution is clearly $y_p(t) = \frac{50}{3}$, a constant. Thus, in general,

$$y(t) = c_1 e^{-(1/50)t} + c_2 e^{-(6/50)t} + \tfrac{50}{3}.$$

Using the second equation of the system to write $x(t)$ in terms of $y(t)$, we find

$$x(t) = \tfrac{50}{2}(D + \tfrac{3}{50})y(t)$$

$$= c_1 e^{-(1/50)t} - \tfrac{3}{2}c_2 e^{-(6/50)t} + \tfrac{50}{2}.$$

Thus the general solution is

$$x(t) = c_1 e^{-(1/50)t} - \tfrac{3}{2}c_2 e^{-(6/50)t} + \tfrac{50}{2},$$

$$y(t) = c_1 e^{-(1/50)t} + c_2 e^{-(6/50)t} + \tfrac{50}{3}.$$

From these equations, we see immediately that

$$\lim_{t \to \infty} x(t) = \tfrac{50}{2},$$

$$\lim_{t \to \infty} y(t) = \tfrac{50}{3}.$$

In other words, the concentration, in pounds per gallon, in the left tank approaches $\frac{1}{2}$, and in the right tank approaches $\frac{1}{3}$.

The constants c_1 and c_2 depend on the initial values $x(0)$ and $y(0)$. Thus the equations

$$x(0) = c_1 - \tfrac{3}{2}c_2 + \tfrac{50}{2},$$

$$y(0) = c_1 + c_2 + \tfrac{50}{3}$$

determine c_1 and c_2 when $x(0)$ and $y(0)$ are known. In particular, if $x(0) - y(0) - 0$, subtracting these equations shows that $c_2 = \frac{10}{3}$, from which it follows by substitution in either equation that $c_1 = -20$. The graphs of x and y are plotted in Figure 6.2 using the same vertical axis for both x and y.

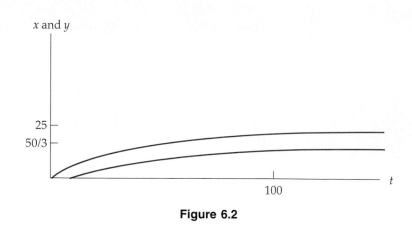

Figure 6.2

The constants c_1, c_2 are determined if we know $x(t_1)$ and $y(t_1)$ at any time t_1. We'll leave this detail as an exercise.

In Example 1, the choice $c_1 = c_2 = 0$ yields the constant equilibrium solution $x = \frac{50}{2}$, $y = \frac{50}{3}$. (This could have been found directly, without solving the entire system.) If these values are taken as the initial values, then the salt content in the two tanks remains at those values forever. Since the exponential terms tend to zero as t tends to infinity, the amounts of salt in the two tanks tend to those values as $t \to \infty$.

e x a m p l e 2   Two tanks, one of capacity 100 gallons and the other of capacity 200 gallons, are initially half-full of pure water. Mixed fluid flows through a pipe from the 100-gallon tank to the other tank at 1 gallon per minute. Mixed fluid flows in the opposite direction through another pipe at 2 gallons per minute. In addition, salt solution at a concentration of 1 pound per gallon runs from an external source into the 100-gallon tank at the rate of 1 gallon per minute. The entire process is stopped as soon as either tank starts to overflow or becomes empty.

To analyze what happens, we first determine the stopping time. The 100-gallon tank gains 3 gallons per minute and loses 1 gallon per minute for a total gain of 2 gallons per minute; thus this tank will start to overflow after $\frac{50}{2} = 25$ minutes. The 200-gallon tank loses 2 gallons per minute and gains 1 gallon per minute for a total loss of 1 gallon per minute; thus this tank would be effectively empty as a mixing tank after 100 minutes. It follows that we need only consider the shorter time interval $0 \leq t \leq 25$ minutes. Let $x(t)$ and $y(t)$ be the respective amounts in the smaller and larger tanks. The amount of solution in the small tank at time t is $50 + 2t$ gallons and in the large tank is $100 - t$ gallons. Then

$$\frac{dx}{dt} = -\frac{x}{50 + 2t} + 2\frac{y}{100 - t} + 1,$$

$$\frac{dy}{dt} = \frac{x}{50 + 2t} - 2\frac{y}{100 - t}.$$

The corresponding initial conditions are $x(0) = y(0) = 0$, since there is initially no salt in either tank. There are no elementary solution formulas, so the solution depends in practice on the numerical methods of the type discussed in Section 7. Since salt is being added to the system at 1 pound per minute, after 25 minutes the total amount of salt in the system will be 25 pounds. Numerical calculation will show that $x(25) \approx 22.2$, $y(25) \approx 2.8$.

EXERCISES

1. Suppose that two initially full 100-gallon tanks of salt solution contain amounts of salt $y(t)$ and $z(t)$ at time t. Suppose that the solution in the y tank is flowing to the z tank at a rate of 1 gallon per minute and that the solution in the z tank is flowing to the y tank at the rate of 4 gallons per minute. Suppose also that the overflow from the y tank goes down the drain, whereas the z tank is kept full by the addition of freshwater. Assume that each tank is kept thoroughly mixed at all times.

   (a) Find a linear system satisfied by y and z.

   (b) Find the general solution of the system in part (a) and then determine the constants in it so that the initial values will be $y(0) = 10$ and $z(0) = 20$.

   (c) Draw the graphs of the particular solutions found in part (b), and interpret the results.

2. In Example 1 of the text, the general solution to a system of differential equations is found to be
$$x(t) = c_1 e^{-(1/50)t} - \tfrac{3}{2}c_2 e^{-(6/50)t} + \tfrac{50}{2},$$
$$y(t) = c_1 e^{-(1/50)t} + c_2 e^{-(6/50)t} + \tfrac{50}{3}.$$

   (a) Find values for the constants c_1 and c_2 so that the initial conditions $x(0) = 25$, $y(0) = \tfrac{2}{3}$ are satisfied.

   (b) Show that it is possible to choose c_1 and c_2 so that an arbitrary initial condition $(x(0), y(0)) = (x_0, y_0)$ is satisfied. Is such a condition always meaningful for the application we have in mind here?

   (c) Show that it is possible to choose c_1 and c_2 so that an arbitrary condition $(x(t_1), y(t_1)) = (x_1, y_1)$ is satisfied. Can you conclude from this result that it is possible to find the initial condition $(x(0), y(0)) = (x_0, y_0)$ that would lead to these values at $t = t_1$? Explain.

3. Two mixing vats of the same 1-gallon size containing salt solution are maintained at full capacity by the following process. Solution is pumped from vat X to vat Y at rate $a > 0$ and also in the opposite direction at rate a.

   (a) Find the system that the respective amounts $x(t)$, $y(t)$ of chemical in solution in the two vats satisfy.

   (b) Show that every solution of the system in part (a) is either a constant equilibrium solution or else tends to one exponentially as t tends to infinity, without exhibiting any oscillatory behavior. [*Hint:* Show that the relevant characteristic roots are nonpositive.]

4. Three mixing vats of the same 1-gallon size containing salt solution stay at full capacity because there is no fluid gain or loss in any of the vats. Solution is pumped cyclically from vat X to vat Y at rate a, from Y to Z at rate a, and from Z to X at rate a.

   (a) Find the system that the respective amounts $x(t), y(t), z(t)$ of chemical in solution in the vats satisfy.

   (b) Show that $\dot{x} + \dot{y} + \dot{z} = 0$, and interpret this equation.

   (c) Show that $x(t) = y(t) = z(t) =$ const is always a solution of the system.

   (d) By examining the relevant characteristic roots, show that the nonconstant solutions exhibit strongly damped oscillatory behavior.

5. Two tanks, one of capacity 100 gallons and the other of capacity 200 gallons, are initially full of liquid. The 100-gallon tank starts with nothing but pure water, but the other tank starts out with 10 pounds of salt dissolved in the water. Solution flows through a pipe from the 100-gallon tank to the other tank at 1 gallon per minute. Solution flows in the opposite direction through another pipe at 2 gallons per minute. Both tanks are allowed to overflow their thoroughly mixed solutions into a drain if necessary, and the entire process is stopped if either tank becomes empty.

   (a) Which tank empties first, and at what time?

   (b) Write down a system of differential equations and initial conditions whose solution describes this process as a function of time.

   (c) Show that the rate of decrease of the total amount of salt in the two tanks is equal to the concentration of salt in the 100-gallon tank.

6. Two tanks, one of capacity 100 gallons and the other of capacity 200 gallons, are each initially half-full of liquid. The 100-gallon tank starts with nothing but pure water, but the other tank starts out with 10 pounds of salt dissolved in the water. Solution flows through a pipe from the 100-gallon tank to the other tank at 1 gallon per minute. Solution flows in the opposite direction through another pipe at 2 gallons per minute. The entire process is stopped if either tank becomes empty or either tank overflows.

   (a) How long does it take for the process to stop?

   (b) Write down a system of differential equations and initial conditions whose solution describes the process as a function of time.

   (c) During the process what is the total change in salt content of the two tanks together?

7. Two tanks, one of capacity 100 gallons and the other of capacity 200 gallons, are each initially half-full of liquid. The 100-gallon tank starts with nothing but pure water, but the other tank starts out with 10 pounds of salt dissolved in the water. Solution flows through a pipe from the 100-gallon tank to the other tank at 2 gallons per minute. Solution flows in the opposite direction through another pipe at 1 gallon per minute. The entire process is stopped if either tank becomes empty or either tank overflows.

   (a) How long does it take for the process to stop?

(b) Write down a system of differential equations and initial conditions whose solution describes the process as a function of time.

(c) How much liquid is in each tank when the process stops?

8. Three tanks, all of capacity 100 gallons, are initially full of liquid. There is an equal exchange of solution between tanks Z and Y at the rate of 2 gallons per minute each way. There is also an equal exchange between tanks X and Y at the rate 2 gallons per minute but no *direct* exchange between tanks X and Z.

(a) Find a system of differential equations whose solution describes this process as a function of time.

(b) Show that the system has equilibrium solutions with all three salt concentrations equal.

*(c) Solve the system by elimination, and assume tanks X and Y start out with nothing but pure water while tank Z starts out with 1 pound of salt dissolved in water.

9. Two 100-gallon tanks X and Y are initially full of pure water. Salt solution is added to X from an external source at 1 gallon per minute, each gallon containing 1 pound of salt. Mixed solution is pumped from X to Y at 2 gallons per minute and drains from Y at 3 gallons per minute. Let $x = x(t)$ and $y = y(t)$ be the respective amounts of salt in X and Y at time $t \geq 0$.

(a) At what time t_1 will X or Y first be empty?

(b) Find a system of differential equations satisfied by $x(t)$ and $y(t)$ for $0 \leq t \leq t_1$.

(c) Solve the system found in part (b).

10. Two 100-gallon tanks X and Y contain initially 50 and 100 gallons, respectively, of pure water. From an external source, salt solution is added to Y at 1 gallon per minute, each gallon containing 1 pound of salt. Mixed solution flows from Y to X at 2 gallons per minute and from X to Y at 1 gallon per minute. Let $x = x(t)$ and $y = y(t)$ be the respective amounts of salt in X and Y at time $t \geq 0$. (*Note:* You are not asked to solve any differential equations for this question.)

(a) At what time t_1 will X begin to overflow? Express the total amount of salt in the two tanks as a function of t while $0 \leq t \leq t_1$.

(b) Find a system of differential equations satisfied by $x(t)$ and $y(t)$ for $0 \leq t \leq t_1$.

(c) Find a system of differential equations satisfied by $x(t)$ and $y(t)$ for $t_1 \leq t$, while X is overflowing.

11. Two 100-gallon tanks X and Y are initially full of salt solution, with x_0 pounds of salt in X and y_0 pounds of salt in Y. Mixed solution is pumped from X to Y at 2 gallons per minute and from Y to X at 3 gallons per minute. Pure water evaporates from X at 2 gallons per minute. Let $x = x(t)$ and $y = y(t)$ be the respective amounts of salt in X and Y at time $t \geq 0$.

(a) At what time t_1 will one of the tanks first overflow or become empty?

(b) Find, but don't solve, a system of differential equations satisfied by $x(t)$ and $y(t)$ for $0 \le t \le t_1$.

(c) Show that $x(t) + y(t)$ remains constant and that the amount of salt in each tank is in equilibrium whenever $x_0 = \frac{3}{2}y_0$.

(d) Assume that $x(0) = 10$ and $y(0) = 20$. Use the equation $x(t) + y(t) = 30$ to solve the system you found in part (b).

12. Two tanks, one of capacity 100 gallons and the other of capacity 200 gallons, are each initially half-full of liquid. The 100-gallon tank starts with nothing but pure water, but the other tank starts out with 10 pounds of salt dissolved in the water. Solution flows from the 100-gallon tank to the other tank at 2 gallons per minute. Solution flows in the opposite direction at 1 gallon per minute. Pure water is added to the 100-gallon tank at 1 gallon per minute. The entire process is stopped if either tank becomes empty or either tank overflows.

(a) How long does it take for the process to stop?

(b) Write down the system of differential equations and initial conditions whose solution describes the process as a function of time. As a check, notice that the total amount of salt present in the system remains unchanged.

(c) Use the check in part (b) to find a first-order linear initial-value problem for $x(t)$ alone.

(d) Solve the initial-value problem in part (c) for $x(t)$. Then find $y(t)$, and estimate the amount of salt in each tank when the process stops.

2   INTERACTING POPULATIONS

The simple population growth equation $dP/dt = kP$, $k > 0$, was modified in Chapter 2, Section 1, to take account of the inhibiting effect of larger population size on the growth rate. One such modification resulted in the nonlinear equation $dP/dt = kP(L - P)$. The factor $L - P$ approaches zero as P approaches L and tends to reduce the growth rate as $P(t)$ approaches L from below. (If $P(t) > L$, the growth rate is negative, and the population declines.) When populations of varying size interact mutually, the possibilities for expressing the interactions are more interesting. The general qualitative behavior of these models is much more significant than specific numerical calculations. The latter are useful largely for illustrative purposes.

---

e x a m p l e **1**

Linear Model

The system

$$\frac{dP}{dt} = kP - mQ, \qquad k > 0, \ m > 0,$$

$$\frac{dQ}{dt} = -nP + lQ, \qquad l > 0, \ n > 0,$$

includes growth inhibition of both populations P, Q by simply subtracting

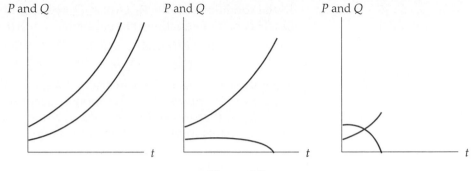

Figure 6.3

from the growth rate of each population a term proportional to the size of the other population. Thus having Q large inhibits the growth of P, and vice versa. In operator form the system is

$$(D - k)P + mQ = 0,$$
$$nP + (D - l)Q = 0.$$

To eliminate Q, operate on the first equation by $D - l$, multiply the second by m, and subtract to get

$$(D - l)(D - k)P - mnP = 0 \quad \text{or} \quad (D^2 - (k + l)D + (kl - mn))P = 0.$$

The characteristic equation $r^2 - (k + l)r + (kl - mn) = 0$ has roots

$$r_1, r_2 = \tfrac{1}{2}((k + l) \pm \sqrt{(k + l)^2 - 4kl + 4mn}) = \tfrac{1}{2}((k + l) \pm \sqrt{(k - l)^2 + 4mn}).$$

Since $(k - l)^2 + 4mn > 0$, the roots are unequal real numbers, so there is no oscillatory behavior, and P has the form $P(t) = c_1 e^{r_1 t} + c_2 e^{r_2 t}$. Solving the first equation in the original system for $Q = (kP - \dot{P})/m$ shows that $Q(t) = ac_1 e^{r_1 t} + bc_2 e^{r_2 t}$, where a and b are constants depending on k, l, m, n: $a = (k - r_1)/m$, $b = (k - r_2)/m$. Figure 6.3 shows typical behaviors of $P(t)$ and $Q(t)$ plotted using the same vertical axis.

The significance of solution formulas for population models lies mainly in the information they provide about general behavior given assumptions about parameters in the model and about initial conditions.

---

e x a m p l e 2

If the effect of each population on the other's growth is essentially the same, we may assume $k = l$, $m = n$ in Example 1. This means that the differential equations are

$$\frac{dP}{dt} = kP - mQ,$$

$$\frac{dQ}{dt} = -mP + kQ, \qquad k > 0, \ m > 0.$$

Also it's easy to see from the previous calculations that the characteristic roots are $r_1 = k + m$ and $r_2 = k - m$. It follows that

$$P(t) = c_1 e^{(k+m)t} + c_2 e^{(k-m)t}, \qquad Q(t) = -c_1 e^{(k+m)t} + c_2 e^{(k-m)t}.$$

Initial conditions $P(0) = P_0$, $Q(0) = Q_0$ require that $c_1 + c_2 = P_0$ and $-c_1 + c_2 = Q_0$. This pair of equations has solutions $c_1 = \frac{1}{2}(P_0 - Q_0)$, $c_2 = \frac{1}{2}(P_0 + Q_0)$. Thus

$$P(t) = \tfrac{1}{2}(P_0 - Q_0)e^{(k+m)t} + \tfrac{1}{2}(P_0 + Q_0)e^{(k-m)t},$$
$$Q(t) = \tfrac{1}{2}(Q_0 - P_0)e^{(k+m)t} + \tfrac{1}{2}(P_0 + Q_0)e^{(k-m)t}.$$

From these equations you can see that a gradual exponential decline in either population is impossible in this model except in the critical circumstance that $P_0 = Q_0$; in that unlikely event, the relation $k < m$ between the growth and inhibition rates implies exponential decline in both P and Q.

The best-known nonlinear population model involves host and parasite populations in an asymmetrical relationship.

---

example 3

Lotka-Volterra Model

Let $P = P(t)$ denote the size at time t of a parasite population that preys on a host population of size $H = H(t)$. In the absence of the parasites, the host population might grow for some time according to $dH/dt = aH$, $a > 0$. However, the growth of H might be inhibited both by the increasingly destructive effect of a larger parasite population P and by the effect of its own increasing demands on its environment. These negative effects can take the form $-bPH$, $b > 0$, when added to the right side of the expression for dH/dt. This leads to the first displayed equation below. On the other hand, if the host population is not present, the parasites can be expected to die out rapidly according to $dP/dt = -dP$, $d > 0$, with additional growth provided by $H > 0$ in the form of the term cHP, $c > 0$. This leads to the second equation below. The equations

$$\frac{dH}{dt} = (a - bP)H,$$

$$\frac{dP}{dt} = (cH - d)P$$

are nonlinear unless $b = c = 0$. We assume here that a, b, c, d are positive constants, though they could in principle depend on t. The factor $a - bP$ acts to increase the size of H if $P(t)$ is below the critical level a/b, and to decrease H in the opposite case. Similarly, the factor $cH - d$ acts to increase the size of P if $H(t)$ is above the critical level d/c, and to decrease P in the opposite case.

The equilibrium solutions satisfy $(a - bP)H = 0$ and $(cH - d)P = 0$, and the solutions are $(H, P) = (0, 0)$ and $(H, P) = (d/c, a/b)$. The $(0, 0)$ solution is not interesting from the point of view of population dynamics, since it represents a total absence of both species and is not a limit of other solutions. To understand the significance of the equilibrium at $H = d/c$, $P = a/b$, we examine the behavior near that point. To that end, let $H = h + d/c$ and $P = p + a/b$, where h and p are assumed to be small. Substitution into the system for H and P gives

$$\dot{h} = (a - bp - a)\left(h + \frac{d}{c}\right) = -\frac{bd}{c}p - bph,$$

$$\dot{p} = (ch + d - d)\left(p + \frac{a}{b}\right) = \frac{ac}{b}h + chp.$$

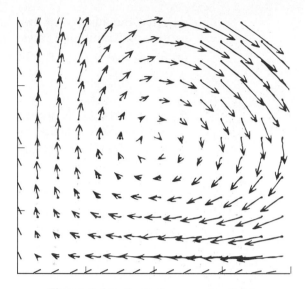

Figure 6.4 Lotka-Volterra vector field.

This system is no simpler than the original one, but since p and h are assumed small, say less than 1 in some scale, the product ph will be even smaller. Removal of the last two terms in the modified version of the system results in the linear system

$$\dot{h} = -\frac{bd}{c}\,p, \qquad \dot{p} = \frac{ac}{b}\,h.$$

This system is easily solved by elimination, and you can check that

$$h(t) = b\sqrt{d}\,A\cos\left(\sqrt{ad}\,t + \phi\right), \qquad p(t) = c\sqrt{a}\,A\sin\left(\sqrt{ad}\,t + \phi\right),$$

where the constants A and ϕ are to be determined by initial conditions. The point $(h(t), p(t))$ traces a family of ellipses centered at $(0, 0)$ in the hp plane. Shifting back to the HP plane, we get ellipses centered at the equilibrium point $(d/c, a/b)$.

It is a remarkable fact that the closed orbits exhibited by the linearized system have as their counterparts closed, convex orbits for the original nonlinear system. You are asked to show in a series of exercises that this is true for every trajectory of the nonlinear system in the positive HP quadrant. The vector field displayed in Figure 6.4 is a scaled version of the one associated with the choice $a = d = 1$, $b = c = \frac{1}{2}$, and it suggests that trajectories will circulate in some way about the equilibrium point $(2, 2)$, but falls well short of establishing that trajectories return to their initial points.

EXERCISES

1. In Example 2 of the text the equations
$$P(t) = \tfrac{1}{2}(P_0 - Q_0)e^{(k+m)t} + \tfrac{1}{2}(P_0 + Q_0)e^{(k-m)t},$$
$$Q(t) = \tfrac{1}{2}(Q_0 - P_0)e^{(k+m)t} + \tfrac{1}{2}(P_0 + Q_0)e^{(k-m)t}$$

describe the history of two competing populations that have common growth and inhibition rates k and m.

(a) Show that if $Q_0 > P_0$, then P will reach extinction, $P(t_e) = 0$, at time

$$t_e = \frac{1}{2m} \ln \frac{Q_0 + P_0}{Q_0 - P_0}.$$

(b) Show that if $Q_0 > P_0$ and $k > m$, then $Q(t)$ increases steadily while $P(t)$ is heading for extinction as shown in part (a).

(c) Show that if $Q_0 > P_0$ and $k < m$, then $Q(t)$ decreases steadily up to time

$$t_1 = \frac{1}{2m} \ln \frac{(m - k)(Q_0 + P_0)}{(m + k)(Q_0 - P_0)}$$

and increases thereafter.

2. The general solution of the linear model in Example 1 of the text is shown to be

$$P(t) = c_1 e^{r_1 t} + c_2 e^{r_2 t},$$

$$Q(t) = c_1 \frac{k - r_1}{m} e^{r_1 t} + c_2 \frac{k - r_2}{m} e^{r_2 t},$$

where $r_1 = \frac{1}{2}((k + l) + \sqrt{(k - l)^2 + 4mn}) > r_2 = \frac{1}{2}((k + l) - \sqrt{(k - l)^2 + 4mn})$.

(a) Show that $mn = (r_1 - k)(r_1 - l) = (r_2 - k)(r_2 - l)$.

(b) Use the result of part (a) to show that the solutions can be written in the more symmetrical form

$$P(t) = -c_1 \frac{k - l + \sqrt{d}}{2n} e^{r_1 t} + c_2 e^{r_2 t},$$

$$Q(t) = c_1 e^{r_1 t} + c_2 \frac{k - l + \sqrt{d}}{2m} e^{r_2 t},$$

where the discriminant d is $(k - l)^2 + 4mn$. (Remember that c_1 and c_2 are *arbitrary* constants.)

(c) Use the result of part (b) to show that if $k = l$, then the initial-value problem $P(0) = P_0$, $Q(0) = Q_0$ has solution

$$P(t) = \frac{1}{2} \left(P_0 - \sqrt{\frac{m}{n}} \, Q_0 \right) e^{(k + \sqrt{mn})t} + \frac{1}{2} \left(\sqrt{\frac{m}{n}} \, Q_0 + P_0 \right) e^{(k - \sqrt{mn})t},$$

$$Q(t) = \frac{1}{2} \left(Q_0 - \sqrt{\frac{n}{m}} \, P_0 \right) e^{(k + \sqrt{mn})t} + \frac{1}{2} \left(Q_0 + \sqrt{\frac{n}{m}} \, P_0 \right) e^{(k - \sqrt{mn})t}.$$

(d) Use the result of part (c) to show that the two populations decline exponentially to zero if and only if both $\sqrt{n} \, P_0 = \sqrt{m} \, Q_0$ and $k^2 < mn$ are true.

3. The Lotka-Volterra equations

$$\frac{dH}{dt} = (a - bP)H,$$

$$\frac{dP}{dt} = (cH - d)P,$$

where a, b, c, d are positive constants, are used to describe the size relationship of parasite $P(t)$ and host $H(t)$ populations at time t.

(a) Show that if $P(t) > a/b$, then $H(t)$ decreases, and that if $H(t) < d/c$, then $P(t)$ decreases. Describe the implications of these results for the host and parasite populations.

(b) Show that the parameterized solution curves $(H, P) = (H(t), P(t))$ satisfy
$$\frac{dH}{dP} = \frac{(a - bP)H}{(cH - d)P}.$$
Then solve this equation by separation of variables to get
$$(H^d e^{-cH})(P^a e^{-bP}) = k, \qquad k \text{ const.}$$

(c) Show that $f(H) = H^d e^{-cH}$ has the unique local maximum value $f(d/c) = (d/ec)^d$ and tends to zero as H tends to infinity. Similarly, show that $g(P) = P^a e^{-bP}$ has unique local maximum $g(P) = (a/eb)^a$ and tends to zero as P tends to infinity.

(d) Show graphically that for $k > 0$ the curves found in part (b) are closed circuits in the first quadrant of the PH plane. [*Hint:* For each value of $H \neq d/c$ there are two corresponding values of P. An analogous statement holds with P and H exchanged.]

*4. Show that the closed circuits identified as orbits of the Lotka-Volterra equations are concave up in the H direction if $H < d/c$ and concave down if $H > d/c$ by showing that
$$\frac{d^2 H}{dP^2} = \frac{1}{d - cH}\left[\frac{aH}{P^2} + \frac{d}{H}\left(\frac{dH}{dP}\right)^2\right].$$
Thus, the sign of the second derivative is determined by the first factor: negative when $H > d/c$, positive when $H < d/c$.

5. The Lotka-Volterra equations can be refined to take account of fixed limits $L > H(t)$ and $M > P(t)$ to the growth of the host and parasite populations as follows:
$$\frac{dH}{dt} = (a - bP)H(L - H),$$
$$\frac{dP}{dt} = (cH - d)P(M - P),$$
where a, b, c, d are positive constants, with $L > d/c > 1$ and $M > a/b > 1$.

(a) Show that if $P(t) > a/b$, then $H(t)$ decreases, and that if $H(t) < d/c$, then $P(t)$ decreases. Describe the implications of these results for the host and parasite populations.

(b) Show that the parameterized solution curves $(H, P) = (H(t), P(t))$ satisfy
$$\frac{dH}{dP} = \frac{(a - bP)H(L - H)}{(cH - d)P(M - P)}.$$
Then solve this equation by separation of variables to get
$$(H^{d/L}(L - H)^{c - d/L})(P^{a/M}(M - P)^{b - a/M}) = f(H)g(P) = k, \qquad k \text{ const.}$$

(c) Show that $f(H)$ in part (b) has its unique local maximum value at $H = d/c$ and is equal to zero at $H = 0$ and $H = L$. Similarly, show that $g(P)$ has its unique local maximum at $P = a/b$ and is zero at $P = 0$ and $P = M$.

*(d) Show that for $k > 0$ the curves found in part (b) are closed circuits in the first quadrant of the PH plane. [*Hint:* For each value of $H \neq c/d$ there are two corresponding values of P. An analogous statement holds with P and H exchanged.]

6. If $H(t)$ and $K(t)$ are the sizes of two populations that are preyed upon by a common parasite population of size $P(t)$, an analog of the Lotka-Volterra system extended to cover the three species is

$$\frac{dH}{dt} = (a_1 - b_1 P)H,$$

$$\frac{dP}{dt} = (c_1 H + c_2 K - d)P,$$

$$\frac{dK}{dt} = (a_2 - b_2 P)K,$$

where the a_k, b_k, c_k, and d are positive.

(a) Verify that if either the H population or the K population is absent, the system reduces to the form of the two-species Lotka-Volterra system. Then find the corresponding constant equilibrium solutions.

(b) The system for three species given above has equilibrium points with strictly positive coordinates only under certain conditions. Find these conditions, and find all the equilibrium points with positive coordinates. [*Hint:* There may, for some choice of a_k, b_k, be infinitely many equilibrium points.]

7. The three-species system described in the previous exercise can be further modified to take into account the direct effect of the growth or decline of $K(t)$ on $H(t)$, and vice versa:

$$\frac{dH}{dt} = (a_1 - b_1 P - b_3 K)H,$$

$$\frac{dP}{dt} = (c_1 H + c_2 K - d)P,$$

$$\frac{dK}{dt} = (a_2 - b_2 P - b_4 H)K.$$

(a) Show that if the constants a_k, b_k, c_k, and d are positive, then the system has finitely many equilibrium solutions.

(b) Find conditions on the constants a_k, b_k, c_k, and d under which all equilibrium solutions have nonnegative coordinates.

3   ELECTRIC NETWORKS

We can analyze an electric network by using a combination of differential and algebraic equations. To understand the analysis it will be enough for our purposes to become familiar with the following ideas. An *LRC* **network** consists of inductors (with inductance L_j), resistors (with resistance R_j), and capacitors

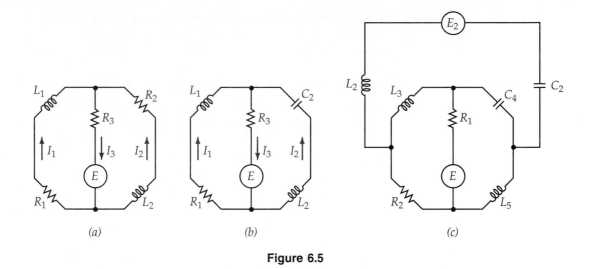

Figure 6.5

(with capacitance C_j) joined by conductors in such a way that any two points of the network can be included in a closed **loop,** or **circuit,** contained in the network. (Such circuits are considered in isolation from one another in Chapter 4, Section 4.) The numbers L_j, R_j, and C_j describe the essential character of a network. In addition a network typically contains voltage sources characterized by voltages E_j together with the direction in which each source E_j would by itself cause current to flow. A point of the network at which entering current can flow out on more than one conductor is called a **junction.** Two junctions are shown in Figure 6.5*a* and *b* and four in Figure 6.5*c*. The segment of a network joining two successive junctions is called a **branch.** Six branches are shown in Figure 6.5*c*, while there are only three each in Figure 6.5*a* and *b*. Each branch will contain at most one voltage source and at most one of each type of network element. In principle there is some resistance, however small, present in every branch, but it may sometimes be small enough that it is neglected.

The problem usually posed is to find the current flowing in the jth branch at time t; we'll denote this current by $I_j(t)$, measured in amperes. We follow the convention that if current is flowing in an arbitrary fixed direction in the jth branch, then $I_j > 0$, while flow in the opposite direction corresponds to $I_j < 0$. In analyzing a network we assign an arbitrary direction, designated *positive*, to each branch and indicate our choices by arrows in the network diagram.

Relations between the currents $I_j(t)$ sufficient to determine their values can be derived from initial conditions together with the two Kirchhoff laws:

■ JUNCTION LAW

The sum of currents directed toward a junction equals the sum of the currents directed away from it.

■ LOOP LAW

The sum of voltage differences across the elements in a closed loop equals the sum of the source voltages in the loop.

The voltage differences in a loop, referred to in the loop law, are caused by the presence of one or more of inductance, resistance, or capacitance in the loop and are computed for the current I_j in the branch containing such a network element as follows.

Voltage difference $L_j \, dI_j/dt$ is due to **inductance** L_j.

Voltage difference $R_j I_j$ is due to **resistance** R_j (Ohm's law).

Voltage difference Q_j/C_j is due to **capacitance** C_j, where Q_j is the charge on the capacitor, and charge is related to current by $I_j = dQ_j/dt$.

In computing the sum of voltage differences in a loop, we assign arbitrarily a fixed direction of traversal to the loop, either clockwise or counterclockwise, and attach a minus sign to a difference for which this direction of traversal is counter to our previously assigned direction of the branch that contains the element causing the difference. A voltage source is designated positive if by itself it would cause current to flow in the direction of traversal of a loop and negative otherwise.

Let us consider first what happens in a network that doesn't contain any capacitors; this gives rise immediately to a system of equations to be satisfied by the currents $I_j(t)$. If the network doesn't contain any inductors either, the system will consist of purely algebraic relations between the currents $I_j(t)$ and whatever voltage sources the network contains. If one or more inductors are also present, the system will consist of a combination of algebraic equations and of differential equations of order one. The following example is of this type.

---

e x a m p l e **1**

Figure 6.5a shows a network that contains just resistors and inductors in addition to a single voltage source. Applying the junction law gives the same equation regardless of which of the two junctions it is applied to. Either way we get

$$I_1 + I_2 = I_3.$$

Suppose that the voltage source causes current to flow in the direction shown for I_3 when $E > 0$. The loop law applied to the left-hand loop then gives

$$L_1\dot{I}_1 + R_1I_1 + R_3I_3 = E.$$

From the right-hand loop we get an equation of the same form relating I_2 and I_3, with L_2 replacing L_1 and R_2 replacing R_1:

$$L_2\dot{I}_2 + R_2I_2 + R_3I_3 = E.$$

Replacing I_3 by $I_1 + I_2$ gives two differential equations for I_1 and I_2:

$$L_1\dot{I}_1 + (R_1 + R_3)I_1 + R_3I_2 = E,$$
$$L_2\dot{I}_2 + (R_2 + R_3)I_2 + R_3I_1 = E.$$

The unit of inductance L is the **henry,** and that of resistance R is the **ohm.** Suppose that $L_1 = L_2 = 0.1$ and that $R_1 = 10$, $R_2 = 20$, and $R_3 = 30$. Suppose also that at the time we first observe the network the voltage source has been switched off and replaced by a conductor. Then $E = 0$, and our differential equations can be written

$$\dot{I}_1 + 400I_1 + 300I_2 = 0,$$
$$300I_1 + \dot{I}_2 + 500I_2 = 0.$$

In operator form these equations are

$$(D + 400)I_1 + 300I_2 = 0,$$
$$300I_1 + (D + 500)I_2 = 0.$$

The elimination method leads to the characteristic equation

$$r^2 + 900r + 110000 = 0,$$

with roots (to the nearest integer) $r_1 = -754$, $r_2 = -146$. An approximation to the general solution then has the form

$$I_1(t) \approx 150c_1e^{-754t} + 150c_2e^{-146t},$$
$$I_2(t) \approx 177c_1e^{-754t} - 127c_2e^{-146t}.$$

Initial values $I_1(0)$ and $I_2(0)$ can be used to determine acceptable values for c_1 and c_2.

We've already seen the relation $dQ/dt = I$ that relates the current in a branch to the charge on a capacitor in the branch. Equations containing a charge Q_j are usually differentiated once with respect to t to eliminate Q_j and to get equations entirely in terms of currents I_j. If this is done, the equation for a loop containing both an inductor and a capacitor will become a second-order differential equation for I_j, because it already contains a term of the form $L_j\, dI_j/dt$, whose derivative is $L_j\, d^2I_j/dt^2$.

e x a m p l e 2

The network shown in Figure 6.5b contains circuit elements of all three types: R, L, and C. Applying the junction law to either junction gives the same equation:

$$I_1 + I_2 = I_3.$$

Suppose that the voltage source causes current to flow in the direction shown for I_3 when $E > 0$. Applying the loop law to the left-hand loop then gives

$$L_1\dot{I}_1 + R_1I_1 + R_3I_3 = E.$$

If $Q_2(t)$ represents the charge on the capacitor so that $\dot{Q}_2 = I_2$, the right-hand loop yields

$$L_2\dot{I}_2 + \frac{Q_2}{C_2} + R_3I_3 = E.$$

If we differentiate this last equation with respect to t and replace $\dot{Q}_2$ by I_2, we get

$$L_2\ddot{I}_2 + \frac{1}{C_2}I_2 + R_3\dot{I}_3 = \dot{E}.$$

Finally we can replace I_3 by $I_1 + I_2$ to get a pair of differential equations for I_1 and I_2:

$$L_1\dot{I}_1 + (R_1 + R_3)I_1 + R_3I_2 = E,$$
$$L_2\ddot{I}_2 + R_3\dot{I}_2 + \frac{1}{C_2}I_2 + R_3\dot{I}_1 = \dot{E}.$$

The form of these two differential equations suggests that specifying initial values for I_1, I_2, and $\dot{I}_2$ will be enough to determine $I_1(t)$ and $I_2(t)$, and hence also their sum $I_3(t)$. (Note that $\dot{I}_2(0)$ is determined by the equation for the right-hand loop if we know $Q_2(0)$, the initial charge on the capacitor, as well as $E(0)$ and $I_3(0)$.) If E represented a constant voltage source of size V_0 that caused current to flow up in the central branch, then E should be replaced by $-V_0$, because the upward direction is counter to the way the two loops were traversed. For a downward-directed source, we would set $E = V_0$. In either case $\dot{E} = 0$, because E is constant. For a variable voltage source, similar remarks apply. For example, if $E(t) = \sin t$, then $\dot{E}(t) = \cos t$, and in the time interval between 0 and π the source would be causing current to flow down in the central branch, followed by an upward flow in the next time interval of length π, and so forth.

EXERCISES

The three networks shown in Figure 6.6 differ only in the elements they contain; each one results from the one to its left by including a new element. In these diagrams let the symbol E stand for a constant voltage source of E volts that would cause current to flow down relative to the diagram. (Such a source could be provided by a battery with its positive $(+)$ terminal attached to the junction below it and its negative $(-)$ terminal attached to the resistor above it.)

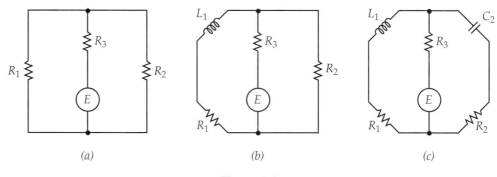

(a)   (b)   (c)

Figure 6.6

1. In the network shown in Figure 6.6a, let $R_1 = 5$, $R_2 = 10$, $R_3 = 15$, and $E = 110$.

   (a) Find a system of equations that determines the currents in the three branches, and solve the system.

   (b) What happens to the solution in part (a) if the voltage source is applied upward instead of downward?

   (c) What happens to the solution if $R_3 = 0$ instead of 15?

   (d) What happens to the solution if $R_1 = R_3 = 0$?

2. In the network shown in Figure 6.6b, let $R_1 = 5$, $R_2 = 10$, $R_3 = 15$, $L_1 = 0.1$, and $E = 110$.

   (a) Find a system of equations that determines the currents in the three branches.

   (b) Find the general solution of the system of equations found in part (a).

   (c) Determine the constant in the general solution by imposing the condition that at time $t = 0$ a current of 3 amperes is flowing upward in the left branch.

   (d) Find the limit as t tends to infinity of the solutions found in part (c).

3. In the network shown in Figure 6.6c let $R_1 = 5$, $R_2 = 10$, $R_3 = 15$, $L_1 = 0.1$, $C_2 = \frac{1}{250} = 0.004$, $E = $ const.

   (a) Find a system of equations that determines the currents in the three branches.

   (b) Find the general solution of the system of equations found in part (a).

   (c) Determine the constants in the general solution by imposing the condition that at time $t = 0$ currents of 2 and 3 amperes, respectively, are flowing upward in the right and left branches and that the initial charge on the capacitor is $Q_2(0) = 5$.

4. In the network in Figure 6.6c let $R_1 = 5$, $R_2 = 10$, $R_3 = 15$, $L_1 = 0.1$, $C_2 = \frac{1}{1250} = 8 \times 10^{-4}$, $E = $ const.

   (a) Find a system of equations that determines the currents in the three branches.

   (b) Find the general solution of the system of equations found in part (a), and show that they exhibit damped oscillation.

   (c) Determine the constants in the general solution by imposing the condition that at time $t = 0$ currents of 2 and 3 amperes, respectively, are flowing upward in the right and left branches and that the initial charge on the capacitor is $Q_2(0) = 5$.

5. Find a system of equations for the currents in the six branches of the network shown in Figure 6.5c, assuming general constant values for the voltages and also for the characteristics of the circuit elements. Note that positive current directions have not been assigned.

6. Suppose two completely independent LRC circuits get joined by a single conductor connecting a point in each circuit. Explain why the resulting configuration is not a network in the sense of our definition.

4 MECHANICAL OSCILLATIONS

4A *Spring-Linked Masses in Linear Motion*

Consider weights of mass m_1 and m_2 tied by springs to each other and also to fixed walls. Suppose the springs have stiffness constants h_1, h_2, and h_3 as shown in Figure 6.7. By **Hooke's law,** the magnitude of the restoring force for

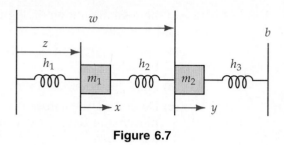

Figure 6.7

the jth spring is proportional to the absolute value of its displacement from rest, with proportionality constant h_j. Let x and y be displacements from equilibrium of the first and second bodies so that equilibrium occurs when $x = y = 0$. The force acting on the first body is by definition equal to $m_1\, d^2x/dt^2$. In calculating the force induced by the springs on m_1 or m_2 at any instant, we have to take account of the relative positions of both bodies. Note that the force exerted by h_3 on m_1 and h_1 on m_3 is transmitted through the middle spring. Thus

$$m_1 \frac{d^2x}{dt^2} = -h_1 x + h_2(y - x).$$

The choice of sign in each term is dictated by whether a displacement x or $y - x$ causes an increase or decrease in the velocity dx/dt. For example, if $y - x > 0$, the middle spring is extended and so acts to increase dx/dt, making $d^2x/dt^2 > 0$. Similarly,

$$m_2 \frac{d^2y}{dt^2} = -h_2(y - x) - h_3 y.$$

In deriving both equations, we have neglected frictional forces. We can rewrite the system in the form

$$\frac{d^2x}{dt^2} = -\frac{h_1 + h_2}{m_1} x + \frac{h_2}{m_1} y,$$

■ **4.1**

$$\frac{d^2y}{dt^2} = \frac{h_2}{m_2} x - \frac{h_2 + h_3}{m_2} y.$$

---

e x a m p l e **1**

If the weights are equal, say $m_1 = m_2 = 1$, and also $h_1 = h_2 = h_3 = 1$, then the system 4.1 becomes

$$\frac{d^2x}{dt^2} = -2x + y,$$

$$\frac{d^2y}{dt^2} = x - 2y.$$

In operator form this is

$$(D^2 + 2)x - y = 0$$
$$-x + (D^2 + 2)y = 0.$$

Operating on the second equation with $D^2 + 2$ and adding gives
$$(D^4 + 4D^2 + 3)y = (D^2 + 1)(D^2 + 3)y = 0.$$
We can read off the characteristic roots at a glance ($\pm i$, $\pm i\sqrt{3}$), so
$$y(t) = c_1 \cos t + c_2 \sin t + c_3 \cos \sqrt{3}\, t + c_4 \sin \sqrt{3}\, t.$$
Using the second of the pair of equations to find x gives
$$\begin{aligned} x(t) &= (D^2 + 2)y(t) \\ &= c_1 \cos t + c_2 \sin t - c_3 \cos \sqrt{3}\, t - c_4 \sin \sqrt{3}\, t. \end{aligned}$$
The constants c_1, c_2, c_3, and c_4 are determined by initial displacements and velocities. The special choice $x(0) = -1$, $y(0) = 1$, $\dot{x}(0) = -\sqrt{3}$, $\dot{y}(0) = \sqrt{3}$ gives

$$\begin{aligned} x(t) &= -\cos \sqrt{3}\, t - \sin \sqrt{3}\, t & y(t) &= \cos \sqrt{3}\, t + \sin \sqrt{3}\, t \\ &= -\sqrt{2} \cos\left(\sqrt{3}\, t - \frac{\pi}{4}\right), & &= \sqrt{2} \cos\left(\sqrt{3}\, t - \frac{\pi}{4}\right). \end{aligned}$$

If a homogeneous system has oscillatory solutions expressible as a sum of terms with distinct frequencies, then a term with one of these frequencies is called a **normal mode** of oscillation for the system. Thus the oscillations $\cos \sqrt{3}\, t$ and $\sin \sqrt{3}\, t$ with circular frequency $\sqrt{3}$ in Example 1 express the same normal mode of the system. Other normal modes in the same example, for instance, $\cos t$ or $2 \sin t$, have circular frequency 1. The typical normal mode of a constant-coefficient system contains a factor $\cos \mu t$ or $\sin \mu t$ with circular frequency μ. The normal modes are clearly important characteristics of an oscillatory system, and an efficient way to find them using eigenvalue methods appears in Chapter 7. In the special case $m\ddot{x} + hx = 0$ of dimension 1 that we treated in detail in Chapter 4, the normal modes all have circular frequency $\omega_0 = \sqrt{h/m}$.

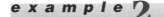

example 2

Suppose the third spring is removed altogether from Example 1, so that $h_3 = 0$. We're left with
$$\begin{aligned} \ddot{x} &= -2x + y, \\ \ddot{y} &= x - y, \end{aligned}$$
or, in operator form,
$$\begin{aligned} (D^2 + 2)x - y &= 0, \\ -x + (D^2 + 1)y &= 0. \end{aligned}$$
Elimination of x proceeds as in Example 1, but this time the differential equation for y is $(D^4 + 3D^2 + 1)y = 0$, with characteristic equation $r^4 + 3r^2 + 1 = 0$. Regarding the left side as quadratic as a function of r^2, we find $r^2 = (-3 \pm \sqrt{5})/2$. Both values are negative, so the characteristic roots are $\pm i\sqrt{(3 + \sqrt{5})/2} \approx \pm 1.62i$ and $\pm i\sqrt{(3 - \sqrt{5})/2} \approx \pm 0.62i$. The normal modes from which solutions are constructed have circular frequencies
$$\mu_1 = \sqrt{\frac{3 + \sqrt{5}}{2}} \quad \text{and} \quad \mu_2 = \sqrt{\frac{3 - \sqrt{5}}{2}}.$$

Note that for Hooke's law to remain valid in practice, the condition $x(t) < y(t)$ has to be maintained with some leeway. Nevertheless, it's quite possible to have this condition violated under certain purely mathematical initial conditions.

In Example 2 it was essential to know in advance the equilibrium positions of the two masses to establish the precise location of each mass relative to the other and to the spring supports. Thus our additional problem is to find an **equilibrium solution** to the appropriate equations of motion, that is, a constant solution, for which all time derivatives are zero. For the equations derived in Example 2, it's easy to check that the unique equilibrium solution is $x(t) = 0$, $y(t) = 0$. Indeed we chose our coordinates so that these would be the equilibrium solutions. The next example shows how to avoid making any assumptions about the coordinates.

Instead of measuring the locations of the two masses shown in Figure 6.7 from their equilibrium positions we can measure both displacements from the same point. If we know the unstressed (i.e., relaxed) lengths l_1, l_2, l_3 of the three springs and the distance b between the supports, this approach allows us to determine the precise location of the equilibrium positions. (This information had to be assumed in our earlier analysis.) Let z and w be the respective distances of masses m_1 and m_2 from the left end, as shown in Figure 6.7. It's now the (positive) extension or (negative) compression of the springs beyond their unstressed lengths that is of critical interest; for the first spring this is $z - l_1$; for the second it is $(w - z) - l_2$; for the third it is $(b - w) - l_3$. Taking into account the force direction of each spring action on the adjacent mass, we find

$$m_1 \frac{d^2z}{dt^2} = -h_1(z - l_1) + h_2((w - z) - l_2),$$

$$m_2 \frac{d^2w}{dt^2} = -h_2((w - z) - l_2) + h_3((b - w) - l_3).$$

These equations simplify to the system

$$m_1 \frac{d^2z}{dt^2} = -(h_1 + h_2)z + h_2 w + h_1 l_1 - h_2 l_2,$$

■ 4.2

$$m_2 \frac{d^2w}{dt^2} = h_2 z - (h_2 + h_3)w + h_2 l_2 - h_3 l_3 + h_3 b.$$

These equations are almost the same as the ones derived in Example 2 except for the presence of additional constant terms on the right side. Thus they constitute a nonhomogeneous system rather than a homogeneous one. Since equilibrium solutions are constant, the second derivatives $\ddot{z}$ and $\ddot{w}$ are identically zero. Consequently, to find the equilibrium positions, all we have to do is set the right sides of the differential equations equal to zero and solve for z and w.

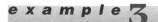 **example 3**

If springs governed by Equations 4.2 have equal stiffness $h_1 = h_2 = h_3$ and equal relaxed lengths $l_1 = l_2 = l_3$, then the h_k and l_k all cancel from the purely algebraic system obtained by setting $\ddot{z} = \ddot{w} = 0$. The system to be solved is then

$$-2z + w = 0,$$
$$z - 2w = -b.$$

The unique solution is the particular constant solution $z_e = b/3$, $w_e = 2b/3$. (A moment's reflection shows that this result is what you would expect on physical grounds: if all three springs have the same properties, then the equilibrium points will be equally spaced between the fixed supports.) Moreover, the general solution to the system can be derived from the solution to the homogeneous system by adding a particular solution. Since the homogeneous system has the same form as the system in Example 2, we have $z(t) = x(t) + z_e$, $w(t) = y(t) + w_e$.

Taking linear damping into account is a simple matter in principle. For example, the two-body system we started with becomes

$$m_1 \frac{d^2x}{dt^2} = -(h_1 + h_2)x + h_2y - k_1\dot{x},$$

$$m_2 \frac{d^2y}{dt^2} = h_2x - (h_2 + h_3)y - k_2\dot{y}.$$

The major technical difference is that the introduction of first-order terms means that the equation for the characteristic roots r can no longer be regarded as a quadratic equation in r^2.

4B   Nonlinear Spring Analysis

Analysis of linear motion of physical bodies linked by springs satisfying Hooke's law leads to a system of second-order linear differential equations with dimension equal to the number n of bodies under consideration. If the motion takes place in a plane, or else in 3-dimensional space, it takes, respectively, two or three coordinates to specify the position of each body. Hence the relevant system of differential equations will then involve $2n$, or else $3n$, coordinates if we're keeping track of n bodies. At first sight it might seem that just setting up the system of differential equations is a formidable problem. There is a simple way to proceed, however, using vector ideas and notation. Figure 6.8 shows two objects with masses m_1, m_2 linked by three springs to each other and to two points **a** and **b.** Depending on how the motion starts, it might in fact be confined to the line joining **a** and **b,** or to some plane containing those two points, or in the most general circumstances the motion of each body might be 3-dimensional in an essential way. Our derivation of the differential equations

Figure 6.8

will include all these cases at once. Along with the stiffness constants h_1, h_2, h_3 we need to take into account the relaxed (i.e., unstressed) lengths of the springs, which we'll denote by l_1, l_2, l_3, respectively. Denoting the position at time t of the body of mass m_1 by $\mathbf{x} = \mathbf{x}(t)$, the total force acting on the body can be expressed as $m_1\ddot{\mathbf{x}}$, that is, mass times the acceleration vector. On the other hand, this same total force can be expressed as the sum of the two forces acting on the body via the two springs attached to it. The first spring will have extended, or maybe compressed, length $|\mathbf{a} - \mathbf{x}|$. The difference $|\mathbf{a} - \mathbf{x}| - l_1$, if it is positive, measures the extension of the spring beyond its relaxed length, and, if this difference is negative, it measures the amount of compression of the spring below its relaxed length. The numerical value of the force due to the first spring is then $h_1(|\mathbf{a} - \mathbf{x}| - l_1)$. Since an extended spring pulls the body toward $\mathbf{a}$, we can express the force vector by its numerical value times the vector of length 1 pointing from the body toward the point $\mathbf{a}$. This unit vector can be written as the vector $\mathbf{a} - \mathbf{x}$ divided by its length: $(\mathbf{a} - \mathbf{x})/|\mathbf{a} - \mathbf{x}|$. The resulting force vector is then the first of the two vectors

$$h_1(|\mathbf{a} - \mathbf{x}| - l_1)\frac{\mathbf{a} - \mathbf{x}}{|\mathbf{a} - \mathbf{x}|}, \qquad h_2(|\mathbf{y} - \mathbf{x}| - l_1)\frac{\mathbf{y} - \mathbf{x}}{|\mathbf{y} - \mathbf{x}|}.$$

Denoting the position of the second body by $\mathbf{y} = \mathbf{y}(t)$, we similarly get the second of these two vectors for the other spring force acting on the first body. The differential equation expressing $m_1\ddot{\mathbf{x}}$ as the sum of these forces is then

$$m_1\ddot{\mathbf{x}} = h_1(|\mathbf{a} - \mathbf{x}| - l_1)\frac{\mathbf{a} - \mathbf{x}}{|\mathbf{a} - \mathbf{x}|} + h_2(|\mathbf{y} - \mathbf{x}| - l_2)\frac{\mathbf{y} - \mathbf{x}}{|\mathbf{y} - \mathbf{x}|}.$$

A parallel derivation yields the equation for the force on the second body:

$$m_2\ddot{\mathbf{y}} = h_2(|\mathbf{x} - \mathbf{y}| - l_2)\frac{\mathbf{x} - \mathbf{y}}{|\mathbf{x} - \mathbf{y}|} + h_3(|\mathbf{b} - \mathbf{y}| - l_3)\frac{\mathbf{b} - \mathbf{y}}{|\mathbf{b} - \mathbf{y}|}.$$

For motion in 3 dimensions, each of these two vector equations is equivalent to three real coordinate equations, making up a system of six equations governing the motion of the entire mass-spring system. One striking feature of this system is that it is, in general, nonlinear. The nonlinearity is due to the presence of the expressions $|\mathbf{b} - \mathbf{y}|$, $|\mathbf{y} - \mathbf{x}|$, $|\mathbf{a} - \mathbf{x}|$. For example, letting $\mathbf{y} = (y_1, y_2, y_3)$ and $\mathbf{b} = (b_1, b_2, b_3)$, we note that

$$|\mathbf{b} - \mathbf{y}| = \sqrt{(b_1 - y_1)^2 + (b_2 - y_2)^2 + (b_3 - y_3)^2},$$

an obviously nonlinear function of the coordinates y_1, y_2, y_3.

The two-body mass-spring system is, in general, nonlinear, though we've already seen that for motion of two bodies on a line the system is linear. The next example explains the connection between the two situations.

---

e x a m p l e **4**

For two bodies moving on a line we may as well take the line of motion to be a coordinate axis. We place the fixed points $\mathbf{a} = a$ and $\mathbf{b} = b$ and the variable points $\mathbf{x} = x$ and $\mathbf{y} = y$ that designate the locations of the two masses in order on the line: $a < x < y < b$. Then the vector lengths can all be interpreted as absolute values, and we can use the order relations among the points to compute their values. For example, $|a - x| = x - a$, since $a < x$. In particular, the unit vectors that we use to determine direction of force all reduce to either 1 or -1. For example, $(a - x)/|a - x| = -1$, since $a < x$ implies $a - x < 0$. Using such relations, the two-mass vector system becomes

$$m_1\ddot{x} = -h_1(x - a - l_1) + h_2(y - x - l_2),$$
$$m_2\ddot{y} = -h_2(y - x - l_2) + h_3(b - y - l_3).$$

Rearranging terms yields the more normal-looking pair

$$m_1\ddot{x} = -(h_1 + h_2)x + h_2y + h_1l_1 - h_2l_2 + h_1a,$$
$$m_2\ddot{y} = h_2x - (h_2 + h_3)y + h_2l_2 - h_3l_3 + h_3b.$$

The system is clearly linear. The reason these equations don't look just like the ones we derived earlier for the special case of motion restricted to a line is that then we used two different locations on the line as the respective zero values for x and y. However, the new system contains more information than the one we found earlier; now we can find the equilibrium values for x and y from the new equations by setting their right-hand sides equal to zero and solving for x and y. The analog for nonlinear systems requires numerical analysis, for example, a multidimensional Newton's method.

Equilibrium solutions of the above pair of second-order equations for two masses are those for which there is no motion; that is, $x(t)$ and $y(t)$ are both constant. It follows that $m_1\ddot{x} = m_1\ddot{y} = 0$, so the pair of differential equations reduces to the pair of purely algebraic equations

$$-(h_1 + h_2)x + h_2y + h_1(a + l_1) - h_2l_2 = 0,$$
$$h_2x - (h_2 + h_3)y + h_2l_2 + h_3(b - l_3) = 0.$$

For example, with $h_1 = h_2 = 1, h_3 = 2, l_1 = 2, l_2 = 3, l_3 = 1$, and $a = 0, b = 8$ we get

$$2x - y = -1,$$
$$-x + 3y = 15.$$

The unique solution is $x = \frac{12}{5} = 2.4, y = \frac{29}{5} = 5.8$. Consequently, starting the system with initial conditions $x(0) = \frac{12}{5}, \dot{x}(0) = 0, y(0) = \frac{29}{5}, \dot{y}(0) = 0$ yields the constant solution we just computed. In physical terms, the forces are precisely balanced at these positions, so there is no motion.

---

example 6

The linear case treated in Example 5 is the only one that has solutions that can be expressed in terms of elementary functions. If we want to resort to numerical approximations to the true solutions, it's sometimes desirable to display the differential equations in coordinate form rather than vector form. Written out, the six real coordinate equations for a two-body mass-spring system might use the following coordinates: $\mathbf{x} = (x_1, x_2, x_3), \mathbf{y} = (y_1, y_2, y_3), \mathbf{a} = (a_1, a_2, a_3), \mathbf{b} = (b_1, b_2, b_3)$. Each of the two vector equations becomes three coordinate equations, making six in all. To avoid repeating expressions for vector length such as $|\mathbf{x} - \mathbf{y}| = \sqrt{(x_1 - y_1)^2 + (x_2 - y_2)^2 + (x_3 - y_3)^2}$ more often than necessary, we rearrange the equations slightly as follows:

$$m_1 \ddot{x}_j = h_1 \left(1 - \frac{l_1}{|\mathbf{a} - \mathbf{x}|}\right)(a_j - x_j) + h_2 \left(1 - \frac{l_2}{|\mathbf{y} - \mathbf{x}|}\right)(y_j - x_j), \qquad j = 1, 2, 3,$$

$$m_2 \ddot{y}_j = h_2 \left(1 - \frac{l_2}{|\mathbf{x} - \mathbf{y}|}\right)(x_j - y_j) + h_3 \left(1 - \frac{l_3}{|\mathbf{b} - \mathbf{y}|}\right)(b_j - y_j), \qquad j = 1, 2, 3.$$

Linear damping would simply add terms of the form $-k_1 \dot{x}_j$ and $-k_2 \dot{y}_j$, respectively, to the two right sides of our equations for the undamped motion, with positive damping constants k_1, k_2.

Since the space in which each of n bodies moves is 3-dimensional, the **configuration space** is $3n$-dimensional, and the **state space** that includes the $3n$ time derivatives is $6n$-dimensional.

---

e x a m p l e 7

There is nothing in the derivation of the vector equations of motion that requires $\mathbf{a}$ and $\mathbf{b}$ to be fixed in space, as we tacitly regarded them at the time. For example, we could hold one of them fixed at $\mathbf{a}(t) = (0, 0, 0)$ and specify a periodic motion for the other one using three coordinates, for example, by $\mathbf{b}(t) = (2 \sin t, \cos t, \sin 2t)$. However, the **configuration space** for a two-body system is 6-dimensional, and the **state space,** or **phase space,** including velocities, is 12-dimensional. At Web site http://math.dartmouth.edu/~rewn/ the Applets SPRINGS2-1, for two springs attached to one mass, SPRING3-1, for three springs attached to one mass, and SPRINGS3-2, for two masses linked in series by three springs, allow you to choose masses m, spring constants h, and spring lengths l for perspective graphic simulations of nonlinear mass-spring systems in 3-dimensional space.

EXERCISES

1. In Example 1 of the text, the system

$$(D^2 + 2)x - y = 0,$$
$$-x + (D^2 + 2)y = 0$$

is shown to have the general solution

$$x(t) = c_1 \cos t + c_2 \sin t - c_3 \cos \sqrt{3}\, t - c_4 \sin \sqrt{3}\, t,$$
$$y(t) = c_1 \cos t + c_2 \sin t + c_3 \cos \sqrt{3}\, t + c_4 \sin \sqrt{3}\, t,$$

where $x(t)$ and $y(t)$ are interpreted as the displacements at time t of two masses in a mass-spring physical system.

(a) Show that the initial conditions $x(0) = 0$, $\dot{x}(0) = 1$ and $y(0) = 1$, $\dot{y}(0) = 0$ can be satisfied by choosing the constants properly in the general solution.

(b) Show that initial conditions of the form $x(0) = x_0$, $y(0) = y_0$, $\dot{x}(0) = u_0$, $\dot{y}(0) = v_0$ can always be satisfied.

Normal Modes. Calculate the circular frequencies of the various constituent oscillations associated with the system 4.1 of the text under the following assumptions.

2. $m_1 = m_2 = 1$, $h_1 = h_2 = h_3 = 1$.

3. $m_1 = m_2 = 1$, $h_1 = h_2 = 1$, $h_3 = 2$.

4. $m_1 = 1$, $m_2 = 2$, $h_1 = 1$, $h_2 = h_3 = 4$.

5. How do the normal modes of the system 4.1 of the text compare with those of the system 4.2?

6. Suppose that the middle spring is removed from the system governed by Equations 4.2.

   (a) Show that the system becomes uncoupled.

   (b) What are the normal modes?

7. Double pendulum. An ideal pendulum of mass m_1 and length l_1 with another one of mass m_2 and length l_2 attached to its free end moves in a single plane with respective displacement angles θ_1, θ_2 shown in Figure 6.23b of Section 9. The linearized system for small oscillations is

$$\ddot{\theta}_1 = -(\rho + 1)\frac{g}{l_1}\,\theta_1 + \rho\frac{g}{l_1}\,\theta_2, \qquad \ddot{\theta}_2 = (\rho + 1)\frac{g}{l_2}\,\theta_1 - (\rho + 1)\frac{g}{l_2}\,\theta_2,$$

where $\rho = m_2/m_1$. Show that if $m_1 = m_2$ and $l_1 = l_2$, then the normal modes have frequencies $((2 - \sqrt{2})g/l_1)^{1/2} \approx 0.765\sqrt{g/l_1}$ and $((2 + \sqrt{2})g/l_1)^{1/2} \approx 1.848\sqrt{g/l_1}$.

8. A 2-dimensional mechanical system $m_1\ddot{x} = f(x, y)$, $m_2\ddot{y} = g(x, y)$ is called **conservative** if there is a **potential function** $U(x, y)$ such that

$$\frac{\partial U(x, y)}{\partial x} = -f(x, y) \qquad \text{and} \qquad \frac{\partial U(x, y)}{\partial y} = -g(x, y).$$

   (a) Show that the general two-body system
$$m_1\ddot{x} = -(h_1 + h_2)x + h_2 y,$$
$$m_2\ddot{y} = h_2 x - (h_2 + h_3)y$$
   is conservative by computing a potential.

   (b) The **kinetic energy** of the system is $T = \frac{1}{2}(m_1\dot{x}^2 + m_2\dot{y}^2)$. Show that for a general conservative system of the type considered here the **total energy** $T + U$ is constant. [*Hint:* Multiply the $\ddot{x}$ equation by $\dot{x}$, the $\ddot{y}$ equation by $\dot{y}$, add the two equations, and integrate.]

9. Suppose two bodies at $x(t) < y(t)$ on a line are joined to each other and to two fixed points l units apart by springs of equal stiffness, so $h_1 = h_2 = h_3$. Find the equilibrium points for the system in terms of the unstressed lengths l_1, l_2, l_3 of the springs and the distance l. Assume no forces are acting except that of the springs. (The answer is intuitively obvious in case $l_1 = l_2 = l_3$, and your general answer should be consistent with this special case.)

10. Suppose two bodies of mass 1 at $0 < x(t) < y(t) < l$ on a line are joined to each other and to two fixed points l units apart by springs of equal stiffness

h and unstressed lengths $l_1 = l_2 = l_3$. Here x and y are both measured from the same origin, whereas in Example 1 of the text each was measured from a different point. Assume no forces are acting except that of the springs.

(a) Show that x and y satisfy

$$\ddot{x} = -2hx + hy,$$
$$\ddot{y} = hx - 2hy + hl.$$

(b) Find the constant equilibrium solutions for the system in terms of the distance l. These solutions remain at the fixed points from which the mass displacements are measured in Example 1 of the text.

11. Suppose two bodies of mass m_1, m_2 at $0 < x(t) < y(t) < l$ on a line are joined to each other and to two fixed points l units apart by springs of respective stiffness h_1, h_2, h_3 and unstressed lengths l_1, l_2, l_3. Here x and y are both measured from the same origin, as in Equation 4.2, rather than from different points measured from a different point. Assume no forces are acting except that of the springs.

   (a) Assume the middle spring is both twice as strong and half as long as the other two. Show that x and y satisfy a system of the form

   $$m_1\ddot{x} = -3hx + 2hy,$$
   $$m_2\ddot{y} = 2hx - 3hy + hl,$$

   where h is some positive constant.

   (b) Find the constant equilibrium solutions for the system in terms of h.

   (c) Find the natural circular frequencies of the system, assuming $m_1 = m_2 = 1$.

12. Derive the vector equation of motion for two bodies of mass m_1, m_2, respectively, if they are joined sequentially by springs to two fixed points and to each other, and assuming a constant gravitational acceleration of magnitude g in the direction of a fixed unit vector $\mathbf{u}$.

13. Find a first-order system in standard form that is equivalent to the second-order system for two bodies and three springs given in Equations 4.2.

14. **(a)** Derive the vector equation of motion for a single body of mass m at $\mathbf{x}(t)$ attached by springs with lengths l_1, l_2 and respective Hooke constants h_1, h_2 to two fixed points $\mathbf{a}, \mathbf{b}$.

   (b) Express the equation of part (a) in xyz coordinates assuming $\mathbf{a} = (0, 0, 0)$ and $\mathbf{b} = (b, 0, 0)$.

   (c) Show that if the motion remains on the segment joining $(0, 0, 0)$ and $(b, 0, 0)$, the system is linear.

15. Derive the vector equations of motion for a single body of mass m at $\mathbf{x}(t)$ attached by springs with lengths l_1, l_2, l_3 and respective Hooke constants h_1, h_2, h_3 to three fixed points $\mathbf{a}, \mathbf{b}, \mathbf{c}$.

*16. On physical grounds the linear system in Equations 4.1 should, in general, have all its solutions representable as linear combinations of undamped sinusoidal oscillations; equivalently, all characteristic roots arising in their solution should be purely imaginary. Prove this as follows.

(a) Note that Equations 4.1 have the form

$$\ddot{x} = -px + qy,$$

$$\ddot{y} = rx - sy, \qquad p, q, r, s \text{ positive constants.}$$

Use the elimination method to show that the squares of the characteristic roots of this more general system are always real.

(b) Show that assuming $p = (h_1 + h_2)/m_1$, $q = h_2/m_1$, $r = h_2/m_2$, $s = (h_2 + h_3)/m_2$ makes the squares of the characteristic roots negative, so the roots themselves are imaginary.

5 INVERSE-SQUARE LAW

Let $\mathbf{x}_1 = \mathbf{x}_1(t)$ and $\mathbf{x}_2 = \mathbf{x}_2(t)$ represent the positions at time t of two bodies in space such that each acts on the other by the inverse-square law of gravitational attraction, with no other forces considered. If m_1 and m_2 are the respective masses of the two bodies, the magnitude of the mutually attractive force is then

$$F = \frac{Gm_1m_2}{r^2},$$

where $r = |\mathbf{x}_1 - \mathbf{x}_2|$ is the distance between $\mathbf{x}_1$ and $\mathbf{x}_2$ (i.e., the length of the vector between them). The **gravitational constant** G is about 6.673×10^{-11} if the relevant units are meters, kilograms, and seconds. The normalized vectors

$$\mathbf{u}_2 = \frac{\mathbf{x}_1 - \mathbf{x}_2}{|\mathbf{x}_1 - \mathbf{x}_2|}, \qquad \mathbf{u}_1 = -\frac{\mathbf{x}_1 - \mathbf{x}_2}{|\mathbf{x}_1 - \mathbf{x}_2|}$$

have length 1 and point, respectively, from the second body to the first, and vice versa. Thus the vectors that describe the force acting on each body are the product of magnitude F and a normalized direction unit vector $\mathbf{u}$; the vector $F\mathbf{u}_1$ acts on the first body, and $F\mathbf{u}_2$ acts on the second body. Since these forces can also be described by Newton's second law as mass times acceleration, we have

$$m_1\ddot{\mathbf{x}}_1 = F\mathbf{u}_1, \qquad m_2\ddot{\mathbf{x}}_2 = F\mathbf{u}_2.$$

Figure 6.9 shows the position and force vectors. The acceleration vectors, which actually govern the motion, are depicted as if m_1 is much larger than m_2. Writ-

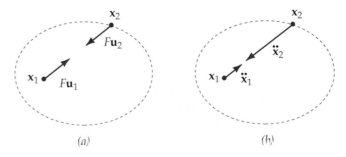

(a)   (b)

Figure 6.9 (a) Equal force magnitudes. (b) Unequal accelerations: $|\ddot{\mathbf{x}}_1| = F/m_1 < |\ddot{\mathbf{x}}_2| = F/m_2$.

ten out in more detail these **Newton equations** are

■ **5.1** $$\ddot{\mathbf{x}}_1 = -\frac{Gm_2}{r^3}(\mathbf{x}_1 - \mathbf{x}_2), \qquad \ddot{\mathbf{x}}_2 = -\frac{Gm_1}{r^3}(\mathbf{x}_2 - \mathbf{x}_1),$$

where m_1 has been canceled from the first equation and m_2 from the second. Subtracting the second equation from the first gives

$$\ddot{\mathbf{x}}_1 - \ddot{\mathbf{x}}_2 = -\frac{G(m_1 + m_2)(\mathbf{x}_1 - \mathbf{x}_2)}{r^3}.$$

Equations 5.1 form a system of vector equations for the motions of the two bodies relative to some coordinate system. If a moving coordinate system has its origin maintained at the center of mass of one of the bodies, say the second, we can let $\mathbf{x} = \mathbf{x}_1 - \mathbf{x}_2$ and consider only the equation of relative motion for the first one:

■ **5.2** $$\ddot{\mathbf{x}} = -\frac{G(m_1 + m_2)}{|\mathbf{x}|^3}\mathbf{x}.$$

Writing $\mathbf{x} = (x, y, z)$ and $k = G(m_1 + m_2)$, we get three numerical equations:

■ **5.3**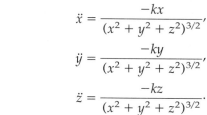
$$\ddot{x} = \frac{-kx}{(x^2 + y^2 + z^2)^{3/2}},$$
$$\ddot{y} = \frac{-ky}{(x^2 + y^2 + z^2)^{3/2}},$$
$$\ddot{z} = \frac{-kz}{(x^2 + y^2 + z^2)^{3/2}}.$$

A solution of this nonlinear system will describe a trajectory of the first body relative to the second, or vice versa. We can eliminate the third equation from consideration by choosing (x, y, z) coordinates so that initial conditions on z are $z(0) = \dot{z}(0) = 0$. By Theorem 1.1, Chapter 5, Section 1E, if $\mathbf{x}(0) \neq 0$ the system then has a unique solution with third coordinate $z = z(t)$ identically zero. Thus the system can be viewed as 2-dimensional (see also Exercise 18a):

$$\ddot{x} = \frac{-kx}{(x^2 + y^2)^{3/2}}, \qquad \ddot{y} = \frac{-ky}{(x^2 + y^2)^{3/2}}.$$

There are no simple formulas for the solution of the remaining two equations of 5.3. The classic approach to the problem is to derive characteristic properties of the solutions. These properties are usually stated as **Kepler's laws** of planetary motion, laws that were discovered empirically for closed orbits before Newton's work.

1. The path described by a solution $(x(t), y(t))$ is an ellipse with the sun (fixed body) at one focus.
2. The radius from the sun to the planet sweeps out equal areas in equal time periods.
3. If P is the time required to complete one orbit and a is the mean distance from the planet to the sun, then $P^2 = 4\pi^2 a^3/(G(m_1 + m_2))$.

Kepler's second law is illustrated for not only an elliptic orbit but also a hyperbolic orbit in Figure 6.10. These beautiful laws are derived in many calculus

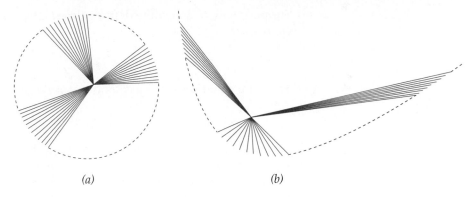

Figure 6.10 Equal areas in equal times. (*a*) Elliptic orbit. (*b*) Hyperbolic orbit.

and physics texts; one approach is outlined in the exercises at the end of the present section. In such derivations it is often assumed that one of the bodies has negligible mass relative to the other so that only the larger mass appears in the equation. For example, our sun is about 333,434 times as massive as the earth, so for some purposes the mass of the earth can be neglected.

---

e x a m p l e 1

Escape Speed

The closed elliptic orbits predicted by the first Kepler law occur only if the speed of separation of the two bodies is not too large. Nonreturning comets follow unbounded hyperbolic orbits for which the second Kepler law holds. To find the relative speed of separation required for a hyperbolic trajectory, suppose the relative motion of the two bodies is restricted to a fixed line, which we take to be an x axis. Thus y and z are always zero, so all that survives of Equations 5.3 is

$$\ddot{x} = -\frac{k}{x^2}, \qquad k = G(m_1 + m_2).$$

To solve this equation (see also Chapter 4, Section 6) we can multiply both sides by $\dot{x}$ to get

$$\dot{x}\ddot{x} = -kx^{-2}\dot{x}.$$

Integrating the left side with respect to t gives

$$\int \dot{x}\ddot{x}\, dt = \int \frac{d}{dt}\left(\tfrac{1}{2}\dot{x}^2\right) dt = \tfrac{1}{2}\dot{x}^2.$$

For the integral of the right side we get

$$-k\int x^{-2}\frac{dx}{dt}\, dt = -k\int x^{-2}\, dx = \frac{k}{x} + C.$$

Combining the preceding three equations we get

$$\tfrac{1}{2}\dot{x}^2 = \frac{k}{x} + C.$$

Suppose now that at some definite time t_0 the values for x and $\dot{x} = v$ are x_0 and v_0, respectively. Then

$$C = \frac{v_0^2}{2} - \frac{k}{x_0}.$$

Thus the equation relating v and x can be written

$$\frac{v^2}{2} - \frac{v_0^2}{2} = \frac{k}{x} - \frac{k}{x_0}.$$

This relation between speed and distance shows that speed decreases as the separation distance x increases; it also allows us to find the **escape speed** of two bodies relative to each other, that is, the speed v_0 that must be attained at distance x_0 so that the speed v will always remain strictly positive thereafter. To achieve this speed we must have

$$0 < v^2 = v_0^2 + \frac{2k}{x} - \frac{2k}{x_0}.$$

Since $k/x \to 0$ as $x \to \infty$, the only way the inequality can hold forever is to have

$$v_0^2 - \frac{2k}{x_0} > 0.$$

Thus the critical escape speed that must be exceeded at distance x_0 must satisfy $v_0^2 = 2k/x_0$. Since $k = G(m_1 + m_2)$, we find that the escape speed is

$$v_0 = \sqrt{\frac{2G(m_1 + m_2)}{x_0}}.$$

The conclusions we have reached are valid even if the relative motion is not restricted to a linear path. A way of showing this is described in Section 6.

It is useful to know what the constant G is in terms of various units of measurement.

Metric Units. Using meters, kilograms, and seconds, it has been calculated from careful observation that $G \approx 6.67 \cdot 10^{-11}$.

Planetary Units. Using the radius of the earth (about 6378 kilometers, or 3963 miles) as the distance unit, the mass of the earth (about $5.98 \cdot 10^{24}$ kilograms) as the unit of mass, and 1 hour as the time unit, it turns out that $G \approx 19.91$.

example 2

Using the results of Example 1 we can compute escape speed relative to the earth. We'll use planetary units and assume the mass of the other body is negligible compared with the earth's mass of $5.98 \cdot 10^{24}$ kilograms. Thus we assume $m_1 + m_2 \approx 1$ in planetary units. With $G = 19.91$, we find $\sqrt{2G(m_1 + m_2)} = \sqrt{2 \cdot 19.91} \approx 6.31$. Escape speed is then about $v_0 = 6.31 x_0^{-1/2}$.

It is correct to measure the distance x_0 between the centers of mass of the two spherically homogeneous bodies, and for an object starting at the surface of the nearly spherical earth it is approximately correct to take x_0 to be 1 earth radius. In this case $v_0 \approx 6.31$ earth radii per second, or about 25,000 miles per hour. Starting at distance $x_0 = 2$ earth radii, which would be about 4000 miles above the earth's surface, we could get away with $v_0 = 6.31/\sqrt{2} \approx 4.46$ earth radii per hour, or about 17,700 miles per hour.

---

example 3

A circular orbit of one body about another is a theoretical possibility but in practice is a very unlikely state of affairs. Nevertheless, some orbits are nearly circular, so understanding circular ones is helpful for getting a qualitative feeling for what the nearby ones are like. You are asked to show in Exercise 14 that the orbital speed v_1 of one body in a circular orbit of radius x_0 about another is

$$v_1 = \sqrt{\frac{G(m_1 + m_2)}{x_0}}.$$

For example, since the moon's mass is about 1.2% of the earth's, for the moon to have a circular orbit of radius x_0 the orbital speed would have to be about $\sqrt{19.91 \cdot 1.012/x_0} \approx 4.49\sqrt{x_0}$, where the orbital radius x_0 is measured in earth radius units. The mean distance of the moon from the earth in its actual elliptical orbit is about 384,404 kilometers, or $384{,}404/6378 \approx 60.27$ earth radii. Since the moon's orbit is not very eccentric, its mean orbital velocity would be about $4.49/\sqrt{60.27} \approx 0.5783$ earth radius per hour, or about 3688 kilometers per hour.

The same sort of argument that leads to Equations 5.3 for the motion of two bodies can be applied to the case of three or more bodies with masses m_k and centers of mass located at $\mathbf{x}_k(t)$, $k = 1, 2, 3, \ldots, n$. Let $r_{ij} = r_{ji} = |\mathbf{x}_i - \mathbf{x}_j|$. Acceleration vectors are added, just as forces are, so the second-order vector equations for three bodies turn out to be

$$\ddot{\mathbf{x}}_1 = \frac{Gm_2}{r_{21}^3}(\mathbf{x}_2 - \mathbf{x}_1) + \frac{Gm_3}{r_{31}^3}(\mathbf{x}_3 - \mathbf{x}_1),$$

■ **5.4**
$$\ddot{\mathbf{x}}_2 = \frac{Gm_1}{r_{12}^3}(\mathbf{x}_1 - \mathbf{x}_2) + \frac{Gm_3}{r_{32}^3}(\mathbf{x}_3 - \mathbf{x}_2),$$

$$\ddot{\mathbf{x}}_3 = \frac{Gm_1}{r_{13}^3}(\mathbf{x}_1 - \mathbf{x}_3) + \frac{Gm_2}{r_{23}^3}(\mathbf{x}_2 - \mathbf{x}_3).$$

To convert these into equations among forces rather than accelerations we would multiply the kth equation by the mass m_k. The analogous set of equations for n bodies, with $6n$-dimensional **state space** vectors for position and velocity $(\mathbf{x}_1, \ldots, \mathbf{x}_n, \dot{\mathbf{x}}_1, \ldots, \dot{\mathbf{x}}_n)$, consists of the n equations

■ **5.5** $\displaystyle \ddot{\mathbf{x}}_k = \sum_{j=1, j \neq k}^{n} \frac{Gm_j}{|\mathbf{x}_j - \mathbf{x}_k|^3}(\mathbf{x}_j - \mathbf{x}_k), \qquad k = 1, 2, \ldots, n.$

To solve such systems we need to know $2n$ initial vectors $\mathbf{x}_k(0)$ and $\dot{\mathbf{x}}_k(0)$. Since each of these vectors is 3-dimensional, the **state space** $\mathscr{S}$ has dimension $6n$; in particular, the three-body problem governed by Equations 5.4 is 18-dimensional.

Notice that the term excluded from the kth sum is $Gm_k(\mathbf{x}_k - \mathbf{x}_k)/|\mathbf{x}_k - \mathbf{x}_k|^3$ and that this omitted term would not be well-defined because the denominator is zero. As an alternative we could have a formal agreement that this ill-defined term should stand for zero, the value that makes the equation correct. (This is, in fact, a useful definition to make when writing the equations for insertion into a numerical solution routine.)

Except for rather special cases, there is no known set of characteristic properties such as the Kepler laws for describing the solutions of Equations 5.5, even for the three-body case governed by Equations 5.4. The numerical methods described in Section 7 can often be used to provide adequate graphical and numerical descriptions of solutions to the initial-value problems. However, an arbitrary choice of velocity coordinates can lead to nonzero velocity for the center of mass of the entire system, causing it eventually to go off to infinity. To avoid this, we can determine the velocity vector of the most massive body from the other $n - 1$ by using the **conservation of momentum equation** (see Exercise 6)

■ 5.6
$$\sum_{k=1}^{n} m_k \ddot{\mathbf{x}}_k(t) = 0.$$

Web site http://math.dartmouth.edu/~rewn/ contains the Applets 2BODY, 3BODY, 4BODY, and PLANETS: these will plot 2-dimensional simulations for the respective planetary motion systems, and they can all be initiated under the assumption of an inverse pth power law of gravitational attraction. The Applet PLANETS contains solar system data for the sun and all nine planets.

EXERCISES

1. If position as a function of time is given by the vector $\mathbf{x}(t) = (x(t), y(t), z(t))$, then the magnitude of the acceleration vector is $a = \sqrt{\ddot{x}^2 + \ddot{y}^2 + \ddot{z}^2}$. Use Equations 5.3 to show that $a = k/r^2$, where $r = \sqrt{x^2 + y^2 + z^2}$, for a single planet orbiting a star.

2. The radius of the earth's atmospheric shell is about 6500×10^3 meters, and the earth's mass is about 5976×10^{21} kilograms. With $G = 6.673 \times 10^{-11}$, estimate the escape speed required near the surface of the shell for a projectile of mass 100 kilograms. How is your answer affected if the projectile mass becomes 1000 kilograms? How about 10^{22} kilograms?

3. Letting $z = 0$ in Equations 5.3, we get a system of two second-order equations for x and y. Find the equivalent first-order system obtained by letting $\dot{x} = z$, $\dot{y} = w$.

4. Equations 5.4 for three bodies are equivalent to nine second-order equations for the coordinates (x_1, y_1, z_1), (x_2, y_2, z_2), (x_3, y_3, z_3) of the three bodies. Write out these equations in terms of x_k, y_k, z_k, and $r_{ij} = \sqrt{(x_i - x_j)^2 + (y_i - y_j)^2 + (z_i - z_j)^2}$.

5. Write the four second-order vector equations for four bodies of mass m_k at $\mathbf{x}_k(t)$, $k = 1, 2, 3, 4$.

6. The **conservation of momentum** law states that, taking into account all forces acting on a physical system of n masses, the sum

$$\sum_{k=1}^{n} m_k \dot{\mathbf{x}}_k(t)$$

of the **momentum vectors** $m_k \dot{\mathbf{x}}_k(t)$ of the system is constant. Verify that conservation of momentum holds for a system governed by:

(a) Equations 5.1.

(b) More generally, for Equations 5.4.

(c) Most generally, for Equations 5.5. [*Hint:* Show that the time derivative of the sum is zero as in Equation 5.6.]

Remark. Conservation of momentum may seem to fail in Equation 5.2, since $\ddot{\mathbf{x}} \neq 0$, but here $\mathbf{x}$ describes only the relative positions of two bodies, not their actual motion.

7. Derive the analog of Equation 5.2 assuming an inverse pth-power gravitation law.

8. Let g be the acceleration of gravity at the surface of a homogeneous spherical body of mass M and radius R. Use the inverse-square law to show that $g = GM/R^2$, where G is the gravitational constant in appropriate units of measurement.

9. (a) Use the equation established in the previous exercise to estimate the gravitational constant using a measured value of 9.8 meters per second per second for the acceleration of gravity near the surface of the earth. (Use the values for the mass and radius of the earth given in the text.)

(b) Estimate the acceleration of gravity near the surface of the earth using the value $6.67 \cdot 10^{-11}$ for the gravitational constant G.

10. Falling-body problems in the absence of air resistance are often modeled by the equation $\ddot{x} = -g$, with $\dot{v}_0 = 0$ and the initial height of the body x_0.

(a) Show that $\dot{x} = \sqrt{2g(x_0 - x)}$ under these assumptions, with distance measured down from the initial position x_0.

(b) Assuming instead that acceleration $\ddot{x} = -k/x^2$, show that $\dot{x} = \sqrt{2k(x_0 - x)/(x_0 x)}$.

11. Suppose at some time that two bodies subject only to their mutual gravitational attraction are at distance r_0 apart and are receding from each other along a fixed line at a certain fraction q of escape velocity, where $0 < q < 1$. Show that their separation velocity reaches zero, and the bodies start to "fall" toward each other, when their distance apart becomes $r_0/(1 - q^2)$.

12. This exercise is a reminder that there would be no such thing as escape velocity for a body of constant mass if the acceleration of gravity were really constant. For linear motion away from the attracting body, we would have $\ddot{x} = -g$, for some positive constant g. Show that no matter how large $x_0 = x(0) > 0$ and $v_0 = \dot{x}(0) > 0$ are, $x(t)$ has a finite maximum.

13. **(a)** Use Kepler's second law, equal areas swept out in equal times, to show that a circular orbit must have constant speed.

(b) Use Kepler's third law, $P^2 = 4\pi^2 a^3/(G(m_1 + m_2))$, together with the result of part (a), to show that a circular orbit of radius a has constant orbital speed $v = \sqrt{G(m_1 + m_2)/a}$. [*Hint:* Express v in terms of the period P.]

14. The **uniform orbital speed** of a satellite of mass m_1 at distance x_0 from an attracting body of mass m_2 is the speed v_1 that the satellite must attain to keep it in a uniform circular orbit.

(a) Show that the orbit $\mathbf{x} = x_0 (\cos (v/x_0)t, \sin (v/x_0)t)$ represents circular motion of radius x_0 with uniform speed v and acceleration $\ddot{\mathbf{x}}$ toward the origin of magnitude v^2/x_0. This acceleration vector is called **centripetal acceleration.** [*Hint:* Compute $|\mathbf{x}|$, $|\dot{\mathbf{x}}|$, and $|\ddot{\mathbf{x}}|$.]

(b) Show that if gravitational acceleration $G(m_1 + m_2)/x_0^2$ is to provide precisely the centripetal acceleration of the circular orbit found in part (a), then the uniform orbital speed will be $v_1 = \sqrt{G(m_1 + m_2)/x_0}$.

(c) How is uniform orbital speed related to escape speed?

(d) How many days would there be in a month if the earth's moon had a uniform circular orbit of radius equal to 384,404 kilometers, the mean distance of the moon from the earth? (The actual number is about 27.32 days.)

*15. The **synchronous orbit** of a body of mass m about a uniformly rotating body of mass $M > m$ is the one that maintains the orbiting body directly over one point on the rotating one. Assume the mass of each body to be concentrated at its center. (See Exercise 33.)

(a) Show that a synchronous orbit must lie in the plane of the rotating body's equator. Then use the first two Kepler laws and the polar area formula $\int \frac{1}{2} r^2 \, d\theta$ to show that a synchronous orbit must be circular.

(b) Use the third Kepler law to show that if P is the period of rotation of the larger body, then the radius of the synchronous orbit is $R = KP^{2/3}$, where $K = \sqrt{G(M + m)/4\pi^2}$.

(c) Show that the synchronous orbit about the earth for a small satellite has radius approximately 6.622846 times the radius of the earth, or 26,360 miles. (Continuing orbital correction of communication satellites is required because of uneven mass concentrations on earth and the influence of other bodies such as the sun and the moon.)

16. The Newton equations for orbits of a single planet of mass m_2 relative to a fixed sun of mass m_1 have the form

$$\ddot{x} = \frac{-kx}{(x^2 + y^2)^{3/2}}, \qquad \ddot{y} = \frac{-ky}{(x^2 + y^2)^{3/2}},$$

where $k = G(m_1 + m_2)$.

(a) Find the relationship that must hold between the positive constants a and ω so that these differential equations will have solutions with circular orbits described by $x(t) = a \cos \omega t$, $y(t) = a \sin \omega t$.

(b) Show that the relationship described in part (a) expresses the third Kepler law.

(c) Show that the orbit

$$\mathbf{x} = (a \cos \omega t, a \sin \omega t), \qquad \omega = \text{const} > 0$$

obeys the second Kepler law.

17. A vector system $\ddot{\mathbf{x}} = -\mathbf{F}(\mathbf{x})$ is called **conservative** if there is a real-valued **potential energy function** $U(\mathbf{x})$ such that $\mathbf{F}(\mathbf{x}) = \nabla U(\mathbf{x})$. (For a 1-dimensional vector field the relation is just $F(x) = U'(x)$; note that a potential function is determined only up to an additive constant.)

(a) Verify that the Newtonian vector field

$$-\mathbf{F}(x, y) = \left(\frac{-kx}{(x^2 + y^2)^{3/2}}, \frac{-ky}{(x^2 + y^2)^{3/2}} \right)$$

has $U(x, y) = -k(x^2 + y^2)^{-1/2}$ as potential; that is, $\nabla U = \mathbf{F}$.

(b) The **kinetic energy** of a body of mass 1 following a path $(x, y) = (x(t), y(t))$ is $T = \frac{1}{2}(\dot{x}^2 + \dot{y}^2)$, and the **total energy** of motion in the Newtonian field is

$$E = T + U = \tfrac{1}{2}(\dot{x}^2 + \dot{y}^2) - \frac{k}{\sqrt{x^2 + y^2}}$$

Verify that the total energy E is constant for the motion in the vector field of part (a). [*Hint:* Show that $dE/dt = 0$, and use the Newton equations of motion.]

(c) Verify that for motion governed by the equation $\ddot{\mathbf{x}} = -\mathbf{F}(\mathbf{x})$ the total energy E is constant if the vector field is conservative: $\mathbf{F}(\mathbf{x}) = \nabla U(\mathbf{x})$.

The next five exercises establish the validity of Kepler's laws.

18. We've seen that the orbit of one body relative to a second always lies in a fixed plane containing both bodies. This is often shown as follows.

(a) Show that if a body of mass m has a path of motion that obeys the inverse-square law $m\ddot{\mathbf{x}} = -(k/|\mathbf{x}|^3)\mathbf{x}$, then the motion is confined to a plane through the center of attraction determined by the initial position and the velocity vectors. [*Hint:* Establish the relation $(d/dt)/(\mathbf{x} \times \dot{\mathbf{x}}) = \mathbf{x} \times \ddot{\mathbf{x}}$ to show that the plane containing $\mathbf{x}$ and $\dot{\mathbf{x}}$ is perpendicular to a fixed vector.]

(b) A **central force** law is one such that motion is governed by an equation of the form $\ddot{\mathbf{x}} = G\mathbf{x}$, where G is a real-valued function of some unspecified variables, for example, $G = G(\mathbf{x})$. Show that motion subject to a central force law is confined to a plane.

19. The **angular momentum** of a planet at position $\mathbf{x}$ in its plane orbit about the sun is the vector $\mathbf{l} = \mathbf{x} \times m\dot{\mathbf{x}}$; that is, $\mathbf{l}$ is the cross product of the position vector $\mathbf{x}$ with the linear momentum vector $m\dot{\mathbf{x}}$.

(a) Introduce rectangular coordinates x, y in the plane of motion so that $\mathbf{x} = (x, y, 0)$ to show that the length $l = |\mathbf{l}|$ of angular momentum can be expressed as $l = m|x\dot{y} - y\dot{x}|$.

(b) Show that in terms of polar coordinates $x = r \cos \theta$, $y = r \sin \theta$ the angular momentum is $mr^2\dot{\theta}$, if $\dot{\theta} > 0$.

(c) Kepler's second law of planetary motion decrees that the radius joining a planet to the sun sweeps out equal areas in equal times. Use the Kepler law together with the formula

$$A = \frac{1}{2} \int_{\theta_1}^{\theta_2} r^2 \, d\theta$$

for area in polar coordinates to show that the angular momentum $mr^2\dot{\theta}$ is constant on an orbit. [*Hint:* Express area swept out along an orbit as an integral with respect to time t between t and $t + \tau$.]

20. Kepler's second law (radius vector from sun to planet sweeps out equal areas in equal times) holds for any **central force law,** that is, a force law expressible in the form $\ddot{\mathbf{x}} = G\mathbf{x}$, where G is some real-valued function. This includes as special cases the inverse-square law of attraction, where $G(\mathbf{x}) = -k|\mathbf{x}|^{-2}$, $k > 0$, and the **Coulomb repulsion law,** where $G(\mathbf{x}) = k|\mathbf{x}|^{-2}$, $k > 0$; the latter governs interaction of particles bearing electric charges of the same sign.

(a) Assuming planar motion and using rectangular coordinates (x, y) for $\mathbf{x}$, show that a central force law can be written

$$\ddot{x} = Gx, \qquad \ddot{y} = Gy,$$

and conclude that $x\ddot{y} - y\ddot{x} = 0$ for a motion governed by a central force law.

(b) Use integration by parts with respect to t in the conclusion of part (a) to show that $x\dot{y} - y\dot{x} = h$ for some constant h.

(c) Change to polar coordinates by $x = r \cos \theta$, $y = r \sin \theta$ to show that $x\dot{y} - y\dot{x} = r^2\dot{\theta}$ and hence, using part (b), show that $r^2\dot{\theta} = h$ for some constant h. (This result can be interpreted as saying that for motion in a central force field, the angular velocity $\dot{\theta}$ is inversely proportional to the square of the distance from the center of the field.)

(d) Use the result of part (c) and a computation of area in polar coordinates to prove Kepler's second law for a central force field by showing that, as a function of time t, area swept out has the form $A = \frac{1}{2}ht + c$. Explain why this proves Kepler's second law under the given assumptions.

*(e) Apply Green's theorem to the equation $x\dot{y} - y\dot{x} = h$ derived in part (b) to show directly, without using polar coordinates, that Kepler's second law holds.

21. If $a \geq b$, an ellipse $x^2/a^2 + y^2/b^2 = 1$ has its **focus** points at $(\pm c, 0)$, where $c^2 = a^2 - b^2$. Parametrize the ellipse by $(x, y) = (a \cos u, b \sin u)$, $0 \leq u \leq 2\pi$, to show that the semimajor axis a is also the average, or mean, distance from points on the ellipse to a focus.

*22. A single planet with position $\mathbf{x} = \mathbf{x}(t)$ obeying $\ddot{\mathbf{x}} = -(k/|\mathbf{x}|^3)\mathbf{x}$ follows an elliptic, parabolic, or hyperbolic path. Here is an outline of a way to show this by deriving a linear differential equation from the vector equation.

(a) Use $\mathbf{x} = (r \cos \theta, r \sin \theta)$ to express the vector equation of motion in the two polar coordinate equations $\ddot{r} - r\dot{\theta}^2 = -k/r^2$, $r\ddot{\theta} + 2\dot{r}\dot{\theta} = 0$.

(b) Show that the second equation derived in part (a) implies $r^2\dot\theta = h$ for some constant h, and use this to write the other equation in the form $\ddot r = h^2 r^{-3} - kr^{-2}$. In particular, show that if $h = 0$, the motion is confined to a line and results in either collision or escape.

(c) Use the results of part (b) to show that if $h \neq 0$,

$$\frac{1}{r^2}\frac{d^2r}{d\theta^2} - 2\frac{1}{r^3}\left(\frac{dr}{d\theta}\right)^2 = \frac{1}{r} - \frac{k}{h^2}.$$

[*Hint:* Use the chain rule to express $\dot r$ and $\ddot r$ in terms of θ derivatives.]

(d) Let $r = 1/u$ so that

$$\frac{dr}{d\theta} = -u^{-2}\frac{du}{d\theta} \quad \text{and} \quad \frac{d^2r}{d\theta^2} = -u^{-2}\frac{d^2}{d\theta^2} + 2u^{-3}\left(\frac{du}{d\theta}\right)^2$$

to show that the equation in part (c) can be written as the second-order linear equation $d^2u/d\theta^2 + u = k/h^2$, sometimes called **Binet's equation.**

(e) Show that the solution $u = 1/r = A\cos(\theta + \alpha) + k/h^2$ to the previous equation represents an ellipse, parabola, or hyperbola in polar coordinates according as $|A| < k/h^2$, $|A| = k/h^2$, or $|A| > k/h^2$. [*Hint:* Change to xy coordinates.]

(f) Each **focus** of an ellipse lies on the major axis at distance c from the center, where $c^2 = a^2 - b^2$ and a and b are the semiaxes. The **eccentricity** is $e - c/a$. Show that for an elliptic orbit the center of attraction is at one focus and the eccentricity is $|A|h^2/k$. Then show that the polar equation for an orbit can be written $kr(1 + e\cos\theta) = h^2$. [*Hint:* For the first part, convert the polar equation, with $\alpha = 0$, to rectangular coordinates.]

(g) Assume that the orbit in part (f) is elliptic, with $0 \leq e < 1$. Show that the time for one complete revolution is $P = 2\pi ab/h$. Then show that $h^2/k = b^2/a$ to derive the third Kepler law $P^2 = 4\pi^2 a^3/k$. [*Hint:* The sum of the maximum and minimum values for r is equal to $2a$.]

23. Suppose a projectile is fired directly away from and at distance x_0 from the center of mass of a planet with initial speed z_0. If z_0 is less than the escape speed, show that the maximum additional distance attained from the center of mass of the planet is

$$\frac{x_0^2 z_0^2}{2GM - x_0 z_0^2},$$

where M is the sum of the masses of the two bodies.

24. A simplified model of a **black hole** is a body such that the escape speed from its surface exceeds c, the speed of light.

(a) How small should the radius r of a solid homogeneous ball of mass m be if it is to become a black hole?

(b) Estimate the radius r of part (a) for the earth and for the sun. Assume that the earth and sun are homogeneous solid balls, that the earth has mass $6 \cdot 10^{24}$ kilograms, and that the sun is 334,000 times as massive as the earth; in units of meters, kilograms, and seconds, $G \approx 6.67 \cdot 10^{-11}$. Also $c \approx 300,000$ kilometers per second.

25. Equations 5.1 with $\mathbf{x}_1 = (x_1, y_1, z_1)$ and $\mathbf{x}_2 = (x_2, y_2, z_2)$ are equivalent to six second-order scalar equations. Write out these equations in terms of coordinates x_k, y_k, z_k and $r = \sqrt{(x_1 - x_2)^2 + (y_1 - y_2)^2 + (z_1 - z_2)^2}$.

26. The distance $x(t)$ between two bodies moving on a fixed line obeys a repelling Coulomb law $\ddot{x} = k/x^2, k > 0$. Show that $\dot{x}(t)$ always increases if $\dot{x}(0) > 0$.

27. Suppose two stars of identical mass m at $\mathbf{x}_1$ and $\mathbf{x}_2$ form a binary system with a fixed center of mass at $\frac{1}{2}(\mathbf{x}_1 + \mathbf{x}_2)$. Show that they can move in the same circular orbit of radius r and that their common orbital speed will then be $v = \frac{1}{2}\sqrt{Gm/r}$. [*Hint:* Use Equations 5.1.]

28. Explain the derivation of Equations 5.4 for three bodies. Be sure to include an explanation of why the differences of vectors $\mathbf{x}_i$ and $\mathbf{x}_j$ appear in the order they do.

29. Explain the derivation of Equations 5.5 for n bodies. Be sure to include an explanation of why the differences of vectors $\mathbf{x}_i$ and $\mathbf{x}_j$ appear in the order they do.

***30.** The principal result in Example 1 of the text was established under the assumption that the relative distance separating two bodies subject only to the forces of mutual gravitational attraction could always be measured along the same fixed line. The purpose of this problem is to show that this last assumption isn't needed, and that in fact the relative speed v and distance r are always related by

$$\frac{v^2}{2} - \frac{v_0^2}{2} = \frac{k}{r} - \frac{k}{r_0}.$$

Here the constant is $k = Gm$, where m is the sum of the two masses, G is the gravitational constant, and v_0 and r_0 are initial speed and distance.

Starting with the Newton vector equation $\ddot{\mathbf{x}} = -k\mathbf{x}/|\mathbf{x}|^3$, we form the dot product of both sides by $\dot{\mathbf{x}}$ to get the scalar equation

$$\dot{\mathbf{x}} \cdot \ddot{\mathbf{x}} = -k\frac{\mathbf{x} \cdot \dot{\mathbf{x}}}{|\mathbf{x}|^3}.$$

(a) Show that the left side of the previous equation is equal to

$$\frac{d}{dt}\frac{v^2}{2},$$

where $v = \sqrt{\dot{\mathbf{x}} \cdot \dot{\mathbf{x}}} = |\dot{\mathbf{x}}|$.

(b) Show that the right side of that same equation is equal to

$$k\left(\nabla\frac{1}{|\mathbf{x}|}\right) \cdot \dot{\mathbf{x}},$$

where ∇ is the gradient operator: $\nabla f = (\partial f/\partial x, \partial f/\partial y, \partial f/\partial z)$.

(c) Show that the result of part (b) can be written $k(d/dt)|\mathbf{x}|^{-1}$, and conclude that

$$\frac{d}{dt}\frac{v^2}{2} = k\frac{d}{dt}\frac{1}{r}.$$

(d) Integrate the previous equation between 0 and an arbitrary positive time t to get the equation relating v, v_0, r, and r_0.

31. Write out the notational details of the derivation of the vector rocket equation, which follows the scalar equation derivation given at the end of Section 4 of Chapter 2. This is mainly a matter of replacing scalar notation by vector notation, but you should verify that these replacements are meaningful and correct.

32. (The results of this exercise are used in numerical solution of the planetary motion equations.) The **center of mass c** of a system of n bodies with masses m_k concentrated respectively at $\mathbf{x}_k$ is

$$\mathbf{c} = \frac{1}{m} \sum_{k=1}^{n} m_k \mathbf{x}_k, \qquad \text{where } m = \sum_{k=1}^{n} m_k.$$

(a) Show that if **conservation of momentum** holds, that is, $\dot{\mathbf{c}}$ is constant, then the center of mass stays on a line. Show also that by subtracting **c** from all vectors $\mathbf{x}_k$, the center of mass can be placed at the origin.

(b) Show that if conservation of momentum holds, then Equation 5.6 holds. Show that if also the center of mass is to stay at the origin, then the initial velocity vectors $\dot{\mathbf{x}}_k(0)$ can't be chosen arbitrarily but must satisfy

$$\sum_{k=1}^{n} m_k \dot{\mathbf{x}}_k(0) - 0.$$

*33. **When can you assume that the mass of a ball is concentrated at its center?** A solid ball B_R of radius R is **spherically homogeneous** if its density is constant on every spherical shell with center at the center of B_R. It's important for the application of Newton's inverse-square law to planetary motions that each of the most significant celestial bodies is fairly near to being a spherically homogeneous ball. (Among bodies that diverge widely from this norm, erratic motion has been observed that is quite uncharacteristic of the orbital motion of a typical planet.) The purpose of this exercise is to establish Newton's result that the gravitational attraction of a spherically homogeneous ball acting at a point **a** is the same as it would be if all the mass of the ball were concentrated at its center, unless **a** is inside the ball, in which case the part of the ball at distance from the center greater than $|\mathbf{a}|$ can be neglected. More specifically, if B_R is centered at the origin and has density $\rho(|\mathbf{x}|)$ at **x**, **Newton's formula** for the attracting force vector acting on a particle of mass 1 at **a** is given by G times the 3-dimensional vector integral

$$\int_{B_R} \rho(|\mathbf{x}|) \frac{\mathbf{x} - \mathbf{a}}{|\mathbf{x} - \mathbf{a}|^3} \, dV_{\mathbf{x}} = -\frac{M_{\mathbf{a}}}{|\mathbf{a}|_3} \mathbf{a},$$

where $M_{\mathbf{a}}$ is the mass of the part of B_R that lies within distance $|\mathbf{a}|$ of its center.

(a) Choose perpendicular xyz axes with origin at the center of B_R and positive z axis passing through $\mathbf{a} = (0, 0, a)$. Show, without computing any indefinite integrals, that the x and y coordinates of the vector integral are zero and that the z coordinate is given in spherical coordinates by

$$2\pi \int_0^R r^2\rho(r) \left[\int_0^\pi \frac{r \cos \phi - a}{(r^2 - 2ar \cos \phi + a^2)^{3/2}} \sin \phi \, d\phi \right] dr.$$

(b) Let $\cos \phi = u$ and integrate by parts to show that the inner integral in part (a) is

$$\int_{-1}^1 (ru - a)(r^2 + a^2 - 2aru)^{-3/2} \, du = \begin{cases} -2/a^2, & a > r, \\ 0, & a < r. \end{cases}$$

(c) Show that the mass of B_S is

$$4\pi \int_0^S r^2\rho(r) \, dr,$$

and then use the previous results to prove Newton's formula for the attracting force.

(d) Use the results of parts (a) and (b) to show that matter distributed in a spherically homogeneous way between two concentric spheres exerts no gravitational attraction at points inside the inner sphere.

(e) Specialize the results of parts (a) and (b) to the case of a homogeneous ball with *constant* density ρ to show that the gravitational attraction of the ball on a unit point mass inside the ball and a units from the center has magnitude proportional to a with constant of proportionality $\frac{4}{3}\pi\rho G$.

6   ENERGY

Kinetic and potential energy are discussed in Chapter 4, Section 6, for 1-dimensional motions, sometimes referred to as systems with *one degree of freedom*. Speaking very generally, the ideas introduced there are contained in the present discussion, but while it is helpful to know about the earlier material, we won't assume here any knowledge of the special case, nor will we dwell on the details of the 1-dimensional examples.

We'll be dealing with vector equations, many of which can be written in the form

$$\ddot{\mathbf{x}} = -\mathbf{F}(\mathbf{x}),$$

where $F(\mathbf{x})$ is a continuous vector-valued function of the vector variable $\mathbf{x}$ in some region D of $\mathcal{R}^2$, $\mathcal{R}^3$ or some higher-dimensional space in which we refer to *two*, *three*, or *more* **degrees of freedom**. More specifically, we'll start out by assuming that $F(\mathbf{x})$ is a **gradient** vector field defined by

$$\nabla U(\mathbf{x}) = \left(\frac{\partial U}{\partial x_1}(\mathbf{x}), \ldots, \frac{\partial U}{\partial x_n}(\mathbf{x}) \right)$$

for some real-valued differentiable function $U(\mathbf{x})$ called a **potential function** of the vector field $F(\mathbf{x})$. Positive constant factors m, denoting mass, will sometimes be explicitly applied to $\ddot{\mathbf{x}}$, even though they can always be absorbed into the right side; the physical interpretation of m as mass sometimes makes it natural to leave these factors on the left, giving both sides of the equation the dimensions of mass times acceleration or force, that is, $\mathbf{F} = m\ddot{\mathbf{x}}$.

The autonomous equation $\ddot{\mathbf{x}} = -\nabla U(\mathbf{x})$ can be solved explicitly only under special assumptions about $U(\mathbf{x})$.

---

example 1

If $\mathbf{x} = (x, y)$ and $U(x, y) = \frac{1}{2}ax^2 + bxy + \frac{1}{2}dy^2$ with a, b, and d constant, then

$$\nabla U(x, y) = \left(\frac{\partial U(x, y)}{\partial x}, \frac{\partial U(x, y)}{\partial y} \right)$$
$$= (ax + by, bx + dy).$$

The system $\ddot{\mathbf{x}} = -\nabla U(\mathbf{x})$ is then written in coordinate form as

$$\ddot{x} = -ax - by,$$
$$\ddot{y} = -bx - dy;$$

this can be solved in terms of elementary functions using the elimination method.

---

example 2

Let $U(x, y) = -k(x^2 + y^2)^{-1/2}$ for $(x, y) \neq (0, 0)$, where k is some positive constant. The gradient of $U(x, y)$ is

$$\nabla U(x, y) = \left(\frac{\partial U}{\partial x}(x, y), \frac{\partial U}{\partial y}(x, y) \right)$$
$$= \left(\frac{kx}{(x^2 + y^2)^{3/2}}, \frac{ky}{(x^2 + y^2)^{3/2}} \right).$$

The resulting field turns out to be the one associated with the Newton equations for the motion of a single planet about a fixed sun:

$$\ddot{x} = \frac{-kx}{(x^2 + y^2)^{3/2}}, \qquad \ddot{y} = \frac{-ky}{(x^2 + y^2)^{3/2}}.$$

The constant k is $G(m_1 + m_2)$, where G is a universal gravitational constant and m_1 and m_2 are the masses of the two bodies. Using the notation $\mathbf{x} = (x, y)$, the system can be written as a single equation, $\ddot{\mathbf{x}} = -k\mathbf{x}|\mathbf{x}|^{-3}$. These equations are derived in Section 5.

---

example 3

The system

$$m_1 \frac{d^2 x}{dt^2} = -(h_1 + h_2)x + h_2 y,$$

$$m_2 \frac{d^2 y}{dt^2} = h_2 x - (h_2 + h_3)y$$

that governs the displacements over time of a linear mass-spring mechanism is derived in Section 4. The associated vector field is given by

$$F(x, y) = ((h_1 + h_2)x - h_2 y, -h_2 x + (h_2 + h_3)y)$$

If we let

$$U(x, y) = \frac{1}{2}(h_1 + h_2)x^2 - h_2 xy + \frac{1}{2}(h_2 + h_3)y^2,$$

then it's easy to check that $\nabla U(x, y) = (U_x(x, y), U_y(x, y)) = -F(x, y)$, where the subscripts denote partial differentiation using the respective variables. The original second-order system can then be written as a vector equation:

$$(m_1\ddot{x}, m_2\ddot{y}) = -(U_x(x, y), U_y(x, y)).$$

Even if a system $\ddot{x} = -\nabla U(\mathbf{x})$ is amenable to explicit solution, that route may turn out to be technically very complicated and may not readily produce the information we want. A more general approach is to find an equivalent first-order system, having twice the dimension of the original system, and from which it may be possible to draw conclusions about the behavior of solutions. The usual choice for such a first-order system is

$$\dot{\mathbf{x}} = \mathbf{z}, \qquad \dot{\mathbf{z}} = -\nabla U(\mathbf{x}).$$

If the vector $\mathbf{x}$ has dimension n, the $2n$-dimensional vectors $(\mathbf{x}, \mathbf{z}) = (x_1, \ldots, x_n, z_1, \ldots, z_n)$ constitute the **phase space** of the system. The **trajectory space** for $\mathbf{x}$ itself, called the **configuration space,** has dimension n; it is traditional to say that the system with an n-dimensional configuration space has n **degrees of freedom.**

---

e x a m p l e **4**

The second-order system of Example 3 can be written using velocity coordinates $z = \dot{x}$, $w = \dot{y}$ as

$$\dot{x} = z,$$
$$\dot{y} = w,$$
$$m_1\dot{z} = -(h_1 + h_2)x + h_2 y,$$
$$m_2\dot{w} = h_2 x - (h_2 + h_3)y.$$

The state space vectors are (x, y, z, w), and the configuration vectors are (x, y).

To fully justify the introduction of phase space at this point we define the **kinetic energy**

$$T(\dot{\mathbf{x}}) = \frac{1}{2} \sum_{k=1}^{n} m_k \dot{x}_k^2 = \frac{1}{2} \sum_{k=1}^{n} m_k z_k^2$$

associated with motion trajectory $\mathbf{x}(t) = (x_1(t), \ldots, x_n(t))$, where the mass m_k is attached to coordinate x_k. In addition, we define a **potential energy function** for such a motion to be a real-valued continuously differentiable function $U(\mathbf{x})$ such that on a path $\mathbf{x} = \mathbf{x}(t)$ of motion

$$m_k\ddot{x}_k = -\frac{\partial U}{\partial x_k}(x_1, \ldots, x_k), \qquad k = 1, \ldots, n.$$

The **total energy** $E(\mathbf{x}, \dot{\mathbf{x}}) = T(\dot{\mathbf{x}}) + U(\mathbf{x})$ turns out to be constant for solutions of such systems (Theorem 6.1 below), and for this reason second-order systems of this particular form are called **conservative.** A special case is a single point mass m, with $m\ddot{\mathbf{x}} = -\nabla U(\mathbf{x})$.

■ 6.1 ENERGY-CONSERVATION THEOREM

Let U be real-valued and continuously differentiable on a region in $\mathcal{R}^n$, and suppose that $(x_1(t), x_2(t), \ldots, x_n(t))$ is a solution of the system

$$m_1 \ddot{x}_1 = -\frac{\partial U}{\partial x_1}, \, m_2 \ddot{x}_2 = -\frac{\partial U}{\partial x_2}, \ldots, m_n \ddot{x}_n = -\frac{\partial U}{\partial x_n},$$

where $m_1, \ldots, m_n$ are positive constants. Then the total energy

$$T + U = \frac{1}{2} \sum_{k=1}^{n} m_k \dot{x}_k^2 + U(x, \ldots, x_n)$$

is constant along the phase curve of the solution.

Proof. Multiply the kth equation of the system by $\dot{x}_k$ and add the resulting equations over k from 1 to n:

$$\sum_{k=1}^{n} m_k \dot{x}_k \ddot{x}_k = - \sum_{k=1}^{n} \frac{\partial U}{\partial x_k} \dot{x}_k.$$

By the chain rule the right side is $-dU/dt$. The integral with respect to t of the kth term on the left is $\frac{1}{2} m_k \dot{x}_k^2$, so the previous equation can be rewritten

$$\sum_{k=1}^{n} m_k \dot{x}_k \ddot{x}_k = -\frac{dU}{dt}, \quad \text{or} \quad \frac{d}{dt} \left[\frac{1}{2} \sum_{k=1}^{n} m_k \dot{x}_k^2 + U(x_1, \ldots, x_n) \right] = 0.$$

It follows that the expression in brackets, namely, the kinetic energy plus the potential energy, is constant on the phase path of the given solution.   ■

---

example 5

In the case $n = 1$ of a 1-dimensional trajectory space, the equation $T + U = C$ very nearly determines the motion. If initial values $x(0) = x_0$ and $\dot{x}(0) = z_0$ are specified, the relation $\frac{1}{2} \dot{x}^2 + U(x) = C$ allows us to solve for $\dot{x}$ and then, in principle at least, find $x(t)$ by integration with respect to t. Several examples are developed in detail in Sections 5 and 6 of Chapter 4.

---

example 6

Let $m_1 = m_2 = 1$ and $h_1 = h_2 = h_3 = 1$ in the mass-spring system of Example 3. Since we want $U_x(x, y) = 2x - y$ and $U_y(x, y) = -x + 2y$, we can use for a potential the function

$$U(x, y) = x^2 - xy + y^2 = \tfrac{1}{2}(x^2 + y^2) + \tfrac{1}{2}(x - y)^2.$$

This identity shows that $U(x, y) \geq 0$. Since $T(\dot{x}, \dot{y}) + U(x, y)$ must be constant for a solution of the system, we know that a solution with initial values $x(0) = x_0, y(0) = y_0, \dot{x}(0) = z_0, \dot{y}(0) = w_0$ must satisfy

$$\tfrac{1}{2}(\dot{x}^2 + \dot{y}^2) + x^2 - xy + y^2 = \tfrac{1}{2}(z_0^2 + w_0^2) + x_0^2 - x_0 y_0 + y_0^2.$$

In particular, suppose for example that $x_0 = 1, y_0 = -1$, and $z_0 = w_0 = 0$. Then

$$\tfrac{1}{2}(\dot{x}^2 + \dot{y}^2) + x^2 - xy + y^2 = 3.$$

As we observed above $x^2 - xy + y^2 \geq 0$, so subtracting it from the left side gives

$$T(\dot{x}, \dot{y}) = \tfrac{1}{2}(\dot{x}^2 + \dot{y}^2) \leq 3.$$

This last inequality places rough restrictions on the velocity coordinates: $|\dot{x}| \leq \sqrt{6}$, $|\dot{y}| \leq \sqrt{6}$. Since in this example the system can be solved explicitly, consequences of this kind can be arrived at directly, though not quite so easily. Indeed, it's easy to check that for initial conditions $x(0) = 1$, $y(0) = -1$, $\dot{x}(0) = \dot{y}(0) = 0$ the solution to the system is

$$x(t) = \cos \sqrt{3}\, t, \qquad y(t) = -\cos \sqrt{3}\, t.$$

(See also Example 1 in Section 4.) From this we find directly that $\dot{x} = -\sqrt{3} \sin \sqrt{3}\, t$ and $\dot{y} = \sqrt{3} \sin \sqrt{3}\, t$. It follows that $T(\dot{x}, \dot{y}) \leq 3$. This last conclusion followed from the energy equation derived above, but without having to solve the system.

---

example 7

The **Newton equations** for the orbit of a single planet attracted by a fixed sun,

$$\ddot{x} = \frac{-kx}{(x^2 + y^2)^{3/2}}, \qquad \ddot{y} = \frac{-ky}{(x^2 + y^2)^{3/2}},$$

have for a potential $U(x, y) = -k(x^2 + y^2)^{-1/2}$, as shown in Example 2. Setting $m_1 = m_2 = 1$ in Theorem 6.1, we conclude that

$$T(\dot{x}, \dot{y}) + U(x, y) = \tfrac{1}{2}(\dot{x}^2 + \dot{y}^2) - k(x^2 + y^2)^{-1/2} = E, \qquad \text{const.}$$

This can be written in vector form as $\tfrac{1}{2}|\dot{\mathbf{x}}|^2 - k/|\mathbf{x}| = E$, where $|\dot{\mathbf{x}}|$ is the orbital speed of the planet and $|\mathbf{x}|$ is the distance of the planet from the sun. Suppose that at some time the orbital speed and radial distance of the planet are, respectively, $|\dot{\mathbf{x}}_0|$ and $|\mathbf{x}_0|$. The total energy $E = T + U$ is a constant, so we have $T(\dot{\mathbf{x}}) + U(\mathbf{x}) = T(\dot{\mathbf{x}}_0) + U(\mathbf{x}_0)$; in other words,

$$\tfrac{1}{2}|\dot{\mathbf{x}}|^2 - \frac{k}{|\mathbf{x}|} = \tfrac{1}{2}|\dot{\mathbf{x}}_0|^2 - \frac{k}{|\mathbf{x}_0|}, \qquad \text{so} \qquad 0 \leq \tfrac{1}{2}|\dot{\mathbf{x}}|^2 = \tfrac{1}{2}|\dot{\mathbf{x}}_0|^2 - \frac{k}{|\mathbf{x}_0|} + \frac{k}{|\mathbf{x}|}.$$

If the trajectory is unbounded, in which case it is hyperbolic or parabolic, the term $k/|\mathbf{x}|$ on the right tends to zero. It follows that the constant on the right side is nonnegative; that is, $E_0 = \tfrac{1}{2}|\dot{\mathbf{x}}_0|^2 - k/|\mathbf{x}_0| \geq 0$. In other words, $|\dot{\mathbf{x}}_0| \geq \sqrt{2k/|\mathbf{x}_0|}$. It can be shown conversely that if $E_0 \geq 0$, then the corresponding orbits are necessarily unbounded, that is, parabolic in case $E_0 = 0$ or hyperbolic in case $E_0 > 0$. For this reason the critical speed $|\dot{\mathbf{x}}_0| = \sqrt{2k/|\mathbf{x}_0|}$ is called the **escape speed** at distance $|\mathbf{x}_0|$.

---

example 8

The differential equations of motion for a two-body system displaying the absolute rather than relative motions of the bodies are

$$m_1\ddot{\mathbf{x}}_1 = -\frac{Gm_1 m_2}{|\mathbf{x}_1 - \mathbf{x}_2|^3}(\mathbf{x}_1 - \mathbf{x}_2), \qquad m_2\ddot{\mathbf{x}}_2 = -\frac{Gm_1 m_2}{|\mathbf{x}_1 - \mathbf{x}_2|^3}(\mathbf{x}_2 - \mathbf{x}_1).$$

It is left as an exercise to show that $U(\mathbf{x}_1, \mathbf{x}_2) = -Gm_1 m_2/|\mathbf{x}_1 - \mathbf{x}_2|$ is a potential function for the system. The total energy is thus

$$E = \tfrac{1}{2}(m_1|\dot{\mathbf{x}}_1|^2 + m_2|\dot{\mathbf{x}}_2|^2) - \frac{Gm_1 m_2}{|\mathbf{x}_1 - \mathbf{x}_2|}.$$

Since E is constant on a given phase path, we could theoretically conclude that $|\dot{\mathbf{x}}_1|$ and $|\dot{\mathbf{x}}_2|$ can't both remain bounded if the distance $|\mathbf{x}_1 - \mathbf{x}_2|$ between x_1 and

x_2 becomes arbitrarily small. Given the prohibition of relativity theory against unbounded velocities, it is apparent that the inverse-square law must be abandoned under certain circumstances.

Finding a potential $U(\mathbf{x})$ for a given gradient field $\nabla U(\mathbf{x})$ is essentially an integration problem that can sometimes be solved by making an educated guess. As a practical matter more elaborate integration techniques may be used, as illustrated in the following example.

example 9

Suppose for the partial derivatives of $U(x, y, z)$ with respect to x, y, and z we have $U_x(x, y, z) = yz$, $U_y(x, y, z) = xz$, $U_z(x, y, z) = xy$. A shrewd guess might be $U(x, y, z) = xyz$, which is easily checked as valid. If guessing fails, we could proceed as follows.

$$U(x, y, z) = \int U_x(x, y, z)\, dx = \int yz\, dx = xyz + C(y, z),$$

where the "constant" of integration may depend on y and z. Differentiating the end result of this equation with respect to y and using the given expression for U_y gives

$$U_y(x, y, z) = xz + C_y(y, z) = xz.$$

We conclude that $C_y(y, z) = 0$, so $C(y, z) = D(z)$ is a function of z alone. Thus $U(x, y, z) = xyz + D(z)$. Now differentiate with respect to z to get

$$U_z(x, y, z) = xy + D'(z) = xy.$$

Thus $D'(z) = 0$, and D must be a constant, $U(x, y, z) = xyz + D$ being the most general potential of the given 3-dimensional vector field.

We conclude the section with a brief look at some special **dissipative systems,** that is, systems for which there is an energy $E(\mathbf{x}, \dot{\mathbf{x}}) = T(\dot{\mathbf{x}}) + U(\mathbf{x})$ that is nonincreasing, but not constant, as t increases.

example 10

The vector equation

$$m\ddot{\mathbf{x}} = -\nabla U(\mathbf{x}) - k\dot{\mathbf{x}}, \qquad k > 0,$$

contains a negative linear term $-k\dot{\mathbf{x}}$ of the sort that is often introduced in mechanical systems to represent a frictional damping force. To get information about solutions, we form the dot product of both sides with $\dot{\mathbf{x}}$ and then integrate from t_0 to t:

$$m \int_{t_0}^{t} \dot{\mathbf{x}} \cdot \ddot{\mathbf{x}}\, dt = -\int_{t_0}^{t} \nabla U(\mathbf{x}) \cdot \dot{\mathbf{x}}\, dt - k \int_{t_0}^{t} |\dot{\mathbf{x}}|^2\, dt.$$

By the chain rule, $d(\tfrac{1}{2}\dot{\mathbf{x}} \cdot \dot{\mathbf{x}})/dt = \dot{\mathbf{x}} \cdot \ddot{\mathbf{x}}$ and $dU(\mathbf{x})/dt = \nabla U(\mathbf{x}) \cdot \dot{\mathbf{x}}$. Hence we can perform the first two integrations and rearrange terms to get

$$\tfrac{1}{2}m|\dot{\mathbf{x}}(t)|^2 + U(\mathbf{x}(t)) = -k \int_{t_0}^{t} |\dot{\mathbf{x}}(t)|^2\, dt + \tfrac{1}{2}m|\dot{\mathbf{x}}(t_0)|^2 + U(\mathbf{x}(t_0)).$$

This can be written in terms of energy as

$$E(\mathbf{x}(t), \dot{\mathbf{x}}(t)) = -k \int_{t_0}^{t} |\dot{\mathbf{x}}(t)|^2 \, dt + E(\mathbf{x}(t_0), \dot{\mathbf{x}}(t_0)).$$

Note that the first term on the right is negative and decreasing at points for which $\dot{\mathbf{x}}(t) \neq 0$. Far from being constant, the energy will thus be decreasing in general unless $\dot{\mathbf{x}} = 0$, as would be the case for an equilibrium solution.

A constant equilibrium solution $\mathbf{x}_0$ is **stable** if, for some t_0, a phase path $(\mathbf{x}(t), \dot{\mathbf{x}}(t))$ can be made to stay arbitrarily close to the point $(\mathbf{x}_0, 0)$ for all $t \geq t_0$ simply by making the initial point $(\mathbf{x}(t_0), \dot{\mathbf{x}}(t_0))$ close enough to $(\mathbf{x}_0, 0)$ in phase space. The following theorem can be viewed as a special case of Liapunov's theorem in Section 8, but we can give a simpler proof of the restricted result state here.

■ 6.2 STABILITY THEOREM

Suppose that $G(\mathbf{x}, \dot{\mathbf{x}})$ is continuous near $(\mathbf{x}_0, \dot{\mathbf{x}})$ and that $\ddot{\mathbf{x}} = G(\mathbf{x}, \dot{\mathbf{x}})$ has $\mathbf{x}(t) = \mathbf{x}_0$ for an equilibrium solution. If there is a continuous function $D(\mathbf{x}, \dot{\mathbf{x}})$ that is nonincreasing on all phase curves in some neighborhood of $(\mathbf{x}_0, 0)$ and has a strict minimum there, then $(\mathbf{x}_0, 0)$ is a stable equilibrium point for the second-order vector equation.

Proof. Assume for simplicity that $\mathbf{x}_0 = 0$, and choose δ so that, while $|\mathbf{x}|^2 + |\dot{\mathbf{x}}|^2 \leq \delta^2$, $D(\mathbf{x}, \dot{\mathbf{x}})$ has its sole minimum at $(\mathbf{x}, \dot{\mathbf{x}}) = (0, 0)$. Let d_0 be the minimum of $D(\mathbf{x}, \dot{\mathbf{x}})$ on the sphere $|\mathbf{x}|^2 + |\dot{\mathbf{x}}|^2 = \delta^2$. Now suppose a phase curve $(\mathbf{x}(t), \dot{\mathbf{x}}(t))$ enters the sphere and satisfies $D(\mathbf{x}, \dot{\mathbf{x}}) < d_0$ for some value $t = t_0$. For this curve to leave the region where $D(\mathbf{x}, \dot{\mathbf{x}}) < d_0$, the value of $D(\mathbf{x}, \dot{\mathbf{x}})$ would have to become at least as large as d_0. But this is impossible, since $D(\mathbf{x}, \dot{\mathbf{x}}) < d_0$ at one point of the path and is nonincreasing thereafter. Hence the curve stays inside the sphere of radius δ for $t \geq t_0$. ■

If a system $\ddot{\mathbf{x}} = -F(\mathbf{x})$ is conservative, with energy $E(\mathbf{x}, \dot{\mathbf{x}})$, then the energy is constant on each phase path and so meets the requirement of the stability theorem for a function D that is nonincreasing on phase curves. If the minimum energy requirement is also met at $(\mathbf{x}_0, 0)$, Theorem 6.2 says that equilibrium is stable. In the next example, kinetic energy plus a formal "potential energy" is still used to construct the function D on phase space, even though the system is not conservative and D is not constant.

---

e x a m p l e 11   A special case of Example 10 is the system

$$\ddot{x} = -2x + y - k\dot{x},$$
$$\ddot{y} = x - 2y - k\dot{y}.$$

If $k = 0$, this system has an energy function

$$E(x, y, \dot{x}, \dot{y}) = T(\dot{x}, \dot{y}) + U(x, y)$$
$$= \tfrac{1}{2}(\dot{x}^2 + \dot{y}^2) + x^2 - xy + y^2.$$

The equilibrium point $(x, y) = (0, 0)$ gives a strict minimum for E, since $E = 0$ there, and the identity

$$x^2 - xy + y^2 = \tfrac{1}{2}(x^2 + y^2) + \tfrac{1}{2}(x - y)^2$$

shows that E is positive except at equilibrium. From Example 10 with $\mathbf{x}(t_0) = 0$, we see that

$$E(x, y, \dot{x}, \dot{y}) = -\int_0^t (\dot{x}^2 + \dot{y}^2)\, dt,$$

so the system is dissipative. Taking $D = E$, we can conclude from the stability theorem that the equilibrium solution $(x(t), y(t)) = (0, 0)$ is stable.

EXERCISES

1. Find a potential U and the total energy $T + U$ for the following systems under the given conditions.

   (a) $\ddot{x} = -x$, $\ddot{y} = -y$; $U(0, 0) = 0$, $x(0) = y(0) = \dot{x}(0) = \dot{y}(0) = 1$.

   (b) $\ddot{x} = -y$, $\ddot{y} = -x$; $U(0, 0) = 0$, $x(0) = y(0) = \dot{x}(0) = \dot{y}(0) = 1$.

   (c) $\ddot{x} = -x - y$, $\ddot{y} = -x + y$; $U(0, 0) = 0$, $x(0) = y(0) = \dot{x}(0) = \dot{y}(0) = 1$.

   (d) $\ddot{x} = -x$, $\ddot{y} = -y$, $\ddot{z} = -z$; $U(0, 0, 0) = 0$, $x(0) = y(0) = z(0) = \dot{x}(0) = \dot{y}(0) = \dot{z}(0) = 1$.

2. Some vector fields are not gradient fields. In particular, if the gradient vector field

   $$\nabla U(\mathbf{x}) = (U_{x_1}(x_1, x_2, \ldots), U_{x_2}(x_1, x_2, \ldots), \ldots, U_{x_n}(x_1, x_2, \ldots))$$

   itself has continuous partial derivatives, then $U_{x_1 x_2}$ must equal $U_{x_2 x_1}$, and in general all second-order mixed partial derivatives with respect to the same two variables must be equal. (This criterion is used for dimension 2 in Chapter 1, Section 4.) Apply the mixed-partial criterion to determine which of the following systems are not conservative. Then find a potential if the system is conservative.

   (a) $\ddot{x} = -y$, $\ddot{y} = x$.

   (b) $\ddot{x} = y$, $\ddot{y} = z$, $\ddot{x} = x$.

   (c) $\ddot{x} = x$, $\ddot{y} = z$, $\ddot{z} = y$.

   (d) $\ddot{x} = xy^4 z^6$, $\ddot{y} = 2x^2 y^3 z^6$, $\ddot{z} = 3x^2 y^4 z^5$.

3. Recall that an **equilibrium solution** of a system is simply a constant solution $\mathbf{x}(t) = \mathbf{x}_0$. Since for such a solution $\dot{\mathbf{x}} = \ddot{\mathbf{x}} = 0$, finding equilibrium solutions of a second-order system $\ddot{\mathbf{x}} = -\nabla U(\mathbf{x})$ amounts to finding vector solutions of $\nabla U(\mathbf{x}) = 0$. But finding such a solution is equivalent to finding a **critical point** of the real-valued function U of the sort that is routinely examined in a search for relative maximum and minimum values of U. (This connection is pursued further in Section 8 on stability of equilibrium solutions.) For each of the following potentials $U(\mathbf{x})$, find all equilibrium solutions of $\ddot{\mathbf{x}} = -\nabla U(\mathbf{x})$.

   (a) $U(x, y) = x^2 + 3xy + 2y^2 - x$.

   (b) $U(x, y) = \cos(x - y)$.

(c) $U(x, y, z) = xyz$.

(d) $U(x, y, z) = (x^2 + y^2 + z^2 - 1)^2$.

4. The conservative system $\ddot{x} = -x$, $\ddot{y} = -y$ has $U(x, y) = \frac{1}{2}(x^2 + y^2)$ for a potential. Consider a solution $x = x(t)$, $y = y(t)$ with $x(0) = x_0$, $y(0) = y_0$, $\dot{x}(0) = \dot{y}(0) = 0$. Use the energy equation for the system to show the following.

   (a) The xy trajectory of the solution remains in the disk $x^2 + y^2 \leq x_0^2 + y_0^2$.

   (b) The speed $v(t) = \sqrt{\dot{x}(t)^2 + \dot{y}(t)^2}$ satisfies $v \leq \sqrt{x_0^2 + y_0^2}$.

   (c) Can you get the above information from an explicit solution?

5. The conservative system $\ddot{x} = -y$, $\ddot{y} = -x$ has $U(x, y) = xy$ for a potential. Consider solution $x = x(t)$, $y = y(t)$ with initial values $x(0) = y(0) = 0$, $\dot{x}(0) = z_0$, $\dot{y}(0) = w_0$. Use the energy equation for the system to show that the xy trajectory of the solution remains in the region determined by $xy \leq \frac{1}{2}(z_0^2 + w_0^2)$. Sketch this region for $z_0 = w_0 = 1$.

6. The conservative nonlinear system $\ddot{x} = -xy^2$, $\ddot{y} = -x^2y$ has $U(x, y) = \frac{1}{2}x^2y^2$ for a potential. Consider a solution $x = x(t)$, $y = y(t)$ with initial values $x(0) = x_0$, $y(0) = y_0$, $\dot{x}(0) = \dot{y}(0) = 0$. Use the energy equation for the system to show the following.

   (a) The xy trajectory of the solution remains in the region determined by $|xy| \leq |x_0 y_0|$. Sketch this region for $x_0 = y_0 = 1$.

   (b) The speed $v(t) = \sqrt{\dot{x}(t)^2 + \dot{y}(t)^2}$ satisfies $v(t) \leq |x_0 y_0|$.

7. The real-valued system $\ddot{x} = -f(x)$, with $f(x)$ continuous on some interval, is always conservative, having potentials $U(y) = \int f(x)\, dx$. Show this by deriving the appropriate constant-energy equation.

8. Show that the system

$$\ddot{x} = ax + by,$$
$$\ddot{y} = cx + dy$$

is conservative if and only if $b = c$. Find a potential if $b = c$.

Oscillation

9. The system

$$m_1\ddot{x} = -(h_1 + h_2)x + h_2 y,$$
$$m_2\ddot{y} = h_2 x - (h_2 + h_3)y$$

governs the oscillations of a two-body mass-spring system. Here the h_k are positive constants.

   (a) Verify that $U(x, y) = \frac{1}{2}(h_1 + h_2)x^2 - h_2 xy + \frac{1}{2}(h_2 + h_3)y^2$ is a potential for the system.

   (b) Use the positivity of the h_k to show that the potential $U(x, y)$ of part (a) is strictly positive except at $(x, y) = (0, 0)$. [*Hint:* Show that the discriminant of a quadratic polynomial is negative.]

10. Choose $m_1 = m_2 = 1$, $h_2 = 1$, $h_2 = 2$, $h_3 = 3$ in the system of the previous exercise.

   (a) Show that a potential for the system is $\frac{3}{2}x^2 - 2xy + \frac{5}{2}y^2$.

(b) Use the constant-energy equation to show that a solution of the system with initial values $x(0) = y(0) = 0$, $\dot{x}(0) = z_0$, $\dot{y}(0) = w_0$ has its xy trajectory confined to the elliptic region $3x^2 - 4xy + 5y^2 \leq z_0^2 + w_0^2$. Sketch this region for the case $z_0 = w_0 = 1$.

11. Show that the system $\ddot{x} = -2x + y - k_1\dot{x}$, $\ddot{y} = x - 2y - k_2\dot{y}$ is dissipative and that at time t the total energy of a solution $(x(t), y(t))$ is decreasing at the rate $k_1\dot{x}(t)^2 + k_2\dot{y}(t)^2$.

12. Find a potential for the system

$$m_1\ddot{x} = -(h_1 + h_2)x + h_2y + h_2l_1 - h_2l_2,$$
$$m_1\ddot{y} = h_2x - (h_2 + h_3)y + h_2l_2 - h_3l_3 + h_3b.$$

Then write down the corresponding total energy $E(x, y, \dot{x}, \dot{y})$.

13. Show that the two-body nonlinear mass-spring system

$$m_1\ddot{\mathbf{x}} = h_1(|\mathbf{a} - \mathbf{x}| - l_1)\frac{\mathbf{a} - \mathbf{x}}{|\mathbf{a} - \mathbf{x}|} + h_2(|\mathbf{y} - \mathbf{x}| - l_2)\frac{\mathbf{y} - \mathbf{x}}{|\mathbf{y} - \mathbf{x}|},$$

$$m_2\ddot{\mathbf{y}}\mathbf{1} = h_2(|\mathbf{x} - \mathbf{y}| - l_2)\frac{\mathbf{x} - \mathbf{y}}{|\mathbf{x} - \mathbf{y}|} + h_3(|\mathbf{b} - \mathbf{y}| - l_3)\frac{\mathbf{b} - \mathbf{y}}{|\mathbf{b} - \mathbf{y}|}$$

has

$$U(\mathbf{x}, \mathbf{y}) = \tfrac{1}{2}(h_1|\mathbf{a} - \mathbf{x}|^2 + h_2|\mathbf{y} - \mathbf{x}|^2 + h_3|\mathbf{b} - \mathbf{y}|^2)$$
$$- h_1l_1|\mathbf{a} - \mathbf{x}| - h_2l_2|\mathbf{y} - \mathbf{x}| - h_3l_3|\mathbf{b} - \mathbf{y}|$$

for a potential. [*Hint:* Write out U in terms of $\mathbf{x} = (x_1, x_2, x_3)$ and $\mathbf{y} = (y_1, y_2, y_3)$.]

14. Consider a **spherical pendulum** for which θ measures the angle from the downward vertical z direction and ϕ measures the rotation angle relative to a horizontal xy plane. The equations of motion can be written, as in Exercise 18 of Chapter 5, Section 1C,

$$\ddot{\theta} = \dot{\phi}^2 \sin\theta\cos\theta - \frac{g}{l}\sin\theta,$$

$$\ddot{\phi} = -2\dot{\theta}\dot{\phi}\cot\theta, \qquad \theta \neq k\pi, \ k \text{ integer}.$$

(a) Show that $x = l\sin\theta\cos\phi$, $y = l\sin\theta\sin\phi$, $z = -l\cos\theta$.

(b) Find the potential energy $U(x, y, z) = gz$ and kinetic energy $T(\dot{x}, \dot{y}, \dot{z}) = \tfrac{1}{2}v^2$ in terms of θ and ϕ.

(c) Use the equations of motion to show directly that the total energy $T + U$ is constant on a phase path.

(d) Separate variables in the equation $\ddot{\phi} = -2\dot{\phi}\dot{\theta}\cot\theta$, and integrate to show that $\dot{\phi}\sin^2\theta = M$, where M is a constant depending on initial conditions for $\dot{\phi}$ and θ. (The angular momentum of the pendulum about its vertical axis is $(l\sin\theta)^2\dot{\phi}$ and is therefore constant.)

*15 **Dynamic equilibrium for the spherical pendulum.** This exercise shows that equilibrium solutions need not always be static as is the case with constant solutions. We'll consider the system of the previous exercise, and we'll assume throughout that $\dot{\phi}(0) \neq 0$ and $\theta \neq 0$.

(a) Rewrite the second equation of the system as $\ddot{\phi} + 2\dot{\phi}\dot{\theta}\cot\theta = 0$, and show that this is equivalent to saying that $\dot{\phi}\sin^2\theta = M$, where M is a constant depending on initial conditions for $\dot{\phi}$ and θ. [*Hint:* Operate by d/dt on the equation containing M.]

(b) Use the result of part (a) in the first equation of the system to show that

$$\ddot{\theta} = -\frac{g}{l}\sin\theta + M^2\frac{\cos\theta}{\sin^3\theta}.$$

(c) Show that the equation of part (b) has exactly one constant equilibrium solution θ^* in $0 < \theta < \pi$, that in fact $0 < \theta^* < \pi/2$, and that the pendulum bob rotates in a circular orbit of radius $l\sin\theta^*$ with constant angular speed $\dot{\phi} = \sqrt{(g/l)}\sec\theta^*$.

(d) Use the Lagrange test (Chapter 4, Section 6B) to show θ^* stable for the equation of part (b).

Inverse-Square Law

16. Show directly that there are no constant equilibrium solutions to the Newton system in Example 2 of the text. Explain how you can draw the same conclusion by looking at the graph of the potential $U(x, y) = (x^2 + y^2)^{-1/2}$.

17. The conservative system $\ddot{x} = -x(x^2 + y^2)^{-3/2}$, $\ddot{y} = -y(x^2 + y^2)^{-3/2}$ has potential $U(x, y) = -(x^2 + y^2)^{-1/2}$. Consider a solution $(x(t), y(t))$ satisfying initial conditions $(x(0), y(0)) = (1, 0)$, $(\dot{x}(0), \dot{y}(0)) = (0, 2)$. Use the energy conservation principle to show:

(a) Speed $v = \sqrt{\dot{x}^2 + \dot{y}^2}$ satisfies $v > \sqrt{2}$.

(b) Distance $r = \sqrt{x^2 + y^2}$ is $2/(v^2 - 2)$.

18. The vector field of the conservative system $\ddot{x} = x(x^2 + y^2)^{-3/2}$, $\ddot{y} = y(x^2 + y^2)^{-3/2}$ is a repelling **Coulomb field** and has potential $U(x, y) = (x^2 + y^2)^{-1/2}$. Consider a solution $(x(t), y(t))$ satisfying initial conditions $(x(0), y(0)) = (1, 0)$, $(\dot{x}(0), \dot{y}(0)) = (0, 2)$. Use the energy conservation principle to show:

(a) Distance $r = \sqrt{x^2 + y^2}$ satisfies $r \geq \frac{1}{3}$.

(b) Speed $v = \sqrt{\dot{x}^2 + \dot{y}^2}$ satisfies $v \leq \sqrt{6}$.

19. Show that the energy equation $\frac{1}{2}|\dot{\mathbf{x}}|^2 - k/|\mathbf{x}| = E$, $k > 0$, for the orbit of a planet about a fixed sun implies arbitrarily large speed for a sufficiently close approach. What principle of relativity theory does this conclusion violate?

20. Verify that $U(\mathbf{x}_1, \mathbf{x}_2) = -Gm_1m_2/|\mathbf{x}_1 - \mathbf{x}_2|$ is a potential function for the two-body system 5.1 of Section 5:

$$m_1\ddot{\mathbf{x}}_1 = -\frac{Gm_1m_2}{|\mathbf{x}_1 - \mathbf{x}_2|^3}(\mathbf{x}_1 - \mathbf{x}_2), \qquad m_2\ddot{\mathbf{x}}_2 = -\frac{Gm_1m_2}{|\mathbf{x}_1 - \mathbf{x}_2|^3}(\mathbf{x}_2 - \mathbf{x}_1).$$

Note that U is a function of six real variables that can be written, for example, $U(\mathbf{x}_1, \mathbf{x}_2) = U(x_1, y_1, z_1, x_2, y_2, z_2)$. Hence ∇U is a vector-valued function of dimension 6.

21. Show that

$$U(\mathbf{x}_1, \mathbf{x}_2, \mathbf{x}_3) = -\frac{Gm_1m_2}{|\mathbf{x}_1 - \mathbf{x}_2|} - \frac{Gm_1m_3}{|\mathbf{x}_1 - \mathbf{x}_3|} - \frac{Gm_2m_3}{|\mathbf{x}_2 - \mathbf{x}_3|}$$

is a potential function for the three-body system 5.4 of Section 5:

$$m_1\ddot{\mathbf{x}}_1 = \frac{Gm_1m_2}{|\mathbf{x}_2 - \mathbf{x}_1|^3}(\mathbf{x}_2 - \mathbf{x}_1) + \frac{Gm_1m_3}{|\mathbf{x}_3 - \mathbf{x}_1|^3}(\mathbf{x}_3 - \mathbf{x}_1),$$

$$m_2\ddot{\mathbf{x}}_2 = \frac{Gm_1m_2}{|\mathbf{x}_2 - \mathbf{x}_1|^3}(\mathbf{x}_1 - \mathbf{x}_2) + \frac{Gm_2m_3}{|\mathbf{x}_3 - \mathbf{x}_2|^3}(\mathbf{x}_3 - \mathbf{x}_2),$$

$$m_3\ddot{\mathbf{x}}_3 = \frac{Gm_1m_3}{|\mathbf{x}_3 - \mathbf{x}_1|^3}(\mathbf{x}_1 - \mathbf{x}_3) + \frac{Gm_2m_3}{|\mathbf{x}_2 - \mathbf{x}_3|^3}(\mathbf{x}_2 - \mathbf{x}_3).$$

Note that U is a function of nine real variables and can be written, for example, $U(\mathbf{x}_1, \mathbf{x}_2, \mathbf{x}_3)$ or $U(x_1, y_1, z_1, x_2, y_2, z_2, x_3, y_3, z_3)$.

22. Suppose that a planet is in a circular orbit of radius a about its sun and that the orbital period is P. Let the potential $U(x, y)$ for the Newton equations of the system be normalized so that as x or y tend to infinity, $U(x, y)$ tends to zero. Use Kepler's third law of Section 5 to show that the total energy $T + U$ of the orbit is equal to $-T = -2\pi^2a^2/P^2$.

23. When the inverse-square law $F = Gm_1m_2r^{-2}$ is replaced by $F = Gm_1m_2r^{-p}$, where $0 < p$, the Newton equations for the relative orbit of a planet about a star are replaced by

$$\ddot{x} = \frac{-kx}{(x^2 + y^2)^{(p+1)/2}},$$

$$\ddot{y} = \frac{-ky}{(x^2 + y^2)^{(p+1)/2}}.$$

(a) Show that this system is conservative for $p \neq 1$ and that a potential is

$$U_p(x, y) = \frac{-k}{p - 1}(x^2 + y^2)^{-(p-1)/2}.$$

(b) Show that in case $p = 1$ the function $U_1(x, y) = (k/2)\ln(x^2 + y^2)$ is a potential.

24. The potentials for the cases $p = 1$ and $0 < p \neq 1$ in the previous exercise appear at first sight to be unrelated. However, recalling that a potential is determined only up to an additive constant, define for $(x, y) \neq (0, 0)$, $p \neq 1$, the potential

$$V_p(x, y) = \frac{1}{p - 1}(1 - (x^2 + y^2)^{-(p-1)/2}).$$

(a) Show that

$$\lim_{p \to 1} V_p(x, y) = \tfrac{1}{2}\ln(x^2 + y^2) = V_1(x, y).$$

(b) Sketch the graphs of $V_{1/2}(x, y)$, $V_1(x, y)$, and $V_2(x, y)$ with respect to 3-dimensional axes.

*25. Exercise 22 of Section 5 shows that for a planet or comet of mass m_1 orbiting a fixed sun of mass m_2 at the origin, the polar coordinate equations

$$r^2\dot\theta = h \quad \text{and} \quad r = \frac{h^2/k}{1 + e\cos\theta}$$

are valid, where the constant h is angular momentum, $k = G(m_1 + m_2)$, and $e \geq 0$ is the eccentricity of the orbit. (Eccentricity $e = 0$ gives a circle, $0 < e < 1$ an ellipse, $e = 1$ a parabola, and $1 < e$ a hyperbola.)

(a) Show that the energy conservation equation for this system takes the polar coordinate form

$$\tfrac{1}{2}(\dot r^2 + r^2\dot\theta^2) - \frac{k}{r} = E.$$

(b) Show that when $\theta = 0$, then $r(0) = h^2/(k(1 + e))$ and $\dot r(0) = 0$ and hence that

$$\frac{1}{2}\frac{h^2}{r(0)^2} - \frac{k}{r(0)} = E.$$

(c) Eliminate $r(0)$ from the equations in part (b) to show that

$$e = \sqrt{1 + 2\left(\frac{h^2}{k^2}\right)E}.$$

(d) Show that an orbit is a circle or an ellipse if $-\tfrac{1}{2}(k^2/h^2) \leq E < 0$, a parabola if $E = 0$, and a hyperbola if $E > 0$.

7   NUMERICAL METHODS

The numerical methods illustrated here apply to a first-order vector equation $\dot{\mathbf{x}} = \mathbf{F}(t, \mathbf{x})$ with initial condition $\mathbf{x}(t_0) = \mathbf{x}_0$ and include a wide variety of systems, both linear and nonlinear. As in Chapters 2 and 6, the methods approximate continuous-time systems by discrete-time systems.

7A   Euler's Method

We choose a step of size h and find successive approximations $\mathbf{x}_k$ to the true values $\mathbf{x}(t_0 + kh)$ of the solution $\mathbf{x}(t)$. The idea is to multiply the difference-quotient approximation to the vector derivative by h:

$$\frac{\mathbf{x}(t + h) - \mathbf{x}(t)}{h} \approx \mathbf{F}(t, \mathbf{x}) \quad \text{becomes} \quad \mathbf{x}(t + h) \approx \mathbf{x}(t) + h\mathbf{F}(t, \mathbf{x}).$$

Thus having found $\mathbf{x}_k$ corresponding to $t_k = t_0 + kh$, we define the approximation $\mathbf{x}_{k+1}$ at $t_{k+1} = t_0 + (k + 1)h$ by

$$\mathbf{x}_{k+1} = \mathbf{x}_k + h\mathbf{F}(t_k, \mathbf{x}_k).$$

For a 2-dimensional system with $\mathbf{x} = (x, y)$,

$$\dot x = F(t, x, y), \qquad x(t_0) = x_0,$$
$$\dot y = G(t, x, y), \qquad y(t_0) = y_0,$$

the zeroth step starts with x_0 and y_0. Then

$$x_1 = x_0 + hF(t_0, x_0, y_0),$$
$$y_1 = y_0 + hG(t_0, x_0, y_0).$$

Next, with $t_1 = t_0 + h$,

$$x_2 = x_1 + hF(t_1, x_1, y_1),$$
$$y_2 = y_1 + hG(t_1, x_1, y_1).$$

In general, with $t_k = t_{k-1} + h = t_0 + kh$, we get

$$x_{k+1} = x_k + hF(t_k, x_k, y_k),$$
$$y_{k+1} = y_k + hG(t_k, x_k, y_k).$$

The basic loop for sequential software would save the initial x with **SET S = X**. Then

```
SET X = X + H * F(T, X, Y)

SET Y = Y + H * G(T, S, Y)

SET T = T + H

PRINT T, X, Y
```

where the saved value **s** is used only in the second step to avoid using the newly computed **x** from the first step.

---

e x a m p l e 1

The system

$$\dot{x} = y, \qquad x(0) = 0, \qquad \dot{y} = tx + 1, \qquad y(0) = 1$$

is equivalent to the second-order equation $\ddot{x} = tx + 1$. Together with step size $H = 0.01$, initial values $x(0) = 0$, $y(0) = 1$, and definitions

```
DEFINE F(T, X, Y) = Y

DEFINE G(T, X, Y) = T * X + 1
```

the loop above will produce a table of approximate values of t, $x(t)$, and the derivative $y(t) = \dot{x}(t)$ at t intervals of 0.01. Replacing the "print" command by **PLOT T, X** produces a graph of $x(t)$. Similarly, **PLOT T, Y** produces a graph of $y(t) = \dot{x}(t)$. See Figure 6.11.

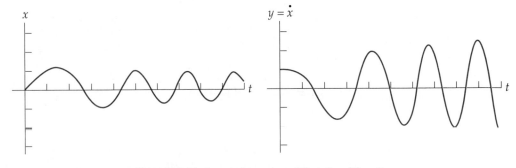

Figure 6.11 $\dot{x} = y$, $\dot{y} = -tx$, $x(0) = 0$, $y(0) = 1$.

7B *Improved Euler Method*

The next simplest numerical method for a vector equation is a modification of the Euler method, sometimes called the Heun method, in which, instead of using the tangent vector $\mathbf{F}(t_k, \mathbf{x}_k)$ to find the next value, we use the vector average

$$\tfrac{1}{2}[\mathbf{F}(t_k, \mathbf{x}_k) + \mathbf{F}(t_k + h, \mathbf{p}_{k+1})],$$

where $\mathbf{p}_{k+1}$ is the value that the Euler method would have predicted, that is,

$$\mathbf{p}_{k+1} = \mathbf{x}_k + h\mathbf{F}(t_k, \mathbf{x}_k).$$

Using the predicted value $\mathbf{p}_{k+1}$, the **improved Euler approximation** to $\mathbf{x}(t_k, x_k)$ is given by

$$\mathbf{x}_{k+1} = \mathbf{x}_k + \frac{h}{2}\,[\mathbf{F}(t_k, \mathbf{x}_k) + \mathbf{F}(t_k + h, \mathbf{p}_{k+1})].$$

For a 2-dimensional system,

$$\dot{x} = F(t, x, y), \qquad x(t_0) = x_0,$$
$$\dot{y} = G(t, x, y), \qquad y(t_0) = y_0,$$

we start with (x_0, y_0). Then letting $\mathbf{p} = (p, q)$, we compute the Euler approximation

$$p_1 = x_0 + hF(t_0, x_0, y_0),$$
$$q_1 = y_0 + hG(t_0, x_0, y_0),$$

followed by the corrected approximation

$$x_1 = x_0 + \frac{h}{2}\,[F(t_0, x_0, y_0) + F(t_0 + h, p_1, q_1)],$$

$$y_1 = y_0 + \frac{h}{2}\,[G(t_0, x_0, y_0) + G(t_0 + h, p_1, q_1)].$$

At the $(k + 1)$th step, we compute $t_k = t_{k-1} + h = t_0 + kh$, and

$$p_{k+1} = x_k + hF(t_k, x_k, y_k),$$
$$q_{k+1} = y_k + hG(t_k, x_k, y_k),$$

$$x_{k+1} = x_k + \frac{h}{2}\,[F(t_k, x_k, y_k) + F(t_{k+1}, p_{k+1}, q_{k+1})],$$

$$y_{k+1} = y_k + \frac{h}{2}[G(t_k, x_k, y_k) + G(t_{k+1}, p_{k+1}, q_{k+1})].$$

One should be careful to avoid using the value x_{k+1} in the fourth equation instead of x_k; in sequential implementation x_k should be stored for use later. An algorithmic loop might look like this:

```
SET P = X + H * F(T, X, Y)

SET Q = Y + H * G(T, X, Y)

SET S = X

SET X = X + (H/2) * (F(T, X, Y) + F(T + H, P, Q))

SET Y = Y + (H/2) * (G(T, S, Y) + G(T + H, P, Q))
```

For three equations $\dot{x} = F_1(t, x, y, z)$, $\dot{y} = F_2(t, x, y, z)$, $\dot{z} = F_3(t, x, y, z)$, recursive formulas for an algorithm can be written in similar style, but for these, and still higher-dimensional systems, it's more efficient to introduce vector arrays in which a vector variable **X** is assigned an appropriate dimension n along with coordinates X(1), . . . , X(n). Coordinate functions F_k can be similarly assigned to a vector function **F**, with the result that the entire computational loop may be given the form **SET X = X + H * F(T, X)**.

An instance of equations of Lotka-Volterra type for three species is $\dot{x} = x(3 - y)$, $\dot{y} = y(x + z - 3)$, $\dot{z} = z(2 - y)$. We can just plot the relationship between two species at a time, winding up with three related pictures. Or we can plot the trajectory of a solution relative to 3-dimensional axes as shown in Figure 6.12. If either $x = 0$ or $z = 0$, this system reduces to dimension 2, for example, $\dot{x} = x(3 - y)$, $\dot{y} = y(x - 3)$ if $z = 0$. The associated trajectories are closed loops as discussed in Section 2.

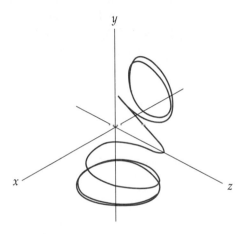

Figure 6.12 Three-species competition.

---

example 3

Recall that Newton's equations of planetary motion for a star with one planet are

$$\ddot{x} = -\frac{GMx}{(x^2 + y^2)^{3/2}}, \qquad x(0) = x_0, \ \dot{x}(0) = z_0,$$

$$\ddot{y} = -\frac{GMy}{(x^2 + y^2)^{3/2}}, \qquad y(0) = y_0, \ \dot{y}(0) = w_0.$$

These equations are derived in Section 5, where we remarked that the second-order system is equivalent to the first-order system

$$\dot{x} = z, \qquad x(0) = x_0,$$
$$\dot{y} = w, \qquad y(0) = y_0,$$
$$\dot{z} = -\frac{GMx}{(x^2 + y^2)^{3/2}}, \qquad z(0) = z_0,$$
$$\dot{w} = -\frac{GMy}{(x^2 + y^2)^{3/2}}, \qquad w(0) = w_0.$$

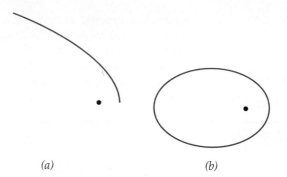

(a)   (b)

Figure 6.13 Hyperbolic and elliptic orbits.

Sometimes we choose units of measurement so that $GM = 4\pi^2$, although for some purposes we could just as well choose them so that GM has some other value. In choosing initial values for position and velocity, recall that we get an elliptic orbit only if the orbital speed is less than the escape speed, $V_e = \sqrt{2GM/r}$. In other words, we want

$$z_0^2 + w_0^2 < \frac{2GM}{\sqrt{x_0^2 + y_0^2}}.$$

Thus, if $GM = 2$ and $(x_0, y_0) = (1, 0)$, we should choose (z_0, w_0) so that $z_0^2 + w_0^2 < 4$ to get a closed trajectory. The orbit in Figure 6.13a was made using the improved Euler method, having chosen $GM = 2$, initial position $(x_0, y_0) = (1, 0)$, and initial velocity $(z_0, w_0) = (0, 2.01)$. The step size was $h = 0.01$, but the result is plotted only for every 10 steps. The closed orbit in Figure 6.13b is the result of changing w_0 to 1.8.

If we were to choose constants $M = 1/G$, $x_0 = 1$, $y_0 = 0$, $z_0 = 0$, $w_0 = 1$ in this example, then the equations would be satisfied by the solution $(x(t), y(t)) = (\cos t, \sin t)$, which has a circular orbit for its trajectory. We could then use this elementary solution as a check on the accuracy of our method of numerical approximation.

The remarks about error accumulation at the end of Chapter 2, Section 6, apply to systems also. However, one outstanding source of error in any numerical method for the approximate solution of differential equations is the evaluation of functions. If, for a single first-order equation $y' = F(x, y)$, the Euler method requires a single evaluation of $F(x, y)$ at each step, the improved Euler method applied to the same equation requires two evaluations, $F(x_k, y_k)$ and $F(x_{k+1}, p_{k+1})$, at the kth step. For an n-dimensional system, the improved Euler method typically requires $2n$ evaluations at the kth step:

$$F_j(t_k, \mathbf{x}_k), F_j(t_{k+1}, \mathbf{p}_{k+1}), \qquad j = 1, 2, \ldots, n,$$

where $F_j(t, \mathbf{x})$ is the jth coordinate function of the system. Since errors made at each step are then fed back into the functions F_j at the next step, it is clear that the possibility for error accumulation will be significantly enhanced for a large system. Nevertheless, it can be shown that the local formula errors for the Euler and improved Euler methods have order h^2 and h^3, respectively; it is only proportionality factors that may increase as the dimension increases.

The distinction between error at a single step and accumulated error is such that the two can be treated separately. If $F(t, x)$ is continuously differentiable in x, then the one-step error using the Euler method has order h^2, where h is the step size. Using the improved Euler method, the one-step error has order h^3 provided $F(t, x)$ is twice continuously differentiable in x. These ideas are introduced here only as a rough guide; deeper understanding requires a thorough study of numerical analysis.

The matter of existence and uniqueness of solutions is particularly pertinent to finding approximate solutions of systems for which analytic solution formulas don't exist. In more theoretical treatments it is shown, for example, that $\dot{x} = F(t, x)$, $x(t_0) = x_0$ has a unique solution if F and its partial derivatives with respect to the x variables are bounded and continuous on an open set in $\mathcal{R}^{n+1}$ containing (t_0, x_0).

EXERCISES

Web site http://math.dartmouth.edu/~rewn/ contains general-purpose approximate methods for systems. These are the Applets DESYS2, DESYS2PLOT, DESYS3, DESYS3PLOT, DESYS4, DESYSN, all designed to approximate the solution trajectories of first-order normal-form systems.

1. The first-order autonomous system $\dot{x} = y$, $\dot{y} = x$ is equivalent to the single equation $\ddot{y} = y$ via the relation $\dot{y} = x$.

 (a) Show that the system has solutions $x(t) = c_1 e^t + c_2 e^{-t}$, $y(t) = c_1 e^t - c_2 e^{-t}$.

 (b) Find the particular solution satisfying $x(0) = 1$, $y(0) = 2$.

 (c) Compute a table of approximations to the particular solution found in part (b). Do this computation on the interval $0 \le t \le \frac{1}{2}$ in steps of size $h = 0.1$, both by computing values from the explicit exponential solution formula and by a direct numerical solution of the system using either the Euler method or its improvement.

2. Find a table of approximations to the solution $x(t)$, $y(t)$ of the system
$$\dot{x} = x + y^2, \qquad \dot{y} = x^2 + y + t,$$
with initial conditions $x(0) = 1$, $y(0) = 2$. Use a step of size $h = 0.1$ on the interval $0 \le t \le \frac{1}{2}$, and make the approximation with (a) Euler method, and (b) improved Euler.

3. (a) Show that the second-order equation $\ddot{y} + 2\dot{y} + y = t$, with initial conditions $y(0) = 1$, $\dot{y}(0) = 2$, is equivalent to the first-order system
$$\dot{x} = -2x - y + t, \qquad x(0) = 2,$$
$$\dot{y} = x, \qquad\qquad\qquad y(0) = 1.$$

 (b) Find a numerical approximation to the solution of the system in part (a) for the interval $0 \le t \le 1$.

 (c) Solve the given second-order equation by using its characteristic equation, and compare the solution with the numerical results of part (b).

4. (a) Find a first-order system equivalent to $y'' - ty' + y = t$, $y(0) = 1$, $y'(0) = 2$.

(b) Find a numerical approximation to the solution of the system in part (a) for $0 \le t \le 1$.

5. Apply the improved Euler method to the equation $dy/dt = t^2 + y^2$, $y(0) = 1$.

*6. A basic result of multivariable calculus says that the gradient vector $\nabla f(x, y) = (f_x(x, y), f_y(x, y))$ is perpendicular to the level curve of f that contains (x, y); consequently, $(-f_y(x, y), f_x(x, y))$ is tangent to a level set, which is then a trajectory of the system $\dot{x} = -f_y(x, y)$, $\dot{y} = f_x(x, y)$. Suppose below that $f(x, y) = x^2 + \frac{1}{2}y^2$.

(a) Use these ideas to make a computer-graphics plot of the elliptic level curves of f.

(b) Make a computer-graphics plot of some **orthogonal trajectories,** that is, curves perpendicular to the level sets of the function f.

(c) Identify the well-known family of orthogonal trajectory curves by solving the relevant uncoupled system analytically.

7. Make computer plots of the solution to the following problems in Chapter 5, Section 1C.

(a) The **bug-pursuit** problem of Exercise 20 with $v = 1$.

(b) The **airplane** problem of Exercise 21 with $v = 3$ and both $\mathbf{w} = (1, 2)$ and $\mathbf{w} = (-1, 2)$.

8. The 4th-order **Runge-Kutta** approximations to the solution of $\dot{\mathbf{x}} = \mathbf{F}(t, \mathbf{x})$, $\mathbf{x}(t_0) = \mathbf{x}_0$, are computed as follows:

$$\mathbf{x}_{k+1} = \mathbf{x}_k + \frac{h}{6}(\mathbf{m}_1 + 2\mathbf{m}_2 + 2\mathbf{m}_3 + \mathbf{m}_4),$$

where

$$\mathbf{m}_1 = \mathbf{F}(t_k, \mathbf{x}_k), \qquad\qquad \mathbf{m}_2 = \mathbf{F}(t_k + \tfrac{1}{2}h, \mathbf{x}_k + \tfrac{1}{2}h\mathbf{m}_1),$$
$$\mathbf{m}_3 = \mathbf{F}(t_k + \tfrac{1}{2}h, \mathbf{x}_k + \tfrac{1}{2}h\mathbf{m}_2), \qquad \mathbf{m}_4 = \mathbf{F}(t_k + h, \mathbf{x}_k + h\mathbf{m}_3).$$

(The local formula error is of order h^5.) Apply a computer algorithm to implement the routine. Then compare the results obtained from it for the problem

$$\dot{x} = -y, \qquad x(0) = 1, \qquad \dot{y} = x, \qquad y(0) = 0$$

with the results obtained from the improved Euler method, as well as with the solution formula $x = \cos t$, $y = \sin t$. Try step sizes $h = 0.1$, 0.01, and 0.001.

9. Apply the Runge-Kutta method of the previous exercise to a 2-dimensional system equivalent to $\ddot{x} = -tx$, $x(0) = 0$, $\dot{x}(0) = 1$, using $h = 0.01$. Compare with the improved Euler method.

Mixing

10. Tank 1 at capacity 100 gallons and tank 2 at 200 gallons are initially full of salt solution. Tank 1 has 5 gallons per minute of salt solution at 1 pound of salt per gallon running in while mixed solution is drawn off, also at 5 gallons per minute, with an additional 3 gallons per minute flowing out to tank 2. Tank 2 has 2 gallons per minute of pure water running in and 3 gallons per minute being drawn off, while 2 gallons per minute flow out to tank 1.

(a) Find a system of differential equations satisfied by the salt contents of the tanks up to the time when one is empty.

(b) Make a computer plot that compares the graphs of the components of the solution to part (a), assuming tank 1 has initially 10 pounds of salt and tank 2 has 20 pounds. Estimate the maximum amounts of salt in each tank and the times when these are attained.

11. Tank 1 at capacity 100 gallons and tank 2 at 200 gallons are initially half-full of salt solution. Tank 1 has 5 gallons per minute of pure water running in while mixed solution is drawn off at 4 gallons per minute, with an additional 3 gallons per minute pumped to tank 2. Tank 2 has 2 gallons per minute of salt solution at 1 pound of salt per gallon running in and 1 gallon per minute being drained off, while 3 gallons per minute are pumped to tank 1.

(a) Find a system of differential equations satisfied by the salt contents of the tanks up to the time when one is full.

(b) Make a computer plot that compares the graphs of the components of the solution to part (a), assuming tank 1 has initially 10 pounds of salt and tank 2 has 20 pounds. Estimate the minimum amount of salt in tank 1 and the time when this is attained.

Gravitational Attraction

12. The system of Newton equations $\ddot{x} = -x(x^2 + y^2)^{-3/2}$, $\ddot{y} = -y(x^2 + y^2)^{-3/2}$ with initial conditions $x(0) = 1$, $y(0) = 0$, $\dot{x}(0) = 0$, $\dot{y}(0) = v_0$ has a solution with a closed trajectory if $v_0 < \sqrt{2}$. Use the improved Euler method to make an approximate computation of the trajectory of a single orbit if:

(a) $v_0 = 0.35$.   **(b)** $v_0 = 0.7$.   **(c)** $v_0 = 1.4$.

13. To implement a routine that applies the improved Euler method to the three-body problem, it's helpful to start with the center of mass of the physical system at the origin:

$$m_1 \mathbf{x}_1 + m_2 \mathbf{x}_2 + m_3 \mathbf{x}_3 = 0.$$

Show this equation valid for all time since the **conservation of momentum** Equation 5.6 holds for the equations for three bodies if also

$$m_1 \dot{\mathbf{x}}_1(0) + m_2 \dot{\mathbf{x}}_2(0) + m_3 \dot{\mathbf{x}}_3(0) = 0.$$

Thus if m_1 is the largest mass, it is appropriate to let the initial values for $\mathbf{x}_1(0)$ and $\dot{\mathbf{x}}_1(0)$ be determined by the corresponding initial values for the other two bodies. Because of the law of conservation of momentum, the previous two equations will be maintained forever.

14. When the inverse-square law $F = m_1 m_2 r^{-2}$ is replaced by $F = m_1 m_2 r^{-p}$, where $0 < p$, the Newton equations for the orbit of a single planet about a fixed sun take the form

$$\ddot{x} = -kx(x^2 + y^2)^{-(p+1)/2}, \qquad \ddot{y} = -ky(x^2 + y^2)^{-(p+1)/2}.$$

Assume $k = 1$ and make a pictorial comparison of the orbits with initial conditions $x(0) = 1$, $y(0) = 0$, $\dot{x}(0) = 0$, $\dot{y}(0) = 0.5$ for the choices $p = 1.9$, $p = 2$, and $p = 2.1$. Discuss the differences among the three cases.

Oscillatory Systems

15. The equations $\ddot{\theta} = -(g/l) \sin\theta + \dot{\phi}^2 \sin\theta \cos\theta$, $\ddot{\phi} = -2\dot{\theta}\dot{\phi} \cot\theta$, $\theta \neq k\pi$, k integer, govern the **spherical pendulum** described in Exercise 18 of Chapter 5, Section 1C. Let $g/l = 1$.

(a) Plot the trajectory in $\theta\phi$ space for a solution if $\theta(0) = \dot{\theta}(0) = 1$, $\phi(0) = 0$, $\dot{\phi}(0) = 1$.

(b) Make a 3-dimensional perspective plot of the xyz path of the bob if the initial conditions are as in part (a). Use the coordinate relations given in Exercise 18 of Chapter 5, Section 1C.

16. The **Van der Pol equation** is $\ddot{x} - \alpha(1 - x^2)\dot{x} + x = 0$, where α is a positive constant.

(a) Let $y = \dot{x}$ and write down the first-order system in x and y that is equivalent to the Van der Pol equation.

(b) Use the improved Euler method to plot numerical solutions to the system found in part (a), using initial values $x(0) = 2$, $y(0) = 0$, while successively letting $\alpha = 0.1$, 1.0, and 2.0. Plot for $0 < t < t_1$, where t_1 is in each plot large enough that the trajectory of $(x(t), y(t))$ appears to be a closed loop.

(c) Repeat the three experiments in part (b) with initial values $x(0) = 1$, $y(0) = 0$ and with $x(0) = 3$, $y(0) = 0$. The closed loops being approximated in part (b) are each an example of a **limit cycle.**

In the next three exercises, use a numerical system solver to plot graphical solutions of the nonlinear equations of motion for mass-spring systems in dimension 3. For two masses and three springs the vector equations were derived in Section 4:

$$m_1\ddot{\mathbf{x}} = h_1(|\mathbf{a} - \mathbf{x}| - l_1)\frac{\mathbf{a} - \mathbf{x}}{|\mathbf{a} - \mathbf{x}|} + h_2(|\mathbf{y} - \mathbf{x}| - l_2)\frac{\mathbf{y} - \mathbf{x}}{|\mathbf{y} - \mathbf{x}|},$$

$$m_2\ddot{\mathbf{x}} = h_2(|\mathbf{x} - \mathbf{y}| - l_2)\frac{\mathbf{x} - \mathbf{y}}{|\mathbf{x} - \mathbf{y}|} + h_3(|\mathbf{b} - \mathbf{y}| - l_3)\frac{\mathbf{b} - \mathbf{y}}{|\mathbf{b} - \mathbf{y}|}.$$

Run your solver with graphic output and with parameter and initial values chosen to show typical nonlinear and, if possible, linear behavior.

17. (a) Plot typical solutions for masses m_1, m_2 joined to each other and to two fixed points $\mathbf{a}$, $\mathbf{b}$ by springs with unstressed lengths l_1, l_2, l_3, and Hooke constants h_1, h_2, h_3.

(b) What is the effect of making $h_3 = 0$ in part (a)? Run your system solver with that assumption included.

(c) Assume initial conditions that will restrict the motion in part (a) to a line and run your system solver under that assumption.

18. Derive the equations for a single mass m joined to two fixed points $\mathbf{a}$, $\mathbf{b}$ by springs with unstressed lengths l_1, l_2, and Hooke constants h_1, h_2. (See Exercise 14 in Section 4B.)

19. (a) Derive the equations for a single mass m joined to three fixed points **a**, **b**, **c** by springs with unstressed lengths l_1, l_2, l_3, and Hooke constants h_1, h_2, h_3.

(b) Then plot typical graphical solutions for some specific constants and initial values.

20. The general **Lorenz system** is $\dot{x} = \sigma(y - x), \dot{y} = \rho x - y - xz, \dot{z} = -\beta z + xy$, where β, ρ, σ are positive constants. For certain values of the parameters, in particular $\beta = \frac{8}{3}, \rho = 28, \sigma = 10$, solution trajectories exhibit an often-studied type of "chaotic" oscillation. Plot the orbits with these parameter choices and initial value $(x, y, z) = (2, 2, 21)$. A particularly good view is obtained by projecting on the plane through the origin perpendicular to the vector $(-2, 3, 1)$. See Chapter 7, Section 5, for a partial explanation of the observed behavior. Observe the effect of small changes in the initial vector on the successive numbers of circuits in each spiral configuration. This is another manifestation of chaotic behavior.

Interacting Populations

21. The special Lotka-Volterra system $\dot{H} = (3 - 2P)H, \dot{P} = (\frac{1}{2}H - 1)P$ is discussed in general terms in Section 2.

(a) Using the initial conditions $H(0) = 3, P(0) = \frac{3}{2}$, compute sufficiently many approximations to the solutions $(H(t), P(t))$ so that the values nearly return to the initial values.

(b) Using the approximate data obtained in part (a), sketch the graphs of $H(t)$ and $P(t)$ using the same vertical axis and the same horizontal t axis.

(c) Sketch an approximate trajectory in the HP plane for the solution found in part (a).

22. The Lotka-Volterra system $\dot{H} = (3 - 2P)H, \dot{P} = (\frac{1}{2}H - 1)P$ has solutions $(H(t), P(t))$ that are periodic, because a given orbit always returns to its initial point in some finite time t_1; thus $H(t + t_1) = H(t)$ and $P(t + t_1) = P(t)$ for all t. The time period t_1 depends on the particular orbit, however. Make numerical estimates of t_1 for the system on orbits with:

(a) $H(0) = 3, P(0) = \frac{1}{2}$.

(b) $H(0) = 3, P(0) = \frac{3}{2}$.

(c) $H(0) = 3, P(0) = 2$.

23. A refinement $\dot{H} = (a - bP)H(L - H), \dot{P} = (cH - d)P(M - P), a, b, c, d$ positive constants, of the Lotka-Volterra equations takes account of fixed limits $L > H(t)$ and $M > P(t)$ to the growth of the host and parasite populations. Assume $L > d/c > 1$ and $M > a/b > 1$. The restrictions L and M are typically imposed by lack of sufficient habitat or food supply. Use $a = 3$, $b = c = 2, d = 1$ and $L = 4, M = 3$ to plot the trajectories for:

(a) $H(0) = 3, P(0) = \frac{1}{2}$.

(b) $H(0) = 3, P(0) = \frac{3}{2}$.

(c) $H(0) = 3, P(0) = 2$.

24. Estimate the time it takes for an orbit to close under each of the three sets of initial conditions proposed in the previous exercise.

25. The three-species system $\dot{x} = x(3 - y)$, $\dot{y} = y(x + z - 3)$, $\dot{z} = z(2 - y)$ discussed in Example 3 of the text has equilibrium solutions at $(3, 3, 0)$ and $(0, 2, 3)$, as well as the trivial one at $(0, 0, 0)$. Use initial conditions $(x, y, z) = (0.001, 3, 2)$ to plot the following:

(a) The orbit of $(x(t), y(t))$.

(b) The orbit of $(y(t), z(t))$.

(c) The orbit of $(x(t), z(t))$.

8   STABILITY FOR AUTONOMOUS SYSTEMS

When an explicit solution formula $\mathbf{x}(t)$ is available for a vector differential equation

$$\frac{d\mathbf{x}}{dt} = \mathbf{F}(t, \mathbf{x}), \qquad t_0 \le t < \infty,$$

it may be possible to determine whether $\lim_{t \to \infty} \mathbf{x}(t)$ exists or whether $\mathbf{x}(t)$ exhibits some other kind of behavior. Even when solution formulas are not available, we can sometimes still get information about the long-term behavior of solutions. We'll make the simplifying assumption that $\mathbf{F}(t, \mathbf{x}) = \mathbf{F}(\mathbf{x})$ does not depend explicitly on t; such a system is called **autonomous.** We'll deal at first only with the case in which $\mathbf{x} = (x, y)$ is 2-dimensional and the system is linear. For nonlinear and higher-dimensional systems see Sections 8B and 8C and also Chapter 7, Section 5.

8A   Linear Systems

We'll assume throughout this subsection that we have a normal-form autonomous system

■ **8.1**

$$\frac{dx}{dt} = ax + by,$$

$$\frac{dy}{dt} = cx + dy, \qquad a, b, c, d \text{ real numbers.}$$

Solving such a system can be reduced by the elimination method to the solution of a single second-order equation. (In practice the system might have arisen initially from an order reduction of second-order linear equation to a first-order system.) We have seen that the solutions are all derived from characteristic roots λ_1, λ_2 and have the form

■ **8.2**

$$x(t) = c_1 e^{\lambda_1 t} + c_2 e^{\lambda_2 t},$$

$$y(t) = c_3 e^{\lambda_1 t} + c_4 e^{\lambda_2 t},$$

unless $\lambda_1 = \lambda_2$, in which case $e^{\lambda_2 t}$ will be replaced by $te^{\lambda_1 t}$. (There will in general be some relation between the constants c_k.) The long-term behavior of a constant solution $x(t) = x_0$, $y(t) = y_0$ is obvious: it simply remains constant, and such a solution is called an **equilibrium solution.** Viewed as a single point, (x_0, y_0) is called an **equilibrium point.** The linear system 8.1 always has $x(t) = 0$, $y(t) = 0$ for an equilibrium solution. The following theorem describes conditions under which all solution trajectories that start close enough to $(0, 0)$ will remain close to the equilibrium point as $t \to \infty$, in which case $(0, 0)$ is called a **stable equilibrium.** If instead there are trajectories starting arbitrarily close to $(0, 0)$ that are infinitely often a fixed distance away from $(0, 0)$, then $(0, 0)$ is called an **unstable equilibrium.**

■ 8.3 THEOREM

Let λ_1 and λ_2 be the characteristic roots that occur in solution 8.2 to system 8.1. We have the following possibilities:

(a) If the real parts of both λ_1 and λ_2 are negative, then $\lim_{t \to \infty} (x(t), y(t)) = (0, 0)$ for all choices of the constants c_k, and $(0, 0)$ is called an **asymptotically stable** equilibrium.

(b) If λ_1 and λ_2 are conjugate and purely imaginary ($\lambda_1 = iq$, $\lambda_2 = -iq$, $q \neq 0$), then all solutions 8.2 are periodic of period $2\pi/q$, and $(0, 0)$ is a **stable** equilibrium.

(c) If either λ_1 or λ_2 has a positive real part, then there are choices of the constants c_k for which $x(t)$ and $y(t)$ are unbounded, and $(0, 0)$ is an **unstable** equilibrium.

(d) If either or both of λ_1 and λ_2 is zero, all solution trajectories are confined to straight lines and may even reduce to single points.

Proof. The proof consists of examining the form that the solutions take in the four cases:

(a) With $\lambda_1 = -p + iq$, $\lambda_2 = -p - iq$, $p > 0$, the solutions are linear combinations of $e^{-pt} \cos qt$ and $e^{-pt} \sin qt$ or of e^{-pt} and te^{-pt}. Since these basic functions tend to zero as t tends to infinity, all solutions tend to zero also.

(b) With $\lambda_1 = iq$, $\lambda_2 = -iq$, $q \neq 0$, the solutions are linear combinations of $\cos qt$ and $\sin qt$ and so are either periodic or zero.

(c) With $\lambda_1 = p + iq$, $\lambda_2 = p - iq$, $p > 0$, the solutions are linear combinations of $e^{pt} \cos qt$ and $e^{pt} \sin qt$ or of e^{pt} and te^{pt}. Thus nonzero linear combinations can assume arbitrarily large values as $t \to \infty$; if the coefficients are zero, the solutions are identically zero.

(d) With $\lambda_1 = 0$ and $\lambda_2 = r \neq 0$, the solutions are of the form $x = c_1 + c_2 e^{rt}$, $y = c_3 + c_4 e^{rt}$. If $r = 0$, we get instead $x = c_1 + c_2 t$, $y = c_3 + c_4 t$. If both c_2 and c_4 are zero, the trajectory is a single point. Otherwise, if, for instance, $c_4 \neq 0$, we can eliminate e^{rt} or t and find straight-line trajectories.   ■

The description that Theorem 8.3 gives for the behavior of solutions of system 8.1 fails to capture the interesting variety of behaviors that solutions can exhibit; the best way to show this is to look at some pictures of what can happen. Since everything depends on the roots λ_1, λ_2 of the characteristic equation, it is convenient to derive the characteristic equation once and for all. Writing the system in operator form, we have

$$(D - a)x - by = 0,$$
$$cx + (D - d)y = 0.$$

Eliminating y gives

$$(D - d)(D - a)x - bcx = 0,$$

or

$$(D^2 - (a + d)D + (ad - bc))x = 0.$$

Thus the **characteristic roots** λ_1, λ_2 are the roots of the **characteristic equation**

$$\lambda^2 - (a + d)\lambda + (ad - bc) = 0;$$

this equation is easy to recall by writing it using a 2 by 2 determinant:

$$\begin{vmatrix} a - \lambda & b \\ c & d - \lambda \end{vmatrix} = (a - \lambda)(d - \lambda) - bc$$

$$= \lambda^2 - (a + d)\lambda + (ad - bc) = 0.$$

example 1

In the system

$$\frac{dx}{dt} = -3x + 2y,$$

$$\frac{dy}{dt} = -4x + 3y$$

$a + d = 0$ and $ad - bc = -1$, so the characteristic equation is

$$\lambda^2 - 1 = 0.$$

The roots are $\lambda_1 = -1$, $\lambda_2 = 1$. The general solution is

$$x(t) = c_1 e^{-t} + c_2 e^t,$$

$$y(t) = c_1 e^{-t} + 2c_2 e^t.$$

The choice $c_1 = 0$, $c_2 = 1$ gives us

$$x(t) = e^t, \qquad y(t) = 2e^t.$$

The trajectory satisfies $2x = y$, which is the equation of a line. Since $x(t) \to \infty$ and $y(t) \to \infty$ as $t \to \infty$, we are on a trajectory in the first quadrant, shown in Figure 6.14a. On the other hand, choosing $c_1 = c_2 = 1$ gives us

$$x(t) = e^{-t} + e^t, \qquad y(t) = e^{-t} + 2e^t.$$

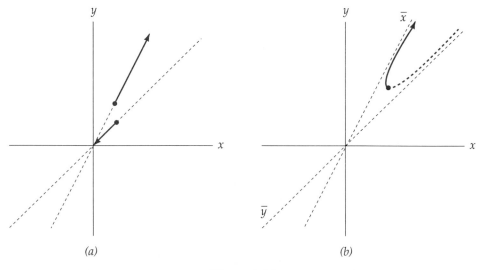

(a)   (b)

Figure 6.14

Since $y(t) - 2x(t) = -e^{-t}$ and $y(t) - x(t) = e^t$, the trajectory satisfies

$$(y - x)(y - 2x) = -1,$$

which is a hyperbola. Again $x(t) \to \infty$ and $y(t) \to \infty$ as $t \to \infty$. Introducing new coordinates $(\bar{x}, \bar{y})$ by $\bar{x} = y - x$, $\bar{y} = y - 2x$, the hyperbola becomes $\bar{x}\bar{y} = -1$, shown in Figure 6.14b. The directed trajectory is shown for $t > 0$. If we choose $c_1 = 1$ and $c_2 = 0$, the trajectory satisfies $x = y$ and is directed toward the origin, as shown in Figure 6.14a.

A **trajectory portrait** for a system consists of directed sketches of enough trajectories to convey a good idea of the variety of possible solution behavior. What follows is a file of portraits organized by type of characteristic root. The first sketch in each case shows the simplest possible behavior relative to xy axes. Also included in most cases is a more typical portrait.

Type I. *Unstable Node:* $0 < \lambda_1 < \lambda_2$

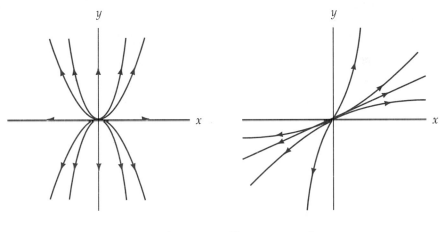

Example: $x = 2e^t, \quad y = e^{3t}; \quad 8y = x^3.$

Type II. *Saddle (Unstable):* $\lambda_1 < 0 < \lambda_2$

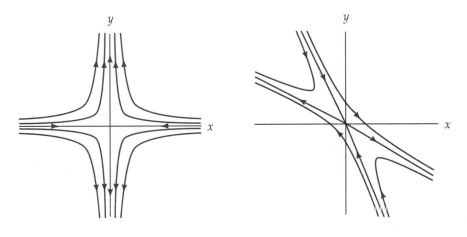

Example: $x = e^{-t}, \quad y = 3e^t; \quad xy = 3.$

Type III. *Asymptotically Stable Node:* $\lambda_1 < \lambda_2 < 0$

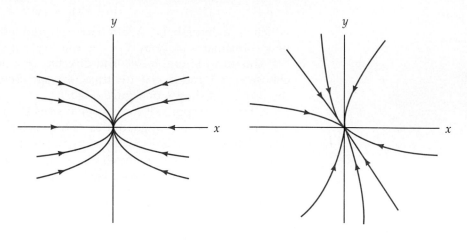

Example: $\qquad x = 2e^{-3t}, \qquad y = e^{-t}; \qquad x = 2y^3.$

Type IV. *Unstable Spiral:* $\lambda_1 = p + iq$, $\lambda_2 = p - iq$, $p > 0$, $q \neq 0$

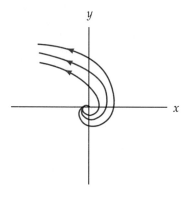

Example: $\quad x = e^t \cos 2t, \qquad y = e^t \sin 2t; \qquad x^2 + y^2 = e^{\arctan y/x}.$

Type V. *Stable Center:* $\lambda_1 = iq$, $\lambda_2 = -iq$, $q \neq 0$

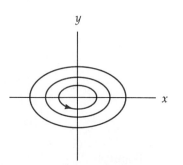

Type VI. *Asymptotically Stable Spiral:* $\lambda_1 = -p + iq$, $\lambda_2 = -p - iq$, $p > 0$, $q \neq 0$

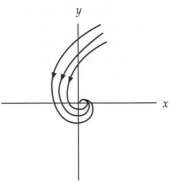

Type VII. *Unstable Star:* $\lambda_1 = \lambda_2 > 0$

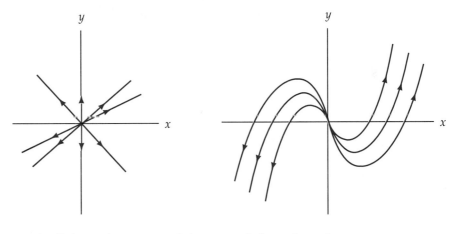

Example (left case): $x = 2e^t$, $y = 3e^t$; $3x = 2y$.
Example (right case): $x = e^t$, $y = te^t$; $y = x \ln x$.

Type VIII. *Asymptotically Stable Star:* $\lambda_1 = \lambda_2 < 0$

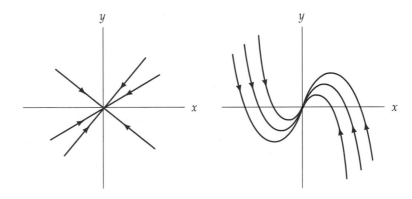

Example (left case): $x = e^{-t}$, $y = 3e^{-t}$; $3x = y$.
Example (right case): $x = e^{-t}$, $y = te^{-1}$; $y = -x \ln x$.

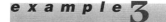

Consider the trajectory portrait in Figure 6.15. All trajectories appear to be spiraling toward the origin, so we might conclude tentatively that we're looking at an asymptotically stable spiral. If the underlying system were linear, the characteristic roots would have the form $\lambda_1 = -p + iq$, $\lambda_2 = -p - iq$, $p > 0$, $q \neq 0$. Indeed, the portrait belongs to the system $\dot{x} = -x + y$, $\dot{y} = -x - y$, which has roots $-1 \pm i$.

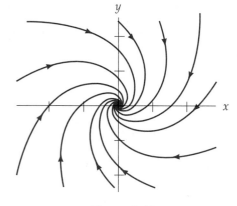

Figure 6.15

The damped oscillator equation $\ddot{x} + k\dot{x} + hx = 0$ is equivalent to the system $\dot{x} = y$, $\dot{y} = -hx - ky$. We find characteristic roots $\lambda_{1,2} = -(k/2) \pm \sqrt{(k/2)^2 - h}$. Assuming as usual that $k > 0$, we see that if $(k/2)^2 > h$, then both roots are negative, so the origin is an asymptotically stable node in the xy phase space. If $(k/2)^2 \leq h$, we get an asymptotically stable spiral or a star.

The characteristic equation
$$\lambda^2 - (a + d)\lambda + (ad - bc) = 0$$
has $\lambda_1 = 0$ for a root precisely when $ad - bc = 0$. This is case (d) in Theorem 8.3 and is not covered by the above stability classification; this case is called **degenerate** for two reasons. One is that the linear combination $c_1 + c_2 e^{\lambda_2 t}$ contains at most one nonconstant function (or none if $a + d = 0$, so that $\lambda_2 = 0$ also). The other reason is that $ad - bc = 0$ precisely when there are infinitely many equilibrium points for the system. The equilibrium points (x, y) occur when
$$ax + by = 0,$$
$$cx + dy = 0$$
both hold. It is left as an exercise to show that the equilibrium points cover at least an entire line when $ad - bc = 0$. (If a, b, c, and d are all zero, every point is an equilibrium point.)

example 4   The system

$$\frac{dx}{dt} = x + y,$$

$$\frac{dy}{dt} = 2x + 2y$$

has equilibrium points (x, y) satisfying

$$x + y = 0,$$
$$2x + 2y = 0.$$

The solutions (x, y) lie on the line $x + y = 0$. The characteristic roots satisfy $\lambda^2 - 3\lambda = 0$ and are $\lambda_1 = 0$, $\lambda_2 = 3$. The solutions of the system are

$$x(t) = c_1 + c_2 e^{3t},$$

$$y(t) = -c_1 + 2c_2 e^{3t}.$$

Elimination of e^{3t} between these two equations shows that the trajectories of the nonequilibrium solutions lie on parallel straight lines with slope 2.

EXERCISES

1. Find the characteristic roots λ_1, λ_2 associated with each of the following systems. Then classify each system according to the types categorized in I through VIII of the text.

   (a) $dx/dt = -3x + 2y,$
   $dy/dt = -4x + 3y.$

   (b) $dx/dt = x + 4y,$
   $dy/dt = 5y.$

   (c) $dx/dt = 2x - y,$
   $dy/dt = x + 2y.$

   (d) $dx/dt = x,$
   $dy/dt = 3x + y.$

   (e) $dx/dt = 2x - 3y,$
   $dy/dt = 2x - 2y.$

   (f) $dx/dt = -x + y,$
   $dy/dt = -4x - y.$

   (g) $dx/dt = x,$
   $dy/dt = 2x.$

   (h) $dx/dt = x + y,$
   $dy/dt = x + y.$

2. Each of the following second-order equations is equivalent to a first-order system obtained by letting $dx/dt = y$. Classify each equation (or equivalently each system) by finding the characteristic roots of the given second-order equation directly.

   (a) $d^2x/dt^2 - x = 0.$

   (b) $d^2x/dt^2 + dx/dt - x = 0.$

   (c) $d^2x/dt^2 + dx/dt + x = 0.$

   (d) $d^2x/dt^2 + x = 0.$

3. The following second-order equations can be thought of as determining the development $x(t)$ of a spring system subject to a frictional force $k\, dx/dt$, where $k \geq 0$ is constant. Find conditions on k under which the equation belongs to any or all of the classes in the standard list.

(a) $d^2x/dt^2 + k\,dx/dt + x = 0$.

(b) $d^2x/dt^2 + k\,dx/dt + 3x = 0$.

(c) $2d^2x/dt^2 + k\,dx/dt + x = 0$.

(d) $m\,d^2x/dt^2 + k\,dx/dt + 3x = 0$, $m > 0$.

4. Sketch typical phase plots in the xy plane for each of the systems in Exercise 1.

5. Sketch typical phase plots in the xy plane, where $y = dx/dt$, for each of the second-order equations in Exercise 2.

6. Show that the system

$$\frac{dx}{dt} = ax + by, \qquad \frac{dy}{dt} = cx + dy$$

has infinitely many constant equilibrium solutions if and only if $ad - bc = 0$. [*Hint:* Try to solve $ax + by = cx + dy = 0$.]

7. Consider the general solution (Equations 8.2 of the text) of a linear system in which $\lambda_1 < \lambda_2 < 0$, and for which $(0, 0)$ is an asymptotically stable node.

   (a) Show that, if c_2 and c_4 are not both zero in Equations 8.2, then as $t \to \infty$ the slope $dy/dx = (dy/dt)/(dx/dt)$ of the associated trajectory approaches c_4/c_2, interpreted as a vertical slope if $c_2 = 0$.

   (b) Show that if $c_2 = c_4 = 0$, but c_1 and c_3 are not both zero, then all trajectories are straight lines with slope c_3/c_1.

8. **(a)** Show that the nonlinear system

$$\frac{dx}{dt} = A - Bx - x + x^2y, \qquad A \neq 0,$$

$$\frac{dy}{dt} = Bx - x^2y$$

has a single constant solution $x(t) = A$, $y(t) = B/A$. Recall that the one-point trajectory of a constant solution is called an **equilibrium point.**

   (b) Assume that $A > 0$ and $B > A^2 + 1$. Trajectories starting sufficiently near, but not at, the equilibrium point exhibit **limit cycle** behavior in that they approach a closed trajectory. Investigate this claim by using a graphical-numerical method.

9. What difference, if any, is there between the trajectory portraits of the system $\dot{x} = f(x, y)$, $\dot{y} = g(x, y)$ and the system $\dot{x} = -f(x, y)$, $\dot{y} = -g(x, y)$?

10. Match the portraits A, B, C, D in the accompanying figure with the systems:

   (a) $\dot{x} = x$, $\dot{y} = x - y$.

   (b) $\dot{x} = x$, $\dot{y} = x + y$.

   (c) $\dot{x} = x - y$, $\dot{y} = x + y$.

   (d) $\dot{x} = 2y$, $\dot{y} = -x$.

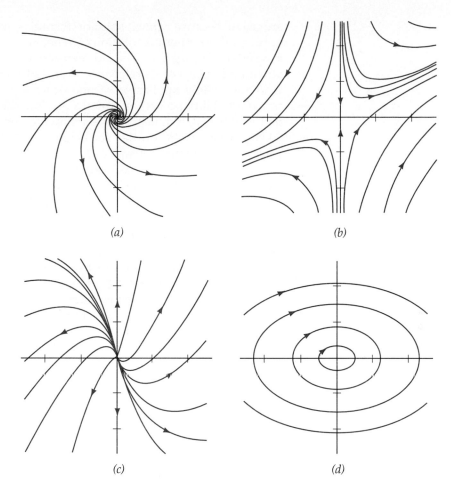

(a) *(b)*

(c) *(d)*

11. Show that the system $\dot{x} = ax + by$, $\dot{y} = cx + dy$ has an asymptotically stable equilibrium (both roots negative or with negative real parts) at the origin if and only if $a + d < 0$ and $ad - bc > 0$.

12. (a) Solve the system $\dot{x} = y$, $\dot{y} = -x - y$, $\dot{z} = x - z$.
   (b) Show that every solution tends to the equilibrium point $(0, 0, 0)$, which is thus asymptotically stable.
   (c) What changes if the last equation is replaced by $\dot{z} = x + z$?

13. (a) Show that Equations 8.1 of the text have $(0, 0)$ as their only equilibrium point precisely when $ad - bc \neq 0$.
   (b) Show that if $ad - bc = 0$ in Equations 8.1, then there are infinitely many eqilibrium points, covering an entire line or even the entire plane.

8B *Nonlinear Systems and Linearization*

Methods for determining the long-term ($t \rightarrow \infty$) stability or instability of solutions of the 2-dimensional nonlinear system

■■ **8.4**
$$\frac{dx}{dy} = f(x, y), \qquad \frac{dy}{dt} = g(x, y)$$

are useful because explicit solution formulas are often impossible to find, and approximate numerical solutions are necessarily always limited to some finite time interval $t_0 \le t \le t_1$. The 2-dimensional case is singled out first for detailed attention because of its relevance to the system $\dot{x} = y$, $\dot{y} = f(t, x, y)$ associated with the widely applicable second-order equation $\ddot{x} = f(x, \dot{x})$. The most extreme instability occurs for systems of dimension more than 2, and systems of higher dimension appear in the following Section 8C and also in Chapter 7, Section 5. First we'll consider the behavior of solution trajectories near an **equilibrium point** of the system 8.4, that is, a point (x_0, y_0) having the property that $f(x_0, y_0) = g(x_0, y_0) = 0$. Associated with each such point is a constant **equilibrium solution** $x(t) = x_0$, $y(t) = y_0$. We will be concerned with finding out whether or not a solution that passes near such a point tends asymptotically to the point or exhibits some other kind of behavior. Given the nature of the general criteria we'll develop, we'll mainly restrict ourselves to considering an **isolated equilibrium point,** that is, an equilibrium point that does not have other equilibrium points arbitrarily close to it. Thus the kind of solution behavior described as "degenerate" in Section 8A will not be included.

Let (x_0, y_0) be an isolated equilibrium point of system 8.4. We are interested in solution behavior near (x_0, y_0), but we change variable, replacing x by $x + x_0$ and y by $y + y_0$, to reduce to the case $(x_0, y_0) = (0, 0)$ of Section 8A. The system then looks like

■■ **8.5** $$\frac{dx}{dt} = f(x + x_0, y + y_0), \qquad \frac{dy}{dt} = g(x + x_0, y + y_0),$$

and the equilibrium point of interest is now at $x = 0$, $y = 0$. Assume f and g are continuously differentiable, and use a first-degree **Taylor expansion:** thus we replace system 8.5 by the **linearized system** in terms of partial derivatives f_x, f_y, g_x, g_y of f and g, namely,

■■ **8.6**
$$\frac{dx}{dt} = f_x(x_0, y_0)x + f_y(x_0, y_0)y,$$

$$\frac{dy}{dt} = g_x(x_0, y_0)x + g_y(x_0, y_0)y.$$

It can be shown that system 8.6 is in a precise sense the best of all linear systems to examine near (x_0, y_0) instead of the nonlinear system 8.5. (Of course, if system 8.5 were itself a homogeneous linear system, then system 8.6 would be exactly the same system as system 8.5. For instance, if $f(x, y) = ax + by$, then $f_x(x_0, y_0) = a$ and $f_y(x_0, y_0) = b$, for all x_0, y_0.)

---

e x a m p l e 1   The equilibrium points of the nonlinear system

$$\frac{dx}{dt} = 2x - y^2,$$

$$\frac{dy}{dt} = x + y$$

are found by solving the algebraic system

$$2x - y^2 = 0,$$

$$x + y = 0.$$

The only solutions are $(x_0, y_0) = (0, 0)$ and $(x_1, y_1) = (2, -2)$, so we have two isolated equilibrium points. Near (x_0, y_0) we consider the linearized system

$$\frac{dx}{dt} = 2x - (2y_0)y,$$

$$\frac{dy}{dt} = x + y.$$

When $y_0 = 0$, the system has solutions

$$x(t) = c_1 e^{2t},$$

$$y(t) = c_1 e^{2t} + c_2 e^t.$$

The two positive characteristic roots show that the linear system has an unstable node at $(0, 0)$. Near $(x_1, y_1) = (2, -2)$ the linearization has the distinct characteristic roots $\lambda_1 = (3 + \sqrt{17})/2$, $\lambda_2 = (3 - \sqrt{17})/2$. Thus the solutions exhibit saddle-type behavior as $t \to \infty$, since $\lambda_1 > 0$ and $\lambda_2 < 0$.

The next important step in Example 1 would be to draw some conclusion about the behavior of solutions of the given nonlinear system from what we know about the solutions of the linearized system at each equilibrium point. To state the connection clearly and succinctly, we will use some standard terminology. Let $x_0(t) = x_0$, $y_0(t) = y_0$ be an equilibrium solution of an autonomous system of differential equations, linear or nonlinear. Then (x_0, y_0) is **asymptotically stable** if every solution $(x(t), y(t))$ of the system that has a trajectory point close enough to (x_0, y_0) satisfies

$$\lim_{t \to \infty} (x(t), y(t)) = (x_0, y_0).$$

For the linear systems in Section 8A, example types III, VI, and VIII show asymptotically stable behavior at $(0, 0)$. Special types of asymptotic stability are the **spiral,** in which the trajectory winds around the equilibrium point as it approaches it, and the **node,** in which the tangent to each trajectory approaches a limiting position. A less strict condition on the system at (x_0, y_0) is that (x_0, y_0) be simply **stable,** which means that every trajectory that has a point close enough to (x_0, y_0) will remain within a preassigned distance of (x_0, y_0). Clearly, if (x_0, y_0) is asymptotically stable for a system, then it is also stable. Type V in Section 8A is stable without being asymptotically stable. All other equilibrium points are called **unstable;** these include types I, II, IV, and VII. In particular, type II, the **saddle point,** can be characterized for the nonlinear system as having two lines through it, one of which is tangent to a trajectory approaching the point and the other of which is tangent to a trajectory that moves away from the point.

■ 8.7 THEOREM

Let $f(x, y)$ and $g(x, y)$ be continuously differentiable in a region R of the xy plane, and suppose the point (x_0, y_0) in R is an isolated equilibrium point of the system

$$\frac{dx}{dt} = f(x, y), \qquad \frac{dy}{dt} = g(x, y).$$

Suppose that the associated linearized system at (x_0, y_0),

$$\frac{dx}{dt} = f_x(x_0, y_0)x + f_y(x_0, y_0)y, \qquad \frac{dy}{dt} = g_x(x_0, y_0)x + g_y(x_0, y_0)y,$$

is **nondegenerate:**

$$f_x(x_0, y_0)g_y(x_0, y_0) - f_y(x_0, y_0)g_x(x_0, y_0) \neq 0.$$

Then, at (x_0, y_0):

(a) Asymptotic stability for the linearized system at $(0, 0)$ implies asymptotic stability for the given system at (x_0, y_0), with nodal and spiral behavior corresponding similarly.

(b) Instability for the linearized system at $(0, 0)$ implies instability for the given system at (x_0, y_0). In particular, if $(0, 0)$ is a saddle point for the linearized system, it is a saddle point for the given system, while nodal and spiral behavior correspond similarly.

(c) If $(0, 0)$ is a stable center for the linearized system, then (x_0, y_0) may be an unstable spiral, an asymptotically stable spiral, or perhaps merely stable for the nonlinear system.

The proof of parts (a) and (b) involves some technical analysis that we will not undertake. Proving part (c) just amounts to looking at some appropriate examples, and this is done in Exercise 4 at the end of the section.

---

e x a m p l e **2**   Continuing with Example 1, we conclude from Theorem 8.7b that the nonlinear system has an unstable equilibrium at $(x_0, y_0) = (0, 0)$, since $(0, 0)$ is an unstable node for the linearized system. The same conclusion holds for the point $(2, -2)$, at which the linearized system has a saddle point. More specifically, $(2, -2)$ is a saddle point for the nonlinear system, since $\lambda_1 = (3 - \sqrt{17})/2 < 0 < \lambda_2 = (3 + \sqrt{17})/2$.

---

e x a m p l e **3**   The undamped nonlinear pendulum equation $\ddot{\theta} = -(g/l)\,\theta$ is derived in Section 2 of Chapter 4. Here $\theta = \theta(t)$ is the displacement angle from the downward-oriented vertical. Letting $x = \theta$ and $y = \dot{\theta}$, the equivalent system is

$$\frac{dx}{dt} = y, \qquad \frac{dy}{dt} = -\frac{g}{l}\sin x.$$

Mechanical intuition suggests that the equilibrium solutions are those for which $y = \dot{\theta} = 0$ and $x = \theta = n\pi$, where n is an integer. When n is even, the pendulum hangs straight down, and when n is odd, it is precisely balanced in a straight-up position, the latter obviously being mechanically unstable. Solving the equations $y = 0$, $-(g/l) \sin x = 0$ for $(x, y) = (0, n\pi)$ confirms some of our intuition. For example, consider the linearized system at $(x_n, y_n) = (0, n\pi)$. We find

$$\frac{dx}{dt} = y,$$

$$\frac{dy}{dt} = -\frac{g}{l} (\cos n\pi) x.$$

The characteristic roots satisfy

$$\lambda^2 + (-1)^n \frac{g}{l} = 0,$$

since $\cos n\pi = (-1)^n$. If n is even (pendulum down), the roots are $\pm i\sqrt{g/l}$. If n is odd, the roots are $\pm\sqrt{g/l}$. In the first case, we have a center for the linear system, so Theorem 8.7 fails to confirm or deny our intuition. It takes a more detailed analysis (see Exercise 3) to show that the expected stability does occur near $(x_0, y_0) = (2m\pi, 0)$. When n is odd, the linear system has a saddle point at $((2m + 1)\pi, 0)$. This tells us that, while the vertical positions with zero velocity are unstable, there is in principle a path in the xy phase space leading to a straight-up balanced position. Figure 4.14a shows the result of some numerical computations of phase-space trajectories. The effect of frictional damping on the nature of equilibrium points is taken up in Exercise 5 and in Section 6 of Chapter 4.

There may be an equilibrium point of an autonomous system that is isolated from every other specific trajectory of the system, though we've seen in Section 8A that this situation is far from typical. (Indeed the stable center numbered V in the display is the only one in the entire list of 2-dimensional linear types that has that property.) The following theorem gives conditions that imply the complementary case, in which an equilibrium point x_0 has at least one other trajectory with points arbitrarily close to x_0.

■ **8.8 THEOREM** Let $F(x)$ be a continuous vector field and let $x(t)$ be a solution of the autonomous system $\dot{x} = F(x)$. If either of the limits

$$\lim_{t \to \infty} x(t) \qquad \text{or} \qquad \lim_{t \to -\infty} x(t)$$

exists then the limit vector is an equilibrium point for the system.

Proof. Denoting the limit of $x(t)$ as $t \to \infty$ by x_0, we just need to show that $F(x_0) = 0$ to guarantee that x_0 is an equilibrium point. Since $F(x)$ is continuous at x_0,

$$\lim_{t \to \infty} F(x(t)) = F(\lim_{t \to \infty} x(t)) = F(x_0).$$

If $F(x_0) \neq 0$, it must have some coordinate $a_k \neq 0$. Then for all large enough t,

$$|F_k(x(t)) - a_k| < \tfrac{1}{2}|a_k|, \qquad \text{that is,} \qquad \tfrac{1}{2}|a_k| - a_k < F_k(x(t)) < \tfrac{1}{2}|a_k| - a_k.$$

If $a_k > 0$, then $\tfrac{1}{2}a_k < F_k(x(t)) = \dot{x}_k(t)$ for all large enough t, so $x_k(t)$ is unbounded above.

If $a_k < 0$, then $\frac{1}{2}a_k > F_k(\mathbf{x}(t)) = \dot{x}_k(t)$ for all large enough t, so $x_k(t)$ is unbounded below.

In either case $\mathbf{x}(t)$ can't have a limit, so $F(\mathbf{x}_0) = 0$ is the only possibility. If $t \to -\infty$, the argument is essentially the same. ∎

e x a m p l e 4

Figure 6.16a is a scaled-down sketch of the vector field of the system $\dot{x} = -x(x^2 + y^2 - 1)$, $\dot{y} = -y(x^2 + y^2 + 1)$. When $y \neq 0$ the second coordinate of the field can't be zero, so all equilibrium points must be on the x axis, where $y = 0$. In that case we just have to find solutions of $-x(x^2 - 1) = 0$; these are $x = -1, 0, 1$, so the equilibrium points are at $(-1, 0)$, $(0, 0)$, and $(1, 0)$. The sketch suggests that the points $(\pm 1, 0)$ are asymptotic equilibrium points, because the nearby arrows point at them. Similarly, $(0, 0)$ appears to be a saddle point. Establishing the truth of these statements using Theorem 8.7 is left as an exercise. Our main interest here is in illustrating the conclusion of Theorem 8.8. It appears from the sketch of the vector field that there is a trajectory lying on the y axis that approaches $(0, 0)$ from above and also one from below. In addition there appears to be a trajectory on the x axis extending from $(0, 0)$ to $(1, 0)$ and another from $(0, 0)$ to $(-1, 0)$. The precise meaning of this last statement is that a solution assuming a value $\mathbf{x}_0$ on the line segment between $(-1, 0)$ and $(0, 0)$ tends along the segment to $(-1, 0)$ as $t \to \infty$ and to $(0, 0)$ as $t \to -\infty$. This, as well as our other assertions about trajectories, can be verified by direct computation or by appealing to Theorem 5.1 of Chapter 2, Section 5.

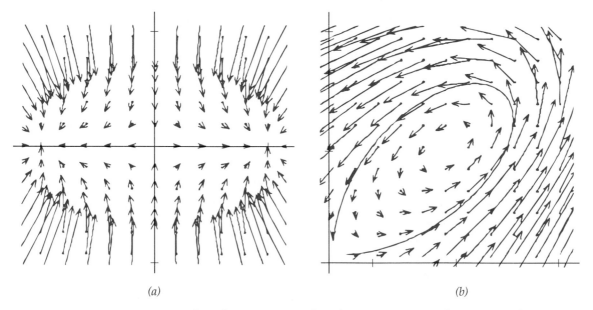

(a)   (b)

Figure 6.16 (a) $\dot{x} = -x(x^2 + y^2 - 1)$, $\dot{y} = -y(x^2 + y^2 + 1)$. (b) $\dot{x} = x - y^2$, $\dot{y} = -y + x^2$.

example 5

The system $\dot{x} = x - y^2$, $\dot{y} = -y + x^2$ has a scaled vector field shown for the first quadrant in Figure 6.16b, along with a trajectory extending counterclockwise away from the origin and back again. The equilibrium point at the origin is evidently of saddle type, since the linearization there is $\dot{x} = x$, $\dot{y} = -y$. The trajectories of the nonlinear system satisfy

$$\frac{dy}{dx} = \frac{\dot{y}}{\dot{x}} = \frac{-y + x^2}{x - y^2}.$$

In differential form this equation is $(y - x^2)\, dx + (x - y^2)\, dy = 0$ and is exact with solution $3xy - x^3 - y^3 = C$. The choice $C = 0$ is an algebraic curve called the folium of Descartes, which in addition to the equilibrium point $(0, 0)$ and the loop in the first quadrant contains unbounded trajectories in the second and fourth quadrants.

If a solution trajectory fails to converge to an equilibrium point as described in Theorem 8.8, it remains for us to describe other possibilities for relatively stable limiting behavior as $t \to +\infty$ or $-\infty$. Let $\mathbf{x} = \mathbf{x}(t)$ be a solution of an autonomous system $\dot{\mathbf{x}} = F(\mathbf{x})$. A point $\mathbf{z}$ is said to be an ω-**limit point** of the solution trajectory if there is a sequence $\{t_k\}$ of times, $\{t_k\} \to +\infty$, such that $\lim_{k \to \infty} \mathbf{x}(t_k) = \mathbf{z}$. The set of all ω-limit points of $\mathbf{x}$ is called the ω-**limit set** of $\mathbf{x}$ and is denoted by $\omega(\mathbf{x})$. The α-**limit set** of $\mathbf{x}$ is defined analogously with $\{t_k\} \to -\infty$. Thus if a solution $\mathbf{x} = \mathbf{x}(t)$ has a unique limit $\mathbf{z}_0$ as $t \to \infty$, then $\omega(\mathbf{x})$ consists of the single point $\mathbf{z}_0$. The next two examples are typical of the other possible limit sets for 2-dimensional systems.

example 6

The system $\dot{x} = -y$, $\dot{y} = x$ has counterclockwise circular trajectories except for a single equilibrium point at the origin. The system

$$\dot{x} = x(1 - x^2 - y^2) - y, \qquad \dot{y} = x + y(1 - x^2 - y^2)$$

is a modification that preserves the single equilibrium point and just one circular trajectory, the one with radius 1. All other trajectories either spiral into the circle from outside it or else spiral out to the circle from inside it, winding around infinitely many times while approaching arbitrarily close to the circle but never touching it. Thus the circular trajectory is the ω-limit set of every trajectory except the equilibrium point. The α-limit set of every trajectory strictly inside the circle is the equilibrium point $(0, 0)$. Figure 6.17a shows a sketch of the vector field and some trajectories that strongly suggests that these statement are valid. The analytical details are left as an exercise.

The circular trajectory in Example 6 is an instance of a **limit cycle,** which is by definition a closed, nonconstant trajectory that is also a limit set of some other trajectory. The Van der Pol equation as treated in Example 8 of Chapter 4, Section 5, provides another example of limit-cycle behavior. The next example

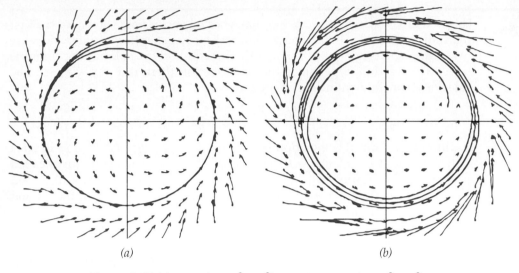

(a)   (b)

Figure 6.17 (a) $\dot{x} = x(1 - x^2 - y^2) - y$; $\dot{y} = x + y(1 - x^2 - y^2)$.
(b) $\dot{x} = x(1 - x^2 - y^2)^3 - y(1 - x^2 - y^2)^2 - y^3$;
$\dot{y} = x(1 - x^2 - y^2)^2 + y(1 - x^2 - y^2)^3 + xy^2$.

is similar to the previous one, because it exhibits the circle $x^2 + y^2 = 1$ as a limit set. While this time the limit set is still a circle, it is not a single cyclic trajectory but decomposes into two semicircular trajectories and two equilibrium points.

e x a m p l e 7   The vector field of the system

$$\dot{x} = x(1 - x^2 - y^2)^3 - y(1 - x^2 - y^2)^2 - y^3,$$
$$\dot{y} = x(1 - x^2 - y^2)^2 + y(1 - x^2 - y^2)^3 + xy^2$$

is obtained by multiplying the one in Example 6 by $(1 - x^2 - y^2)^2$ and then adding the vector $y^2(-y, x)$. Comparing Figure 6.16a and b shows some geometric similarities. In particular, the modified system turns out to have the circle $x^2 + y^2 = 1$ as an ω-limit set for all but a few of its trajectories. One striking difference is that, in addition to a common equilibrium point at $(0, 0)$, the modified system has equilibrium points at $(-1, 0)$ and $(1, 0)$ that aren't shared by the simpler system. Since these additional equilibrium points lie on the unit circle, the circle is not a limit cycle as in Example 6. Indeed, this limit set consists of two semicircular trajectories separated by equilibrium points. The semicircle $x^2 + y^2 = 1$, $y > 0$, is a trajectory with α-limit point $(1, 0)$ and ω-limit point $(-1, 0)$. The semicircle $x^2 + y^2 = 1$, $y < 0$, is a trajectory with α-limit point $(-1, 0)$ and ω-limit point $(1, 0)$. The circle $x^2 + y^2 = 1$ is an ω-limit set. Verifying some of the details is left for the exercises.

The next theorem gives conditions under which an ω-limit set of a 2-dimensional system must be "tame," restricted to being either an equilibrium point, a limit cycle, or the type considered in Example 7. The core of the theorem is called the **Poincaré-Bendixon theorem.** For a proof see S. Wiggins, *Applied*

Nonlinear Dynamical Systems and Chaos, Springer-Verlag, 1990. Higher-dimensional systems satisfying the same conditions may have a much wider range of behavior, in particular, chaotic behavior. See Chapter 4, Section 7C, where the third dimension is time t.

■ 8.9 THEOREM

Assume a 2-dimensional autonomous system has a continuously differentiable vector field, and let R be a region with at most finitely many equilibrium points of the system such that if $\mathbf{x}_0$ is in R, then the trajectory $\mathbf{x}(t)$ with $\mathbf{x}(0) = \mathbf{x}_0$ remains in R for $t \geq 0$. Then the only possible ω-limit set for a trajectory starting in R is either (a) an equilibrium point, (b) a limit cycle, or (c) a connected set consisting of equilibrium points joined pairwise by trajectories.

A region R such that trajectories starting there remain there for $t \geq 0$ is called **positively invariant.** Verifying positive invariance may be difficult to do precisely in general, but for 2-dimensional systems a reasonably accurate sketch of the relevant vector field may be convincing enough. For instance, in the vector fields shown in Figure 6.17, it's clear that the annular region $\frac{1}{2} < x^2 + y^2 < 2$ around the circle $x^2 + y^2 = 1$ is positively invariant for both systems. In Example 6 this region contains no equilibrium points at all, so the only possible ω-limit set is a limit cycle of type (b) in Theorem 8.9. In Example 7 the unit circle is the prime candidate to be a limit set. Since it contains equilibrium points, the circle can't be a limit cycle, so it must be of type (c).

EXERCISES

1. Find all the equilibrium points (x_0, y_0) for each of the following systems. Then find the associated linearized system for each such point. Use Theorem 8.7 to draw what conclusions you can about the behavior of solution trajectories near the equilibrium point. *Warning:* If the linearized system is degenerate $(ad = bc)$, Theorem 8.7 does not apply.

(a) $\dot{x} = y,$
$\quad \dot{y} = x + y^2.$

(b) $\dot{x} = y,$
$\quad \dot{y} = y + x^2.$

(c) $\dot{x} = -y,$
$\quad \dot{y} = x + x^2.$

(d) $\dot{x} = x + y^2,$
$\quad \dot{y} = x^2 - y.$

(e) $\dot{x} = x - xy^2,$
$\quad \dot{y} = x - y^3.$

(f) $\dot{x} = e^x - 1,$
$\quad \dot{y} = x + y.$

(g) $\dot{x} = x(1 - x^2 - y^2) - y, \ \dot{y} = x + y(1 - x^2 - y^2).$

(h) $\dot{x} = x(1 - x^2 - y^2)^3 - y(1 - x^2 - y^2)^2 - y^3,$
$\quad \dot{y} = x(1 - x^2 - y^2)^2 + y(1 - x^2 - y^2)^3 + xy^2.$

2. Each of the following systems has infinitely many equilibrium points. In each case, single out all isolated equilibrium points for which the linearized system in nondegenerate, and draw what conclusions you can about solution trajectories near these equilibrium points.

(a) $\dot{x} = 1 - y,$
$\quad \dot{y} = \sin x.$

(b) $\dot{x} = x(1 - x^2 - y^2),$
$\quad \dot{y} = y(1 - x^2 - y^2).$

3. In Example 3 of the text we looked at the system

$$\frac{dx}{dt} = y, \qquad \frac{dy}{dt} = -\frac{g}{l}\sin x,$$

which is derived from the nonlinear pendulum equation. Since $dy/dx = (dy/dt)/(dx/dt)$, the solutions are related by

$$\frac{dy}{dx} = \frac{-g\sin x}{ly}.$$

(a) Use separation of variables to show that solution trajectories satisfy

$$y^2 = \frac{2g}{l}\cos x + c, \qquad \text{where } c \text{ is const.}$$

(b) Show that if the constant c in part (a) satisfies $|c| < 2g/l$, then the trajectories are closed circuits about the equilibrium points $(x, y) = (2m\pi, 0)$, corresponding to stable oscillations.

(c) Show that if $c > 2g/l$, the trajectories are periodic graphs, unbounded in the x direction. These trajectories correspond to a perpetual winding motion of the pendulum.

(d) Show that if $c = 2g/l$, the trajectories represent critical paths of motion that extend from one unstable equilibrium point to another. (Hence one such trajectory fits, for two distinct equilibrium points, the limit behavior described in Theorem 8.8.)

4. Consider the family of nonlinear systems

$$\frac{dx}{dt} = y + \alpha x(x^2 + y^2),$$

$$\frac{dy}{dt} = -x + \alpha y(x^2 + y^2),$$

where α is constant. Linearized analysis is inadequate for this system.

(a) Show that the only equilibrium point is $(x_0, y_0) = (0, 0)$, regardless of the value of α.

(b) Show that the linearized system associated with $(0, 0)$ is $\dot{x} = y$, $\dot{y} = -x$, and that the origin is a stable center as $t \to \infty$. Note that $(0, 0)$ is also a stable center for the given system when $\alpha = 0$.

(c) Show that in polar coordinates the given system takes the form $\dot{r} = \alpha r^3$, $\dot{\theta} = -1$. [Hint: Apply d/dt to the equations $x = r\cos\theta$, $y = r\sin\theta$.]

(d) Solve the polar-form system in part (c), and show that if $\alpha > 0$, the equilibrium point is unstable, and if $\alpha < 0$, it is stable. Thus the parameter value $\alpha = 0$ is called a **bifurcation point** for the system, because the stability of the system at the equilibrium point changes in a fundamental way as α passes through zero.

5. A nonlinear pendulum with frictional damping has a displacement angle $\theta = \theta(t)$ that satisfies $\ddot{\theta} + (k/(lm))\dot{\theta} + (g/l)\sin\theta = 0$, where $k > 0$ is constant.

(a) With $x = \theta$, show the equation for θ equivalent to the first-order system

$$\dot{x} = y, \qquad \dot{y} = -\frac{g}{l}\sin x - \frac{k}{lm}y.$$

(b) Show that the equilibrium points of the system in part (a) are independent of k.

(c) Show that the unstable equilibrium points are all saddles.

(d) Show that the stable equilibrium points are nodes if $k^2 > 4glm^2$ and asymptotic spirals if $k^2 < 4glm^2$. Why does this last distinction make sense physically?

6. The **Lotka-Volterra equations** are discussed in Section 2:

$$\frac{dH}{dt} = (a - bP)H, \qquad \frac{dP}{dt} = (cH - d)P, \qquad a, b, c, d > 0.$$

(a) Show that for $H > 0$ and $P > 0$, the only equilibrium point is $(H_0, P_0) = (d/c, a/b)$, and find the associated linearized system.

(b) Show that the equilibrium solution of the linearized system is a stable center.

(c) Explain why the result of part (b) is consistent with the results of text Example 3 of Section 2.

(d) Discuss the equilibrium solution of the nonlinear system at $(H, P) = (0, 0)$. Do your conclusions make sense, given the interpretation of P and H as sizes of parasite and host populations, respectively?

7. Discuss the nature of the equilibrium point of the system in Exercise 8 of Section 8A, under the assumptions $A > 0$, $B > A^2 + 1$.

8. The nonlinear system $\dot{x} = 2ye^{x^2+y^2}$, $\dot{y} = -2xe^{x^2+y^2}$ is shown in Example 4 of the next section to have a stable equilibrium at $(0, 0)$. Show that an analysis of the linearized system at $(0, 0)$ fails to lead to this conclusion.

9. Find all equilibrium points of $\dot{x} = -y(1 - x^2 - y^2)$, $\dot{y} = x(1 - x^2 - y^2)$. Then show that all other trajectories are circles and that no trajectory converges to an equilibrium point. [*Hint:* $x\,dx + y\,dy = 0$.]

10. Find all equilibrium points of $\dot{x} = x(2 - x - y)$, $\dot{y} = y(x - 1)$ and discuss their stability. Then, by hand or using computer graphics, make a sketch of some typical trajectories near the equilibrium points.

11. Compute the equilibrium solutions of the system $\dot{x} = -x(x^2 + y^2 - 1)$, $\dot{y} = -y(x^2 + y^2 + 1)$ and discuss their stability.

12. The system $\dot{x} = -x(x^2 + y^2 - 1)$, $\dot{y} = -y(x^2 + y^2 + 1)$ discussed in Example 4 of the text reduces to $\dot{x} = -x(x^2 - 1)$ on the x axis and to $\dot{y} = -y(y^2 + 1)$ on the y axis.

(a) Solve the y equation explicitly to show that a trajectory starting at y_0 on the y axis converges to the origin as $t \to \infty$.

(b) Solve the x equation explicitly to show that if $|x_0| < 1$, a trajectory starting at x_0 on the x axis converges to either $(1, 0)$ or $(-1, 0)$ as $t \to \infty$ and converges to the origin as $t \to -\infty$.

13. The system $\dot{x} = -x(x^2 + y^2 - 1)$, $\dot{y} = -y(x^2 + y^2 + 1)$ discussed in Example 4 of the text reduces to $\dot{x} = -x(x^2 - 1)$ on the x axis and to $\dot{y} = -y(y^2 + 1)$ on the y axis. Use Theorem 5.1 of Chapter 2, Section 5, to establish the following.

(a) Show that a trajectory starting at y_0 on the y axis converges to the origin as $t \to \infty$.

(b) Show that if $|x_0| < 1$, a trajectory starting at x_0 on the x axis converges to either $(1, 0)$ or $(-1, 0)$ as $t \to \infty$ and converges to the origin as $t \to -\infty$.

14. In Example 6 of the text we considered the system
$$\dot{x} = x(1 - x^2 - y^2) - y, \quad \dot{y} = x + y(1 - x^2 - y^2).$$

(a) Show the system has a circular orbit traced by the solution $x = \cos t$, $y = \sin t$.

(b) Show that all solutions satisfy $(d/dt)(y/x) = 1 + (y/x)^2$.

(c) Use part (b) to show the polar angle θ of a point on a nonconstant trajectory satisfies $\theta = \arctan (y/x) = t + c$ and hence that all such trajectories wind counterclockwise infinitely often around the origin.

(d) Show that all solutions satisfy $x\dot{x} + y\dot{y} = (x^2 + y^2)(1 - x^2 - y^2)$.

(e) Use part (d) to show that the polar radius r of a point on a trajectory satisfies $dr/dt = r(1 - r^2)$, with solutions $r = ke^t/\sqrt{k^2 e^{2t} \pm 1}$, the sign depending on whether $0 < r < 1$ or $r > 1$. Show that these solutions have as an ω-limit set the circular trajectory $x^2 + y^2 = 1$.

15. In Example 6 of the text we considered the system
$$\dot{x} = x(1 - x^2 - y^2) - y, \qquad \dot{y} = x + y(1 - x^2 - y^2).$$

(a) Show that all solutions satisfy $x\dot{x} + y\dot{y} = (x^2 + y^2)(1 - x^2 - y^2)$ and hence that the polar radius r of a point on a trajectory satisfies $dr/dt = r(1 - r^2)$.

(b) Show that the xy plane with $(0, 0)$ deleted is positively invariant for the system by integrating the result of part (a) from 0 to t.

(c) Explain how to conclude from Theorem 8.9 that the circle $x^2 + y^2 = 1$ is a limit cycle.

16. In Example 7 of the text we considered the system
$$\dot{x} = x(1 - x^2 - y^2)^3 - y(1 - x^2 - y^2)^2 - y^3,$$
$$\dot{y} = x(1 - x^2 - y^2)^2 + y(1 - x^2 - y^2)^3 + xy^2.$$

(a) Show that all solutions satisfy $x\dot{x} + y\dot{y} = (x^2 + y^2)(1 - x^2 - y^2)^3$ and hence that the polar radius r of a point on a trajectory satisfies $dr/dt = r(1 - r^2)^3$.

(b) Show that the xy plane with $(0, 0)$ deleted is positively invariant for the system by integrating the result of part (a) from 0 to t.

(c) Use Theorem 8.9 to show that the circle $x^2 + y^2 = 1$ is an ω-limit set.

(d) Show that the system has trajectories on the unit circle satisfying $\dot{x} = -y^3, \dot{y} = xy^2$. Explain why the unit semicircle with $y > 0$ is a trajectory with $(1, 0)$ as an α-limit point and $(-1, 0)$ as an ω-limit point. What can you say about the semicircle with $y < 0$?

17. The **Van der Pol equation** $\ddot{x} + \alpha(x^2 - 1)\dot{x} + x = 0$ is equivalent to the system $\dot{x} = y, \dot{y} = -x - \alpha(x^2 - 1)y$.

(a) Find the linearization of the system near $(x_0, y_0) = (0, 0)$.

(b) Discuss the behavior of solutions near $(x_0, y_0) = (0, 0)$ and their dependence on the constant $\alpha > 0$. What happens if $\alpha = 0$ or if $\alpha < 0$?

(c) Make a computer-aided sketch of the vector field of the system for $\alpha = 0.5$, use it to identify a positively invariant region R of the xy plane that excludes $(0,0)$, and conclude that this Van der Pol oscillator has a limit cycle. For a sketch of some trajectories see Figure 4.15 in Chapter 4.

18. It can be shown that if a 2-dimensional system has a closed (i.e., periodic) solution trajectory and the system's vector field is also defined and continuous everywhere inside this trajectory, then the system has an equilibrium point inside the trajectory curve.

   (a) Illustrate this principle with an example.

   (b) Find an example of a 2-dimensional system with an equilibrium point but which has no closed trajectories.

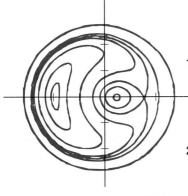

19. (a) Show that the **Van der Pol equation** $\ddot{x} + \alpha(x^2 - 1)\dot{x} + x = 0$ is equivalent to the system $\dot{x} = -\alpha(x^3/3 - x) + y$, $\dot{y} = -x$.

   (b) Find the linearization at the equilibrium point $(x_0, y_0) = (0,0)$ of the system in part (a), and compare it with the linearization found in the previous exercise.

20. (a) $\dot{x} = \frac{1}{2}y(\frac{39}{16} - \frac{3}{4}(x^2 + y^2))$,
   $\dot{y} = -\frac{1}{2}x(\frac{39}{16} - \frac{3}{4}(x^2 + y^2)) + \frac{9}{16}$
   has three equilibrium solutions. One is $(-2, 0)$; what are the others?

   (b) Use a numerical system plotter to plot typical solution curves for the system in part (a), paying particular attention to direction of traversal and behavior near equilibrium points. Before adding direction arrows, your picture should look something like the one in the margin.

21. Use **Green's theorem** in the form $\int_R (\partial f/\partial x + \partial g/\partial x)\, dx\, dy = \int_C (f\, dy - g\, dx)$ to prove the **Bendixon criterion**: If $\partial f/\partial x + \partial g/\partial y$ is either always negative or always positive in a region R, then R can't contain a closed trajectory C of the system $\dot{x} = f(x, y)$, $\dot{y} = g(x, y)$. Assume the partial derivatives are continuous on R. [*Hint:* Show the line integral is zero.]

8C Liapunov's Method

The method of Section 8B for determining trajectory behavior near an equilibrium point x_0 depends on comparing the behavior of a nonlinear system to that of a related linear system. The Liapunov method described below is sometimes called *direct*, because it avoids reference to an intervening auxiliary system. Furthermore, the method allows us to determine a specific region of initial points such that solutions starting there tend to a given equilibrium point x_0; the set of *all* such points is called the **basin of attraction** of x_0. The geometric idea behind the Liapunov method is quite simple. Given a first-order system $\dot{x} = F(x)$ with an equilibrium point x_0, suppose you can find a real-valued function $V(x)$ on $\mathcal{R}^n$ that has a strict local minimum at $x = x_0$ and such that $-\nabla V(x) \cdot F(x) > 0$ in some neighborhood of x_0. Recall that the *negative* of the gradient vector $\nabla V(x)$ points in the direction of maximum *decrease* of V at x. On a solution trajectory $x = x(t)$, we have $\dot{x} = F(x)$, so the vector $F(x)$ points in the direction of the trajectory. But the relations

$$-\nabla V(x) \cdot F(x) = -\nabla V(x) \cdot \dot{x}$$
$$= |\nabla V(x)||\dot{x}| \cos \theta > 0$$

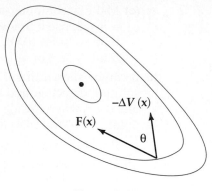

Figure 6.18

show that the cosine of the angle θ between the trajectory and the direction of maximum decrease satisfies $\cos \theta > 0$. In other words, we must have $-\pi/2 < \theta < \pi/2$, so the trajectory points in a "downhill" direction relative to the graph of V. It is reasonable to expect that maintaining this general direction near a strict minimum $V(\mathbf{x}_0)$ for V will keep the trajectory from wandering away from $\mathbf{x}_0$; this is stable behavior, which may even result in asymptotically stable behavior with $\mathbf{x}(t)$ tending to $\mathbf{x}_0$. Figure 6.18 shows level curves of $V(\mathbf{x})$ for an example in 2 dimensions, with $-\nabla V(x)$ perpendicular to a level curve and $\mathbf{F}(\mathbf{x})$ pointing in the general direction of the minimizing point for $V(\mathbf{x})$. The expression $\nabla V \cdot \mathbf{F}$ occurs often enough that the abbreviation $\dot{V}$ is often used for it:

$$\dot{V}(\mathbf{x}) = \nabla V(\mathbf{x}) \cdot \mathbf{F}(\mathbf{x}).$$

The notation is fairly natural; if $\mathbf{x}(t)$ is a trajectory, the chain rule then gives us

$$\frac{d}{dt} V(\mathbf{x}) = \nabla V(\mathbf{x}) \cdot \dot{\mathbf{x}} = \nabla V(\mathbf{x}) \cdot \mathbf{F}(\mathbf{x}) = \dot{V}(\mathbf{x}).$$

■ 8.10 LIAPUNOV STABILITY THEOREM

Let $\dot{\mathbf{x}} = \mathbf{F}(\mathbf{x})$ have an equilibrium solution $\mathbf{x}_0$ in a region R in which $\mathbf{F}(\mathbf{x})$ is continuous. Let $V(\mathbf{x})$ be real-valued and continuously differentiable on a neighborhood $B_r \subset R$ centered at $\mathbf{x}_0$ such that $V(\mathbf{x})$ has a strict minimum at $\mathbf{x} = x_0$. Let m be the minimum of $V(\mathbf{x})$ on the boundary of B_r, and let S be the subset of B_r for which $V(\mathbf{x}) < m$.

(a) If $\dot{V}(\mathbf{x}) \leq 0$ on B_r, then $\mathbf{x}_0$ is a stable equilibrium point, and a trajectory that enters S stays in S.

(b) If $\dot{V}(\mathbf{x}) < 0$ on B_r except at $\mathbf{x}_0$, then $\mathbf{x}_0$ is an asymptotically stable equilibrium point, and S is contained in the basin of attraction of $\mathbf{x}_0$.

A function V with an isolated minimum at $\mathbf{x}_0$ is called a **Liapunov function** for $\dot{x} = \mathbf{F}(\mathbf{x})$ if it satisfies condition (a) and a **strict Liapunov function** if it satisfies (b).

Proof. By a shift of coordinates we can assume that $\mathbf{x}_0 = 0$ and that $V(0) = 0$. Suppose r is chosen so that $V(\mathbf{x})$ is continuous and positive for $0 < |\mathbf{x}| \leq r$. Let m be the minimum of $V(\mathbf{x})$ on the sphere $|\mathbf{x}| = r$ in $\mathfrak{R}^n$. Now suppose a solution $\mathbf{x}(t)$ satisfies $0 < V(\mathbf{x}) < m$ for some value $t = t_0$. For the trajectory to leave the region S, the value of $V(\mathbf{x})$ would have to become at least as large as m. But this is impossible. The reason is that $V(\mathbf{x}) < m$ at one point of the trajectory, and since

$$\frac{d}{dt} V(\mathbf{x}(t)) = \nabla V(\mathbf{x}(t)) \cdot \dot{\mathbf{x}}(t)$$

$$= V(\mathbf{x}(t)) \cdot \mathbf{F}(\mathbf{x}) = \dot{V}(\mathbf{x}(t)) \leq 0,$$

$V(\mathbf{x})$ is nonincreasing thereafter along the trajectory. Hence $\mathbf{x}_0$ is stable, and the trajectory stays inside S. This proves conclusion (a).

To prove (b), observe that since now $\dot{V}(\mathbf{x}) < 0$, we must have $V(\mathbf{x})$ strictly decreasing along a trajectory unless the trajectory reaches $\mathbf{x}_0 = 0$. We'll first prove that $\dot{V}(\mathbf{x})$ tends to zero on a trajectory starting in S. Suppose not, and that there is a solution $\mathbf{x}(t)$ such that $0 < b < V(\mathbf{x}(t))$ for $t \geq 0$. Choose a positive $s < r$ such that $V(\mathbf{x}) < b$ for $|\mathbf{x}| < s$. It follows that $|\mathbf{x}(t)| > s$ for $t \geq t_0$. Since $\dot{V}(\mathbf{x})$ has an isolated zero value at 0, there is a $\mu > 0$ such that $\dot{V}(\mathbf{x}) \leq -\mu < 0$ when $s \leq |\mathbf{x}| \leq r$. Applying the fundamental theorem of calculus to the equation displayed above then gives

$$V(\mathbf{x}(t)) - V(\mathbf{x}(t_0)) = \int_{t_0}^{t} \dot{V}(\mathbf{x}(u)) \, du \leq -\mu(t - t_0).$$

The left side of the inequality is bounded below as t increases, but the right side is not. This contradiction negates our assumption, so $V(\mathbf{x})$ must tend to zero along the trajectory. To show that $\mathbf{x}(t)$ tends to $\mathbf{x}_0$, let ρ be an arbitrarily small positive number. We'll show that the trajectory stays in the neighborhood B_ρ of 0. Let m_ρ be the minimum of $V(\mathbf{x})$ on the boundary of B_ρ. Since $V(\mathbf{x}(t))$ tends to 0, we can find a t_ρ such that $V(\mathbf{x}(t_\rho)) < m_\rho$. It follows from part (a) that the trajectory stays in B_ρ for $t > t_\rho$. ∎

---

e x a m p l e **1**

The **Van der Pol equation**

$$\ddot{x} + \alpha(1 - x^2)\dot{x} + x = 0$$

is equivalent to the system

$$\dot{x} = y,$$
$$\dot{y} = -x - \alpha y(1 - x^2).$$

Thus we're dealing with $\dot{\mathbf{x}} = \mathbf{F}(\mathbf{x})$, where $\mathbf{F}(x, y) = (y, -x - \alpha y(1 - x^2))$. (The equilibrium point at $(x_0, y_0) = (0, 0)$ can be characterized for $\alpha \neq 0$ using the method of Section 8B, but using the Liapunov method will give us some additional information about the region of stability.) A Liapunov function should have a strict minimum at $(x, y) = (0, 0)$. A choice of the form $V(x, y) = ax^2 + by^2$, with positive a, b has the required minimum and allows us some latitude to choose a and b so that $\dot{V}(x, y) = \nabla V(x, y) \cdot \mathbf{F}(x, y) < 0$. We find that

$$\dot{V}(x, y) = \nabla V(x, y) \cdot \mathbf{F}(x, y)$$

$$= (2ax, 2by) \cdot (y, -x - \alpha(1 - x^2)y)$$

$$2axy \quad 2bxy \quad 2b\alpha y^2(1 \quad x^2).$$

We choose $a = b$ to make the first two terms cancel, since those terms could contribute positive values. Assume $\alpha > 0$. Then $\dot{V}(x, y) = -2b\alpha y^2(1 - x^2) \leq 0$ for $(x, y) \neq (0, 0)$ and for any $b > 0$, whenever $|x| \leq 1$. Let $b = 1$. Evidently

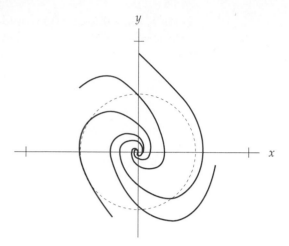

Figure 6.19 Asymptotically stable equilibrium.

$\dot{V}(x, y) \leq 0$ when $x^2 + y^2 \leq 1$, so a solution trajectory starting in or on a circle of radius 1 about the origin must stay in that disk. Also $\dot{V}(x, y) < 0$ when $x^2 + y^2 < 1$. Thus a solution trajectory starting in the interior of the disk must tend to $(0, 0)$. What the Liapunov theorem guarantees is that every trajectory starting inside the circle not only stays there but tends to $(0, 0)$. The actual region of asymptotic stability is somewhat larger, as the computer plot in Figure 6.19 shows. In the case $\alpha = 0$, $\dot{V}(x, y) = 0 \leq 0$ for all (x, y), so it follows that $(0, 0)$ is stable, though this follows independently from the elementary solution of the system.

---

example 2

The system
$$\dot{x} = y(z + 1), \qquad \dot{y} = -x(z + 1), \qquad \dot{z} = -z^3$$
has a single equilibrium point $(x_0, y_0, z_0) = (0, 0, 0)$. We try to find a Liapunov function of the form $V(x, y, z) = ax^2 + by^2 + cz^2$ with a, b, c positive. Then
$$\dot{V}(x, y, z) = \nabla V(x, y, z) \cdot \mathbf{F}(x, y, z)$$
$$= (2ax, 2by, 2cz) \cdot (y(z + 1), x(z + 1), -z^3)$$
$$= 2axy(z + 1) - 2bxy(z + 1) - 2cz^4.$$

Choosing $a = b = c = 1$, we have a Liapunov function with $\dot{V}(x, y, z) = -2z^4 \leq 0$; this function is not strict because $\dot{V}(x, y, 0) = 0$ not only at $(0, 0, 0)$ but also at every point of the xy plane. However, we can conclude that the origin is a stable equilibrium and that every solution trajectory that enters the neighborhood $x^2 + y^2 + z^2 \leq r^2$ will stay there from that time on.

The previous examples show that finding an appropriate Liapunov function is not necessarily an automatic routine. However, there are two important classes of system for which there are sometimes "ready-made" Liapunov functions. One of these classes consists of systems of the form $\dot{\mathbf{x}} = -\nabla V(\mathbf{x})$, where $V(\mathbf{x})$ is a continuously differentiable function; such a system is called a **gradi-**

ent system. It is a standard result of multivariable calculus (see Exercise 18) that the solution trajectories of the system are perpendicular to the level sets of V and point in the direction of maximum decrease of V at each point $\mathbf{x}$. An equilibrium point of a gradient system is a critical point $\mathbf{x}_0$ of V, and if $V(\mathbf{x}_0)$ is an isolated local minimum value for V, then V itself will be a Liapunov function for the system at $\mathbf{x}_0$. The reason is that

$$\dot{V}(\mathbf{x}) = \nabla V(\mathbf{x}) \cdot (-\nabla V(\mathbf{x}))$$
$$= -|\nabla V(\mathbf{x})|^2 \le 0.$$

To say that $\mathbf{x}_0$ is an *isolated* minimum point for V means that there is a neighborhood B of $\mathbf{x}_0$ such that $\nabla V(\mathbf{x}) \ne 0$ throughout B except at $\mathbf{x}_0$ itself. Thus $\dot{V}(\mathbf{x}) < 0$ except at $\mathbf{x} = \mathbf{x}_0$, so $\mathbf{x}_0$ will be an asymptotically stable equilibrium point.

---

example 3

Suppose $V(x, y) = e^{x^2+y^2}$. The associated gradient system is

$$\dot{x} = -2xe^{x^2+y^2}, \qquad \dot{y} = -2ye^{x^2+y^2}.$$

Taking $V(x, y)$ itself for a Liapunov function at the unique critical point $(0, 0)$ we find

$$\dot{V}(\mathbf{x}) = -|\nabla V(\mathbf{x})|^2 = -4(x^2 + y^2)e^{2(x^2+y^2)}.$$

Thus $\dot{V}(x, y) < 0$ except at the critical equilibrium point. It follows from part (b) of the Liapunov theorem that $(0, 0)$ is asymptotically stable, but the theorem says even more. The minimum of $V(x, y) = e^{x^2+y^2}$ on the circle $x^2 + y^2 = r^2$ centered at $(0, 0)$ is $m = e^{r^2}$. But $V(x, y) < m$ on the disk S inside the circle. Thus every such disk is in the basin of attraction of $(0, 0)$, so the basin consists of all points (x, y) in $\mathcal{R}^2$.

Apart from gradient systems, there is another kind of system that sometimes has a ready-made Liapunov function. These systems necessarily have even dimension $2n$. We start with an arbitrary continuously differentiable function $H(\mathbf{x}, \mathbf{y}) = H(x_1, \ldots, x_n, y_1, \ldots, y_n)$ or $2n$ real variables. The system of $2n$ first-order equations

$$\dot{x}_k = \frac{\partial H}{\partial y_k}, \qquad k = 1, \ldots, n$$

$$\dot{y}_k = -\frac{\partial H}{\partial x_k}, \qquad k = 1, \ldots, n$$

is called a **Hamiltonian system** with **Hamiltonian** H. What is important for us here is that the gradient at $(\mathbf{x}, \mathbf{y})$ of H as a function of the $2n$ variables is perpendicular to the $2n$-dimensional vector field of the Hamiltonian system. To see this, take as a Liapunov function at an equilibrium point the Hamiltonian function: $V(\mathbf{x}, \mathbf{y}) = H(\mathbf{x}, \mathbf{y})$. Then

$$\dot{V}(\mathbf{x}, \mathbf{y}) = \left(\frac{\partial H}{\partial x_1}, \ldots, \frac{\partial H}{\partial x_n}, \frac{\partial H}{\partial y_1}, \ldots, \frac{\partial H}{\partial y_n}\right) \cdot \left(\frac{\partial H}{\partial y_1}, \ldots, \frac{\partial H}{\partial y_n}, -\frac{\partial H}{\partial x_1}, \ldots, -\frac{\partial H}{\partial x_n}\right)$$

$$= \frac{\partial H}{\partial x_1}\frac{\partial H}{\partial y_1} + \cdots + \frac{\partial H}{\partial x_n}\frac{\partial H}{\partial y_n} - \frac{\partial H}{\partial y_1}\frac{\partial H}{\partial x_1} - \cdots - \frac{\partial H}{\partial y_n}\frac{\partial H}{\partial v_n} = 0.$$

Thus if the critical point in question gives an isolated minimum $H(x_0, y_0)$, then part (a) of the Liapunov theorem applies, and we conclude that (x_0, y_0) is a stable equilibrium.

---

e x a m p l e 4

Let $H(x, y) = e^{x^2 + y^2}$. The associated Hamiltonian system is
$$\dot{x} = 2ye^{x^2 + y^2}, \qquad \dot{y} = -2xe^{x^2 + y^2}.$$
Try $V(x, y) = H(x, y)$ for a Liapunov function. Since $\dot{V}(x, y) = 0$ for all (x, y), we just have to note that V has an isolated minimum at $(0, 0)$ to conclude from part (a) of the Liapunov theorem that $(0, 0)$ is a stable equilibrium. Every disk centered at $(0, 0)$ is a stability region. The reason is that the minimum of $V(x, y) = e^{x^2 + y^2}$ on the circle $x^2 + y^2 = r^2$ centered at $(0, 0)$ is $m = e^{r^2}$. But $V(x, y) < m$ on the disk S inside the circle. Thus every trajectory that enters such a disk stays there from that time on.

Every second-order system of dimension n is equivalent to a $2n$-dimensional first-order system and so meets the even-dimensionality requirement for Hamiltonian systems. Given the usual method for reducing the order of $\ddot{\mathbf{x}} = \mathbf{F}(\mathbf{x})$ by setting $\dot{\mathbf{x}} = \mathbf{y}$, the only additional requirement is that $\mathbf{F}(\mathbf{x}) = -\nabla U(\mathbf{x})$ for some continuously differentiable real-valued function $U(\mathbf{x})$. (In other words, the original second-order system should be $\ddot{\mathbf{x}} = -\nabla U(\mathbf{x})$, a **conservative** system as discussed in Section 6.) It follows that the equivalent first-order system $\dot{\mathbf{x}} = \mathbf{y}$, $\dot{\mathbf{y}} = -\nabla U(\mathbf{x})$ is Hamiltonian with Hamiltonian function
$$H(\mathbf{x}, \mathbf{y}) = \tfrac{1}{2}|\mathbf{y}|^2 + U(\mathbf{x})$$
$$= \tfrac{1}{2}(y_1^2 + \cdots + y_n^2) + U(x_1, \ldots, x_n).$$
Clearly the Hamiltonian function has a strict minimum at $(\mathbf{x}_0, 0)$ if and only if U has a strict minimum at $\mathbf{x}_0$. This observation together with the Liapunov theorem proves over again the stability test of Section 6 to the effect that $\ddot{\mathbf{x}} = -\nabla U(\mathbf{x})$ *has a stable equilibrium at a strict minimum of* $U(\mathbf{x})$.

---

e x a m p l e 5

Let $U(x, y) = x^2 y^2$. The second-order conservative system $\ddot{x} = -2xy^2$, $\ddot{y} = -2x^2 y$ is equivalent to the first-order system
$$\dot{x} = z,$$
$$\dot{y} = w,$$
$$\dot{z} = -2xy^2,$$
$$\dot{w} = -2x^2 y.$$
This system is Hamiltonian with Hamiltonian function
$$H(x, y, z, w) = x^2 y^2 + \tfrac{1}{2}z^2 + \tfrac{1}{2}w^2.$$
Since the function H has an isolated minimum at the equilibrium point, $(x, y, z, w) = (0, 0, 0, 0)$. By the previous remark $\dot{H}$ is identically zero, so H is a Liapunov function for the system at the equilibrium, and we can conclude that the equilibrium is stable.

If an attempt to show that a function $V(\mathbf{x})$ is a Liapunov function fails because $\dot{V}(\mathbf{x}) > 0$ near an equilibrium point $\mathbf{x}_0$, it may be possible to use V to show that $\mathbf{x}_0$ is unstable.

■ 8.11 INSTABILITY TEST

Let $\dot{\mathbf{x}} = \mathbf{F}(\mathbf{x})$ have an equilibrium solution $\mathbf{x}_0$ in a region R in which $\mathbf{F}(\mathbf{x})$ is continuous. Let $V(\mathbf{x})$ be real-valued and continuously differentiable on a neighborhood $B_r \subset R$ centered at $\mathbf{x}_0$ such that $V(\mathbf{x})$ has a strict minimum at $\mathbf{x} = \mathbf{x}_0$ and such that $\dot{V}(\mathbf{x}) > 0$ on B_r except at $\mathbf{x}_0$. Then $\mathbf{x}_0$ is an unstable equilibrium.

Proof. Let B_a, with $0 < a < r$, be an arbitrarily small neighborhood of $\mathbf{x}_0$. We will show that if $\mathbf{x}_1 \neq \mathbf{x}_0$ is in B_a, the system trajectory starting at $\mathbf{x}_1$ must leave B_a. Let $0 < \epsilon < |\mathbf{x}_1|$, and let $\mu > 0$ be the minimum of V on the boundary of B_ϵ. Then

$$V(\mathbf{x}(t)) - V(\mathbf{x}(0)) = \int_0^t \dot{V}(\mathbf{x}((u))\, du \geq \mu t.$$

It follows that $V(\mathbf{x}(t))$ is unbounded on this trajectory, and in particular becomes larger than the maximum of V on B_a. Hence the trajectory leaves B_a. ■

---

e x a m p l e **6**

The system

$$\dot{x} = 2xe^{x^2+y^2}, \qquad \dot{y} = 2ye^{x^2+y^2}$$

is the gradient system associated with the function $-e^{x^2+y^2}$. However, the latter function has its maximum rather than its minimum at the unique equilibrium point $(0, 0)$. If we try letting $V(x, y) = e^{x^2+y^2}$, we find

$$\dot{V}(x, y) = (2xe^{x^2+y^2}, 2ye^{x^2+y^2}) \cdot (2xe^{x^2+y^2}, 2ye^{x^2+y^2})$$
$$= 4(x^2 + y^2)e^{x^2+y^2} > 0,$$

except at $(x, y) = (0, 0)$. By the instability test, the equilibrium is unstable.

EXERCISES

1. For each of the following systems, determine whether the constants a and b can be chosen so that $V(x, y) = ax^2 + by^2$ is (i) a Liapunov function for the system at $(0, 0)$ or (ii) a strict Liapunov function.
   (a) $\dot{x} = -x^3,\ \dot{y} = -y^3$.
   (b) $\dot{x} = -x^2,\ \dot{y} = -y^2$.
   (c) $\dot{x} = -\sin x,\ \dot{y} = -\sin y$.
   (d) $\dot{x} = -\sin y,\ \dot{y} = -\sin x$.
   (e) $\dot{x} = -xy^2,\ \dot{y} = -x^2y$.
   (f) $\dot{x} = -x^2y,\ \dot{y} = -xy^2$.

2. For each of the following systems, determine whether the constants a, b, and c can be chosen so that $V(x, y, z) = ax^2 + by^2 + cz^2$ is (i) a Liapunov function for the system at $(0, 0, 0)$ or (ii) a strict Liapunov function.
   (a) $\dot{x} = -x^3,\ \dot{y} = -y^3,\ \dot{z} = -z^3$.
   (b) $\dot{x} = -x,\ \dot{y} = -y,\ \dot{z} = -z$.
   (c) $\dot{x} = -yz,\ \dot{y} = -xz,\ \dot{z} = -xy$.
   (d) $\dot{x} = -xe^y,\ \dot{y} = -ye^z,\ \dot{z} = -ze^x$.
   (e) $\dot{x} = -x,\ \dot{y} = -y,\ \dot{z} = 0$.
   (f) $\dot{x} = -x,\ \dot{y} = -y^3,\ \dot{z} = x^2z$.

3. (a) Show that the system $\dot{x} = -x - 3y^2$, $\dot{y} = xy - y^3$ has a unique equilibrium point at $(x, y) = (0, 0)$.

(b) Choose a and b so that $V(x, y) = ax^2 + by^2$ is a strict Liapunov function at $(0, 0)$. Then deduce that $(0, 0)$ is asymptotically stable for the system, with the entire xy plane as basin of attraction.

4. The system $\dot{x} = y(z + 1)$, $\dot{y} = -x(z + 1)$, $\dot{z} = -z^3$ is shown in Example 2 of the text to have a stable equilibrium at $(0, 0, 0)$ using the Liapunov method. The question is left open there as to whether the equilibrium is in fact asymptotically stable.

(a) Solve the third equation of the system for $z = z(t)$.

(b) Show that the trace of a nonequilibrium solution $(x(t), y(t), z(t))$ in the xy plane satisfies $x^2 + y^2 = x_0^2 + y_0^2$ and so is either a circle or point. [*Hint:* $dy/dx = -x/y$.]

(c) Explain why the equilibrium at $(0, 0, 0)$ is not asymptotically stable.

5. The system $\dot{x} = yz$, $\dot{y} = -xz$, $\dot{z} = -z$ is similar to the one treated in Example 2 of the text using the Liapunov method.

(a) Find all equilibrium points of the system.

(b) Solve the third equation of the system for $z = z(t)$.

(c) Show that the trace of a nonequilibrium solution $(x(t), y(t), z(t))$ in the xy plane satisfies $x^2 + y^2 = x_0^2 + y_0^2$ and so is either a circle or point. [*Hint:* $dy/dx = -x/y$.]

(d) Show that $(0, 0, 0)$ is a stable equilibrium and that all others are unstable.

6. The system $\dot{x} = yz$, $\dot{y} = -xz$, $\dot{z} = -z^3$ is similar to the one treated in Example 2 of the text using the Liapunov method.

(a) Find all equilibrium points of the system.

(b) Solve the third equation of the system for $z = z(t)$.

(c) Show that the trace of a nonequilibrium solution $(x(t), y(t), z(t))$ in the xy plane satisfies $x^2 + y^2 = x_0^2 + y_0^2$ and so is either a circle or point. [*Hint:* $dy/dx = -x/y$.]

(d) Show that $(0, 0, 0)$ is a stable equilibrium, but is not asymptotically stable, and that all other equilibrium points are unstable.

7. The Hamiltonian system

$$\dot{x} = 2ye^{x^2+y^2}, \qquad \dot{y} = -2xe^{x^2+y^2}$$

is shown in Example 4 of the text to have a stable equilibrium at $(0, 0)$. Show that an analysis of the linearized system at $(0, 0)$ fails to lead to this conclusion.

8. Show that the solution trajectories of a two-dimensional Hamiltonian system $\dot{x} = \partial H/\partial y$, $\dot{y} = -\partial H/\partial x$ are tangent to level sets $H(\mathbf{x}, \mathbf{y}) = C$ at nonequilibrium points.

9. Consider the **Lotka-Volterra system** $\dot{x} = x(a - by)$, $\dot{y} = y(cx - d)$, where a, b, c, d are positive constants, and let $H_0(x, y) = cx - d \ln x + by - a \ln y$.

(a) Show that each solution trajectory satisfies an implicit relation $H_0(x, y) = C$. [*Hint:* Compute dy/dx and observe that the variables separate.]

(b) Show that $H_0(x, y)$ has a strict minimum at the equilibrium $(d/c, a/b)$.

(c) Compute $\dot{H}_0(x, y)$ to show that the equilibrium is stable.

(d) Show that if the vector field of the Lotka-Volterra system is multiplied by $\rho(x, y) = -(xy)^{-1}$, the result is the Hamiltonian system with Hamiltonian $H_0(x, y)$.

10. Let $H(x, y) = x^d y^a e^{-cx-by}$, a, b, c, d positive, and consider the Hamiltonian system $\dot{x} = H_y(x, y)$, $\dot{y} = -H_x(x, y)$.

(a) Show that the Hamiltonian system has the same solution trajectories as the Lotka-Volterra system of the previous exercise but that the solutions themselves are not the same.

(b) Show that $H(x, y)$ has a strict maximum at the equilibrium $(d/c, a/b)$, making it unsuitable as a Liapunov function for the system.

11. Let $H(x, y)$ be continuously differentiable. Suppose that for each solution trajectory of a given 2-dimensional system $\dot{x} = f(x, y)$, $\dot{y} = g(x, y)$ there is a constant C such that $H(x, y) = C$ on the trajectory. Show that the Hamiltonian system

$$\dot{x} = H_y(x, y), \qquad \dot{y} = -H_x(x, y)$$

has the same trajectories as the given system.

12. Let $U(x)$ be a continuously differentiable function of one variable, and consider the differential equation $\ddot{x} = -U'(x)$.

(a) Show that the second-order equation is equivalent to the first-order system

$$\dot{x} = y, \dot{y} = -U'(x).$$

(b) Show that the system in part (a) is Hamiltonian with Hamiltonian function $H(x, y) = U(x) + \frac{1}{2}y^2$, with the conclusion that if $U(x_0)$ is an isolated minimum for $U(x)$, then $(x_0, 0)$ is a stable equilibrium point for the system and hence for $\ddot{x} = -U'(x)$.

(c) Noting that the system in part (a) is equivalent to the 1-dimensional conservative system $\ddot{x} = -U'(x)$, show that the Hamiltonian $H(x, y)$ is constant on trajectories $(x(t), y(t))$ of the given system.

13. Consider the second-order linear system

$$\ddot{x} = -ax - by,$$
$$\ddot{y} = -bx - cy,$$

where a, b, c, d are constants.

(a) Show that the system is equivalent to a first-order Hamiltonian system, with Hamiltonian function

$$H(x, y, z, w) = \tfrac{1}{2}(ax^2 + 2bxy + cy^2 + z^2 + w^2).$$

(b) Show that the system has a stable equilibrium at the origin if $a > 0$ and $ac > b^2$.

14. (a) If the vector field of a 2-dimensional Hamiltonian system $\dot{x} = f(x, y)$, $\dot{y} = g(x, y)$ is continuously differentiable, show that the **divergence** div $(f, g) = \partial f/\partial x + \partial g/\partial y$ is identically zero.

(b) Verify that the vector field of the system $\dot{x} = x^2 y$, $\dot{y} = -xy^2$ has divergence identically zero, and find a Hamiltonian function for the system. [*Hint:* Solve $H_y(x, y) = x^2 y$.]

15. Consider the nonlinear oscillator equation $\ddot{x} + k\dot{x} + hx + \alpha x^2 + \beta x^3 = 0$ with linear damping and h, α, β positive. Let
$$V(x, y) = ax^2 + bx^3 + cx^4 + dy^2.$$
    (a) By letting $\dot{x} = y$, find an equivalent first-order system.

    (b) Show that if $k \geq 0$, the constants a, b, c, d can be chosen so that V is a Liapunov function for the equilibrium solution $x = 0$, $\dot{x} = 0$ and hence that the equilibrium is stable.

16. Consider the family of nonlinear systems
$$\dot{x} = y + \alpha x(x^2 + y^2),$$
$$\dot{y} = -x + \alpha y(x^2 + y^2),$$
    where α is constant. Linearized analysis is inadequate for this system. Let $V(x, y) = x^2 + y^2$.

    (a) Show that the only equilibrium point is $(x_0, y_0) = (0, 0)$, regardless of the value of α.

    (b) Show that V is a Liapunov function if $\alpha < 0$ and that the origin is asymptotically stable.

    (c) Use V to show that the origin is unstable if $\alpha > 0$.

    (d) What can you say about stability if $\alpha = 0$?

17. Suppose $\mathbf{x} = \mathbf{x}(t)$ satisfies the gradient system $\dot{\mathbf{x}} = -\nabla V(\mathbf{x})$. Use the chain rule to show that the tangent vectors $\dot{\mathbf{x}}(t)$ (i) are perpendicular to the level sets of V, defined implicitly by equations $V(\mathbf{x}) = c$, and (ii) point in the direction of maximum decrease of V at $\mathbf{x}(t)$. [*Hint:* $\nabla V(\mathbf{x}) \cdot \dot{\mathbf{x}} = |\nabla V(\mathbf{x})||\dot{\mathbf{x}}| \cos \theta$.]

18. Suppose $\mathbf{x} = \mathbf{x}(t)$ traces the solution trajectory γ of a system $\dot{\mathbf{x}} = \mathbf{F}(\mathbf{x})$ and that $V(\mathbf{x})$ is a real-valued, continuously differentiable function defined on γ.

    (a) Show that if $\nabla V(\mathbf{x}) \cdot \mathbf{F}(\mathbf{x}) = 0$ for all $\mathbf{x}$ on γ, then γ is a level set for V; i.e., there is a constant c such that $V(\mathbf{x}) = c$ for all $\mathbf{x}$ on γ.

    (b) Apply the result of part (a) to the system $\dot{x} = -y$, $\dot{y} = x$ with $V(x, y) = x^2 + y^2$ to show that all nonequilibrium trajectories are circular.

    (c) Apply the result of part (a) to the system $\dot{x} = yz$, $\dot{y} = -2xz$, $\dot{z} = xy$ with $V(x, y, z) = x^2 + y^2 + z^2$ to show that each nonequilibrium trajectory lies on a sphere.

    (d) What conclusions can you draw from part (a) about trajectories of Hamiltonian systems in general?

9 CALCULUS OF VARIATIONS

First encounters with differential equations are often motivated by an underlying physical dynamical system. In physical terms, the question posed is "Given the mechanism, how does it run?" Formulated in this way, a theoretical model of the mechanism is often described by a system of one or more differential equations, and the answer to our question is provided by a solution that we think of as a dynamical system in the mathematical sense. In Section 9A we make precise a way in which solutions to certain problems in geometry can be regarded as "best possible" in comparison with "nearby" alternatives. In Section 9B we extend these ideas to Lagrange's powerful technique for deriving dynamical models.

9A   Euler-Lagrange Equations

Consider the problem of finding a function f such that the integral

$$I(f) = \int_{t_1}^{t_2} F(t, f(t), f'(t)) \, dt$$

has a minimum (or maximum) value. It is understood that the function $F(t, x, \dot{x})$ is given and that the function f is to be sought from among some reasonable alternatives, for example, some set of real-valued, continuously differentiable functions on the interval $t_1 \leq t \leq t_2$.

---

example 1

Suppose we want to find the function $f\colon [t_1, t_2] \to \mathscr{R}$ whose graph has the shortest length subject to conditions of the form $f(t_1) = a$, $f(t_2) = b$, as well as the assumption that f' is continuous. The arc-length integral is

$$I_L(f) = \int_{t_1}^{t_2} \sqrt{1 + (f'(t))^2} \, dt,$$

so the problem is a special case of the one posed above with $F(t, x, \dot{x}) = \sqrt{1 + \dot{x}^2}$. The solution should, of course, be a function $x = f(t)$ with linear graph, namely,

$$f(t) = (b - a)t + a, \qquad t_1 \leq t \leq t_2,$$

joining (t_1, a) and (t_2, b), and compared with a curved graph in Figure 6.20a. Note that it is not $F(t, x, \dot{x})$ that must be minimized as a function of three variables but a certain definite integral with a function taking on the role of

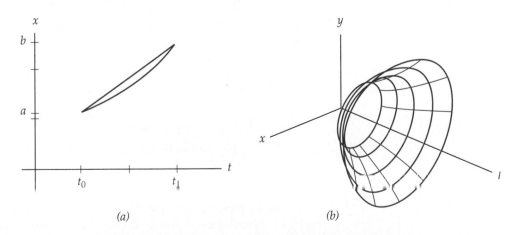

(a)   (b)

Figure 6.20 (a) Shortest length. (b) Smaller surface area.

independent variable. Hence a direct application of the ordinary calculus methods for minimization will not suffice. After considering another example we'll describe an appropriate extension of those methods.

---

e x a m p l e **2**

If the graph of the type of function $f(t)$ considered in Example 1 is rotated around the t axis, the result is a surface of revolution with area given by

$$I_A(f) = \int_{t_1}^{t_2} 2\pi f(t)\sqrt{1 + (f'(t))^2}\, dt.$$

It is no longer intuitively obvious how f should be chosen to minimize the surface area, given that $f(t_1) = a$ and $f(t_2) = b$ are fixed. We might guess, incorrectly as it turns out, that a conical surface gives the least area. Note, however, that the integral I_A is only a simple modification of the arc-length integral, with integrand now determined by

$$F(t, x, \dot{x}) = 2\pi x\sqrt{1 + \dot{x}^2}.$$

The method described below will apply as well to area as to length, and it will show that we may be able to minimize area with a surface like the one in Figure 6.20*b*.

Recall that to minimize or maximize a real-valued function of one or many variables, we derive certain necessary conditions that must be satisfied in order for an extreme value to exist. These conditions usually take the form of setting certain derivatives equal to zero and solving the resulting equations. Such conditions are, of course, not in general *sufficient* for the existence of an extreme value, so it may be necessary to examine further the results of our calculations to see which, if any, of the points we find do indeed give extreme values. We proceed in a similar way to derive a *necessary* condition for f to give an extreme value to the integral

$$I(f) = \int_{t_1}^{t_2} F(t, f(t), f'(t))\, dt.$$

Denote by $\mathscr{A}$ the set of all twice continuously differentiable functions $f(t)$ on $[t_1, t_2]$ and satisfy $f(t_1) = a$, $f(t_2) = b$. Suppose that f_0 is a member of $\mathscr{A}$. To say that f_0 minimizes I on $\mathscr{A}$ is to say that $I(f_0) \le I(f)$ for all functions f in $\mathscr{A}$. Now if h is a twice continuously differentiable function on $[t_1, t_2]$ satisfying $h(t_1) = h(t_2) = 0$, then for real ε, $f_0(t) + \varepsilon h(t)$ defines a function $f_0 + \varepsilon h$ that is also in $\mathscr{A}$, that is, is twice continuously differentiable for $t_1 \le t \le t_2$ with $f(t_1) = a$, $f(t_2) = b$. See Figure 6.21. Since f_0 minimizes I on $\mathscr{A}$, it follows that

$$I(f_0) \le I(f_0 + \varepsilon h).$$

Thus the real-valued function $J(\varepsilon)$ defined by

$$J(\varepsilon) = I(f_0 + \varepsilon h)$$
$$= \int_{t_1}^{t_2} F(t, f_0(t) + \varepsilon h(t), f_0'(t) + \varepsilon h'(t))\, dt$$

assumes its minimum value at $\varepsilon = 0$. We can now apply ordinary calculus techniques to the real-valued function J of the real variable ε and conclude that

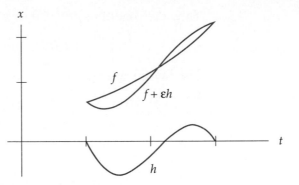

Figure 6.21 Graph of f and a nearby comparison.

if F is sufficiently well-behaved so that J is differentiable, then $J'(0) = 0$. This analysis can be interpreted as saying that minimizing the integral requires "directional derivatives with respect to h" to be zero at a minimizing function f_0. To explain what the condition $J'(0) = 0$ says in practical terms, we have the following theorem.

■ 9.1 THEOREM

If $F(t, x, \dot{x})$ is twice continuously differentiable and

$$I(f) = \int_{t_1}^{t_2} F(t, f(t), f'(t))\, dt$$

has a minimum or a maximum for $f = f_0$ in $\mathscr{S}$, then setting $x = f_0$ and $\dot{x} = f_0'$ satisfies the differential equation

$$\frac{\partial F}{\partial x}(t, x, \dot{x}) - \frac{d}{dt}\frac{\partial F}{\partial \dot{x}}(t, x, \dot{x}) = 0,$$

called the **Euler-Lagrange differential equation** associated with $I(f)$.

Proof. Let $J(\varepsilon) = I(f_0 + \varepsilon h)$ be as defined above with f_0 in $\mathscr{S}$ and h continuously differentiable, with $h(t_1) = h(t_2) = 0$. Apply $\partial/\partial\varepsilon$ under the integral sign, using the chain rule to get

$$J'(\varepsilon) = \int_{t_1}^{t_2} \left[\frac{\partial F}{\partial x}(t, f_0(t) + \varepsilon h(t), f_0'(t) + \varepsilon h'(t))h(t) \right.$$
$$\left. + \frac{\partial F}{\partial \dot{x}}(t, f_0(t) + \varepsilon h(t), f_0'(t) + \varepsilon h'(t))h'(t) \right] dt.$$

Integration by parts shows that the integral of the second term is

$$\left[\frac{\partial F}{\partial \dot{x}} h \right]_{t_1}^{t_2} - \int_{t_1}^{t_2} h\, \frac{d}{dt}\frac{\partial F}{\partial \dot{x}}\, dt.$$

But because $h(t_1) = h(t_2) = 0$, the integrated term is zero, so

$$J'(\varepsilon) = \int_{t_1}^{t_2} \left[\frac{\partial F}{\partial x} - \frac{d}{dt}\frac{\partial F}{\partial \dot{x}} \right] h\, dt.$$

Thus the condition $J'(0) = 0$ says that

$$\int_{t_1}^{t_2} \left[\frac{\partial F}{\partial x}(t, f_0(t), f_0'(t)) - \frac{d}{dt} \frac{\partial F}{\partial \dot{x}}(t, f_0(t), f_0'(t)) \right] h(t) \, dt = 0.$$

The Euler-Lagrange equation now follows immediately from the following.

■ LEMMA

Let G be continuous for $t_1 \le t \le t_2$ and suppose that

$$\int_{t_1}^{t_2} G(t) h(t) \, dt = 0$$

for all continuously differentiable h such that $h(t_1) = h(t_2) = 0$. Then $G(t) = 0$ for all t in $[t_1, t_2]$.

Proof of Lemma. The proof proceeds by contradiction, supposing that there is a t_0 with $t_1 < t_0 < t_2$ such that $G(t_0) \ne 0$. It would follow that there is an interval (a, b) on which G is not zero and such that $t_1 < a < t_0 < b < t_2$. If we construct a continuously differentiable function h that is zero outside (a, b) and positive on (a, b) where G is positive, then

$$\int_{t_1}^{t_2} G(t) h(t) \, dt = \int_a^b G(t) h(t) \, dt \ne 0.$$

This contradiction establishes the theorem.   ■

Functions h with the properties required in the lemma can be constructed in many ways; here is an explicit example:

$$h(t) = \begin{cases} 0, & \text{for } t \le a \text{ and } b \le t, \\ (t - a)^2 (t - b)^2, & \text{for } a < t < b. \end{cases}$$

---

e x a m p l e 3

Continuing now from the arc-length problem in Example 1, we see that for f to minimize the arc-length integral

$$I(f) = \int_{t_1}^{t_2} \sqrt{1 + (f'(t))^2} \, dt$$

among twice continuously differentiable functions, we must have

$$\frac{\partial F}{\partial x} - \frac{d}{dt} \frac{\partial F}{\partial \dot{x}} = 0$$

where $F(t, x, \dot{x}) = \sqrt{1 + \dot{x}^2}$. Since $\partial F / \partial x = 0$ and $\partial F / \partial \dot{x} = \dot{x} / \sqrt{1 + \dot{x}^2}$, we get

$$-\frac{d}{dt} \frac{\partial F}{\partial \dot{x}} = -\frac{d}{dt} \left(\frac{\dot{x}}{\sqrt{1 + \dot{x}^2}} \right) = -\frac{\ddot{x}}{(1 + \dot{x}^2)^{3/2}} = 0.$$

Thus $\ddot{x} = f''(t) = 0$ must hold for $t_1 \le t \le t_2$, so $f(t) = \alpha t + \beta$ for some constants α and β. If endpoint conditions $f(t_1) = a$ and $f(t_2) = b$ are prescribed, then

$$f(t) = \frac{b - a}{t_2 - t_1} (t - t_1) + a,$$

is the only solution. The graph of f is, of course, the expected minimizing straight line.

The following theorem provides an alternative way to solve an Euler-Lagrange equation if the associated integrand is not explicitly dependent on the function $x(t)$ itself. The criterion can in that case be reduced to satisfying a 1-parameter family of first-order equations, called collectively a **first integral** of the Euler-Lagrange equation.

■ 9.2 THEOREM

If $x = x(t)$ satisfies the Euler-Lagrange equation of the function $F(t, \dot{x})$, then it must also satisfy $\partial F / \partial \dot{x} = C$ for some constant C.

Proof. Converting to subscript notation for partial derivatives, the Euler-Lagrange equation is $F_x - d(F_{\dot{x}})/dt = 0$. Since $F(t, \dot{x})$ is independent of x, the first term is zero, so $d(F_{\dot{x}})/dt = 0$. It follows that $F_{\dot{x}}$ is independent of t. In other words, $F_{\dot{x}}$ has a constant value C for all admissible x.   ■

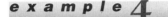

Applying Theorem 9.2 to the arc-length problem of Example 1, we note that $F(\dot{x}) = \sqrt{1 + \dot{x}^2}$ is explicitly independent of x. Hence a solution of the Euler-Lagrange equations must also satisfy

$$\frac{\partial F(\dot{x})}{\partial \dot{x}} = \frac{\partial}{\partial \dot{x}} \sqrt{1 + \dot{x}^2} = \frac{\dot{x}}{\sqrt{1 + \dot{x}}} = C.$$

Squaring both sides of the last equation and simplifying gives $(1 - C^2)\dot{x}^2 = C^2$. (The value $C = 1$ is clearly inconsistent.) It follows that $\dot{x} = \alpha$ for some constant α and hence that $x = \alpha t + \beta$, as we discovered in Example 3.

As a practical matter, the virtue of Theorem 9.2 is that is saves us the trouble of computing a second derivative in finding the Euler-Lagrange equation and then reversing right away to solve a second-order equation. A similar efficiency is possible if the integrand $F(t, x, \dot{x})$ does not explicitly depend on t as one of its three variables. Thus we have to deal with a function $F(x, \dot{x})$ that is called an **autonomous** integrand. Such integrands arise often enough that they deserve some special attention; the Euler-Lagrange equation for an autonomous problem can be reduced to a particularly simple form.

■ 9.3 THEOREM

If $x = f(t)$ satisfies the Euler-Lagrange equation of the function $F(x, \dot{x})$, then it must also satisfy $\dot{x}\partial F / \partial \dot{x} - F = C$ for some constant C.

Proof. Direct application of the chain rule to $F(x, \dot{x})$ shows that

$$\frac{d}{dt}\left(\dot{x}\frac{\partial F}{\partial \dot{x}} - F \right) = \dot{x}\frac{d}{dt}\frac{\partial F}{\partial \dot{x}} + \ddot{x}\frac{\partial F}{\partial \dot{x}} - \dot{x}\frac{\partial F}{\partial x} - \ddot{x}\frac{\partial F}{\partial \dot{x}}$$

$$= \dot{x}\frac{d}{dt}\frac{\partial F}{\partial \dot{x}} - \dot{x}\frac{\partial F}{\partial x} = \dot{x}\left[\frac{d}{dt}\frac{\partial F}{\partial x} - \frac{\partial F}{\partial x} \right].$$

It follows that whenever the Euler-Lagrange equation

$$\frac{d}{dt}\frac{\partial F}{\partial \dot{x}} - \frac{\partial F}{\partial x} = 0$$

is satisfied, we must also have

$$\frac{d}{dt}\left(\dot{x}\frac{\partial F}{\partial \dot{x}} - F\right) = 0.$$

Thus it is enough to consider solutions $x = f(t)$ of the first-integral equation

$$\dot{x}\frac{\partial F}{\partial \dot{x}}(x, \dot{x}) - F(x\,\dot{x}) = C,$$

for constant C.   ∎

---

e x a m p l e 5

The surface area problem of Example 2 requires us to minimize

$$\int_{t_1}^{t_2} x\sqrt{1 + \dot{x}^2}\, dt$$

given that $x(t_1) = a$, and $x(t_2) = b$. The solutions of the Euler-Lagrange equation are solutions of the first-integral equation of Theorem 9.3, where $F(x, \dot{x}) = x\sqrt{1 + \dot{x}^2}$. Thus we try to solve

$$\frac{x\dot{x}^2}{\sqrt{1 + \dot{x}^2}} - x\sqrt{1 + \dot{x}^2} = C$$

for constant C. Multiplying through by $\sqrt{1 + \dot{x}^2}$ gives $x = C\sqrt{1 + \dot{x}^2}$ or $x^2 - C^2 = C^2\dot{x}^2$. The identity $\cosh^2 u - \sinh^2 u = 1$ shows that the previous equation has solutions

$$x(t) = C \cosh\frac{1}{C}(t + c),$$

where c is another constant. (See Exercise 10 for a direct derivation.) Some curves of this family are shown in Figure 6.22.

It can be shown that it is sometimes, but not always, possible to satisfy both conditions of the form $x(t_1) = a$, $x(t_2) = b$ with this family of solutions. Furthermore, even when there is a curve of the family through two given points, the surface obtained by rotation about the t axis doesn't necessarily give the surface of least area satisfying the conditions placed upon it, but only a *relative* minimum as compared with nearby smooth surfaces of revolution.

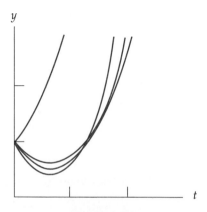

Figure 6.22

The inconclusive state of affairs described at the end of Example 5 is a more complicated analog of what happens when we try to find a minimum value for a real function of one variable by setting the first derivative equal to zero: (1) There may not be any points where the derivative equals zero; (2) a point where the derivative is zero may not provide even a local minimum, let alone a global one. For a real-valued function $f(x)$, it's customary to call a solution of $f'(x) = 0$ on some interval a **critical point**. A solution of an Euler-Lagrange equation is by analogy called a **critical function** or a **stationary function**.

9B   Vector Equations

To extend the range of problems that we can treat, we consider vector-valued functions $\mathbf{x} = g(t)$ and integrals of the form

$$I(g) = \int_{t_1}^{t_2} F(t, g(t), g'(t))\, dt,$$

where F is a function from $\mathcal{R}^{2n+1}$ to $\mathcal{R}$. We assume that g is twice continuously differentiable from $[t_1, t_2]$ to $\mathcal{R}^n$. Because we can apply the basic variational technique to a single coordinate function at a time relative to $g(t) = (g_1(t), \ldots, g_n(t))$, we get a system of n Euler-Lagrange equations

▮ 9.4 $$\frac{\partial F}{\partial x_k} - \frac{d}{dt}\frac{\partial F}{\partial \dot{x}_k} = 0, \qquad k = 1, \ldots, n,$$

where $F(t, \mathbf{x}, \dot{\mathbf{x}}) = F(t, x_1, \ldots, x_n, \dot{x}_1, \ldots, \dot{x}_n)$ is treated as a function of $2n + 1$ variables. A solution of the system 9.4 is a vector-valued function $\mathbf{x} = g(t)$ that describes a path in $\mathcal{R}^n$, called a **critical** or **stationary** path, for the integral $I(g) = \int_{t_1}^{t_2} F(t, g, g')\, dt$.

***e x a m p l e* 6**   Suppose that a conservative force field in $\mathcal{R}^3$ has **potential** $U(x, y, z)$; in other words, suppose that the field is given by $(-\partial U/\partial x, -\partial U/\partial y, -\partial U/\partial z)$. The function $U(x, y, z)$ is called a **potential energy** function of the field at (x, y, z). A particle of mass m in moving on a path in such a way that its position at time t is $\mathbf{x}(t)$ is said to have **kinetic energy** $T = \frac{1}{2}m|\dot{\mathbf{x}}(t)|^2$. Elementary arguments with line integrals show that $\frac{1}{2}m|\dot{\mathbf{x}}(t)|^2 + U(g(t))$ must be a constant provided that Newton's law $\mathbf{F} = m\mathbf{a}$ holds. (It is total energy $T + U$ that is conserved in a conservative field.) We can conversely derive Newton's law from the following assumption about the energy:

▮ 9.5 HAMILTON'S PRINCIPLE

The path followed by a particle in a conservative field is stationary for the integral

$$\int_{t_1}^{t_2} L(\mathbf{x}(t), \dot{\mathbf{x}}(t))\, dt$$

where $L(g(t), g'(t)) = T(\dot{\mathbf{x}}) - U(\mathbf{x}(t))$ is called the **Lagrange function** or **Lagrangian** of the conservative system.

The Euler-Lagrange equations associated with

$$L(\mathbf{x}, \dot{\mathbf{x}}) = \frac{m}{2}(\dot{x}^2 + \dot{y}^2 + \dot{z}^2) - U(x, y, z)$$

are

$$-\frac{\partial U}{\partial x} - \frac{d}{dt}m\dot{x} = 0, \quad \text{or} \quad m\ddot{x} = -\frac{\partial U}{\partial x},$$

$$-\frac{\partial U}{\partial y} - \frac{d}{dt}m\dot{y} = 0, \quad \text{or} \quad m\ddot{y} = -\frac{\partial U}{\partial y},$$

$$-\frac{\partial U}{\partial z} - \frac{d}{dt}m\dot{z} = 0, \quad \text{or} \quad m\ddot{z} = -\frac{\partial U}{\partial z}.$$

These equations are summarized in vector form by $m\ddot{\mathbf{x}} = -\nabla U(\mathbf{x})$; that is, mass times acceleration equals a force expressed as the gradient of a potential energy function $U(\mathbf{x})$.

A major advantage of the Lagrangian formulation of mechanics is that it simplifies the derivation of the differential equations of motion for conservative systems. Even if the spatial configuration of the system is constrained to lie on a curve or surface, it is still true that the solution chosen from the spatially restricted set of solutions will be stationary for the constrained motion. Furthermore, we are free to use any convenient system of coordinates. The next example shows this advantage very clearly.

---

e x a m p l e 7

A **spherical** pendulum is one that is allowed to move freely in 3-dimensional space about its fixed pivot. Let l be the length of such a pendulum, and let m be its mass, concentrated at its center of gravity. We assume the system has no forces acting on it other than the downward gravitational force $F = (0, 0, -mg)$, expressed in rectangular xyz coordinates. In rectangular coordinates there is a potential energy $U(x, y, z) = mgz$. Thus the Lagrangian is

$$L = T - U = \tfrac{1}{2}m(\dot{x}^2 + \dot{y}^2 + \dot{z}^2) - mgz.$$

To adapt this Lagrangian to the spherical pendulum, with 2-dimensional motion shown in Figure 6.23a, we introduce spherical $\phi\theta$ coordinates on a fixed sphere of radius l:

$$x = l\cos\phi\sin\theta, \quad y = l\sin\phi\sin\theta, \quad z = -l\cos\theta,$$

where θ is measured from the downward vertical. We compute

$$\dot{x} = l(\dot{\theta}\cos\phi\cos\theta - \dot{\phi}\sin\phi\sin\theta),$$

$$\dot{y} = l(\dot{\theta}\sin\phi\cos\theta + \dot{\phi}\cos\phi\sin\theta),$$

$$\dot{z} = l\dot{\theta}\sin\theta.$$

An elementary computation shows that $L = \tfrac{1}{2}ml^2(\dot{\theta}^2 + \dot{\phi}^2\sin^2\theta) + mgl\cos\theta$. The Euler-Lagrange equations are

$$\frac{\partial L}{\partial\phi} - \frac{d}{dt}\frac{\partial L}{\partial\dot{\phi}} = -\frac{d}{dt}(ml^2\dot{\phi}\sin^2\theta) = 0,$$

$$\frac{\partial L}{\partial\theta} - \frac{d}{dt}\frac{\partial L}{\partial\dot{\theta}} = (ml^2\dot{\phi}^2\sin\theta\cos\theta - mgl\sin\theta) - \frac{d}{dt}(ml^2\dot{\theta}) = 0.$$

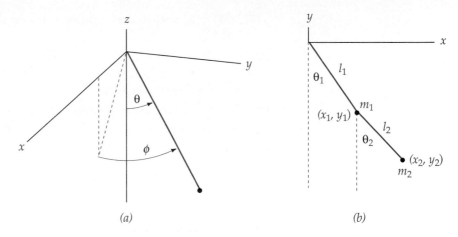

(a) *(b)*

Figure 6.23 (*a*) Spherical pendulum. (*b*) Double pendulum.

These equations simplify to

$$(\sin \theta)\ddot{\phi} + 2\dot{\theta}\dot{\phi} \cos \theta = 0, \qquad \ddot{\theta} - \dot{\phi}^2 \sin \theta \cos \theta + \frac{g}{l} \sin \theta = 0.$$

If there is no rotation about the vertical axis, then $\dot{\phi} = 0$, and the second of these equations becomes just $\ddot{\theta} = -(g/l) \sin \theta$, the usual form for the equation of a pendulum swinging in a fixed plane. (See Exercise 18 in Section 1C, Chapter 5, for another derivation of this system.)

There is an obvious technical difficulty in our derivation of the Euler-Lagrange equations: we needed to assume the existence and continuity of the second derivative f'' even though f'' does not enter explicitly in the integral $I(f) = \int_{t_1}^{t_2} F(t, f(t), f'(t)) \, dt$ to be minimized or maximized. Fortunately, there is a theorem asserting that if f is stationary for I, and if F is twice continuously differentiable, with $\partial^2 F/\partial \dot{x}^2 \neq 0$, then continuous differentiability of f automatically implies the existence of two continuous derivatives for f. The theorem is called **duBois-Reymond's lemma.** The theorem allows the domain of I to be enlarged, and this enlargement implies a strengthening of the conclusions that can be drawn from the Euler-Lagrange equations: a solution of these equations is stationary not only among twice-differentiable functions but also among once-differentiable functions. For a proof of the theorem see R. Courant and D. Hilbert, *Methods of Mathematical Physics*, vol. 1, Interscience, 1953, p. 200.

EXERCISES

1. (a) Show directly that $\int_0^1 \sqrt{1 + (f'(t))^2} \, dt$ has a minimum subject to the condition $f(0) = 1$ for some f continuously differentiable on $[0, 1]$.

   (b) The set of functions over which the minimum was sought in part (a) is reduced in size if we add the additional restriction $f(1) = 2$. What is the value of the integral for a critical function f under the conditions $f(0) =$

1, $f(1) = 2$, and f continuously differentiable, and for what f is it attained?

(c) Show that subject to any given fixed boundary conditions $f(0) = a$, $f(1) = b$, the integral $\int_0^1 \sqrt{1 + (f'(t))^2}\, dt$ will assume arbitrarily large values for suitable functions f.

2. Find the Euler-Lagrange equation for each of the following integrands:

   (a) $F(t, x, \dot{x}) = tx + \dot{x}^2$.   **(d)** $F(t, \dot{x}) = t\dot{x} - \dot{x}^2$.

   (b) $F(x, \dot{x}) = x^2 + \dot{x}^2$.   **(e)** $F(t, x, \dot{x}) = x + t\dot{x}^2$.

   (c) $F(\dot{x}) = \dot{x}/\sqrt{1 + \dot{x}^2}$.   **(f)** $\dot{x}^2 + \dot{x}^4$.

3. Find the complete solution of each equation derived in the previous exercise.

4. Find the system of Euler-Lagrange equations for each of the following integrands:

   (a) $F(x, y, \dot{x}, \dot{y}) = \sqrt{x^2 + y^2 + \dot{x}^2 + \dot{y}^2}$.   **(d)** $F(\dot{x}, \dot{y}) = \dot{x}\dot{y}$.

   (b) $F(x, y, \dot{x}, \dot{y}) = x^2 + \dot{y}^2$.   **(e)** $F(x, y, \dot{x}, \dot{y}) = x\dot{y} + y\dot{x}$.

   (c) $F(t, \dot{x}, \dot{y}) = t^2 + \dot{x}^2 - \dot{y}^2$.   **(f)** $F(x, y, \dot{x}, \dot{y}) = \dot{x}^2 + \dot{y}^2 + 2xy$.

5. Find the complete solution of each system derived in the previous exercise.

6. Show that for an integrand of the form $F(\dot{x})$ the Euler-Lagrange equation can be written in the form $\ddot{x}F''(\dot{x}) = 0$. Does this follow from either Theorem 9.2 or Theorem 9.3?

7. **(a)** Use the previous exercise to show that the stationary functions for a twice continuously differentiable integrand $F(\dot{x})$ for which $F''(\dot{x}) \neq 0$ have the form $x(t) = \alpha t + \beta$.

   (b) Find all stationary functions if $F(\dot{x}) = \dot{x}^3$. What about $F(\dot{x}) = \dot{x}$? (Be careful!)

8. The solution of the first integral of the Euler-Lagrange equation in Example 5 of the text was found by judicious guessing. Solve the equation $x^2 - C^2 = C^2\dot{x}^2$ by computing an appropriate indefinite integral.

9. Show that there may not be a value for C such that the graph of $x(t) = C \cosh(t + c)/C$ contains two given points (t_1, x_1) and (t_2, x_2), with $t_1 \neq t_2$.

10. Derive the equation $\ddot{\theta} = -(g/l) \sin \theta$ for an ideal **simple pendulum** using Hamilton's principle and the relations $x = l \sin \theta$, $y = -l \cos \theta$.

11. Find Newton's equations for a particle in a field with potential $U(\mathbf{x}) = |\mathbf{x}|^{-1}$.

12. **Double pendulum.** Figure 6.23b shows an ideal pendulum of mass m_1 and length l_1, with another one of mass m_2 and length l_2 attached to its free end. The kinetic energy of the system is $T = \frac{1}{2}m_1(\dot{x}_1^2 + \dot{y}_1^2) + \frac{1}{2}m_2(\dot{x}_2^2 + \dot{y}_2^2)$, and a potential energy function is $U(x_1, y_1, x_2, y_2) = m_1gy_1 + m_2gy_2$.

   (a) Derive in terms of (θ_1, θ_2) the Euler-Lagrange equations for the Lagrangian $T - U$:

$$(m_1 + m_2)l_1\ddot{\theta}_1 + m_2l_2\ddot{\theta}_2 \cos(\theta_1 - \theta_2) + m_2l_2\dot{\theta}_2^2 \sin(\theta_1 - \theta_2)$$
$$+ (m_1 + m_2)g \sin \theta_1 = 0,$$
$$l_1\ddot{\theta}_1 \cos(\theta_1 - \theta_2) + l_2\ddot{\theta}_2 - l_1\dot{\theta}_1^2 \sin(\theta_1 - \theta_2) + g \sin \theta_2 = 0.$$

(b) Solve the equations in part (a) for $\ddot\theta_1$ and $\ddot\theta_2$, with $\mu = 1 + (m_1/m_2)$, to get

$$\ddot\theta_1 = \frac{g(\sin\theta_2\cos(\theta_1-\theta_2) - \mu\sin\theta_1) - (l_2\dot\theta_2^2 + l_1\dot\theta_1^2\cos(\theta_1-\theta_2))\sin(\theta_1-\theta_2)}{l_1(\mu - \cos^2(\theta_1-\theta_2))},$$

$$\ddot\theta_2 = \frac{g\mu(\sin\theta_1\cos(\theta_1-\theta_2) - \sin\theta_2) + (\mu l_1\dot\theta_1^2 + l_2\dot\theta_2^2\cos(\theta_1-\theta_2))\sin(\theta_1-\theta_2)}{l_2(\mu - \cos^2(\theta_1-\theta_2))}.$$

(c) Make computer plots of the motions of both pendulums, one in $\theta_1\dot\theta_1$ phase space and a separate plot for the other in $\theta_2\dot\theta_2$ phase space; for both plots use $g = 32$, $m_1 = m_2$, $l_1 = l_2 = 1$, and initial values $(\theta_1(0),$ $\dot\theta_1(0)) = (2, 1)$ and $(\theta_2(0), \dot\theta_2(0)) = (8, -3)$. Then observe the extreme sensitivity of the global form of these pictures to minute changes in initial values, which is a characteristic of **chaotic** behavior

Geodesics. A curve $\mathbf{x} = g(t)$ that is stationary for the arc-length integral $\int_{t_1}^{t_2}|\dot{\mathbf{x}}(t)|\,dt$ is called a **geodesic**. A geodesic may or may not be a curve of shortest length joining two points that lie on it. The following exercises are about geodesics.

13. Find the Euler-Lagrange equations for the arc-length integral $\int_{t_1}^{t_2}\sqrt{\dot x^2 + \dot y^2 + \dot z^2}\,dt$ in 3-dimensional space, and show that the stationary paths lie on line segments.

14. Curves parametrized by $x = a\cos u(t)$, $y = a\sin u(t)$, $z = v(t)$ lie on a cylinder of radius $a > 0$ with the z axis as the axis of symmetry. Assume that $u(t)$ and $v(t)$ are continuously differentiable.

(a) Show that the arc-length integral for curves on the cylinder in terms of u and v is equal to $\int_{t_1}^{t_2}\sqrt{a^2\dot u^2 + \dot v^2}\,dt$.

(b) Find the Euler-Lagrange equations for the integral in part (a); then show that together they imply that a stationary path on the cylinder lies on either a circle, a helix, or a line parallel to the z axis.

(c) Find two stationary paths on the cylinder that have only the endpoints $(a, 0, 0)$ and $(a, 0, a)$ in common.

7 CHAPTER

Matrix Methods

1 EIGENVALUES AND EIGENVECTORS

The facts about matrices used in this chapter are reviewed in Appendix B. In particular, we'll use the products of matrices, including matrices and column vectors, inverse matrices, and the determinant criterion $\det A \neq 0$ for the existence of the inverse A^{-1} of a square matrix A. Our focus in Sections 1 and 2 will be on **homogeneous** systems in normal form such as

$$\begin{aligned} \dot{x} &= 2x + 3y, \\ \dot{y} &= 4x - y, \end{aligned} \qquad \text{and} \qquad \begin{aligned} \dot{x} &= -x + 2y - 3z, \\ \dot{y} &= 4x - 4y - 3z, \\ \dot{z} &= 2x - 2y + 3z. \end{aligned}$$

Written in matrix form, these systems look like

$$\begin{pmatrix} \dot{x} \\ \dot{y} \end{pmatrix} = \begin{pmatrix} 2 & 3 \\ 4 & -1 \end{pmatrix} \begin{pmatrix} x \\ y \end{pmatrix} \qquad \text{and} \qquad \begin{pmatrix} \dot{x} \\ \dot{y} \\ \dot{z} \end{pmatrix} = \begin{pmatrix} -1 & 2 & -3 \\ 4 & -4 & -3 \\ 2 & -2 & 3 \end{pmatrix} \begin{pmatrix} x \\ y \\ z \end{pmatrix}.$$

In abbreviated form the two systems can be written $\dot{\mathbf{x}} = A\mathbf{x}$ and $\dot{\mathbf{x}} = B\mathbf{x}$; the expanded form above follows from the "row-by-column" definition of matrix product once it is understood that

$$A = \begin{pmatrix} 2 & 3 \\ 4 & -1 \end{pmatrix} \qquad \text{and} \qquad B = \begin{pmatrix} -1 & 2 & -3 \\ 4 & -4 & -3 \\ 2 & -2 & 3 \end{pmatrix}.$$

The dimension of the column vector $\mathbf{x}$ and its derivative $\dot{\mathbf{x}}$ in each equation is dictated by the dimensions of A, which is 2 by 2, and B, which is 3 by 3.

434

1A   Exponential Solutions

The vector differential equation $\dot{\mathbf{x}} = A\mathbf{x}$ appears in many examples in Chapter 6. These examples show that if A is a constant n by n matrix, then exponentials arise naturally in the solutions. For example, in the case $n = 1$, we would have $\dot{x} = ax$, with solutions of the form $x(t) = ce^{at}$. Consequently, we try solutions

■ 1.1 $$\mathbf{x}(t) = (u_1 e^{\lambda t}, u_2 e^{\lambda t}, \ldots, u_n e^{\lambda t}) = e^{\lambda t}\mathbf{u},$$

where $\mathbf{u}$ is a constant vector in $\mathcal{R}^n$. Differentiation of $\mathbf{x}(t)$ gives

$$\frac{d\mathbf{x}}{dt} = (u_1 \lambda e^{\lambda t}, u_2 \lambda e^{\lambda t}, \ldots, u_n \lambda e^{\lambda t}) = \lambda e^{\lambda t}\mathbf{u}.$$

The justification for this differentiation is that differentiation of vector-valued functions is defined so that it is carried out one entry at a time. For example, if

$$\mathbf{x}(t) = e^{\lambda t}\begin{pmatrix} u_1 \\ u_2 \end{pmatrix} = \begin{pmatrix} e^{\lambda t}u_1 \\ e^{\lambda t}u_2 \end{pmatrix},$$

then $$\frac{d\mathbf{x}(t)}{dt} = \begin{pmatrix} \lambda e^{\lambda t}u_1 \\ \lambda e^{\lambda t}u_2 \end{pmatrix} = \lambda e^{\lambda t}\begin{pmatrix} u_1 \\ u_2 \end{pmatrix} = \lambda e^{\lambda t}\mathbf{u}.$$

Since $e^{\lambda t}$ is a scalar factor, it can be taken past the matrix A, so

$$A\mathbf{x} = A(e^{\lambda t}\mathbf{u}) = e^{\lambda t}A\mathbf{u}.$$

To solve the differential equation, we equate $A\mathbf{x}$ and $d\mathbf{x}/dt$ and so require

■ 1.2 $$A\mathbf{u} = \lambda\mathbf{u}.$$

The choice $\mathbf{u} = 0$ is too trivial to be interesting. We ignore that possibility and define a *nonzero* vector $\mathbf{u}$ to be an **eigenvector** of A with **eigenvalue** λ if $\mathbf{u}$ and λ satisfy Equation 1.2. Thus Equation 1.1 provides a nontrivial solution to the vector differential equation whenever $\mathbf{u}$ is an eigenvector of A with corresponding eigenvalue λ. The problem of solving the vector differential equation has been reduced to an algebraic problem to which we can apply the ideas and techniques of matrix algebra. More specifically, we want to find all real, or complex, numbers λ such that the **eigenvalue-eigenvector equation**

■ 1.3 $$(A - \lambda I)\mathbf{u} = 0$$

has a nonzero solution vector $\mathbf{u}$. While Equation 1.2 is equivalent to Equation 1.3, the latter equation has the technical advantage that it displays the vector equation in homogeneous form. If $A - \lambda I$ should happen to be invertible, we could apply its inverse to both sides of Equation 1.3 and conclude that $\mathbf{u}$ must be 0. Since this is too trivial to be useful, we choose λ so that $(A - \lambda I)^{-1}$ fails to exist. But Theorem B.9 of Appendix B tells us that the matrix $(A - \lambda I)$ fails to have an inverse precisely when λ satisfies the **characteristic equation**

■ 1.4 $$\det(A - \lambda I) = 0.$$

The determinant $P(\lambda) = \det(A - \lambda I)$ is a polynomial, called the **characteristic polynomial** of A. The roots of $P(\lambda)$ are the eigenvalues of A. For example, if $A = \begin{pmatrix} 2 & 1 \\ 3 & 2 \end{pmatrix}$, then we want

$$\det\left[\begin{pmatrix} 2 & 1 \\ 3 & 2 \end{pmatrix} - \lambda\begin{pmatrix} 1 & 0 \\ 0 & 1 \end{pmatrix}\right] = \det\begin{pmatrix} 2-\lambda & 1 \\ 3 & 2-\lambda \end{pmatrix} = 0,$$

or $$(2-\lambda)^2 - 3 = \lambda^2 - 4\lambda + 1 = 0.$$

This characteristic equation is a condition on λ alone. Having found all values of λ that satisfy it, we can then go ahead to find corresponding eigenvectors $\mathbf{u}$ from Equation 1.3. Note that if $\mathbf{u}$ is a nonzero solution of Equation 1.2 or 1.3, then so is $c\mathbf{u}$ for any nonzero constant c. If we have solutions of the form $e^{\lambda_k t}\mathbf{u}_k$, it is easy to verify that a **linear combination**

$$\mathbf{x}(t) = c_1 e^{\lambda_1 t}\mathbf{u}_1 + \cdots + c_n e^{\lambda_n t}\mathbf{u}_n$$

is also a solution.

---

e x a m p l e 1

To solve the system

$$\begin{pmatrix} dx_1/dt \\ dx_2/dt \end{pmatrix} = \begin{pmatrix} x_1 + x_2 \\ 4x_1 + x_2 \end{pmatrix}$$

$$= \begin{pmatrix} 1 & 1 \\ 4 & 1 \end{pmatrix}\begin{pmatrix} x_1 \\ x_2 \end{pmatrix},$$

try to find nonzero vectors $\mathbf{u} = (u_1, u_2)$ that satisfy the eigenvector equation

$$\begin{pmatrix} 1 & 1 \\ 4 & 1 \end{pmatrix}\begin{pmatrix} u_1 \\ u_2 \end{pmatrix} = \lambda \begin{pmatrix} u_1 \\ u_2 \end{pmatrix}$$

for some number λ. In other words, we find numbers λ such that the equation

$$\begin{pmatrix} 1 - \lambda & 1 \\ 4 & 1 - \lambda \end{pmatrix}\begin{pmatrix} u_1 \\ u_2 \end{pmatrix} = \begin{pmatrix} 0 \\ 0 \end{pmatrix}$$

has nonzero solutions. If the 2 by 2 matrix is invertible, then only the solution $(u_1, u_2) = (0, 0)$ exists, so we must have

$$\det \begin{pmatrix} 1 - \lambda & 1 \\ 4 & 1 - \lambda \end{pmatrix} = 0.$$

In other words,

$$(1 - \lambda)^2 - 4 = \lambda^2 - 2\lambda - 3$$
$$= (\lambda - 3)(\lambda + 1) = 0.$$

The only solutions are $\lambda = 3$ and $\lambda = -1$.

Case 1: $\lambda = 3$. We want nonzero vectors (u_1, u_2) such that

$$\begin{pmatrix} -2 & 1 \\ 4 & -2 \end{pmatrix}\begin{pmatrix} u_1 \\ u_2 \end{pmatrix} = \begin{pmatrix} 0 \\ 0 \end{pmatrix}.$$

The two numerical equations

$$-2u_1 + u_2 = 0$$
$$4u_1 - 2u_2 = 0$$

are equivalent to $u_2 = 2u_1$, so we can choose $u_1 = 1$, $u_2 = 2$. Thus, since $\lambda = 3$,

$$\mathbf{x}_1(t) = e^{3t}\begin{pmatrix} 1 \\ 2 \end{pmatrix}$$

is a solution, along with any numerical multiple.

Case 2: $\lambda = -1$. We want nonzero vectors (v_1, v_2) such that

$$\begin{pmatrix} 2 & 1 \\ 4 & 2 \end{pmatrix}\begin{pmatrix} v_1 \\ v_2 \end{pmatrix} = \begin{pmatrix} 0 \\ 0 \end{pmatrix}.$$

Clearly, $v_1 = 1$, $v_2 = -2$ will do, so

$$\mathbf{x}_2(t) = e^{-t}\begin{pmatrix} 1 \\ -2 \end{pmatrix}$$

is a solution, as well as any numerical multiple. Adding our two solutions we get a more general solution of the vector differential equation:

$$\mathbf{x}(t) = c_1 e^{3t}\begin{pmatrix} 1 \\ 2 \end{pmatrix} + c_2 e^{-t}\begin{pmatrix} 1 \\ -2 \end{pmatrix}$$

$$= \begin{pmatrix} c_1 e^{3t} + c_2 e^{-t} \\ 2c_1 e^{3t} - 2c_2 e^{-t} \end{pmatrix}.$$

To see the geometric significance of the computation in Example 1, we'll do it over again, with reference to Figure 7.1. Note that neither of the two vectors

$$\mathbf{u} = \begin{pmatrix} 1 \\ 2 \end{pmatrix}, \qquad \mathbf{v} = \begin{pmatrix} 1 \\ -2 \end{pmatrix}$$

is a multiple of the other. Thus any vector $\mathbf{x}$ can be written using $y_1 y_2$ coordinates as

$$\mathbf{x} = y_1\mathbf{u} + y_2\mathbf{v}.$$

By the rules for multiplication by a matrix A,

$$A\mathbf{x} = Ay_1\begin{pmatrix} 1 \\ 2 \end{pmatrix} + Ay_2\begin{pmatrix} 1 \\ -2 \end{pmatrix}$$

$$= y_1 A\begin{pmatrix} 1 \\ 2 \end{pmatrix} + y_2 A\begin{pmatrix} 1 \\ -2 \end{pmatrix}.$$

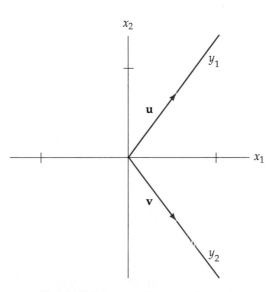

Figure 7.1 Eigenvector coordinates.

Since **u** and **v** are eigenvectors, with eigenvalues 3 and -1, respectively, we have $A\mathbf{u} = 3\mathbf{u}$ and $A\mathbf{v} = -\mathbf{v}$. In other words, A has the effect of multiplying the first vector **u** by 3 and the second vector **v** by -1. Hence

$$A\mathbf{x} = y_1 A\mathbf{u} + y_2 A\mathbf{v}$$
$$= 3y_1 \mathbf{u} - y_2 \mathbf{v}.$$

We also have

$$\frac{d\mathbf{x}}{dt} = \frac{dy_1}{dt}\mathbf{u} + \frac{dy_2}{dt}\mathbf{v}.$$

To make $d\mathbf{x}/dt = A\mathbf{x}$, we equate the coefficients of **u** and **v** on the right sides of the last two equations to get

$$\frac{dy_1}{dt} = 3y_1,$$

$$\frac{dy_2}{dt} = -y_2.$$

The reason for introducing eigenvectors is that the new system is particularly simple to solve, because each involves only one unknown function. In other words, the new system is *uncoupled*. In fact, we see immediately that

$$y_1(t) = c_1 e^{3t},$$
$$y_2(t) = c_2 e^{-t}.$$

Thus we arrive again at our original solution in terms of $x_1 x_2$ coordinates:

$$\mathbf{x}(t) = c_1 e^{3t}\mathbf{u} + c_2 e^{-t}\mathbf{v} = c_1 e^{3t}\begin{pmatrix} 1 \\ 2 \end{pmatrix} + c_2 e^{-t}\begin{pmatrix} 1 \\ -2 \end{pmatrix}.$$

The procedure in Example 1 can be tried in any number of dimensions, as outlined below.

Summary of the Method. To solve the n-dimensional constant-coefficient equation

$$\frac{d\mathbf{x}}{dt} = A\mathbf{x},$$

proceed as follows:

1. Find the eigenvalues of A by solving the characteristic equation
$$\det(A - \lambda I) = 0.$$

2. For each eigenvalue λ_k, find an eigenvector $\mathbf{u}_k$ by solving
$$(A - \lambda_k I)\mathbf{u} = 0.$$

3. Form the linear combination
$$\mathbf{x}(t) = c_1 e^{\lambda_1 t}\mathbf{u}_1 + \cdots + c_n e^{\lambda_n t}\mathbf{u}_n.$$

If the eigenvectors $\mathbf{u}_1, \ldots, \mathbf{u}_n$ are **linearly dependent** (i.e., one of them is replaceable by a linear combination of the others), the procedure outlined above produces solutions but not the most general one. In this case, we can use the elimination method explained in Chapter 5, or the general methods of the following Sections 2 and 3 in this chapter. However, it can be shown that if det $(A - \lambda I) = 0$ has no multiple roots, then the corresponding eigenvectors will always be **linearly independent** (i.e., none of them is replaceable by a linear combination of the others), and step 3 produces the most general solution.

If A has some complex eigenvalues, the procedure outlined above is the same, with complex exponentials, or their trigonometric equivalent, in the final result.

---

e x a m p l e 2

The system

$$\frac{dx_1}{dt} = x_1 - x_2,$$

$$\frac{dx_2}{dt} = x_1 + x_2$$

has matrix

$$A = \begin{pmatrix} 1 & -1 \\ 1 & 1 \end{pmatrix}.$$

The eigenvalues are solutions of

$$\det \begin{pmatrix} 1 - \lambda & -1 \\ 1 & 1 - \lambda \end{pmatrix} = 0;$$

this is the same as

$$(1 - \lambda)^2 + 1 = \lambda^2 - 2\lambda + 2 = 0,$$

having solutions $\lambda_1 = 1 + i$ and $\lambda_2 = 1 - i$.

Case 1: $\lambda = 1 + i$. The eigenvectors are the nonzero solutions of

$$\begin{pmatrix} -i & -1 \\ 1 & -i \end{pmatrix} \begin{pmatrix} u_1 \\ u_2 \end{pmatrix} = \begin{pmatrix} 0 \\ 0 \end{pmatrix},$$

that is, of

$$-iu_1 - u_2 = 0,$$
$$u_1 - iu_2 = 0.$$

A simple nontrivial solution, found by elimination, is $\begin{pmatrix} u_1 \\ u_2 \end{pmatrix} = \begin{pmatrix} 1 \\ -i \end{pmatrix}.$

Case 2: $\lambda = 1 - i$. The eigenvectors are the nonzero solutions of

$$\begin{pmatrix} i & -1 \\ 1 & i \end{pmatrix} \begin{pmatrix} v_1 \\ v_2 \end{pmatrix} = \begin{pmatrix} 0 \\ 0 \end{pmatrix},$$

that is, of

$$iv_1 - v_2 = 0$$
$$v_1 + iv_2 = 0.$$

Again using elimination, we find the nontrivial solution

$$\begin{pmatrix} v_1 \\ v_2 \end{pmatrix} = \begin{pmatrix} 1 \\ i \end{pmatrix}.$$

The general solution of the differential equation is

$$\mathbf{x}(t) = c_1 e^{(1+i)t} \begin{pmatrix} 1 \\ -i \end{pmatrix} + c_2 e^{(1-i)t} \begin{pmatrix} 1 \\ i \end{pmatrix}$$

$$= c_1 e^t \begin{pmatrix} \cos t + i \sin t \\ -i \cos t + \sin t \end{pmatrix} + c_2 e^t \begin{pmatrix} \cos t - i \sin t \\ i \cos t + \sin t \end{pmatrix}$$

$$= \begin{pmatrix} (c_1 + c_2)e^t \cos t + i(c_1 - c_2)e^t \sin t \\ -i(c_1 - c_2)e^t \cos t + (c_1 + c_2)e^t \sin t \end{pmatrix}.$$

If we rename the constants so that $c_1 + c_2 = d_1$ and $i(c_1 - c_2) = d_2$, then

$$\mathbf{x}(t) = \begin{pmatrix} d_1 e^t \cos t + d_2 e^t \sin t \\ -d_2 e^t \cos t + d_1 e^t \sin t \end{pmatrix}$$

$$= d_1 \begin{pmatrix} e^t \cos t \\ e^t \sin t \end{pmatrix} + d_2 \begin{pmatrix} e^t \sin t \\ -e^t \cos t \end{pmatrix}.$$

Because $c_1 = (d_1 - id_2)/2$ and $c_2 = (d_1 + id_2)/2$, the constants c_1 and c_2 can always be chosen so that d_1 and d_2 have any preassigned values; in particular, they can be chosen to be real numbers.

1B   *Eigenvector Matrices and Initial Conditions*

As usual, we may want to choose arbitrary constants in general solutions such as those preceding so as to satisfy prescribed initial conditions. This calculation can be reduced to a routine as follows. We denote by U the matrix whose *columns* are the vectors $\mathbf{u}_1, \ldots, \mathbf{u}_n$ in some fixed order. The matrix $U = (\mathbf{u}_1, \ldots, \mathbf{u}_n)$ is called the **eigenvector matrix** of the system. We denote by Λ_t the diagonal matrix

$$\Lambda_t = \begin{pmatrix} e^{\lambda_1 t} & 0 & \cdots & 0 \\ 0 & e^{\lambda_2 t} & \cdots & 0 \\ \vdots & \vdots & & \vdots \\ 0 & 0 & \cdots & e^{\lambda_n t} \end{pmatrix},$$

with corresponding eigenvalues λ_k in the *same order* as the eigenvectors. If $\mathbf{c}$ is some constant column vector with entries $c_1, \ldots, c_n$, we can form the vector-valued function

$$\mathbf{x}(t) = U\Lambda_t c.$$

Clearly, $\mathbf{x}(t)$ has the form

$$\mathbf{x}(t) = U \begin{pmatrix} c_1 e^{\lambda_1 t} \\ \vdots \\ c_n e^{\lambda_n t} \end{pmatrix} = c_1 e^{\lambda_1 t}\mathbf{u}_1 + \cdots + c_n e^{\lambda_n t}\mathbf{u}_n,$$

and so is a solution of $d\mathbf{x}/dt = A\mathbf{x}$. But if U is **invertible,** that is, if U^{-1} exists, we can let $c = U^{-1}\mathbf{x}_0$ for some $\mathbf{x}_0$ in $\mathcal{R}^n$. Thus

■ 1.5 $$\mathbf{x}(t) = U\Lambda_t U^{-1}\mathbf{x}_0.$$

Since $\Lambda_0 = I$, we have $\mathbf{x}(0) = UU^{-1}\mathbf{x}_0$; so $\mathbf{x}(0) = \mathbf{x}_0$. Thus $\mathbf{x}(t)$ is the solution of the vector differential equation that satisfies $\mathbf{x}(0) = \mathbf{x}_0$. To satisfy an initial condition at $t = t_0$, we let

$$\mathbf{c} = \Lambda_{-t_0} U^{-1}\mathbf{x}_0,$$

so that

$$\mathbf{x}(t) = U\Lambda_{(t-t_0)} U^{-1}\mathbf{x}_0$$

is the desired solution.

---

example 3

In Example 1, we found for

$$A = \begin{pmatrix} 1 & 1 \\ 4 & 1 \end{pmatrix}$$

the eigenvalues $\lambda_1 = 3$, $\lambda_2 = -1$ with eigenvectors

$$\mathbf{u}_1 = \begin{pmatrix} 1 \\ 2 \end{pmatrix}, \qquad \mathbf{u}_2 = \begin{pmatrix} 1 \\ -2 \end{pmatrix}.$$

We have

$$\Lambda_t = \begin{pmatrix} e^{3t} & 0 \\ 0 & e^{-t} \end{pmatrix},$$

$$U = \begin{pmatrix} 1 & 1 \\ 2 & -2 \end{pmatrix},$$

$$U^{-1} = \begin{pmatrix} \frac{1}{2} & \frac{1}{4} \\ \frac{1}{2} & -\frac{1}{4} \end{pmatrix}.$$

Thus, if $\mathbf{x}_0 = (2, 3)$, the solution

$$\mathbf{x}(t) = \begin{pmatrix} 1 & 2 \\ 2 & -2 \end{pmatrix}\begin{pmatrix} e^{3t} & 0 \\ 0 & e^{-t} \end{pmatrix}\begin{pmatrix} \frac{1}{2} & \frac{1}{4} \\ \frac{1}{2} & -\frac{1}{4} \end{pmatrix}\begin{pmatrix} 2 \\ 3 \end{pmatrix}$$

$$= \begin{pmatrix} \frac{7}{4}e^{3t} + \frac{1}{4}e^{-t} \\ \frac{7}{2}e^{3t} - \frac{1}{2}e^{-t} \end{pmatrix}$$

satisfies the differential equation $d\mathbf{x}/dt = A\mathbf{x}$ and also the initial condition $\mathbf{x}(0) = (2, 3)$.

To satisfy instead the initial condition $\mathbf{x}(1) = (3, 4)$, we form

$$\mathbf{x}(t) = \begin{pmatrix} 1 & 1 \\ 2 & -2 \end{pmatrix}\begin{pmatrix} e^{3(t-1)} & 0 \\ 0 & e^{-(t-1)} \end{pmatrix}\begin{pmatrix} \frac{1}{2} & \frac{1}{4} \\ \frac{1}{2} & -\frac{1}{4} \end{pmatrix}\begin{pmatrix} 3 \\ 4 \end{pmatrix}$$

$$= \begin{pmatrix} \frac{5}{2}e^{3(t-1)} + \frac{1}{2}e^{-(t-1)} \\ 5e^{3(t-1)} - e^{-(t-1)} \end{pmatrix}.$$

EXERCISES

1 The matrix $A = \begin{pmatrix} 1 & 12 \\ 3 & 1 \end{pmatrix}$ has eigenvalues 7 and -5. Which of the following vectors is an eigenvector of A? For those that are, what is the corresponding eigenvalue?

(a) $\begin{pmatrix} 2 \\ 1 \end{pmatrix}$.

(d) $\begin{pmatrix} -2 \\ 2 \end{pmatrix}$.

(b) $\begin{pmatrix} -2 \\ 1 \end{pmatrix}$.

(e) $\begin{pmatrix} 1 \\ 1 \end{pmatrix}$.

(c) $\begin{pmatrix} -4 \\ -2 \end{pmatrix}$.

2. Find all the eigenvalues of each of the following matrices.

(a) $\begin{pmatrix} 1 & 4 \\ 1 & 1 \end{pmatrix}$.

(c) $\begin{pmatrix} 2 & 4 \\ 1 & 2 \end{pmatrix}$.

(b) $\begin{pmatrix} 0 & 4 \\ 1 & 0 \end{pmatrix}$.

(d) $\begin{pmatrix} 1 & 0 & 0 \\ 2 & 1 & 0 \\ 0 & 1 & 2 \end{pmatrix}$.

3. For each matrix in Exercise 2, find an eigenvector corresponding to each eigenvalue.

4. Each of the following 2-dimensional systems of differential equations has the form

$$\frac{d\mathbf{x}}{dt} = A\mathbf{x}$$

in which the matrix A has constant entries. In each example find the eigenvalues of A, and for each eigenvalue find a corresponding eigenvector.

(a) $\begin{pmatrix} \dot{x} \\ \dot{y} \end{pmatrix} = \begin{pmatrix} -3 & 2 \\ -4 & 3 \end{pmatrix}\begin{pmatrix} x \\ y \end{pmatrix}$.

(c) $\dot{x} = x + 4y,$
$\dot{y} = 5y.$

(b) $\dot{x} = 3x,$
$\dot{y} = 2y.$

(d) $\begin{pmatrix} \dot{x} \\ \dot{y} \end{pmatrix} = \begin{pmatrix} 2 & -1 \\ 1 & 2 \end{pmatrix}\begin{pmatrix} x \\ y \end{pmatrix}$.

5. Use the eigenvalues and eigenvectors of each of the systems in Exercise 4 to write the general solution of the system in the form

$$\mathbf{x}(t) = c_1 e^{\lambda_1 t}\mathbf{u}_1 + c_2 e^{\lambda_2 t}\mathbf{u}_2.$$

In the case of complex eigenvalues, convert the solution to real form.

6. For the systems of differential equations in Exercise 4, find the particular solution that satisfies the corresponding condition listed.

(a) $(x(0), y(0)) = (1, 3)$.

(c) $(x(0), y(0)) = (1, 1)$.

(b) $(x(0), y(0)) = (-1, 0)$.

(d) $(x(0), y(0)) = (0, 0)$.

7. (a) Find the general solution of the system $\dot{\mathbf{x}} = A\mathbf{x}$, where

$$A = \begin{pmatrix} 1 & -1 & 4 \\ 3 & 2 & -1 \\ 2 & 1 & -1 \end{pmatrix}.$$

(b) Find a constant vector $\mathbf{x}_p$ that is a particular solution of the system $\dot{\mathbf{x}} = A\mathbf{x} + \mathbf{c}$, where $\mathbf{c}$ is the column vector with entries 1, 0, 2, top to bottom.

(c) Find, by any method, the particular solution $\mathbf{x}$ of the system in part (b) that satisfies the condition $\mathbf{x}(0) = (1, 1, 2)$.

8. (a) Find, by any method, the general solution of the system

$$\frac{dx}{dt} = -x + y, \qquad \frac{dy}{dt} = -y.$$

(b) Show that the eigenvectors of the related matrix

$$\begin{pmatrix} -1 & 1 \\ 0 & -1 \end{pmatrix}$$

are insufficient for the application of the method of Section 1B.

9. (a) Find the general solution of the system

$$\frac{dx}{dt} = x, \qquad \frac{dy}{dt} = y + z, \qquad \frac{dz}{dt} = z.$$

(b) Show that the eigenvectors of the related matrix

$$\begin{pmatrix} 1 & 0 & 0 \\ 0 & 1 & 1 \\ 0 & 0 & 1 \end{pmatrix}$$

are insufficient for the application of the method of Section 1B.

10. Show that, if A is n by n with invertible eigenvector matrix U, and the eigenvalues of A have negative real parts, then all solutions of $dx/dt = Ax$ tend to zero as $t \to +\infty$.

11. The second-order constant-coefficient differential equation

$$\frac{d^2y}{dt^2} + a\frac{dy}{dt} + by = 0$$

is equivalent to the first-order system

$$\frac{dx}{dt} = -ax - by, \qquad \frac{dy}{dt} = x.$$

Show that the eigenvalues of the matrix

$$\begin{pmatrix} -a & -b \\ 1 & 0 \end{pmatrix}$$

are the same as the characteristic roots of the second-order differential equation.

12. Let U be an invertible n by n matrix with columns $\mathbf{u}_1, \ldots, \mathbf{u}_n$. Let D be the diagonal matrix with diagonal entries $\lambda_1, \ldots, \lambda_n$, and define the n by n matrix A by $A = UDU^{-1}$.

(a) Show that $A\mathbf{u}_k = \lambda_k \mathbf{u}_k$, for $k = 1, \ldots, n$.

(b) Find the 2 by 2 matrix that has eigenvectors $\mathbf{u}_1 = (2, 3)$, $\mathbf{u}_2 = (1, 1)$, and corresponding eigenvalues $\lambda_1 = 3$, $\lambda_2 = 1$.

(c) Show that the system $dx/dt = Ax$ has solutions $e^{\lambda_k t}\mathbf{u}_k$, for $k = 1, \ldots, n$.

(d) Find a 2-dimensional system having

$$\mathbf{x}(t) = c_1 e^{3t} \begin{pmatrix} 2 \\ 3 \end{pmatrix} + c_2 e^{t} \begin{pmatrix} 1 \\ 1 \end{pmatrix}$$

as its general solution.

13. In applications of linear systems, we're sometimes concerned mainly with the general behavior of solutions. By finding the eigenvalues $\lambda = a + ib$ of the matrix A, find out whether the solutions of $dx/dt = Ax$ oscillate ($b \neq 0$), tend to 0 as $t \to +\infty$ ($a < 0$), or become unbounded ($a > 0$). Of course a single equation may have solutions exhibiting more than one kind of behavior.

(a) $\dfrac{d}{dt}\begin{pmatrix} x \\ y \end{pmatrix} = \begin{pmatrix} -1 & -2 \\ 1 & 0 \end{pmatrix}\begin{pmatrix} x \\ y \end{pmatrix}.$

(c) $\dfrac{d}{dt}\begin{pmatrix} x \\ y \end{pmatrix} = \begin{pmatrix} 2x - y \\ 3x - y \end{pmatrix}.$

(b) $\dfrac{d}{dt}\begin{pmatrix} x \\ y \end{pmatrix} = \begin{pmatrix} 0 & -1 \\ 1 & 0 \end{pmatrix}\begin{pmatrix} x \\ y \end{pmatrix}.$

(d) $dx/dt = 2x + y,$
$dy/dt = 3x + y.$

14. The second-order system

$$\ddot{x} = ax + by, \qquad \ddot{y} = cx + dy$$

is equivalent to the first-order system

$$\begin{pmatrix} \dot{x} \\ \dot{y} \\ \dot{z} \\ \dot{w} \end{pmatrix} = \begin{pmatrix} 0 & 0 & 1 & 0 \\ 0 & 0 & 0 & 1 \\ a & b & 0 & 0 \\ c & d & 0 & 0 \end{pmatrix}\begin{pmatrix} x \\ y \\ z \\ w \end{pmatrix}.$$

(a) Verify this statement and show that the characteristic equation of the first-order system is $\lambda^4 - (a + d)\lambda^2 + (ad - bc) = 0$.

(b) Assume that a, b, c, d are real numbers. Show that all eigenvalues of the system are purely imaginary, and hence that all solutions are sums of periodic oscillations, if and only if both $a + d < 0$ and $ad - bc > 0$. [*Hint:* Find out when $\lambda^2 < 0$.]

15. Verify that if $x_1(t)$ and $x_2(t)$ are solutions of $\dot{x} = Ax$, then so is the vector function $c_1x_1(t) + c_2x_2(t)$ for arbitrary constants c_1 and c_2.

16. Let A be an n by n matrix that has 0 for an eigenvalue. Explain why A can't have an inverse.

17. **Finding a matrix with given eigenvalues and eigenvectors.** Suppose we want an n by n matrix A that has $\lambda_1, \ldots, \lambda_n$ and $u_1, \ldots, u_n$ as eigenvalues and corresponding eigenvectors: $Au_k = \lambda_k u_k$. Let Λ be the square matrix with the λ's on the main diagonal, and let U be the square matrix with the vectors u_k as columns, in the same order as the λ's. Show that if U^{-1} exists, then $A = U\Lambda U^{-1}$ is the desired matrix.

18. Use the method described in the previous exercise to find a matrix A with the following eigenvalues and corresponding eigenvectors.

(a) $\lambda_1 = 2,\ \lambda_2 = 3,\ u_1 = \begin{pmatrix} 1 \\ 2 \end{pmatrix},\ u_2 = \begin{pmatrix} 2 \\ 5 \end{pmatrix}.$

(b) $\lambda_1 = 2,\ \lambda_2 = 3,\ \lambda_3 = 3,\ u_1 = \begin{pmatrix} 1 \\ 2 \\ 3 \end{pmatrix},\ u_2 = \begin{pmatrix} 2 \\ 5 \\ 0 \end{pmatrix},\ u_3 = \begin{pmatrix} -1 \\ 0 \\ -14 \end{pmatrix}.$

19. The matrix

$$A = \begin{pmatrix} a & 0 \\ 0 & b \end{pmatrix}\begin{pmatrix} \cos\theta & -\sin\theta \\ \sin\theta & \cos\theta \end{pmatrix} = \begin{pmatrix} a\cos\theta & -a\sin\theta \\ b\sin\theta & b\cos\theta \end{pmatrix}$$

represents a counterclockwise rotation about the origin in $\mathcal{R}^2$ through angle θ, followed by dilation in two perpendicular directions.

  (a) Assume $a = b = 1$. Show then that A has a real eigenvalue if and only if $\theta = k\pi$ for some integer k. What are the possible eigenvalues and eigenvectors?

  (b) Show that A has distinct real eigenvalues if and only if $(a + b)^2 \cos^2 \theta > 4ab$.

20. Let $A = \begin{pmatrix} a & b \\ c & d \end{pmatrix}$, where the entries are real numbers and $bc \geq 0$.

  (a) Show that all eigenvalues of A are real.

  (b) Show that both eigenvalues of A are positive when $a + d > 0$ and when $ad - bc \geq 0$, that both are negative when $a + d < 0$ and $ad - bc \geq 0$, and that they have opposite signs when $ad - bc < 0$.

21. The purpose of this exercise is to show that square matrices of the form $U\Lambda_t U^{-1}$ behave in some respects like the exponential function e^t. Let U be an invertible n by n matrix and let $\lambda_1, \ldots, \lambda_n$ be n arbitrarily chosen complex numbers. Let Λ_t be the n by n matrix with entries $e^{\lambda_1 t}, \ldots, e^{\lambda_n t}$ on the main diagonal and zeros elsewhere. Define $E(t) = U\Lambda_t U^{-1}$. Prove the following properties of $E(t)$.

  (a) $E(0) = I$.

  (b) $E(s + t) = E(s)E(t) = E(t)E(s)$.

  (c) $E(t)E(-t) = I$.

22. Show that the independent solutions obtained at the end of Example 2 of the text can be obtained just from case 1 by taking real and imaginary parts of $e^{(1+i)t} \begin{pmatrix} 1 \\ -i \end{pmatrix}$.

2 MATRIX EXPONENTIALS

The exponential e^A of a square matrix A is fundamental for describing solutions of linear differential systems with constant coefficients. The ordinary numerical exponential e^x can be defined in several ways, and the simplest one on which to model our matrix definition is the infinite sum

$$e^x = 1 + \frac{x}{1!} + \frac{x^2}{2!} + \frac{x^3}{3!} + \cdots + \frac{x^k}{k!} \cdots$$

This formula is valid for all real or complex values of x, and while it is not usual to do so, all the properties of e^x can be derived directly from it. For example, notice that the formal derivative of the infinite sum is equal to the sum itself, which suggests that $de^x/dx = e^x$, as we know. Also, letting $x - 0$ shows that $e^0 = 1$. From these two properties alone it follows that e^x has all the familiar properties of an exponential function. Using similar arguments, we will be able to establish analogous properties for e^A, where A is a square matrix. Our initial approach will use an infinite series of matrices, because this allows us to see most directly the general properties of the solutions we are looking for. For the purpose of computing explicit solution formulas, however, the exponential series is effective only in special cases. Thus other methods will be introduced to find the desired formulas. The state of affairs is somewhat like that associated with the definition and computation of values for the ordinary exponential function e^x. One payoff for the work we do, in both the matrix and numerical case, is that we have a notation that strongly suggests the properties of the mathematical quantity that it represents.

If A has dimension n by n and I is the n by n identity matrix, we consider the sum

$$I + A + \frac{A^2}{2!} + \cdots + \frac{A^k}{k!} = \sum_{j=0}^{k} \frac{A^j}{j!}.$$

This sum of n by n matrices is, of course, also an n by n matrix. We define the **exponential** of A by

$$e^A = \lim_{k \to \infty} \sum_{j=0}^{k} \frac{A^j}{j!} = \sum_{j=0}^{\infty} \frac{A^j}{j!},$$

where the existence of the matrix limit is understood to mean that the limit exists in each of the n^2 entries in the matrix. It is sometimes convenient to use the notation $\exp A$ for e^A. For example, if

$$A = \begin{pmatrix} 2 & 0 \\ 0 & 3 \end{pmatrix}, \ A^2 = \begin{pmatrix} 2^2 & 0 \\ 0 & 3^2 \end{pmatrix}, \ \ldots, \ A^j = \begin{pmatrix} 2^j & 0 \\ 0 & 3^j \end{pmatrix},$$

then

$$\exp \begin{pmatrix} 2 & 0 \\ 0 & 3 \end{pmatrix} = \lim_{k \to \infty} \sum_{j=0}^{k} \frac{1}{j!} \begin{pmatrix} 2^j & 0 \\ 0 & 3^j \end{pmatrix}$$

$$= \begin{pmatrix} \sum_{j=0}^{\infty} \dfrac{2^j}{j!} & 0 \\ 0 & \sum_{j=0}^{\infty} \dfrac{3^j}{j!} \end{pmatrix} = \begin{pmatrix} e^2 & 0 \\ 0 & e^3 \end{pmatrix}.$$

It is a remarkable and useful fact that the exponential of a square matrix always exists and has many of the properties of the ordinary real or complex exponential function. The justification is provided by the following fundamental theorem, in which the derivative of a matrix is defined to be simply the matrix obtained by differentiating the individual entries; thus

$$\frac{d}{dt} \begin{pmatrix} 1 & t \\ 0 & e^t \end{pmatrix} = \begin{pmatrix} 0 & 1 \\ 0 & e^t \end{pmatrix}.$$

■ 2.1 THEOREM

If A is an n by n real or complex matrix, the matrix series

$$\sum_{j=0}^{\infty} \frac{t^j A^j}{j!}$$

converges to an n by n matrix e^{tA} having properties:

(a) $e^{(t+s)A} = e^{tA}e^{sA} = e^{sA}e^{tA}$, for all numbers t and s.

(b) e^{tA} is invertible, and $e^{-tA}e^{tA} = e^{tA}e^{-tA} = I$.

(c) $(d/dt)e^{tA} = Ae^{tA} = e^{tA}A$.

Proof. If $A = (a_{ij})$, choose a positive number b such that $|a_{ij}| \leq b$ for i, $j = 1, \ldots, n$. Since the entries in A^2 are of the form

$$a_{i1}a_{1j} + \cdots + a_{in}a_{nj},$$

it follows that they are all at most nb^2 in absolute value. Proceeding inductively, the entries in A^k are at most $n^{k-1}b^k$ in absolute value. It follows that each entry in e^A is defined by an absolutely convergent infinite series dominated by the convergent series

$$1 + b + \frac{nb^2}{2!} + \cdots + \frac{1}{j!}n^{j-1}b^j + \cdots.$$

Hence all entries exist, and e^A is defined. These estimates show that if the entries a_{ij} in A are replaced by the entries ta_{ij} in tA, then the convergence is uniform on every bounded interval $c \leq t \leq d$.

To prove property (a), we apply the binomial theorem to $(t+s)^j$ to get

$$e^{(t+s)A} = \sum_{j=0}^{\infty} \frac{(t+s)^j A^j}{j!} = \sum_{j=0}^{\infty} \frac{1}{j!}\left[\sum_{l=0}^{j} \binom{j}{l} t^l s^{j-l}A^j\right].$$

Since $\binom{j}{l} = \frac{j!}{l!(j-l)!}$, we can cancel $j!$ to get

$$e^{(t+s)A} = \sum_{j=0}^{\infty}\left[\sum_{l=0}^{j} \frac{t^l A^l}{l!}\frac{s^{j-l}A^{j-l}}{(j-l)!}\right].$$

On the other hand, this last sum is just the product of the two absolutely convergent series that represent e^{tA} and e^{sA}, respectively. Note that the product can be formed in either order.

Property (b) follows from (a) on taking $t = 1$ and $s = -1$. Clearly, $e^{0A} = I$, so $I = e^A e^{-A} = e^{-A}e^A$.

Formally, the derivative of e^{tA} can be computed from the definition by

$$\frac{d}{dt}e^{tA} = \frac{d}{dt}\sum_{j=0}^{\infty} \frac{t^j A^j}{j!}$$

$$= \sum_{j=1}^{\infty} \frac{jt^{j-1}A^j}{j!}$$

$$= A\sum_{j=1}^{\infty} \frac{t^{j-1}A^{j-1}}{(j-1)!} = Ae^{tA}.$$

Note that the factor A could be taken out on the right just as well as on the left. This computation can be justified using term-by-term differentiation of series. The reason is that the differentiated series in each entry is uniformly convergent on every bounded t interval because of the estimate made in the first part of the proof.   ∎

2A   *Solving Systems*

The quickest way to see the importance of e^{tA} is to show that

$$\mathbf{x}(t) = e^{tA}\mathbf{x}_0$$

defines the solution of

$$\frac{d\mathbf{x}}{dt} = A\mathbf{x}, \qquad \mathbf{x}(0) = \mathbf{x}_0.$$

First, to differentiate the vector $e^{tA}\mathbf{x}_0$, we need only differentiate each entry in the matrix e^{tA}, because $\mathbf{x}_0$ has constant entries. Hence

$$\frac{d}{dt}e^{tA}\mathbf{x}_0 = Ae^{tA}\mathbf{x}_0$$

by part (c) of Theorem 2.1. Thus the differential equation is satisfied. Second, to show that the initial condition is satisfied, note that

$$\mathbf{x}(0) = I\mathbf{x}_0 = \mathbf{x}_0.$$

More generally,

$$\mathbf{x}(t) = e^{(t-t_0)A}\mathbf{x}_0$$

satisfies the initial condition $\mathbf{x}(t_0) = \mathbf{x}_0$.

■ 2.2 THEOREM

The unique solution of

$$\frac{d\mathbf{x}}{dt} = A\mathbf{x}, \qquad \mathbf{x}(0) = \mathbf{x}_0,$$

is given by $\mathbf{x}(t) = e^{tA}\mathbf{x}_0$.

Proof. Multiply the differential equation by e^{-tA} to get

$$e^{-tA}\frac{d\mathbf{x}}{dt} - e^{-tA}A\mathbf{x} = 0.$$

The product rule for derivatives (see Exercise 14) shows that this is the same as $d/dt(e^{-tA}\mathbf{x}) = 0$, so $e^{-tA}\mathbf{x} = \mathbf{c}$ for some constant vector $\mathbf{c}$. Since $e^{tA}e^{-tA} = I$, multiplying by e^{tA} gives $\mathbf{x}(t) = e^{tA}\mathbf{c}$. Finally $\mathbf{c} = e^{0A}\mathbf{c} = \mathbf{x}(0) = \mathbf{x}_0$.   ■

---

e x a m p l e 1

The system

$$\frac{dx}{dt} = x,$$

$$\frac{dy}{dt} = -y$$

can be written in matrix form as $d\mathbf{x}/dt = A\mathbf{x}$, where

$$A = \begin{pmatrix} 1 & 0 \\ 0 & -1 \end{pmatrix}.$$

Hence the system has general solution

$$\mathbf{x}(t) = e^{tA}\mathbf{c},$$

where **c** is the 2-dimensional column vector with constant entries c_1, c_2. To compute the exponential matrix, write

$$e^t \begin{pmatrix} 1 & 0 \\ 0 & -1 \end{pmatrix} = e \begin{pmatrix} t & 0 \\ 0 & -t \end{pmatrix}$$

$$= \sum_{j=0}^{\infty} \frac{1}{j!} \begin{pmatrix} t & 0 \\ 0 & -t \end{pmatrix}^j$$

$$= \sum_{j=0}^{\infty} \frac{1}{j!} \begin{pmatrix} t^j & 0 \\ 0 & (-t)^j \end{pmatrix}$$

$$= \begin{pmatrix} \sum_{j=0}^{\infty} \frac{1}{j!} t^j & 0 \\ 0 & \sum_{j=0}^{\infty} \frac{1}{j!} (-t)^j \end{pmatrix} = \begin{pmatrix} e^t & 0 \\ 0 & e^{-t} \end{pmatrix}.$$

Hence the general solution is

$$\mathbf{x}(t) = \begin{pmatrix} e^t & 0 \\ 0 & e^{-t} \end{pmatrix} \begin{pmatrix} c_1 \\ c_2 \end{pmatrix} = \begin{pmatrix} c_1 e^t \\ c_2 e^{-t} \end{pmatrix}.$$

Of course, this solution can easily be found by a glance at the original system; we derive it using the exponential matrix because it provides a very simple example of the definition and use of the exponential.

e x a m p l e 2

The system

$$\frac{dx}{dt} = x + y,$$

$$\frac{dy}{dt} = y$$

can be written in matrix form $d\mathbf{x}/dt = A\mathbf{x}$ as

$$\begin{pmatrix} dx/dt \\ dy/dt \end{pmatrix} = \begin{pmatrix} 1 & 1 \\ 0 & 1 \end{pmatrix} \begin{pmatrix} x \\ y \end{pmatrix}.$$

We compute

$$A = \begin{pmatrix} 1 & 1 \\ 0 & 1 \end{pmatrix}, \; A^2 = \begin{pmatrix} 1 & 2 \\ 0 & 1 \end{pmatrix}, \dots, \; A^k = \begin{pmatrix} 1 & k \\ 0 & 1 \end{pmatrix}, \dots$$

Then

$$e^{tA} = \sum_{k=0}^{\infty} \frac{t^k}{k!} \begin{pmatrix} 1 & k \\ 0 & 1 \end{pmatrix} = \begin{pmatrix} \sum_{k=0}^{\infty} \frac{t^k}{k!} & \sum_{k=1}^{\infty} \frac{t^k}{(k-1)!} \\ 0 & \sum_{k=0}^{\infty} \frac{t^k}{k!} \end{pmatrix}$$

$$= \begin{pmatrix} e^t & te^t \\ 0 & e^t \end{pmatrix}.$$

Hence the solution with initial conditions $x(0) = 2$, $y(0) = -3$ is

$$\begin{pmatrix} x(t) \\ y(t) \end{pmatrix} = \begin{pmatrix} e^t & te^t \\ 0 & e^t \end{pmatrix} \begin{pmatrix} 2 \\ -3 \end{pmatrix} = \begin{pmatrix} 2e^t - 3te^t \\ -3e^t \end{pmatrix}.$$

This system could also be solved by the elimination method, but not by the method of Section 1.

2B   Relationship to Eigenvectors

The connection between matrix exponential solutions and the eigenvector method is as follows. Let the eigenvectors of the square matrix A be $\mathbf{u}_1, \mathbf{u}_2, \ldots, \mathbf{u}_n$, and suppose that the eigenvector matrix

$$U = (\mathbf{u}_1, \mathbf{u}_2, \ldots, \mathbf{u}_n)$$

with these vectors as columns is invertible. We also form the diagonal matrix

$$\Lambda_t = \begin{pmatrix} e^{\lambda_1 t} & \cdots & 0 \\ \vdots & & \vdots \\ 0 & \cdots & e^{\lambda_n t} \end{pmatrix},$$

where λ_k is the eigenvalue of $\mathbf{u}_k$. Then we have seen that

$$\mathbf{x}(t) = U\Lambda_t U^{-1}\mathbf{x}_0$$

solves the initial-value problem for the equation $d\mathbf{x}/dt = A\mathbf{x}$. Since

$$\mathbf{x}(t) = e^{tA}\mathbf{x}_0$$

solves the same problem, we are faced with the question of whether the two solutions are the same. It turns out by Theorem 2.2 that there is in fact only one solution satisfying $\mathbf{x}(0) = \mathbf{x}_0$. Hence $e^{tA}\mathbf{x}_0 = U\Lambda_t U^{-1}\mathbf{x}_0$ for all t. It follows, since $\mathbf{x}_0$ is arbitrary, that

■ **2.3** $$e^{tA} = U\Lambda_t U^{-1}.$$

example 3

In Example 1 of Section 1 we solved the system

$$\begin{pmatrix} dx_1/dt \\ dx_2/dt \end{pmatrix} = \begin{pmatrix} 1 & 1 \\ 4 & 1 \end{pmatrix} \begin{pmatrix} x_1 \\ x_2 \end{pmatrix}$$

by finding the eigenvalues $\lambda_1 = 3$, $\lambda_2 = -1$ and corresponding eigenvectors $(1, 2)$, $(1, -2)$ of the 2 by 2 matrix A of the system. Thus

$$U = \begin{pmatrix} 1 & 1 \\ 2 & -2 \end{pmatrix}, \quad \Lambda_t = \begin{pmatrix} e^{3t} & 0 \\ 0 & e^{-t} \end{pmatrix}, \quad U^{-1} = \begin{pmatrix} \frac{1}{2} & \frac{1}{4} \\ \frac{1}{2} & -\frac{1}{4} \end{pmatrix}$$

and $$e^{tA} = U\Lambda_t U^{-1} = \begin{pmatrix} \frac{1}{2}e^{3t} + \frac{1}{2}e^{-t} & \frac{1}{4}e^{3t} - \frac{1}{4}e^{-t} \\ e^{3t} - e^{-t} & \frac{1}{2}e^{3t} + \frac{1}{2}e^{-t} \end{pmatrix}.$$

As a check on the computation, notice that $e^{0A} = I$. This example shows that if the eigenvector matrix U of A is invertible, it may be easier to use U and Λ to compute e^{tA} than to use the matrix power-series definition.

Finally, suppose that the eigenvector matrix U of A is invertible. By Equation 2.3 and Theorem 2.2 the general solution of $dx/dt = Ax$ can be written

$$x(t) = U\Lambda_t U^{-1}x_0$$
$$= U\Lambda_t c,$$

where c is the column vector with entries $c_1, \ldots, c_n$. Multiplying this last expression out gives the usual form

$$x(t) = c_1 e^{\lambda_1 t}\mathbf{u}_1 + \cdots + c_n e^{\lambda_n t}\mathbf{u}_n$$

for the solution when the eigenvector matrix is invertible.

EXERCISES

1. Find the exponential e^{tA} of each of the following matrices A by first computing the successive terms $I, tA, t^2 A^2/2!, \ldots$ in the series definition.

   (a) $A = \begin{pmatrix} -1 & 0 \\ 0 & 1 \end{pmatrix}$.

   (c) $A = \begin{pmatrix} i & 0 \\ 0 & -i \end{pmatrix}$.

   (b) $A = \begin{pmatrix} 1 & 0 \\ 1 & 0 \end{pmatrix}$.

   (d) $A = \begin{pmatrix} 1 & 0 & 0 \\ 0 & 2 & 0 \\ 0 & 0 & 4 \end{pmatrix}$.

2. In Example 2 of the text, it was shown that

$$e^{t\left(\begin{smallmatrix} 1 & 1 \\ 0 & 1 \end{smallmatrix}\right)} = \begin{pmatrix} e^t & te^t \\ 0 & e^t \end{pmatrix}.$$

   Verify directly, by using this equation, that

   (a) $e^{t\left(\begin{smallmatrix} 1 & 1 \\ 0 & 1 \end{smallmatrix}\right)}$ and $e^{-t\left(\begin{smallmatrix} 1 & 1 \\ 0 & 1 \end{smallmatrix}\right)}$ are inverse to one another.

   (b) $e^{t\left(\begin{smallmatrix} 1 & 1 \\ 0 & 1 \end{smallmatrix}\right)}e^{s\left(\begin{smallmatrix} 1 & 1 \\ 0 & 1 \end{smallmatrix}\right)} = e^{(t+s)\left(\begin{smallmatrix} 1 & 1 \\ 0 & 1 \end{smallmatrix}\right)}$.

   (c) $\dfrac{d}{dt} e^{t\left(\begin{smallmatrix} 1 & 1 \\ 0 & 1 \end{smallmatrix}\right)} = \begin{pmatrix} 1 & 1 \\ 0 & 1 \end{pmatrix}e^{t\left(\begin{smallmatrix} 1 & 1 \\ 0 & 1 \end{smallmatrix}\right)}$.

   (d) What is the solution of the system

$$\begin{pmatrix} dx/dt \\ dy/dt \end{pmatrix} = \begin{pmatrix} 1 & 1 \\ 0 & 1 \end{pmatrix}\begin{pmatrix} x \\ y \end{pmatrix} \quad \text{that satisfies} \quad \begin{pmatrix} x(0) \\ y(0) \end{pmatrix} = \begin{pmatrix} -1 \\ 2 \end{pmatrix}?$$

3. If A is a 1 by 1 matrix with entry a, what is e^{tA}?

4. Let

$$A = \begin{pmatrix} -1 & -2 & -2 \\ 0 & 1 & -1 \\ 0 & 0 & 2 \end{pmatrix}.$$

   Using methods to be developed in Section 3, we will be able to show quite easily that

$$e^{tA} = \begin{pmatrix} e^{-t} & e^{-t} - e^t & e^{-t} - e^t \\ 0 & e^t & e^t - e^{2t} \\ 0 & 0 & e^{2t} \end{pmatrix}.$$

(a) Replace t by $-t$ in e^{tA} and verify directly that $e^{-tA}e^{tA} = I$ and that $e^{0A} = I$.

(b) If $B(t) = (d/dt)e^{tA}$, verify that $B(0) = A$.

5. (a) Find the general solution of the system

$$\begin{pmatrix} \dot{x} \\ \dot{y} \end{pmatrix} = \begin{pmatrix} 9 & -4 \\ 4 & 1 \end{pmatrix} \begin{pmatrix} x \\ y \end{pmatrix}$$

by the method of elimination.

(b) Use the result of part (a) to compute the matrix e^{tA}, where

$$A = \begin{pmatrix} 9 & -4 \\ 4 & 1 \end{pmatrix}.$$

$$\left[\textit{Hint: } \text{Find solutions such that } \mathbf{x}_1(0) = \begin{pmatrix} 1 \\ 0 \end{pmatrix} \text{ and } \mathbf{x}_2(0) = \begin{pmatrix} 0 \\ 1 \end{pmatrix}. \right]$$

6. Let A be an n by n matrix with real entries. The matrix e^{itA} is defined by

$$e^{itA} = \sum_{k=0}^{\infty} \frac{1}{k!} (i)^k t^k A^k.$$

Define $\cos tA$ to be the real part of the series and $\sin tA$ to be the imaginary part so that

$$e^{itA} = \cos tA + i \sin tA.$$

Show that the matrices $\cos tA$ and $\sin tA$ satisfy the following:

(a) $\cos (-t)A = \cos tA$, $\sin (-t)A = -\sin tA$.

(b) $(d/dt) \cos tA = -A \sin tA$, $\quad (d/dt) \sin tA = A \cos tA$. [*Hint:* Express $\cos tA$ and $\sin tA$ in terms of e^{itA}.]

(c) $(\cos tA)^2 + (\sin tA)^2 = I$, where I is the n by n identity matrix.

7. Let A be the 2 by 2 matrix

$$\begin{pmatrix} 1 & 1 \\ 0 & 1 \end{pmatrix}$$

Define $\cos tA$ and $\sin tA$ as in Exercise 6, and verify the formulas given in parts (a), (b), and (c).

8. Show that if A is an n by n matrix, then a system of the form

$$\frac{d^2\mathbf{x}}{dt^2} + A^2\mathbf{x} = 0$$

has solutions of the form

$$\mathbf{x}(t) = (\cos tA)\mathbf{c}_1 + (\sin tA)\mathbf{c}_2,$$

where $\mathbf{c}_1$ and $\mathbf{c}_2$ are constant n-dimensional vectors, whereas $\cos tA$ and $\sin tA$ are the n by n matrices defined in Exercise 6. Is this always the most general solution?

9. For each of the following matrices A, compute e^{tA} by using Equation 2.3. Then find the inverse matrix e^{-tA}, and check your original computation by showing that the derivative of e^{tA} at $t = 0$ is equal to A.

(a) $\begin{pmatrix} 0 & 1 \\ -6 & 5 \end{pmatrix}$.

(c) $\begin{pmatrix} -1 & 0 & 0 \\ 0 & \frac{3}{2} & -\frac{1}{2} \\ 0 & -\frac{1}{2} & \frac{3}{2} \end{pmatrix}$.

(b) $\begin{pmatrix} -4 & 4 \\ -6 & 6 \end{pmatrix}$.

10. Use the answers to Exercise 4 of Section 1 to find e^{tA} for each of the following matrices A.

(a) $A = \begin{pmatrix} -3 & 2 \\ -4 & 3 \end{pmatrix}$.

(c) $A = \begin{pmatrix} 1 & 4 \\ 0 & 5 \end{pmatrix}$.

(b) $A = \begin{pmatrix} 3 & 0 \\ 0 & 2 \end{pmatrix}$.

(d) $A = \begin{pmatrix} 2 & -1 \\ 1 & 2 \end{pmatrix}$.

11. Let A be the n by n matrix with all entries equal to 1.

   (a) Show that $A^2 = nA$ and, more generally, that $A^k = n^{k-1}A$ for integer $k \geq 1$.

   (b) Show that $e^{tA} = I + (1/n)(e^{nt} - 1)A$.

   (c) Find the four entries in e^{tA} when $n = 2$.

12. It can be shown that if A and B are n by n matrices, then $e^A e^B = e^{A+B}$ if and only if $AB = BA$. Find 2 by 2 matrices A and B for which the exponential equation fails to hold.

13. It can be shown that $e^{tA} e^{sB} = e^{sB} e^{tA}$ if A and B commute, that is, if $AB = BA$. Find 2 by 2 matrices A and B such that $e^{tA} e^{sB} \neq e^{sB} e^{tA}$.

14. Prove that if $B(t)$ is an n by n matrix and $\mathbf{x}(t)$ is an n-dimensional column vector, both with differentiable entries, then $(d/dt)B(t)\mathbf{x}(t) = B'(t)\mathbf{x}(t) + B(t)\dot{\mathbf{x}}(t)$.

2C Independent Solutions

Theorem 2.2 in Section 2A shows that for every equation $d\mathbf{x}/dt = A\mathbf{x}$ with A a constant square matrix, there is an exponential solution formula $\mathbf{x}(t) = e^{tA}\mathbf{c}$. In Example 2 of that section we had

$$e^{tA}\mathbf{c} = \begin{pmatrix} e^t & te^t \\ 0 & e^t \end{pmatrix}\begin{pmatrix} c_1 \\ c_2 \end{pmatrix}$$

$$= c_1\begin{pmatrix} e^t \\ 0 \end{pmatrix} + c_2\begin{pmatrix} te^t \\ e^t \end{pmatrix}.$$

Since c_1, c_2 are arbitrary constants, it's important to avoid the redundancy that would occur in the formula in case one of the two columns in the matrix is a constant multiple of the other. We will see that this cannot happen in general, but to state the general result we need to look more closely at what is meant by linear independence of vector functions. Let $\mathbf{x}_1(t), \mathbf{x}_2(t), \ldots, \mathbf{x}_m(t)$ be n-dimensional column vectors whose entries are functions on some common interval $a < t < b$. (It is not ruled out of course that some or all of the entries may happen to be constant.) Vector functions $\mathbf{x}_k(t)$, $k = 1, \ldots, m$, defined on a t-

interval are said to be **linearly independent** if whenever

$$c_1\mathbf{x}_1(t) + c_2\,\mathbf{x}_2(t) + \cdots + c_m\mathbf{x}_m(t) = 0$$

for all t, then the constant coefficients c_k are all zero. When we have only two functions ($m = 2$), asserting linear independence is the same as saying that neither function is a constant multiple of the other. The reason is that if either c_1 or c_2 is not zero, we could divide by it and express one vector as a multiple of the other. Similarly, if we have $m > 2$ vector functions, their linear independence means that none of them can be replaced by a sum of scalar multiples, or **linear combination,** of the others. The negation of linear independence of a set of vectors is called **linear dependence,** and it means simply that at least one of the vectors is a linear combination of the others.

e x a m p l e 1

The vector functions

$$\mathbf{x}_1(t) = \begin{pmatrix} e^t \\ 0 \end{pmatrix}, \qquad \mathbf{x}_2 = \begin{pmatrix} te^t \\ e^t \end{pmatrix}$$

that form the exponential matrix in Example 1, Section 2A, are linearly independent. For

$$c_1 \begin{pmatrix} e^t \\ 0 \end{pmatrix} + c_2 \begin{pmatrix} te^t \\ e^t \end{pmatrix} = \begin{pmatrix} 0 \\ 0 \end{pmatrix}, \qquad -\infty < t < \infty,$$

is the same as

$$c_1 e^t + c_2 t e^t = 0$$
$$c_2 e^t = 0.$$

It follows that $c_2 = 0$. Hence $c_1 = 0$. This conclusion holds for any fixed value of t, so in particular the constant vectors $\begin{pmatrix} 1 \\ 0 \end{pmatrix}$ and $\begin{pmatrix} 0 \\ 1 \end{pmatrix}$ are linearly independent. Just set $t = 0$. Theorem 2.4 below shows that the columns of a matrix e^{tA} are always independent vector functions and are also independent for each t. (But see Exercise 8 and the next example.)

e x a m p l e 2

Consider the vector functions

$$\mathbf{x}_1(t) = \begin{pmatrix} e^t \\ 0 \\ 0 \end{pmatrix}, \qquad \mathbf{x}_2(t) = \begin{pmatrix} 0 \\ e^t \\ te^t \end{pmatrix}, \qquad \mathbf{x}_3(t) = \begin{pmatrix} 0 \\ e^t \\ t^2 e^t \end{pmatrix}.$$

The check for independence for $-\infty < t < \infty$ is to solve

$$c_1 \begin{pmatrix} e^t \\ 0 \\ 0 \end{pmatrix} + c_2 \begin{pmatrix} 0 \\ e^t \\ te^t \end{pmatrix} + c_3 \begin{pmatrix} 0 \\ e^t \\ t^2 e^t \end{pmatrix} = \begin{pmatrix} 0 \\ 0 \\ 0 \end{pmatrix}$$

for c_1, c_2, c_3. This is the same as

$$c_1 e^t = 0,$$
$$c_2 e^t + c_3 e^t = 0,$$
$$c_2 te^t + c_3 t^2 e^t = 0.$$

The first equation shows that $c_1 = 0$. The middle equation implies $c_2 = -c_3$, so the last equation says $c_2 t - c_2 t^2 = 0$ for all t. Thus $c_2 = c_3 = 0$, so the vector functions are independent as defined on an interval $a < t < b$. Note, however, that when $t = 0$, we get

$$\mathbf{x}_1(0) = \begin{pmatrix} 1 \\ 0 \\ 0 \end{pmatrix}, \qquad \mathbf{x}_2(0) = \begin{pmatrix} 0 \\ 1 \\ 0 \end{pmatrix}, \qquad \mathbf{x}_3(0) = \begin{pmatrix} 0 \\ 1 \\ 0 \end{pmatrix},$$

and these constant vectors are linearly dependent. This shows that functions may be linearly independent while their restrictions to some smaller domain (in this example a single point) may be linearly dependent.

Here is the theorem that guarantees independence of the columns of an exponential matrix for any and all values of t.

■ 2.4 THEOREM

Let A be an n by n matrix with constant entries and let $\mathbf{x}_k(t)$ be the kth column of the exponential matrix e^{tA}. Then the vector functions $\mathbf{x}_1(t), \ldots, \mathbf{x}_n(t)$ are linearly independent over any set of t-values.

Proof. Apply the matrix e^{-tA} to both sides of the vector equation

$$c_1 \mathbf{x}_1(t) + \cdots + c_n \mathbf{x}_n(t) = 0.$$

Using the distributivity of matrix multiplication, we get

$$c_1 e^{-tA} \mathbf{x}_1(t) + \cdots + c_n e^{-tA} \mathbf{x}_n(t) = 0.$$

But e^{-tA} is the inverse of the matrix whose kth column is $\mathbf{x}_k(t)$. Hence $e^{-tA} \mathbf{x}_k(t)$ is the kth column of the identity matrix I. Thus our equation becomes

$$c_1 \mathbf{e}_1 + \cdots + c_k \mathbf{e}_k + \cdots + c_n \mathbf{e}_n = 0,$$

where $\mathbf{e}_k$ is the column vector with 1 in the kth entry and 0 elsewhere. Adding up the linear combination gives

$$\begin{pmatrix} c_1 \\ \vdots \\ c_k \\ \vdots \\ c_n \end{pmatrix} = \begin{pmatrix} 0 \\ \vdots \\ 0 \\ \vdots \\ 0 \end{pmatrix}.$$

This proves that the vectors $\mathbf{x}_k(t)$ are linearly independent. ■

EXERCISES

1. Each of the given matrices is the exponential matrix of some constant matrix A. In each case, express the vector function $\mathbf{y}(t) = e^{tA}\mathbf{c}$ as a linear combination of the columns of e^{tA}.

   (a) $e^{tA} = \begin{pmatrix} 1 & 0 \\ 0 & e^t \end{pmatrix}$.

   (b) $e^{tA} = \begin{pmatrix} e^t & 0 \\ 0 & e^{2t} \end{pmatrix}$.

(c) $e^{tA} = \begin{pmatrix} e^t & 0 & 0 \\ 0 & e^{2t} & 0 \\ 0 & 0 & e^{3t} \end{pmatrix}$.   **(d)** $e^{tA} = \begin{pmatrix} e^t & te^t & 0 \\ 0 & e^t & 0 \\ 0 & 0 & e^{2t} \end{pmatrix}$.

2. For each of the exponential matrices in Exercise 1 find A by computing the derivative of e^{tA} at $t = 0$. Then verify that each column of e^{tA} is a solution of the differential equation $\dot{\mathbf{x}} = A\mathbf{x}$.

3. Not every square matrix with linearly independent columns is an exponential matrix. For example, an exponential matrix e^{tA} must equal I when $t = 0$. Show that each of the following matrices has linearly independent columns, but is not an exponential matrix.

   (a) $\begin{pmatrix} e^t & 1 \\ 0 & e^t \end{pmatrix}$.   **(b)** $\begin{pmatrix} e^{2t} & t^2 e^t \\ 0 & e^t \end{pmatrix}$.   **(c)** $\begin{pmatrix} e^t & te^{2t} \\ 0 & e^t \end{pmatrix}$.

4. Theorem 2.4 is a simple consequence of the following more general theorem: If $A(t)$ is an invertible square matrix for each t in some interval $a < t < b$, then the columns of $A(t)$ are linearly independent vector functions on this interval. Show how to prove this theorem using the ideas in the proof of Theorem 2.4.

5. Let D be the n by n diagonal matrix with entries $d_1, \ldots, d_n$ on the main diagonal and zeros elsewhere. Show that e^{tD} is the diagonal matrix with entries $e^{d_1 t}, \ldots, e^{d_n t}$.

6. Prove that a set $\{\mathbf{x}_1(t), \mathbf{x}_2(t), \ldots, \mathbf{x}_m(t)\}$ of vector-valued functions of the same dimension is linearly independent on a t interval if and only if no one of them is a linear combination of the others.

7. A system $\dot{\mathbf{x}} = A\mathbf{x}$ has solution $\mathbf{x} = (c_1 e^{2t} + c_2 e^{3t}, c_1 e^{2t} - c_2 e^{3t})$. Find the matrix A by first finding e^{tA}.

8. Theorem 2.4 shows that if $M(t)$ is an exponential matrix, then not only are its columns independent vector functions of t, but in addition the columns are independent vectors for each *fixed* t.

   (a) Show that the matrix $\begin{pmatrix} e^t & te^t \\ te^t & e^t \end{pmatrix}$ has independent vector functions for its columns.

   (b) Show that the matrix in part (a) can't be an exponential matrix, because there is a value $t = t_0$ for which the columns are *not* independent vectors.

3   COMPUTING e^{tA} IN GENERAL

The method described below is usually simpler than appealing directly to the definition of e^{tA} or first finding an eigenvector matrix as described in Section 2C. Furthermore it works just as well whether the matrix A has distinct eigenvectors or not. The idea behind it is that an n by n exponential matrix series can be expressed in the compressed form

■■ **3.1**
$$e^{tA} = \sum_{j=0}^{n-1} b_j(t)A^j,$$

where the coefficient functions $b_k(t)$ contain the eigenvalues $\lambda_1, \ldots, \lambda_n$ of A explicitly. The next theorem shows how to compute the coefficients by solving a system of linear equations. The complete proof of the theorem is complicated, so we'll just sketch it.

■■ **3.2 THEOREM**

The coefficient functions $b_k(t)$ in the expansion 3.1 satisfy the equations

$$e^{t\lambda_k} = \sum_{j=0}^{n-1} b_j(t)\lambda_k^j, \qquad k = 1, \ldots, n. \tag{1}$$

If some m of the eigenvalues λ_k are equal, say $\lambda_1 = \cdots = \lambda_m$, then the following $m-1$ additional relations hold just at the repeated eigenvalue $\lambda = \lambda_1$:

$$\frac{d^\mu}{d\lambda^\mu} e^{t\lambda} = \frac{d^\mu}{d\lambda^\mu} \sum_{j=0}^{n-1} b_j(t)\lambda^j, \qquad \text{at } \lambda = \lambda_1 \text{ for } \mu = 1, \ldots, m-1. \tag{2}$$

Note that the coefficients $b_j(t)$ depend only on the eigenvalues of A, not on the specific entries in A.

Sketch of Proof. We will assume that Equation 3.1 holds for some choice of the coefficients $b_j(t)$. Let $\mathbf{v}_k$ be an eigenvector of A corresponding to eigenvalue λ_k: $A\mathbf{v}_k = \lambda_k\mathbf{v}_k$, $\mathbf{v}_k \neq 0$. Apply the matrix sum on the right side of Equation 3.1 to $\mathbf{v}_k$:

$$\sum_{j=0}^{n-1} b_j(t)A^j\mathbf{v}_k = \left(\sum_{j=0}^{n-1} b_j(t)\lambda_k^j \right)\mathbf{v}_k.$$

Similarly apply the matrix e^{tA} to $\mathbf{v}_k$ to get another expression for the same thing:

$$e^{tA}\mathbf{v}_k = \lim_{N\to\infty} \sum_{j=0}^{N} \frac{t^j}{j!} A^j\mathbf{v}_k = \sum_{j=0}^{\infty} \frac{t^j}{j!} \lambda_k^j\mathbf{v}_k = e^{t\lambda_k}\mathbf{v}_k.$$

Since $\mathbf{v}_k$ is an eigenvector, it is not zero. Hence the coefficients of $\mathbf{v}_k$ at the ends of the previous two displayed equations must be equal. Equations (1) follow immediately.

If there are multiple eigenvalues, we alter the entries of A slightly so as to produce matrices A_h with distinct eigenvalues whose limit as $h \to 0$ is A. Equations (2) follow by calculating an appropriate limit of a difference quotient as $h \to 0$. ■

---

e x a m p l e **1**

The matrix $A = \begin{pmatrix} 1 & 1 \\ 4 & 1 \end{pmatrix}$ was shown in Example 1 of Section 1 to have eigenvalues $\lambda_1 = -1$, $\lambda_2 = 3$. Equations (1) of Theorem 3.2 are then

$$e^{-t} = b_0(t) - b_1(t),$$
$$e^{3t} = b_0(t) + 3b_1(t).$$

Solve for $b_0(t)$ and $b_1(t)$ to get $b_1(t) = -\frac{1}{4}e^{-t} + \frac{1}{4}e^{3t}$, $b_0(t) = \frac{3}{4}e^{-t} + \frac{1}{4}e^{3t}$. Plugging these coefficient functions into Equation 3.1 gives

$$e^{tA} = (\tfrac{3}{4}e^{-t} + \tfrac{1}{4}e^{3t})\begin{pmatrix} 1 & 0 \\ 0 & 1 \end{pmatrix} + (-\tfrac{1}{4}e^{-t} + \tfrac{1}{4}e^{3t})\begin{pmatrix} 1 & 1 \\ 4 & 1 \end{pmatrix}$$

$$= \begin{pmatrix} \tfrac{1}{2}e^{-t} + \tfrac{1}{2}e^{3t} & -\tfrac{1}{4}e^{-t} + \tfrac{1}{4}e^{3t} \\ -e^{-t} + e^{3t} & \tfrac{1}{2}e^{-t} + \tfrac{1}{2}e^{3t} \end{pmatrix}.$$

As a partial check on the accuracy of our computation we can verify that our expression for e^{tA} equals the identity matrix I when $t = 0$.

---

e x a m p l e 2

In Example 2 of Section 1 we saw that the matrix $\begin{pmatrix} 1 & -1 \\ 1 & 1 \end{pmatrix}$ has eigenvalues $\lambda_1 = 1 + i$ and $\lambda_2 = 1 - i$. Equations (1) of Theorem 3.2 are then

$$e^{(1+i)t} = b_0(t) + b_1(t)(1 + i),$$
$$e^{(1-i)t} = b_0(t) + b_1(t)(1 - i).$$

Subtracting the second equation from the first gives

$$2b_1(t) = e^t(e^{it} - e^{-it}), \qquad \text{so} \qquad b_1(t) = \frac{1}{2i}e^t(e^{it} - e^{-it}) = e^t \sin t.$$

Substituting for $b_1(t)$ in the first equation gives

$$b_0(t) = -(1 + i)e^t \sin t + e^t(\cos t + i \sin t) = e^t(\cos t - \sin t).$$

Then

$$e^{t\begin{pmatrix} 1 & -1 \\ 1 & 1 \end{pmatrix}} = e^t(\cos t - \sin t)\begin{pmatrix} 1 & 0 \\ 0 & 1 \end{pmatrix} + e^t \sin t\begin{pmatrix} 1 & -1 \\ 1 & 1 \end{pmatrix} = \begin{pmatrix} e^t \cos t & -e^t \sin t \\ e^t \sin t & e^t \cos t \end{pmatrix}.$$

---

e x a m p l e 3

The matrix $A = \begin{pmatrix} 0 & 1 \\ -4 & -4 \end{pmatrix}$ has characteristic equation $\lambda^2 + 4\lambda + 4 = 0$ with eigenvalues $\lambda_1 = \lambda_2 = -2$. Equations (1) of Theorem 3.2 are identical:

$$e^{t\lambda} = b_0(t) + b_1(t)\lambda, \qquad \lambda = -2, \qquad \text{or} \qquad e^{-2t} = b_0(t) - 2b_1(t).$$

Hence we need another equation satisfied by the coefficients. We differentiate with respect to λ in the first equation above and then set $\lambda = -2$ to get

$$te^{t\lambda} = b_1(t), \qquad \lambda = -2, \qquad \text{or} \qquad te^{-2t} = b_1(t).$$

We see right away that $b_0(t) = e^{-2t} + 2te^{-2t}$. Equation 3.1 then becomes

$$e^{tA} = b_0(t)\begin{pmatrix} 1 & 0 \\ 0 & 1 \end{pmatrix} + b_1(t)\begin{pmatrix} 0 & 1 \\ -4 & -4 \end{pmatrix}$$

$$= (1 + 2t)e^{-2t}\begin{pmatrix} 1 & 0 \\ 0 & 1 \end{pmatrix} + te^{-2t}\begin{pmatrix} 0 & 1 \\ -4 & -4 \end{pmatrix}$$

$$= e^{-2t}\begin{pmatrix} 1 + 2t & t \\ -4t & 1 - 2t \end{pmatrix}.$$

example 4

Let $A = \begin{pmatrix} 2 & 1 & 0 \\ 0 & 2 & 1 \\ 0 & 0 & 2 \end{pmatrix}$. The characteristic equation is $(2 - \lambda)^3 = 0$, so there is a triple root $\lambda_1 = 2$. Equations (1) of Theorem 3.2 reduce to

$$e^{t\lambda} = b_0(t) + b_1(t)\lambda + b_2(t)\lambda^2, \qquad \lambda = 2,$$

or

$$e^{2t} = b_0(t) + 2b_1(t) + 4b_2(t).$$

We get two additional relations among the $b_k(t)$ by differentiating the first equation above twice with respect to λ and then setting $\lambda = 2$ after each differentiation. First we get $te^{t\lambda} = b_1(t) + 2b_2(t)\lambda$ and $t^2e^{t\lambda} = 2b_2(t)$; then

$$te^{2t} = b_1(t) + 4b_2(t), \qquad t^2e^{2t} = 2b_2(t).$$

Solving the last three displayed equations for the $b_k(t)$ gives

$$b_2(t) = \tfrac{1}{2}t^2e^{2t}, \qquad b_1(t) = (t - 2t^2)e^{2t}, \qquad b_0(t) = (1 - 2t + 2t^2)e^{2t}.$$

Equation 3.1 is then

$$e^{tA} = (1 - 2t + 2t^2)e^{2t}\begin{pmatrix} 1 & 0 & 0 \\ 0 & 1 & 0 \\ 0 & 0 & 1 \end{pmatrix} + (t - 2t^2)e^{2t}\begin{pmatrix} 2 & 1 & 0 \\ 0 & 2 & 1 \\ 0 & 0 & 2 \end{pmatrix} + \tfrac{1}{2}t^2e^{2t}\begin{pmatrix} 4 & 4 & 1 \\ 0 & 4 & 4 \\ 0 & 0 & 4 \end{pmatrix}$$

$$= e^2\begin{pmatrix} 1 & t & \tfrac{1}{2}t^2 \\ 0 & 1 & t \\ 0 & 0 & 1 \end{pmatrix}.$$

As we indicated earlier, a detailed explanation of why Theorem 3.2 holds is difficult. However, it seems appropriate to consider in some detail at least the purely algebraic core of Equation 3.1, which is contained in the following remarkable theorem that allows us to express high powers of A in terms of lower powers. (See also Theorem B.10 in Appendix B.)

■ 3.3 CAYLEY-HAMILTON THEOREM

If A is an n by n matrix with characteristic polynomial

$$P(\lambda) = \det (A - \lambda I),$$

then the matrix polynomial obtained by substituting A^k for λ^k in $P(\lambda)$ satisfies $P(A) = 0$, with the understanding that $A^0 = I$ replaces $\lambda^0 = 1$ in the substitution.

example 5

Suppose $A = \begin{pmatrix} a & b \\ c & d \end{pmatrix}$. The characteristic polynomial of A is

$$\det \begin{pmatrix} a & \lambda & b \\ c & d - \lambda \end{pmatrix} = \lambda^2 - (a + d)\lambda + (ad - bc).$$

The Cayley-Hamilton theorem asserts that $A^2 - (a + d)A + (ad - bc)I = 0$, or $A^2 = (a + d)A - (ad - bc)I$. Multiplying this last equation by A gives

$$A^3 = (a + d)A^2 - (ad - bc)A = (a + d)((a + d)A - (ad - bc)I) - (ad - bc)A$$
$$= ((a + d)^2 - (ad - bc))A - (a + d)(ad - bc)I.$$

Continuing as in Example 5, it becomes clear that any power of A, and hence any polynomial in A, can be expressed as a first-degree polynomial in A. The main difficulty in proving Theorem 3.2 is that we're dealing with an infinite sum that defines e^{tA}. Nevertheless it can be shown that the successive additions to coefficients terms in $I, A, \ldots, A^{n-1}$ accumulate to finite sums that obey the conclusions of Theorem 3.2.

It's worth pointing out that if a square matrix A is invertible, then the Cayley-Hamilton theorem can be used to write A^{-1} as a polynomial in A. For if

$$a_0I + a_1A + \cdots + a_{n-1}A^{n-1} \pm A^n = 0$$

is the characteristic equation of A with λ replaced by A, we can multiply by A^{-1} to get

$$a_0A^{-1} + a_1I + \cdots + a_{n-1}A^{n-2} \pm A^{n-1} = 0.$$

Now solve for A^{-1}, noting that $a_0 = \det A \neq 0$.

example 6

The matrix $A = \begin{pmatrix} 2 & 3 & 1 \\ 0 & 2 & 2 \\ 0 & 0 & 2 \end{pmatrix}$ has characteristic equation

$$(2 - \lambda)^3 = 8 - 12\lambda + 6\lambda^2 - \lambda^3 = 0,$$

so $8I - 12A + 6A^2 - A^3 = 0$. Multiply by A^{-1} to get

$$8A^{-1} - 12I + 6A - A^2 = 0, \quad \text{or} \quad 8A^{-1} = 12I - 6A + A^2 = 0.$$

Hence we need only divide by 8 after computing

$$8A^{-1} = 12\begin{pmatrix} 1 & 0 & 0 \\ 0 & 1 & 0 \\ 0 & 0 & 1 \end{pmatrix} - 6\begin{pmatrix} 2 & 3 & 1 \\ 0 & 2 & 2 \\ 0 & 0 & 2 \end{pmatrix} + \begin{pmatrix} 4 & 12 & 10 \\ 0 & 4 & 8 \\ 0 & 0 & 4 \end{pmatrix} = \begin{pmatrix} 4 & -6 & 4 \\ 0 & 4 & -4 \\ 0 & 0 & 4 \end{pmatrix}.$$

EXERCISES

1. Solve $\dot{\mathbf{x}} = A\mathbf{x}$ for the matrices A listed below by finding e^{tA}; the general solution is then a linear combination of the columns of e^{tA}.

(a) $\begin{pmatrix} 8 & -3 \\ 10 & -3 \end{pmatrix}$.

(b) $\begin{pmatrix} 8 & 9 \\ -4 & -4 \end{pmatrix}$.

(c) $\begin{pmatrix} 1 & 1 & 2 \\ 0 & 1 & -1 \\ 0 & 0 & 2 \end{pmatrix}$.

(d) $\begin{pmatrix} 1 & 1 & 1 \\ 0 & 1 & 0 \\ 0 & 0 & 2 \end{pmatrix}$.

(e) $\begin{pmatrix} 1 & 1 & 1 & 0 \\ 0 & 0 & -1 & 0 \\ 1 & 1 & 2 & 0 \\ 1 & 0 & -1 & 1 \end{pmatrix}$.

(f) $\begin{pmatrix} 1 & 0 & 0.5 & -0.5 \\ 1 & 0 & -1 & 0 \\ 0 & 2 & 2.5 & 0.5 \\ -1 & 1 & 0.5 & 1.5 \end{pmatrix}$.

2. Solve the system

$$\frac{dx}{dt} = 2x \qquad + z,$$

$$\frac{dy}{dt} = -x + 3y + z,$$

$$\frac{dz}{dt} = -x \qquad + 4z.$$

(*Hint:* $\lambda = 3$ is a triple eigenvalue.)

3. Solve the system

$$\frac{dx}{dt} = 2x \qquad + z,$$

$$\frac{dy}{dt} = \qquad y \qquad + w,$$

$$\frac{dz}{dt} = \qquad 2z + w,$$

$$\frac{dw}{dt} = \qquad -y \qquad + w.$$

4. Find the solution to the equation in Exercise 1 that satisfies the corresponding initial condition given here.

(a) $\mathbf{x}(0) = \begin{pmatrix} 2 \\ -3 \end{pmatrix}$.

(c) $\mathbf{x}(0) = \begin{pmatrix} 0 \\ 1 \\ 0 \end{pmatrix}$.

(b) $\mathbf{x}(1) = \begin{pmatrix} 1 \\ 1 \end{pmatrix}$.

(d) $\mathbf{x}(0) = \begin{pmatrix} 2 \\ 1 \\ 1 \end{pmatrix}$.

5. Let $A = \begin{pmatrix} \alpha & 1 \\ 0 & \beta \end{pmatrix}$ where α, β are real numbers.

(a) Show that if $\alpha \neq \beta$, then A has two linearly independent eigenvectors.

(b) Show that if $\alpha = \beta$, then the only eigenvectors of A are of the form

$$\mathbf{u} = \begin{pmatrix} c \\ 0 \end{pmatrix}, \quad c \neq 0.$$

(c) Compute e^{tA} in each of the two cases.

6. Show by using the definition of e^{tA} as a matrix power series that $e^{tI} = e^{t}I$.

7. Theorem 3.2 shows that the coefficients $b_k(t)$ in an expansion

$$e^{tA} = \sum_{k=0}^{n-1} b_k(t) A^k$$

are completely determined by the eigenvalues of the n by n matrix A. For example, the matrices $\begin{pmatrix} 1 & \beta \\ 0 & 2 \end{pmatrix}$, $\begin{pmatrix} 2 & \beta \\ 0 & 1 \end{pmatrix}$ both have characteristic polynomials with 1 and 2 as roots. Hence the $b_k(t)$ are the same for all these matrices regardless of the value of the entry b.

(a) Compute $b_0(t)$ and $b_1(t)$ for the two matrices above, and use them to find the corresponding exponential matrices, each depending on the parameter β.

(b) Compute the exponential matrix for $\begin{pmatrix} \alpha & \beta \\ 0 & \alpha \end{pmatrix}$.

8. Use Theorem 3.3 to compute A^2 and A^3 if $A = \begin{pmatrix} 1 & 2 \\ 3 & 4 \end{pmatrix}$.

9. Determine which of the following matrices have inverses and then use the Cayley-Hamilton theorem to find the inverses of the ones that are invertible.

(a) $\begin{pmatrix} 1 & 0 & 0 \\ 3 & 1 & 5 \\ -2 & 0 & 1 \end{pmatrix}$.

(b) $\begin{pmatrix} 1 & 2 & 3 \\ -1 & 1 & 0 \\ 0 & 3 & 3 \end{pmatrix}$.

(c) $\begin{pmatrix} 2 & 4 & 8 \\ 1 & 0 & 0 \\ 1 & -3 & -7 \end{pmatrix}$.

(d) $\begin{pmatrix} t & 0 & 0 \\ 0 & 2 & 0 \\ 0 & 0 & 1 \end{pmatrix}$, t real.

(e) $\begin{pmatrix} 1 & 2 & 1 \\ 0 & 0 & 1 \\ 0 & 0 & 3 \end{pmatrix}$.

(f) $\begin{pmatrix} 1 & -1 & 1 \\ 0 & -1 & 1 \\ 0 & 0 & 1 \end{pmatrix}$.

(g) $\begin{pmatrix} 1 & 2 & -1 & 3 \\ 0 & 2 & 0 & 1 \\ 0 & 0 & 1 & 1 \\ 0 & 0 & 0 & 4 \end{pmatrix}$.

(h) $\begin{pmatrix} 1 & 0 & 1 & 0 \\ 0 & 2 & 0 & 0 \\ 0 & 0 & 3 & 0 \\ 0 & 0 & 0 & 4 \end{pmatrix}$.

(i) $\begin{pmatrix} 1 & 0 & 0 \\ 0 & e^t & te^t \\ 0 & 0 & e^t \end{pmatrix}$, t real.

10. Show that equations (2) of Theorem 3.2 can be written more directly for computing, though less memorably, as

$$t^\mu e^{t\lambda} = \sum_{j=\mu}^{n-1} \frac{j!}{(j-\mu)!} b_j(t) \lambda^{j-\mu}, \qquad \text{at } \lambda = \lambda_1 \text{ for } \mu = 1, \ldots, m-1.$$

4 NONHOMOGENEOUS SYSTEMS

4A *Solution Formula*

To develop efficient methods for solving nonhomogeneous systems, we need a formula for the derivative of a matrix product or, more particularly, the product of a matrix and a vector. The rule is similar to the usual product rule for real derivatives and is easily proved using that special case:

■ **4.1** $$\frac{d}{dt}[A(t)B(t)] = \left[\frac{d}{dt}A(t)\right]B(t) + A(t)\left[\frac{d}{dt}B(t)\right].$$

Recall that the derivative of a matrix is just the matrix obtained by differentiating the entries. However, the order of the factors on the right is important because it involves matrix multiplication, which is not in general commutative. To prove the formula, we differentiate one entry at a time on the left side; the derivative of the ijth entry is

$$\frac{d}{dt}\sum_{k=1}^{n} a_{ik}(t)b_{kj}(t) = \sum_{k=1}^{n}\frac{da_{ik}}{dt}(t)b_{kj}(t) + \sum_{k=1}^{n}a_{ik}(t)\frac{db_{kj}}{dt}(t).$$

But this is just the ijth entry in the sum of products of matrices on the right, so the formula is proved.

To solve the nonhomogeneous vector equation

$$\frac{d\mathbf{x}}{dt} = A\mathbf{x} + \mathbf{b}(t),$$

where A is a constant matrix, we write

$$\frac{d\mathbf{x}}{dt} - A\mathbf{x} = \mathbf{b}(t)$$

and multiply by the exponential matrix integrating factor e^{-tA} to get

$$e^{-tA}\frac{d\mathbf{x}}{dt} - e^{-tA}A\mathbf{x} = e^{-tA}\mathbf{b}(t).$$

But by the product rule for differentiation of matrices, this is the same as

$$\frac{d}{dt}(e^{-tA}\mathbf{x}) = e^{-tA}\mathbf{b}(t).$$

Integration of both sides gives, with constant vector **c**,

$$e^{-tA}\mathbf{x}(t) = \int e^{-tA}\mathbf{b}(t)\,dt + \mathbf{c}.$$

Recalling that e^{tA} is the inverse of e^{-tA}, we multiply by it to get the **general solution**

■ **4.2** $$\mathbf{x}(t) = e^{tA}\int e^{-tA}\mathbf{b}(t)\,dt + e^{tA}\mathbf{c}$$

A review of the steps used in deriving Equation 4.2 shows that the term "general solution" is justified. if $\mathbf{x}(t)$ is some solution of the given equation, then it must have the form of Equation 4.2 for some choice of the n-dimensional constant vector **c**.

e x a m p l e 1

In Example 2 of Section 2, we saw that the system

$$\begin{pmatrix} dx/dt \\ dy/dt \end{pmatrix} = \begin{pmatrix} 1 & 1 \\ 0 & 1 \end{pmatrix} \begin{pmatrix} x \\ y \end{pmatrix}$$

had associated with it the fundamental matrix exponential

$$e^{tA} = \begin{pmatrix} e^t & te^t \\ 0 & e^t \end{pmatrix}.$$

As an alternative to using Equation 4.2, we can find a particular solution of

$$\begin{pmatrix} dx/dt \\ dy/dt \end{pmatrix} = \begin{pmatrix} 1 & 1 \\ 0 & 1 \end{pmatrix} \begin{pmatrix} x \\ y \end{pmatrix} + \begin{pmatrix} e^t \\ e^{-t} \end{pmatrix}.$$

we compute the particular solution

$$e^{tA} \int e^{-tA} \begin{pmatrix} e^t \\ e^{-t} \end{pmatrix} dt = \begin{pmatrix} e^t & te^t \\ 0 & e^t \end{pmatrix} \int \begin{pmatrix} e^{-t} & -te^{-t} \\ 0 & e^{-t} \end{pmatrix} \begin{pmatrix} e^t \\ e^{-t} \end{pmatrix} dt$$

$$= \begin{pmatrix} e^t & te^t \\ 0 & e^t \end{pmatrix} \int \begin{pmatrix} 1 - te^{-2t} \\ e^{-2t} \end{pmatrix} dt$$

$$= \begin{pmatrix} e^t & te^t \\ 0 & e^t \end{pmatrix} \begin{pmatrix} t + \frac{1}{2}te^{-2t} + \frac{1}{4}e^{-2t} \\ -\frac{1}{2}e^{-2t} \end{pmatrix}$$

$$= \begin{pmatrix} te^t + \frac{1}{4}e^{-t} \\ -\frac{1}{2}e^{-t} \end{pmatrix}.$$

Adding the particular solution found to the general homogeneous solution we already had gives

$$\mathbf{x}(t) = \begin{pmatrix} e^t & te^t \\ 0 & e^t \end{pmatrix} \begin{pmatrix} c_1 \\ c_2 \end{pmatrix} + \begin{pmatrix} te^t + \frac{1}{4}e^{-t} \\ -\frac{1}{2}e^{-t} \end{pmatrix}$$

$$= \begin{pmatrix} c_1 e^t + c_2 te^t + te^t + \frac{1}{4}e^{-t} \\ c_2 e^t - \frac{1}{2}e^{-t} \end{pmatrix}$$

for the general solution.

4B   Linearity

Matrix multiplication acts linearly on n-dimensional columns $\mathbf{x}$ in the sense that

$$A(c_1\mathbf{x}_1 + c_2\mathbf{x}_2) = c_1 A\mathbf{x}_1 + c_2 A\mathbf{x}_2.$$

Similarly, d/dt acts linearly:

$$\frac{d}{dt}(c_1\mathbf{x}_1 + c_2\mathbf{x}_2) = c_1 \frac{d}{dt}\mathbf{x}_1 + c_2 \frac{d}{dt}\mathbf{x}_2.$$

Now suppose $A = A(t)$ has functions of t for entries and define the operator $D - A(t)$ by

$$[D - A(t)]\mathbf{x} = \frac{d\mathbf{x}}{dt} - A(t)\mathbf{x}.$$

It is an easy exercise to show that $D - A(t)$ is a linear operator. This is helpful to know here because of what Theorem 3.1 of Chapter 3, Section 3A, says in this instance about the general solution $\mathbf{x}(t)$ of

$$\frac{d\mathbf{x}}{dt} - A(t)\mathbf{x} = \mathbf{b}(t):$$

find a particular solution $\mathbf{x}_p(t)$ and add it to the general solution $\mathbf{x}_h(t)$ of the **homogeneous equation**

$$\frac{d\mathbf{x}}{dt} - A(t)\mathbf{x} = 0$$

to get $\mathbf{x}(t) = \mathbf{x}_p(t) + \mathbf{x}_h(t)$. Equation 4.2 of this section has just this form for the case of constant entries in A. Section 4C shows how to find $\mathbf{x}_p(t)$ if we have $\mathbf{x}_h(t)$. General methods for finding $\mathbf{x}_h(t)$ when $A = A(t)$ has nonconstant entries are not available. We can in that case resort to the numerical methods of Chapter 6, Section 7.

4C Variation of Parameters

It is an important fact that even in the case of a nonconstant, continuous n by n matrix $A(t)$ there is a formula for a particular solution of

$$\frac{d\mathbf{x}}{dt} = A(t)\mathbf{x} + \mathbf{b}(t)$$

in terms of solutions of the associated homogeneous equation. Suppose $\mathbf{x}_1(t), \ldots, \mathbf{x}_n(t)$ is a set of n linearly independent solutions of

$$\frac{d\mathbf{x}}{dt} = A(t)\mathbf{x}.$$

Solutions would be independent, for example, if their values for some $t = t_0$ were independent vectors. We now form the n by n **fundamental matrix**

$$X(t) = (\mathbf{x}_1(t) \cdots \mathbf{x}_n(t))$$

whose columns are these independent vector solutions. If A has constant entries, the columns of e^{tA} will suffice; just set $\mathbf{x}_k(t) = e^{tA}\mathbf{e}_k$.

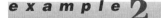

example 2

If A is a constant matrix, then the matrix $X(t) = e^{tA}$ is an example of a fundamental matrix, because its columns are linearly independent solutions of $d\mathbf{x}/dt = A\mathbf{x}$. In Example 1, a fundamental matrix is

$$X_1(t) = e^{tA} = \begin{pmatrix} e^t & te^t \\ 0 & e^t \end{pmatrix}.$$

Another fundamental matrix for the same system is

$$X_2(t) = \begin{pmatrix} te^t & e^t \\ e^t & 0 \end{pmatrix},$$

although $X_2(t)$ can't be an exponential matrix, because $X_2(0) \neq I$.

Next we try to find a vector-valued function $\mathbf{v}(t)$ such that

$$\mathbf{x}_p(t) = X(t)\mathbf{v}(t)$$

is a solution of the nonhomogeneous equation. It turns out that this can always be done as follows. Using the product rule for differentiation, we substitute $X(t)\mathbf{v}(t)$ into the nonhomogeneous equation to get

$$\frac{dX(t)}{dt}\mathbf{v}(t) + X(t)\frac{d\mathbf{v}(t)}{dt} = A(t)X(t)\mathbf{v}(t) + \mathbf{b}(t).$$

Since each column of $X(t)$ is a solution of the homogeneous equation, we have

$$\frac{dX(t)}{dt} = A(t)X(t).$$

Therefore, the first term cancels on each side, leaving

$$X(t)\frac{d\mathbf{v}(t)}{dt} = \mathbf{b}(t).$$

If $X(t)$ is invertible, we can multiply both sides of this last equation by $X^{-1}(t)$ to get

$$\frac{d\mathbf{v}(t)}{dt} = X^{-1}(t)\mathbf{b}(t).$$

Integration gives a formula for $\mathbf{v}(t)$. (We need only one solution, so we omit the constant of integration.)

$$\mathbf{v}(t) = \int X^{-1}(t)\mathbf{b}(t)\,dt.$$

Finally,

■ 4.3
$$\begin{aligned}\mathbf{x}_p(t) &= X(t)\mathbf{v}(t)\\ &= X(t)\int X^{-1}(t)\mathbf{b}(t)\,dt.\end{aligned}$$

Notice that this formula is the same as the one previously derived in the constant-coefficient case, with e^{tA} now replaced by the more general $X(t)$. This process for finding $\mathbf{x}_p$ is called **variation of parameters,** because to find it we allow the constant vector $\mathbf{v}_0$ in the homogeneous vector solution $X(t)\mathbf{v}_0$ to vary as a function of t.

It can be shown that a fundamental matrix $X(t)$ is always invertible; in specific examples, we will always verify this simply by finding the inverse matrix. We already know that in the special case of exponential matrices we have invertibility.

---

e x a m p l e 3

It is easy to verify that the homogeneous system associated with

$$\frac{d\mathbf{x}}{dt} = \begin{pmatrix} 1 & e^t \\ 0 & 1 \end{pmatrix}\mathbf{x} + \begin{pmatrix} e^t \\ e^{2t} \end{pmatrix}$$

has independent solutions

$$\mathbf{x}_1(t) = \begin{pmatrix} e^t \\ 0 \end{pmatrix}, \qquad \mathbf{x}_2(t) = \begin{pmatrix} e^{2t} \\ e^t \end{pmatrix}.$$

We form a fundamental matrix $X(t)$ and its inverse:

$$X(t) = \begin{pmatrix} e^t & e^{2t} \\ 0 & e^t \end{pmatrix}, \quad X^{-1}(t) = \begin{pmatrix} e^{-t} & -1 \\ 0 & e^{-t} \end{pmatrix}.$$

Formula 4.3 gives the particular solution

$$\mathbf{x}_p(t) = \begin{pmatrix} e^t & e^{2t} \\ 0 & e^t \end{pmatrix} \int \begin{pmatrix} e^{-t} & -1 \\ 0 & e^{-t} \end{pmatrix} \begin{pmatrix} e^t \\ e^{2t} \end{pmatrix} dt$$

$$= \begin{pmatrix} e^t & e^{2t} \\ 0 & e^t \end{pmatrix} \int \begin{pmatrix} 1 - e^{2t} \\ e^t \end{pmatrix} dt$$

$$= \begin{pmatrix} e^t & e^{2t} \\ 0 & e^t \end{pmatrix} \begin{pmatrix} t - \frac{1}{2}e^{2t} \\ e^t \end{pmatrix} = \begin{pmatrix} te^t + \frac{1}{2}e^{3t} \\ e^{2t} \end{pmatrix}.$$

The general solution is then

$$\mathbf{x}(t) = c_1\mathbf{x}_1(t) + c_2\mathbf{x}_2(t) + \mathbf{x}_p(t).$$

If the matrix A in $\dot{\mathbf{x}} = A\mathbf{x} + \mathbf{b}$ is constant, the practical significance of the fundamental matrix $X(t)$ as compared with the more special exponential matrix e^{tA} is that the exponential matrix must equal the identity matrix when $t = 0$, whereas the columns of $X(t)$ are required only to be independent solutions of the homogeneous equation $\dot{\mathbf{x}} = A\mathbf{x}$.

4D   *Summary of Methods*

For linear systems in the standard form $dx/dt = Ax + b$, and hence for the systems and equations reducible to this form, we usually proceed as follows:

1. Find the general solution of the homogeneous equation $dx/dt = Ax$, either by elimination, by the eigenvector method in Section 1A, or by finding e^{tA} directly. In the constant-coefficient case the homogeneous solution is always of the form $\mathbf{x}_h(t) = e^{tA}\mathbf{c}$, where $\mathbf{c}$ is a constant vector. If A is not constant, there is no general method for finding $\mathbf{x}_h(t)$, and we will very likely have to use numerical methods.
2. Find a particular solution to the nonhomogeneous equation, either as a by-product of the elimination method, by undetermined coefficient, if applicable, or by Formula 4.2 or 4.3.
3. Write the general solution as $\mathbf{x}(t) = \mathbf{x}_h(t) + \mathbf{x}_p(t)$.

EXERCISES

1. Use Formula 4.2 to solve the following equations of the form $dx/dt = Ax + b(t)$. The associated homogeneous equations $dx/dt = Ax$ were found in Exercise 10 of Section 2D to have exponential matrices e^{tA} as shown

(a) $\begin{pmatrix} \dot{x} \\ \dot{y} \end{pmatrix} = \begin{pmatrix} -3 & 2 \\ -4 & 3 \end{pmatrix} \begin{pmatrix} x \\ y \end{pmatrix} + \begin{pmatrix} e^{2t} \\ 1 \end{pmatrix}; \quad e^{tA} = \begin{pmatrix} -e^t + 2e^{-t} & e^t - e^{-t} \\ -2e^t + 2e^{-t} & 2e^t - e^{-t} \end{pmatrix}.$

(b) $\begin{pmatrix} \dot{x} \\ \dot{y} \end{pmatrix} = \begin{pmatrix} 3 & 0 \\ 0 & 2 \end{pmatrix}\begin{pmatrix} x \\ y \end{pmatrix} + \begin{pmatrix} e^t + 1 \\ e^{-t} \end{pmatrix}; \; e^{tA} = \begin{pmatrix} e^{3t} & 0 \\ 0 & e^{2t} \end{pmatrix}.$

(c) $\begin{pmatrix} \dot{x} \\ \dot{y} \end{pmatrix} = \begin{pmatrix} 1 & 4 \\ 0 & 5 \end{pmatrix}\begin{pmatrix} x \\ y \end{pmatrix} + \begin{pmatrix} 1 \\ e^t \end{pmatrix}; \; e^{tA} = \begin{pmatrix} e^t & e^{5t} - e^t \\ 0 & e^{5t} \end{pmatrix}.$

(d) $\begin{pmatrix} \dot{x} \\ \dot{y} \end{pmatrix} = \begin{pmatrix} 2 & -1 \\ 1 & 2 \end{pmatrix}\begin{pmatrix} x \\ y \end{pmatrix} + \begin{pmatrix} e^{2t} \\ 2e^{2t} \end{pmatrix}; \; e^{tA} = \begin{pmatrix} e^{2t}\cos t & -e^{2t}\sin t \\ e^{2t}\sin t & e^{2t}\cos t \end{pmatrix}.$

2. Find a particular solution of each of the differential equations in Exercise 1 that satisfies the following corresponding initial condition:

   (a) $x(0) = (-1, -1)$.   **(c)** $x(0) = (0, 1)$.

   (b) $x(0) = (0, 0)$.   **(d)** $x(0) = (-1, -1)$.

3. **(a)** Show that for a solution of the form
$$x(t) = e^{tA}c, \quad c \text{ const,}$$
   to satisfy the condition $x(t_0) = x_0$, we must have $c = e^{-t_0 A}x_0$.

   (b) Show that if $X(t)$ is an n by n matrix with linearly independent columns, in particular if $X(t)$ is a fundamental matrix, then for
$$x(t) = X(t)c, \quad c \text{ const,}$$
   to satisfy $x(t_0) = x_0$ we must have $c = X^{-1}(t_0)x_0$.

4. Let $X(t)$ be a fundamental matrix such that $X(t)c$ is the general solution to the homogeneous equation
$$\dot{x} = A(t)x.$$

   (a) Show that if $x_p(t)$ is a particular solution of the nonhomogeneous system
$$\dot{x} = A(t)x + b(t),$$
   then the general solution of the nonhomogeneous system can be written $x(t) = x_p(t) + X(t)c$, where c is a constant.

   (b) Show that, if the general solution in part (a) is to satisfy an initial condition $x(t_0) = x_0$, then c should be chosen so that $c = X^{-1}(t_0)(x_0 - x_p(t_0))$.

5. The systems in Exercise 1 can be solved by the method of **undetermined coefficients** as follows: Form linear combinations of the terms, and their derivatives, that occur in each entry of the nonhomogeneous part of the differential equation, taking care to include appropriate multiples by powers of t for terms that are also homogeneous solutions. Then substitute into the equation to determine the coefficients of combination. Use this method on each of the parts of Exercise 1.

6. Each of the following systems has the *homogeneous* solutions shown. Verify that these are linearly independent solutions. Find a particular solution of the nonhomogeneous equation, using Equation 4.3.

   (a) $\begin{pmatrix} \dot{x} \\ \dot{y} \end{pmatrix} = \begin{pmatrix} 3/(2t) & -1/2 \\ -1/(2t^2) & 1/(2t) \end{pmatrix}\begin{pmatrix} x \\ y \end{pmatrix} + \begin{pmatrix} t^3 \\ 3t^2 \end{pmatrix}; \begin{pmatrix} t \\ 1 \end{pmatrix}, \begin{pmatrix} -t^2 \\ t \end{pmatrix}.$

   (b) $\begin{pmatrix} \dot{x} \\ \dot{y} \end{pmatrix} = \begin{pmatrix} t/(t-1) & -1/(t-1) \\ 1 & 0 \end{pmatrix}\begin{pmatrix} x \\ y \end{pmatrix} + \begin{pmatrix} 1 - t \\ 1 - t^2 \end{pmatrix}; \begin{pmatrix} 1 \\ t \end{pmatrix}, \begin{pmatrix} e^t \\ e^t \end{pmatrix}.$

7. (a) Let $\mathbf{x}_1(t), \ldots, \mathbf{x}_n(t)$ be linearly independent, continuously differentiable functions taking values in $\mathscr{R}^n$. Let $X(t)$ be the n by n matrix with columns $\mathbf{x}_1(t), \ldots, \mathbf{x}_n(t)$. Show that if we define

$$A(t) = X'(t)X^{-1}(t),$$

then the system $d\mathbf{x}/dt = A(t)\mathbf{x}$ has $\mathbf{x}_1(t), \ldots, \mathbf{x}_n(t)$ as solutions and thus has $X(t)$ as a fundamental matrix.

(b) Find a first-order homogeneous linear system of the form $d\mathbf{x}/dt = A(t)\mathbf{x}$ having

$$\mathbf{x}_1(t) = \begin{pmatrix} e^t \\ 2e^{2t} \end{pmatrix}, \qquad \mathbf{x}_2(t) = \begin{pmatrix} 1 \\ e^t \end{pmatrix}$$

as solutions. Are these two solutions linearly independent?

8. Let A be an n by n invertible matrix of constants, and let $\mathbf{b}$ be a fixed vector in $\mathscr{R}^n$. Show that the equation

$$\frac{d\mathbf{x}}{dt} = A\mathbf{x} + \mathbf{b}$$

always has $\mathbf{x}_p = -A^{-1}\mathbf{b}$ for a particular solution.

9. Show that if A is a constant n by n matrix, then the equation

$$\frac{d\mathbf{x}}{dt} = A\mathbf{x}$$

has $X(t) = e^{tA}$ for its fundamental matrix of independent column solutions with $X(0) = I$.

10. Let $A(t)$ be a square matrix with entries that are differentiable on some interval.

(a) Show that

$$\frac{dA^2}{dt} = 2A\frac{dA}{dt}$$

when A and dA/dt commute.

(b) Show that

$$\frac{de^A}{dt} = e^A\frac{dA}{dt}$$

when A and dA/dt commute.

11. (a) Write out in complete detail the derivation of an analog to Equation 4.2 to show that, for A constant and $\mathbf{b}(t)$ continuous, the initial-value problem

$$\frac{d\mathbf{x}}{dt} = A\mathbf{x} + (t), \qquad \mathbf{x}(t_0) = \mathbf{x}_0$$

has a unique solution of the form

$$\mathbf{x}(t) = e^{tA}\int_{t_0}^{t} e^{-uA}\mathbf{b}(u)\, du + e^{(t-t_0)A}\mathbf{x}_0.$$

[*Hint:* Integrate from t_0 to t instead of using an indefinite integral.]

(b) Use the result of part (a) to show that the solution to the initial-value problem with the special choice $\mathbf{x}(t_0) = 0$ is

$$\mathbf{x}(t) = \int_{t_0}^{t} e^{(t-u)A}\mathbf{b}(u) \, du.$$

The matrix factor $e^{(t-u)A}$ is a **Green's function** for the initial-value problem and plays a role analogous to that of the Green's function introduced in Chapter 3, Section 3C.

*12. It's not true that every square matrix $M(t)$ with entries that are differentiable functions of t arises as an e^{tA}. For example, it must be the case that (i) $M(0) = I$, (ii) $M(t + s) = M(t)M(s)$, and (iii) $\dot{M}(t) = AM(t)$. The purpose of this exercise is to characterize those differentiable matrices $M(t)$ that are exponential matrices by showing that if (i) and (ii) hold for all real t and s, then there is a constant matrix A such that $M(t) = e^{tA}$.

(a) Show that if (ii) holds and $M(t_0)$ is invertible for some t_0, then (i) holds also.

(b) Show that if (i) and (ii) both hold and $M(t)$ has differentiable entries, then (iii) holds for some constant matrix A. [*Hint:* Apply (i) and (ii) to $(1/h)(M(t + h) - M(t))$, and let $h \to 0$.]

(c) Show that if (i) and (iii) hold for some constant matrix A, then $M(t) = e^{tA}$. [*Hint:* Multiply both sides of (iii) by e^{-tA}, and show that the matrix $d/dt(e^{tA}M(t))$ is the zero matrix.]

13. Prove the product rule 4.1 for square matrices with differentiable functions as entries, under the following assumptions.

(a) A and B are 2 by 2 matrices.

(b) $A = (a_{ij}(t))$ and $B = (b_{ij}(t))$ are n by n matrices with product defined by

$$AB = \sum_{k=1}^{n} a_{ik}(t)b_{kj}(t).$$

5 STABILITY FOR AUTONOMOUS SYSTEMS

In Section 8 of Chapter 6 there is a detailed examination of possible behaviors for 2-dimensional autonomous systems depending on the nature of the associated eigenvalues, assuming that these are nonzero. A similar analysis of higher-dimensional systems quickly leads to an overwhelmingly large number of possibilities. What we'll do here instead is just establish conditions that distinguish among various kinds of stable and unstable behavior (Figure 7.2) of solutions $\mathbf{x}(t)$ to $\dot{\mathbf{x}} = \mathbf{F}(\mathbf{x})$ near an equilibrium point $\mathbf{x}_0$:

1. $\mathbf{x}_0$ is **asymptotically stable** if there is a number $d_0 > 0$ such that every solution starting within distance d_0 of $\mathbf{x}_0$ tends to $\mathbf{x}_0$ as t tends to infinity.
2. $\mathbf{x}_0$ is **stable** if there is a $d_0 > 0$ such that all solutions starting within some distance $d_1 < d_0$ from $\mathbf{x}_0$ remain within distance d_0 of $\mathbf{x}_0$. Clearly asymptotic stability near an equilibrium point $\mathbf{x}_0$ implies stability near $\mathbf{x}_0$.
3. An equilibrium point $\mathbf{x}_0$ that is not stable is called **unstable**.

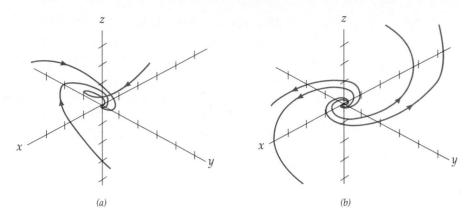

Figure 7.2 (*a*) Asymptotically stable. (*b*) Unstable.

5A   Linear Systems

▇▇ 5.1 THEOREM

Let A be an n by n constant matrix. An equilibrium solution $\mathbf{x}_0 = 0$ for the system $\dot{\mathbf{x}} = A\mathbf{x}$ is asymptotically stable if every eigenvalue of A has a negative real part and is unstable if A has at least one eigenvalue with a positive real part.

Proof. The method for computing the exponential matrix described in Section 3 shows that every entry in the fundamental solution matrix e^{tA} has the form $e^{\lambda t} Q(t)$, where $Q(t)$ is a polynomial and λ is an eigenvalue of A. If every such λ has a negative real part, then the entries, and hence all solutions, tend to zero as $t \to +\infty$. On the other hand, if the eigenvalue λ_+, with eigenvector $\mathbf{v}_+$, has a positive real part, then the solution $\mathbf{x}(t) = \delta e^{\lambda_+ t} \mathbf{v}_+$ is unbounded in every non-zero coordinate as $t \to +\infty$, regardless of how small the positive number δ is chosen. Finally, if $\lambda_- < 0$, then $\delta e^{\lambda_- t} \mathbf{v}_- \to 0$.   ■

If all eigenvalues of the matrix A have negative real parts, then all solutions tend exponentially to 0, and the origin is accordingly called a **sink** for the system $\dot{\mathbf{x}} = A\mathbf{x}$. On the other hand, if all eigenvalues have positive real parts, the origin is called a **source,** with all nonzero solutions diverging from the origin. If some eigenvalues have positive real parts and all others have negative real parts, the equilibrium is unstable and is called a **saddle point.**

e x a m p l e 1

The 3-dimensional system $\dot{x} = y$, $\dot{y} = -x - y$, $\dot{z} = x - z$ has a unique equilibrium point at the origin $(0, 0, 0)$. In matrix form the system is

$$\dot{\mathbf{x}} = \begin{pmatrix} 0 & 1 & 0 \\ -1 & -1 & 0 \\ 1 & 0 & -1 \end{pmatrix}$$

with characteristic equation

$$\det \begin{pmatrix} -\lambda & 1 & 0 \\ -1 & -1-\lambda & 0 \\ 1 & 0 & -1-\lambda \end{pmatrix} = 0.$$

This last equation is just $\lambda^3 + 2\lambda^2 + 2\lambda + 1 = 0$. By inspection we see that $\lambda_1 = -1$ is a root. Factoring out $\lambda + 1$ leaves the quadratic $\lambda^2 + \lambda + 1$ with roots $\lambda_2 = (-1 + \sqrt{3}i)/2$, $\lambda_3 = (-1 - \sqrt{3}i)/2$. Since all roots have a negative real part, either -1 or $-\frac{1}{2}$, the equilibrium is an asymptotically stable sink by Theorem 5.1. The general solution can be computed without much trouble, but it isn't needed if we're only checking equilibrium stability.

Showing the existence of one positive eigenvalue is enough to guarantee instability, as in the next example.

---

e x a m p l e 2

If we replace the third equation in the system of Example 1 by $\dot{z} = x + z$, the characteristic equation becomes

$$P(\lambda) = \det \begin{pmatrix} -\lambda & 1 & 0 \\ -1 & -1-\lambda & 0 \\ 1 & 0 & 1-\lambda \end{pmatrix} = 0 \qquad \text{or} \qquad P(\lambda) = -\lambda^3 + 1 = 0.$$

Clearly $\lambda_1 = 1$ is a root. Thus there is a positive eigenvalue, and by Theorem 5.1 the equilibrium at the origin is unstable for this system. Dividing $\lambda^3 - 1$ by $\lambda - 1$ leaves the quadratic $\lambda^2 + \lambda + 1$, which has roots $-\frac{1}{2} \pm \sqrt{3}i$. Since all eigenvalues have either positive or negative real parts, the origin is an example of a saddle point.

5B   Nonlinear Systems

Building on the eigenvalue analysis for linear systems, we consider equilibrium points $\mathbf{x}_0 = (a_1, \ldots, a_n)$ of a nonlinear system $\dot{\mathbf{x}} = \mathbf{F}(\mathbf{x})$, that is, points such that $\mathbf{F}(\mathbf{x}_0) = 0$. The first step is to *linearize* the system, replacing each real-valued equation $\dot{x}_k = F_k(x_1, \ldots, x_n)$ in the system by its own linearization

$$\dot{x}_k = \sum_{j=1}^{n} \frac{\partial F_k}{\partial x_j}(\mathbf{x}_0)(x_j - a_j), \qquad k = 1, \ldots, x_n.$$

The resulting linear autonomous system is called the **linearization** of $\dot{\mathbf{x}} = \mathbf{F}(\mathbf{x})$ at $\mathbf{x}_0$. The linearization can be written $\dot{\mathbf{x}} = \mathbf{F}'(\mathbf{x}_0)(\mathbf{x} - \mathbf{x}_0)$ using the **derivative matrix** or **Jacobian matrix**

$$\mathbf{F}'(\mathbf{x}_0) = \begin{pmatrix} \dfrac{\partial F_1}{\partial x_1}(\mathbf{x}_0) \cdots \dfrac{\partial F_1}{\partial x_n}(\mathbf{x}_0) \\ \vdots \qquad\qquad \vdots \\ \dfrac{\partial F_n}{\partial x_1}(\mathbf{x}_0) \cdots \dfrac{\partial F_n}{\partial x_n}(\mathbf{x}_0) \end{pmatrix}.$$

We'll see that the eigenvalues of $\mathbf{F}'(\mathbf{x}_0)$ are the key to our criteria, and for this reason we can think of working with the simpler homogeneous equation $\dot{\mathbf{x}} = \mathbf{F}'(\mathbf{x}_0)\mathbf{x}$ as is done in Chapter 6, Section 8B.

---

e x a m p l e 3

The system

$$\dot{x} = -x + y,$$
$$\dot{y} = 2x - y - xz,$$
$$\dot{z} = xy - z$$

belongs to a family called the **Lorenz system.** The equilibrium solutions are just the solutions to the system obtained by setting the right-hand sides equal to zero:

$$-x + y = 0,$$
$$2x - y - xz = 0,$$
$$xy - z = 0.$$

Noting that $x = y$, it's easy to see that there are just three solutions $(1, 1, 1)$, $(0, 0, 0)$, and $(-1, -1, 1)$. The derivative matrix $\mathbf{F}'(x, y, z)$ for the linearization at (x, y, z) is

$$\begin{pmatrix} \dfrac{\partial(-x+y)}{\partial x} & \dfrac{\partial(-x+y)}{\partial y} & \dfrac{\partial(-x+y)}{\partial z} \\ \dfrac{\partial(2x-y-xz)}{\partial x} & \dfrac{\partial(2x-y-xz)}{\partial y} & \dfrac{\partial(2x-y-xz)}{\partial z} \\ \dfrac{\partial(xy-z)}{\partial x} & \dfrac{\partial(xy-z)}{\partial y} & \dfrac{\partial(xy-z)}{\partial z} \end{pmatrix} = \begin{pmatrix} -1 & 1 & 0 \\ 2-z & -1 & -x \\ y & x & -1 \end{pmatrix}.$$

Evaluating this matrix at $(x, y, z) = (1, 1, 1)$ gives

$$\mathbf{F}'(1, 1, 1) = \begin{pmatrix} -1 & 1 & 0 \\ 2-z & -1 & -x \\ y & x & -1 \end{pmatrix}_{(1, 1, 1)} = \begin{pmatrix} -1 & 1 & 0 \\ 1 & -1 & -1 \\ 1 & 1 & -1 \end{pmatrix}.$$

Similarly, the linearization matrices at $(0, 0, 0)$ and $(-1, -1, 1)$ are obtained by evaluating the same derivative matrix at these two additional points:

$$\mathbf{F}'(0, 0, 0) = \begin{pmatrix} -1 & 1 & 0 \\ 2 & -1 & 0 \\ 0 & 0 & -1 \end{pmatrix}; \quad \mathbf{F}'(-1, -1, 1) = \begin{pmatrix} -1 & 1 & 0 \\ 1 & -1 & 1 \\ -1 & -1 & -1 \end{pmatrix}.$$

The proof of the next theorem is omitted.

■ 5.2 THEOREM

Assume that the real-valued coordinate functions F_k of $\mathbf{F}$ are continuously differentiable and that the system $\dot{\mathbf{x}} = \mathbf{F}(\mathbf{x})$ has an equilibrium point at $\mathbf{x}_0$. The equilibrium solution $\mathbf{x}_0$ for the system is asymptotically stable if every eigenvalue of the derivative matrix $\mathbf{F}'(\mathbf{x}_0)$ has a negative real part and is unstable if $\mathbf{F}'(\mathbf{x}_0)$ has at least one eigenvalue with a positive real part. If there are eigenvalues with a real part zero, $\mathbf{x}_0$ may be either stable or unstable.

---

e x a m p l e 4

The special Lorenz system treated in Example 3 has an equilibrium at $(1, 1, 1)$ with characteristic polynomial $P(\lambda) = \det(\mathbf{F}'(1, 1, 1) - \lambda I)$. From Example 3 we see that

$$P(\lambda) = \det \begin{pmatrix} -1 - \lambda & 1 & 0 \\ 1 & -1 - \lambda & -1 \\ 1 & 1 & -1 - \lambda \end{pmatrix}.$$

Computing the determinant, we get $P(\lambda) = (\lambda + 1)^3 + 1$, and we see by inspection that $\lambda_1 = -2$ is a root. Division by $\lambda + 2$ gives

$$P(\lambda) = -(\lambda + 2)(\lambda^2 + \lambda + 1).$$

The roots of the quadratic factor are

$$\lambda_2 = \frac{-1 + \sqrt{3}\, i}{2} \quad \text{and} \quad \lambda_3 = \frac{-1 - \sqrt{3}\, i}{2}.$$

Thus the real parts of all three eigenvalues are negative, so we conclude from Theorem 5.2 that $(1, 1, 1)$ is an asymptotically stable equilibrium solution.

---

e x a m p l e 5

Continuing with the special Lorenz system, we examine the equilibrium at $(0, 0, 0)$. The relevant characteristic polynomial is $P(\lambda) = \det(F'(0, 0, 0) - \lambda I)$, or

$$\det \begin{pmatrix} -1 - \lambda & 1 & 0 \\ 2 & -1 - \lambda & 0 \\ 0 & 0 & -1 - \lambda \end{pmatrix}.$$

The determinant is computed to be

$$-(\lambda + 1)^3 + 2\lambda + 2 = -(\lambda + 1)(\lambda^2 + 2\lambda - 1).$$

The roots are $\lambda_1 = -1$, $\lambda_2 = -1 - \sqrt{2}$, and $\lambda_3 = -1 + \sqrt{2}$. Since $\lambda_3 > 0$, we conclude from Theorem 5.2 that $(0, 0, 0)$ is an unstable equilibrium. This point is in fact a saddle point, since there are two negative eigenvalues that contribute to making the other basic solutions tend to zero. Checking out the equilibrium at $(-1, -1, 1)$ is left as an exercise.

---

e x a m p l e 6

The general **Lorenz system** is

$$\dot{x} = \sigma(y - x), \qquad \dot{y} = \rho x - y - xz, \qquad \dot{z} = -\beta z + xy, \qquad \beta, \rho, \sigma \text{ positive const.}$$

This system has been studied intensively with the aim of understanding trajectories such as the one shown in Figure 7.3. With the choice of parameters shown there, the equilibrium points, aside from the obvious one at the origin, are at $(\pm 6\sqrt{2}, \pm 6\sqrt{2}, 27)$. The trajectory shown in Figure 7.3 has initial point

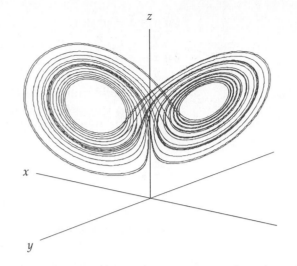

Figure 7.3 A Lorenz orbit: $\beta = \frac{8}{3}$, $\rho = 28$, $\sigma = 10$; view from $(x, y, z) = (-2, 3, 1)$.

$(2, 2, 21)$. It winds around in the area of one equilibrium an apparently random number of times, then switches to the other equilibrium with similar behavior. The eigenvalues of the linearizations are the same at the two equilibrium points; they are approximately as follows: $\lambda_1 \approx -13.85$, $\lambda_2, \lambda_3 \approx 0.09 \pm 10.19i$. Thus these two points are saddle points, and it can be shown that each one has a surface containing it on which all trajectories gradually spiral away from it, as well as a trajectory transverse to the surface that converges to the point. The typical trajectory behavior lies somewhere between these extremes, winding away from one equilibrium until it is attracted by the other, then reversing. The number of circuits about each point, and the path taken, is very sensitive to minute changes in the initial conditions. This issue is taken up numerically in Exercise 20 of Chapter 6, Section 7.

EXERCISES

1. Find the eigenvalues of the following square matrices A. Then use that information to decide about the stability or instability of the zero solution to the associated linear system $\dot{\mathbf{x}} = A\mathbf{x}$.

    (a) $\begin{pmatrix} 2 & 1 \\ 1 & -3 \end{pmatrix}$.

    (b) $\begin{pmatrix} 4 & -3 \\ 3 & -4 \end{pmatrix}$.

    (c) $\begin{pmatrix} 1 & 1 & -1 \\ 0 & 1 & 3 \\ 0 & 0 & -1 \end{pmatrix}$.

    (d) $\begin{pmatrix} -6 & -1 & -2 \\ -2 & -4 & -2 \\ 2 & -1 & -2 \end{pmatrix}$.

2. Find all equilibrium points of the following systems $\dot{u} = \mathbf{F}(u)$. For each equilibrium $\mathbf{x}_0$, compute the derivative matrix $\mathbf{F}'(\mathbf{x}_0)$. Using the eigenvalues of $\mathbf{F}'(\mathbf{x}_0)$ try to determine the stability or instability of each equilibrium.

(a) $\dot{x} = x - y^2 + 1,$
$\dot{y} = x^2 - 1.$

(c) $\dot{x} = y,$
$\dot{y} = e^x - 1.$

(b) $\dot{x} = x - y,$
$\dot{y} = x^2 + y^2 - 1.$

(d) $\dot{x} + \dot{y} = y,$
$\dot{x} - \dot{y} = x.$

3. Find all equilibrium points of the following systems $\dot{x} = F(x)$. For each equilibrium x_0, compute the derivative matrix $F'(x_0)$. Using the eigenvalues of $F'(x_0)$ try to determine the stability or instability of each equilibrium.

(a) $\dot{x} = -x - y^2,$
$\dot{y} = -y,$
$\dot{z} = -z - x^2.$

(b) $\dot{x} = x - y,$
$\dot{y} = x^2 + y^2 - 1,$
$\dot{z} = z.$

4. Find the characteristic polynomial of the derivative matrix $F'(-1, -1, 1)$ associated with the equilibrium $(-1, -1, 1)$ of the special Lorenz system in Example 3 of the text. Then find the eigenvalues of $F'(-1, -1, 1)$ and use the information to decide about the stability of $(-1, -1, 1)$.

5. The general **Lorenz system** is

$$\dot{x} = \sigma(y - x),$$
$$\dot{y} = \rho x - y - xz,$$
$$\dot{z} = -\beta z + xy, \qquad \beta, \rho, \sigma \text{ positive const.}$$

(a) Show that (i) if $0 < \rho \le 1$, the only equilibrium point is $(0, 0, 0)$, and (ii) if $\rho > 1$, there are three equilibrium points: $(0, 0, 0)$ and $(\pm\sqrt{\beta(\rho - 1)}, \pm\sqrt{\beta(\rho - 1)}, \rho - 1)$.

(b) Show that at the equilibrium points $(\pm\sqrt{\beta(\rho - 1)}, \pm\sqrt{\beta(\rho - 1)}, \rho - 1)$, $\rho > 1$, the characteristic equation of the linearized system is

$$\lambda^3 + (\beta + \sigma + 1)\lambda^2 + \beta(\rho + \sigma)\lambda + 2\sigma\beta(\rho - 1) = 0.$$

Explain why there must be at least one negative eigenvalue.

(c) Use Newton's method or a careful graphical analysis to show that with the choices $\beta = \frac{8}{3}$, $\rho = 28$, $\sigma = 10$, $P(\lambda)$ has a negative root λ_1 at about -13.85.

(d) Use the result of part (c) to show that $P(\lambda)$ has complex conjugate roots with a positive real part about 0.09. Conclude that these two nonzero equilibrium points are saddle points. [*Hint:* Divide $P(\lambda)$ by $\lambda - \lambda_1$.]

6. Consider the second-order system

$$\ddot{x} = -4x + y^2, \qquad \ddot{y} = -y + 2xy.$$

(a) Show that the assumption $y(t) = 0$ for all t allows us to infer the existence of periodic solutions for some initial conditions.

(b) Let $\dot{x} = z$, $\dot{y} = w$ to get an equivalent first-order system. Find all equilibrium solutions $(x(t), y(t))$ of the second-order system.

(c) Use linearization to show instability at nonzero equilibrium points.

(d) Show that the origin is stable for the first-order system. Do this by showing that the system is Hamiltonian, with Hamiltonian function

$$H(x, y, z, w) = \tfrac{1}{2}(4x^2 + y^2 + z^2 + w^2) - xy^2,$$

having the origin as an isolated minimum. (See Chapter 6, Section 8C on Liapunov's method for this part.)

7. The second-order **Hénon-Heilas system** may be written
$$\ddot{x} = -x - 2xy, \qquad \ddot{y} = -y - x^2 + y^2.$$

   (a) Show that the assumption $x(t) = 0$ for all t allows us to infer the existence of periodic solutions for some initial conditions.

   (b) Let $\dot{x} = z$, $\dot{y} = w$ to get an equivalent first-order system. Find all equilibrium solutions $(x(t), y(t))$ of the second-order system.

   (c) Use linearization to show instability of the nonzero equilibrium points.

   (d) Show that the origin is stable for the first-order system. Do this by showing that the system is Hamiltonian, with Hamiltonian function
$$H(x, y, z, w) = \tfrac{1}{2}(x^2 + y^2 + z^2 + w^2) + x^2 y - \tfrac{1}{3}y^3,$$
   having the origin as an isolated minimum. (See Chapter 6, Section 8C on Liapunov's method for this part.)

8. Do Exercise 4 in Chapter 6, Section 8B, confirming the last statement in Theorem 5.2.

9. An equilibrium point $\mathbf{x}_0$ of a system $\dot{\mathbf{x}} = \mathbf{F}(\mathbf{x})$ is called **hyperbolic** if the eigenvalues of the linearization at $\mathbf{x}_0$ all have nonzero real parts; if both signs occur, $\mathbf{x}_0$ is a **saddle point**.

   (a) Which of the types I through VIII in Chapter 6, Section 8, are of hyperbolic type and which are of saddle type?

   (b) Which of the types I through VIII in Chapter 6, Section 8, are sources and which are sinks?

8

Laplace
Transforms

1 BASIC PROPERTIES

The techniques described in Chapter 3 can be used to find formulas for the solution of initial-value problems such as

$$y'' + ay' + by = f(t), \qquad y(0) = y_0, \qquad y'(0) = y_1.$$

If we are in a position to assume that $f(t)$ is defined for all $t \geq 0$, and that $f(t)$ and the solution $y(t)$ don't grow too rapidly as $t \to \infty$, we can use an alternative method that has achieved widespread popularity in electrical engineering and control theory. Experience with exponential integrating factors shows that it is natural to multiply a solution $y(t)$ by a factor e^{-st}, where s is some real or complex number, to get a product $e^{-st}y(t)$. What seems less natural, but nevertheless turns out to be effective, is to integrate with respect to t between 0 and ∞. In general this gives us an improper integral. In particular, the equation

■ **1.1**
$$\int_0^\infty e^{-st}e^{at}\, dt = \frac{1}{s-a}$$

holds when $s - a > 0$, and is used repeatedly. To verify this equation, we compute the partial integral

$$\int_0^T e^{-(s-a)t}\, dt = \left[-\frac{1}{s-a}e^{-(s-a)t} \right]_0^T$$

$$= -\frac{1}{s-a}e^{-(s-a)T} + \frac{1}{s-a}.$$

478

When $T \to \infty$, the exponential factor tends to zero, so letting $T \to \infty$ on both sides verifies the formula. If for some real or complex numbers s an integral of the form

$$\mathcal{L}[f](s) = \int_0^\infty e^{-st}f(t)\,dt$$

converges to a function $\mathcal{L}[f]$ depending on s, then $\mathcal{L}[f]$ is called the **Laplace transform** of the function f. The key to using the Laplace transform to solve differential equations is the following formula:

■ **1.2** $$\int_0^\infty e^{-st}y'(t)\,dt = -y(0) - s\int_0^\infty e^{-st}y(t)\,dt,$$

which we prove under the assumptions (a) the improper integral on the left is convergent, and (b) $\lim_{t\to\infty} e^{-st} y(t) = 0$. Indeed, integration by parts of the partial integral on the left gives

$$\int_0^T e^{-sT}y'(t)\,dt = [e^{-st} y(t)]_0^T + s\int_0^T e^{-st}y(t)\,dt$$

$$= e^{-sT}y(T) - y(0) + s\int_0^T e^{-st}y(t)\,dt.$$

Because of assumption (b), the first term on the right tends to 0 as $T \to \infty$. Equation 1.2 follows by letting $T \to \infty$ in the preceding equation.

---

e x a m p l e **1**

The example

$$y' + 2y = 0, \qquad y(0) = 3$$

is too simple to show the real advantages of using Laplace transforms, but it does illustrate the general principles involved. To do this, convert to the Laplace transform of both sides of the differential equation by multiplying both sides by e^{-st} and then integrating from 0 to ∞ with respect to t. The result can be written

$$\int_0^\infty e^{-st}y'\,dt + 2\int_0^\infty e^{-st}y\,dt = 0.$$

Equation 1.2 allows us to rewrite the first integral, obtaining

$$-3 + s\int_0^\infty e^{-st}y\,dt + 2\int_0^\infty e^{-st}y\,dt = 0.$$

Here we have used the assumption that $y(0) = 3$. We also rely on our knowledge of the exponential nature of the solution to justify the assumptions (a) and (b) needed for the application of Equation 1.2. The previous equation can now be solved for the Laplace transform of the solution $y(t)$ in the form

$$\int_0^\infty e^{-st}y\,dt = \frac{3}{s+2}. \tag{1}$$

Thus we have found not the solution $y(t)$ but its Laplace transform. However, if we set $a = -2$ in Equation 1.1 and then multiply by 3, we get

$$\int_0^\infty e^{-st}3e^{-2t}\,dt = \frac{3}{s+2}. \tag{2}$$

Since we already know from the general theory of Chapter 1 that $y(t)$ must be an exponential solution, there remains only the question of the constants involved, and we see by comparing equations (1) and (2) that the solution

$$y(t) = 3e^{-2t}$$

satisfies our requirements. Even though we really do not need the fact here, and it's some work to prove, it is worth commenting that a function $f(t)$ defined for $t > 0$ is determined at its points of continuity by its Laplace transform $\mathscr{L}[f]$.

To apply the Laplace transform to differential equations with order higher than one, we need a simple extension of Equation 1.2. To simplify the notation, we write Equation 1.2 in the form

$$\mathscr{L}[y'](s) = -y(0) + s\mathscr{L}[y](s).$$

Applying this equation to $\mathscr{L}[y''](s)$, the Laplace transform of y'', gives

$$\mathscr{L}[y''](s) = -y'(0) + s\mathscr{L}[y'].$$

Applying this new equation gives

$$\mathscr{L}[y''](s) = -y'(0) + s\{-y(0) + s\mathscr{L}[y](s)\}$$
$$= -y'(0) - sy(0) + s^2\mathscr{L}[y](s).$$

The same formal routine leads after n steps to the formula

■ **1.3** $\mathscr{L}[y^{(n)}](s) = -y^{(n-1)}(0) - sy^{(n-2)}(0) - \cdots - s^{n-1}y(0) + s^n\mathscr{L}[y](s).$

Of course, the assumptions (a) and (b) needed for Equation 1.2 have to be increased. We assume

(A) The integrals $\int_0^\infty e^{-st} y^{(k)}(t)\, dt$ are convergent for $k = 1, 2, \ldots, n$.
(B) $\lim_{t \to \infty} e^{-st}y^{(k)}(t) = 0$ for $k = 0, 1, \ldots, n - 1$.

For simplicity in solving equations, we denote the Laplace transform of y by Y, of f by F, and so on. Thus $Y(s) = \mathscr{L}[y](s)$ and $F(s) = \mathscr{L}[f](s)$

---

e x a m p l e **2**

The initial-value problem

$$y'' - y' - 2y = 3e^t, \qquad y(0) = 1, \qquad y'(0) = 0,$$

can be solved by applying the Laplace transform to both sides. Using Equation 1.3 for $n = 1$ and 2, together with the initial conditions, we let $Y(s) = \mathscr{L}[y](s)$ to get

$$\mathscr{L}[y'] = -y(0) + sY(s)$$
$$= -1 + sY(s),$$
$$\mathscr{L}[y''] = -y'(0) - sy(0) + s^2Y(s)$$
$$= -s + s^2Y(s).$$

Because integration from 0 to ∞ and multiplication by e^{-st} both act linearly on functions $y(t)$, the equation

$$\mathscr{L}[y'' - y' - 2y] = \mathscr{L}[3e^t]$$

can be written

$$\mathscr{L}[y''] - \mathscr{L}[y'] - 2\mathscr{L}[y] = 3\mathscr{L}[e^t].$$

The expressions for $\mathcal{L}[y]$, $\mathcal{L}[y']$, and $\mathcal{L}[y'']$ in terms of Y, together with Equation 1.1, allow us to write this last equation as

$$[-s + s^2Y(s)] - [-1 + sY(s)] - 2[Y(s)] = 3\frac{1}{s-1}.$$

Rearrangement gives

$$(s^2 - s - 2)Y(s) = \frac{3}{s-1} + s - 1 = \frac{s^2 - 2s + 4}{s-1}$$

or

$$Y(s) = \frac{s^2 - 2s + 4}{(s-1)(s^2 - s - 2)}.$$

Having found an expression for $Y(s)$, our problem is now to identify precisely the solution $y(t)$ that satisfies $\mathcal{L}[y](s) = Y(s)$. Because $Y(s)$ is a rational function, it can theoretically always be broken down according to the partial-fraction decomposition associated with the computation of indefinite integrals. In our example, the decomposition can be carried out because the denominator of $Y(s)$ can be factored. We need to determine the coefficients A, B, and C in

$$\frac{s^2 - 2s + 4}{(s-1)(s+1)(s-2)} = \frac{A}{s-1} + \frac{B}{s+1} + \frac{C}{s-2}.$$

Multiplying through by $s - 1$ and then setting $s = 1$ gives $A = -\frac{3}{2}$. Similarly, we multiply by $s + 1$ and then set $s = -1$ to get $B = \frac{7}{6}$. Finally, we multiply by $s - 2$ and then set $s = 2$ to get $C = \frac{4}{3}$. As a result, we have

$$Y(s) = -\frac{\frac{3}{2}}{s-1} + \frac{\frac{7}{6}}{s+1} + \frac{\frac{4}{3}}{s-2}.$$

Equation 1.1 now allows us to identify $y(t)$ as

$$y(t) = -\tfrac{3}{2}e^t + \tfrac{7}{6}e^{-t} + \tfrac{4}{3}e^{2t}.$$

As a check on the computation, we can verify that $y(0) = 1$ and $y'(0) = 0$.

In the examples given previously, we used the linearity of $\mathcal{L}$, the Laplace transform operator. The property is formally expressed by the two equations

■■■ **1.4**
$$\mathcal{L}[y_1 + y_2] = \mathcal{L}[y_1] + \mathcal{L}[y_2],$$
$$\mathcal{L}[cy] = c\mathcal{L}[y], \qquad c \text{ const.}$$

These equations, together with Equation 1.3 for the Laplace transform of a derivative, need to be supplemented in practice by the calculation of Laplace transforms of specific functions, as is done in Equation 1.1, which asserts that $\mathcal{L}[e^{at}](s) = 1/(s - a)$. Table 1 contains more than enough entries to do all the problems in this section.

It is important to note that the tables of Laplace transforms intended for solving differential equations are usually meant to be used in both directions, from function $f(t)$ to transform $\mathcal{L}[f](s)$ and also back the other way. Furthermore, since the transform $\mathcal{L}$ acts linearly, so does the inverse transform $\mathcal{L}^{-1}$ that takes us back from $Y(s)$ to $f(t)$. Thus, for example, we might have $\mathcal{L}^{-1}[c_1Y_1(s) + c_2Y_2(s)] = c_1f_1(t) + c_2f_2(t)$.

Table 1: Laplace transformations

$f(t)$	$\mathcal{L}[f](s) = \int_0^\infty e^{-st}f(t)\,dt$
1. 1	$\dfrac{1}{s}$
2. t	$\dfrac{1}{s^2}$
3. t^n	$\dfrac{n!}{s^{n+1}}$, $n = 0, 1, 2, \ldots$
4. e^{at}	$\dfrac{1}{s-a}$
5. te^{at}	$\dfrac{1}{(s-a)^2}$
6. $t^n e^{at}$	$\dfrac{n!}{(s-a)^{n+1}}$, $n = 0, 1, 2, \ldots$
7. $\sin bt$	$\dfrac{b}{s^2 + b^2}$
8. $\cos bt$	$\dfrac{s}{s^2 + b^2}$
9. $t \sin bt$	$\dfrac{2bs}{(s^2 + b^2)^2}$
10. $t \cos bt$	$\dfrac{s^2 - b^2}{(s^2 + b^2)^2}$
11. $e^{at} \sin bt$	$\dfrac{b}{(s-a)^2 + b^2}$
12. $e^{at} \cos bt$	$\dfrac{s-a}{(s-a)^2 + b^2}$
13. $e^{at} - e^{bt}$	$\dfrac{a-b}{(s-a)(s-b)}$
14. $ae^{at} - be^{bt}$	$\dfrac{(a-b)s}{(s-a)(s-b)}$
15. $(b-c)e^{at} + (c-a)e^{bt} + (a-b)e^{ct}$	$\dfrac{(a-b)(b-c)(a-c)}{(s-a)(s-b)(s-c)}$
16. $\sin bt - bt \cos bt$	$\dfrac{2b^3}{(s^2 + b^2)^2}$

More elaborate tables may contain several hundred entries. Such tables are meant to be used in both directions, so while the entry

$$\mathscr{L}[t^n] = \frac{n!}{s^{n+1}}, \qquad n = 0, 1, 2, \ldots$$

provides the transform of $f(t) = t^n$, it also provides, after division by $n!$, the **inverse transform,** denoted $\mathscr{L}^{-1}[s^{-(n+1)}] = t^n/n!, n = 0, 1, 2, \ldots$. For the proof that for every Laplace transform $Y(s)$ there is a *unique* function y such that $\mathscr{L}[y](s) = Y(s)$, we can refer to more theoretical accounts of the subject. All entries in Table 1 can be computed using elementary integration techniques

In finding inverse transforms, it is sometimes essential to decompose a rational function $P(s)/Q(s)$, with the degree of P less than the degree of Q, according to the following rules:

1. If the denominator $Q(s)$ has the factor $(s - a)^m$ as the highest power of $s - a$ that divides $Q(s)$, then include in the decomposition of $P(s)/Q(s)$ the fractions of the form

$$\frac{A_j}{(s - a)^j}, \qquad j = 1, 2, \ldots, m.$$

2. If the denominator $Q(s)$ has the factor $(s^2 + ps + q)^n$ as the highest power of $s^2 + ps + q$, irreducible over the reals, that divides $Q(s)$, then include in the decomposition of $P(s)/Q(s)$ all fractions of the form

$$\frac{B_k s + C_k}{(s^2 + ps + q)^k}, \qquad k - 1, 2, \ldots, n.$$

e x a m p l e   3

To find the function $f(t)$ having Laplace transform

$$F(s) = \frac{s + 1}{(s - 1)^2(s^2 + 1)},$$

we decompose the function into a sum of fractions as follows:

$$\frac{s + 1}{(s - 1)^2(s^2 + 1)} = \frac{A}{s - 1} + \frac{B}{(s - 1)^2} + \frac{Cs + D}{s^2 + 1}.$$

To compute B, we can multiply through by $(s - 1)^2$ and then set $s = 1$. We get $B = 1$. The same kind of trick does not apply directly to the other coefficients, but if we subtract $1/(s - 1)^2$ from both sides, we find we can cancel $s - 1$ on the left to get

$$\frac{-s^2 + s}{(s - 1)^2(s^2 + 1)} = \frac{-s}{(s - 1)(s^2 + 1)} = \frac{A}{s - 1} + \frac{Cs + D}{s^2 + 1}.$$

Now multiply by $s - 1$ and then set $s = 1$ to get $A = -\frac{1}{2}$. As a result,

$$\frac{-s}{(s - 1)(s^2 + 1)} = \frac{-\frac{1}{2}}{s - 1} + \frac{Cs + D}{s^2 + 1}.$$

To find C and D, we can multiply through by $(s - 1)(s^2 + 1)$ to get

$$-s = -\tfrac{1}{2}(s^2 + 1) + (s - 1)(Cs + D).$$

Rearranging the powers of s gives

$$\tfrac{1}{2}s^2 - s + \tfrac{1}{2} = Cs^2 + (D - C)s - D.$$

We equate coefficients of like powers on both sides and find that $C = \frac{1}{2}$, whereas $D = -\frac{1}{2}$. The result is that

$$\frac{s + 1}{(s - 1)^2(s^2 + 1)} = \frac{-\frac{1}{2}}{(s - 1)} + \frac{1}{(s - 1)^2} + \frac{-\frac{1}{2}s}{s^2 + 1} - \frac{\frac{1}{2}}{s^2 + 1}.$$

From Table 1 we conclude that

$$\mathscr{L}^{-1}[F(s)] = -\tfrac{1}{2}e^t + te^t + \tfrac{1}{2}\cos t - \tfrac{1}{2}\sin t.$$

Laplace transforms are applicable to linear systems, as in the next example.

example 4

We'll solve the initial-value problem

$$\ddot{x} + \dot{y} = 4x, \qquad x(0) = 0, \qquad \dot{x}(0) = 1,$$
$$4\dot{x} - \ddot{y} = 9y, \qquad y(0) = -1, \qquad \dot{y}(0) = 2.$$

Denote $\mathscr{L}[x]$ and $\mathscr{L}[y]$ by X and Y, respectively. Then

$$(s^2X - sx(0) - \dot{x}(0)) + (sY - y(0)) = 4X,$$
$$4(sX - x(0)) - (s^2Y - sy(0) - \dot{y}(0)) = 9Y.$$

Inserting the specific initial values gives

$$(s^2X - 1) + (sY + 1) = 4X,$$
$$4(sX - x(0)) - (s^2Y + s - 2) = 9Y.$$

Now rearrange into a form suitable for solving for X and Y:

$$(s^2 - 4)X + sY = 0,$$
$$4sX - (s^2 + 9)Y = s - 2.$$

Suppose we choose to eliminate Y. Multiply the first equation by $s^2 + 9$, the second by s, and add to get

$$(s^4 + 9s^2 - 36)X = s^2 - 2s, \qquad \text{or} \qquad X(s) = \frac{s^2 - 2s}{s^4 + 9s^2 - 36}.$$

The denominator in the expression for X is quadratic in s^2 with roots $s^2 = 3$, -12, so we can factor and find the partial-fraction decomposition

$$X(s) = \frac{s^2 - 2s}{(s^2 - 3)(s^2 + 12)}$$

$$= \frac{2}{15}\frac{s}{s^2 + 12} + \frac{4}{5}\frac{1}{s^2 + 12} - \frac{2 + \sqrt{3}}{30}\frac{1}{s + \sqrt{3}} - \frac{2 - \sqrt{3}}{30}\frac{1}{s - \sqrt{3}}.$$

Reading from right to left in Table 1, we find

$$x(t) = \frac{2}{15}\cos\sqrt{12}t + \frac{4}{5\sqrt{12}}\sin\sqrt{12}t - \frac{2 + \sqrt{3}}{30}e^{-\sqrt{3}t} - \frac{2 - \sqrt{3}}{30}e^{\sqrt{3}t}.$$

To find $Y(s)$, and then $y(t)$ we can return to the pair of linear equations for X and Y. Substituting the factored expression above for $X(s)$ in the first of those two equations gives

$$(s^2 - 4)\frac{s^2 - 2s}{(s^2 - 3)(s^2 + 12)} + sY(s) \qquad \text{or} \qquad Y(s) = -\frac{(s - 2)(s^2 - 4)}{(s^2 - 3)(s^2 + 12)}.$$

Partial-fraction decomposition yields

$$Y(s) = -\frac{16}{15}\frac{s}{s^2 + 12} + \frac{32}{15}\frac{1}{s^2 + 12} + \frac{2 + \sqrt{3}}{30\sqrt{3}}\frac{1}{s + \sqrt{3}} - \frac{2 - \sqrt{3}}{30\sqrt{3}}\frac{1}{s - \sqrt{3}}.$$

Using Table 1 gives

$$y(t) = -\frac{16}{15}\cos\sqrt{12}t + \frac{32}{15\sqrt{12}}\sin\sqrt{12}t + \frac{2 + \sqrt{3}}{30\sqrt{3}}e^{-\sqrt{3}t} - \frac{2 - \sqrt{3}}{30\sqrt{3}}e^{\sqrt{3}t}.$$

Transforms of Periodic Functions. Table 1 lists the transforms of two periodic functions, $\mathcal{L}[\sin bt](s) = b/(s^2 + b^2)$ and $\mathcal{L}[\cos bt](s) = s/(s^2 + b^2)$, and these are routinely computed using integration by parts. For many functions of period $T > 0$, it's much more efficient to reduce the computation to an integral over the interval $0 \le t \le T$. In the context of the Laplace transform, to say that $f(t)$ has **period** T usually means that $f(t + T) = f(t)$ for $t > 0$. For all such functions having a Laplace transform, we have the following computational aid.

■ 1.5 THEOREM

If $f(t)$ has a Laplace transform and has period T, then

$$\mathcal{L}[f](s) = \frac{1}{1 - e^{-Ts}}\int_0^T e^{-st}f(t)\, dt.$$

Proof.

$$\mathcal{L}[f](s) = \int_0^T e^{-st}f(t)\, dt + \int_T^\infty e^{-st}f(t)\, dt \qquad \text{[Breaking up integral.]}$$

$$= \int_0^T e^{-st}f(t)\, dt + \int_0^\infty e^{-s(u+T)}f(u + T)\, du \qquad \text{[Let } t = u + T.\text{]}$$

$$= \int_0^T e^{-st}f(t)\, dt + \int_0^\infty e^{-s(u+T)}f(u)\, du \qquad \text{[Use periodicity of } f.\text{]}$$

$$= \int_0^T e^{-st}f(t)\, dt + e^{-sT}\mathcal{L}[f](s). \qquad \text{[Pull out factor } e^{-Ts}.\text{]}$$

Now solve for $\mathcal{L}[f](s)$.   ■

---

example 5

Consider the "square-wave" function

$$f(t) = \begin{cases} 1, & 2k \le t < 2k + 1, \\ -1, & 2k + 1 \le t < 2k + 2; \ k = 0, 1, 2, \ldots. \end{cases}$$

This function has period 2, and its graph is sketched in Figure 8.1. Setting $T = 2$ in the right side of the equation in Theorem 1.5, we note that the integral decomposes into a sum of two pieces:

$$\int_0^1 1 \cdot e^{-st}\, dt + \int_1^2 (-1) \cdot e^{-st}\, dt = \left[\frac{e^{-st}}{-s}\right]_0^1 - \left[\frac{e^{-st}}{-s}\right]_1^2$$

$$\frac{-1}{s}(e^{-s} - 1) \qquad \frac{-1}{s}(e^{-2s} - e^{-s})$$

$$= \frac{1}{s}(1 - 2e^{-s} + e^{-2s}).$$

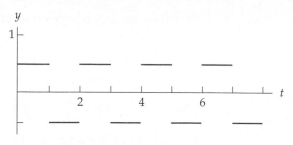

Figure 8.1 Square wave.

According to Theorem 1.5, the transform of $f(t)$ equals this last expression divided by $1 - e^{-2s}$:

$$\mathcal{L}[f](s) = \frac{1 - 2e^{-s} + e^{-2s}}{s(1 - e^{-2s})}.$$

e x a m p l e 6   Suppose we want the Laplace transform of the function

$$f(t) = \begin{cases} \sin bt, & 0 \le t \le 2\pi/b,\ b > 0, \\ 0, & 2\pi/b < t. \end{cases}$$

The graph of this function consists of one complete cycle of $\sin bt$ on $0 \le t \le 2\pi/b$ completed by the t axis elsewhere. The transform can be computed using integration by parts, but it's simpler first to note that

$$\mathcal{L}[f](s) = \int_0^\infty e^{-st} f(t)\, dt = \int_0^{2\pi/b} e^{-st} \sin bt\, dt.$$

Theorem 1.5 together with entry 7 in Table 1 implies

$$\frac{b}{s^2 + b^2} = \frac{1}{1 - e^{-2\pi s/b}} \mathcal{L}[f](s).$$

Solving for $\mathcal{L}[f]$ gives us

$$\mathcal{L}[f](s) = \frac{b(1 - e^{-2\pi s/b})}{s^2 + b^2}.$$

EXERCISES

1. (a) Verify that

$$\mathcal{L}[e^{at}](s) = \int_0^\infty e^{-st} e^{at}\, dt = \frac{1}{s - a}, \qquad \text{for } s > a.$$

(b) Use integration by parts to verify that

$$\mathcal{L}[t](s) = \int_0^\infty t e^{-st}\, dt = \frac{1}{s^2}, \qquad \text{for } s > 0.$$

2. Use Equation 1.3 to show that, if $y(0) = 2$ and $y'(0) = 3$, then

$$\int_0^\infty e^{-st} y''(t)\, dt = s^2 Y(s) - 2s - 3,$$

where $Y(s)$ is the Laplace transform of $y(t)$.

3. By computing the appropriate integral, or by using the table of Laplace transforms (Table 1), compute $\mathcal{L}[f](s)$, where $f(t)$ is as follows:

(a) $t \sin 2t$.

(b) $\cos t + 2 \sin t$.

(c) $t^2 + 2t - 1$.

(d) $\cos (t + a)$.

(e) $(2t + 1)e^{3t}$.

(f) $e^t + e^{-t}$.

4. Use Table 1 to find the inverse Laplace transforms of the following functions $F(s)$; that is, find $y(t)$ such that $\mathcal{L}[y](s) = F(s)$.

(a) $1/(s^2 - 1)$.

(b) $s/(s^2 - 4)$.

(c) $2/(s^2 + 4)$.

(d) $4s/(s^2 + 4)^2$.

(e) $1/((s - 2)^2 + 9)$.

(f) $1/s^2 - 1/(s - 1)^2$.

5. Use the Laplace transform to solve the initial-value problems. Check by substitution.

(a) $y' - y = t$, $y(0) = 2$.

(b) $y' + 2y = 1$, $y(0) = 1$.

(c) $y' + 3y = \cos 2t$, $y(0) = 0$.

(d) $y'' + y = e^{-t} + 1$,
$y(0) = -1$, $y'(0) = 1$.

(e) $2y'' - y' = 2 \cos 3t$,
$y(0) = 0$, $y'(0) = 2$.

(f) $y''' = t$, $y(0) = y'(0) = 0$,
$y''(0) = 1$.

(g) $\dot{x} = 6x + 8y$, $x(0) = 0$.
$\dot{y} = -4x - 6y$, $y(0) = 1$.

(h) $\dot{x} = x + 2y$, $x(0) = 0$,
$\dot{y} = -2x + y$, $y(0) = 1$.

(i) $dx/dt = x + 2y$, $x(0) = 1$,
$dy/dt = x + y + t$, $y(0) = -1$.

(j) $dx/dt = -y - t$, $x(0) = 2$,
$dy/dt = x + t$, $y(0) = 0$.

(k) $\ddot{x} - \dot{y} = -t + 1$, $x(0) = 0$,
$\dot{x}(0) = 1$,
$\dot{x} - x + 2\dot{y} = 4e^t$, $y(0) = 0$.

(l) $\ddot{x} = y$, $x(0) = \dot{x}(0) = 0$.
$\ddot{y} = x$, $y(0) = 0$, $\dot{y}(0) = 1$.

6. Define $H_a(t) = \begin{cases} 0, & \text{if } t < a \\ 1, & \text{if } a \le t \end{cases}$.

(a) Show that $\mathcal{L}[H_a](s) = (1/s)e^{-as}$.

(b) Show that if $g(t) = H_a(t)f(t)$ for $0 < t$ and $a \ge 0$, then $\mathcal{L}[g](s) = e^{-as}\mathcal{L}[f(t + a)](s)$.

(c) Sketch the graph of $H_a(t)$ for $a = 1$.

(d) Solve the differential equation $y'' = H_a(t)$, $0 < a$, with the conditions $y(0) = 1$, $y'(0) = 0$.

7. Let $P(D)$ be an nth-order linear constant-coefficient differential operator. Show that

$$\mathcal{L}[P(D)y](s) = P(s)\mathcal{L}[y](s) + Q(s),$$

where $Q(s)$ is some polynomial of degree $n - 1$. Use induction.

8. (a) Show that if $f(t)$ is defined and differentiable only for $t > 0$ (instead of $t \ge 0$), then

$$\mathcal{L}[f'](s) = -f(0+) + s\mathcal{L}[f](s), \qquad \text{where } f(0+) = \lim_{t \to 0+} f(t).$$

(b) Show that, if the limits $f^k(0+)$ all exist, then Formula 1.3 can be generalized to

$$\mathcal{L}[f^{(n)}](s) = -f^{(n-1)}(0+) - \cdots - s^{(n-1)}f(0+) + s^n\mathcal{L}[f](s).$$

Transforms of Periodic Functions. Sketch the graph of each of the functions in the next four exercises; then use Theorem 1.5 to verify the correctness of the given transform.

9. Graph the function of period 1 defined for $0 \le t < 1$ by $f(t) = t$; verify

$$\mathcal{L}[f](s) = \frac{1 - se^{-s} - e^{-s}}{s^2(1 - e^{-s})}.$$

10. Graph the function $f_2(t)$ that agrees with $f(t)$ in the previous exercise for $0 \le t \le 2$ and is equal to zero for $2 \le t$; verify

$$\mathcal{L}[f_2](s) = \frac{(1 - se^{-s} - e^{-s})(1 - e^{-2s})}{s^2(1 - e^{-s})} = \frac{(1 - e^{-s} - se^{-s})(1 + e^{-s})}{s^2}.$$

11. Graph the function of period 2 defined for $0 \le t < 2$ by

$$g(t) = \begin{cases} 1, & 0 \le t < 1 \\ 0, & 1 \le t < 2. \end{cases}$$

Show that

$$\mathcal{L}[g](s) = \frac{1 - e^{-s}}{s(1 - e^{-2s})}.$$

12. Graph the function $g_4(t)$ that agrees with $g(t)$ in the previous exercise for $0 \le t \le 4$ and is equal to zero for $4 \le t$; verify

$$\mathcal{L}[g_4](s) = \frac{(1 - e^{-s})(1 - e^{-4s})}{s(1 - e^{-2s})} = \frac{1 - e^{-s} + e^{-2s} - e^{-3s}}{s}.$$

2 CONVOLUTION

Let us review the solution of the second-order differential equation

$$y'' + py' + qy = f(t), \qquad y(0) = y_0, \qquad y'(0) = y_1.$$

Taking the Laplace transform of both sides gives

$$(s^2 + ps + q)Y(s) = F(s) + y'(0) + (s + p)y(0).$$

The polynomial factor $P(s) = s^2 + ps + q$ on the left is the **characteristic polynomial** of the operator $D^2 + pD + q$. The reciprocal $Q(s) = 1/P(s)$ is called the **transfer function** of the operator, and if we multiply by $Q(s)$, or divide by $P(s)$, we get the formula

$$Y(s) = \frac{F(s) + y'(0) + (s + p)y(0)}{P(s)}$$

for the Laplace transform $Y = \mathcal{L}[y]$. The remaining step is to find the inverse transform $y(t) = \mathcal{L}^{-1}[y](t)$. The essence of the method is to use the Laplace transform to reduce the solution of the problem to some routine algebraic manipulations.

In addition to the table of specific Laplace transforms in Section 1 (Table 1), a number of general formulas, such as Formula 1.3 in Section 1, are useful in solving problems. The most important of these answers the following question: If $F(s)$ and $G(s)$ are the Laplace transforms of $f(t)$ and $g(t)$, respectively, what function has a Laplace transform equal to the product $F(s)G(s)$? It turns out that under rather general hypotheses there is an answer, given by the **convolution integral**

$$f * g(t) = \int_0^t f(u)g(t - u)\, du.$$

The function $f * g(t)$ is called the **convolution** of the functions $f(t)$ and $g(t)$ and is defined for $t \geq 0$, provided that f and g are integrable on every finite interval. The convolution $f * g$ is to be thought of as a kind of product of f and g, and, in fact, it turns out that $f * g = g * f$, although this is not obvious from the definition. The basic information about convolutions is summarized as follows.

■ 2.1 THEOREM

Let $f(u)$ and $g(u)$ be integrable on $0 \leq u \leq t$ for every positive t; then $f * g$ and $g * f$ both exist and are equal; that is, convolution is commutative:

$$f * g = g * f.$$

If $\mathcal{L}[|f|](s)$ and $\mathcal{L}[|g|](s)$ are both finite, then

$$\mathcal{L}[f * g](s) = (\mathcal{L}[f](s))(\mathcal{L}[g](s)).$$

Proof. The first statement is easily proved by changing the variable in the definition of $f * g$. Holding t fixed and replacing u by $t - v$ in the integral, we get

$$f * g(t) = \int_0^t f(u)g(t - u)\, du = -\int_t^0 f(t - v)g(v)\, dv$$

$$= \int_0^t g(v)f(t - v)\, dv = g * f(t).$$

Proving the second statement involves an important technical point for which we will not give a proof, but the rest of the argument is complete. To simplify the writing of limits of integration, we can extend both $f(t)$ and $g(t)$ to have the value 0 for $t < 0$. Then we can write

$$\mathcal{L}[f](s)\mathcal{L}[g](s) = \int_{-\infty}^{\infty} e^{su}f(u)\, du \int_{-\infty}^{\infty} e^{sv}g(v)\, dv$$

$$= \int_{-\infty}^{\infty} f(u)\left[\int_{-\infty}^{\infty} e^{-s(u+v)}g(v)\, dv\right] du.$$

We can make the change of variable $v = t - u$ in the inner integral to get

$$\mathcal{L}[f](s)\mathcal{L}[g](s) = \int_{-\infty}^{\infty} f(u)\left[\int_{-\infty}^{\infty} e^{-st}g(t - u)\, dt\right] du.$$

Under the hypotheses of the theorem, we can interchange the order of integration. (We will assume the required theorem, called Fubini's theorem.) Then we have

$$\mathcal{L}[f](s)\mathcal{L}[g](s) = \int_{-\infty}^{\infty} e^{-st}\left\{\int_{-\infty}^{\infty} f(u)g(t - u)\, du\right\} dt.$$

Because we have assumed $f(t)$ and $g(t)$ are zero for $t < 0$, the inner integral is zero for $t < 0$. It follows that the t integration is needed only for $0 \leq t \leq \infty$.

Similarly, the u integration is needed only for $0 \le u \le t$. Hence

$$\mathscr{L}[f](s)\mathscr{L}[g](s) = \int_0^\infty e^{-st}\left[\int_0^t f(u)g(t-u)\,du\right]dt$$

$$= \mathscr{L}[f*g](s),$$

which is what we wanted to prove. ∎

e x a m p l e **1**

1. From Table 1 in Section 1, we see that $\mathscr{L}[t](s) = 1/s^2$ and $\mathscr{L}[\sin t](s) = 1/(s^2 + 1)$. It follows from Theorem 2.1 that

$$\frac{1}{s^2} \cdot \frac{1}{s^2 + 1} = \mathscr{L}\left[\int_0^t (t-u)\sin u\,du\right](s).$$

But holding t fixed, we can use integration by parts to show that

$$\int_0^t (t-u)\sin u\,du = [-(t-u)\cos u]_0^t - \int_0^t \cos u\,du$$

$$= t - \sin t.$$

We could have obtained the same result by computing a partial-fraction decomposition of the form

$$\frac{1}{s^2} \cdot \frac{1}{s^2 + 1} = \frac{A}{s} + \frac{B}{s^2} + \frac{Cs + d}{s^2 + 1},$$

and then finding the inverse transform of each term.

Table 2 lists the most frequently used general properties of the Laplace transform. The entries that have not already been discussed can easily be proved using elementary calculus techniques. The table omits the precise conditions under which each formula holds. The distinction between formulas 2

Table 2: Laplace transform properties

1. $\mathscr{L}[af + bg] = a\mathscr{L}[f] + b\mathscr{L}[g]$, a, b const.

2. $\mathscr{L}[f'](s) = s\mathscr{L}[f](s) - f(0+)$, $f(0+) = \lim\limits_{t\to 0+} f(t)$.

3. $\mathscr{L}[f^{(n)}](s) = s^n\mathscr{L}[f](s) - s^{n-1}f(0+) - s^{n-2}f'(0+) - \cdots - f(n-1)(0+)$.

4. $\mathscr{L}\left[\int_0^t f(u)\,du\right](s) = \dfrac{1}{s}\mathscr{L}[f](s)$.

5. $\mathscr{L}[e^{at}f(t)](s) = \mathscr{L}[f](s-a)$.

6. $\mathscr{L}[H_a(t)f(t-a)](s) = e^{-as}\mathscr{L}[f](s)$, $H_a(t) = \begin{cases} 0, & t < a, \\ 1, & a \le t. \end{cases}$

7. If $F(s) = \mathscr{L}[f(t)](s)$, then $F^{(n)}(s) = (-1)^n\mathscr{L}[t^nf(t)](s)$.

8. $\mathscr{L}[f*g](s) = \mathscr{L}[f](s)\mathscr{L}[g](s)$.

and 3 in the table and Equations 1.2 and 1.3 of Section 1 occurs because in the latter we assumed that we were dealing with solutions of differential equations and that these solutions had continuous derivatives at $t = 0$. The corresponding formulas in Table 2 are valid under the weaker assumption that $\lim_{t \to 0+} f^{(k)}(t)$ exists but is not necessarily equal to $f^{(k)}(0)$. (See Exercise 8 of Section 1.)

---

e x a m p l e 2

From Table 1, we find that

$$\mathcal{L}[\sin t](s) = \frac{1}{s^2 + 1}.$$

Taking $f(t) = \sin t$ in formula 4 of Table 2 gives

$$\frac{1}{s(s^2 + 1)} = \frac{1}{s} \mathcal{L}[\sin t](s)$$

$$= \mathcal{L}\left[\int_0^t \sin u \, du\right](s) = \mathcal{L}[-\cos t + 1].$$

Hence

$$\mathcal{L}^{-1}\left[\frac{1}{s(s^2 + 1)}\right] = -\cos t + 1.$$

Repeating the application of formula 4 gives

$$\frac{1}{s^2(s^2 + 1)} = \mathcal{L}\left[\int_0^t (-\cos u + 1) \, du\right]$$

$$= \mathcal{L}[-\sin t + t].$$

This establishes the formula that was derived in Example 1 using convolution.

---

e x a m p l e 3

Starting with the formula

$$\mathcal{L}[\cos t](s) = \frac{s}{s^2 + 1},$$

we can apply formula 7 in Table 2 to get

$$\mathcal{L}[t \cos t](s) = -\frac{d}{ds} \frac{s}{s^2 + 1}$$

$$= \frac{s^2 - 1}{(s^2 + 1)^2}.$$

Another application of the same formula gives

$$\mathcal{L}[t^2 \cos t](s) = -\frac{d}{ds} \frac{s^2 - 1}{(s^2 + 1)^2}$$

$$= \frac{2s^3 - 6s}{(s^2 + 1)^3}.$$

e x a m p l e **4**

To apply formula 6 of Table 2 to the function $f(t) = t$, we define $f(t) = 0$ for $t < 0$. The graphs of $f(t)$ and $f(t - a)$ are indicated in Figure 8.2 for $a = 1, a = 2$, and $a = 3$. Each function is zero where it is not positive. From formula 6 in Table 2 we find

$$\mathscr{L}[f(t - 1)](s) = e^{-s}\mathscr{L}[f(t)](s)$$

$$= e^{-s}\mathscr{L}[t](s) = e^{-s}\frac{1}{s^2}.$$

Similarly

$$\mathscr{L}[f(t - 2)](s) = e^{-2s}\mathscr{L}[f(t)](s)$$

$$= e^{-2s}\mathscr{L}[t](s) = e^{-2s}\frac{1}{s^2}.$$

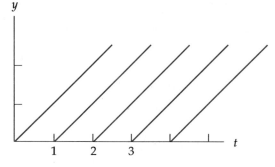

Figure 8.2 f and z shift to the right.

EXERCISES

1. Find the convolution $f * g$ of the following pairs of functions:
   (a) $f(t) = t$, $g(t) = e^{-t}$, $t \geq 0$.
   (b) $f(t) = t^2$, $g(t) = (t^2 + 1)$, $t \geq 0$.
   (c) $f(t) = 1$, $g(t) = t$, $t \geq 0$.

2. Use the convolution of two functions to find the inverse Laplace transform of each of the following products of Laplace transforms:

   (a) $\dfrac{1}{s^2(s + 1)}$.   (b) $\dfrac{1}{(s - 1)(s - 2)}$.   (c) $s^{-3}e^{-2s}$.

3. By using the formula in Tables 1 and 2, find the inverse Laplace transform of each of the following functions.

   (a) $\dfrac{1}{s(s + 3)^2}$.   (c) $\dfrac{1}{s^2 + 2s + 2}$.   (e) $\dfrac{e^{-s} + 1}{s}$.

   (b) $\dfrac{e^{-2s}}{s(s^2 + 4)}$.   (d) $\dfrac{1}{s^2 + 1}$.   (f) $\dfrac{s}{s^2 + 10}$.

4. Solve the following initial-value problems. Check by substitution.
   (a) $y'' - y = \sin 2t + 1$, $y(0) = 1$, $y'(0) = 1$.
   (b) $y'' + 2y = t$, $y(0) = 0$, $y'(0) = 1$.
   (c) $y'' + y' = t + e^{-t}$, $y(0) = 2$, $y'(0) = 1$.

5. Solve the equation

$$y' + y = \int_0^t y(u)\, du + t,$$

given that $y(0) = 1$.

6. Use integration by parts to show that

$$\int_0^t (t - u)^n f(u)\, du = n! \int_0^t \left(\int_0^{t_1} \left(\cdots \int_0^{t_n} f(t_{n+1})\, dt_{n+1} \right) \cdots dt_2 \right) dt_1.$$

7. The gamma function, denoted by Γ, can be defined by

$$\Gamma(\nu) = \int_0^\infty t^{\nu-1} e^{-t}\, dt, \qquad \nu > 0.$$

   (a) Use integration by parts to show that
   $$\Gamma(\nu + 1) = \nu \Gamma(\nu).$$
   (b) Deduce from part (a) that $\Gamma(n + 1) = n!$, for $n = 0, 1, 2, \ldots$.
   (c) Show that if $\nu > -1$, then

   $$\mathscr{L}[t^\nu](s) = \frac{\Gamma(\nu + 1)}{s^{\nu+1}}.$$

8. Derive formula 7 in Table 2 by formal differentiation with respect to s.

3 GENERALIZED FUNCTIONS

3A *The Delta Function*

The solution of the initial-value problem
$$y'' - ay' + by = f(t), \qquad y(0) = y_0, \qquad y'(0) = z_0,$$
is often wanted when $f(t)$ is an **impulse function,** that is, a function that is identically zero outside some small interval and that assumes relatively large values on the interval itself. Figure 8.3 shows the graphs of just such functions assuming the value $1/h$ on the intervals $t_0 < t < t_0 + h$ and $0 < t < h$. The integral of the first one is

$$\int_{t_0}^{t_0+h} \frac{1}{h}\, dt = 1.$$

As h tends to zero, the value of the function on the drcreasing interval $t_0 \le t \le t_0 + h$ tends to infinity while the value of the integral remains 1. When $t_0 = 0$, we denote the function by $\Delta_h(t)$. Thus the shifted version, different from zero on $(t_0, t_0 + h)$, is $\Delta_h(t - t_0)$.

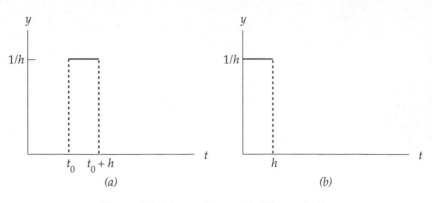

Figure 8.3 (a) $y = \Delta_h(t - t_0)$. (b) $y = \Delta_h(t)$.

We now compute

$$\lim_{h \to 0} \int_0^\infty \Delta_h(t - t_0) g(t)\, dt$$

under the assumption that g is continuous at t_0. If we define

$$G(t) = \int_0^t g(u)\, du,$$

then

$$\int_0^\infty \Delta_h(t - t_0) g(t)\, dt = \int_{t_0}^{t_0 + h} \frac{1}{h} g(t)\, dt$$

$$= \frac{1}{h} \left[\int_0^{t_0 + h} g(t)\, dt - \int_0^{t_0} g(t)\, dt \right]$$

$$= \frac{1}{h} (G(t_0 + h) - G(t_0)).$$

As $h \to 0$, this last expression tends to $G'(t_0)$. Since $G(t)$ is defined as an integral of $g(t)$, by one of the fundamental theorems of calculus, $G'(t_0) = g(t_0)$. Hence

$$\lim_{h \to 0} \int_0^\infty \Delta_h(t - t_0) g(t)\, dt = g(t_0).$$

If we were to take the limit under the integral sign, we would get

$$\lim_{h \to 0} \Delta_h(t - t_0) = \begin{cases} 0, & t \neq t_0, \\ +\infty, & t = t_0. \end{cases}$$

The resulting function is not Riemann integrable, but the way we arrived as it does suggest that we might extend the definition of integral to the **Dirac delta** δ times a function $g(t)$ to be such that

$$\int_0^\infty \delta(t - t_0) g(t)\, dt = g(t_0)$$

whenever $g(t)$ is continuous at $t = t_0$. We define the notation δ_a by $\delta_a(t) = \delta(t - a)$ and think of it as representing **unit point mass** concentrated at a.

e x a m p l e **1**

Using the definition of integral suggested above, we have $\int_0^\infty \delta(t - 2)e^{-t} dt = e^{-2}$. More generally, we can compute the Laplace transform of δ_{t_0} by

$$\mathcal{L}[\delta_{t_0}](s) = \int_0^\infty \delta(t - t_0)e^{-st} dt = e^{-t_0 s}.$$

In particular, with $t_0 = 0$, we get $\mathcal{L}[\delta](s) = 1$. The convolution of δ with f turns out to be f:

$$\int_0^t f(t - u)\delta(u) \, du = f(t), \qquad t > 0.$$

Thus the rule $\mathcal{L}[f * \delta](s) = \mathcal{L}[f](s)\mathcal{L}[\delta](s)$, shown already when δ is replaced by an ordinary function, still holds for convolution of δ with a continuous function.

The formal definition of $\delta_{t_0}(t) = \delta(t - t_0)$, given previously, can be interpreted as providing a large instantaneous impulse at $t = t_0$. To give some meaning to the word "large" in this context, consider the following. We define the **Heaviside function** H_{t_0} by

$$H_{t_0}(t) = \begin{cases} 0, & t < t_0, \\ 1, & t \geq t_0. \end{cases}$$

Its graph for a particular value of t_0 is shown in Figure 8.4, followed by the graph of the special case for which $t_0 = 0$. Note that H_{t_0} is a perfectly well defined function in the ordinary sense of the word "function." It does have the property, however, that its derivative H'_{t_0} fails to exist at t_0, even though $H'_{t_0}(t)$ exists for all values of t except t_0. Given the abruptness of the jump in H_{t_0} at t_0, we might guess that $H'_{t_0}(t) = \delta(t - t_0)$. To justify this equation, we need to show that it is consistent with the usual computations of calculus. Thus let $g(t)$ be a differentiable function tending rapidly enough to zero as $t \to \infty$ so that the improper integral

$$\int_0^\infty g(t) \, dt$$

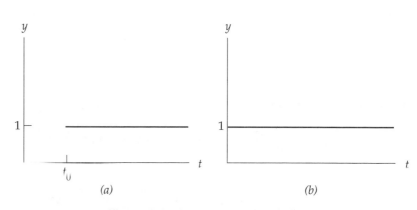

(a)   (b)

Figure 8.4 (a) $y = H_{t_0}(t)$. (b) $y = H_0(t)$.

exists. For example, $g(t) = e^{-t}$ is such a function. We now do the following formal integration by parts, assuming $t_0 > 0$:

$$\int_0^\infty H'_{t_0}(t)g(t)\,dt = H_{t_0}(t)g(t)\Big|_0^\infty - \int_0^\infty H_{t_0}(t)g'(t)\,dt$$

$$= H_{t_0}(\infty)g(\infty) - H_{t_0}(0)g(0) - \int_{t_0}^\infty g'(t)\,dt.$$

Here we have used the property that $H_{t_0}(t) = 0$ for $t < t_0$. But $H_{t_0}(\infty) = \lim_{t \to \infty} H_{t_0}(t) = 1$, and $g(\infty) = \lim_{t \to \infty} g(t) = 0$. Also $H_{t_0}(0) = 0$ since $t_0 > 0$. Hence

$$\int_0^\infty H'_{t_0}(t)g(t)\,dt = -\int_{t_0}^\infty g'(t)\,dt = -g(\infty) + g(t_0) = g(t_0).$$

Comparing this result with our previous definition,

$$\int_0^\infty \delta(t - t_0)g(t)\,dt = g(t_0),$$

we see that $H'_{t_0}(t)$ behaves just like $\delta(t - t_0)$. So we define the **derivative of the Heaviside function** by $H'_{t_0} = \delta_{t_0}$ so that

$$H'_{t_0}(t) = \delta_{t_0}(t) = \delta(t - t_0).$$

---

e x a m p l e **2**

$$\int_0^\infty H'_{t_0}(t)e^{-st}\,dt = \int_0^\infty \delta(t - t_0)e^{-st}dt = e^{-t_0 s}.$$

Thus

$$\mathscr{L}[H'_{t_0}](s) = e^{-t_0 s}.$$

On the other hand,

$$\mathscr{L}[H_{t_0}](s) = \int_0^\infty H_{t_0}(t)e^{-st}\,dt = \int_{t_0}^\infty e^{-st}\,dt = -\frac{1}{s}e^{-st}\Big|_{t_0}^\infty$$

$$= \frac{1}{s}e^{-t_0 s}.$$

It follows that

$$\mathscr{L}[H'_{t_0}](s) = s\mathscr{L}[H_{t_0}](s) - H_{t_0}(0+) = s\mathscr{L}[H_{t_0}](s).$$

This rule has already been shown to hold with H_{t_0} replaced by an honestly differentiable function f for which $f(0+) = 0$. It is this kind of consistency with the rules of ordinary calculus that facilitates operating with $\delta(t - t_0)$ or its equivalent, $H'_{t_0}(t)$.

3B   *Solution of Equations*

Here is a direct solution of $y'' = \delta(t - t_0)$. We recall that $H'_{t_0}(t) = \delta(t - t_0)$, so

$$y' = H_{t_0}(t) + c_1,$$

$$y = \int_0^t H_{t_0}(u)\,du + c_1 t + c_2,$$

$$= \begin{cases} 0 + c_1 t - c_2, & t < t_0, \\ (t - t_0) + c_1 t - c_2, & t \geq t_0. \end{cases}$$

Since $H_{t_0}(t) = 0$ for $t < t_0$, $y(t)$ can be written
$$y = (t - t_0)H_{t_0}(t) + c_1 t + c_2.$$
If we also require $y(0) = y'(0) = 0$, we get the solution
$$y = (t - t_0)H_{t_0}(t).$$
As a check, we calculate, using the product rule,
$$y' = H_{t_0}(t) + (t - t_0)H_{t_0}'(t)$$
$$= H_{t_0}(t) + (t - t_0)\delta(t - t_0).$$
But the second term simply acts like 0, because
$$\int_0^\infty (t - t_0)\delta(t - t_0)g(t)\, dt = (t_0 - t_0)g(t_0) = 0,$$
for every g. It follows that $y'' = H_{t_0}'(t) = \delta(t - t_0)$, as we wanted to show.

---

e x a m p l e 3

We can use the Laplace transform to solve the problem $y'' = \delta(t - t_0)$, $y(0) = 0$, $y'(0) = 0$, more quickly. We find
$$\mathcal{L}[y''](s) = s^2 \mathcal{L}[y](s) - sy(0+) - y'(0+) = s^2 \mathcal{L}[y](s).$$
Since $\mathcal{L}[\delta(t - t_0)](s) = e^{-t_0 s}$, we have
$$\mathcal{L}[y](s) = \frac{1}{s^2} e^{-t_0 s}.$$
Since $\mathcal{L}[t](s) = 1/s^2$, reference to formula 6 in Table 2 shows that
$$y(t) = (t - t_0)H_{t_0}(t).$$
The interpretation of $\delta(t - t_0)$ in the equation $y'' = \delta(t - t_0)$ follows from the formula for y' we derived previously:
$$y' = H_{t_0}(t) + c_1.$$
Thus the presence of $\delta(t - t_0)$ produces a jump of height $+1$ in y' at the point t_0.

---

e x a m p l e 4

To solve the initial-value problem
$$y'' + 2y' + y = 2\delta(t - 1), \qquad y(0) = 1, \qquad y'(0) = 1,$$
apply $\mathcal{L}$ to both sides:
$$s^2 \mathcal{L}[y](s) + 2s\mathcal{L}[y] + \mathcal{L}[y] - 3 - s = 2e^{-s}.$$
Then
$$\mathcal{L}[y](s) = \frac{2}{(s + 1)^2} + \frac{1}{s + 1} + 2\frac{e^{-s}}{(s + 1)^2}.$$
From Tables 1 and 2, we find
$$y(t) = 2te^{-t} + e^{-t} + 2(t - 1)e^{-(t-1)}H_1(t).$$
Notice that $y(t)$ is continuous at $t = 1$. However, its derivative $y'(t)$ will jump up by 2 at $t = 1$.

e x a m p l e 5

It may seem that the advantage of using the Laplace transform in Example 4 lies partly in the ease with which it handles the δ function. The real advantage, however, still lies in the way the initial conditions are handled. To see this, note that we can easily find a particular solution of the differential equation using a Green's function. The roots of the characteristic equation $r^2 + 2r + 1 = 0$ are $r_1 = r_2 = -1$. Hence the Green's function is $g(t - u) = (t - u)e^{-(t-u)}$. The solution satisfying $y(0) = y'(0) = 0$ is

$$y_p(t) = \int_0^t (t - u)e^{-(t-u)}2\delta(u - 1)\, du.$$

The integral can be written

$$y_p(t) = \int_0^\infty (t - u)e^{-(t-u)} H_u(t) \cdot 2\delta(u - 1)\, du,$$

since $H_u(t) = 1$ when $u < t$ and 0 otherwise. It follows right away from the definition of δ that

$$y_p(t) = 2(t - 1)e^{-(t-1)}H_1(t).$$

To finish the problem, we would still have to use the general homogeneous solution, $c_1e^{-t} + c_2te^{-t}$, to satisfy the initial conditions.

EXERCISES

1. Sketch the graphs of $\Delta_1(t)$, $\Delta_1(t - 2)$, and $H_2(t - 4)$.

2. Show that if $\Delta_h(t) = 1/h$ for $0 \le t < h$ and $\Delta_h(t) = 0$ otherwise, then

$$\Delta_h(t) = \frac{H_0(t) - H_h(t)}{h}.$$

   Conclude then that

$$\Delta_h(t - t_0) = \frac{H_0(t - t_0) + H_h(t - t_0)}{h}.$$

3. Consider the convolution integral with $g(u) = 0$ if $u < 0$:

$$\Delta_h * g(t) = \int_0^t \delta_h(u)g(t - u)\, du.$$

   (a) By making simple changes of variable in the integral, show that this convolution can be written

$$\Delta_h * g(t) = \frac{1}{h} \int_{t-h}^h g(u)\, du, \qquad \text{if } t \ge h \ge 0.$$

   (b) Show that the convolution can also be written

$$\Delta_h * g(t) = \frac{1}{h} \int_0^t g(u)\, du, \qquad \text{if } h \ge t \ge 0.$$

4. (a) Solve $y' = \Delta_h(t - t_0)$ under the initial condition $y(0) = 0$.

   (b) Sketch the graph of the solution $y_p(t)$ found in part (a) when $t_0 = 1$.

   (c) Find $\lim_{h \to 0+} y_p(t)$, and compare the result with $H_{t_0}(t)$.

5. Express the solution of the initial-value problem

$$y'' = \delta_{t_0}(t), \qquad y(0) = y'(0) = 0,$$

in terms of $H_{t_0}(t)$ if $t_0 > 0$.

6. (a) Solve the initial-value problems

$$y'' + y = \delta_\pi(t), \qquad y(0) = y'(0) = 0,$$

and
$$y'' + y = \delta_{2\pi}(t), \qquad y(0) = y'(0) = 0.$$

(b) Use the results of part (a) to solve

$$y'' + y = c\delta_\pi(t) + d\delta_{2\pi}(t), \qquad y(0) = y'(0) = 0, \qquad \text{where } c, d \text{ const.}$$

(c) Show that if $c = d$ in part (b), then the solution is not only 0 for $0 < t < \pi$ but also for $2\pi < t$.

7. Solve the following initial-value problems using transforms or Green's functions.

(a) $y'' + y' = \delta(t - 1), y(0) = 0, y'(0) = 1$.

(b) $y'' + y' = \delta(t - 1), y(0) = 1, y'(0) = 1$.

(c) $y'' - y = 2\delta(t - 2), y(0) = 0, y'(0) = 0$.

(d) $y'' + 4y = \delta(t - \pi), y(0) = 0, y'(0) = 0$.

8. Verify that

$$\mathcal{L}[\delta(t - t_0)f(t)] = e^{-t_0 s}f(t_0).$$

Then use this result to solve

$$y'' + 2y' + y = \delta(t - 1)t, \qquad y(0) = 0, \qquad y'(0) = 0.$$

9. The Heaviside function H_a is useful for representing functions that are defined piecewise on intervals. In (a) through (e) sketch the graphs of the function $f(t)$. In (f) through (i) sketch the graph of $g(t)$ and then write the function as sum of multiples of Heaviside functions H_a, using coefficient multipliers that may depend on t.

(a) $f(t) = H_1(t) - H_2(t)$.

(b) $f(t) = H_0(t) + 2H_1(t)$.

(c) $f(t) = H_0(t) + H_1(t) - 2H_2(t)$.

(d) $f(t) = t[H_0(t) - H_1(t)] + H_1(t)$.

(e) $f(t) = t[H_0(t) + H_1(t)] - t[H_1(t) - H_2(t)]$.

(f) $g(t) = \begin{cases} 1, & 1 \le t < 2, \\ 0, & \text{elsewhere.} \end{cases}$

(h) $g(t) = \begin{cases} t, & 0 \le t < 2, \\ 4 - t, & 2 \le t < 4, \\ 0, & \text{elsewhere.} \end{cases}$

(g) $g(t) = \begin{cases} 1, & 1 \le t < 2, \\ -2, & 2 \le t < 3, \\ 0, & \text{elsewhere.} \end{cases}$

(i) $g(t) = \begin{cases} t, & 0 \le t < 2, \\ t - 2, & 2 \le t < 4, \\ 0, & \text{elsewhere.} \end{cases}$

10. (a) Derive $\mathcal{L}[H_a(t)](s) = s^{-1}e^{-as}$, $a > 0$.

(b) Derive $\mathcal{L}[tH_a(t)](s) = (1 + as)s^{-2}e^{-as}$, $a > 0$.

11. Use the formulas in Exercise 10 to compute the Laplace transforms of the functions in Exercise 9.

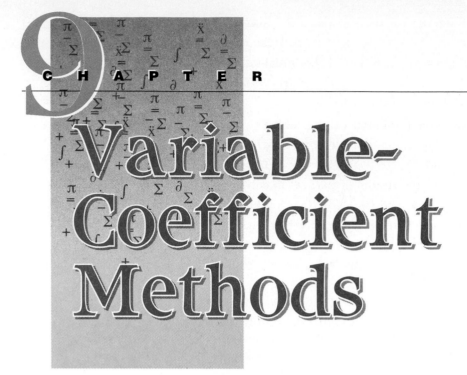

9 CHAPTER

Variable-Coefficient Methods

The discussion in Chapter 3 shows that if $f(x)$ is a continuous function and a and b are constants, the linear differential equation

$$y'' + ay' + by = f(x)$$

is such that all its solutions can be obtained by specializing the constants c_1 and c_2 in a formula that in general looks like

$$y = c_1 y_1(x) + c_2 y_2(x) + y_p(x).$$

Much of this chapter is devoted to seeing in detail that a similar solution formula is valid even if $a = a(x)$ and $b = b(x)$ are allowed to be continuous functions of x. The most obvious difference will be that we can no longer expect $y_1(x)$ and $y_2(x)$ to have the simple forms $x^l e^{\alpha x} \cos \beta x$ and $x^l e^{\alpha x} \sin \beta x$ that they have when a and b are constant. The search for the analogous solutions to the general linear equation

■ **1.1** $$y'' + a(x)y' + b(x)y = f(x)$$

will first be reduced to finding just one such solution to the corresponding **homogeneous** equation, for which $f(x)$ is assumed to be identically zero.

Section 4 of the present chapter gives some graphic indication of how very different the structure of the family of solutions to a nonlinear second-order equation can be from the straightforward formula given above for a linear equation. In addition there are a number of interesting applications of the equations considered there.

1   INDEPENDENT SOLUTIONS

Recall that two functions y_1, y_2 defined on the same interval are **linearly independent** if whenever a linear combination $c_1 y_1 + c_2 y_2$ is identically zero, then the constant coefficients c_1 and c_2 must both be zero. It is a routine exercise to show that y_1 and y_2 are linearly independent if and only if neither one is a constant multiple of the other; correspondingly, y_1 and y_2 are **linearly dependent** if one of them *is* a constant multiple of the other, say $y_2 = ky_1$. When y_1 and y_2 are linearly dependent, the general linear combination $c_1 y_1 + c_2 y_2$ then reduces to a multiple of a single function. For example, if $y_2 = ky_1$, then $c_1 y_1 + c_2 y_2 = (c_1 + c_2 k)y_1$.

e x a m p l e 1

A glance at the graphs of $y_1(x) = e^x$ and $y_2(x) = e^{-x}$ in Figure 9.1a shows that neither of these functions is a multiple of the other on any interval, because one of them increases, the other decreases, and both are positive. Hence the two functions are linearly independent. The same is true of any two exponentials $e^{\alpha x}$, $e^{\beta x}$, where $\alpha \neq \beta$. For if $e^{\alpha x} = ce^{\beta x}$, then $e^{(\alpha - \beta)x} = c$. But the exponential is a constant only if $\alpha = \beta$. Hence $e^{\alpha x}$, $e^{\beta x}$ are independent unless they are equal.

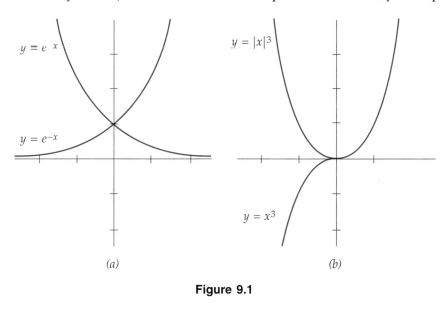

(a)   (b)

Figure 9.1

e x a m p l e 2

The formulas $y_1(x) = x^3$, $y_2(x) = |x|^3$ define independent functions on some intervals and dependent functions on other intervals. On any interval consisting of only positive numbers, the two functions are equal and so are dependent. On a negative interval, $y_2(x) = -y_1(x)$, as shown in Figure 9.1b. This is again a dependence relation. However, over an entire interval $a < x < b$ with $a < 0 < b$, neither function is a unique constant multiple of the other, so the functions thus defined are linearly independent.

It often happens that it is fairly easy to find one **nontrivial** (i.e., not identically zero) solution y_1 of a homogeneous equation

$$y'' + a(x)y' + b(x)y = 0.$$

For example, power series, discussed in Sections 6 and 7, may provide a solution, or we may be fortunate enough to guess one. But then to be able to write the general solution in the form

$$y = c_1 y_1 + c_2 y_2,$$

we need to be able to find a linearly independent solution y_2. It turns out that this can be done just by finding a suitable factor $u(x)$ that makes

$$y_2(x) = y_1(x)u(x)$$

a solution also. The reason the method works is that having one factor as a solution of the differential equation makes several terms drop out when the product $y_1 u$ is substituted into the equation. To illustrate the method most simply, we start with a constant-coefficient equation for which we already know how to write down the general solution.

It is easy to check that

$$y'' - 2y' + y = 0$$

has $y_1(x) = e^x$ for a solution. To find another, try

$$\begin{aligned} y(x) &= y_1(x)u(x) \\ &= e^x u(x). \end{aligned}$$

We find

$$\begin{aligned} y &= e^x u, \\ y' &= e^x u + e^x u', \\ y'' &= e^x u + 2e^x u' + e^x u''. \end{aligned}$$

When we substitute these expressions into the differential equation, we can collect the terms having u as a factor at the beginning to get

$$(e^x u - 2e^x u + e^x u) + (e^x u'' + 2e^x u' - 2e^x u') = 0.$$

The first three terms add up to zero, and the second three boil down to $e^x u''$. We are then left with $e^x u'' = 0$; so divide out e^x, leaving $u'' = 0$. The solution is

$$u(x) = c_1 x + c_2.$$

So, since $y = u y_1$, and $y_1 = e^x$, we get

$$y(x) = y_1(x)(c_1 x + c_2) = (c_1 x + c_2)e^x.$$

This is in fact the general solution, but we can obviously pick out a solution independent of $y_1(x) = e^x$ by setting $c_1 = 1$ and $c_2 = 0$: $y_2(x) = xe^x$. If this method for finding a second independent solution seems shorter than what we did in Chapter 3, it is only because this time we started with one solution. Furthermore, the present method does not by itself prove that we have found the most general solution.

The routine we applied in Example 3 works quite generally, and rather than go through it repeatedly we can condense the general result into a single theorem. Because the technique that leads to the next theorem involves solving only a first-order equation, it is often described as an **order-reduction method**.

■ 1.2 THEOREM

Assume the coefficient functions $a(x)$ and $b(x)$ are continuous on an interval I. If $y_1(x)$ satisfies $y'' + a(x)y' + b(x)y = 0$, and $y_1(x) \neq 0$ on an I, then

$$y_2(x) = y_1(x) \int \frac{\exp\left(-\int a(x)\, dx\right)}{y_1^2(x)}\, dx$$

defines another solution on I, independent of the solution y_1. *Note:* For this to work the differential equation must be normalized so that y'' has coefficient 1.

Proof. We let

$$y = y_1 u,$$
$$y' = y_1' u + y_1 u',$$
$$y'' = y_1'' u + 2y_1' u' + y_1 u''.$$

Substitution into the differential equation allows us to collect the terms containing u at the end:

$$y_1 u'' + (2y_1' + a(x)y_1)u' + (y_1'' + a(x)y_1' + b(x)y_1)u = 0.$$

The factor $(y_1'' + a(x)y_1' + b(x)y_1)$ is zero, because y_1 is a solution of the original equation. The remaining terms form a differential equation of first order for u',

$$(u')' + \left(\frac{2y_1'}{y_1} + a(x)\right)u' = 0,$$

valid on any interval where $y_1(x) \neq 0$. An exponential integrating factor for this first-order linear equation for u' is the exponential of

$$\int \left(\frac{2y_1'}{y_1} + a(x)\right) dx = 2\log|y_1| + \int a(x)\, dx.$$

We find
$$u' = c_1 \exp\left(-2\log|y_1| - \int a(x)\, dx\right)$$
$$= c_1 \frac{\exp\left(-\int a(x)\, dx\right)}{y_1^2}.$$

Then
$$u(x) = c_1 \int \frac{\exp\left(-\int a(x)\, dx\right)}{y_1^2(x)}\, dx + c_2.$$

We take $c_1 = 1$ and $c_2 = 0$ to get the solution $y_2 = y_1 u$ in the simplest form, namely,

$$y_2(x) = y_1(x) \int \frac{\exp\left(-\int a(x)\, dx\right)}{y_1^2(x)}\, dx.$$

This formula defines a solution independent of y_1, because for the integral factor to be a constant, the integrand $y_1^{-2} \exp\left(-\int a\, dx\right)$ would have to be identically zero on the interval in question. But this is impossible, given the nature of the two factors: one is an exponential and the other is assumed to be nonzero on I.   ■

Because of the specific role played by $a(x)$, direct application of Theorem 1.2 should be made only when the differential equation to be solved has its coefficient of y'' **normalized** to be 1.

e x a m p l e 4

This example illustrates how the apparent problem with zero values of y_1 in the denominator can be resolved in practice. The **Bessel equation** of index ν can be written

$$x^2 y'' + x y' + (x^2 - \nu^2) y = 0.$$

Suitably normalized, the equation is

$$y'' + \frac{1}{x} y' + \left(1 - \frac{\nu^2}{x^2}\right) y = 0.$$

When $\nu = \frac{1}{2}$, direct substitution shows that the formula

$$y_1(x) = \frac{\sin x}{\sqrt{x}}, \qquad x > 0,$$

provides a solution. See Exercise 17 for a derivation. Since $a(x) = 1/x$, Theorem 1.2 gives

$$y_2(x) = \frac{\sin x}{\sqrt{x}} \int \frac{x}{\sin^2 x} e^{-\ln x} \, dx$$

$$= \frac{\sin x}{\sqrt{x}} \int \frac{dx}{\sin^2 x}$$

$$= -\frac{\sin x}{\sqrt{x}} \cot x = -\frac{\cos x}{\sqrt{x}}.$$

The solutions y_1 and y_2 are independent, because $\cos x$ and $\sin x$ are independent.

The order-reduction method produces a formula for a second, independent solution of a second-order differential equation if we already have one solution. Finding that first solution may be fairly easy, involving only a little intelligent guessing. At the other extreme, we may begin with no particularly handy formula for a solution and have to rely on an existence theorem to tell us that there is a solution. From there we may proceed to study some particular solution, using a combination of numerical methods and the form of the differential equation to deduce properties of the solution. Once we are comfortable with this first solution, we can then proceed to derive an independent solution by the order-reduction method. The solutions arrived at in this way constitute the great majority of the so-called special functions, also called *higher transcendental functions*, examples of which are the Bessel functions of Section 8.

We've seen by direct computation that for the constant-coefficient case, two independent solutions to a homogeneous second-order equation play an important role in determining the most general solution to the equation. If the coefficients are continuous functions, the same result still holds.

■ 1.3 THEOREM

Let $a(x)$ and $b(x)$ be continuous on an interval I. If y_1 and y_2 are linearly independent solutions of the homogeneous equation $y'' + a(x)y' + b(x)y = 0$, the general solution is given by the 2-parameter family $y = c_1 y_1(x) + c_2 y_2(x)$. Arbitrary initial conditions $y(x_0) = y_0$, $y'(x_0) = z_0$ at a point x_0 in I can be satisfied by a unique choice of the constants c_1 and c_2.

Our earlier Theorem 1.2 established only that the homogeneous equation has two independent solutions if it has a nontrivial solution. The proof of Theorem 1.3 depends heavily on proving the existence of one solution, something that we were able to prove by successive integration for the constant-coefficient case in Chapter 3. The argument that replaces this elementary one is beyond the scope of our treatment, and we omit the proof.

EXERCISES

1. Decide whether the following functions are linearly independent on their indicated intervals. Sketch the relevant graphs.

    (a) $\cos x$, $\sin x$; $-\infty < x < \infty$.

    (b) x, x^2; $-\infty < x < \infty$.

    (c) $\tan x$, $\sin x$; $-\pi/2 < x < \pi/2$.

    (d) x^5, $|x|^5$; $-\infty < x < \infty$.

    (e) x^5, $|x|^5$; $-\infty < x < 0$.

    (f) e^x, xe^{2x}; $-\infty < x < \infty$.

2. Two functions $y_1(x)$ and $y_2(x)$ are **linearly independent** on an x interval if the identity $c_1 y_1(x) + c_2 y_2(x) = 0$ holds only for coefficients $c_1 = c_2 = 0$. Show that y_1 and y_2 are linearly independent on an interval if and only if neither one is a constant multiple of the other.

3. Each of the following equations has the given function as a solution. Verify this, and find a linearly independent solution using Theorem 1.2.

    (a) $y'' - 4y' + 4y = 0$; $y_1(x) = e^{2x}$.

    (b) $xy'' - (2x + 1)y' + (x + 1)y = 0$; $y_1(x) = e^x$.

    (c) $(x - 1)y'' - xy' + y = 0$; $y_1(x) = e^x$.

    (d) $x^2 y'' - 2xy' + 2y = 0$; $y_1(x) = x$, $x > 0$.

    (e) $(x^2 - 1)y'' - 2xy' + 2y = 0$; $y_1(x) = x$.

4. Each of the following equations is of the kind that can be treated using the order-reduction method of this section, but because the term in y is already absent, it is simpler to solve each equation directly by treating it as a first-order equation in the unknown function y', and then find y by integration. For each equation, find the general solution in the form $c_1 y_1 + c_2 y_2$. Use either method.

    (a) $y'' + xy' = 0$.

    (b) $y'' + \dfrac{1}{x} y' = 0$.

5. Use Theorem 1.2 to derive a solution independent of $y_1(x)$ for each equation, assuming $\alpha \neq 0$ is constant.

    (a) $y'' - 2\alpha y' + \alpha^2 y = 0$; $y_1(x) = e^{\alpha x}$.

    (b) $y'' + \alpha^2 y = 0$; $y_1(x) = \cos \alpha x$.

6. The **Legendre equation** of index 1 is

$$(1 - x^2)y'' - 2xy' + 2y = 0$$

    and has a solution $y_1(x) = x$.

    (a) Use Theorem 1.2 to find an independent solution in terms of elementary functions.

(b) Find the solution $y(x)$ to the Legendre equation such that $y(0) = 1$ and $y'(0) = 2$.

7. The **Hermite equation** of index 1 is

$$y'' - 2xy' + 2y = 0$$

and has a solution $y_1(x) = x$. Find an independent solution in power series form by using the expansion $e^x = \sum_{k=0}^{\infty} x^k/k!$ and integrating term by term.

8. Let $L(y) = x^2 y'' - 2xy' + 2y$. Verify that $y_1(x) = x$ is a solution of $L(y) = 0$. Then find a solution of the nonhomogeneous equation $L(y) = x^2$ in the form $y(x) = y_1(x)u(x) = xu(x)$. [*Hint:* Substitute $y = xu(x)$ into the equation to be solved and then solve for u.]

9. Theorem 1.2 contains the formula $e^{-\int a(x)\,dx}$ which is reminiscent of the integrating factor used to solve first-order linear equations. Here is a mathematical connection. Suppose y_1 and y_2 are solutions of the same homogeneous equation:

$$y_1'' + a(x)y_1' + b(x)y_1 = 0,$$
$$y_2'' + a(x)y_2' + b(x)y_2 = 0.$$

(a) Multiply the first equation by y_2, the second by y_1, and subtract to show that

$$w = y_1 y_2' - y_2 y_1' \qquad \text{satisfies} \qquad w' + a(x)w = 0.$$

(b) Solve the first-order equation in part (a) to get

$$y_1 y_2' - y_2 y_1' = ce^{-\int a(x)\,dx}.$$

This result is called **Abel's formula**.

(c) Show that Abel's formula can be written using determinants as

$$\begin{vmatrix} y_1(x) & y_2(x) \\ y_1'(x) & y_2'(x) \end{vmatrix} = ce^{-\int a(x)\,dx}.$$

The determinant is called the **Wronskian determinant** of y_1 and y_2.

(d) Show that the Wronskian of solutions y_1, y_2 of $y'' + b(x)y = 0$ is constant. (This case is less special than it looks; see Exercise 17.)

(e) Show that the Wronskian of the two solutions y_1, y_2 is either identically zero or else never zero.

10. Verify that, if $L = D^2 + a(x)D + b(x)$ on some interval $x_1 < x < x_2$, then L acts linearly on twice-differentiable functions y_1, y_2 on that interval. That is, show that

$$L(c_1 y_1 + c_2 y_2) = c_1 L(y_1) + c_2 L(y_2), \qquad c_1, c_2 \text{ const.}$$

11. **(a)** Show that the **Euler differential equation** (see also Section 7A)

$$x^2 y'' + axy' + by = 0, \qquad a, b \text{ real const,}$$

can be solved as follows, assuming $x > 0$. Let $y = x^\mu$, so that $y' = \mu x^{\mu-1}$, and $y'' = \mu(\mu - 1)x^{\mu-2}$. Show that for y to solve the differential equation, μ must satisfy the **indicial equation** $\mu^2 + (a - 1)\mu + b = 0$.

(b) Show that if the indicial equation has real roots $\mu_1 \neq \mu_2$, then $y_1 = x^{\mu_1}$, $y_2 = x^{\mu_2}$ are solutions.

(c) Show that if the indicial equation has complex conjugate roots $\mu_1 = \alpha + i\beta$, $\mu_2 = \alpha - i\beta$, then $y_1 = x^\alpha \cos(\beta \ln x)$, $y_2 = x^\alpha \sin(\beta \ln x)$ are solutions. Note that, by definition, $x^{\alpha+i\beta} = x^\alpha e^{i\beta \ln x}$ for $x > 0$.

(d) Show that if μ_1 is a double root of the indicial equation, then $y_1 = x^{\mu_1}$, $y_2 = x^{\mu_1} \ln x$ are solutions. Use Theorem 1.2 for this.

12. Use the results of the previous exercise to solve the following Euler equations.

(a) $x^2 y'' + xy' - y = 0$.   **(c)** $x^2 y'' + 3xy' + y = 0$.

(b) $x^2 y'' + 4xy' + y = 0$.   **(d)** $x^2 y'' + xy' + y = 0$.

13. Find independent solutions of the Euler equation of Exercise 11 for $x > 0$ as follows.

(a) Let $t = \ln x$ and define $u(t) = y(e^t)$. Use the chain rule to show that $dy/dx = x^{-1}\dot{u}$ and $d^2y/dx^2 = x^{-2}(\ddot{u} - \dot{u})$.

(b) Show that $x^2 y'' + axy' + by = \ddot{u} + (a - 1)\dot{u} + bu$ and hence that the Euler equation has solution $e^{\mu_0 t} = x^{\mu_0}$, where μ_0 is a root of

$$\mu^2 + (a - 1)\mu + b = 0.$$

(c) Use Theorem 1.2 to find a solution independent of x^{μ_0}. Note particularly the case $\mu_0 = \frac{1}{2}(1 - a)$.

14. The following systems have solutions that can be found by first eliminating one of the unknown functions of t. Find the general solution.

(a) $\dot{x} = x - (t - 1)y$, $\dot{y} = 0$.

(b) $\dot{x} = -2t^{-2}y$, $t > 0$, $\dot{y} = x$. [*Hint:* Use Exercise 11.]

15. Let $f(x)$ and $g(x)$ be twice-differentiable functions on an interval $a < x < b$ on which $f(x)g'(x) - f'(x)g(x) \neq 0$.

(a) Show that the 3 by 3 determinant equation

$$\begin{vmatrix} y & f & g \\ y' & f' & g' \\ y'' & f'' & g'' \end{vmatrix} = 0$$

is a second-order homogeneous linear equation on the interval (a, b) having $f(x)$ and $g(x)$ as solutions.

(b) Find a second-order homogeneous linear equation having $f(x) = \sin x$ and $g(x) = x \sin x$ as solutions for $0 < x < \pi$.

(c) Find a second-order homogeneous linear equation having $f(x) = x$ and $g(x) = e^x$ as solutions for all $x \neq 1$.

16. It is sometimes erroneously inferred from circumstantial evidence that the **Wronskian determinant**

$$w[f, g](x) = \begin{vmatrix} f(x) & g(x) \\ f'(x) & g'(x) \end{vmatrix}$$

of two linearly independent functions f and g can't equal 0. Put this idea to test by computing $w[f, g](x)$ for all real x when $f(x) = x^2$, $g(x) = x|x|$. Note that you need to use the definition of derivative as a limit of a difference quotient to compute $g'(0)$.

17. Solving the general second-order equation $y'' + a(x)y' + b(x)y = 0$ can be reduced to the problem of solving $u'' + g(x)u = 0$, with

$$g(x) = -\tfrac{1}{2}a'(x) - \tfrac{1}{4}a^2(x) + b(x) \qquad \text{and} \qquad y(x) = u(x)e^{-(1/2)\int a(x)\,dx}.$$

(a) Verify the assertion above by substituting the expression for $y(x)$ into the general equation. [*Hint:* Denote the exponential factor by $v(x)$ and note that $v' = -\tfrac{1}{2}av$ and $v'' = -\tfrac{1}{2}a'v + \tfrac{1}{4}a^2v$.]

(b) The **Bessel equation** of index $\tfrac{1}{2}$ can be written for $x > 0$ as

$$y'' + \frac{1}{x}y' + \left(1 - \frac{1}{4x^2}\right)y = 0.$$

Show that for this equation $g(x)$ is constant. Then use the method outlined above to find two independent solutions.

2 NONHOMOGENEOUS EQUATIONS

2A *Variation of Parameters*

The technique used in Section 1 to find an independent solution of a homogeneous differential equation can be applied just as well to solving a nonhomogeneous equation as long as we have in hand a single nontrivial solution y_1 of the associated homogeneous equation. Because the multiplicative parameter u in the trial solution

$$y(x) = y_1(x)u(x)$$

is allowed to be a function of x, the procedure is sometimes called **variation of parameters,** although this term is often reserved for a refinement to be described later. The substitution works for the same reason it works for the homogeneous equation: it leaves us with a first-order linear equation that can be solved for u'.

---

e x a m p l e 1

The equation

$$x^2y'' - 2xy' + 2y = x^4$$

has $y_1(x) = x$ for a *homogeneous* solution. Letting $y = y_1(x)u(x) = xu(x)$ we compute

$$y = xu, \qquad y' = xu' + u, \qquad y'' = xu'' + 2u'.$$

Substitution followed by simplification of the differential equation yields

$$x^2(xu'' + 2u') - 2x(xu' + u) + 2xu = x^3u'' = x^4.$$

This last equation is easily solved for u to get $u = \tfrac{1}{6}x^3 + c_1x + c_2$. So our solution is

$$y = y_1(x)u(x) = xu(x)$$
$$= \tfrac{1}{6}x^4 + c_1x^2 + c_2x.$$

Finding the homogeneous solution $y_1(x) = x$ was done by guessing, but lacking a correct guess we could have tried to find a solution in power series form using methods to be described in Sections 6 and 7. (See also Exercise 11 of Section 1.)

The number of integrations required can be reduced to one if we already know two independent solutions y_1, y_2 of the associated homogeneous equation. The routine is complicated enough that rather than repeat it for every application we'll standardize it, along with the final result, as follows. Thus suppose $a(x)$, $b(x)$ are continuous on some interval and that $f(x)$ is continuous. Then consider the **normalized equation**

$$y'' + a(x)y' + b(x)y = f(x),$$

with coefficient of y'' equal to 1. If y_1, y_2 are homogeneous solutions, we form the linear combination

$$y(x) = y_1(x)u_1(x) + y_2(x)u_2(x),$$

where $u_1(x)$ and $u_2(x)$ are to be determined so that $y(x)$ is a solution of the nonhomogeneous equation. (Note that if u_1 and u_2 were constant, $y(x)$ would be a solution of the associated homogeneous equation.) What we do now is much like what we did in Example 1: compute the derivatives y', y'' and substitute into the nonhomogeneous equation. Then rearrange the terms as follows:

$$(y_1'' + ay_1' + by_1)u_1 + (y_2'' + ay_2' + by_2)u_2 + (y_1u_1' + y_2u_2')'$$
$$+ a(y_1u_1' + y_2u_2') + (y_1'u_1' + y_2'u_2') = f.$$

The first two collections of terms are zero, because y_1 and y_2 are homogeneous solutions. What remains of the equation will be satisfied if we can choose u_1 and u_2 so that the two equations

■ 2.1
$$y_1u_1' + y_2u_2' = 0,$$
$$y_1'u_1' + y_2'u_2' = f$$

hold identically for all x on the interval in question. This little system of equations can be solved for u_1' and u_2' by elimination, with the result that

$$u_1'(x) = \frac{-y_2(x)f(x)}{y_1(x)y_2'(x) - y_2(x)y_1'(x)},$$

$$u_2'(x) = \frac{y_1(x)f(x)}{y_1(x)y_2'(x) - y_2(x)y_1'(x)}.$$

The expression in the denominators is the same in both formulas and can be written as the 2 by 2 determinant

$$w(x) = \begin{vmatrix} y_1(t) & y_2(x) \\ y_1'(x) & y_2'(x) \end{vmatrix} = y_1(x)y_2'(x) - y_2(x)y_1'(x),$$

called the **Wronskian determinant** of y_1, y_2. It is shown in the proof of Theorem 2.6 in Section 2C that $w(x)$ is never zero if y_1, y_2 are linearly independent solutions, so the examples we consider here will have that property. To complete the solution, integrate the formulas for u_1', u_2' to find u_1 and u_2, and then combine with y_1, y_2 to get a particular solution

■ 2.2 $y_p(x) = y_1(x)u_1(x) + y_2(x)u_2(x),$

$$= y_1(x) \int \frac{-y_2(x)f(x)}{w(x)} \, dx + y_2(x) \int \frac{y_1(x)f(x)}{w(x)} \, dx.$$

Because Equations 2.1 are easier to remember than Equation 2.2, people often prefer to start with them in each problem and carry out the rest of the computation to arrive at Equation 2.2. The next example will be done that way. Because of the way that f enters Equations 2.1 and 2.2, the equation to be solved must be in normalized form to make these formulas valid.

---

e x a m p l e **2**

The equation $x^2y'' - 2xy' + 2y = x^3$ is normalized, for example, for positive x, to

$$y'' - \frac{2}{x}y' + \frac{2}{x^2}y = x, \qquad x > 0,$$

and has homogeneous solutions $y_1(x) = x$, $y_2(x) = x^2$. The latter can be discovered by the order-reduction method of Section 1. For this example, Equations 2.1 are

$$xu_1' + x^2u_2' = 0,$$
$$u_1' + 2xu_2' = x.$$

Multiplying the second equation by x and then subtracting the first equation from it gives $x^2u_2' = x^2$, or $u_2' = 1$. It then follows from the first equation that $u_1' = -x$. Integrating to find u_1, u_2 gives

$$u_1(x) = -\tfrac{1}{2}x^2, \qquad u_2(x) = x.$$

A particular solution is

$$y_p = y_1u_1 + y_2u_2$$
$$= x \cdot (-\tfrac{1}{2}x^2) + x^2 \cdot x = \tfrac{1}{2}x^3.$$

Adding constants of integration to u_1 and u_2 would only add a linear combination of homogeneous solutions to y_p. In any case, we have the general solution

$$y = c_1x + c_2x^2 + \tfrac{1}{2}x^3.$$

We will now solve a generalization of Example 1 but using Equation 2.2 instead of Equation 2.1.

---

e x a m p l e **3**

In normalized form, we consider

$$y'' - \frac{2}{x}y' + \frac{2}{x^2}y = f(x), \qquad x > 0.$$

Homogeneous solutions are $y_1(x) = x$, $y_2(x) = x^2$. The Wronskian determinant of y_1, y_2 is

$$w(x) = \begin{vmatrix} x & x^2 \\ 1 & 2x \end{vmatrix} = 2x^2 - x^2 = x^2.$$

Equation 2.2 reduces to

$$y_p(x) = x \int \frac{-x^2f(x)}{x^2}\,dx + x^2 \int \frac{xf(x)}{x^2}\,dx$$
$$= -x \int f(x)\,dx + x^2 \int \frac{f(x)}{x}\,dx.$$

To make the integration fairly easy, we can use the example $f(x) = x \cos x$. For this choice, we use integration by parts to get

$$y_p = -x \int x \cos x \, dx + x^2 \int \cos x \, dx$$

$$= -x\left(x \sin x - \int \sin x \, dx\right) + x^2 \int \cos x \, dx$$

$$= -x(x \sin x + \cos x) + x^2 \sin x$$

$$= -x \cos x.$$

2B   Green's Functions

A slight modification of Equation 2.2 gives us the analog of the Green's function formula that we found in Chapter 3 for the initial-value problem $y(x_0) = y'(x_0) = 0$ associated with a constant-coefficient operator. All we need to do to get this representation is to convert the integrals in Equation 2.2 into definite integrals over the interval from x_0 to x:

■ **2.3** $\displaystyle y_p(x) = y_1(x) \int_{x_0}^{x} \frac{-y_2(t)f(t)}{w(t)} \, dt + y_2(x) \int_{x_0}^{x} \frac{y_1(t)f(t)}{w(t)} \, dt,$

$$-\int_{x_0}^{x} \frac{y_1(t)y_2(x) - y_2(t)y_1(x)}{w(t)} f(t) \, dt,$$

where $w(t) = y_1(t)y_2'(t) - y_2(t)y_1'(t)$ is the Wronskian determinant of $y_1(t)$, $y_2(t)$.

Setting $x = x_0$ in Equation 2.3 gives $y_p(x_0) = 0$. It is left as an exercise to show that $y_p'(x_0) = 0$ also. Thus the **Green's function**

$$G(x, t) = \frac{y_1(t)y_2(x) - y_2(t)y_1(x)}{w(t)}$$

generates the solution

$$y_p(x) = \int_{x_0}^{x} G(x, t)f(t) \, dt$$

to the initial-value problem

$$y'' + a(x)y' + b(x)y = f(x), \qquad y(x_0) = y'(x_0) = 0.$$

Quite apart from the neatness with which Equation 2.3 displays a solution, it is convenient for calculating solutions when the forcing function $f(x)$ is discontinuous.

e x a m p l e 4   The equation

$$y'' - \frac{2x}{x^2 - 1} y' + \frac{2}{x^2 - 1} y = f(x)$$

 has homogeneous solutions $y_1(x) = x$, $y_2(x) = x^2 + 1$. Their Wronskian is

$$w(x) = \begin{vmatrix} x & x^2 + 1 \\ 1 & 2x \end{vmatrix} = x^2 - 1.$$

The Green's function for an initial-value problem is then

$$G(x, t) = \frac{t(x^2 + 1) - (t^2 + 1)x}{t^2 - 1}.$$

Suppose we want the solution with forcing function

$$f(x) = \begin{cases} 0, & -1 < x < 0, \\ 1, & 0 \le x \le \frac{1}{2}, \\ 0, & \frac{1}{2} < x < 1, \end{cases}$$

and satisfying $y(0) = y'(0) = 0$. This is

$$y_p(x) = \int_0^x G(x, t) f(t) \, dt$$

$$= \begin{cases} 0, & -1 < x < 0, \\ \int_0^x G(x, t) \, dt, & 0 \le x \le \frac{1}{2}, \\ \int_0^{1/2} G(x, t) \, dt, & \frac{1}{2} < x < 1. \end{cases}$$

We compute, for $0 \le x < 1$,

$$\int_0^x G(x, t) \, dt = (x^2 + 1) \int_0^x \frac{t}{t^2 - 1} \, dt - x \int_0^x \frac{t^2 + 1}{t^2 - 1} \, dt$$

$$= \tfrac{1}{2}(x^2 + 1) \ln (1 - x^2) - x\left(x + \ln \frac{1 - x}{1 + x}\right).$$

The complete solution is then

$$y_p(x) = \begin{cases} 0 & -1 < x < 0, \\ \tfrac{1}{2}(x^2 + 1) \ln (1 - x^2) - x \ln \frac{1 - x}{1 + x} - x^2, & 0 \le x \le \frac{1}{2}, \\ \tfrac{1}{2} (\ln \tfrac{3}{4})(x^2 + 1) - (\tfrac{1}{2} - \ln 3)x, & \frac{1}{2} < x < 1. \end{cases}$$

The first and third lines are of course solutions of the homogeneous equation on their respective intervals, because $f(x) = 0$ there. Finding the solution that satisfies more general initial conditions, $y(0) = y_0$, $y'(0) = z_0$, is just a matter of solving for c_1 and c_2 using

$$y(x) = c_1 x + c_2(x^2 + 1) + y_p(x),$$

and

$$y'(x) = c_1 + 2c_2 x + y_p'(x).$$

But $y_p(0) = y_p'(0) = 0$, so we find c_1, c_2 from equations $c_2 = y_0$, $c_1 = z_0$.

2C   *Summary*

This section and the previous one show how, given one solution to the homogeneous equation $y'' + a(x)y' + b(x)y = 0$, we can find solution formulas of the form

▬ 2.4 $$y(x) = c_1 y_1(x) + c_2 y_2(x) + y_p(x)$$

for differential equations of the form

▬ 2.5 $$y'' + a(x)y' + b(x)y = f(x).$$

We already proved in Chapter 3 that if $a(x)$ and $b(x)$ are constant, then Equation 2.5 has its most general solution of the form 2.4. Furthermore, we saw there that by choosing c_1 and c_2 properly we could satisfy arbitrary initial conditions of the form $y(x_0) = y_0$, $y'(x_0) = z_0$. This last possibility followed simply from the meaning of c_1 and c_2 as arbitrary constants of integration. If $a(x)$ and $b(x)$ are continuous functions, the analogous theorem still holds. The general results can be summarized as follows.

■ 2.6 THEOREM

Let $a(x)$, $b(x)$, and $f(x)$ be continuous on an interval $a < x < b$. Then the general solution of $y'' + a(x)y' + b(x)y = f(x)$ can be written in the form

$$y = c_1 y_1 + c_2 y_2 + y_p,$$

where y_1 and y_2 are independent solutions of the associated homogeneous equation and y_p is some particular solution of the nonhomogeneous equation. Every pair of initial conditions $y(x_0) = y_0$, $y'(x_0) = z_0$ can be satisfied by a unique choice of c_1 and c_2.

Proof. We need to appeal to Theorem 1.3 of Section 1 for the existence and form of the general solution to the homogeneous equation. Theorem 3.1 of Chapter 3, Section 3, proved there for an arbitrary linear equation, guarantees the form of the solution to the nonhomogeneous equation with y_p given by Equation 2.3 or by variation of parameters. To prove that the Wronskian $w(x)$ appearing there is never zero, suppose on the contrary that for some x_0 in the interval we have $w(x_0) = 0$. But $w(x_0)$ is the determinant of the system

$$c_1 y_1(x_0) + c_2 y_2(x_0) = 0,$$
$$c_1 y_1'(x_0) + c_2 y_2'(x_0) = 0.$$

We know from Theorem 1.3 that the system has a unique solution for c_1 and c_2. Hence the determinant $w(x_0)$ can't be zero. Initial conditions for the nonhomogeneous equation can be written

$$c_1 y_1(x_0) + c_2 y_2(x_0) = y_0 - y_p(x_0),$$
$$c_1 y_1'(x_0) + c_2 y_2'(x_0) = z_0 - y_p'(x_0).$$

Since the determinant $w(x_0) \neq 0$, there is a unique solution for c_1 and c_2 here also. ■

e x a m p l e **5**

It's easy to verify that $x^2 y'' - 2xy' + 2y = 0$ has independent solutions $y_1(x) = x$, $y_2(x) = x^2$. The Wronskian is $w(x) = x^2 \neq 0$ except at $x = 0$. The Green's function is $G(x, t) = (x^2/t) - x$. We have the general solution to the normalized nonhomogeneous equation $y'' - (2/x)y' + (2/x^2)y = f(x)$ given by

$$y(x) = c_1 x + c_2 x^2 + \int_{x_0}^x \left(\frac{x^2}{t} - x \right) f(t)\, dt, \qquad x_0 \neq 0.$$

It's left as an exercise to show, for instance, that with $f(x) = x$ the integral term works out to 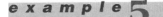 $y_p(x) = \frac{1}{2}x^3 - x_0 x^2 + \frac{1}{2}x_0^2 x$. The last two terms can be combined with the homogeneous solution to give $y(x) = c_1 x + c_2 x^2 + \frac{1}{2}x^3$.

e x a m p l e 6

The equation $(x - 1)y'' - xy' + y = 1$ has a particular solution $y_p(x) = 1$, and the associated homogeneous equation has independent solutions $y_1(x) = x$ and $y_2(x) = e^x$. Initial conditions $y(x_0) = y_0$, $y'(x_0) = z_0$ are satisfied by solving for c_1, c_2 in

$$y_0 = c_1 y_1(x_0) + c_2 y_2(x_0) + 1,$$
$$z_0 = c_1 y_1'(x_0) + c_2 y_2'(x_0).$$

If $x_0 = 0$, it turns out that $c_1 = z_0 - y_0 + 1$ and $c_2 = y_0 - 1$. However, Theorem 2.6 fails to apply at $x_0 = 1$, because the coefficient $x - 1$ of y'' is zero there. You are asked to show in Exercise 10 that the initial-value problem at $x_0 = 1$ has a solution only if $z_0 = y_0 - 1$ and that in that case the solution is not unique.

EXERCISES

1. For each of the following differential equations, try to find or guess a solution y_1 of the associated homogeneous equation. Then determine $u(x)$ so that $y(x) = y_1(x)u(x)$ is the general solution of the given equation.

   (a) $y'' - 4y' + 4y = e^x$.

   (b) $y'' + (1/x)y' = x$, $x > 0$.

   (c) $x^2 y'' - 3xy' + 3y = x^4$, $x > 0$. [*Hint:* Try $y_1 = x^n$, for some n].

   (d) $xy'' - (2x + 1)y' + (x + 1)y = 3x^2 e^x$, $x > 0$. [*Hint:* Try $y_1(x) = e^{rx}$.]

2. Find a particular solution y_p by solving Equation 2.1 for $u_1(x)$, $u_2(x)$ to get $y_p(x) = y_1(x)u_1(x) + y_2(x)u_2(x)$. If suitable y_1 and y_2 are not given, find them first. Then write down the general solution.

   (a) $y'' + y' - 2y = e^{2x}$.

   (b) $y'' + y = \tan x$, $-\pi/2 < x < \pi/2$.

   (c) $y'' + y = \sec x$, $-\pi/2 < x < \pi/2$.

   (d) $y'' - y = xe^x$.

   (e) $x^2 y'' - 2xy' + 2y = 1$; $y_1(x) = x$, $y_2(x) = x^2$, $x > 0$.

3. Use Equation 2.2 to find a formula for a solution $y_p(x)$ to each of the following equations, with homogeneous solution $y_1(x)$ given in some cases. Don't forget to make sure that the equation is normalized.

   (a) $y'' - 2y' + y = e^x$.   **(d)** $2y'' + 8y = e^x$.

   (b) $y'' + 3y' + 2y = 1 + e^x$.   **(e)** $xy'' - y' = x^2$; $y_1(x) = 1$.

   (c) $y'' + 3y' + 2y = (1 + e^x)^{-1}$.   **(f)** $x^2 y'' - 2xy' + 2y = x^4$; $y_1(x) = x$.

4. Find the Green's function $G(x, t)$ for an initial-value problem for each of the following equations. Then express the solution y_p satisfying $y_p(x_0) = y_p'(x_0) = 0$ in terms of $g(x)$. Be sure to normalize $g(x)$.

   (a) $y'' + 3y' + 2y = g(x)$.   **(c)** $y'' + (1/x)y' = g(x)$, $x > 0$.

   (b) $2y'' + 4y = g(x)$.   **(d)** $x^2 y'' - 2xy' + 2y = g(x)$.

[*Hints:* Part (c) is first-order in y'; part (d) has homogeneous solutions x^n for some n.]

5. Using the same equations as in Exercise 4 and the solutions of the initial-value problems found there, solve the following corresponding initial-value problems with the given choice for $g(x)$.

   (a) $g(x) = \begin{cases} 0, & x < 1, \\ 1, & 1 \le x; \end{cases} y(0) = 1, \ y'(0) = 2.$

   (b) $g(x) = \begin{cases} 1, & x < 1, \\ 0, & 1 \le x; \end{cases} y(0) = -1, \ y'(0) = 1.$

   (c) $g(x) = 1/x; \ y(1) = 0, \ y'(1) = 2.$

   (d) $g(x) = \begin{cases} 0, & 0 < x < 2, \\ 3, & 2 \le x \le 4, \ y(3) = 0, \ y'(3) = 0, \\ 1, & 4 < x. \end{cases}$

6. Show that the derivative of the Green's function formula 2.3 for $y_p(x)$ can be written

$$y_p'(x) = \int_{x_0}^x \frac{y_1(t)y_2'(x) - y_2(t)y_1'(x)}{w(t)} f(t) \, dt.$$

Show then that $y_p'(x_0) = 0$. [*Hint:* Separate the integral in Equation 2.3 into two integrals, and then apply the product rule for differentiation.]

7. The **Leibnitz rule** for differentiating an integral states that

$$\frac{d}{dx} \int_{a(x)}^{b(x)} F(x, t) \, dt = \int_{a(x)}^{b(x)} \frac{\partial F}{\partial x}(x, t) \, dt + b'(x)F(x, b(x)) - a'(x)F(x, a(x)),$$

if $\partial F/\partial x$ is continuous in x and integrable in t. Use this result to establish the formula for $y_p'(x)$ in Exercise 6.

8. The constant-coefficient equation $y'' + ay' + by = f(x)$ has homogeneous solutions $y_1(x) = e^{r_1 x}$, $y_2(x) = e^{r_2 x}$. These solutions are independent if $r_1 \ne r_2$; otherwise we consider $y_1(x) = e^{r_1 x}$, $y_2(x) = xe^{r_1 x}$. Find the Green's function for the equation in the following cases by using the approach of the present section.

   (a) $r_1 \ne r_2$.   **(b)** $r_1 = r_2$.

   (c) $r_1 = \alpha + i\beta, \ r_2 = \alpha - i\beta, \ \beta \ne 0$.

9. Do the calculation of the Green's function integral in Example 5 of the text for the following choices:

   (a) $f(x) = x$.   **(b)** $f(x) = x^2$.   **(c)** $f(x) = 1/x$.

10. Consider the differential equation $(x - 1)y'' - xy' + y = 1$ of Example 6 of the text.

    (a) Show that the initial-value problem $y(1) = y_0$, $y'(1) = z_0$ has a solution only if $z_0 = y_0 - 1$, and that in that case there are infinitely many solutions. [*Hint:* x and e^x are homogeneous solutions.]

    (b) Explain why the results of part (a) don't contradict Theorem 2.6.

3   BOUNDARY PROBLEMS

3A   *Two-Point Boundaries*

Initial-value problems for second-order linear differential equations are those in which we specify for a solution $y = y(x)$ the values $y(a)$ and $y'(a)$ at a single point $x = a$. We have seen that, under reasonable hypotheses, such problems always have a unique solution. If instead we try to specify the solution values $y(a)$ and $y(b)$ at two distinct points $x = a$ and $x = b$, then the variety of outcomes if we ask about existence of a solution is much wider: There may be a unique solution, there may be infinitely many solutions, or there may be no solution at all. It is customary to call the search for a solution with values given at two points a **boundary-value problem**. The problem we're considering has an important physical interpretation for the oscillator equation

$$\ddot{y} = -k(t)\dot{y} - h(t)y$$

with specified boundary values $y(t_0) = \alpha$ and $y(t_1) = \beta$. The solutions $y(t)$ of such an equation often represent the displacements of an oscillating mechanism as a function of time t. Our question then becomes the following: Over the time interval from t_0 to t_1 is it possible to go from displacement $y = \alpha$ to displacement $y = \beta$, and if so, is there a unique way to do this? The next example illustrates the various possibilities.

---

example 1

The differential equation $y'' + y = 0$ has the general solution

$$y(x) = c_1 \cos x + c_2 \sin x.$$

To satisfy the boundary conditions

$$y(0) = 1, \qquad y\left(\frac{\pi}{2}\right) = 2,$$

we must have

$$y(0) = c_1 = 1, \qquad y\left(\frac{\pi}{2}\right) = c_2 = 2.$$

Thus c_1 and c_2 are uniquely determined, and the resulting solution is

$$y(x) = \cos x + 2 \sin x.$$

The graph of the solution is shown in Figure 9.2a.

If we keep the same differential equation, but change the boundary conditions to

$$y(0) = 1, \qquad y(2\pi) = 1,$$

then the constants c_1 and c_2 in the general solution are subject to the requirement

$$y(0) = c_1 = 1, \qquad y(2\pi) = c_1 = 1,$$

while c_2 can evidently be chosen arbitrarily. The resulting 1-parameter family of solutions is infinite and is made up of all functions of the form

$$y(x) = \cos x + c_2 \sin x.$$

The graphs of three of these solutions are shown in Figure 9.2b.

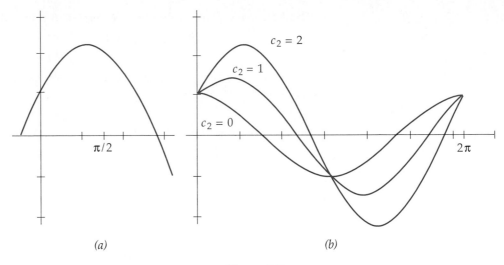

(a)   *(b)*

Figure 9.2

Still keeping the same differential equation, we can change the boundary conditions to

$$y(0) = 1, \qquad y(\pi) = 1.$$

These conditions require of c_1 that

$$y(0) = c_1 = 1, \qquad y(\pi) = -c_1 = 1.$$

Since we cannot have both $c_1 = 1$ and $c_1 = -1$, it follows that there is no solution satisfying this third set of conditions.

It turns out that the case of nonexistence and the case of multiple existence can together be characterized for a linear differential equation in terms of the equation and the boundary points, the boundary values themselves being irrelevant. The statement is as follows:

■ 3.1 THEOREM

Let the linear homogeneous differential equation

$$y'' + p(x)y' + q(x)y = 0$$

have continuous coefficients $p(x)$ and $q(x)$ on the interval $a \le x \le b$, and let α and β be arbitrary numbers. Then the differential equation has a unique solution satisfying both $y(a) = \alpha$ and $y(b) = \beta$ if and only if either one or the other of the following equivalent statements holds:

(a) The only solution satisfying $y(a) = y(b) = 0$ is the identically zero solution.

(b) For two independent solutions, we have the 2 by 2 determinant condition

$$y_1(a)y_2(b) - y_2(a)y_1(b) = \begin{vmatrix} y_1(a) & y_2(b) \\ y_1(b) & y_2(b) \end{vmatrix} \neq 0.$$

Proof. Let the general solution of the differential equation be

$$y(x) = c_1 y_1(x) + c_2 y_2(x).$$

The boundary conditions $y(a) = \alpha$, $y(b) = \beta$ amount to

$$c_1 y_1(a) + c_2 y_2(a) = \alpha,$$
$$c_1 y_1(b) + c_2 y_2(b) = \beta.$$

It is easy to show (see Exercise 12) that these equations have a unique solution for c_1 and c_2 if and only if

$$y_1(a) y_2(b) - y_2(a) y_1(b) \neq 0.$$

But the condition that the determinant be nonzero is precisely the condition that the equations

$$c_1 y_1(a) + c_2 y_2(a) = 0,$$
$$c_1 y_1(b) + c_2 y_2(b) = 0$$

have only the trivial solution $c_1 = c_2 = 0$. ∎

Theorem 3.1 shows that, given the differential equation, the existence of a unique boundary-value solution depends on the location of the boundary points. The term **conjugate points** is used to denote a pair of boundary points for which there *fails* to exist a unique solution. Thus a pair of points is conjugate relative to a homogeneous differential equation if there is a nontrivial solution that is zero at the two points. From the statement of the theorem, we see that, equivalently, a pair of points a and b is conjugate if and only if, for two linearly independent solutions y_1 and y_2, we have the relation

$$\begin{vmatrix} y_1(a) & y_2(a) \\ y_1(b) & y_2(b) \end{vmatrix} = 0.$$

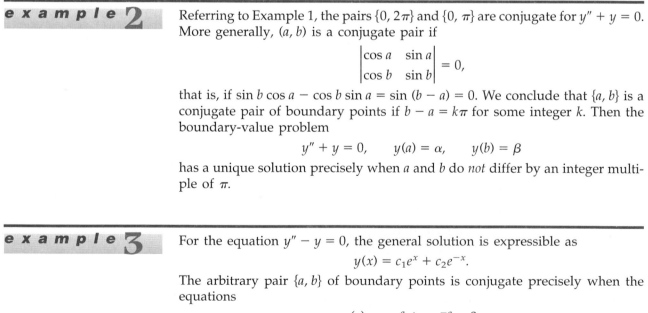

e x a m p l e 2

Referring to Example 1, the pairs $\{0, 2\pi\}$ and $\{0, \pi\}$ are conjugate for $y'' + y = 0$. More generally, (a, b) is a conjugate pair if

$$\begin{vmatrix} \cos a & \sin a \\ \cos b & \sin b \end{vmatrix} = 0,$$

that is, if $\sin b \cos a - \cos b \sin a = \sin(b - a) = 0$. We conclude that $\{a, b\}$ is a conjugate pair of boundary points if $b - a = k\pi$ for some integer k. Then the boundary-value problem

$$y'' + y = 0, \qquad y(a) = \alpha, \qquad y(b) = \beta$$

has a unique solution precisely when a and b do *not* differ by an integer multiple of π.

e x a m p l e 3

For the equation $y'' - y = 0$, the general solution is expressible as

$$y(x) = c_1 e^x + c_2 e^{-x}.$$

The arbitrary pair $\{a, b\}$ of boundary points is conjugate precisely when the equations

$$y(a) = c_1 e^a + c_2 e^{-a} = 0,$$
$$y(b) = c_1 e^b + c_2 e^{-b} = 0$$

are satisfied by some c_1 and c_2 not both zero. But for real a and b,

$$\begin{vmatrix} e^a & e^{-a} \\ e^b & e^{-b} \end{vmatrix} = e^{(a-b)} - e^{-(a-b)} = 0$$

only when $a = b$. So there is always a unique solution to the boundary-value problem

$$y'' - y = 0, \qquad y(a) = \alpha, \qquad y(b) = \beta, \qquad \text{if } a \neq b.$$

3B   Nonhomogeneous Equations

We consider boundary-value problems for the nonhomogeneous equation with continuous coefficients $p(x)$, $q(x)$, and $f(x)$,

$$y'' + p(x)y' + q(x)y = f(x).$$

The key is first to find a solution satisfying the special **homogeneous** boundary conditions

$$y(a) = 0, \qquad y(b) = 0.$$

It turns out that this can always be done provided a and b are not conjugate points for the differential equation. Recall from Section 2 the variation of parameters formula,

$$y(x) = -y_1(x) \int \frac{y_2(x)f(x)}{w(x)}\, dx + y_2(x) \int \frac{y_1(x)f(x)}{w(x)}\, dx,$$

where y_1 and y_2 are independent solutions and w is their Wronskian. By choosing the constants of integration properly in the two integrals, we can get the most general solution of the nonhomogeneous differential equation. In particular, we can specify these constants by using integrals with limits as follows:

▆ **3.2**
$$y_p(x) = -y_1(x) \int_b^x \frac{y_2(t)f(t)}{w(t)}\, dt + y_2(x) \int_a^x \frac{y_1(t)f(t)}{w(t)}\, dt$$

$$= \int_a^x \frac{y_2(x)y_1(t)}{w(t)} f(t)\, dt + \int_x^b \frac{y_1(x)y_2(t)}{w(t)} f(t)\, dt.$$

Now suppose that the linearly independent solutions y_1 and y_2 have been chosen so that $y_1(a) = 0$ and $y_2(b) = 0$. It then follows by setting $x = a$ and $x = b$, respectively, that

$$y_p(a) = 0 \qquad \text{and} \qquad y_p(b) = 0;$$

thus we have found a particular solution satisfying the preceding boundary condition.

The assumption that a and b are not conjugate points ensures that the desired independent solutions y_1 and y_2 can always be found, as stated in the following theorem.

▆ **3.3 THEOREM**

If the equation $y'' + p(x)y' + q(x)y = 0$ has continuous coefficients and $x = a$, $x = b$ are a pair of nonconjugate boundary points for the equation, then there are linearly independent solutions y_1 and y_2 satisfying $y_1(a) = 0$, $y_2(b) = 0$.

Proof. All we have to do for y_1 is pick a nontrivial initial-value solution, with $y_1(a) = 0$, $y_1'(a) \neq 0$. Then automatically $y_1(b) \neq 0$, because otherwise the fact that a and b are not conjugate would imply that y_1 is necessarily identically zero. Similarly, choose y_2 so that $y_2(b) = 0$, $y_2'(b) \neq 0$, with the result that $y_2(a) \neq 0$. Then y_1 and y_2 are linearly independent, because if one were a multiple of the other, each would have to be zero at both boundary points a and b. But $y_1(a)y_2(b) \neq y_1(b)y_2(a)$, so this is impossible.   ∎

We denote by G the function of two variables defined for $a \leq x \leq b$ and $a \leq t \leq b$ by

$$G(x, t) = \begin{cases} \dfrac{y_2(x)y_1(t)}{w(t)}, & a \leq t \leq x, \\[2mm] \dfrac{y_1(x)y_2(t)}{w(t)}, & x \leq t \leq b, \end{cases}$$

where $y_1(a) = y_2(b) = 0$ and $w(t) = y_1(t)y_2'(t) - y_2(t)y_1'(t)$ is the Wronskian of $y_1(t)$ and $y_2(t)$. Then G is called a **Green's function** for the boundary-value problem, and by Equation 3.2 we can write

◼ **3.4**
$$y_p(x) = \int_a^b G(x, t)f(t)\, dt$$

for the solution of

$$y'' + p(x)y' + q(x)y = f(x), \qquad y(a) = y(b) = 0.$$

It is important to understand that while the independent homogeneous solutions y_1 and y_2 must satisfy the conditions $y_1(a) = 0$ and $y_2(b) = 0$, respectively, there is still some arbitrariness in the choice that we make of them. We can illustrate this point with a very simple example.

e x a m p l e 4

Suppose we want to solve the equation $y'' = f(x)$, subject to $y(0) = 0$, $y(1) = 0$. (This can in principle be done by repeated integration.) The most general solution of the associated homogeneous equation is

$$y(x) = c_1 + c_2 x.$$

To satisfy $y_1(0) = 0$ we need to have $c_1 = 0$. Hence we can choose for simplicity $c_2 = 1$. At $x = 1$ we must have $c_1 + c_2 = 0$, so we can choose this time $c_1 = 1$, $c_2 = -1$. Thus we have selected

$$y_1(x) = x, \qquad y_2(x) = 1 - x$$

for our independent homogeneous solutions. The Wronskian $w[y_1 y_2](x)$ is

$$\begin{vmatrix} x & 1 - x \\ 1 & -1 \end{vmatrix} = -1,$$

which happens to be a constant. Thus the Green's function is given by

$$G(x, t) = \begin{cases} \dfrac{(1 - x)t}{-1}, & 0 \leq t \leq x, \\[2mm] \dfrac{x(1 - t)}{-1}, & x \leq t \leq 1. \end{cases}$$

The formula for the particular solution of $y'' = 1$ is obtained by taking $f(t) = 1$ in the general Green's formula 3.4. Thus

$$y_p(x) = -(1-x)\int_0^x t\, dt - x\int_x^1 (1-t)\, dt$$

$$= -(1-x)\frac{x^2}{2} - x\left(\frac{1}{2} - x + \frac{x^2}{2}\right) = \frac{x^2}{2} - \frac{x}{2}.$$

Clearly, $y_p(0) = y_p(1) = 0$ and $y_p''(x) = 1$, so we have found the correct solution, though by a more elaborate method than necessary.

3C   *Nonhomogeneous Boundary Conditions*

To solve the nonhomogeneous equation

$$y'' + p(x)y' + q(x)y = f(x)$$

with general **nonhomogeneous boundary conditions**

$$y(a) = \alpha, \qquad y(b) = \beta, \qquad \alpha, \beta \text{ not both zero,}$$

we continue to assume that a and b are not conjugate points of the associated homogeneous equation. Then first solve the equation with homogeneous boundary conditions

$$y(a) = 0, \qquad y(b) = 0,$$

for example, by the method just outlined in Section 3B, calling this solution y_p. Now the general solution of the nonhomogeneous equation is

$$y(x) = c_1 y_1(x) + c_2 y_2(x) + y_p(x),$$

where y_1 and y_2 are independent homogeneous solutions and c_1 and c_2 are constants. Since $y_p(a) = y_p(b) = 0$, all we have to do to satisfy $y(a) = \alpha, y(b) = \beta$ is choose c_1 and c_2 so that

$$c_1 y_1(a) + c_2 y_2(a) = \alpha,$$
$$c_1 y_1(b) + c_2 y_2(b) = \beta.$$

This can always be done if a and b are not conjugate, because in that case the determinant of the system of linear equations for c_1 and c_2 is not zero. See Exercise 12a.

e x a m p l e **5**   The differential equation

$$y'' + y = 1$$

has linearly independent homogeneous solutions $\cos x$ and $\sin x$. To satisfy boundary conditions of the form

$$y(0) = \alpha, \qquad y\left(\frac{\pi}{2}\right) = \beta,$$

we can check that the points $x = 0$ and $x = \pi/2$ are not conjugate by observing that

$$\begin{vmatrix} \cos 0 & \cos \pi/2 \\ \sin 0 & \sin \pi/2 \end{vmatrix} = 1.$$

Rather than construct a Green's function, we can observe by inspection that $y(x) = 1$ is a particular solution, so

$$y(x) = c_1 \cos x + c_2 \sin x + 1$$

is the most general solution of the differential equation. To satisfy $y(0) = \alpha$, $y(\pi/2) = \beta$, we solve the equations

$$c_1 \cos 0 + c_2 \sin 0 + 1 = \alpha,$$

$$c_1 \cos \frac{\pi}{2} + c_2 \sin \frac{\pi}{2} + 1 = \beta$$

for c_1 and c_2 to get $c_1 = \alpha - 1$, $c_2 = \beta - 1$. Thus

$$y(x) = (\alpha - 1) \cos x + (\beta - 1) \sin x + 1$$

is the unique solution satisfying the boundary conditions. On the other hand, the points $x = 0$, $x = \pi$ are conjugate for $y'' + y = 0$, because $\sin x$ is a nontrivial solution that is zero at both points. Thus an attempt to solve

$$c_1 \cos 0 + c_2 \sin 0 + 1 = \alpha$$

$$c_1 \cos \pi + c_2 \sin \pi + 1 = \beta$$

leads to $c_1 = \alpha - 1$ *and* $-c_1 = \beta - 1$. Thus there is no solution to the boundary-value problem with boundary conditions

$$y(0) = \alpha, \qquad y(\pi) = \beta$$

unless $\alpha - 1 = -\beta + 1$ or $\alpha + \beta = 2$. In that case, there are infinitely many solutions, all of the form

$$y(x) = (\alpha - 1) \cos x + c_2 \sin x + 1,$$

where c_2 is arbitrary. Clearly, these solutions all satisfy

$$y(0) = \alpha, \qquad y(\pi) = 2 - \alpha.$$

EXERCISES

1. For each of the following differential equations, determine whether (i) there is a unique solution, (ii) there are infinitely many solutions, or (iii) there is no solution satisfying the given boundary conditions.

   (a) $y'' + y = 0$, $y(0) = 1$, $y(\pi/4) = 0$.

   (b) $y'' + y = 0$, $y(0) = 1$, $y(2\ \pi) = 0$.

   (c) $y'' + y = 0$, $y(0) = 1$, $y(2\pi) = 1$.

   (d) $y'' + 2y = 0$, $y(1) = 1$, $y(2) = 0$.

2. Find all pairs of conjugate points for the following differential equations of the form $L[y] = 0$.

   (a) $y'' + 2y = 0$.   (c) $y'' + y' + y = 0$.

   (b) $y'' - 2y = 0$.

   (d) $y'' + xy' + y = 0$. [*Hint:* $e^{-x^2/2}$ is a solution; show that $e^{-x^2/2} \int_2^x e^{t^2/2}\, dt$ is an independent one.]

3. A pair of nonconjugate boundary points is given for each of the differential equations in Exercise 2. Find in each case a pair of linearly independent solutions y_1 and y_2 that satisfy $y_1(a) = 0$ and $y_2(b) = 0$. Then use these solutions to construct a Green's function for the equation on the interval $[a, b]$.

(a) $a = 0$, $b = \pi/(2\sqrt{2})$.

(c) $a = 0$, $b = \pi/\sqrt{3}$.

(b) $a = 1$, $b = 2$.

(d) $a = 0$, $b = 1$.

4. Use the Green's function found in each part of Exercise 3 to represent the solution $y(x)$ of the corresponding differential equation $L[y] = f$, where f is defined as follows, and where $y(a) = y(b) = 0$. Compute the solution by any method.

(a) $f(x) = x$, $0 \leq x \leq \pi/(2\sqrt{2})$.

(c) $f(x) = \cos x$, $0 \leq x \leq \pi/\sqrt{3}$.

(b) $f(x) = \begin{cases} -1, & 1 \leq x < \frac{1}{2}, \\ 1, & \frac{1}{2} \leq x \leq 2. \end{cases}$

(d) $f(x) = e^{-x}$, $0 \leq x \leq 1$.

5. Modify the solutions found in Exercise 4 so that instead of satisfying the homogeneous conditions $y(a) = y(b) = 0$, they satisfy the following corresponding nonhomogeneous conditions.

(a) $y(0) = 1$, $y(\pi/(2\sqrt{2})) = 0$.

(c) $y(0) = -1$, $y(\pi/\sqrt{3}) = 0$.

(b) $y(1) = 2$, $y(2) = -1$.

(d) $y(0) = 1$, $y(1) = 1$.

6. Show that the points $a = 0$, $b = \pi\sqrt{2}$ are conjugate for the differential equation $y'' + 2y = 0$, and find conditions on α and β under which there are solutions of this equation that satisfy $y(0) = \alpha$, $y(\pi\sqrt{2}) = \beta$. Find all such solutions. Does one of these solutions also satisfy $y'(0) = 1$?

7. For a homogeneous linear boundary problem, the only way to get a nontrivial (i.e., not identically zero) solution is to have the solution nonunique. In particular, consider the problem of finding a number λ and a nontrivial solution $y(x)$ to the equation

$$L[y] = \lambda y,$$

where L is a second-order linear operator. (Such a number λ is called an **eigenvalue,** and a corresponding nontrivial solution is called an **eigenfunction.**) Show that if $a < b$, the problem

$$y'' = \lambda y, \qquad y(a) = y(b) = 0$$

has nontrivial solutions if and only if $\lambda = -k^2\pi^2/(b - a)^2$, where $k = 1$, $2, \ldots$. What are the corresponding solutions $y_k(x)$?

8. Consider the oscillator equation

$$my'' + ky' + hy = 0,$$

where m, k, and h are constant and $m > 0$.

(a) Show that if $k^2 \geq 4mh$, then there are no conjugate pairs a, b. (The case $k^2 = 4mh$ requires special attention.)

(b) Explain the physical significance of the conclusion in part (a).

(c) Find all conjugate pairs a, b in case $k^2 < 4mh$.

9. Buckling rod. A uniform horizontal rod is subjected to a horizontal compressive force of magnitude F. The rod buckles an amount $y = y(x)$ at distance x from one end, where $0 \le x \le l$, in such a way that $y'' + (F/k)y = 0$; here k is a positive constant depending on the elasticity, the cross-sectional shape, and the density of the rod.

(a) Assuming boundary conditions $y(0) = y(l) = 0$ at the ends, show that actual displacement takes place only when the applied force has one of the critical values $F_n = kn^2\pi^2/l^2$, where n is a positive integer.

(b) Since the differential equation is homogeneous, a constant multiple of a solution is also a solution, and the boundary-value problem doesn't have a unique solution. Sketch the general shapes of the critical displacement functions $y_n(x)$ corresponding to the three smallest critical forces.

10. (a) Show that the boundary conditions

$$A[y] = a_1y(0) + a_2y'(0) = 0,$$
$$B[y] = b_1y(\pi) + b_2y'(\pi) = 0,$$

when applied to the differential equation $y'' + y = 0$, can be satisfied by a solution $y(x) \ne 0$ precisely when $a_1b_2 = a_2b_1$.

(b) Show that $y'' + y = f(x)$ has a unique solution for each continuous f on $0 \le x \le \pi$ and satisfying $A[y] = \alpha$, $B[y] = \beta$, precisely when $a_1b_2 \ne a_2b_1$.

11. Beam deflection. The downward deflection $y = y(x)$ of a uniform horizontal beam of length L satisfies the 4th-order differential equation

$$y^{(4)} = R, \qquad 0 \le x \le L,$$

where R is an appropriate constant and x represents distance from one end of the beam. If the beam has a simple support at an end $x = x_0$, then $y(x_0) = y''(x_0) = 0$. If the beam is held rigidly horizontal at x_0, say by embedding it in a concrete wall, then $y(x_0) = y'(x_0) = 0$. If the beam is unsupported at $x = x_0$, then $y''(x_0) = y'''(x_0) = 0$. Solve the differential equation under the following sets of conditions and arbitrary $R > 0$.

(a) $y(0) = y''(0) = 0$, $y(L) = y''(L) = 0$.

(b) $y(0) = y'(0) = 0$, $y(L) = y''(L) = 0$.

(c) $y(0) = y'(0) = 0$, $y''(L) = y'''(L) = 0$.

(d) $y(0) = y''(0) = 0$, $y''(L) = y'''(L) = 0$. Give a physical explanation for what goes wrong under these conditions.

12. Consider the algebraic equations

$$Az + Bw = K,$$
$$Cz + Dw = L.$$

(a) Show that these equations have a unique solution for z, w if $AD - BC \ne 0$.

(b) Show that if $AD - BC = 0$, then these equations have either infinitely many solutions or else no solutions at all. [*Hint:* If A, B, C, D are all zero, the conclusion is obvious. If, for example, $B \ne 0$, then $C = AD/B$.]

13. In certain 1-dimensional diffusion processes the concentration $y(x)$ of the diffusing substance depends only on the distance x from one end of a medium of width 1. Then $y(x)$ can be shown to satisfy the equation

$$y'' = ky \exp \frac{\gamma(1 - y)}{1 + \delta(1 - y)},$$

with boundary conditions $y(0) = 0$, $y(1) = 1$. Here k, γ, and δ are nonnegative constants.

(a) Solve the equation under the assumption $\gamma = 0$ and arbitrary $k \geq 0$.

(b) Sketch the graph of the solution found in part (a) for $k = 1$.

(c) For general constants γ and δ see Exercise 7 of Section 4 on numerical methods.

14. Under circumstances different from those assumed in the previous exercise, a diffusion concentration may satisfy an equation of the form

$$y'' = py' - qy^2,$$

where p and a are nonnegative constants and there are boundary conditions of the form $y(0) = 0$, $y(1/p) = 1$.

(a) Solve the equation with $q = 0$ and arbitrary $p \geq 0$.

(b) Sketch the graph of the solution found in part (a).

(c) For other constants p and q, use numerical methods as in Exercise 8 of Section 4.

4   SHOOTING

Shooting is a method for applying numerical solution of initial-value problems to solution of boundary-value problems. Assuming that there is a solution to

$$y'' + a(x)y' + b(x)y = F(x), \qquad y(a) = \alpha, \qquad y(b) = \beta,$$

we can try to approximate the solution as follows. Replace the condition $y(b) = \beta$ at b by an initial condition $y'(a) = z_0$ at a. Suppose we are very lucky in our choice of z_0; it might then happen that the solution of the problem with initial value $y(a) = \alpha$, $y'(a) = z_0$ turns out to satisfy $y(b) = \beta$ nearly enough that we are satisfied with the result, and so we accept this for our approximate solution. If the discrepancy $y(b) - \beta$ is too great, we can then use this difference to guide us in adjusting z so as to reduce $|y(b) - \beta|$ to an acceptable size. The method applies just as well to the general second-order problem

$$y'' = f(x, y, y'), \qquad y(a) = \alpha, \qquad y(b) = \beta.$$

The term "shooting" suggests the idea that from $y(a)$ we "aim" at $y(b)$ by adjusting $y'(a)$ and making a succession of "shots" after each adjustment. The values of each approximate solution should be recorded numerically or graphically in case one of these solutions turns out to be acceptable. Figure 9.3a shows the graphs of six shots for a pendulum equation, the last two of which are graphically indistinguishable at the scale used there.

To start the method, we make two trial shots, with different "elevations" $y'(a) = z_1$ and $y'(a) = z_2$, and note the corresponding approximate values y_1, y_2 that we get at $x = b$. Of course, $y(b) = \beta$ is what we want, so we try to improve

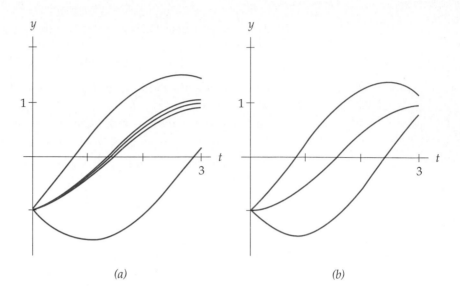

Figure 9.3 (a) $\ddot{y} = -\sin y$, $y(0) = -1$, $y(3) = 1$. (b) $\ddot{y} = -y$, $y(0) = -1$, $y(3) = 1$.

our accuracy by choosing the next slope $y'(a) = z_3$ to satisfy the linear interpolation equation

■ **4.1** $$z_3 = z_2 + \frac{z_1 - z_2}{y_1(b) - y_2(b)}(\beta - y_2(b)).$$

This is just the equation in the yz plane of the line through (y_1, z_1) and (y_2, z_2) but evaluated at $y = \beta$. If the shot aimed with the choice $z = z_3$ has an unacceptable value at $x = b$, then try again, replacing z_1 by z_2 and z_2 by z_3 in Equation 4.1 to get z_4. It is a good idea to try to choose z_1, z_2 so that the corresponding y values $y_1(b), y_2(b)$ straddle the desired value $y = \beta$ and are not too close together. Aim each shot after that using a z value from Equation 4.1 along with a previous one that will lead to a straddle. Exercise 5 shows that three shots should be enough for a linear equation.

The central part of a computing routine to implement the method is a subroutine to approximate the solution of an initial-value problem with $y(a) = \alpha$, $y'(a) = z$. The subroutine is called repeatedly until $|y(b) - \beta|$ is small enough.

example 1

The boundary-value problem $\ddot{y} = -\sin y$, $y(0) = -1$, $y(3) = 1$ for an undamped pendulum equation requires us to pick two initial velocities; we try $y'(0) = z_1 = 1$, which swings the pendulum toward the position $y = 1$ that we want to attain at time $t = 3$, and $y'(0) = z_2 = -1$, which initially swings the pendulum away from $y = 1$. Taking 200 steps from $t = 0$ to $t = 3$ gives the two extreme solution curves shown in Figure 9.3a, where $y_1(3) \approx 1.46487$ and $y_2(3) \approx 0.173326$. Since these are quite far from the desired value 1, we apply the interpolation formula 4.1 to get a new slope

$$z_3 = -1 + \frac{1 - (-1)}{1.46487 - 0.173326}(1 - 0.173326) \approx 0.280133.$$

Taking another shot with this slope gives an approximate solution y_3 with $y_3(3) \approx 1.04513$. To get more improvement, we note that y_2 and y_3 straddle the target value 1 at $x = 3$, so we interpolate their initial slopes to get a slope $z_4 = 0.213875$ for the next shot. Continuing in this way gives $z_5 \approx 0.152445$, $z_6 \approx 0.150604$, with $y_6(3) = 1.00001$. We may choose to accept this solution without further improvement; its graph is the lowest of the intermediate ones in Figure 9.3a. If we apply the same process to the linearized boundary-value problem $\ddot{y} = -y$, $y(0) = -1$, $y(3) = 1$, we discover that the third shot already gives $y_3(3) = 1.00000$ with $y_3'(0) = z_3 = 0.070893$, as shown in Figure 9.3b.

example 2

The equation

$$\ddot{y} - 0.1t\dot{y} + y = 0$$

is linear with a single nonconstant coefficient. If t is restricted to be fairly small, the solutions of the equation can be expected to behave something like the solutions of

$$\ddot{y} + y = 0;$$

these are $y = c_1 \cos t + c_2 \sin t$. For the given equation, consider the boundary-value problem of finding a solution satisfying

$$y(0) = 1, \qquad y(3) = -1.$$

To estimate a trial value for $y'(0)$, we solve the boundary problem for the constant-coefficient equation, with the result

$$y_c(t) = \cos t - 0.07 \sin t,$$
$$y_c'(t) = \sin t - 0.07 \cos t.$$

Thus we could try $z_1 = -0.07$ for a first shot. A conservative approach would be to try $z_1 = 2$ and $z_2 = -2$ for the first two shots. With this choice, and with 500 steps of the improved Euler method, the result for the third shot turns out to give the value $y(3) = -1.00000$, along with $y'(0) = -1.1728$. The graphs of the three solutions are shown in Figure 9.4a.

It is easy (see Exercise 5) to show that the method described when applied to a *linear* equation, whether homogeneous or not, will in principle lead to the desired solution at the third shot. Furthermore, it's not necessary in a linear example for the terminal values of the first two shots to straddle the desired terminal value. This straddle requirement is typically quite important for a nonlinear equation.

example 3

The nonlinear boundary problem $\ddot{y} = -y^3$, $y(0) = 1$, $y(2.5) = -1$ requires us to straddle the value -1 with the terminal values of the first two shots from $y(0) = 1$. It turns out that using $y'(0) = 1$ and $y'(0) = -1$ will give us the required straddle, and that on the sixth shot, having taken care to maintain a straddle each time, we get $y(2.5) = -1.00000$ with $y'(0) = -0.517353$. See Figure 9.4b for a graphical representation of the results.

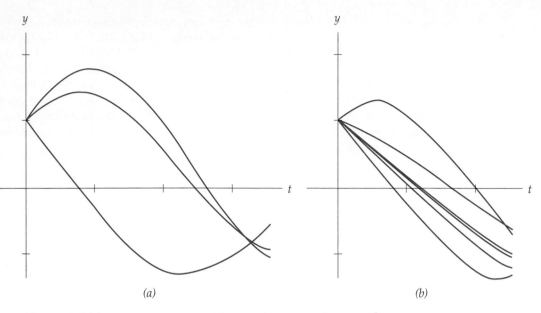

(a)   (b)

Figure 9.4 (a) $\ddot{y} - 0.1t\dot{y} + y = 0$, $y(0) = 1$, $y(3) = -1$. (b) $\ddot{y} = -y^3$, $y(0) = 1$, $y(2.5) = -1$.

The Web site http://math.dartmouth.edu/~rewn/ contains the prepared Applet BOUNDARY2 that finds graphic and numeric approximations to 2nd-order two-point boundary problems by iterated shooting.

EXERCISES

1. (a) The differential equation $y'' - y = 1$ has the solution
$$y = (e + 1)^{-1}(e^x + e^{-x+1}) - 1$$
satisfying $y(0) = y(1) = 0$. Find solutions y_1, y_2 satisfying
$$y_1(0) = 0, \qquad y_1'(0) = z_1 = 1,$$
$$y_2(0) = 0, \qquad y_2'(0) = z_2 = -1.$$
Then find a third solution y_3 such that $y_3(0) = 0$, $y_3'(0) = z_3$, where z_3 is determined from Equation 4.1 with $b = 1$ and $\beta = 0$. Compare the result with the preceding solution.

(b) Derive the solution given at the beginning of part (a) directly or by using a Green's function.

2. (a) The differential equation $y'' + y = 1$ has the solution
$$y = -\cos x + \sin x + 1$$
satisfying $y(0) = y(2\pi) = 0$. Find solutions y_1, y_2 satisfying
$$y_1(0) = 0, \qquad y_1'(0) = 1,$$
$$y_2(0) = 0, \qquad y_2'(0) = -1.$$
Then explain why you cannot use Equation 4.1 to find a third solution.

(b) Find all solutions of the given boundary-value problem.

3. (a) Find graphical or numerical data for the solution of the boundary-value problem

$$y'' + \frac{2x}{1 + x^2} y' = F(x), \qquad y(0) = -1, \qquad y(10) = 1.5$$

for the case $F(x) = \frac{1}{5}$.

(b) Show that the general solution of the differential equation in part (a) has the form

$$y = c_1 + c_2 \arctan x + \int_0^x (1 + t^2)(\tan^{-1} x - \tan^{-1} t)F(t)\, dt.$$

(c) Explain how to determine c_1 and c_2 in part (b) to find a solution to the boundary-value problem in part (a).

4. Consider the nonlinear pendulum problem

$$\ddot{y} = -\sin y - 0.1\dot{y}, \qquad y(0) = -1, \qquad y(3) = 1.$$

(a) Find a numerical solution for this problem, estimate the value of $y'(0)$ for your solution, and compare this value with the one given for the corresponding nonlinear problem in Example 1 of the text.

(b) Give a physical interpretation of your comparison results in part (a).

5. If we apply the shooting method to a solvable linear, perhaps nonhomogeneous, boundary problem

$$y'' + u(x)y' + b(x)y = f(x), \qquad y(a) = \alpha, \qquad y(b) - \beta,$$

we will in principle arrive at a solution after just three shots. This exercise shows why, showing that Equation 4.1 of the text is just right for linear equations.

(a) Suppose that the first two shots give us solutions $y_1(x)$ and $y_2(x)$ that satisfy the initial conditions

$$y_1(a) = \alpha, \qquad y_2'(a) = z_1,$$
$$y_2(a) = \alpha, \qquad y_2'(a) = z_2,$$

where $z_1 \neq z_2$. Form the 1-parameter family

$$y_3(x) = \lambda y_1(x) + (1 - \lambda)y_2(x),$$

where λ is a constant to be determined. (Note that for $0 \leq \lambda \leq 1$, $y_3(x)$ ranges from $y_2(x)$ if $\lambda = 0$ to $y_1(x)$ if $\lambda = 1$.) Show that, regardless of the value of λ, $y_3(x)$ satisfies the given differential equation and also satisfies the single boundary condition $y_3(a) = \alpha$.

(b) Assuming $y_1(b) \neq y_2(b)$, solve the equation $\beta = \lambda y_1(b) + (1 - \lambda)y_2(b)$ for λ to get

$$\lambda = \frac{\beta - y_2(b)}{y_1(b) - y_2(b)}.$$

(c) Differentiate the equation for $y_3(x)$ in part (a), and show that using the value of λ found in part (b) yields Equation 4.1 with $z_3 = y_3'(a)$.

(d) Show that the choice of λ in part (b) yields $y_3(b) = \beta$.

6. Using the boundary conditions $y(0) = 0$, $y(10) = 0$, find a graphical solution of the **nonhomogeneous Airy equation**

$$y'' + xy = 1.$$

7. Apply the shooting method to the boundary problem described in Exercise 13 of Section 3 under the assumptions $k = 1$, $\gamma = 0.3$, $\delta = 5$. Compare your solution numerically with the exact solution for the case $k = 1$, $\gamma = 0$, $\delta = 5$.

8. Apply the shooting method to the boundary problem described in Exercise 14 of Section 3 under the assumptions $p = 1$, $q = 2$. Compare your solution numerically with the exact solution for the case $p = 1$, $q = 0$.

9. The purpose of this exercise is to see what can happen if you try to apply numerical solution to a boundary-value problem that has no solution. The last part of Example 1 of Section 3 shows that the problem $y'' = -y$, $y(0) = 1$, $y(\pi) = 1$ fails to have a solution. However, note that any attempt to solve the problem numerically forces us to approximate π by a rational number p, perhaps in decimal or binary form.

(a) Show that if $p \neq \pi$ is a close approximation to π, the boundary-value problem to satisfy $y'' = -y$, $y(0) = 1$, $y(p) = 1$ has the unique solution

$$y(x) = \cos x + \frac{1 - \cos p}{\sin p} \sin x.$$

(b) Show that for the solution proposed in part (a), the slope at $x = p$, namely, $y'(p)$, tends to $-\infty$ if p approaches π from below and tends to $+\infty$ if p approaches π from above.

(c) Explain what the result will be of a naive attempt to find a nonexistent solution to the boundary-value problem by shooting. You may want to conduct some computer experiments.

10. Find a graphical solution to the boundary problem $\ddot{y} = -3y^3$, $y(1) = 1$, $y(2.5) = -1$, noting that some care is needed to straddle the value -1 with the terminal values of the first two shots.

11. Find a graphical solution to the boundary problem $\ddot{y} = -3y^3$, $y(1) = 1$, $y(6) = -1$, noting that some care is needed to straddle the value -1 with the terminal values of the first two shots.

12. The **catenary equation** $y'' = k\sqrt{1 + (y')^2}$ is satisfied by the shape $y = y(x)$ of a chain suspended between two fixed supports at $y(a) = \alpha$ and $y(b) = \beta$. The positive constant k depends on the length of the chain. Plot the solution assuming $k = \frac{1}{2}$ and $y(0) = 1$, $y(6) = 2$.

5   POWER SERIES SOLUTIONS

A differential equation of the form $y' = f(x, y)$, if it has a solution $y = y(x)$ near $x = x_0$, determines the derivative $y'(x_0)$ by $y'(x_0) = f(x_0, y(x_0))$. If f is sufficiently differentiable, even higher derivatives are similarly determined at x_0. Our aim is to find polynomial approximations to solutions of differential equations and, if possible, extend the approximations to power series representations

$$y(x) = \sum_{k=0}^{\infty} c_k(x - x_0)^k$$

for solutions convergent in some open interval symmetric about x_0. Functions that can be so represented by power series are called **analytic** and have the property that they can be differentiated and integrated term by term as often as you like within the open interval of convergence. In addition, an analytic function $y(x)$ as displayed above is represented by its **Taylor series** at x_0, which simply means that the coefficients c_k can in principle be computed by the **Taylor coefficient formula**

$$c_k = \frac{1}{k!} y^{(k)}(x_0) , \qquad k = 0, 1, 2, \dots .$$

e x a m p l e 1

The equation $y' = y^2$ has $y(x) = (1 - x)^{-1}$ for a solution satisfying $y(0) = 1$. The power series expansion is the familiar geometric series formula

$$y = \frac{1}{1 - x} = 1 + x + x^2 + x^3 + \cdots, \qquad -1 < x < 1.$$

This expansion is also the Taylor series of $y(x)$, so by the Taylor coefficient formulas we have

$$y = y(0) + \frac{y'(0)}{1!} x + \frac{y''(0)}{2!} x^2 + \frac{y'''(0)}{3!} x^3 + \cdots.$$

Comparison of coefficients of x^k in the two expansions shows that $y^{(k)}(0)/k! = 1$, so $y^{(k)}(0) = k!$, $k = 0, 1, 2, \dots$. Suppose, however, that we had no formula for the coefficients to begin with. (Most examples are of this sort.) Starting with the given differential equation, successive derivatives can be computed and then simplified by substitution from the earlier equations:

$$y' = y^2 = 1!y^2,$$
$$y'' = 2yy' = 2y^3 = 2!y^3,$$
$$y''' = 6y^2y' = 6y^4 = 3!y^4,$$
$$y^{(4)} = 24y^3y' = 24y^5 = 4!y^5.$$

The general pattern is evidently $y^{(k)} = k!y^{k+1}$. The formal Taylor expansion for a solution $y = y(x)$ about x_0 with $y(x_0) = y_0 \neq 0$ can then be simplified using a geometric series:

$$y(x) = y_0 + y_0^2(x - x_0) + y_0^3(x - x_0)^2 + \cdots$$
$$= y_0(1 + y_0(x - x_0) + y_0^2(x - x_0)^2 + \cdots)$$
$$= y_0 \sum_{k=0}^{\infty} [y_0(x - x_0)]^k = \frac{y_0}{1 - y_0(x - x_0)}.$$

This expansion contains the one we started with as the special case when $x_0 = 0$ and $y_0 = 1$. However, that expansion was valid just for $-1 < x < 1$, while the general one is valid for $|x - x_0| < 1/|y_0|$, since then $|y_0(x - x_0)| < 1$.

Higher-order equations, for example,

$$y'' = f(x, y, y'),$$

can be treated the same way.

e x a m p l e 2

Suppose we want successive derivatives $y^{(k)}(0)$ to a solution of

$$y'' = yy'$$

given that $y(0) = 1$ and $y'(0) = -1$. First compute from the given equation some formulas for higher derivatives. Then simplify by substituting the given values $y(0) = 1$, $y'(0) = -1$. We find

$$y'' = yy'; \qquad\qquad\qquad y''(0) = -1,$$
$$y''' = yy'' + (y')^2 = y^2 y' + (y')^2; \qquad\qquad y'''(0) = 0,$$
$$y^{(4)} = 2y(y')^2 + y^2 y'' + 2y'y' = 4y(y')^2 + y^3 y'; \qquad y^{(4)}(0) = 3.$$

The first five terms of the Taylor expansion of $y(x)$ about $x = 0$ then add up to

$$y(0) + \frac{y'(0)}{1!} x + \frac{y''(0)}{2!} x^2 + \frac{y'''(0)}{3!} x^3 + \frac{y^{(4)}(0)}{4!} x^4 = 1 - x - \tfrac{1}{2}x^2 + \tfrac{1}{8}x^4.$$

In this example there does not seem to be a simple pattern, but clearly we could compute as many terms as we had time and space for.

e x a m p l e 3

From the differential equation

$$y'' = xy' - y,$$

we compute, using substitution after differentiating,

$$y''' = xy'' = x(xy' - y)$$
$$= x^2 y' - xy,$$
$$y^{(4)} = x^2 y'' + xy' - y = x^2(xy' - y) + xy' - y$$
$$= (x^3 + x)y' - (x^2 + 1)y.$$

If we denote $y(0)$ by c_0 and $y'(0)$ by c_1, then

$$y''(0) = -c_0, \qquad y'''(0) = 0, \qquad y^{(4)}(0) = -c_0.$$

Thus the Taylor expansion of $y(x)$ about $x = 0$ has the form

$$y(x) = c_0 + c_1 x - \frac{c_0}{2!} x^2 + 0 - \frac{c_0}{4!} x^4 - \cdots$$
$$= c_0(1 - \tfrac{1}{2}x^2 - \tfrac{1}{24}x^4 - \cdots) + c_1 x.$$

For comparison, note that if the original differential equation were replaced by

$$y'' = -y,$$

then solutions would have the form

$$y(x) = c_0 \cos x + c_1 \sin x,$$

where $y(0) = c_0$, $y'(0) = c_1$.

EXERCISES

1. Find the first three nonzero terms in the Taylor expansion about $x = 0$ of $y = y(x)$ if y satisfies each of the following relations. Then use the resulting approximation near $x = 0$ to sketch the graph of the solution in a small neighborhood of $x = 0$.

   (a) $y' = y^2 + y$, $y(0) = 1$.

   (b) $y' = y^2 + x$, $y(0) = -1$.

   (c) $y' = xy$, $y(0) = 2$.

   (d) $y'' = xy$, $y(0) = 1$, $y'(0) = 0$.

2. Find the first four nonzero terms in the Taylor expansion of $y = y(x)$ about $x = 0$ if $y'' = yy'$ and $y(0) = y'(0) = 1$.

3. Find the first four nonzero terms in the Taylor expansion of $y = y(x)$ about $x = 1$ if $y''' = y$ and $y(1) = 2$, $y'(1) = 0$, $y''(1) = 1$.

4. Suppose $y'' = x^2 y$ while $y(0) = c_0$ and $y'(0) = c_1$. Show that $y = y(x)$ has the form

$$y = c_0(1 + \tfrac{1}{12}x^4 + \cdots) + c_1(x + \tfrac{1}{20}x^5 + \cdots).$$

5. Show that if $y''' = y^2 y'$ and $y(0) = y'(0) = y''(0) = 1$ then

$$y = 1 + x + \tfrac{1}{2}x^2 + \tfrac{1}{6}x^3 + \tfrac{1}{8}x^4 + \tfrac{3}{40}x^5 + \cdots.$$

6. In Example 2 of the text the computation stops with the 4th-degree term. Find the 5th-degree term.

7. Solve the initial-value problem $y' = e^x y$, $y(0) = 1$ in terms of elementary functions. Then find the first three nonzero terms in the Taylor expansion of that solution about $x = 0$.

8. The initial-value problem $y' = y$, $y(0) = 1$ has a well-known solution. Derive the power series form of that solution by the method of this section.

9. The initial-value problem $y'' = y$, $y(0) = y'(0) = 1$ has a well-known solution. Derive the power series form of that solution by the method of this section.

10. (a) Show that the equation $y' = |x|$ has solutions

$$y(x) = \begin{cases} \dfrac{x^2}{2} + C, & x \geq 0, \\[2mm] -\dfrac{x^2}{2} + C, & x < 0. \end{cases}$$

(b) Sketch the graph of the solution in part (a) for $-2 < x < 2$ assuming $C = 0$. Then show that none of the solutions in part (a) is analytic in an interval containing $x_0 = 0$.

11. Verify using a geometric series the assertion made at the end of Example 1 of the text that the expansion

$$\frac{y_0}{1 - y_0(x - x_0)} = y_0(1 + y_0(x - x_0) + y_0^2(x - x_0)^2 + \cdots$$

is valid for $|x - x_0| < 1/|y_0|$.

6   UNDETERMINED COEFFICIENTS

The solutions of many of the differential equations we have studied can be represented in terms of their Taylor expansions. For example, polynomials, the elementary transcendental functions $\cos x$, $\sin x$, e^x, and linear combinations of all these have Taylor expansions that are valid for all x. In fact, there is a large and important class of differential equations that has solutions that are **analytic,** that is, representable by power series. Even if an analytic solution is not a combination of elementary functions, its power series expansion may serve to define an important new function. Furthermore, the partial sums of a series expansion often give useful approximations to the true solution.

Recall that a Taylor expansion of a function f is a power series

$$\sum_{k=0}^{\infty} c_k(x - x_0)^k = c_0 + c_1(x - x_0) + c_2(x - x_0)^2 + \cdots,$$

where $c_k = f^{(k)}(x_0)/k!$. Such a series, unless it converges only for $x = x_0$, always converges absolutely in an interval $x_0 - r < x < x_0 + r$ that is symmetric about x_0; for functions f that are *analytic,* the series actually converges to $f(x)$ for all x in that interval. Furthermore, Taylor expansions about the same point x_0 can be conveniently added and multiplied, and a Taylor expansion can be differentiated or integrated term by term to produce the Taylor expansion of the derivative or integral of the expanded function within the interval of convergence.

example 1

The Taylor expansions

$$e^x = \sum_{k=0}^{\infty} \frac{x^k}{k!}, \qquad -\infty < x < \infty,$$

$$\cos x = \sum_{k=0}^{\infty} \frac{(-1)^k x^{2k}}{(2k)!}, \qquad -\infty < x < \infty,$$

$$\sin x = \sum_{k=0}^{\infty} \frac{(-1)^k x^{2k+1}}{(2k + 1)!}, \qquad -\infty < x < \infty,$$

$$\frac{1}{1 - x} = \sum_{k=0}^{\infty} x^k, \qquad -1 < x < 1,$$

$$\ln (1 - x) = -\sum_{k=1}^{\infty} \frac{x^k}{k}, \qquad -1 < x < 1,$$

can all be computed directly from the general Taylor formula and can be shown to converge to the value of the function on the left for each x in the indicated interval. Furthermore, we can compute the derivative of a function such as $\cos 2x$ as follows:

$$\frac{d}{dx} \cos 2x = \frac{d}{dx} \sum_{k=0}^{\infty} \frac{(-1)^k (2x)^{2k}}{(2k)!}$$

$$= \sum_{k=0}^{\infty} \frac{d}{dx} \frac{(-1)^k (2x)^{2k}}{(2k)!}$$

$$= \sum_{k=0}^{\infty} \frac{(-1)^k (2x)^{2k-1}(4k)}{(2k)!} \qquad \text{(Term is 0 when } k = 0.)$$

$$= 2 \sum_{k=1}^{\infty} \frac{(-1)^k (2x)^{2k-1}}{(2k - 1)!} \qquad \text{(} 2k \text{ canceled.)}$$

$$= -2 \sum_{k=0}^{\infty} \frac{(-1)^k (2x)^{2k+1}}{(2k + 1)!} = -2 \sin 2x.$$

In the last step we simply replaced k by $k + 1$ throughout, including the summation limits, to make the expansion look more like the corresponding expansion in the preceding examples.

Many of the most important examples of series solutions of differential equations are expressible as power series about the point $x_0 = 0$. Since Taylor expansions about zero are also a little easier to work with, most of our examples will be of that kind. Our main job in this section is efficient computation of coefficients in the expansion of solutions of linear differential equations. We begin with a familiar example for which we already know all solutions. The method involves treating the coefficients in a power series as "undetermined coefficients," to be calculated successively from a **recurrence relation.**

example 2

To solve $y'' + y = 0$, we try to find a solution of the form

$$y(x) = \sum_{k=0}^{\infty} c_k x^k.$$

This form of the expansion is particularly appropriate if we want to solve an initial-value problem with $y(0)$ and $y'(0)$ specified, because then $c_0 = y(0)$ and $c_1 = y'(0)$, if there is a Taylor expansion for the solution about $x - 0$. Proceeding under that assumption for the moment, we compute

$$y'(x) = \sum_{k=1}^{\infty} k c_k x^{k-1},$$

$$y''(x) = \sum_{k=2}^{\infty} (k - 1) k c_k x^{k-2}$$

$$= \sum_{k=0}^{\infty} (k + 1)(k + 2) c_{k+2} x^k.$$

To simplify adding terms in x^k of the series for $y''(x)$ to terms in x^k for $y(x)$, we shifted the index of summation by 2 in the series for $y''(x)$; by analogy with the definite integral, we also shift the summation limits by -2. For $y(x)$ to represent a solution we want

$$y''(x) + y(x) = \sum_{k=0}^{\infty} c_k x^k + \sum_{k=0}^{\infty} (k + 1)(k + 2) c_{k+2} x^k$$

$$= \sum_{k=0}^{\infty} [c_k + (k + 1)(k + 2) c_{k+2}] x^k = 0.$$

Next we use the fundamental fact that for a Taylor expansion to be identically zero, all the coefficients must be zero. Hence we get a recurrence relation for the c_k's:

$$c_{k+2} = -\frac{c_k}{(k + 1)(k + 2)}, \qquad k = 0, 1, 2, \ldots.$$

Recalling that we can specify a particular solution by determining the numbers $c_0 = y(0)$, $c_1 = y'(0)$, it is natural to compute recursively from the previous formula to get

$$c_2 = -\frac{c_0}{1 \cdot 2} = -\frac{c_0}{2!}, \qquad\qquad c_3 = -\frac{c_1}{2 \cdot 3} = -\frac{c_1}{3!},$$

$$c_4 = -\frac{c_2}{3 \cdot 4} = \frac{c_0}{4!}, \qquad\qquad c_5 = -\frac{c_3}{4 \cdot 5} = \frac{c_1}{5!},$$

$$c_6 = -\frac{c_4}{5 \cdot 6} = -\frac{c_0}{6!}, \qquad\qquad c_7 = -\frac{c_5}{6 \cdot 7} = -\frac{c_1}{7!},$$

$$\vdots \qquad\qquad\qquad\qquad\qquad \vdots$$

$$c_{2k} = -\frac{c_{2k-2}}{(2k-1)(2k)} = \frac{(-1)^k c_0}{(2k)!}, \qquad c_{2k+1} = -\frac{c_{2k-1}}{(2k)(2k+1)} = \frac{(-1)^k c_1}{(2k+1)!},$$

If we take $y(0) = c_0$ and $y'(0) = c_1 = 0$, then only the first column can contain nonzero entries, and we get the solution

$$y_0(x) = c_0 - \frac{c_0}{2!}x^2 + \frac{c_0}{4!}x^4 - \cdots + (-1)^k\frac{c_0}{(2k)!}x^{2k} + \cdots$$

$$= c_0 \cos x.$$

On the other hand, the choice $y(0) = c_0 = 0$ and $y'(0) = c_1$ makes the entries in the first column all zero, so we get another solution from the second column:

$$y_1(x) = c_1 - \frac{c_1}{3!}x^3 + \frac{c_1}{5!}x^5 - \cdots + (-1)\frac{c_1}{(2k+1)}x^{2k+1} + \cdots$$

$$= c_1 \sin x.$$

Thus the general solution is $y(x) = c_0 \cos x + c_1 \sin x$ as usual.

---

e x a m p l e 3

Solutions of **Airy's equation** $y'' + xy = 0$ are not obtainable in terms of elementary functions, so we try

$$y(x) = \sum_{k=0}^{\infty} c_k x^k,$$

$$y''(x) = \sum_{k=2}^{\infty} (k-1)kc_k x^{k-2}.$$

Substitution into the differential equation gives

$$y''(x) + xy(x) = \sum_{k=2}^{\infty} (k-1)kc_k x^{k-2} + \sum_{k=0}^{\infty} c_k x^{k+1}$$

$$= \sum_{k=0}^{\infty} (k+1)(k+2)c_{k+2}x^k + \sum_{k=1}^{\infty} c_{k-1}x^k = 0,$$

where this time we shifted the index in the first summation up by 2 and in the second summation down by 1 to make the powers of x agree. Thus we get a single term, the constant $2c_2$, in the first summation that does not correspond to a term in the second summation. We can then write $y'' + xy = 0$ as

$$2c_2 + \sum_{k=1}^{\infty} [(k + 1)(k + 2)c_{k+2} + c_{k-1}]x^k = 0;$$

setting all coefficients equal to zero gives the *recurrence relation*

$$c_2 = 0, \quad \text{and} \quad c_{k+2} = -\frac{c_{k-1}}{(k + 1)(k + 2)}, \quad k = 1, 2, 3, \ldots.$$

We find as a result that the terms are determined in sequences with indices differing by 3 and that

$$0 = c_2 = c_5 = c_8 = \cdots = c_{3k+2} = \cdots.$$

However, if c_0 or c_1 is not zero, we compute as follows:

$$c_3 = -\frac{c_0}{2 \cdot 3},$$

$$c_6 = -\frac{c_3}{5 \cdot 6} \qquad = \frac{c_0}{2 \cdot 3 \cdot 5 \cdot 6},$$

$$c_9 = -\frac{c_6}{8 \cdot 9} \qquad = -\frac{c_0}{2 \cdot 3 \cdot 5 \cdot 6 \cdot 8 \cdot 9},$$

$$\vdots \qquad\qquad \vdots$$

$$c_{3k} = -\frac{c_{3k-3}}{(3k - 1)3k} = \frac{(-1)^k c_0}{2 \cdot 3 \cdot 5 \cdot 6 \cdots (3k - 1)3k},$$

$$c_4 = -\frac{c_1}{3 \cdot 4},$$

$$c_7 = -\frac{c_4}{6 \cdot 7} \qquad = \frac{c_1}{3 \cdot 4 \cdot 6 \cdot 7},$$

$$c_{10} = -\frac{c_7}{9 \cdot 10} \qquad = -\frac{c_1}{3 \cdot 4 \cdot 6 \cdot 7 \cdot 9 \cdot 10},$$

$$\vdots \qquad\qquad \vdots$$

$$c_{3k+1} = -\frac{c_{3k-2}}{3k(3k + 1)} = \frac{(-1)^k c_1}{3 \cdot 4 \cdot 6 \cdot 7 \cdots 3k(3k + 1)},$$

The solution determined by $y(0) = c_0 = 1$, $y'(0) = c_1 = 0$ is

$$y_0(x) = 1 + \sum_{k=1}^{\infty} \frac{(-1)^k x^{3k}}{2 \cdot 3 \cdot 5 \cdot 6 \cdots (3k - 1)3k},$$

and the solution determined by $y(0) = c_0 = 0$, $y'(0) = c_1 = 1$ is

$$y_1(x) = x + \sum_{k=1}^{\infty} \frac{(-1)^k x^{3k+1}}{3 \cdot 4 \cdot 6 \cdot 7 \cdots 3k(3k + 1)}.$$

The graphs of y_0 and y_1 shown in Figure 9.5 can be drawn using the numerical methods of Chapter 4, Section 7, or by plotting partial sums of the corre-

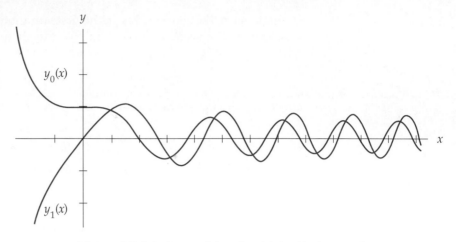

Figure 9.5 Solutions $y_0(x)$ and $y_1(x)$ to Airy's equation.

sponding power series. The power series for y_0 and y_1 both converge for all x because the denominators of the kth terms each contain increasing integer factors, $2k$ in number. Hence each series has terms dominated by those of an everywhere-convergent series, for example,

$$\left| \frac{(-1)^k x^{3k}}{2 \cdot 3 \cdot 5 \cdot 6 \cdots (3k-1)3k} \right| \le \frac{|x|^{3k}}{(2k)!}.$$

In fact, estimates of this kind can be used to test the accuracy of stopping with a specified number of terms in a Taylor expansion. In this example, to get an estimate when $|x| \le 1$, we estimate the tail of the factorial series by a geometric series with ratio $1/4n^2$:

$$(2n)! \sum_{k=n}^{\infty} \frac{1}{(2k)!} = 1 + \frac{1}{(2n+1)(2n+2)} + \frac{1}{(2n+1)\cdots(2n+4)} + \cdots$$

$$\le 1 + \frac{1}{4n^2} + \frac{1}{(4n^2)^2} + \cdots$$

$$= \frac{4n^2}{4n^2 - 1}.$$

Thus the error in stopping after $n-1$ terms on the interval $-1 \le x \le 1$ is at most

$$\frac{4n^2}{4n^2 - 1} \cdot \frac{1}{(2n)!}.$$

For $n = 5$, that is, keeping terms of degree 4, the error is at most $3 \cdot 10^{-7}$.

Since the two solutions y_0 and y_1 are linearly independent, the general solution of $y'' + xy = 0$ has the form

$$y(x) = c_0 y_0(x) + c_1 y_1(x).$$

The solutions of $y'' + ay = 0$ have infinitely many zero values if $a > 0$ and at most one such value when $a < 0$. Analogously it can be shown that a solution of the Airy equation has infinitely many positive zeros, but at most one negative zero.

example 4

The **Legendre equation** of index m is

$$(1 - x^2)y'' - 2xy' + m(m + 1)y = 0,$$

and it is discussed in Chapter 11, Section 2. Equation 2.5 is given there for polynomial solutions in case m is a nonnegative integer. To find other solutions in the form of power series, we let

$$y(x) = \sum_{k=0}^{\infty} c_k x^k, \qquad y'(x) = \sum_{k=1}^{\infty} k c_k x^{k-1}, \qquad y''(x) = \sum_{k=2}^{\infty} (k-1)k c_k x^{k-2},$$

so that the c_k are determined by

$$(1 - x^2)\sum_{k=2}^{\infty} (k-1)k c_k x^{k-2} - 2x \sum_{k=1}^{\infty} k c_k x^{k-1} + m(m+1)\sum_{k=0}^{\infty} c_k x^k = 0.$$

Shifting the index by 2 in the first sum allows us to write

$$[2c_2 + m(m+1)c_0] + [2 \cdot 3c_3 - 2c_1 + m(m+1)c_1]x$$

$$+ \sum_{k=2}^{\infty} [(k+1)(k+2)c_{k+2} - ((k-1)k + 2k - m(m+1))c_k]x^k = 0.$$

Setting the coefficient of each power of x equal to zero gives

$$c_2 = -\frac{m(m+1)}{2}c_0, \qquad c_3 = -\frac{(m-1)(m+2)}{2 \cdot 3}c_1,$$

$$c_{k+2} = \frac{(k+1+m)(k-m)}{(k+1)(k+2)}c_k, \qquad k \geq 2.$$

Since the *recurrence relation* contains a shift by 2, it is natural to split the coefficients into those of even and those of odd index:

$$c_4 = \frac{(3+m)(2-m)}{3 \cdot 4}c_2 = \frac{(m-2)\,m(m+3)(m+1)}{4!}c_0,$$

$$c_6 = \frac{(5+m)(4-m)}{5 \cdot 6}c_4 = -\frac{(m-4)(m-2)m(m+5)(m+3)(m+1)}{6!}c_0,$$

$$\vdots$$

and

$$c_5 = \frac{(4+m)(3-m)}{4 \cdot 5}c_3 = \frac{(m-3)(m-1)(m+4)(m+2)}{5!}c_1,$$

$$c_7 = \frac{(6+m)(5-m)}{6 \cdot 7}c_5 = -\frac{(m-5)(m-3)(m-1)(m+6)(m+4)(m+2)}{7!}c_1,$$

$$\vdots$$

Clearly, if $m = 2l$ is a positive even integer, then all even coefficients are zero beyond $2l$. Thus the series expansion with even powers reduces to an even polynomial. Similarly, if $m = 2l + 1$ is a positive odd integer, the series expansion with odd powers reduces to an odd polynomial. For example, when $m = 4$, the independent solutions we derived are

$$P_4(x) = 1 - \frac{4 \cdot 5}{2!} x^2 + \frac{2 \cdot 4 \cdot 7 \cdot 5}{4!} x^4, \qquad \text{if } c_0 = 1, c_1 = 0,$$

$$Q_4(x) = x - \frac{3 \cdot 6}{3!} x^3 + \frac{1 \cdot 3 \cdot 8 \cdot 6}{5!} x^5 - \frac{-1 \cdot 1 \cdot 3 \cdot 10 \cdot 8 \cdot 6}{7!} x^7 + \cdots,$$

$$\text{if } c_0 = 0, c_1 = 1.$$

The ratio test for convergence (see Exercise 7) shows that the infinite series solution converges for $-1 < x < 1$. These two solutions form the basis for the collection of all solutions of the homogeneous Legendre equation of index 4 on the interval $-1 < x < 1$. In general, the Legendre equation of integer index m evidently has independent solutions of which one is a polynomial and the other is not. The importance of the polynomial solutions P_m is explained in Chapter 11; they are analogous to orthogonal eigenvectors, with corresponding eigenvalues $-m(m + 1)$, of a differential operator L, where

$$L(y) = (1 - x^2)y'' - 2xy'.$$

EXERCISES

1. **(a)** Use the undetermined-coefficient method to derive the general solution to the differential equation $y'' - y = 0$ in terms of a power series expansion about $x_0 = 0$.

   (b) Show that the general series solution derived in part (a) represents $y = c_0 \cosh x + c_1 \sinh x$, where c_0 and c_1 are arbitrary constants.

2. **(a)** Show that if $y = f(x)$ is a solution of $y'' + xy = 0$ for all x, then $y = f(-x)$ is a solution of

$$y'' - xy = 0.$$

   (b) Use the result of part (a) together with the result of Example 3 of the text to find a power series expansion for the general solution of $y'' - xy = 0$.

3. **(a)** Apply the method of power series to solve the first-order differential equation

$$y' + 2xy = 0.$$

   (b) Solve the differential equation in part (a) by finding an exponential multiplier and then integrating.

   (c) Do the results of parts (a) and (b) agree for all x?

4. **(a)** Apply the power series method to find the general solution of the differential equation

$$y'' - xy' = 0$$

   in the form $y(x) = c_0 y_0(x) + c_1 y_1(x)$.

(b) Solve the differential equation in part (a) by solving the equivalent system

$$y' = u,$$
$$u' = xu.$$

(c) Do the results of parts (a) and (b) agree for all x?

5. (a) Apply the power series method to find the general solution of the differential equation

$$y'' + xy' + y = 0$$

in the form $y(x) = c_0 y_0(x) + c_1 y_1(x)$.

(b) Show that a special case of the general solution found in part (a) is the solution $y(x) = e^{-(x^2/2)}$.

(c) Given that $y_0(x) = e^{-(x^2/2)}$, use the order-reduction method of Section 1 to find an independent solution $y_1(x)$ in terms of the function $F(x) = \int_0^x e^{t^2/2}\, dt$.

*(d) Reconcile the result of part (c) with the first three terms of the solution of part (a). [*Hint:* Multiply power series.]

6. (a) Apply the power series method to find a single particular solution of the differential equation

$$y'' + xy' + y = x.$$

(b) Combine the result of part (a) with that of the previous exercise to write the general solution of $y'' + xy' + y = x$.

(c) Starting with the solution $y_0(x) = e^{-(x^2/2)}$ to the associated homogeneous equation, use variation of parameters to find the general solution to the equation in part (a).

7. Apply the ratio test for series convergence to the series solution $Q_4(x)$ found for the Legendre equation in Example 4 of the text. Show that the series converges for $-1 < x < 1$. [*Hint:* The general recurrence relation for this example is helpful here.]

8. The **Bessel equation** of integer index n (see Section 7B) is

$$x^2 y'' + xy' + (x^2 - n^2)y = 0.$$

(a) Show that when $n = 0$, a solution of the form $\sum_{k=0}^{\infty} c_k x^k$ satisfies $(k + 2)^2 c_{k+2} = -c_k$.

(b) Show that if we choose $c_0 = 1$, $c_1 = 0$ in part (a), we get the solution

$$J_0(x) = \sum_{k=0}^{\infty} (-1)^k 2^{-2k}(k!)^{-2} x^{2k},$$

called a Bessel function of index 0.

(c) Show that $J_1(x) = -J_0'(x)$ defines a solution of the Bessel equation of index 1. [*Hint:* Apply d/dx to the differential equation for J_0; then express J_0 in terms of J_0' and J_0''.]

9. A **Bessel function** of integer index n is defined by

$$J_n(x) = \left(\frac{x}{2}\right)^n \sum_{k=0}^{\infty} (-1)^k \frac{x^{2k}}{2^{2k} k!(n + k)!}.$$

(a) Show that J_n satisfies the Bessel equation of index n given in Exercise 8.

(b) Show that if y_n is a solution of the Bessel equation of index n and $u_n(x) = \sqrt{x}y_n(x)$, then u_n satisfies

$$u'' + \left(1 - \frac{4n^2 - 1}{4x^2}\right)u = 0.$$

10. Find a power series expansion about $x_0 = 1$ for the solution to initial-value problem $y' = 3(x - 1)^2 y$, $y(1) = 1$ as follows.

(a) Use the undetermined-coefficient method of this section.

(b) Find the solution to the problem in terms of elementary functions by working with the solution to part (a).

(c) Find some form of the solution to the related problem $y' = 3(x - 1)^2 y$, $y(0) = 1$ by any method you choose.

11. (a) Show that the only power series expansion about $x_0 = 0$ that provides a solution for $x^2 y'' - y = 0$ has all its coefficients equal to zero and so is the identically zero solution.

(b) Show nevertheless that this differential equation has solutions defined for $x > 0$ by determining those exponents r for which $y_r(x) = x^r$ is a solution.

12. (a) Show that the only power series expansion about $x_0 = 0$ that provides a solution for $x^2 y'' + xy' + (x^2 - (\frac{1}{2})^2)y = 0$ has all its coefficients equal to zero and so is the identically zero solution.

(b) Show nevertheless that this differential equation, the Bessel equation of index $\frac{1}{2}$, has solutions defined for $x > 0$ defined by series of the form

$$y = \sum_{k=0}^{\infty} c_k x^{k-1/2}.$$

(c) Show that the series solutions found in part (b) can be combined to give solutions of the form $y = x^{-1/2}(c_0 \cos x + c_1 \sin x)$.

13. Consider the solutions of Airy's equation $y'' = -xy$ of Example 3 of the text.

(a) Show that for $x > 0$, the graph of a solution $y(x)$ must be concave down when $y > 0$ and concave up when $y < 0$.

(b) Show that for $x < 0$, the graph of a solution $y(x)$ must be concave up when $y > 0$ and concave down when $y < 0$.

(c) Explain how the observations of parts (a) and (b) are consistent with the general nature of the graphs in Figure 9.5.

7   SINGULAR POINTS

The power series method of Section 6 is usually applied just to second-order homogeneous equations of the form $a_0(x)y'' + a_1(x)y' + a_2(x)y = 0$, in which we assume that the coefficient functions $a_0(x)$, $a_1(x)$, $a_2(x)$ are analytic at each point of their common domain of definition; in other words, each of these three functions has a power series representation in some neighborhood of each point of that domain. Our special focus in this section will be on isolated points

$x = x_0$ such that $a_0(x_0) = 0$; in that case the differential equation typically fails to define y'' in terms of x, y, and y', with the result that solutions may fail to be defined at x_0. Even if solutions are defined at x_0, satisfying the conventional initial conditions $y(x_0) = y_0$, $y'(x_0) = z_0$ is not possible in general. Nevertheless, solutions that exist in a neighborhood of such a point are sometimes very useful and so deserve special study. We begin by putting the differential equation in **normal form** with leading coefficient 1:

$$y'' + a(x)y' + b(x)y = 0, \qquad \text{where } a(x) = \frac{a_1(x)}{a_0(x)} \text{ and } b(x) = \frac{a_2(x)}{a_0(x)}.$$

If there is sufficient cancellation in the ratios $a(x)$ and $b(x)$ so that these two coefficients are analytic at x_0, then x_0 is called an **ordinary point** for the differential equation, and the method of Section 6 can in principle be applied; otherwise x_0 is called a **singular point,** and a modification of our earlier technique may be necessary. If the coefficient $a_0(x)$ of y'' has not been divided out to produce the normalized form, a singular point may arise from a point $x = x_0$ where $a_0(x) = 0$. A standard bit of terminology is to refer to a point x_0 at which an analytic function has value zero as a **zero** of the function.

---

e x a m p l e 1

The normal form for $x^2 y'' - xy' + x^2 y = 0$ is $y'' - x^{-1}y' + y = 0$. In this example, $a(x) = -x^{-1}$ fails to be analytic just at $x_0 = 0$, even though $b(x) = 1$ is analytic everywhere. Hence $x_0 = 0$ is the equation's only singular point.

Even though solutions $y(x)$ exist in a neighborhood of a singular point x_0 of a differential equation, the behavior of $y(x)$ as x approaches x_0 may be quite unruly; examples are given later. However, most of the examples that come up in practice have solutions expressible at their singularities by relatively mild modifications of ordinary power series. These equations have the property that their singular points are not too "bad" in the following precise sense. A singular point x_0 of the equation $y'' + a(x)y' + b(x)y = 0$ is called a **regular singular point** if the functions $(x - x_0)a(x)$ and $(x - x_0)^2 b(x)$ can both be defined at x_0 so as to be analytic there.

---

e x a m p l e 2

The equation $y'' - x^{-1}y' + y = 0$ has its only singular point at $x_0 = 0$. This point is a regular singular point, since $xa(x) = -xx^{-1} = -1$ and $x^2 b(x) = x^2$ are analytic for all x, in particular at the singular point $x_0 = 0$.

---

e x a m p l e 3

The **Legendre equation** of index 1 is $(1 - x^2)y'' - 2xy' + 2y = 0$. The normal form is $y'' - (2x/(1 - x^2))y' + (2/(1 - x^2))y = 0$, and the normal-form coefficients $a(x) = -2x/(1 - x^2)$ and $b(x) = 2/(1 - x^2)$ have singular points at $x = \pm 1$. At $x_0 = 1$ we have

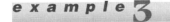

$$(x - 1)a(x) = -2(x - 1)\frac{x}{1 - x^2} = \frac{2x}{1 + x}$$

Also,

$$(x - 1)^2 b(x) = 2(x - 1)^2 \frac{x}{1 - x^2} = -\frac{2(x - 1)}{1 + x}.$$

Thus $(x - 1)a(x)$ and $(x - 1)^2 b(x)$ are both analytic at $x = 1$, with respective values 1 and 0 there. Notice that to pass the regularity test at $x_0 = 1$, these two functions are not required to be analytic at the other singular point at $x_1 = -1$. Similar computations show that at $x_1 = -1$ the functions $(x + 1)a(x)$ and $(x + 1)^2 b(x)$ are analytic.

We'll consider two differential equations in detail, the Euler equation and the Bessel equation. Between them, their solutions exhibit most of the important features associated with equations having regular singular points.

7A   Euler's Equation

The Euler differential equation

$$x^2 y'' + axy' + by = 0, \qquad a, b \text{ const},$$

arises naturally in the solution of the Laplace equation in Chapter 10 and is the simplest example of an equation with a regular singular point. The form of the equation suggests that there may be a solution of the form $y = x^\mu$ for $x > 0$. This is indeed the case, and we proceed to find the correct values for μ by first finding $y' = \mu x^{\mu-1}$ and $y'' = \mu(\mu - 1)x^{\mu-2}$. (The validity of these differentiation formulas for complex values of μ and $x > 0$ follows from the definition $x^{\alpha+i\beta} = x^\alpha x^{i\beta} = x^\alpha e^{i\beta \ln x}$.) Substitution into the differential equation gives

$$\mu(\mu - 1)x^\mu + \mu a x^\mu + b x^\mu = 0, \qquad \text{or} \qquad (\mu(\mu - 1) + \mu a + b)x^\mu = 0.$$

Since $x^\mu \neq 0$, we can divide by it, getting the **indicial equation** for the **index** μ:

$$\mu(\mu - 1) + \mu a + b = 0 \qquad \text{or} \qquad \mu^2 + (a - 1)\mu + b = 0.$$

The roots of the indicial equation determine the solutions of the differential equation, as the roots of the characteristic equation do for a constant-coefficient equation. We'll assume that the coefficients a and b are real numbers. The indices are given by the quadratic formula: $\mu = \frac{1}{2}((1 - a) \pm \sqrt{(a - 1)^2 - 4b})$.

Distinct Real Roots: $\mu_1 \neq \mu_2$. This is the simplest case in every respect. Since the differential equation is linear, it has solutions $y = c_1 x^{\mu_1} + c_2 x^{\mu_2}$ for $x > 0$.

Equal Roots: $\mu_1 = \mu_2 = (1 - a)/2$. There is a solution $y_1 = x^{(1-a)/2}$. To find an independent solution, we resort to Theorem 1.2 of Section 1. The normalized form of the differential equation is $y'' + (a/x)y' + (b/x^2)y = 0$. An independent solution is

$$y_2 = x^{(1-a)/2} \int \frac{e^{-\int (a/x)\,dx}}{x^{1-a}}\,dx$$

$$= x^{(1-a)/2} \int \frac{e^{-a \ln x}}{x^{1-a}}\,dx = x^{(1-a)/2} \ln x.$$

Since the equation is linear, we have solutions $y = x^{(1-a)/2}(c_1 + c_2 \ln x)$ for $x > 0$.

Complex Roots: $4b > (a - 1)^2$. Denoting the distinct complex roots by $\mu_1 = \alpha + i\beta$, $\mu_2 = \alpha - i\beta$, $\beta > 0$, we have

$$x^{\alpha \pm i\beta} = x^\alpha x^{\pm i\beta} = x^\alpha e^{\pm i\beta \ln x} = x^\alpha(\cos(\beta \ln x) \pm \sin(\beta \ln x)).$$

Thus we have the independent real solutions $x^\alpha \cos(\beta \ln x)$ and $x^\alpha \sin(\beta \ln x)$. As x increases from 1 to ∞, $\beta \ln x$ increases from 0 to ∞. Hence the sine and cosine of $\beta \ln x$ oscillate infinitely often, though with decreasing frequency because the rate of increase of $\ln x$ tends to zero. As x decreases from 1 to 0, $\beta \ln x$ decreases from 0 to $-\infty$, so the sine and cosine of $\beta \ln x$ oscillate infinitely often, with increasing frequency as x approaches zero.

example 4

The Euler equation $x^2 y'' - 2xy' + 2y = 0$ has indicial equation $\mu^2 - 3\mu + 2 = (\mu - 1)(\mu - 2) = 0$, with roots $\mu_1 = 1$, $\mu_2 = 2$. The corresponding independent solutions are $y_1(x) = x$ and $y_2(x) = x^2$. The constants in the linear combination $y = c_1 x + c_2 x^2$ can be chosen so as to satisfy arbitrary initial conditions $y(x_1) = y_0$, $y'(x_1) = z_0$, *except* at the singular point $x_1 = 0$, where zero is the only possible value for the solution. Solving for the constants gives $c_1 = (2y_0 - x_1 z_0)/x_1$, $c_2 = (x_1 z_0 - y_0)/x_1^2$.

example 5

The family of equations $x^2 y'' - \nu xy' + \nu y = 0$ with real parameter ν has $\mu^2 - (\nu + 1)\mu + \nu = (\mu - 1)(\mu - \nu) = 0$ for indicial equation with roots $\mu_1 = 1$, $\mu_2 = \nu$. If $\nu \neq 1$, we get x and x^ν for independent solutions. If $\nu = 1$, the double root leads to the solutions x and $x \ln x$. See Figure 9.6 for some graphs.

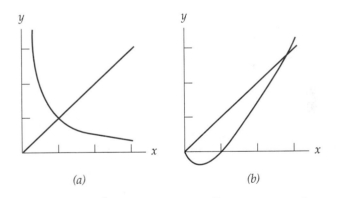

(a) (b)

Figure 9.6 (a) $y_1 = x$, $y_2 = x^{-1}$; $\mu_1 = 1$, $\mu_2 = -1$. (b) $y_1 = x$, $y_2 = x \ln x$; $\mu_1 = \mu_2 = 1$.

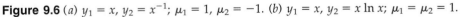

example 6

The family of equations $x^2 y' + m^2 y = 0$ with real parameter $m^2 > \frac{1}{4}$ has the indicial equation $\mu^2 - \mu + m^2 = 0$, with roots $\mu = \frac{1}{2} \pm (i/2)\sqrt{4m^2 - 1}$. With $m^2 = \frac{401}{4}$, we get $x^{1/2} \sin(10 \ln x)$ and $x^{1/2} \sin(10 \ln x)$ for independent solutions. Now change the differential equation by including the first-order term $2y'$, to get $x'y'' + 2y' + m^2 y = 0$. The indicial equation becomes $\mu^2 + \mu + m^2 = 0$, with roots $\mu = -\frac{1}{2} \pm (i/2)\sqrt{4m^2 - 1}$. This just changes the factor $x^{1/2}$ in the solutions to the previous equation to $x^{-1/2}$. Figure 9.7 shows some typical graphs.

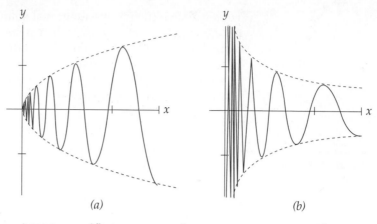

Figure 9.7 (*a*) $y = x^{1/2} \sin (10 \ln x)$; $x^2 y'' + \frac{101}{4} y = 0$. (*b*) $y = x^{-1/2} \sin (10 \ln x)$; $x^2 y + 2y' + \frac{101}{4} y = 0$.

EXERCISES

1. Find all singular points of each of the following differential equations, and state which of these points is regular and which is not.

   (a) $xy'' + xy' + y = 0$.

   (b) $x^2 y'' + x^2 y' + y = 0$.

   (c) $(1 - x)y'' + xy' + y = 0$.

   (d) $(1 - x^2)xy'' + xy' + y = 0$.

   (e) $y'' + xy' + y = 0$.

   (f) $x^3 y'' + xy' + y = 0$.

   (g) $e^x y'' + xy' + y = 0$.

   (h) $x^2 y'' + (\sin x)y' + y = 0$.

2. Find the roots of the indicial equation for each of the following differential equations. Then find a pair of independent real-valued solutions defined for $x > 0$.

   (a) $x^2 y'' + xy' + y = 0$.

   (b) $x^2 y'' - xy' - y = 0$.

   (c) $x^2 y'' + y = 0$.

   (d) $xy'' + y' = 0$.

   (e) $x^2 y'' + 5xy' - 5y = 0$.

   (f) $x^2 y'' - 3xy' + y = 0$.

   (g) $x^2 y'' - 4xy' + y = 0$.

   (h) $x^2 y'' - xy' + 7y = 0$.

3. For what positive values of x does the function $\sin (\ln x)$ in the examples equal zero?

4. For what positive values of x does the function $x \sin (\ln x)$ have relative maximum or minimum values?

5. The most general equation with a regular singular point at $x_0 = 0$ can be written $x^2 y'' + xa(x)y' + b(x)y = 0$. Assume $x > 0$, and let $t = \ln x$ so that $x = e^t$.

   (a) Use the chain rule for differentiation to show that

   $$x \frac{dy}{dx} = \frac{dy}{dt} \quad \text{and} \quad x^2 \frac{d^2 y}{dx^2} = \frac{d^2 y}{dt^2} - \frac{dy}{dt}.$$

   (b) Use the result of part (a) to show that the differential equation is equivalent, for $x > 0$, to $\ddot{y} + (a(e^t) - 1)\dot{y} + b(e^t)y = 0$.

(c) Specialize the result of part (b) to show that the constant-coefficient Euler equation is equivalent to $\ddot{y} + (a - 1)\dot{y} + by = 0$. Show that the characteristic roots of this last equation are the same as the roots of the Euler equation's indicial equation.

6. Use the definition $x^{\alpha + i\beta} = x^{\alpha}x^{i\beta} = x^{\alpha}e^{i\beta \ln x}$ to verify that the derivative of $x^{\alpha + i\beta}$ is given by $(a + i\beta)x^{\alpha + i\beta - 1}$ for $x > 0$.

7B Bessel's Equation

Bessel functions share certain properties with the sine function that allow it to play a part in the representation of oscillatory phenomena, as shown in Chapter 11, Section 3, as well as in many other areas of pure and applied mathematics. From our present point of view the definition of Bessel function starts most naturally with the **Bessel equation** $x^2 y'' + xy' + (x^2 - \nu^2)y = 0$. This differential equation depends on the parameter ν, which for our purposes it will be sufficient to assume is a nonnegative real number. Integer indices ν, particularly $\nu = 0$ and $\nu = 1$, are the ones that occur most often in applications, but other choices for ν also occur in practice. In particular, choosing $\nu = n + \frac{1}{2}$ for integer n leads to "elementary" solutions.

A routine check shows that the Bessel equation never has one-term solutions of the form x^{μ} that work for the Euler equation of Section 7A. Nevertheless there is still an indicial equation to be satisfied by constants μ if we try to find a **Frobenius series** solution, namely, one of the form

∎ **7.1**
$$y(x) = \sum_{k=0}^{\infty} c_k x^{k+\mu}.$$

Since μ need not be an integer, we'll assume for simplicity that $x > 0$ throughout the following discussion to avoid fractional powers of negative numbers. If $\mu = n$ is a nonnegative integer, the Frobenius series is an ordinary power series with first term $c_0 x^n$.

We apply term-by-term differentiation to the series in Equation 7.1 and then substitute into the Bessel equation in the form $x^2 y'' + xy' - \nu^2 y = -x^2 y$ to get

$$\sum_{k=0}^{\infty} [(k + \mu)(k + \mu - 1) + (k + \mu) - \nu^2]c_k x^{k+\mu} = -\sum_{k=0}^{\infty} c_k x^{k+\mu+2}.$$

Now simplify the left side and shift the summation index on the right:

$$\sum_{k=0}^{\infty} ((k + \mu)^2 - \nu^2)c_k x^{k+\mu} = -\sum_{k=2}^{\infty} c_{k-2} x^{k+\mu}.$$

Next, equate coefficients of like powers $x^{k+\mu}$. When $k = 0$, we get $(\mu^2 - \nu^2) \cdot c_0 = 0$, and when $k = 1$, we get $((1 + \mu)^2 - \nu^2)c_1 = 0$. The constant c_0 is designed to be the first nonzero coefficient in the series, so, dividing it out, we're left with the equation $\mu^2 - \nu^2 = 0$. This is the **indicial equation** that requires $\mu = \pm\nu$, where ν is the index of the Bessel equation. We'll assume $\mu = \nu \geq 0$ at first. The equation containing c_1 then reduces to $(1 + 2\nu)c_1 = 0$, so we take $c_1 = 0$. Equating coefficients of index $k \geq 2$ allows us to compute coefficients recursively using the **recurrence relation** $((k + \nu)^2 - \nu^2)c_k = -c_{k-2}$ or $c_k =$

$-c_{k-2}/k(k + 2\nu)$. Since $c_1 = 0$, it follows that $c_k = 0$ if k is odd. Having noted that $c_k = 0$ if k is odd, we write $k = 2j$ for the remaining even indices. The recurrence relation says then that

■ 7.2
$$c_{2j} = \frac{(-1)^j c_{2j-2}}{2j(2j + 2\nu)}.$$

As we mentioned earlier, the cases $\nu = 0$ and $\nu = 1$ are the most important, so we'll first consider nonnegative integer values $\nu = n \geq 0$. With $\nu = n$ in Equation 7.2, we get

$$c_2 = -\frac{c_0}{2^2(n + 1)}, \quad c_4 = \frac{c_0}{2^4 2!(n + 1)(n + 2)},$$

$$c_6 = -\frac{c_0}{2^6 \, 3!(n + 1)(n + 2)(n + 3)},$$

and in general

$$c_{2j} = \frac{(-1)^j c_0}{2^{2j} j!(n + 1)(n + 2) \cdots (n + j)}.$$

We now make an arbitrary choice for c_0, and set $c_0 = 1/(2^n n!)$. Then

■ 7.3
$$c_{2j} = \frac{(-1)^j}{2^{2j+n} j!(n + j)!}.$$

We define the **Bessel function** J_n for $n = 0, 1, 2, \ldots$ by

$$J_n(x) = \sum_{j=0}^{\infty} \frac{(-1)^j x^{2j+n}}{2^{2j+n} j!(n + j)!} = \sum_{j=0}^{\infty} \frac{(-1)^j (x/2)^{2j+n}}{j!(n + j)!}.$$

a series that converges for all x.

---

e x a m p l e 1

Index $\nu = 0$

A computer plot of

$$J_0(x) = \sum_{j=0}^{\infty} \frac{(-1)^j (x/2)^{2j}}{(j!)^2}$$

for $x \geq 0$ is shown in Figure 9.8. Since the expansion contains only even powers of x, J_0 is an even function, and the graph for $x < 0$ is obtained by reflecting in the vertical axis.

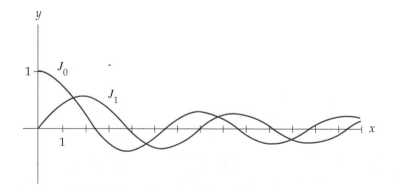

Figure 9.8 Graphs of J_0 and J_1 for $0 \leq x \leq 14$.

It can be proved that $J_0(x) = 0$ for an infinite sequence $\{x_n\}$ of positive numbers with successive differences $x_{n+1} - x_n$ tending to π. (For example, the difference between the locations of the last two zeros shown in Figure 9.8 is about 3.14.) This is just one among many analogies between solutions of the Bessel equation and solutions of damped harmonic oscillator equations. The first four zeros of J_0 are at approximately 2.40, 5.52, 8.65, 11.79.

e x a m p l e 2

Index $\nu = 1$

The Bessel function J_1 is

$$J_1(x) = \sum_{j=0}^{\infty} \frac{(-1)^j (x/2)^{2j+1}}{j!(j+1)!}.$$

As with J_0, the distance between successive zeros tends to π. It is left as an exercise using term-by-term differentiation of the series for J_0 to show that $J_0'(x) = -J_1(x)$ for all x. This last relation is analogous to $d\cos x/dx = -\sin x$, and its simplicity confirms the "correctness" of the choices 1 and $\frac{1}{2}$ for c_0 when $\nu = 0, 1$, respectively. The graphical relationship between J_0 and J_1 is indicated in Figure 9.8, where it appears that the locations of the local maxima and minima of J_0 coincide with the zeros of J_1, the first four positive ones being at approximately 3.85, 7.02, 10.17, 13.32.

Identifying series expansions for solutions of the Bessel equation of integer index n that are independent of J_n is relatively complicated. Any such independent solution is called a Bessel function of the **second kind,** and up to the addition of a constant multiple of J_n, these solutions are given by the formula in Theorem 1.2 of Section 1 in the form

■ **7.4** $$y_n(x) = J_n(x) \int \frac{C_n}{x J_n^2(x)} \, dx, \qquad C_n = \text{const} \neq 0.$$

Fortunately all we usually need know for our later applications is that all these independent solutions y_n are unbounded near zero, as the following theorem shows. It follows from Theorem 2.6 of Section 2 that every solution y_n of the index-n Bessel equation has the form

$$y_n(x) = c_1 J_n(x) + c_2 Z_n(x),$$

where Z_n is called a Bessel function of the **second kind;** Figure 9.9 shows their form for $n = 0$ and $n = 1$. The exercises take up a specific version of Bessel functions of the second kind, the **Weber functions** $Y_n(x)$.

■ **7.5 THEOREM**

A solution y_n given by Equation 7.4 tends to plus or minus infinity as x tends to 0. In particular, for each n there are numbers $b_n > 0$ and $\delta_n > 0$ such that for $0 < |x| \leq \delta_n$,

$$|y_0(x)| \geq b_0 \ln \left| \frac{1}{x} \right| \qquad \text{and} \qquad |y_n(x)| \geq \frac{b_n}{x^n}, \qquad n = 1, 2, 3, \ldots.$$

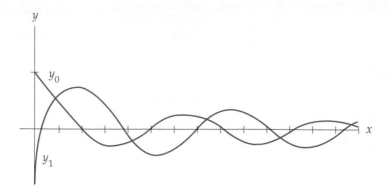

Figure 9.9 Bessel functions of the second kind: $y_0(1) = 1$, $y_0'(1) = -1$; $y_1(1) = y_1'(1) = 1$.

Proof. For simplicity take $C_n = 1/2^n n!$, and assume first that $x > 0$. Note that $y_n(x)$ differs by only a constant multiple of the continuous function J_n from

$$z_n(x) = x^n E_n(x) \int_x^{\delta_n} \frac{dt}{t^{2n+1} E_n^2(t)},$$

where $E_n(x) = x^{-n} J_n(x)/2^n n!$ is a continuous even function such that $E_n(0) = 1$. Choose δ_n between 0 and 1 so that $\frac{1}{2} \le E_n(x) \le 2$ if $|x| \le \delta_n$. Then for $0 < x \le \delta_n$,

$$z_n(x) \ge \tfrac{1}{2} x^{-n} \int_x^{\delta_n} \tfrac{1}{4} t^{-2n-1}\, dt \ge \begin{cases} -\tfrac{1}{8}\ln x + \tfrac{1}{8}\ln \delta_n, & n = 0. \\ (1/2n)x^{-n} - (1/2n)\delta_n^{-n}, & n \ge 1. \end{cases}$$

It follows that z_n plus a constant satisfies the claims made for y_n, so $y_n = z_n + C_n J_n$ also satisfies the claim. A similar argument with an integral over $-\delta_n \le x \le x$ proves the result for $x < 0$.   ∎

Boundary-value problems for a Bessel equation are typically posed over an interval $0 \le x \le a$, in which the requirement at $x = 0$ is simply that the solution should remain bounded as x approaches 0 from the right. At $x = a$ a more conventional condition $y(a) = b$ is imposed.

---

e x a m p l e 3

The index-0 Bessel equation can be written $xy'' + y' + xy = 0$. A condition requiring that $y(x)$ remain bounded near $x = 0$ suggests that we should avoid using an unbounded solution and concentrate on solutions $c_1 J_0(x)$. A condition $y(a) = b$ requires $c_1 J_0(a) = b$, which can be satisfied by choosing $c_1 = b/J_0(a)$ if $J_0(a) \ne 0$. (If $J_0(a) = 0$, then $c_1 J_0(a) = b$ has no solution unless $b = 0$, in which case c_1 is arbitrary; this is one reason why it's important to know the location of the zeros of Bessel functions.)

GAMMA FUNCTION. Finding an independent pair of solutions for $x > 0$ to the Bessel equation is somewhat simpler if ν is not an integer. The only additional complexity arises from the need to choose the initial constant c_0 in Equation 7.1 so that relations among the resulting Bessel functions are fairly simple. Just as in the case $\nu = n \ge 0$, the recurrence relation 7.2 leads to even-index coefficients c_k of the form

$$c_{2j} = \frac{(-1)^j c_0}{2^{2j} j! (\nu + 1)(\nu + 2)\cdots(\nu + j)}.$$

If $\nu \geq 0$ is an integer, the appropriate choice for c_0 turned out to be $c_0 = 1/2^\nu \nu!$, resulting in Equation 7.3. To make this work for noninteger ν, and even $\nu < 0$, it's customary to generalize the factorial function by using the **gamma function,** defined by an improper integral:

$$\Gamma(\nu) = \int_0^\infty t^{\nu-1} e^{-t}\, dt, \qquad \nu > 0.$$

Some properties of $\Gamma(\nu)$ that are pertinent for us are summarized in the following theorem.

■ 7.6 THEOREM

For n a nonnegative integer, $\Gamma(n + 1) = n!$, and if $\nu > 0$ is a real number, $\Gamma(\nu + 1) = \nu\Gamma(\nu)$. This last relation can be used to extend the definition of $\Gamma(\nu)$ to noninteger values $\nu < 0$ so that the relation becomes an identity for the extended function. Defining $1/\Gamma(v) = 0$ at $\nu = 0, -1, -2, -3, \ldots$ generalizes $1/n!$ to $1/\nu\,! = 1/\Gamma(\nu + 1)$ for all ν. In particular, $(-\frac{1}{2})! = \sqrt{\pi}$ and $\frac{1}{2}! = \sqrt{\pi}/2$. (Relevant graphs are shown in Figure 9.10.)

Proof. For $\nu > 0$, integration by parts shows that

$$\Gamma(\nu + 1) = \int_0^\infty t^\nu e^{-t}\, dt$$

$$= \left. t^\nu e^{-t} \right|_0^\infty + \nu \int_0^\infty t^{\nu-1} e^{-t}\, dt = \nu\Gamma(\nu).$$

In particular, if n is a positive integer, $\Gamma(n + 1) = n\Gamma(n) = n(n - 1)\Gamma(n - 2) = \cdots = n(n - 1)(n - 2)\cdots 3 \cdot 2\Gamma(1)$. Direct computation gives $\Gamma(1) = 1$, so $\Gamma(n + 1) = n!$. We define $\Gamma(\nu) = \Gamma(\nu + 1)/\nu$ for $-1 < \nu < 0$ and more generally

$$\Gamma(\nu) = \frac{\Gamma(\nu + n)}{\nu(\nu + 1)\cdots(\nu + n - 1)} \qquad \text{for } -n < \nu < -n + 1.$$

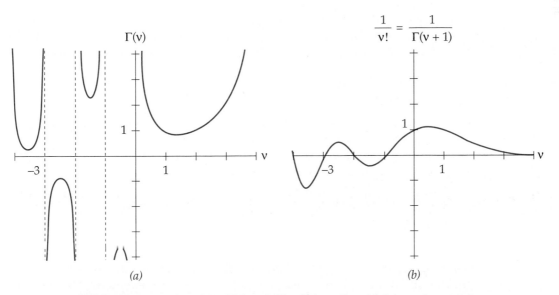

Figure 9.10 Graphs of (*a*) $\Gamma(\nu)$ and (*b*) $1/\Gamma(\nu + 1) = 1/\nu!$ for $-4 < \nu < 4$.

(Thus $\Gamma(\nu)$ for $0 > \nu$ not equal to a negative integer is defined in terms of the original $\Gamma(\nu)$ for $0 < \nu < 1$.) Suppose $-n < \nu < -n + 1$. Then it follows that $-n + 1 < \nu + 1 < -n + 2$, and

$$\Gamma(\nu + 1) = \frac{\Gamma(\nu + 1 + n - 1)}{(\nu + 1) \cdots (\nu + 1 + n - 2)}$$

$$= \frac{\Gamma(\nu + n)}{(\nu + 1) \cdots (\nu + n - 1)} = \nu\Gamma(\nu).$$

The quotient relation shows that $\Gamma(\nu) \to \pm\infty$ as ν approaches 0 or a negative integer from one side or the other, so it is natural to define $1/\nu! = 1/\Gamma(\nu + 1)$ to be 0 at negative integers. Computing values for $\nu = \pm\frac{1}{2}$ is left as an exercise.

∎

Series Expansions for J_ν. At first sight the graphs shown in Figure 9.10 for $\nu < 0$ look like a rather arbitrary attempt to extend the definition of $1/n!$; the justification is that using these definitions allows us to extend the definition of $J_n(x)$ and other important functions in a convincing way to values of n that aren't nonnegative integers. Replacing the integer $n \geq 0$ by the real index ν in the definition of $J_n(x)$ gives

$$J_\nu(x) = \sum_{j=0}^{\infty} \frac{(-1)^j (x/2)^{2j+\nu}}{j!(\nu + j)!} = \sum_{j=0}^{\infty} \frac{(-1)^j (x/2)^{2j+\nu}}{j!\Gamma(\nu + j + 1)}.$$

This series converges for all $x > 0$ and represents an analytic function if ν is an integer. Even if ν isn't an integer, x^ν can be factored out, so we see that $J_\nu(x)$ is x^ν times a function that is analytic for all x and so can be differentiated as often as we like. While it's routine and tedious to verify that $J_\nu(x)$ satisfies the Bessel equation for $x > 0$, this fact follows formally from knowing the result for an arbitrary integer $\nu = n$; we can then just replace n by ν in that term-by-term verification and use the fundamental relation $\nu! = \nu(\nu - 1)!$ as we would for arbitrary positive integer index. The same remarks apply if we make a formal replacement of n by $-\nu < 0$ so that both $J_\nu(x)$ and $J_{-\nu}(x)$ are solutions of the Bessel equation of index ν. (Recall that the roots of the indicial equation are $\pm\nu$.) Here is a summary of some important relations involving Bessel functions.

■ 7.7 THEOREM

The Bessel equation $x^2y'' + xy' + (x^2 - \nu^2)y = 0$ has $J_\nu(x)$ and $J_{-\nu}(x)$ for an independent pair of solutions precisely when $\nu > 0$ is not an integer, and $J_{-\nu}(x)$ is unbounded near $x = 0$. If $\nu = n$ is an integer, then $J_n(x) = (-1)^n J_{-n}(x)$. For each real ν, **Bessel recurrence identities** hold:

$$J_{\nu+1}(x) = \frac{2\nu}{x} J_\nu(x) - J_{\nu-1}(x) = \frac{\nu}{x} J_\nu(x) - J'_\nu(x).$$

Proof. Assume $\nu > 0$. Then $J_\nu(x) = x^\nu C_1(x)$ where $C_1(x)$ is continuous, so $J_\nu(x)$ is bounded near $x = 0$. If $-\nu$ is not a negative integer, then none of the coefficients in $J_{-\nu}(x) = x^{-\nu} C_2(x)$ is zero, so the function is unbounded as x approaches 0 from the right. Hence, J_ν and $J_{-\nu}$ must be independent. On the other

hand, if $-\nu = -n$ is a negative integer, then $1/(-n + j)! = 0$ for $j = 1, 2, \ldots,$
$n - 1$, so

$$J_{-n}(x) = \sum_{j=0}^{\infty} \frac{(-1)^j (x/2)^{2j-n}}{j!(-n+j)!} = \sum_{j=n}^{\infty} \frac{(-1)^j (x/2)^{2j-n}}{j!(-n+j)!}$$

$$= \sum_{j=0}^{\infty} \frac{(-1)^{j+n}(x/2)^{2j+n}}{(j+n)!j!} = (-1)^n J_n(x).$$

The proofs of the recurrence identities are also routine exercises in series ma-
nipulation. ∎

The functions $J_{n+1/2}(x)$, where n is an integer, are exceptional in that they
are "elementary" functions.

---

e x a m p l e 4

From the definition we have

$$J_{-1/2}(x) = \sum_{j=0}^{\infty} \frac{(-1)^j (x/2)^{2j-1/2}}{j!(j-\frac{1}{2})!} = \sqrt{2}x^{-1/2} \sum_{j=0}^{\infty} \frac{(-1)^j x^{2j}}{2^{2j}j!(j-\frac{1}{2})!}.$$

To simplify the terms in the sum, note that

$$(j-\tfrac{1}{2})! = (j-\tfrac{1}{2})(j-\tfrac{3}{2})\cdots(j-(j-\tfrac{1}{2}))(-\tfrac{1}{2})! = 2^{-j}(2j-1)(2j-3)\cdots 3 \cdot 1\sqrt{\pi}.$$

Hence $2^{2j}j!(j-\frac{1}{2})! = (2j)!\sqrt{\pi}$. It follows that the last sum displayed above is
the Taylor expansion of $\cos x$ about 0, so $J_{-1/2}(x) = \sqrt{2/\pi x}\cos x$. A similar
computation with $\nu = \frac{1}{2}$ would show that $J_{1/2}(x) = \sqrt{2/\pi x}\sin x$, but the same
result can be achieved more easily using the expression for $J_{\nu+1}$ in terms of J_ν
and J'_ν in Theorem 7.7.

---

e x a m p l e 5

Using the results of Example 4 and Theorem 7.7, we find for $x > 0$

$$J_{3/2}(x) = \frac{1}{x}J_{1/2}(x) - J_{-1/2}(x)$$

$$= \frac{1}{x}\left(\sqrt{\frac{2}{\pi x}}\sin x\right) - \sqrt{\frac{2}{\pi x}}\cos x = \sqrt{\frac{2}{\pi}}x^{-3/2}(\sin x - x\cos x).$$

EXERCISES

1. Differentiate the series for $J_0(x)$ to show that $J'_0(x) = -J_1(x)$ for all x.

2. Differentiate the series for $J_1(x)$ to show that $J'_1(x) = J_0(x) - J_1(x)/x$ for all x.

3. Show that the Bessel equation of order ν does not have a solution of the
form $y = x^\mu$ for $x > 0$.

4. A standard theorem on convergent alternating series $\sum_{k=0}^{\infty}(-1)^k a_k, \; a_k \geq 0$,
states that the error e_n made in replacing the sum by a partial sum $s_n =$

$\sum_{k=0}^{n} (-1)^k a_k$ satisfies $|e_n| \leq a_{n+1}$. Apply this theorem along with the two previous exercises to show the following.

(a) $J_0(1) \approx 0.7651977.$

(c) $J_1(1) \approx 0.4400506.$

(b) $J_0'(1) \approx -0.4400506.$

(d) $J_1'(1) \approx 0.3251471.$

5. Use the appropriate numerical results from the previous exercise in a numerical differential equation solver to plot the following. Recall that the initial-value problems at 0 are not uniquely determined and that numerical approximation for the initial-value problem starting at $x = 1$ requires a negative step $(h < 0)$ on the interval $0 \leq x \leq 1$.

(a) Plot the graph of $J_0(x)$ for $0 \leq x \leq 10$.

(b) Plot the graph of $J_1(x)$ for $0 \leq x \leq 10$.

6. Use the numerical estimates from Exercise 3 in a numerical differential equation solver to estimate the location of the following to three decimal places.

(a) The first five positive zeros of $J_0(x)$.

(b) The first five positive zeros of $J_1(x)$.

7. The relationship between the Bessel equation and the harmonic oscillator equation can be illuminated as follows.

(a) Let $u(x) = x^{1/2} y(x)$ for $x > 0$, and show that the function $u(x)$ satisfies the equation $u'' + (1 + (\frac{1}{4} - \nu^2)/x^2)u = 0$ if and only if $y(x)$ is a solution of $x^2 y'' + xy + (x^2 - \nu^2)y = 0$ for $x > 0$. Note that for large x the differential equation for u is approximately $u'' + u = 0$.

(b) Use part (a) to show that the Bessel equation of index $\nu = \frac{1}{2}$ has for $x > 0$ the general solution

$$y(x) = c_1 \frac{\cos x}{\sqrt{x}} + c_2 \frac{\sin x}{\sqrt{x}}.$$

(It can be shown that for large values of x, solutions of Bessel equations have approximately this form, which explains the spacing of the zeros for large x.)

8. **Weakening spring.** The harmonic oscillator equation $y'' + h_0 y = 0$, with constant $h_0 > 0$, represents the undamped oscillation of a spring. If the stiffness of the spring undergoes slow exponential decay over time, then h_0 should be replaced by $h(t) = h_0 e^{-at}$, for some small constant $a > 0$. (For a numerical analysis of the more general linear oscillator, see Exercise 21 for Chapter 4, Sections 7A and 7B, titled "Time-dependent linear spring mechanism.")

(a) Use the chain rule to show that the change of variable $u = ce^{-dt}$ transforms the differential equation into

$$d^2 u^2 \frac{d^2 y}{du^2} + d^2 u \frac{dy}{du} + h_0 \left(\frac{u}{c}\right)^{a/d} y = 0.$$

(b) Show that the choices $c = 2\sqrt{h_0}/a$ and $d = a/2$ give the zero-index Bessel equation $u^2 y'' + uy' + u^2 y = 0$.

(c) Show that bounded solutions of the original oscillator equation have the form

$$y(t) = CJ_0\left(\frac{2\sqrt{h_0}}{a}\,e^{-at/2}\right).$$

(d) Show that the number of oscillations of the solution in part (c) is limited by the number of sign changes of $J_0(x)$ in the interval $0 < x \le 2\sqrt{h_0}/a$.

9. The index-ν Bessel functions of the **first kind** are the nontrivial constant multiples of $J_\nu(x)$. For fixed ν these functions all have the same zeros, and it follows from the uniqueness theorem for second-order equations that a (necessarily independent) index-ν Bessel function of the second kind has all its zeros different from those of J_ν. Use the modified Euler method to generate approximate solutions of the appropriate Bessel equation to the right and left of the given point to make computer graphics plots of the following functions of the second kind.

(a) The index-0 function Z_0 such that $Z_0(\frac{1}{2}) = 0$ and $Z_0'(\frac{1}{2}) = 1$.

(b) The index-$\frac{1}{2}$ function $Z_{1/2}$ such that $Z_{1/2}(\pi/2) = 0$ and $Z_{1/2}'(\pi/2) = -\sqrt{2/\pi}$. This is an elementary function; can you identify it?

(c) The index-1 function Z_1 such that $Z_1(\frac{1}{2}) = 0$ and $Z_1'(\frac{1}{2}) = 1$.

(d) The index-$(-\frac{1}{2})$ function $Z_{-1/2}$ such that $Z_{-1/2}(\pi/2) = 0$ and $Z_{-1/2}'(\pi/2) = -(2/\pi)^{3/2}$. This is an elementary function; can you identify it?

10. Find a solution of the boundary problem $y(2) = 3$, $y(x)$ bounded near 0, for the equation $x^2y'' + xy' + (x^2 - 1)y = 0$.

11. Consider the equation $x^2y'' + xy' + k^2x^2y = 0$, where k is constant.

(a) Show that the equation has solutions $y(x) = c_1 J_0(kx)$.

(b) Let $k = 2$, and solve the boundary problem with conditions $y(3) = 4$ and $y(x)$ bounded near 0.

12. Consider the equation $x^2y'' + xy' + (k^2x^2 - 1)y = 0$, where k is constant.

(a) Show that the equation has solutions $y(x) = c_1 J_1(kx)$.

(b) Let $k = 5$, and solve the boundary problem with conditions $y(2) = 1$ and $y(x)$ bounded near 0.

Gamma Function

13. Show that $(-\frac{1}{2})! = \Gamma(\frac{1}{2}) = \sqrt{\pi}$ by first showing that $\int_0^\infty t^{-1/2}e^{-t}\,dt = 2\int_0^\infty e^{-u^2}\,du$. Then compute this last integral by observing that it is the square root of the double integral

$$\int_0^\infty \int_0^\infty e^{-u^2-v^2}\,du\,dv,$$

which can be evaluated by changing to polar coordinates in the first quadrant of the uv plane.

14. Use $\Gamma(\nu + 1) = \nu!\,\Gamma(\nu)$ together with the result of the previous exercise to show that $\frac{1}{2}! = \Gamma(\frac{3}{2}) = \sqrt{\pi}/2$.

15. The reciprocal factorial $1/\nu!$ can also be defined for all ν, consistent with the conventional interpretation of $n!$ for integer n, by

$$\frac{1}{\nu!} = \lim_{k \to \infty} \frac{(\nu + 1)(\nu + 2) \cdots (\nu + k)}{k! k^{\nu}}.$$

It can be proved that this limit exists and is finite for all ν. Under just that assumption, prove the following.

(a) $(\nu + 1) \cdot 1/(\nu + 1)! = 1/\nu!$.

(b) $1/\nu! = 0$ if ν is a negative integer.

(c) $1/\nu!$ has the conventional meaning when ν is a positive integer n. Note that

$$\frac{1}{n!} = \frac{(n + 1) \cdots (n + k)}{n!(n + 1) \cdots (n + k)} = \frac{(n + 1) \cdots (n + k)k^{n}}{k!(k + 1) \cdots (k + n)k^{n}}.$$

(It can be proved that the function $1/\nu!$ defined here is the same as the one defined in the text using $\Gamma(\nu)$.)

General Index ν

16. Use the relevant infinite series to prove:

(a) $J_{\nu+1}(x) = (2\nu/x)J_{\nu}(x) - J_{\nu-1}(x)$.

(b) $J_{\nu+1}(x) = (\nu/x)J_{\nu}(x) - J'_{\nu}(x)$.

17. Show that $J_{\nu-1}(x) = (\nu/x)J_{\nu}(x) + J'_{\nu}(x)$.

18. Show that $J_{1/2}(x) = \sqrt{2/\pi x}\, \sin x$:

(a) Using the appropriate identity in Theorem 7.7 together with $J_{-1/2}(x) = \sqrt{2/\pi x}\, \cos x$.

(b) Directly from the Frobenius series for $J_{1/2}(x)$.

*19. The form of a recurrence identity stated for the Bessel functions J_{ν} in Theorem 7.7 can be used to generate, for fixed ν, a sequence of solutions to the Bessel equations of index $\nu, \nu + 1, \nu + 2, \ldots$, in the following steps. Let Z_{ν} be an arbitrary solution of the index-ν equation so that

$$Z''_{\nu} + \frac{1}{x} Z'_{\nu} + \left(1 - \frac{\nu^2}{x^2}\right) Z_{\nu} = 0.$$

Define $Z_{\nu+1}$ by

$$Z_{\nu+1}(x) = \frac{\nu}{x} Z_{\nu}(x) - Z'_{\nu}(x).$$

(a) Show that $xZ'_{\nu+1} = (\nu + 1)Z'_{\nu} + (x - (\nu^2 + \nu)/x)Z_{\nu}$.

(b) Show that $(xZ'_{\nu+1})' = (x - (\nu + 1)^2/x)Z'_{\nu} - \nu(1 - (\nu + 1)^2/x^2)Z_{\nu}$.

(c) Show that $Z_{\nu+1}$ satisfies the Bessel equation of index $\nu + 1$.

20. For noninteger ν define the **Weber function** Y_{ν} by

$$Y_{\nu}(x) = \frac{J_{\nu}(x) \cos \pi\nu - J_{-\nu}(x)}{\sin \pi\nu}.$$

For integer n it can then be shown that $Y_n(x) = \lim_{\nu \to n} Y_{\nu}(x)$ defines a solution independent of $J_n(x)$ for the index-n Bessel equation. You are asked to use the recurrence identities stated for $J_n(x)$ in Theorem 7.7 to show that $Y_n(x)$ satisfies the same relations.

(a) Show that $Y_{n+1}(x) = (2n/x)Y_n(x) - Y_{n-1}(x)$.

(b) Show that $Y_{n+1}(x) = (n/x)Y_n(x) - Y_n'(x)$.

21. (a) Apply L'Hôpital's rule to the limit definition of $Y_n(x)$ in the previous exercise to show that the Weber function Y_n of integer index satisfies

$$Y_n(x) = \frac{1}{\pi} \lim_{\nu \to n} \frac{\partial}{\partial \nu} (J_\nu(x) - (-1)^n J_{-\nu}(x)).$$

(b) Assuming that limits and order of partial differentiation can be interchanged as needed, use the result of part (a) to verify formally that $Y_n(x)$ satisfies the Bessel equation of index n. (The stated assumption isn't universally valid but can be justified in this case.)

(c) Show that Y_n and J_n are independent by showing that $Y_n(x)$ is unbounded for x near zero. [*Hint:* Consider the series expansions for J_ν and $J_{-\nu}$.]

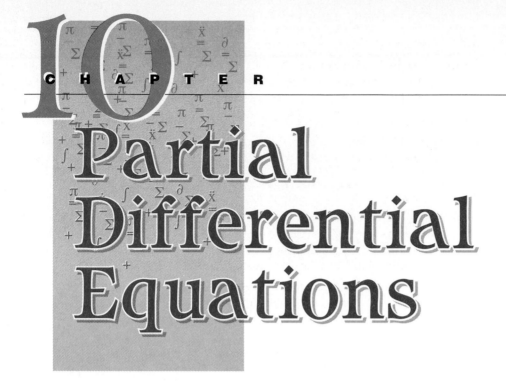

10 CHAPTER
Partial Differential Equations

1 INTRODUCTION

The purpose of this section is to acquaint the reader with some prominent similarities and differences between ordinary and partial differential equations.

1A Equations and Solutions

Differential equations that involve partial derivatives of some real function $u(x, y, \ldots)$ of several variables arise just as often in pure and applied mathematics as do ordinary differential equations for functions of one variable. For example, the height $u(x, t)$ of a moving water wave is usually treated as a function of position x in some direction and time t. The differential equation that $u(x, t)$ satisfies under various physical conditions may contain partial derivatives

$$\frac{\partial u}{\partial x}, \frac{\partial u}{\partial t}, \frac{\partial^2 u}{\partial x^2}, \frac{\partial^2 u}{\partial t \partial x}, \frac{\partial^2 u}{\partial t^2}, \ldots,$$

both with respect to x and t. (To make the formulas easier to write, we will sometimes use the alternative notations

$$u_x, u_t, u_{xx}, u_{xt}, u_{tt}, \ldots$$

for these partials.) Our examples will be restricted to linear equations of at most second order in two independent variables, say x and y; such an equation has the form

■ **1.1** $au_{xx} + bu_{xy} + cu_{yy} + du_x + eu_y + fu = g,$

where the coefficients a, b, c, d, e, f and the function g are assumed to be continuous functions, perhaps constant, of x and y in some common region of the xy plane. In addition, we will look only for solutions $u(x, y)$ all of whose first- and second-order partial derivatives are continuous, so that in particular the relation $u_{xy} = u_{yx}$ will hold for these solutions.

---

e x a m p l e 1

In the region of the xy plane for which $y > 0$, the "upper half-plane," we will solve the equation

$$yu_{xy} = x.$$

Since $y > 0$, we can write the equation as $u_{xy} = x/y$ and integrate both sides with respect to y, for each *fixed* x, getting

$$u_x(x, y) = x \ln y + c(x).$$

The integration "constant" $c(x)$ can be chosen to be a differentiable function of x. Now integrate both sides with respect to x:

$$u(x, y) = \tfrac{1}{2}x^2 \ln y + C(x) + D(y).$$

Here $C'(x) = c(x)$ is constant (as a function of y), and so will be a twice-differentiable function of y. Assuming $D(y)$ is also twice-differentiable, we have just found the most general twice-differentiable solution of the equation for $y > 0$.

For a second-order linear differential equation of ordinary type (i.e., with a single independent variable), we are used to having two arbitrary constants in the general solution formula. Notice, however, that in Example 1 the general solution contains two arbitrary twice-differentiable *functions*, allowing for a wide variety in the choice of a particular solution. This wide freedom of choice is an indication that we can expect to find particular solutions satisfying fairly complicated combinations of boundary and initial conditions. We will refer to boundary and initial conditions together as **side conditions,** and they will be explored later in the chapter in the context of specific physical problems.

---

e x a m p l e 2

With $u = u(x, y)$, the differential equation

$$u_{xx} - y^2 u = 0, \qquad y > 0,$$

is peculiar in that it contains a derivative only with respect to x, even though the solution will be of the form $u = u(x, y)$. (In a sense we're really dealing with an ordinary differential equation here, and solving the equation depends only on techniques for ordinary differential equations.) Thinking of y as held fixed, we find

$$u(x, y) = c_1(y)e^{xy} + c_2(y)e^{-xy}.$$

Some conditions that single out one solution are, for example,

$$u(0, y) = g(y), \qquad u_x(0, y) = h(y),$$

where g and h are preassigned functions of y. (Note that in this example the solution need not be even once-differentiable as a function of y.) Imposing these conditions on $u(x, y)$ leads to the following linear equations for determin-

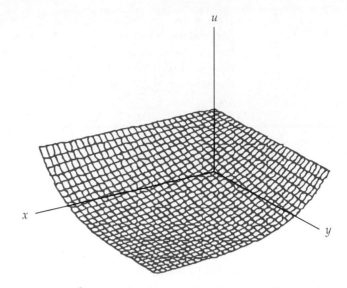

Figure 10.1 $u_{xx} - y^2 u = 0$, $u(0, y) = 1$, $u_x(0, y) = -y$; solution $u(x, y) = e^{-xy}$.

ing $c_1(y)$ and $c_2(y)$:

$$c_1(y) + c_2(y) = g(y), \qquad yc_1(y) - yc_2(y) = h(y).$$

The unique solution is $c_1(y) = \frac{1}{2}(g(y) + h(y)/y)$, $c_2(y) = \frac{1}{2}(g(y) - h(y)/y)$. Thus the particular solution satisfying the side conditions is

$$u(x, y) = \frac{1}{2}\left[g(y) + \frac{h(y)}{y}\right]e^{xy} + \frac{1}{2}\left[g(y) - \frac{h(y)}{y}\right]e^{-xy}.$$

In particular, if $g(y) = u(0, y) = 1$ and $h(y) = u_x(0, y) = -y$, we get $u(x, y) = e^{-xy}$. The graph of this solution appears in Figure 10.1 for $0 < x < 2, 0 < y < 2$.

The partial differential operator

$$L = a\frac{\partial^2}{\partial x^2} + b\frac{\partial^2}{\partial y \partial x} + c\frac{\partial^2}{\partial y^2} + d\frac{\partial}{\partial x} + e\frac{\partial}{\partial y} + f$$

is easily verified to be linear in its action on twice-differentiable functions u and v:

$$L(\alpha u + \beta v) = \alpha Lu + \beta Lv, \qquad \alpha, \beta \text{ const.}$$

Just as for ordinary differential operators, it follows that we can restate linearity as a **superposition principle** for homogeneous equations $Lu = 0$. This principle just says that if u, v are homogeneous solutions, then so is a linear combination $\alpha u + \beta v$. For partial differential operators, some of the most important solution techniques use the extension to an arbitrary number of homogeneous solutions $u_1, \ldots, u_n$. Repeated application of the two-term formula gives

$$L(\alpha_1 u_1 + \cdots + \alpha_n u_n) = \alpha_1 Lu_1 + \cdots + \alpha_n Lu_n.$$

In this way, we start to construct solutions to general boundary problems from more elementary solutions.

1B   *Exponential Solutions*

For a constant-coefficient homogeneous equation

■■ **1.2** $au_{xx} + bu_{xy} + cu_{yy} + du_x + eu_y + fu = 0,$

we can always find solutions of exponential form, just as in the case of a linear ordinary differential equation with constant coefficients. Guided by what worked for those linear differential equations, we assume a solution of the form $u(x, y) = e^{rx+sy}$. We see that $u_x = ru$, $u_y = su$, $u_{xx} = r^2u$, $u_{xy} = rsu$, and $u_{yy} = s^2u$. Thus a solution of the simple exponential form will have to have its parameters r, s chosen so that its **characteristic equation**

■■ **1.3** $ar^2 + brs + cs^2 + dr + es + f = 0$

is satisfied. The graph of this equation is an ellipse, a parabola, or a hyperbola accordingly as $b^2 - 4ac$ is negative, zero, or positive, and the nature of the solutions to the corresponding differential equation depends significantly on which condition holds. Since the type of the differential equation depends just on the quadratic terms, $ar^2 + brs + cs^2$ is called the **characteristic polynomial** of the differential equation. Examples of each of the three types will be taken up in detail later in the chapter.

---

e x a m p l e 3

The equation

$$u_{xx} - u_{yy} = 0$$

has characteristic equation $r^2 - s^2 = 0$ or $r^2 = s^2$. Thus exponential solutions have the form

$$u(x, y) = e^{rx+ry} \qquad \text{or} \qquad u(x, y) = e^{rx-ry},$$

since $s = \pm r$. Corresponding to each complex number r, there is a pair of solutions

$$u(x, y) = \begin{cases} e^{r(x+y)}, \\ e^{r(x-y)}. \end{cases}$$

By forming linear combinations of these basic exponential solutions, it is possible to obtain representations for solutions satisfying many different boundary conditions. For example, the conditions

$$u(x, 0) = \cos \beta x,$$
$$u_y(x, 0) = 0,$$

for $-\infty < x < \infty$, are satisfied by $u(x, y) = \cos \beta x \cos \beta y$. Since

$$\cos \beta x \cos \beta y = \tfrac{1}{2}(e^{i\beta x} + e^{-i\beta y}) \cdot \tfrac{1}{2}(e^{i\beta y} + e^{-i\beta y})$$
$$= \tfrac{1}{4}(e^{i\beta(x+y)} + e^{i\beta(x-y)} + e^{-i\beta(x+y)} + e^{-i\beta(x-y)}),$$

$\cos \beta x \cos \beta y$ is a solution of the differential equation that we get by letting $r = i\beta$ and $r = -i\beta$ in the exponential solutions and forming linear combinations. The techniques of **Fourier expansion** explained in Section 2 will replace such computations by a straightforward routine.

EXERCISES

1. Verify by direct substitution that each of the following differential equations is satisfied by the given function. Then verify that the function itself satisfies the given side condition.

   (a) $u_{xx} - u_y = 0$; $u(x, y) = e^{-y} \sin x$; $u(0, y) = 0$.

   (b) $u_{xx} + u_{yy} = 0$; $u(x, y) = e^y \sin x$; $u(0, y) = u(\pi, y) = 0$.

   (c) $u_{xx} - u_{yy} = 0$; $u(x, y) = \sin x \sin y$; $u(x, 0) = u(x, \pi) = 0$.

   (d) $u_{xx} - 4u_y = 0$; $u(x, y) = y^{-1/2} e^{-x^2/y}$, $y > 0$; $\lim_{x \to \infty} u(x, y) = 0$.

   (e) $x^2 u_{xx} + x u_x + u_{yy} = 0$; $u(x, y) = x \sin y$; $u(x, 0) = u(x, \pi) = 0$.

2. (a) Solve the differential equation $u_{xy} = 0$ by integrating first with respect to y, then x.

   (b) Solve the differential equation $u_{xy} = 1$ by integrating first with respect to y, then x.

   (c) Verify that $u_p(x, y) = xy$ is a particular solution of $u_{xy} = 1$. Then use the linearity of the partial differential operator $L = \partial^2/\partial y \partial x$ to find the general twice-differentiable solution of $u_{xy} = 1$ by using the solution to part (a). [*Hint:* Use Theorem 3.1 of Chapter 3, Section 3.]

3. Find the general twice-differentiable solution $u(x, y)$ of the differential equation
$$u_{xx} + 2u_x + u = y.$$

4. To find the general solution in $u(x, y)$ of $u_{xx} - u_{yy} = 0$ for all x, y, define new variables z, w by
$$z = x + y,$$
$$w = x - y.$$
Define $\bar{u}$ as a function of two variables by
$$\bar{u}(z, w) = u\left(\frac{z + w}{2}, \frac{z - w}{2}\right).$$

   (a) Show that if (z, w) corresponds to (x, y) as above, then $\bar{u}(z, w) = u(x, y)$.

   (b) Assuming that $u_{xx} - u_{yy} = 0$, use the chain rule to show that $\bar{u}_{zw} = 0$. [*Hint:* Compute $u_{xx} - u_{yy}$ in terms of $\bar{u}_{zz}$, $\bar{u}_{zw}$, and $\bar{u}_{ww}$.)

   (c) Show that the general solution of $\bar{u}_{zw} = 0$ is $\bar{u}(z, w) = g(z) + h(w)$, where g, h are arbitrary twice-differentiable functions.

   (d) Conclude that the solution of the original equation is
$$u(x, y) = g(x + y) + h(x - y).$$

5. We consider here the geometric properties of the set of points in the rs plane that satisfy the characteristic equation
$$ar^2 + brs + cs^2 + dr + es + f = 0$$
of the associated operator
$$L = a\frac{\partial^2}{\partial x^2} + b\frac{\partial^2}{\partial y \partial x} + c\frac{\partial^2}{\partial y^2} + d\frac{\partial}{\partial x} + e\frac{\partial}{\partial y} + f;$$

these properties provide an important classification of the operators L. If a, b, c are not all zero, the characteristic equation is that of a conic section, perhaps consisting only of lines or a point, of elliptic, parabolic, or hyperbolic type. The corresponding operator L is then said to be, respectively, **elliptic, parabolic,** or **hyperbolic.** For example, $\partial^2/\partial x^2 + \partial^2/\partial y^2$ is elliptic, because $r^2 + s^2 = 0$ is a degenerate ellipse; $\partial^2/\partial x^2 + \partial/\partial y$ is parabolic, because $r^2 + s = 0$ is a parabola; $\partial^2/\partial x^2 - \partial^2/\partial y^2$ is hyperbolic, because $r^2 - s^2 = 0$ is a degenerate hyperbola. Classify each of the following operators as elliptic, parabolic, or hyperbolic.

(a) $\dfrac{\partial^2}{\partial x^2} + \dfrac{\partial^2}{\partial y^2} - \dfrac{\partial}{\partial x}$.

(c) $\dfrac{\partial^2}{\partial x^2} - 2\dfrac{\partial^2}{\partial y^2} + 4\dfrac{\partial}{\partial y}$.

(b) $\dfrac{\partial^2}{\partial x \partial y}$.

(d) $\dfrac{\partial^2}{\partial y^2} - \dfrac{\partial}{\partial x} - 3$.

6. The partial differential equation $u_{xx} + u_{yy} = u$ has solutions of the form $u(x, y) = e^{rx+sy}$ for certain constants r and s. Find the relationship that must hold between r and s for such a solution to exist. Then write a formula for such a solution under the assumption that $r = 1$.

7. The partial differential equation $u_{xx} + 2u_x + u = u_{yy}$ has solutions of the form $u(x, y) = e^{rx+sy}$ for certain constants r and s. Find the relationship that must hold between r and s for such a solution to exist.

8. The partial differential equation $u_{xx} - u_{yy} + u_y = 0$ has solutions of the form $u(x, y) = e^{rx+sy}$ for certain constants r and s. Find the relationship (i.e., characteristic equation) that must be satisfied by both r and s for such a solution to exist. Then write a formula for all such solutions in terms of x, y, and s.

9. Find all solutions of the **telegraph equation** $u_{tt} + bu_t + cu = au_{xx}$, a, b, c positive constants, that are independent of t; that is, find the *steady-state* solutions. [*Hint:* $u_t = u_{tt} = 0$.]

10. **(a)** Find the characteristic equation of the telegraph equation $u_{tt} + bu_t + cu = au_{xx}$, a, b, c nonnegative constants.

  (b) Show that product solutions $u(x, t) = e^{rx+st}$ of the telegraph equation cannot have r and s both purely imaginary unless $b = 0$.

Partial differential equations have lots of solutions that are not of exponential form; some of these are developed later in the chapter. The following exercises give some indication of the possible variety of solutions.

11. **(a)** Show that the **wave equation** $u_{tt} = a^2 u_{xx}$ has solution $u(x, t) = f(x - at)$, where f is an arbitrary twice-differentiable function of one real variable and a is constant. For example, if $f(z) = 1/(1 + z^2)$, then $f(x - at) = 1/(1 + (x - at)^2)$.

  (b) Use the linearity of the wave equation to show that it has solutions of the form $u(x, t) = f(x - at) + g(x - at)$.

12. **(a)** Show that the **Laplace equation** $u_{xx} + u_{yy} = 0$ has solutions $u(x, y) = f(x \pm iy)$, where $f(z)$ is an arbitrary **analytic function** given by a power series in z. (For example, if $f(z) = z^2$, then $u(x, y) = (x \pm iy)^2 = x^2 - y^2 \pm 2ixy$ are complex solutions.)

(b) Use the linearity of the Laplace equation to show that it has solutions of the form $u(x,y) = f(x + iy) + f(x - iy)$. What solution $u(x, y)$ results from choosing $f(z) = z^2$?

13. The 3rd-order nonlinear **Korteweg-deVries (KdV) wave equation** is

$$\frac{\partial u}{\partial t} + \beta u \frac{\partial u}{\partial x} + \frac{\partial^3 u}{\partial x^3} = 0, \qquad \beta \text{ const.}$$

(a) Show that if $u = u(x, t)$ is a nonzero solution and c is a constant different from 0 or 1, then $cu(x, t)$ is not a solution of the KdV equation with nonlinear term $\beta u u_x$ but satisfies the equation with β replaced by β/c.

(b) Show that the KdV equation with constant $\beta = 12$ has solution $u(x, t) = \text{sech}^2 (x - 2t)$, where $\text{sech}\, w = 1/\cosh w$ is the hyperbolic cosecant.

(c) Sketch the graph of the function $w(x) = u(x, t)$ in part (b) for $t = 0, 1$, and 2, and show that the solitary hump moves to the right above the x axis with velocity 4 as t increases.

2   FOURIER SERIES

2A   *Introduction*

Finite linear combinations of the exponential solutions discussed in Section 1B are not enough to satisfy the wide variety of side conditions that come up in practice. However, if we allow infinite series whose partial sums contain linear combinations of exponentials of the form $e^{i\beta x}$, or what amounts to the same thing, of $\cos \beta x$ and $\sin \beta x$, it turns out that we can produce solutions that satisfy many interesting side conditions. For the moment disregarding any question of convergence, we define a **trigonometric series** to be a series of the form

2.1
$$\frac{a_0}{2} + \sum_{k=1}^{\infty} (a_k \cos kx + b_k \sin kx).$$

In the special circumstance that the coefficients a_k, b_k arise from an integrable function $f(x)$ by means of the **Euler formulas**

2.2 $\qquad a_k = \frac{1}{\pi} \int_{-\pi}^{\pi} f(x) \cos kx \, dx, \qquad b_k = \frac{1}{\pi} \int_{-\pi}^{\pi} f(x) \sin kx \, dx,$

then the trigonometric series is called the **Fourier series** of f. The coefficients a_k, b_k as given by Equations 2.2 are called the **Fourier coefficients** of f. The most fundamental question about a Fourier series is the extent to which the series represents the function. The importance of such series for us right here stems from the fact that the individual terms of the series can be incorporated into a solution of some differential equation.

Our first examples are meant to illustrate the beautiful way in which the partial sums of a Fourier series attempt to mimic the function f that generates them. A partial sum

2.3 $\qquad\qquad S_N(x) = \frac{a_0}{2} + \sum_{k=1}^{N} (a_k \cos kx + b_k \sin kx)$

is called a **trigonometric polynomial.** Note that each term in S_N is a **periodic function** f of period 2π; that is, $f(x + 2\pi) = f(x)$ for all x. It follows that S_N is also periodic:

$$S_N(x + 2\pi) = S_N(x), \qquad \text{for all } x.$$

Note also that the Fourier coefficients a_k, b_k are determined by integral formulas that use the values of $f(x)$ only for $-\pi \le x \le \pi$. This restriction will be removed in Section 3.

2B   Computing Coefficients

The integrals in the Euler formulas deserve special consideration. If we let $f(x)$ equal $\cos lx$ or $\sin lx$ in Equations 2.2, it so happens that for integers k and l we have

■■ 2.4

$$\frac{1}{\pi} \int_{-\pi}^{\pi} \cos kx \sin lx \, dx = 0,$$

$$\frac{1}{\pi} \int_{-\pi}^{\pi} \cos kx \cos lx \, dx = \begin{cases} 0, & k \ne l, \\ 1, & k = l \ne 0. \end{cases}$$

$$\frac{1}{\pi} \int_{-\pi}^{\pi} \sin kx \sin lx \, dx = \begin{cases} 0, & k \ne l \text{ or } k = l = 0, \\ 1, & k = l \ne 0. \end{cases}$$

The three Equations 2.4 are called **orthogonality relations** for reasons explained in Chapter 11, and they follow easily from trigonometric identities. (See Exercise 5.) As a sample application of the orthogonality relations, suppose that a trigonometric series satisfies some condition that allows us to integrate it term by term on the interval $-\pi \le x \le \pi$. For example, the series might just be a finite sum, with all a_k and b_k equal to zero from some point on. We will prove the following theorem.

■■ 2.5 THEOREM

If the trigonometric series 2.1 converges to a function $f(x)$ and can be integrated term by term over the interval $[-\pi, \pi]$, then the coefficients a_k, b_k must be given by the Euler formulas 2.2.

Proof. We have defined $f(x)$ by

$$f(x) = \frac{a_0}{2} + \sum_{k=1}^{\infty} (a_k \cos kx + b_k \sin kx).$$

Then for a fixed integer $l > 0$,

$$\frac{1}{\pi} \int_{-\pi}^{\pi} f(x) \cos lx \, dx = \frac{1}{\pi} \int_{-\pi}^{\pi} \left[\frac{a_0}{2} + \sum_{k=0}^{\infty} (a_k \cos kx + b_k \sin kx) \right] \cos lx \, dx$$

$$\frac{a_0}{2\pi} \int_{-\pi}^{\pi} \cos lx \, dx + \sum_{k=1}^{\infty} \left[\frac{a_k}{\pi} \int_{-\pi}^{\pi} \cos kx \cos lx \, dx \right.$$

$$\left. + \frac{b_k}{\pi} \int_{-\pi}^{\pi} \sin kx \cos lx \, dx \right].$$

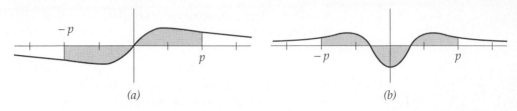

Figure 10.2 (*a*) Odd function. (*b*) Even function.

The first two relations in 2.4 show that all but one of these last terms is zero; the only term that survives is the term containing a_l, and we find, for $l \neq 0$,

$$\frac{1}{\pi} \int_{-\pi}^{\pi} f(x) \cos lx \, dx = \frac{a_l}{\pi} \int_{-\pi}^{\pi} \cos lx \cos lx \, dx = a_l.$$

When $l = 0$, the only nonzero term that survives in the sum is the first one. Since the integral of 1 over the interval is 2π, we get $2a_0$, which accounts for the term $\frac{1}{2}a_0$ in the expansion. A similar computation shows that $(1/\pi)\int_{-\pi}^{\pi} f(x) \sin lx \, dx = b_l$. ∎

Theorem 2.5 shows that, under fairly broad conditions on the coefficients of a trigonometric series, the coefficients must be given by the Euler formulas 2.2. In particular, the Fourier series of a trigonometric polynomial is precisely equal to the polynomial. (For example, $f(x) = \sin x + \cos 2x$ is its own Fourier series.) It follows that Formulas 2.2 for determining the a_k and b_k appear to be quite appropriate if we want to represent integrable functions $f(x)$ by trigonometric series using Formula 2.1.

The first of the three Equations 2.4 follows, without any detailed computation, from the useful observation that the integral of an odd function, $g(x) = \cos kx \sin lx$ in this case, over an interval symmetric about the origin is necessarily zero. Recall that an **odd function** $g(x)$ satisfies $g(-x) = -g(x)$ for $-p \leq x \leq p$ and that an **even function** satisfies $g(-x) = g(x)$ for $-p \leq x \leq p$. Figure 10.2*a* and *b* illustrates the two types. The pictures show geometrically, and it's easy to show analytically, that

■ 2.6

a. $\displaystyle\int_{-p}^{p} g(x) \, dx = 0$ if g is odd,

b. $\displaystyle\int_{-p}^{p} g(x) \, dx = 2\int_{0}^{p} g(x) \, dx$ if g is even.

---

e x a m p l e 1

Let $f(x) = |x|$ for $-\pi \leq x \leq \pi$. Then

$$a_k = \frac{1}{\pi} \int_{-\pi}^{\pi} |x| \cos kx \, dx, \qquad b_k = \frac{1}{\pi} \int_{-\pi}^{\pi} |x| \sin kx \, dx.$$

Clearly, $|x| \sin kx$ has integral zero over $[-\pi, \pi]$, because it is an odd function. Hence $b_k = 0$ for $k = 1, 2, \ldots$. On the other hand, the graph of $|x| \cos kx$ is symmetric about the y axis, so we can settle for twice its integral over $[0, \pi]$. For

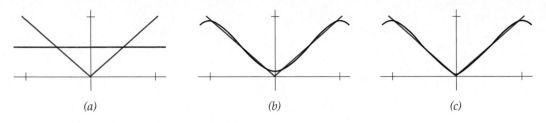

(a)   *(b)*   *(c)*

Figure 10.3 Fourier approximations to $f(x) = |x|$ on $[-\pi, \pi]$. (a) $s_0(x) = \pi/2$;
(b) $s_1(x) = s_0(x) - (4/\pi) \cos x$; (c) $s_3(x) = s_1(x) - (4/9\pi) \cos 3x$.

$k \neq 0$ we integrate by parts, getting

$$a_k = \frac{2}{\pi} \int_0^\pi x \cos kx \, dx$$

$$= \frac{2}{\pi} \left[\frac{x \sin kx}{k} \right]_0^\pi - \frac{2}{\pi k} \int_0^\pi \sin kx \, dx$$

$$= \left[\frac{2}{\pi k^2} \cos kx \right]_0^\pi = \frac{2}{\pi k^2} (\cos k\pi - 1)$$

$$= \frac{2}{\pi k^2} ((-1)^k - 1) = \begin{cases} 0, & k = 2, 4, 6, \ldots, \\ -4/(\pi k^2), & k = 1, 3, 5, \ldots. \end{cases}$$

When $k = 0$, we have $a_0 = (2/\pi) \int_0^\pi x \, dx = \pi$. To summarize,

$$b_k = 0, \quad k = 1, 2, 3, \ldots, \quad a_0 = \pi, \quad a_k = \begin{cases} 0, & k = 2, 4, 6, \ldots, \\ -\dfrac{4}{\pi k^2}, & k = 1, 3, 5, \ldots. \end{cases}$$

Hence, the Nth Fourier approximation is given for $N = 1, 3, 5, \ldots$ by the trigonometric polynomial

$$s_N(x) = \frac{\pi}{2} - \frac{4}{\pi} \cos x - \frac{4}{\pi} \frac{\cos 3x}{3^2} - \cdots - \frac{4}{\pi} \frac{\cos Nx}{N^2}.$$

If N is even, we have $s_N(x) = s_{N-1}(x)$. Figure 10.3 shows how the graphs of s_0, s_1, and s_3 approximate that of $|x|$ on $[-\pi, \pi]$; additional terms improve the approximation as we'll see below.

example 2

Let

$$g(x) = \begin{cases} 1, & 0 \le x \le \pi, \\ -1, & -\pi \le x < 0. \end{cases}$$

To compute a_k and b_k we break the interval of integration $[-\pi, \pi]$ at 0:

$$a_k = \frac{-1}{\pi} \int_{-\pi}^0 \cos kx \, dx + \frac{1}{\pi} \int_0^\pi \cos kx \, dx.$$

Since cosine is an even function, $\cos(-kx) = \cos kx$, the two integrals are equal, so we get $a_k = 0$. Similarly,

$$b_k = \frac{-1}{\pi} \int_{-\pi}^0 \sin kx \, dx + \frac{1}{\pi} \int_0^\pi \sin kx \, dx.$$

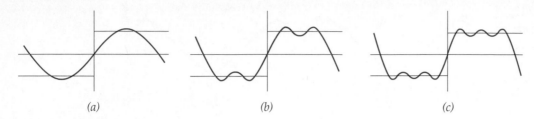

Figure 10.4 Fourier approximations to $g(x)$ on $[-\pi, \pi]$. (a) $s_1(x) = (4/\pi) \sin x$; (b) $s_3(x) = s_1(x) + (4/3\pi) \sin 3x$; (c) $s_5(x) = s_3(x) + (4/5\pi) \sin 5x$.

Since sine is an odd function, $\sin(-kx) = -\sin kx$, the integrals themselves are negatives of each other, and we get, since we can assume $k \neq 0$,

$$b_k = \frac{2}{\pi} \int_0^\pi \sin kx \, dx = \frac{2}{\pi} \left[\frac{-\cos kx}{k} \right]_0^\pi$$

$$= \frac{2}{k\pi} \, (-(-1)^k + 1) = \begin{cases} 0, & k \text{ even,} \\ \dfrac{4}{k\pi}, & k \text{ odd.} \end{cases}$$

In summary,

$$a_k = 0, \quad k = 0, 1, 2, \ldots, \quad b_k = \begin{cases} 0, & k = 2, 4, 6, \ldots, \\ \dfrac{4}{k\pi}, & k = 1, 3, 5, \ldots. \end{cases}$$

Hence, for odd N the Nth Fourier approximation to g is given by

$$s_N(x) = \frac{4}{\pi} \sin x + \frac{4}{\pi} \frac{\sin 3x}{3} + \cdots + \frac{4}{\pi} \frac{\sin Nx}{N}.$$

The graphs of s_1, s_3, and s_5 are shown in Figure 10.4, together with that of $g(x)$.

2C   Convergence

An important question is whether the Fourier approximations $s_N(x)$ converge as $N \to \infty$ to $f(x)$ for some values of x, where f is the function on $[-\pi, \pi]$ from which the Fourier coefficients are computed. Since the partial sums of a Fourier series are themselves periodic functions, a function to which they converge must also be periodic and is called a **periodic extension** of f. This periodic extension may differ from the precise definition of $f(x)$ at some points in the interval $-\pi \leq x \leq \pi$. Indeed, changing a value $f(x_0)$ at a point x_0 has no effect on the integral formulas for the Fourier coefficients, and it is customary to change values of f at some points whenever it's convenient in defining a periodic extension of a function from an interval to the entire real number line. Figure 10.5 sketches a periodic extension from the interval $-\pi \leq x \leq \pi$.

The Fourier series of f is by definition the infinite series

$$\frac{a_0}{2} + \sum_{k=1}^{\infty} (a_k \cos kx + b_k \sin kx),$$

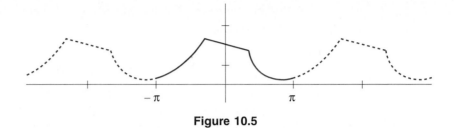

Figure 10.5

where a_k and b_k are given by the Euler formulas 2.2. Theorem 2.7 states conditions on f under which the Fourier series can be used to represent f. Thus we'll assume that the graph of f is not only bounded on $[-\pi, \pi]$ but is **piecewise-monotone,** which means that the interval $[-\pi, \pi]$ can be broken into finitely many subintervals, with endpoints $-\pi = x_1 < x_2 < \cdots < x_n = \pi$, such that f is either nondecreasing or nonincreasing on each open subinterval (x_k, x_{k+1}). It can be proved that the Fourier series of f will then converge to a 2π-periodic extension of f, also denoted by f. At a possible discontinuity at $x = x_0$ the series will converge to the "average" value

$$\tfrac{1}{2}[f(x_0-) + f(x_0+)].$$

Here $f(x_0-)$ stands for the left-hand limit of f at x_0, and $f(x_0+)$ is the right-hand limit. The graph of a typical piecewise-monotone function is shown in Figure 10.5, with the average value indicated by a dot at each jump and the periodic extension shown by dots. A somewhat stronger version of the next theorem is called **Dirichlet's theorem,** in which the conclusion is the same but the hypotheses are weaker.

■ 2.7 THEOREM

Let f be bounded and piecewise-monotone on $[-\pi, \pi]$. Then the 2π-periodic Fourier series of f converges at every point x to a periodic extension of f. In particular, if f is continuous at x, then the series converges to $f(x)$. If f is discontinuous at x_0, the series converges to the average value $\tfrac{1}{2}[f(x_0-) + f(x_0+)]$.

Examples 1 and 2 gave an indication of how partial sums of a Fourier series converge. In each of those examples, the function satisfies the conditions of boundedness and piecewise-monotonicity; hence the series converges to the value claimed in Theorem 2.7.

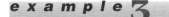 *example* **3**

The function g defined in Example 2 is rather arbitrarily defined to have the value 1 at $x = 0$. In spite of this arbitrariness, Theorem 2.7 allows us to conclude that the Fourier series of g converges as follows:

$$\sum_{k=0}^{\infty} \frac{4}{\pi} \frac{\sin (2k + 1)x}{2k + 1} = \begin{cases} 1, & 0 < x < \pi, \\ 0, & x = 0, \\ -1, & -\pi < x < 0. \end{cases}$$

To be very specific, we can set $x = \pi/2$ and arrive at the alternating series expansion

$$\sum_{k=0}^{\infty} \frac{(-1)^k}{2k + 1} = \frac{\pi}{4}.$$

Theorem 2.7 shows to some extent the reason for choosing the coefficients in a trigonometric polynomial according to Formulas 2.2. The reason is that, under favorable circumstances, the resulting sequence of trigonometric polynomials will converge to the function f. Figures 10.5 and 10.6 show functions f extended periodically, with period 2π, from the interval $[-\pi, \pi]$ to other values of x. Since the partial sums of the Fourier series of f are also periodic with period 2π, whatever convergence takes place on $[-\pi, \pi]$ extends periodically to all values of x. At this point Theorem 2.7 has implications only for functions of period 2π, but we'll see in Section 3 that the statement can easily be modified to be valid for functions with any positive period $2p$.

The coefficient values a_k and b_k that we get from the Euler formulas are quite independent of the finite values assigned to $f(x)$ at isolated points, because the definite integrals in the Euler formulas don't distinguish between two functions that differ at finitely many points. For example, the two functions

$$f(x) = \begin{cases} 0, & -\pi \le x \le 0, \\ 1, & 0 < x \le \pi, \end{cases} \quad \text{and} \quad g(x) = \begin{cases} 0, & -\pi \le x < 0, \\ 1, & 0 \le x \le \pi, \end{cases}$$

differ only at $x = 0$, where $f(0) = 0$ and $g(0) = 1$. Intuitively speaking the areas under the two graphs should be the same, namely, π. (That this is actually the case is a significant property of the integral.) Since the Fourier series of a piecewise-monotone function converges to the average of the right and left limits at each point x, it makes sense simply to redefine such a periodic function to have the average value $\frac{1}{2}(f(x_0+) + f(x_0-))$ at each jump discontinuity x_0 and refer to the end result as the **normalized function.**

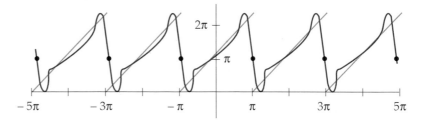

Figure 10.6 $f(x) = x + \pi$ extended periodically from $(-\pi, \pi)$ and normalized at jumps. The fourth Fourier partial sum is superimposed.

e x a m p l e 4   If the function $f(x) = x + \pi$ is extended periodically from $-\pi < x < \pi$ to other values of x, its graph consists of parallel line segments of slope 1; it remains undefined at odd integer multiples of π, since it's initially undefined at $\pm\pi$. To produce a normalized version of the function, defined for all x, all we have to do is define the function to have the average value π at odd integer multiples of π. The Fourier series can be computed for the function as originally defined, and the series will converge to the normalized function for real x. The periodically extended function is shown in Figure 10.6 together with the fourth partial sum of the Fourier expansion. The Fourier coefficients can be computed as follows.

$$a_0 = \frac{1}{\pi} \int_{-\pi}^{\pi} (x + \pi) \, dx$$

$$= \frac{1}{\pi} \int_{-\pi}^{\pi} x \, dx + \frac{1}{\pi} \int_{-\pi}^{\pi} \pi \, dx = 0 + 2\pi = 2\pi.$$

(Note that the integral of x over an interval symmetric about 0 is always 0.) When $k > 0$,

$$a_k = \frac{1}{\pi} \int_{-\pi}^{\pi} (x + \pi) \cos kx \, dx$$

$$= \frac{1}{\pi} \int_{-\pi}^{\pi} x \cos kx \, dx + \frac{1}{\pi} \int_{-\pi}^{\pi} \pi \cos kx \, dx = 0 + 0 = 0.$$

The last integral above is 0, because the indefinite integral is 0 at $\pm\pi$. The previous one is most easily seen to be zero by observing that the integrand, $x \cos kx$, is an odd function so that the integral over $[-\pi, 0]$ is the negative of the integral over $[0, \pi]$. Now for the b_k's,

$$b_k = \frac{1}{\pi} \int_{-\pi}^{\pi} (x + \pi) \sin kx \, dx$$

$$= \frac{1}{\pi} \int_{-\pi}^{\pi} x \sin kx \, dx + \frac{1}{\pi} \int_{-\pi}^{\pi} \pi \sin kx \, dx$$

$$= \frac{2(-1)^{k+1}}{k} + 0 = \frac{2(-1)^{k+1}}{k}.$$

The last integral above is zero, because the integrand is an odd function. The previous one can be computed using integration by parts, with

$$\frac{1}{\pi} \int_{-\pi}^{\pi} x \sin kx \, dx = -\frac{1}{\pi k} x \cos kx \Big|_{-\pi}^{\pi} + \frac{1}{\pi k} \int_{-\pi}^{\pi} \cos kx \, dx$$

$$= -\frac{1}{k} \cos k\pi - \frac{1}{k} \cos (-k\pi) + 0 = \frac{2(-1)^{k+1}}{k}.$$

The full expansion, including the constant $a_0/2$, is then

$$f(x) = \pi + \sum_{k=1}^{\infty} \frac{2(-1)^{k+1}}{k} \sin kx$$

$$= \pi + 2 \sin x - \sin 2x + \tfrac{2}{3} \sin 3x - \tfrac{1}{2} \sin 4x + - \cdots.$$

EXERCISES

The following observations about values of sine and cosine are useful for computing Fourier coefficients; k is always an integer here.

$$\sin k\pi = 0; \quad \cos k\pi = (-1)^k; \quad \sin \left(k + \tfrac{1}{2}\right)\pi = (-1)^k; \quad \cos \left(k + \tfrac{1}{2}\right)\pi = 0.$$

1. Compute the Fourier coefficients of each of the following functions, and write the corresponding Fourier series in the form of Equation 2.1. Sketch the graph of each function extended to have period 2π on the interval $-2\pi \le x \le 2\pi$. Finally, sketch the graphs of the first three partial sums $S_0(x)$, $S_1(x)$, $S_2(x)$ of the Fourier series.

   (a) $f(x) = x, \ -\pi < x < \pi.$

   (b) $f(x) = \begin{cases} -\pi - x, & -\pi < x < 0, \\ \pi - x, & 0 \le x \le \pi. \end{cases}$

   (c) $f(x) = x^2, \ -\pi < x < \pi.$

   (d) $f(x) = |x| + 1, \ -\pi \le x \le \pi.$

   (e) $f(x) = \begin{cases} 0, & -\pi < x \le 0, \\ 1, & 0 < x \le \pi. \end{cases}$

   (f) $f(x) = x + 1, \ -\pi < x < \pi.$

   (g) $f(x) = \begin{cases} -\pi, & -\pi < x < 0, \\ \pi, & 0 \le x \le \pi. \end{cases}$

   (h) $f(x) = 2x + 1, \ -\pi < x < \pi.$

   (i) $f(x) = -|x|, \ -\pi \le x \le \pi.$

   (j) $f(x) = \begin{cases} -1, & -\pi < x \le 0, \\ 2, & 0 < x \le \pi. \end{cases}$

2. According to Theorem 2.7, the Fourier series of each function $f(x)$ in Exercise 1 converges to some function $F(x)$ whose graph may or may not differ at some points from the graph of $f(x)$. In each case, sketch $F(x)$ on the interval $-2\pi \le x \le 2\pi$.

3. Show that if $f(x)$ and $g(x)$ have the Fourier coefficients a_k, b_k and a_k', b_k', respectively, then $\alpha f(x) + \beta g(x)$, α and β constant, has Fourier coefficients $\alpha a_k + \beta a_k'$, $\alpha b_k + \beta b_k'$.

4. The Nth partial sum of a trigonometric series is called a **trigonometric polynomial** of degree N, and a trigonometric polynomial is necessarily the Fourier series of the function it represents. For example, the identity $\cos^2 x = \tfrac{1}{2} + \tfrac{1}{2}\cos 2x$ is the Fourier expansion of $\cos^2 x$. Find the Fourier series of each of the following functions by using appropriate identities, for example the ones in Exercise 6.

   (a) $\sin^2 x.$

   (b) $\cos^3 x.$

   (c) $\sin 2x \cos x.$

   (d) $\sin^3 x.$

   (e) $\cos^4 x - \sin^4 x.$

5. Using elementary properties of integrals, prove Equations 2.6a and b of the text.

6. Establish the three orthogonality relations in Equations 2.4 of the text, as follows:

   (a) Use the trigonometric identities

   $$\cos \alpha \cos \beta = \tfrac{1}{2}\cos (\alpha + \beta) + \tfrac{1}{2}\cos (\alpha - \beta),$$
   $$\sin \alpha \cos \beta = \tfrac{1}{2}\sin (\alpha + \beta) + \tfrac{1}{2}\sin (\alpha - \beta),$$
   $$\sin \alpha \sin \beta = \tfrac{1}{2}\cos (\alpha - \beta) - \tfrac{1}{2}\cos (\alpha + \beta)$$

   to compute the relevant integrals.

(b) Use instead the identities
$$\cos nx = \tfrac{1}{2}(e^{inx} + e^{-inx}), \qquad \sin nx = (1/2i)(e^{inx} - e^{-inx})$$
together with the identity $e^{i(\alpha+\beta)} = e^{i\alpha}e^{i\beta}$ for the exponential function.

7. Carry out the details of the proof that $(1/\pi)\int_{-\pi}^{\pi} f(x) \sin lx\, dx = b_l$, parallel to the computation in Theorem 2.5.

8. Suppose $f(x) = x^2 - x$ for $-\pi \le x \le \pi$. What does the Fourier series of f converge to at $x = \pi$ and at $x = -\pi$?

9. Let $f(x) = \sqrt{|x|}$ for $-\pi \le x \le \pi$. Does f satisfy the hypotheses of Theorem 2.7?

10. Show that $f(x) = x \sin(1/x)$ is not piecewise-monotone for $0 < x < \pi$.

3 ADAPTED FOURIER EXPANSIONS

3A General Intervals

The direct application of Fourier methods to practical problems usually requires some modification of the standard formulation presented in Section 2. In the present section we describe some of these modifications and calculate some examples.

While the interval $[-\pi, \pi]$ is a natural one for Fourier expansions because it is a period interval for the trigonometric functions, it may be that a function encountered in an application needs to be approximated on some other interval. If the function f to be approximated is defined not on the interval $[-\pi, \pi]$ but on $[-p, p]$, a suitable change in the computation of the approximation can be made as follows. With f defined on $[-p, p]$, we define a function f_p restricted to $[-\pi, \pi]$ by

$$f_p(x) = f\left(\frac{px}{\pi}\right), \qquad -\pi \le x \le \pi.$$

Then we can compute the Fourier coefficients of f_p by Formulas 2.2. The resulting Fourier sums $s_N(x)$ will converge to f_p on $[-\pi, \pi]$ under the assumptions of Theorem 2.7 of Section 2C. To approximate f itself on $[-p, p]$, it's then natural to use the partial sums

$$s_N\left(\frac{\pi x}{p}\right) = \frac{a_0}{2} + \sum_{k=1}^{N}\left(a_k \cos\frac{k\pi x}{p} + b_k \sin\frac{k\pi x}{p}\right), \qquad -p \le x \le p.$$

The coefficients a_k and b_k can be computed directly in terms of f by making a change of variable as follows. We replace x by $\pi x/p$ throughout the integral over $[-\pi, \pi]$ to get

$$a_k = \frac{1}{\pi}\int_{-\pi}^{\pi} f_p(x) \cos kx\, dx = \frac{1}{\pi}\int_{-\pi}^{\pi} f\left(\frac{px}{\pi}\right)\cos kx\, dx$$

$$= \frac{1}{p}\int_{-p}^{p} f(x) \cos\frac{k\pi x}{p}\, dx.$$

A similar computation holds for b_k, so the coefficient formulas in terms of f itself are

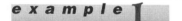 **3.1**
$$a_k = \frac{1}{p} \int_{-p}^{p} f(x) \cos \frac{k\pi x}{p} \, dx, \qquad b_k = \frac{1}{p} \int_{-p}^{p} f(x) \sin \frac{k\pi x}{p} \, dx$$

for the coefficients in the Fourier expansion

$$\frac{a_0}{2} + \sum_{k=1}^{N} \left(a_k \cos \frac{k\pi x}{p} + b_k \sin \frac{k\pi x}{p} \right)$$

for the $2p$-periodic extension of the function f originally defined on $[-p, p]$. We can conclude from the foregoing results that π **can be replaced by** p **in convergence Theorem 2.7 of Section 2C;** the most important assumption is the piecewise monotonicity of the function to be expanded.

---

e x a m p l e **1**

If
$$h(x) = \begin{cases} 1, & 0 \le x \le p, \\ -1, & -p \le x < 0, \end{cases}$$

then
$$a_k = 0, \qquad\qquad\qquad\qquad k = 0, 1, 2, \ldots,$$

$$b_k = \frac{2}{p} \int_0^p \sin \frac{k\pi x}{p} \, dx$$

$$= \frac{2}{\pi} \int_0^{\pi} \sin kx \, dx = \begin{cases} 0, & k = 2, 4, 6, \ldots, \\ 4/(\pi k), & k = 1, 3, 5, \ldots. \end{cases}$$

Hence the Nth Fourier approximation to h is given, for odd N, by

$$s_N(x) = \frac{4}{\pi} \sin \frac{\pi x}{p} + \frac{4}{3\pi} \sin \frac{3\pi x}{p} + \cdots + \frac{4}{N\pi} \sin \frac{N\pi x}{p}, \qquad -p \le x \le p.$$

For a function f defined on an arbitrary interval $a \le x \le b$, it is helpful to think of a periodic extension f_E of f having period $b - a$ and defined for all real numbers x. Such an extension is illustrated in Figure 10.7. We set $2p = b - a$ so that $p = (b - a)/2$ and $-p = -(b - a)/2$. We then compute the Fourier coefficients of f_E over the interval $[-p, p]$ according to Formulas 3.1. Furthermore, because the integrands in Formulas 3.1 have period $2p$, we can use the geometrically obvious fact that the integration can be performed over any interval of length $2p = b - a$ in particular, over $[a, b]$. (See Exercise 10.) Thus Formulas 3.1

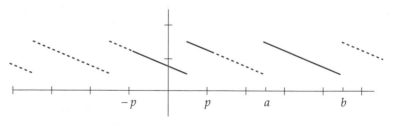

Figure 10.7 $p = (b - a)/2$.

can be rewritten as

$$\blacksquare\ 3.2 \qquad a_k = \frac{2}{b-a} \int_a^b f(x) \cos\frac{2k\pi x}{b-a} dx, \qquad b_k = \frac{2}{b-a} \int_a^b f(x) \sin\frac{2k\pi x}{b-a} dx.$$

The associated trigonometric polynomials are

$$s_N(x) = \frac{a_0}{2} + \sum_{k=1}^{N}\left(a_k \cos\frac{2k\pi x}{b-a} + b_k \sin\frac{2k\pi x}{b-a}\right).$$

Equations 3.2 are useful computationally in part because the definition of $f(x)$ on $[a, b]$ may make for easier computing than integrating over a symmetric interval $[-p, p]$.

---

example 2

Let $f(x) = x$ for $0 < x < 1$. We find, integrating by parts for $k \ne 0$,

$$a_k = 2\int_0^1 x \cos 2k\pi x \, dx$$

$$= 2\left[x\frac{\sin 2k\pi x}{2k\pi}\right]_0^1 - \frac{2}{2k\pi}\int_0^1 \sin 2k\pi x \, dx = 0,$$

$$b_k = 2\int_0^1 x \sin 2k\pi x \, dx$$

$$= 2\left[-x\frac{\cos 2k\pi x}{2k\pi}\right]_0^1 + \frac{2}{2k\pi}\int_0^1 \cos 2k\pi x \, dx$$

$$= -\frac{\cos 2k\pi}{k\pi} = -\frac{1}{k\pi}.$$

Since $a_0 = 2\int_0^1 x \, dx = 1$, then $a_0/2 = \frac{1}{2}$, and the Fourier series is

$$\frac{1}{2} - \frac{1}{\pi}\left(\sin 2\pi x + \frac{\sin 4\pi x}{2} + \frac{\sin 6\pi x}{3} + \cdots\right).$$

3B   Sine and Cosine Expansions

It sometimes happens that an expansion in terms only of cosines or only of sines is more convenient to use than a general Fourier expansion. We begin with the observation that the cosines in a Fourier expansion are even functions (i.e., $\cos(-k\pi x/p) = \cos(k\pi x/p)$) and that the sine terms are odd (i.e., $\sin(-k\pi x/p) = -\sin(k\pi x/p)$). It follows that if f is an even periodic function, the product

$$f(x) \sin\frac{k\pi x}{p}$$

is odd, a simple verification. Therefore, for the Fourier sine coefficient b_k, we have by Equations 3.1,

$$b_k = \frac{1}{p}\int_{-p}^p f(x) \sin\frac{k\pi x}{p} dx = 0.$$

It follows that **an even function has only cosine terms in its Fourier expansion, plus a possible constant.** Similarly, if f is an odd periodic function, the product

$$f(x) \cos \frac{k\pi x}{p}$$

is also odd; so for the Fourier cosine coefficient we have

$$a_k = \frac{1}{p} \int_{-p}^{p} f(x) \cos \frac{k\pi x}{p} \, dx = 0.$$

Thus **an odd function has only sine terms in its Fourier expansion.**

The facts in the preceding paragraph are the key to solving the following problem. Given a function $f(x)$ defined just on the interval $0 \le x \le p$, find a trigonometric series expansion for f consisting only of cosine terms or, alternatively, only of sine terms. The trick is to extend the definition of f from the interval $0 \le x \le p$ to all real x in such a way that the extension is periodic of period $2p$ and is either even or odd. We then compute the Fourier series of the extension. If f_e is an even periodic extension of f, then f_e will have only cosine terms in its Fourier series but is designed to represent f itself just on $0 \le x \le p$. Similarly, if f_o is an odd periodic extension of f, then f_o has only sine terms in an expansion designed to represent f for $0 \le x \le p$. For a **sine expansion** of $f(x)$ on $0 \le x \le p$ we have

■ 3.3 $$\sum_{k=1}^{\infty} b_k \sin \frac{k\pi x}{p}, \qquad b_k = \frac{2}{p} \int_0^p f(x) \sin \frac{k\pi x}{p} \, dx.$$

For a **cosine expansion** of $f(x)$ on $0 \le x \le p$ we have

■ 3.4 $$\tfrac{1}{2}a_0 + \sum_{k=1}^{\infty} a_k \cos \frac{k\pi x}{p}, \qquad a_k = \frac{2}{p} \int_0^p f(x) \cos \frac{k\pi x}{p} \, dx.$$

We illustrate the procedure with two examples.

e x a m p l e 3   We'll compute the cosine expansion for the function defined by $f(x) = 1 - x$ for $0 \le x \le 2$. We consider the even periodic extension shown in Figure 10.8. To find the extension we define f_e by $f_e(x) = f(-x)$ for $-2 \le x < 0$ and then extend periodically, with period 4, to the whole x axis. We can use Formula 3.4 to compute the Fourier cosine expansion of f_e. (Since f_e is even, we know that

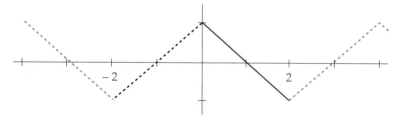

Figure 10.8

$b_k = 0$ for all k.) The coefficient formula in 3.4 allows us to write

$$a_k = \frac{1}{2} \int_{-2}^{2} f_e(x) \cos \frac{k\pi x}{2} \, dx$$

$$= \int_{0}^{2} f_e(x) \cos \frac{k\pi x}{2} \, dx.$$

Since on $0 \le x \le 2$, the function f_e is the same as the given function $f(x) = 1 - x$, we have, for $k > 0$.

$$a_k = \int_{0}^{2} (1 - x) \cos \frac{k\pi x}{2} \, dx$$

$$= \left[\frac{2}{k\pi} (1 - x) \sin \frac{k\pi x}{2} \right]_{0}^{2} + \frac{2}{k\pi} \int_{0}^{2} \sin \frac{k\pi x}{2} \, dx$$

$$= \frac{4}{\pi^2 k^2} (1 - \cos k\pi)$$

$$= \begin{cases} 0, & k \text{ even,} \\ \dfrac{8}{\pi^2 k^2}, & k \text{ odd.} \end{cases}$$

Finally, $a_0 = \int_{0}^{2} (1 - x) \, dx = 0$. Thus the cosine expansion of f on $0 \le x \le 2$ has for its general nonzero term

$$\frac{8}{\pi^2 k^2} \cos \frac{k\pi x}{2}, \qquad k \text{ odd.}$$

Written out, the expansion looks like

$$\frac{8}{\pi^2} \left(\cos \frac{\pi x}{2} + \frac{\cos (3\pi x/2)}{9} + \frac{\cos (5\pi x/2)}{25} + \cdots \right).$$

---

e x a m p l e 4

Starting with the same function as in Example 3, $f(x) = 1 - x$ for $0 \le x \le 2$, we compute a sine expansion by considering the odd periodic extension shown in Figure 10.9. We first define $f_o(x) = -f(-x)$ for $-2 \le x < 0$ and then extend periodically with period 4. (Since f_o is odd, we know that $a_k = 0$ for all k.) Also, by Equations 3.1 or 3.3,

$$b_k = \frac{1}{2} \int_{-2}^{2} f_o(x) \sin \frac{k\pi x}{2} \, dx$$

$$= \int_{0}^{2} f_o(x) \sin \frac{k\pi x}{2} \, dx.$$

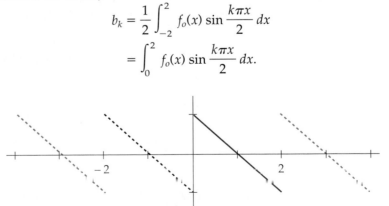

Figure 10.9

But $f_o(x) = 1 - x$ for $0 \le x \le 2$, so

$$b_k = \int_0^2 (1-x) \sin \frac{k\pi x}{2} \, dx$$

$$= \left[-\frac{2}{k\pi} (1-x) \cos \frac{k\pi x}{2} \right]_0^2 - \frac{2}{k\pi} \int_0^2 \cos \frac{k\pi x}{2} \, dx$$

$$= \frac{2}{\pi k} (-1)^k + \frac{2}{k\pi} - \frac{2}{k\pi} \left[\frac{2}{k\pi} \sin \frac{k\pi x}{2} \right]_0^2$$

$$= \begin{cases} 0, & k \text{ odd}, \\ \dfrac{4}{\pi k}, & k \text{ even}. \end{cases}$$

Thus the general nonzero term in the sine expansion is

$$\frac{4}{k\pi} \sin \frac{k\pi x}{2}, \qquad \text{for even } k.$$

The sine expansion is then

$$\frac{2}{\pi} \left(\sin \pi x + \frac{\sin 2\pi x}{2} + \frac{\sin 3\pi x}{3} + \cdots \right).$$

Finally, it's worth noting that Examples 1 and 2 of Section 3A can be regarded as cosine and sine expansions, respectively, of the given functions when they are restricted to $0 \le x \le \pi$.

EXERCISES

Web site http://math.dartmouth.edu/~rewn/ contains the Applet FOURIER for plotting partial sums of Fourier series.

1. Find the Fourier series for the function
$$f(x) = -x, \qquad -2 < x < 2.$$
To what values will the series converge at $x = 2$ and $x = -2$?

2. Find the Fourier series for the function
$$f(x) = 1 + x, \qquad 1 < x < 2.$$
To what values will the series converge at $x = 1$ and $x = 2$?

3. Find the sine expansion of each of the following functions. In each case sketch the graph of the odd periodic extension of f after normalizing f to have the average value $\frac{1}{2}(f(x+) + f(x-))$ at each jump discontinuity. Sketch also the graph of the sum of the first two nonzero terms of the sine expansion; this should exhibit some reasonable relation to your sketch of f.

(a) $f(x) = 1, \, 0 < x < \pi$.

(b) $f(x) = x^2, \, 0 < x < \pi$.

(c) $f(x) = \cos x, \, 0 < x < \pi/2$.

(d) $f(x) = \begin{cases} 1, & 0 < x < 1, \\ 0, & 1 \le x < 2. \end{cases}$

(e) $f(x) = \begin{cases} 0, & 0 < x < 2, \\ x - 2, & 2 \le x < 3. \end{cases}$

4. Find the cosine expansion of each of the following functions. In each case sketch the graph of the even periodic extension of f after normalizing f to have the average value at each jump discontinuity. Sketch also the graph of the sum of the first two nonzero terms of the cosine expansion; this should exhibit some reasonable relation to your sketch of f.

(a) $f(x) = 1$, $0 < x < \pi$.

(b) $f(x) = x^2$, $0 < x < \pi$.

(c) $f(x) = \sin x$, $0 < x < \pi/2$.

(d) $f(x) = \begin{cases} 0, & 0 < x < 1, \\ 1, & 1 \le x < 2. \end{cases}$

(e) $f(x) = \begin{cases} 1, & 0 \le x < 1, \\ 0, & 1 \le x < 3. \end{cases}$

5. For the function

$$f(x) = x, \qquad 0 < x < \pi.$$

(a) Find the Fourier cosine expansion.

(b) Find the Fourier sine expansion.

(c) Compare the results of (a) and (b) with the complete Fourier expansion of

$$g(x) = x, \qquad -\pi < x < \pi.$$

6. Show that if the combinations suggested below are defined, then:

(a) A product of even functions is even.

(b) A product of odd functions is even.

(c) The product of an even function and an odd function is odd.

(d) A linear combination of even functions is even, and of odd functions is odd.

7. Let f be a real-valued function defined on a symmetric interval $-a < x < a$.

(a) Show that f can be written as the sum of two functions, an even part P_e and an odd part P_o. [*Hint:* Let $P_e(x) = \frac{1}{2}(f(x) + f(-x))$.]

(b) What are even and odd parts for a trigonometric polynomial?

(c) What are even and odd parts for $f(x) = e^x$?

8. Show that the even and odd functions P_e and P_o of the previous exercise are uniquely determined by f.

9. Let f be an odd function on $[-\pi, \pi]$ (that is, $f(-x) = -f(x)$), and let g be an even function (that is, $g(-x) = g(x)$). Let a_k, b_k and a_k', b_k' be the Fourier coefficients of f and g, respectively. Show that

$$a_k = 0, \qquad\qquad b_k = \frac{2}{\pi} \int_0^\pi f(x) \sin kx \, dx,$$

$$a_k' = \frac{2}{\pi} \int_0^\pi g(x) \cos kx \, dx, \qquad b_k' = 0.$$

10. Show that if $f(x + 2p) = f(x)$ for all real x, then the integral of f over every interval of length $2p$ is the same; specifically, show that

$$\int_t^{t+2p} f(x) \, dx = \int_0^{2p} f(x) \, dx.$$

(a) Do this by using

$$\int_t^{t+2p} f(x) \, dx = \int_t^0 f(x) \, dx + \int_0^{2p} f(x) \, dx + \int_{2p}^{t+2p} f(x) \, dx.$$

(b) Do this for continuous f by defining $F(t) = \int_t^{t+2p} f(x)\,dx$ and showing that $F'(t) = 0$.

11. (a) Show that if f is periodic and differentiable, then f' is periodic.

(b) Show by example that if f' is periodic, then f need not be periodic.

12. Show that if f is even and differentiable, then f' is odd. [*Hint:* Consider the limit of $(f(-x + h) - f(-x))/h$ as $h \to 0$.]

13. Show that if f is odd and differentiable, then f' is even. See the hint for the previous exercise.

14. Show that the set of functions $\{\sqrt{2/p}\,\sin\,(k\pi x/p)\}_{k=1}^{\infty}$ is an **orthonormal set** on the interval $0 \leq x \leq p$; that is, show

$$\frac{2}{p}\int_0^p \sin\frac{k\pi x}{p}\sin\frac{l\pi x}{p}\,dx = \begin{cases} 0, & k \neq l, \\ 1, & k = l \neq 0. \end{cases}$$

(a) Do this directly, using the identity

$$\sin\alpha\sin\beta = \tfrac{1}{2}\cos\,(\alpha - \beta) - \tfrac{1}{2}\cos\,(\alpha + \beta).$$

(b) Show how the result follows from Equations 2.4 of Section 2.

15. Show that the set of functions $\{\sqrt{2/p}\,\cos\,(k\pi x/p)\}_{k=1}^{\infty}$ is an **orthonormal set** on the interval $0 \leq x \leq p$; that is, show

$$\frac{2}{p}\int_0^p \cos\frac{k\pi x}{p}\cos\frac{l\pi x}{p}\,dx = \begin{cases} 0, & k \neq l, \\ 1, & k = l \neq 0. \end{cases}$$

(a) Do this directly, using the identity $\cos\alpha\cos\beta = \tfrac{1}{2}\cos\,(\alpha - \beta) + \tfrac{1}{2}\cos\,(\alpha + \beta)$.

(b) Show how the result follows from Equations 2.4 of Section 2.

4 HEAT AND WAVE EQUATIONS

In this section we show how a Fourier series can be used to solve some problems in heat conduction and wave motion. Each of the pertinent partial differential equations is an example of an **evolution equation,** in this case one that determines how the solution develops over a whole range of points as time increases. Each of these equations will have solutions described by **dynamical systems** whose **states** are themselves functions of a space variable.

4A One-Dimensional Heat Equation

Suppose we are given a thin homogeneous wire of length p. Let $u(x, t)$ be the temperature at time t at a point x units from one end. Thus $0 \leq x \leq p$, and we'll let $t \geq 0$. We will assume that the only heat transfer is along the direction of the wire and that the temperature at the two ends is held fixed. For this reason we can, without loss of generality, represent the wire as a straight line segment along an x axis and picture the temperature at time t and position x as the

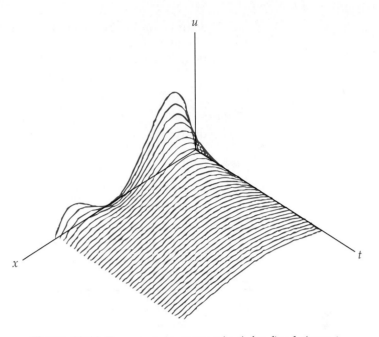

Figure 10.10 Temperature states $u(x, t)$ for fixed times t.

graph of a function $u = u(x, t)$; an example is shown in Figure 10.10 with single temperature states traced as functions of x. Figure IS.9 in the Introductory Survey shows a sketch of the same temperature function with emphasis on the variation of temperature at individual points x.

The basic physical principle of heat conduction is that heat flow is proportional to, and in the direction opposite to, the temperature gradient ∇u. Recall that ∇u is the direction in which the temperature is increasing most rapidly, so it is reasonable that heat should flow in the opposite direction, from hotter to colder. Since the medium is 1-dimensional and is represented by a segment of the x axis, the gradient is represented by $\partial u / \partial x$. The rate of change of total heat in a segment $[x_1, x_2]$ with respect to time t is proportional to the flow rate in at the right end, $u_x(x_2, t)$, plus the flow rate in at the left end, $-u_x(x_1, t)$. Note, for example, that $u_x(x, t) > 0$ implies that heat flows to the left at x and that $u_x(x, t) < 0$ implies that heat flows to the right at x. The rate of heat flow is

$$k \left[-\frac{\partial u}{\partial x}(x_1, t) + \frac{\partial u}{\partial x}(x_2, t) \right], \tag{1}$$

where the positive constant of proportionality k is the **conductivity** of the heat-conducting medium. This constant is arrived at empirically for each material. By the fundamental theorem of calculus, expression (1) can be written as

$$k \int_{x_1}^{x_2} \frac{\partial^2 u}{\partial x^2}(x, t)\, dx \tag{2}$$

It is also true that the rate of change of heat in the segment is

$$\frac{d}{dt} \int_{x_1}^{x_2} \rho c u(x, t)\, dx = \rho c \int_{x_1}^{x_2} \frac{\partial u}{\partial t}(x, t)\, dx. \tag{3}$$

The factors ρ and c are, respectively, the linear **density** and **heat capacity** per unit of material. We can equate expressions (2) and (3) for rate of change of total heat to give

$$a^2 \int_{x_1}^{x_2} \frac{\partial^2 u}{\partial x^2}(x, t)\, dx = \int_{x_1}^{x_2} \frac{\partial u}{\partial t}(x, t)\, dx,$$

where $a^2 = k/(c\rho)$. Allowing x_2 to vary, we can differentiate both sides of this last equation with respect to the upper limit, getting

■ **4.1**
$$a^2 \frac{\partial^2 u}{\partial x^2}(x, t) = \frac{\partial u}{\partial t}(x, t).$$

Equation 4.1 is the one-dimensional **heat equation** or **diffusion equation;** it has the **qualitative interpretation** that if the graph of $u(x, t)$ is concave up as a function of x on some interval so that $\partial^2 u/\partial x^2 > 0$ there, then $\partial u/\partial t > 0$ there also so that temperature is increasing. Similarly, decreasing temperature goes with downward concavity of the graph of u as a function of x. Figure 10.10 supports these observations.

Equation 4.1 is linear in the sense that if u_1 and u_2 are solutions, then so are linear combinations $b_1 u_1 + b_2 u_2$. To single out particular solutions, we first impose **boundary conditions** of the form

$$u(0, t) = 0 \qquad \text{and} \qquad u(p, t) = 0, \tag{4}$$

so temperature is held at zero at both ends of an x interval $[0, p]$. We also suppose we know the initial conditions throughout the interval in terms of a function $h(x)$:

$$u(x, 0) = h(x).$$

The standard method of solution is by **separation of variables,** in which we start by trying to find product solutions of the form

$$u(x, t) = X(x)T(t)$$

with boundary condition $X(0) = X(p) = 0$. Then use the linearity of the differential equation to form sums of multiples of solutions to satisfy the initial condition $u(x, 0) = h(x)$. If such product solutions exist, substitution into $a^2 u_{xx} = u_t$ gives

$$a^2 X''(x)T(t) = X(x)T'(t),$$

for $0 \le x \le p$, $0 < t$. Dividing through by $X(x)T(t)$ gives

$$a^2 \frac{X''(x)}{X(x)} = \frac{T'(t)}{T(t)}. \tag{5}$$

The left side is independent of t, and the right side is independent of x; since the two sides are equal, both must be independent of t *and* x, that is, equal to a constant $-\lambda^2$. (Note that $-\lambda^2 > 0$ if λ is imaginary.) This notation for the constant is not essential but is motivated by the convenient form of the resulting ordinary differential equations. (See Section 4C for an alternative approach.) Indeed, setting both sides of Equation (5) equal to $-\lambda^2$ gives the two equations

$$a^2 X'' + \lambda^2 X = 0, \tag{6}$$

$$T' + \lambda^2 T = 0. \tag{7}$$

Equation (6) has solutions

$$X(x) = c_1 \cos \frac{\lambda x}{a} + c_2 \sin \frac{\lambda x}{a}.$$

But $u(0, t) = 0$ requires $X(0) = 0$, so $c_1 = 0$. Similarly, $u(p, t) = 0$ requires $X(p) = 0$, so $c_2 \sin (\lambda p/a) = 0$. This condition can be achieved, without making $c_2 = 0$, only by choosing λ so that $\lambda p/a = k\pi$, where k is an integer. That is, we must take $\lambda = ka\pi/p$, with the result that $X(x)$ has the form

$$X(x) = c_2 \sin \frac{k\pi x}{p}, \qquad k = 1, 2, \ldots.$$

Equation (7) has solutions $T(t) = de^{-\lambda^2 t}$, which, because now $\lambda^2 = k^2 a^2 \pi^2/p^2$, becomes

$$T(t) = de^{-k^2(a^2\pi^2/p^2)t}.$$

The product solutions $u(x, t)$ are thus given, except for a constant factor $b_k = dc_2$, by

$$u_k(x, t) = e^{-k^2(a^2\pi^2/p^2)t} \sin \frac{k\pi}{p} x, \qquad k = 1, 2, \ldots.$$

A limit of linear combinations $\sum_{k=1}^{N} b_k u_k(x, t)$ as $N \to \infty$ then looks like

■ **4.2**
$$u(x, t) = \sum_{k-1}^{\infty} b_k e^{-k^2(a^2\pi^2/p^2)t} \sin \frac{k\pi x}{p}.$$

Equation 4.2 represents a dynamical system in which, for fixed t, the **states** are functions $U(x) = u(x, t)$ to be determined by an initial state $h(x) = u(x, 0)$.

To satisfy the initial condition $u(x, 0) = h(x)$, we set $t = 0$ in Equation 4.2 and require

$$\sum_{k=1}^{\infty} b_k \sin \frac{k\pi}{p} x = h(x),$$

for suitable choices of the b_k. If $h(x)$ can be expressed in this form, we can expect that a solution to the problem is given by Equation 4.2. The boundary conditions $u(0, t) = u(p, t) = 0$ require that the temperature remain zero at the ends, and the initial condition $u(x, 0) = h(x)$ specifies the initial temperature at each point x of the wire between 0 and p. Thus we are naturally led to the problem of finding a Fourier sine series representation for $h(x)$, as in Equation 3.3 of Section 3. The coefficients in the expansion are

■ **4.3**
$$b_k = \frac{2}{p} \int_0^p h(x) \sin \frac{k\pi x}{p} dx, \qquad k = 1, 2, 3, \ldots.$$

The matter of conditions under which the infinite series 4.2 actually represents a solution of Equation 4.1 is taken up briefly in Exercise 6.

***

e x a m p l e **1**   Let's be more specific about solving the heat equation. For simplicity let $p = \pi$. Recall that to solve $a^2 u_{xx} = u_t$ with boundary condition $u(0, t) = u(\pi, t) = 0$ and initial condition $u(x, 0) = h(x)$, we want in general to be able to represent h

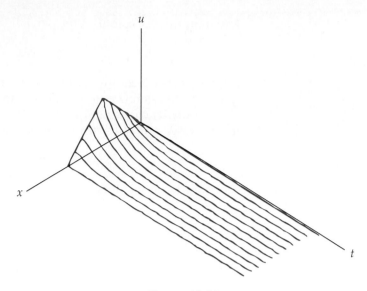

Figure 10.11

by an infinite series of the form

$$h(x) = \sum_{k=1}^{\infty} b_k \sin kx.$$

Suppose, for example, that h is given in $[0, \pi]$ by

$$h(x) = \begin{cases} x, & 0 \le x \le \pi/2, \\ \pi - x, & \pi/2 < x \le \pi. \end{cases}$$

The graph of h is shown in Figure 10.11 in perspective above the x axis.

To find the Fourier sine expansion of h on $[0, \pi]$, we extend h to the interval $[-\pi, \pi]$ in such a way that the cosine terms in the expansion of h will all be zero, leaving only the sine terms to be computed. We do this by extending the graph of h asymmetrically about the vertical axis so as to represent an odd function. Then $a_k = 0$, because $h(x) \cos k(x)$ is an odd function on $[-\pi, \pi]$. Now use Equation 4.3:

$$b_k = \frac{1}{\pi} \int_{-\pi}^{\pi} h(x) \sin kx\, dx$$

$$= \frac{2}{\pi} \int_0^{\pi} h(x) \sin kx\, dx$$

$$= \frac{2}{\pi} \int_0^{\pi/2} x \sin kx\, dx + \frac{2}{\pi} \int_{\pi/2}^{\pi} (\pi - x) \sin kx\, dx$$

$$= \frac{4}{\pi k^2} \sin \frac{k\pi}{2}.$$

Hence

$$b_k = \begin{cases} 0, & k \text{ even,} \\ 4/(\pi k^2), & k = 1, 5, 9, \ldots, \\ -4/(\pi k^2), & k = 3, 7, 11, \ldots. \end{cases}$$

Theorem 2.5 (Section 2) then implies that

$$h(x) = \frac{4}{\pi}\left(\frac{\sin x}{1^2} - \frac{\sin 3x}{3^2} + \frac{\sin 5x}{5^2} - \frac{\sin 7x}{7^2} + \cdots\right)$$

for each x in $[0, \pi]$. Finally, from Equation 4.2 we expect the solution to the equation $a^2 u_{xx} = u_t$ to be given by

$$u(x, t) = \frac{4}{\pi}\left(e^{-a^2 t} \sin x - \frac{1}{3^2} e^{-3^2 a^2 t} \sin 3x + \frac{1}{5^2} e^{-5^2 a^2 t} \sin 5x\right.$$
$$\left. - \frac{1}{7^2} e^{-7^2 a^2 t} \sin 7x + \cdots\right).$$

Verification that $u(0, t) = u(\pi, t) = 0$ follows immediately from setting $x = 0$ and $x = \pi$. By setting $t = 0$, we get the representation of h by its Fourier series, which is guaranteed by Theorem 2.5 (Section 2). That $u(x, t)$ satisfies the equation $a^2 u_{xx} = u_t$ depends on term-by-term differentiation of the series for u. The graph of the solution is sketched in Figure 10.11.

4B   Equilibrium Solutions

A solution $u(x, t) = v(x)$ of the heat equation that is independent of t is called an **equilibrium solution,** because the state $v(x)$ doesn't vary with time. The heat equation for such functions becomes simply $v'' = 0$, and all solutions are necessarily of the form $u(x, t) = v(x) = \alpha + \beta x$, where α and β are constant. Solutions of this type are useful for solving the time-dependent problem when we have **nonhomogeneous boundary conditions** of the form

$$u(0, t) = u_0, \qquad u(p, t) = u_1, \qquad t > 0,$$

where at least one of u_0 and u_1 is a nonzero constant. The idea is to choose α, β in the equilibrium solution $v(x) = \alpha + \beta x$ so that

$$v(0) = u_0, \qquad v(p) = u_1.$$

Then $$u(x, t) = w(x, t) + v(x)$$

will satisfy $u(0, t) = u_0$, $u(p, t) = u_1$ if $w(x, t)$ is a solution of the heat equation satisfying **homogeneous conditions**

$$w(0, t) = w(p, t) = 0, \qquad t > 0.$$

Note that the function $w + v$ is indeed a solution of the heat equation, because both w and v are solutions and because the heat operator $a^2 D_{xx} - D_t$ acts linearly.

To solve the problems of the form

$$a^2 u_{xx} = u_t, \qquad u(0, t) = u_0, \qquad u(p, t) = u_1, \qquad u(x, 0) = h(x),$$

we first find an equilibrium solution $v(x) = \alpha + \beta x$. We need $v(0) = u_0 = \alpha$ and $v(p) = u_1 = \alpha + \beta p$. Thus $\alpha = u_0$, and $\beta = (u_1 - u_0)/p$. Then solve the heat equation $a^2 w_{xx} - w_t$ with boundary conditions $w(0, t) = w(p, t) = 0$ and initial condition $w(x, 0) = h(x) - v(x)$. You can check immediately that the solution to the original problem has the form $u(x, t) = v(x) + w(x, t)$, or

 **4.4**

$$u(x, t) = v(x) + \sum_{k=1}^{\infty} b_k e^{-k^2(\pi^2 a^2/p^2)t} \sin \frac{k\pi x}{p}$$

with

$$b_k = \frac{2}{p} \int_0^p (h(x) - v(x)) \sin \frac{k\pi x}{p} \, dx.$$

---

e x a m p l e **2**

Consider the problem $a^2 u_{xx} = u_t$, with side conditions
$$u(0, t) = 10, \qquad u(5, t) = 30, \qquad u(x, 0) = h(x) = 10 - 2x, \qquad 0 < x < 5.$$
We find the equilibrium solution $v(x) = 10 + 4x$. The desired solutions
$$u(x, t) = (10 + 4x) + w(x, t)$$
come from finding solutions $w(x, t)$ that satisfy homogeneous boundary conditions
$$w(0, t) = w(5, t) = 0, \qquad t \geq 0,$$
and an initial condition
$$u(x, 0) = w(x, 0) + v(x) = h(x);$$
this last condition is just
$$w(x, 0) = h(x) - v(x) = -6x.$$
Thus our solution $u(x, t)$ has the form of Equation 4.4, where the b_k are Fourier sine coefficients
$$b_k = \frac{2}{p} \int_0^5 -6x \sin \frac{k\pi x}{p} \, dx = \frac{60(-1)^k}{k\pi}.$$
The solution to the problem is thus
$$u(x, t) = (10 + 4x) - \frac{60}{\pi} \sum_{k=1}^{\infty} \frac{(-1)^{k+1}}{k} e^{-(k/5)^2 a^2 \pi^2 t} \sin \frac{k\pi x}{5}.$$

Web site http://math.dartmouth.edu/~rewn/ contains the Applet HEAT for plotting solutions to the 1-dimensional heat equation.

EXERCISES

1. Solve the heat equation $a^2 u_{xx} = u_t$ subject to the following boundary and initial conditions.

   (a) $u(0, t) = u(p, t) = 0$; $u(x, 0) = \sin(\pi x/p)$, $0 < x < p$.

   (b) $u(0, t) = u(1, t) = 0$; $u(x, 0) = x$, $0 < x < 1$.

   (c) $u(0, t) = u(1, t) = 0$; $u(x, 0) = 1 - x$, $0 < x < 1$.

   (d) $u(0, t) = u(\pi, t) = 0$; $u(x, 0) = x(\pi - x)$, $0 < x < \pi$.

   (e) $u(0, t) = u(\pi, t) = 0$; $u(x, 0) = \sin x + \frac{1}{2} \sin 2x$, $0 < x < \pi$.

   (f) $u(0, t) = u(2, t) = 0$; $u(x, 0) = \begin{cases} 1, & 0 < x < 1, \\ 0, & 1 < x < 2. \end{cases}$

2. Find equilibrium solutions $u(x, t) = v(x)$ of the heat equation $a^2 u_{xx} = u_t$ that satisfy each of the following conditions:

   (a) $u(0, t) = -1$, $u(2, t) = 1$.

   (b) $u(0, t) = 0$, $u(100, t) = 100$.

   (c) $u_x(0, t) = 1$, $u(1, t) = 2$.

3. Solve the heat equation $a^2 u_{xx} = u_t$ subject to the following boundary and initial conditions:

   (a) $u(0, t) = 1$, $u(p, t) = 3$; $u(x, 0) = \sin(\pi x/p)$, $0 < x < p$.

   (b) $u(0, t) = -1$, $u(2, t) = 1$; $u(x, 0) = x$, $0 < x < 2$.

   (c) $u(0, t) = 0$, $u(1, t) = 1$; $u(x, 0) = 1 - x$, $0 < x < 1$.

4. Verify that the partial differential operator $L = a^2 D_{xx} - D_t$, defined by $L(u) = a^2 u_{xx} - u_t$, is linear in its action on twice-differentiable functions.

5. Find all solutions of the form $u(x, t) = X(x)T(t)$ for each of the following equations:

   (a) $u_{xx} + u_x = u_t$.   **(c)** $x u_x = 2 u_t$.

   (b) $u_{xx} - u_x = u_t$.   **(d)** $u_{xx} = u_{tt}$.

6. (a) Verify that the series expansion for $u(x, t)$ given by Equation 4.2 of the text is formally a solution of $a^2 u_{xx} = u_t$. Use term-by-term differentiation of the series.

   ***(b)** Show that the formal verification in part (a) is valid for $t > 0$ by demonstrating that the differentiated series all converge uniformly for $t > \delta > 0$. *Note:* If f is smooth enough, it can be shown that $\lim_{t \to 0+} u(x, t) = f(x)$.

7. Find the equilibrium solution to each of the following problems:

   (a) $u_{xx} = u_t + 2$; $u(0, t) = 1$, $u(1, t) = 2$.

   (b) $u_{xx} = u_t + u$; $u(0, t) = 0$, $u(2, t) = 0$.

   (c) $u_{xx} = u_t + x$; $u(0, t) = 1$, $u(1, t) = 2$.

Insulated Endpoints. The 1-dimensional heat flow problem

$$u_{xx} = u_t, \qquad u_x(0, t) = u_x(p, t) = 0, \qquad u(x, 0) = f(x)$$

for $0 \le x \le p$ can be interpreted as requiring the conducting medium to have insulated ends at $x = 0$ and $x = p$. The reason is that the temperature gradient u_x is always 0 at the endpoints, so there is no heat flow past those points. The solution involves Fourier cosine series. The next four exercises are about problems of this kind.

8. (a) Show that product solutions $u(x, t) = T(t)X(x)$ of the insulated endpoint problem have the form

$$u_k(x, t) = a_k e^{-k^2 \pi^2 t/p^2} \cos \frac{k\pi}{p} x.$$

   (b) Use the Fourier cosine expansion for $f(x)$ on $0 \le x \le p$ to solve the boundary-value problem for a general initial temperature $f(x)$.

   (c) What is the equilibrium temperature function $u(x, \infty)$ for $0 \le x \le p$?

9. Solve the heat equation $a^2 u_{xx} = u_t$ with insulated end conditions $u_x(0, t) = u_x(1, t) = 0$ and initial condition $u(x, 0) = x$ for $0 < x < 1$.

10. Solve the heat equation $a^2 u_{xx} = u_t$ with insulated end conditions $u_x(0, t) = u_x(1, t) = 0$ and initial condition $u(x, 0) = 1$ for $0 < x < 1$.

11. Solve the heat equation $a^2 u_{xx} = u_t$ with insulated end conditions $u_x(0, t) = u_x(\pi, t) = 0$ and initial condition $u(x, 0) = \cos x$ for $0 < x < \pi$.

12. Suppose that $u(x, t)$ satisfies $u_{xx} = u_t$ as well as $u(0, t) = \alpha$, $u(p, t) = \beta$ for $t \geq 0$. Without doing any detailed computations, make a qualitative sketch of the graph of u relative to axes oriented as in Figures 10.10 and 10.11. Do this with each of the following choices for $p = b - a$, for α and β, and for initial function $u(x, 0)$.

    (a) $u(x, 0) = x - x^2$, $0 \leq x \leq 1$, $\alpha = \beta = 0$.

    (b) $u(x, 0) = x^2 - x^3$, $0 \leq x \leq 1$, $\alpha = \beta = 0$.

    (c) $u(x, 0) = -x^2$, $0 \leq x \leq 1$, $\alpha = 0$, $\beta = -1$.

    (d) $u(x, 0) = |x - 1|$, $0 \leq x \leq 2$, $\alpha = 0$, $\beta = 0$.

    (e) $u(x, 0) = -x$, $0 \leq x \leq 1$, $\alpha = 0$, $\beta = -1$.

    (f) $u(x, 0) = 1 - x$, $0 \leq x \leq 2$, $\alpha = 1$, $\beta = -1$.

13. The partial differential equation $t u_{xx} = u_t$ has product solutions of the form $u(x, t) = X(x)T(t)$.

    (a) Find two ordinary differential equations satisfied by $X(x)$ and $T(t)$, respectively. (The equation for X should not contain t, and the equation for T should not contain x.)

    (b) Solve the ordinary differential equations found in part (a), and use the results to specify the general form of the solutions $X(x)T(t)$.

14. The partial differential equation $txu_x = u_t$ has product solutions of the form $u(x, t) = X(x)T(t)$ for $x > 0$ and $t > 0$. Find the general form of all such solutions.

15. Suppose that $u(x, t)$ satisfies $a^2 u_{xx} = u_t$ and that $u(x, t_0)$ is concave up as a function of x for $a < x < b$. Show that for each x with $a < x < b$, there is a time $t(x)$ such that $u(x, t)$ increases as t increases from t_0 to $t(x)$. What if $u(x, t_0)$ is concave down?

16. The assumption that the separation constant C for the problem $a^2 u_{xx} = u_{tt}$, $u(0, t) = u(p, t) = 0$ has the special form $-\lambda^2$ is convenient but not essential. Derive the product solutions $u_k(x, t) = \exp(-k^2(a^2\pi^2/p^2)t) \sin(k\pi/p)x$ using C instead of $-\lambda^2$.

4C   *One-Dimensional Wave Equation*

Consider for physical motivation a stretched elastic string of length p and uniform length density ρ placed along the x axis in 3-dimensional space. Suppose that the ends of the string are held fixed at $(x, y, z) = (0, 0, 0)$ and $(x, y, z) = (p, 0, 0)$ and that the absolute magnitude of the tension force τ is constant over

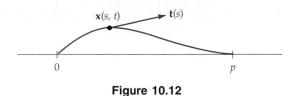

Figure 10.12

the entire length of the string. If the string is somehow made to vibrate, our problem is to determine the position vector $\mathbf{x}(s, t)$ at time t of a point on the string a distance s from the end fixed at $x = 0$. Figure 10.12 shows a possible configuration. We imagine the string subdivided into short pieces of length Δs and then derive two different expressions for the total force acting on a typical segment of the subdivision. If $\mathbf{t}(s)$ is the unit tangent vector to the string at $\mathbf{x}(s)$, then the opposing forces at $\mathbf{x}(s_0)$ and $\mathbf{x}(s_0 + \Delta s)$ are

$$\tau\mathbf{t}(x_0 + \Delta s) \qquad \text{and} \qquad -\tau\mathbf{t}(s_0).$$

Hence the total force acting on the segment is

$$\tau[\mathbf{t}(s_0 + \Delta s) - \mathbf{t}(s_0)].$$

The mass of the short section is $\rho\,\Delta s$, density times length. Also, by Newton's law, the force equals mass, $\rho\,\Delta s$, times acceleration $\mathbf{a}(s_0)$. It follows that

$$\rho\mathbf{a} = \tau\left[\frac{\mathbf{t}(s_0 + \Delta s) - \mathbf{t}(s_0)}{\Delta s}\right].$$

But
$$\mathbf{a} = \frac{\partial^2\mathbf{x}}{\partial t^2}(s, t) \qquad \text{and} \qquad \mathbf{t}(s) = \frac{\partial\mathbf{x}}{\partial s}(s, t). \tag{1}$$

Making these substitutions in the previous equation and letting $\Delta s \to 0$, we get

$$\rho\frac{\partial^2\mathbf{x}}{\partial t^2} = \tau\lim_{\Delta s \to 0}\frac{1}{\Delta s}\left[\frac{\partial\mathbf{x}(s_0 + \Delta s)}{\partial s} - \frac{\partial\mathbf{x}(s_0)}{\partial s}\right] = \tau\frac{\partial^2\mathbf{x}}{\partial s^2}.$$

This vector differential equation is equivalent to a system of three scalar equations in which $\mathbf{x} = (x, y, z)$ and $a^2 = \tau/\rho$:

$$\frac{\partial^2 x}{\partial t^2} = a^2\frac{\partial^2 x}{\partial s^2}, \qquad \frac{\partial^2 y}{\partial t^2} = a^2\frac{\partial^2 y}{\partial s^2}, \qquad \frac{\partial^2 x}{\partial t^2} = a^2\frac{\partial^2 z}{\partial s^2}.$$

Technically, the problem of finding solution functions $x(s, t)$, $y(s, t)$, and $z(s, t)$ is the same for all three equations. In practice, however, the equation for $x(s, t)$ is usually set aside when the longitudinal motion (along the x axis) is slight, and we make this assumption. Between the other two equations there is little difference in physical significance unless some other special assumption is made. From now on we'll use the neutral symbols $u(s, t)$ to stand for $y(s, t)$ or $z(s, t)$ as the case may be. We might in particular suppose that the string has been plucked in such a way that its motion takes place entirely in the xy plane, with $z(s, t) \equiv 0$. Finally, we assume that the displacements are small enough that we can replace $\partial^2 y/\partial s^2$ by $\partial^2 y/\partial x^2$. This last assumption is one that requires experimental justification in actual practice. (See also the last exercise at the end of this section.) Thus, with some loss of generality, we consider, instead of the vector equation, a single scalar **wave equation** for 1-dimensional displacement $u(x, t)$:

■ **4.5**
$$\frac{\partial^2 u}{\partial t^2} = a^2 \frac{\partial^2 u}{\partial x^2}.$$

Equation 4.5 is the 1-dimensional **wave equation;** it has the **qualitative interpretation** that if $u(x, t)$ is concave down as a function of x ($u_{xx} < 0$), then the acceleration $u_{tt}(x, t)$ is negative, or alternatively the velocity $u_t(x, t)$ is decreasing. Similarly, upward concavity in the x direction implies increasing velocity. A moment's reflection shows that this interpretation is physically reasonable, since a curved string under tension will tend to straighten itself.

The differential equation 4.5 doesn't completely specify the vibration of a string unless we impose some initial conditions:

$$u(x, 0) = f(x), \tag{2}$$

$$\frac{\partial u}{\partial t}(x, 0) = g(x). \tag{3}$$

The first condition specifies the initial ($t = 0$) displacement in the u direction as a function of x, rather than s, and the second equation specifies the initial velocity in the u direction, also as a function of x. For string ends held fixed we use boundary conditions

$$u(0, t) = 0 \qquad \text{and} \qquad u(p, t) = 0. \tag{4}$$

The graph of a fairly complicated solution is shown in Figure 10.13; the individual curves on the graph represent string shapes at equally spaced times. In

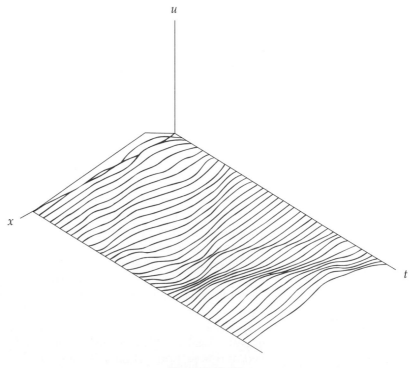

Figure 10.13

this case the string is not only released from an initial shape (i.e., "plucked" as on a harp) but is simultaneously given a downward initial velocity (i.e., "hammered" as on a piano) near its center.

As in the case of the heat equation, we use separation of variables and rely on the linearity of Equation 4.5 and conditions (4) for constructing solutions that satisfy conditions (2) and (3) also. We try

$$u(x, t) = X(x)T(t),$$

upon which Equation 4.5 becomes

$$X(x)T''(t) = a^2 X''(x)T(t) \qquad \text{or} \qquad \frac{T''(t)}{T(t)} = a^2 \frac{X''(x)}{X(x)}.$$

Since T''/T and X''/X are independent of x and t, respectively, both sides of the last equation are constant; if λ is such a constant, then

$$\frac{X''(x)}{X(x)} = \lambda \qquad \text{and} \qquad \frac{T''(t)}{T(t)} = a^2 \lambda.$$

The first equation, $X''(x) = \lambda X(x)$, has solutions

$$X(x) = c_1 e^{\sqrt{\lambda} x} + c_2 e^{-\sqrt{\lambda} x}.$$

The boundary conditions (4) require

$$X(0) = 0 \qquad \text{and} \qquad X(p) = 0;$$

so $\qquad c_1 + c_2 = 0 \qquad \text{and} \qquad c_1 e^{\sqrt{\lambda} p} + c_2 e^{-\sqrt{\lambda} p} = 0.$

Solving for c_1 and c_2, we find that for nonzero solutions to exist, we must have $c_1 = -c_2$ and $e^{-\sqrt{\lambda} p} = e^{\sqrt{\lambda} p}$, so $e^{2\sqrt{\lambda} p} = 1$. Allowing for complex exponents, we observe first that

$$e^{\alpha + i\beta} = e^{\alpha}(\cos \beta + i \sin \beta) = 1$$

precisely when $\alpha = 0$ and $\beta = 2k\pi i$ for some integer k. Since also $e^{2\sqrt{\lambda} p} = 1$, we have $2\sqrt{\lambda} p = 2\pi k i$ for some integer k. We see that λ can't be positive, more specifically that

$$\sqrt{\lambda} = \frac{k\pi i}{p} \qquad \text{and} \qquad \lambda = -\frac{k^2 \pi^2}{p^2}.$$

Solutions $X(x)$ are then of the form

$$X(x) = c_1 e^{(\pi k i/p)x} - c_1 e^{(-\pi k i/p)x}$$

$$= 2c_1 i \sin \frac{\pi k}{p} x.$$

We can let $b_k = 2c_1 i$ and write $X_k(x) = b_k \sin (\pi k/p)x$. The differential equation for T, $T''(t) = a^2 \lambda T(t)$, now takes the form

$$T''(t) + \frac{\pi^2 k^2 a^2}{p^2} T(t) = 0,$$

since we have determined that $\lambda = -\pi k^2/p^2$. Solutions of this equation are of the form

$$T_k(t) = C_k \cos \frac{\pi k a}{p} t + D_k \sin \frac{\pi k a}{p} t.$$

Writing $A_k = C_k b_k$ and $B_k = D_k b_k$, product solutions $u_k = X_k T_k$ take the form

$$u_k(x, t) = \left[A_k \cos \frac{\pi k a}{p} t + B_k \sin \frac{\pi k a}{p} t \right] \sin \frac{\pi k}{p} x.$$

To satisfy initial conditions (2) and (3), we form finite or infinite sums of the type

■ **4.6** $$u(x, t) = \sum_{k=1}^{\infty} \left[A_k \cos \frac{\pi k a}{p} t + B_k \sin \frac{\pi k a}{p} t \right] \sin \frac{\pi k}{p} x.$$

The initial conditions become formally

$$u(x, 0) = \sum_{k=1}^{\infty} A_k \sin \frac{\pi k}{p} x = f(x) \tag{5}$$

and $$u_t(x, 0) = \sum_{k=1}^{\infty} \frac{\pi k a}{p} B_k \sin \frac{\pi k}{p} x = g(x) \tag{6}$$

The coefficients A_k and $(\pi k a / p) B_k$ are then determined so that they are the Fourier sine coefficients of f and g, respectively. Thus Equation 4.6 represents a dynamical system in which, for fixed t, a **state** consists of a pair of functions $(U(x), V(x)) = (u(x, t), u_t(x, t))$, where $U(x)$ describes the string's shape and $V(x)$ is its velocity at x. In particular, $(f(x), g(x)) = (u(x, 0), u_t(x, 0))$ is an initial state for some trajectory in the state space.

---

e x a m p l e 1

For a simple example, we consider a string that is released with no initial velocity so that $u_t(x, 0) = 0$. We assume the initial displacement is given by

$$u(x, 0) = b \sin \frac{\pi x}{p}, \tag{7}$$

and $u(0, t) = u(p, t) = 0$. It happens that f and g are so simple in this case that we can find their Fourier sine coefficients by inspection. We choose $B_k = 0$ in equation (6) to make $u_y(x, t) = 0$. In equation (5), set $A_1 = b$ and $A_k = 0$ for $k \neq 1$. The solution to Equation 4.5 then takes the form

$$u(x, t) = b \cos \frac{\pi a}{p} t \sin \frac{\pi}{p} x.$$

Recall that for Equation 4.5 to be physically realistic the vibrations of the string should be fairly small; in other words, the coefficient b should be small. In Figure 10.13, the graph of our solution is shown with $a = 1$, $p = \pi$, and b chosen unrealistically large in order to bring out some qualitative features of the picture.

In general, the computation of the coefficients A_k, B_k in equation (6) is effected by the Fourier sine formulas

■ **4.7** $$A_k = \frac{2}{p} \int_0^p f(x) \sin \frac{k \pi x}{p} \, dx,$$

■ **4.8** $$B_k = \frac{2}{k \pi a} \int_0^p g(x) \sin \frac{k \pi x}{p} \, dx.$$

example 2

The solution of $a^2 u_{xx} = u_{tt}$ that satisfies

$$u(0, t) = u(1, t) = 0, \qquad u(x, 0) = x(1 - x), \qquad u_t(x, 0) = \sin \pi x$$

is found as follows. We compute from Equations 4.7 and 4.8, with $p = 1$,

$$A_k = 2 \int_0^1 x(1 - x) \sin k\pi x \, dx = \frac{4}{k^3 \pi^3} [1 - (-1)^k].$$

$$B_k = \frac{2}{k\pi a} \int_0^1 \sin \pi x \sin k\pi x \, dx = \begin{cases} 1/(\pi a), & k = 1 \\ 0, & k = 2, 3, \dots \end{cases}$$

The term corresponding to $k = 1$ in the general expansion is just

$$\left(\frac{8}{\pi^3} \cos \pi a t + \frac{1}{\pi a} \sin \pi a t \right) \sin \pi x.$$

All the other nonzero terms have odd index, so the solution formula is

$$u(x, t) = \left(\frac{8}{\pi^3} \cos \pi a t + \frac{1}{\pi a} \sin \pi a t \right) \sin \pi x$$

$$+ \sum_{j=1}^{\infty} \frac{8}{(2j + 1)^3 \pi^3} \cos (2j + 1)\pi a t \sin (2j + 1)\pi x.$$

Web site http://math.dartmouth.edu/~rewn/ contains the Applet WAVE for plotting solutions to the 1-dimensional wave equation.

The method of Fourier series and separation of variables is important for the wave equation not only because it provides solution formulas. (In fact, the d'Alembert method discussed in Exercises 3, 4, and 5 does this also, and in a way that is in some cases more direct.) What the Fourier method allows us to see is an analysis of vibratory motion into simpler vibrations, and it is this analysis, and its flexibility in applying to other problems, that makes the Fourier method so important.

example 3

Suppose $u(x, t)$ is the displacement at time t and position x of a taut string with ends fixed: $u(0, t) = u(p, t) = 0$. To simplify the discussion, suppose the initial velocity is zero: $u_t(x, 0) = 0$. Then the Fourier solution has the form

$$u(x, t) = \sum_{k=1}^{\infty} A_k \cos \frac{k\pi a}{p} t \sin \frac{k\pi}{p} x,$$

where the amplitudes A_k of the individual terms are determined as Fourier sine coefficients of the initial displacement $u(x, 0) = f(x)$. A typical term

$$u_k(x, t) = A_k \cos \frac{k\pi a}{p} t \sin \frac{k\pi}{p} x$$

is periodic in the time variable t with period T determined by $k\pi a T/p = 2\pi$ so that $T = 2p/ka$. The frequency, or number of vibrations per time unit, is thus $\nu = ka/2p$. The **fundamental frequency** is the frequency that corresponds to the first nonzero A_k. If $A_1 \neq 0$, the fundamental vibration has frequency $a/2p$ and

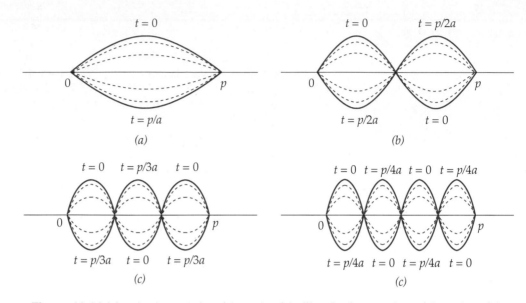

Figure 10.14 (a) $u_1(x, t) = \cos(\pi at/p) \sin(\pi x/p)$. (b) $u_2(x, t) = \cos(2\pi at/p) \sin(2\pi x/p)$. (c) $u_3(x, t) = \cos(3\pi at/p) \sin(3\pi x/p)$. (d) $u_4(x, t) = \cos(4\pi at/p) \sin(4\pi x/p)$.

all others have frequencies that are integer multiples of the fundamental frequency. If we think of each term $u_k(x, t)$ as generating an audible tone, it is usual to speak of the **fundamental** tone and its associated **overtones** or **harmonic tones** of higher frequency.

Some displacements of a fundamental tone and its overtones appear in Figure 10.14 with uniform amplitude 1. However, it is the relative sizes of the amplitudes A_k that give a complex vibration its recognizable quality. A point that remains stationary in one of these oscillations is called a **nodal point.** For example, $u_2(x, t)$ has a nodal point at $x = p/2$. Otherwise, for fixed x each term exhibits vertical motion as time t varies and is called a **"standing" wave.** Exercise 4 motivates use of the term "wave."

EXERCISES

1. Solve the wave equation $a^2 u_{xx} = u_{tt}$ subject to the following boundary and initial conditions.

   (a) $u(0, t) = u(\pi, t) = 0$, $u(x, 0) = \sin x$, $u_t(x, 0) = 0$.

   (b) $u(0, t) = u(\pi, t) = 0$, $u(x, 0) = \begin{cases} x, & 0 < x \le \pi/2, \\ \pi - x, & \pi/2 < x < \pi, \end{cases}$ $u_t(x, 0) = 0$.

   (c) $u(0, t) = u(\pi, t) = 0$, $u(x, 0) = 0$, $u_t(x, 0) = \begin{cases} 0, & 0 < x \le \pi/2, \\ 1, & \pi/2 < x < \pi. \end{cases}$

   (d) $u(0, t) = u(1, t) = 0$, $u(x, 0) = x(1 - x)$, $u_t(x, 0) = \sin \pi x$.

2. The **nonhomogeneous wave equation**

$$a^2 u_{xx} = u_{tt} + g$$

incorporates g, the acceleration of gravity, into the vibrating-string problem.

   (a) Find the equilibrium solutions $u(x, t) = v(x)$ of the nonhomogeneous equation.

   (b) Among the solutions found in part (a), select the solution that satisfies the boundary conditions $u(0, t) = u(p, t) = 0$.

   (c) Explain how to modify the Fourier solution method for $a^2 u_{xx} = u_{tt}$ in order to cover the solution of the nonhomogeneous problem.

3. **The d'Alembert solution to the wave equation.** Let $U(x)$ and $V(x)$ be twice-differentiable functions defined for all real x.

   (a) Show that $u(x, t) = U(x + at) + V(x - at)$ defines a solution of $a^2 u_{xx} = u_{tt}$ that is valid for all real x and all real t.

   (b) Assume that $f''(x)$ exists. Show that

$$u(x, t) = \tfrac{1}{2}[f(x + at) + f(x - at)]$$

   is a solution to the wave equation having the form described in part (a) and also satisfies the initial conditions $u(x, 0) = f(x)$, $u_t(x, 0) = 0$.

   (c) Assume that $g'(x)$ exists. Show that

$$u(x, t) = \frac{1}{2a} \int_{x-at}^{x+at} g(s) \, ds$$

   is a solution to the wave equation having the form described in part (a) and also satisfies the initial $u(x, 0) = 0$, $u_t(x, 0) = g(x)$.

   (d) Combine the results of parts (b) and (c) to find a solution formula for the wave equation subject to the initial conditions $u(x, 0) = f(x)$, $u_t(x, 0) = g(x)$.

4. **(a)** Show that the d'Alembert solution $U(x + at) + V(x - at)$ described in the previous exercise represents the sum of two wave motions, the first moving left with speed $a > 0$, the other moving right with the same speed.

   (b) Show for the general term of Equation 4.6 that

$$\left[A \cos \frac{k\pi a}{p} t + B \sin \frac{k\pi a}{p} t \right] \sin \frac{\pi k}{p} x$$

$$= \sqrt{A^2 + b^2} \cos \left(\frac{k\pi a t}{p} - \theta \right) \sin \frac{k\pi x}{p},$$

   where θ depends on A and B. This term is called a **standing wave.**

   (c) Show that the term in part (b) can be written in the d'Alembert form

$$\tfrac{1}{2}\sqrt{A^2 + B^2} \left[\sin \left(\frac{k\pi}{p} (x + at) - \theta \right) + \sin \left(\frac{k\pi}{p} (x - at) + \theta \right) \right],$$

   where θ depends on A and B. Use an identity from Exercise 5 of Section 2 to prove this.

5. This exercise imposes boundary conditions and initial *velocity* zero on d'Alembert's solution described in Exercise 3.

(a) Let $f(x)$ be twice-differentiable for $0 \leq x \leq p$. Extend $f(x)$ to the interval $-p < x < 0$ by defining

$$f(x) = -f(-x), \quad \text{if } -p < x < 0.$$

Then extend $f(x)$ to $-\infty < x < \infty$ to have period $2p$. Show that $f(x)$ so extended is not only periodic but odd; that is, $f(-x) = -f(x)$ for $-\infty < x < \infty$.

(b) Sketch the odd periodic extension of $f(x) = x(1 - x)$, as described in part (a), from $0 \leq x \leq 1$ to $-\infty < x < \infty$.

(c) Show that if $f(x)$ is odd and has period $2p$ for $-\infty < x < \infty$, then the function

$$u(x, t) = \tfrac{1}{2}[f(x + at) + f(x - at)]$$

is a d'Alembert solution to $a^2 u_{xx} = u_{tt}$ that satisfies the boundary conditions $u(0, t) = u(p, t) = 0$ and initial conditions $u(x, 0) = f(x)$ and $u_t(x, 0) = 0$.

6. This exercise imposes boundary conditions and initial *displacement* zero on d'Alembert's solution described in Exercise 3.

(a) Let $G(x)$ be twice-differentiable for $0 \leq x \leq p$, with $G'(0) = G'(p) = 0$. Extend $G(x)$ to the interval $-p < x < 0$ by defining

$$G(x) = G(-x), \quad \text{if } -p < x < 0.$$

Then extend $G(x)$ to $-\infty < x < \infty$ to have period $2p$. Show that $G(x)$ so extended is not only periodic but even; that is, $G(-x) = G(x)$ for $-\infty < x < \infty$.

(b) Sketch the even periodic extension of $G(x) = x^2(1 - x)^2$, as described in part (a), from $0 \leq x \leq 1$ to $-\infty < x < \infty$.

(c) Show that if $G(x)$ is even and has period $2p$ for $-\infty < x < \infty$, then the function

$$u(x, t) = \frac{1}{2a} [G(x + at) - G(x - at)]$$

is a d'Alembert solution to $a^2 u_{xx} = u_{tt}$ that satisfies the boundary conditions $u(0, t) = u(p, t) = 0$ and initial conditions $u(x, 0) = 0$ and $u_t(x, 0) = G'(x)$. Why did we assume $G'(0) = G'(p) = 0$?

7. The 1-dimensional wave equation with linear frictional drag is

$$a^2 u_{xx} = u_{tt} + k u_t,$$

where k is a positive constant. Find all product solutions $u(x, t) = X(x)T(t)$.

8. The partial differential equation $t u_{xx} = u_{tt}$ has product solutions of the form $u(x, t) = X(x)T(t)$. Find two ordinary differential equations, each dependent on a parameter λ, satisfied by $X(x)$ and $T(t)$, respectively. (The equation for X should be independent of t, and the equation for T should be independent of x.)

*9. Our derivation of the wave equation produced an equation of the form $a^2 u_{ss} = u_{tt}$. This is a linear equation if we keep arc length s as an independent space variable. However, introducing the more convenient coordinate variable x invariably leads to a **nonlinear wave equation** unless we accept a linearization. Here are three different nonlinear alternatives, each depending on a different assumption. The derivations depend on the chain rule for differentiation and the knowledge that arc length s and coordinate distance x are related by $ds/dx = \sqrt{1 + u_x^2}$ so that $dx/ds = (1 + u_x^2)^{-1/2}$. Observe that in each case the acceleration $|u_{tt}|$ decreases as the inclination $|u_x|$ increases.

(a) Replace $\partial^2 u/\partial s^2$ by $\partial(\partial u/\partial x)/\partial s$, and show that the wave equation becomes

$$a^2(1 + u_x^2)^{-1/2} u_{xx} = u_{tt}.$$

(b) Replace $\partial^2 u/\partial s^2$ by $\partial(\partial u/\partial s)/\partial x$, and show that the wave equation becomes

$$a^2(1 + u_x^2)^{-3/2} u_{xx} = u_{tt}.$$

(Note that the left side is $a^2 \kappa(x, t)$, where κ is the curvature of the vibrating filament.)

(c) Compute $\partial^2 u/\partial s^2$ with no special assumptions, and show that the wave equation is

$$a^2(1 + u_x^2)^{-2} u_{xx} = u_{tt}.$$

5   LAPLACE EQUATION

The 2-dimensional **Laplace equation**

■ **5.1** $u_{xx} + u_{yy} = 0,$

arises naturally as the time-independent **equilibrium** versions of the 2-dimensional heat and wave equations

$$a^2(u_{xx} + u_{yy}) = u_t, \qquad a^2(u_{xx} + u_{yy}) = u_{tt}.$$

The derivation of these equations is similar to the ones given in Section 4. If R is a region in the xy plane whose boundary consists of finitely many smooth curves, we let $u(x, y, t)$ be the temperature at (x, y) taken at time t. Let the temperature at the boundary be maintained so that it does not change as time goes on, although the temperature may be different at different points of the boundary. Now let enough time elapse so that there is no detectable change in the function u regarded as a function of t; in other words, wait long enough to be able to assume that $u_t(x, y, t) = 0$ throughout the interior of the region from that time on. In effect, we've assumed that $u(x, y, t) = u(x, y)$ is independent of t, in which case the heat equation reduces to the Laplace equation 5.1, since $u_t = 0$. It is often convenient to abbreviate the Laplace operator by

$$\Delta = \frac{\partial^2}{\partial x^2} + \frac{\partial^2}{\partial y^2}$$

The operator Δ acts linearly on twice-differentiable functions $u(x, y)$, and a solution of the partial differential equation $\Delta u = 0$ is called a **harmonic function.**

Solving the Laplace equation in two independent variables x and y is somewhat different from solving the heat or wave equation in the variables x and t.

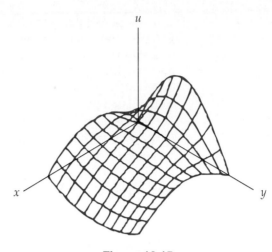

Figure 10.15

One important difference is that the boundary and initial conditions on $u(x, t)$ that are appropriate for the heat and wave equations are replaced by a boundary condition on a solution $u(x, y)$ that is defined on some 2-dimensional region. We will consider only rectangular regions, although techniques exist that are applicable to much more complicated regions.

A solution of the Laplace equation over a rectangle $0 \le x \le p$, $0 \le y \le q$ is shown in Figure 10.15. On each of the four sides there are specified boundary values for $u(x, y)$:

$$u(x, 0) = f_0(x), \qquad u(x, q) = f_q(x),$$
$$u(0, y) = g_0(y), \qquad u(p, y) = g_p(y).$$

To simplify matters for the moment, suppose that $f_0(x)$, $g_0(y)$, and $g_p(y)$ are identically zero; we will see that the general case will follow from this special case in a simple way.

Using the separation-of-variables method, we look for product solutions in the form

$$u(x, y) = X(x)Y(y).$$

Substitution into $u_{xx} + u_{yy} = 0$ gives

$$X''Y + XY'' = 0$$

for $0 < x < p$, $0 < y < q$. Assuming $X \ne 0$, $Y \ne 0$, we divide by XY, separate the variables, and write

$$\frac{Y''}{Y} = -\frac{X''}{X} = \lambda,$$

where λ is constant with respect to x and y because Y''/Y does not depend on x and X''/X does not depend on y. This gives us the two ordinary differential equations

$$X'' + \lambda X = 0, \qquad Y'' - \lambda Y = 0.$$

If $X_\lambda(x)$ and $Y_\lambda(y)$ are the solutions of these equations for the same λ, then $u_\lambda(x, y) = X_\lambda(x)Y_\lambda(y)$ will be a solution of $u_{xx} + u_{yy} = 0$.

If boundary conditions on $u(x, y)$ are $u(x, 0) = 0$, $u(0, y) = 0$, $u(p, y) = 0$, leaving the top edge at $y = q$ for later, then the corresponding implications for X and Y are

$$
\begin{aligned}
u(x, 0) = 0 &\quad \text{requires} \quad X(x)Y(0) = 0, &\quad 0 < x < p, \\
u(0, y) = 0 &\quad \text{requires} \quad X(0)Y(y) = 0, &\quad 0 < y < q, \\
u(p, y) = 0 &\quad \text{requires} \quad X(p)Y(y) = 0, &\quad 0 < y < q.
\end{aligned}
$$

We achieve these three conditions by making

$$X(0) = X(p) = 0, \qquad Y(0) = 0.$$

The resulting two-point boundary problem for X is then

$$X'' + \lambda X = 0, \qquad X(0) = X(p) = 0.$$

The nontrivial solutions of the problem are

$$X_k(x) = C_k \sin \frac{k\pi}{p} x, \qquad k = 1, 2, 3, \ldots .$$

with $\lambda = k^2\pi^2/p^2$ being the only λ values of interest from now on. Similarly, the equation for Y is now

$$Y'' - \frac{k^2\pi^2}{p^2} Y = 0, \qquad Y(0) = 0;$$

this has solutions that can be written as either

$$Y_k(y) = \alpha_k e^{k\pi y/p} + \beta_k e^{-k\pi y/p}$$

or

$$Y_k(y) = A_k \cosh \frac{k\pi y}{p} + B_k \sinh \frac{k\pi y}{p}.$$

(Recall that $\cosh x = (e^x + e^{-x})/2$ and $\sinh x = (e^x - e^{-x})/2$.) Either form for $Y(y)$ will do, but writing the constants is a little simpler in terms of hyperbolic functions, so we will use them. We satisfy $Y(0) = 0$ by making $A_k = 0$. Writing $b_k = C_k B_k$, we have

$$u_k(x, y) = X_k(x)Y_k(y) = b_k \sinh \frac{k\pi y}{p} \sin \frac{k\pi x}{p}.$$

The series representation

■ **5.2**
$$u(x, y) = \sum_{k=1}^{\infty} b_k \sinh \frac{k\pi y}{p} \sin \frac{k\pi x}{p}$$

will satisfy $u(x, q) = f_q(x)$, as well as $u(x, 0) = 0$, if we can choose the b_k so that

$$\sum_{k=1}^{\infty} b_k \sinh \frac{k\pi q}{p} \sin \frac{k\pi x}{p} = f_q(x)$$

So we take $b_k \sinh (k\pi q/p)$ to be the kth Fourier sine coefficient of $f(x) = f_q(x)$:

■ **5.3**
$$b_k = \left(\sinh \frac{k\pi q}{p} \right)^{-1} \frac{2}{p} \int_0^p f(x) \sin \frac{k\pi x}{p} \, dx.$$

If f_q is smooth enough that its Fourier series converges to it, we will have satisfied the boundary condition at $y = q$. It can be shown that the second-order partial derivatives of $u(x, y)$ as defined by Equation 5.2 can be computed using term-by-term differentiation. Since each term satisfies the Laplace equation, the linearity of the equation guarantees that $u(x, y)$ will satisfy it also.

example 1

We want to solve the Laplace equation in the rectangle $0 < x < 2$, $0 < y < 3$ (i.e., $p = 2$, $q = 3$) with boundary conditions shown in Figure 10.16a:

$$u(x, 0) = 0, \qquad u(x, 3) = 1, \qquad 0 < x < 2,$$
$$u(0, y) = 0, \qquad u(2, y) = 0, \qquad 0 < y < 3.$$

Note that no condition is imposed at the corners of the rectangle. Equation 5.3 yields

$$b_k = \left(\sinh \frac{3k\pi}{2} \right)^{-1} \int_0^2 1 \sin \frac{k\pi x}{2}\, dx$$

$$= \frac{4(1 - (-1)^k)}{k\pi \sinh (3k\pi/2)}, \qquad k = 1, 2, 3, \ldots.$$

Notice that all even-order coefficients b_{2j} are zero. The solution formula 5.2 gives us the particular solution

$$u_3(x, y) = \sum_{j=0}^{\infty} \frac{8}{(2j + 1)\pi \sinh 3(2j + 1)\pi/2} \sinh \frac{(2j + 1)\pi y}{2} \sin \frac{(2j + 1)\pi x}{2}.$$

Replacing $\sinh k\pi y/p$ by $\sinh k\pi(q - y)/p$ in Equation 5.2 gives a different solution $v(x, y)$ of the Laplace equation, a solution that satisfies $v(x, q) = 0$. The corresponding coefficients b_k are then determined by Equation 5.3 with $f_0(x) = v(x, 0)$ plugged in for $f(x)$. This simple shift allows us to solve the Laplace equation with zero boundary values everywhere around the rectangle except along the edge where $y = 0$.

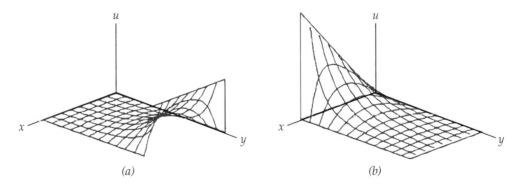

Figure 10.16 (a) $u(x, 3) = 1$, $0 < x < 2$. (b) $u(x, 0) = x$, $0 < x < 2$.

example 2

Using the preceding remark, we can solve the Laplace equation with boundary conditions

$$u(x, 0) = x, \qquad u(x, 3) = 0, \qquad 0 < x < 2,$$
$$u(0, y) = 0, \qquad u(2, y) = 0, \qquad 0 < y < 3.$$

The coefficients b_k in Equation 5.2 are determined by the requirement $u(x, 0) = f_0(x) = x$ along the bottom edge of the rectangle. From Equation 5.3 we compute

$$b_k = \left(\sinh \frac{3k\pi}{2}\right)^{-1} \int_0^2 x \sin \frac{k\pi x}{2} \, dx$$

$$= \left(\sinh \frac{3k\pi}{2}\right)^{-1} \frac{4(-1)^{k+1}}{k\pi}.$$

The particular solution, graphed in Figure 10.16b, is

$$u_0(x, y) = \sum_{k=1}^{\infty} \frac{4(-1)^{k+1}}{k\pi \sinh (3k\pi/2)} \sinh \frac{k\pi(3 - y)}{2} \sin \frac{k\pi x}{2}.$$

The factor $\sinh k\pi(3 - y)/2$ replaces $\sinh (k\pi y/2)$ as remarked earlier, so $u_0(x, 3) = 0$ will hold. But each term in the series is a solution of the Laplace equation, and it only requires justification of term-by-term differentiation of the series to show that u_0 itself satisfies the Laplace equation.

e x a m p l e 3

To solve the Laplace equation with boundary conditions

$$u(x, 0) = 5x, \qquad u(x, 3) = 4, \qquad 0 < x < 2,$$
$$u(0, y) = 0, \qquad u(2, y) = 0, \qquad 0 < y < 3.$$

we just have to form a linear combination of the solutions u_3 and u_0 that we found in Examples 1 and 2. The reason is, first, that $5u_0(x, y) + 4u_3(x, y)$ satisfies the Laplace equation because the Laplace operator Δ is linear:

$$\Delta(5u_0 + u_3) = 5\Delta u_0 + 4\Delta u_3 = 0 + 0.$$

Second, $\qquad\qquad 5u_0(x, 3) + 4u_3(x, 3) = 5x + 0 = 5x,$

and $\qquad\qquad 5u_0(x, 0) + 4u_3(x, 0) = 0 + 4 = 4.$

The reason this works is that each of u_3 and u_0 has a nonzero boundary value only where the other has a zero boundary value.

Examples 1 and 2 have had nonzero boundary values imposed only on the top and bottom edges of the region in which a solution is sought. By reversing the roles of x and y in these examples, and in Equations 5.2 and 5.3, we can solve the Laplace equation with nonzero boundary values on the right and left edges of a rectangle.

The Web site http://math.dartmouth.edu/~rewn/ contains the Applet LAPLACE for plotting solutions to the 2-dimensional Laplace equation on a rectangle.

EXERCISES

1. Each of the following functions $u(x, y)$ is a solution of the Laplace equation $u_{xx} + u_{yy} = 0$. Verify this, and verify that $u(x, y)$ satisfies the indicated boundary conditions. Then sketch the graph of $z = u(x, y)$ for (x, y) inside the indicated boundary.

(a) $u(x, y) = (e^y - e^{-y}) \sin x$; $u(0, y) = u(\pi, y) = 0$ for $0 < y < \ln 3$,
$u(x, 0) = 0$, $u(x, \ln 3) = \frac{8}{3} \sin x$ for $0 < x < \pi$.

(b) $u(x, y) = x^2 - y^2$; $u(0, y) = -y^2$, $u(1, y) = 1 - y^2$ for $0 < y < 1$,
$u(x, 0) = x^2$, $u(x, 1) = x^2 - 1$ for $0 < x < 1$.

(c) $u(x, y) = e^x \cos y$; $u(0, y) = \cos y$, $u(1, y) = e \cos y$ for $0 < y < \pi/2$,
$u(x, 0) = e^x$, $u(x, \pi/2) = 0$ for $0 < x < 1$.

(d) $u(x, y) = (\sinh 2)^{-1} \sinh (2 - y) \sin x$; $u(0, y) = u(\pi, y) = 0$ for $0 < y < 2$,
$u(x, 0) = \sin x$, $u(x, 2) = 0$ for $0 < x < \pi$.

2. As usual, define $\cosh y = (e^y + e^{-y})/2$, $\sinh y = (e^y - e^{-y})/2$. Prove the following:

(a) $d \cosh y/dy = \sinh y$.

(c) $\cosh^2 y - \sinh^2 y = 1$.

(b) $d \sinh y/dy = \cosh y$.

(d) $e^y = \cosh y + \sinh y$.

3. **(a)** Solve the Laplace equation $u_{xx} + u_{yy} = 0$ subject to the boundary conditions $u(0, y) = u(\pi, y) = 0$, $u(x, 0) = 0$, $u(x, \pi) = 0$, $0 < x < \pi, 0 < y < \pi$.

(b) Sketch the graph of the solution for $0 < x < \pi$, $0 < y < \pi$.

4. Find a series solution for the Laplace equation subject to each of the following sets of boundary conditions:

(a) $u(x, 0) = u(x, 1) = 1$, $0 < x < 1$,
$u(0, y) = 0$, $u(1, y) = 0$, $0 < y < 1$.

(b) $u(x, 0) = 0$, $u(x, 1) = x(1 - x)$, $0 < x < 1$,
$u(0, y) = u(1, y) = 0$, $0 < y < 1$.

(c) $u(x, 0) = 1$, $u(x, 1) = 2$, $0 < x < 1$,
$u(0, y) = 0$, $u(1, y) = 0$, $0 < y < 1$.

(d) $u(x, 0) = u(x, 1) = 0$, $0 < x < 1$,
$u(0, y) = y$, $u(1, y) = 0$, $0 < y < 1$.

(e) $u(x, 0) = 1$, $u(x, 2) = -1$, $0 < x < 1$,
$u(0, y) = y$, $u(1, y) = y$, $0 < y < 2$.

5. **(a)** Assume that $f(x + iy)$ and $g(x - iy)$ are twice-differentiable functions of x and y in some region R of the xy plane common to both f and g. Verify that

$$u(x, y) = f(x + iy) + g(x - iy)$$

defines a solution u of the Laplace equation in R. (*Remark:* It can be shown that f and g will in fact turn out to be infinitely often differentiable if R is an **open** set.)

(b) Let $f(x + iy) = g(x + iy) = e^{x+iy}$. What is the function $u(x, y)$ defined in part (a)?

6. Describe all solutions of the Laplace equation $u_{xx} + u_{yy} = 0$ that are independent of the variable y and are defined for (x, y) satisfying $a < x < b$ for some $a < b$.

7. **(a)** The derivation of Equation 5.3 for the coefficients b_k in Equation 5.2 was based on the assumption that $u(x, q)$ was specified for $0 < x < p$. If instead we specify the values of $u_y(x, q)$ for $0 < x < p$ by $u_y(x, q) = f(x)$,

show that Equation 5.3 is replaced by

$$b_k = \left(\cosh \frac{k\pi q}{p}\right)^{-1} \frac{2}{k\pi} \int_0^p f(x) \sin \frac{k\pi x}{p} \, dx.$$

(b) Solve the Laplace equation under the boundary conditions

$$u(x, 0) = 0, \qquad u_y(x, q) = 1, \qquad \text{for } 0 < x < p,$$
$$u(0, y) = u(p, y) = 0, \qquad \text{for } 0 < y < q.$$

8. **Graphs of harmonic functions.** Note that the Laplace equation $\Delta u = 0$ in two dimensions can be written $u_{xx} = -u_{yy}$. Also, recall that a *saddle point* for a function of two variables is one where the graph of the function has opposite concavity in two different directions. See Figure 10.15.

   (a) Show that if $\Delta u(x, y) = 0$ in a region and $u_{xx}(x_0, y_0) \neq 0$, then (x_0, y_0) corresponds to a saddle point of the graph of u. (A typical point on the graph of a nonlinear harmonic function $u(x, y)$ is a saddle point.)

   (b) Show that every point of the graph of the harmonic function $u(x, y) = x^2 - y^2$ is a saddle point, paying special attention to $(x_0, y_0) = (0, 0)$.

9. The **Laplace equation in polar coordinates** (r, θ), where $x = r \cos \theta$, $y = r \sin \theta$, is $r^2 \bar{u}_{rr} + r\bar{u}_r + \bar{u}_{\theta\theta} = 0$. Thus $\bar{u}(r, \theta) = u(r \cos \theta, r \sin \tau\theta)$, where $u(x, y)$ satisfies the xy Laplace equation $u_{xx} + u_{yy} = 0$.

   (a) Compute $\bar{u}_r$, $\bar{u}_{rr}$, and $\bar{u}_{\theta\theta}$ by applying the chain rule to $u(r \cos \theta, r \sin \theta)$. Then verify that $\bar{u}$ satisfies the polar form of the Laplace equation.

   (b) Verify that the polar form of the Laplace equation has series solutions

   $$\bar{u}(r, \theta) = r^n(a_n \cos n\theta + b_n \sin n\theta), \qquad n = 0, 1, 2, \ldots.$$

   *(c) Let $F(\theta) = u(R \cos \theta + a, R \sin \theta + b)$ be the boundary values of a harmonic function on the circular region $(x - a)^2 + (y - b)^2 \leq R^2$. Define $\bar{u}(r, \theta) = \frac{1}{2}a_0 + \sum_{n=1}^{\infty} r^n(a_n \cos n\theta + b_n \sin n\theta)$, where

   $$a_n = \frac{1}{R^n\pi}\int_{-\pi}^{\pi} F(\theta) \cos n\theta \, d\theta, \qquad b_n = \frac{1}{R^n\pi}\int_{-\pi}^{\pi} F(\theta) \sin n\theta \, d\theta.$$

   Show that the series for $\bar{u}(r, \theta)$ and its partial derivatives converge uniformly if $0 \leq r \leq R_0$ for any $R_0 < R$, and hence that $\bar{u}(r, \theta)$ is a solution of the Laplace equation in polar coordinates for $0 \leq r < R$ and $-\pi \leq \theta < \pi$.

   (d) Set $r = 0$ in the expansion of part (c) to show that a value $u(a, b)$ of a harmonic function u equals the average value of u over small circles centered at (a, b).

CHAPTER REVIEW AND SUPPLEMENT

1. What characteristic equation relating the constants r and s must be satisfied in order for $u(x, y) = e^{rx+sy}$ to be a solution of each of the following partial differential equations?

   (a) $u_{xx} - u_{yy} + 2u_x = 0$.

   (b) $u_{xx} + u_{xy} + u = 0$.

   (c) $u_{xx} + u_{yy} = u$.

   (d) $uu_{xxx} - u_y^2 = 0$.

2. Let $f(x) = \begin{cases} 1, & 0 \leq x \leq \pi/2, \\ 0, & \pi/2 < x \leq \pi. \end{cases}$

(a) Sketch the graph of the even periodic extension of $f(x)$ with period 2π.

(b) Sketch the graph of the odd periodic extension of $f(x)$ with period 2π.

(c) Sketch the graph of the periodic extension of $f(x)$ with period π.

3. A function f is defined on the interval $0 \leq x < 4$ by

$$f(x) = \begin{cases} x^2, & 0 \leq x \leq 1, \\ 2 - x, & 1 < x < 2, \\ 1, & 2 \leq x < 3, \\ x - 3, & 3 \leq x < 4. \end{cases}$$

(a) The function f has a unique periodic extension of period 4 to $-\infty < x < \infty$. Sketch the graph of this extension for $0 \leq x \leq 8$, indicating clearly the value of the function at points of discontinuity.

(b) To what respective values does the Fourier series of f converge at $x = 0$, $x = 1$, $x = 2$, $x = 3$, and $x = 4$?

(c) Sketch the graph of the odd function on $-4 \leq x \leq 4$ that agrees with f for $0 \leq x \leq 4$.

(d) Sketch the graph of the even function on $-4 \leq x \leq 4$ that agrees with f for $0 \leq x \leq 4$.

4. (a) Show that the Fourier sine expansion of $f(x) = x$ for $0 \leq x \leq \pi$ is

$$2 \sum_{k=1}^{\infty} (-1)^{k+1} \frac{1}{k} \sin kx.$$

(b) Using the result of part (a), show that $\sum_{n=1}^{\infty} (-1)^{n-1}/(2n - 1) = \pi/4$.

5. (a) Find the Fourier expansion of $f(x) = x + x^2$ for $-\pi \leq x \leq \pi$.

(b) Using the result of part (a), show that $\sum_{n=1}^{\infty} 1/n^2 = \pi^2/6$.

6. The partial differential equation $u_{xx} + 5u_x + u_t = 0$ has product solutions of the form $u(x, t) = X(x)T(t)$. Find two ordinary differential equations, each dependent on a parameter λ, satisfied by $X(x)$ and $T(t)$, respectively. (The equation for X should be independent of t, and the equation for T should be independent of x.)

7. The heat equation $u_{xx} = u_t$ with $u(0, t) = u(1, t) = 0$ has solution

$$u(x, t) = \sum_{k=1}^{\infty} A_k e^{-k^2 \pi^2 t} \sin k\pi x.$$

Find coefficients A_k so that:

(a) $u(x, 0) = 2 \sin 3\pi x$.   (b) $u(x, 0) = x$ for $0 < x < 1$.

8. Mark each of the following statements T (for true) or F (for false), as the case may be. It's possible to arrive logically at correct answers without doing any elaborate computation.

_____ The cosine coefficients a_k of the full Fourier expansion of $f(x) = x/(1 + x^2)$ for $-\pi < x < \pi$ are all zero.

_____ The cosine coefficients a_k of the full Fourier expansion of $f(x) = \cos(1 + x^2)$ for $-1 < x < 1$ are all zero.

_____ The Fourier sine series of $f(x) = 2x^2$, $0 \le x \le 1$, converges to 2 at $x = 1$.

_____ The steady-state solution of $4u_{xx} + 4u_x + u = u_t$ that satisfies $u(0, t) = 1$ and $u(1, t) = 2$ is $u(x, t) = 1 + x$.

9. Match each item in the following list of functions with its Fourier series by writing an identifying letter in the appropriate blank. Do as little calculation as possible in reaching a decision.

(A) $f(x) = \begin{cases} -1, & -\pi < x < 0 \\ 1, & 0 < x < \pi \end{cases}$ _____ $\dfrac{\pi}{2} - \dfrac{4}{\pi} \sum\limits_{k=1}^{\infty} \dfrac{\cos(2k-1)x}{(2k-1)^2}$

(B) $f(x) = x$, $-\pi < x < \pi$ _____ $\dfrac{2}{\pi} - \dfrac{4}{\pi} \sum\limits_{k=1}^{\infty} \dfrac{\cos 2kx}{4k^2 - 1}$

(C) $f(x) = x^2$, $-\pi < x < \pi$ _____ $\dfrac{4}{\pi} \sum\limits_{k=1}^{\infty} \dfrac{\sin(2k-1)x}{2k-1}$

(D) $f(x) = |\sin x|$, $-\pi < x < \pi$ _____ $2 \sum\limits_{k=1}^{\infty} \dfrac{(-1)^{k-1}}{k} \sin kx$

(E) $f(x) = |x|$, $-\pi < x < \pi$ _____ $\dfrac{\pi^2}{3} + 4 \sum\limits_{k=1}^{\infty} (-1)^k \dfrac{\cos kx}{k^2}$

10. Find product solutions $U(x, t) = X(x)T(t)$ for each of the following. Also find, if possible, nonconstant steady-state solutions for each equation, that is, solutions that are independent of t.

   (a) $U_{xx} = k^2 U_{tt}$, k const
   (b) $U_{xx} = U_t + tU$
   (c) $U_{xx} = U_t$
   (d) $U_{xx} + U_{tt} = 0$

11. Solve the wave equation $a^2 y_{xx} = y_{tt}$ subject to boundary conditions $y(0, t) = y(2, t) = 0$ and initial conditions $y(x, 0) = 0$, $y_t(x, 0) = 1 - |x - 1|$ for $0 \le x \le 2$.

12. Solve the heat equation $a^2 u_{xx} = u_t$ subject to boundary conditions $u(0, t) = 0$ and $u(p, t) = 1$, and initial condition $u(x, 0) = x/p$ for $0 \le x \le p$.

13. Let L be a linear operator.

   (a) Explain what "linear operator" means in general.
   (b) If $\mathbf{u}$ and $\mathbf{v}$ are solutions of $L(\mathbf{x}) = 0$, prove that $\mathbf{x} = c\mathbf{u} + d\mathbf{v}$ is also a solution of $L(\mathbf{x}) = 0$ for constant c, d.
   (c) Explain exactly what it means for λ and $\mathbf{u}$ to be a corresponding eigenvalue-eigenvector pair for L.
   (d) Give three examples of linear operators and three of nonlinear operators.

CHAPTER

Sturm-Liouville Expansions

1 ORTHOGONAL FUNCTIONS

Many of the useful properties of Fourier expansions are shared by a large class of similar expansions in which the functions $\sin kx$ and $\cos kx$ are replaced by some sequence $\{\phi_k\}_{k=1,2,\ldots}$ of functions, all of which are mutually orthogonal. Recall that orthogonality of the trigonometric system amounts to having an integral of the product of two distinct functions equal to zero. The same is true for other systems of orthogonal functions, but since the precise integral that is to be computed varies from system to system, it is convenient to introduce a general notation for treating all systems at once. To the extent that the different systems have common features, the uniform notation allows us to display some of their similarities more clearly; furthermore, some formal geometric intuition can then be applied. To be specific, we define the **Fourier inner product** of real-valued functions $f(x)$ and $g(x)$ for $-\pi \leq x \leq \pi$ by

■ 1.1
$$\langle f, g \rangle = \int_{-\pi}^{\pi} f(x)g(x)\, dx.$$

It follows by direct computation (see Chapter 10, Section 2) that if we set $\phi_1(x) = 1/\sqrt{2\pi}$, $\phi_{2n}(x) = (\cos nx)/\sqrt{\pi}$, and $\phi_{2n+1}(x) = (\sin nx)/\sqrt{\pi}$, then
$$\langle \phi_k, \phi_l \rangle = \begin{cases} 0, & k \neq l, \\ 1, & k = l. \end{cases}$$

Thus the sequence $\{\phi\}_{k=1,2,\ldots}$ is not only **orthogonal** in the sense that $\langle \phi_k, \phi_l \rangle = 0$ when $k \neq l$, but is **orthonormal**. The "normal" part of the terminology comes from having $\langle \phi_k, \phi_k \rangle = 1$. This normalization comes, in the case of the trigonometric system, from dividing by $\sqrt{\pi}$.

It turns out that many, although by no means all, consequences of defining the given inner product can be derived from four properties of a **general inner product:**

■ 1.2

Positivity:	$\langle f, f \rangle > 0$	except that $\langle 0, 0 \rangle = 0$.
Symmetry:	$\langle f, g \rangle = \langle g, f \rangle$.	
Additivity:	$\langle f + g, h \rangle = \langle f, h \rangle + \langle g, h \rangle,$	$\langle f, g + h \rangle = \langle f, g \rangle + \langle f, h \rangle.$
Homogeneity:	$\langle cf, g \rangle = c\langle f, g \rangle,$	$\langle f, cg \rangle = c\langle f, g \rangle.$

These properties can easily be verified for the particular inner product defined in Equation 1.1. For example, symmetry is just the statement that

$$\int_{-\pi}^{\pi} f(x)g(x)\,dx = \int_{-\pi}^{\pi} g(x)f(x)\,dx.$$

One reason for singling out these four properties here is that they are shared by many other similar products, including the ordinary dot product of coordinate vectors. In addition, we'll see in Section 2 that inner products of the form

■ 1.3
$$\langle f, g \rangle = \int_{a}^{b} f(x)g(x)r(x)\,dx$$

with **weight function** $r(x) \geq 0$ arise in a natural way.

With any definition of $\langle f, g \rangle$ that satisfies the relation 1.2, we can define a **norm** or **length** of a real-valued function f such that f^2 is integrable by

$$\|f\| = \langle f, f \rangle^{1/2}.$$

For our definition of Fourier inner product, with $r(x) = 1$, this amounts to

$$\|f\| = \left(\int_{-\pi}^{\pi} f^2(x)\,dx \right)^{1/2}.$$

More generally, we may consider functions $f(x)$ and $g(x)$ that are continuous for $a \leq x \leq b$. We denote the set of such functions by $C[a, b]$, and we define

$$\langle f, g \rangle = \int_{a}^{b} f(x)g(x)\,dx, \qquad \|f\| = \left(\int_{a}^{b} f^2(x)\,dx \right)^{1/2}$$

for f and g in $C[a, b]$.

---

e x a m p l e 1

Let $f(x)$ and $g(x)$ be integrable functions. We could compute $\langle f + g, f - g \rangle$ directly from the specific definition of $\langle f, g \rangle$ in Equation 1.1. Alternatively, we can use the symmetry, additivity, and homogeneity properties:

$$\begin{aligned}
\langle f + g, f - g \rangle &= \langle f + g, f \rangle + \langle f + g, -g \rangle \\
&= \langle f, f \rangle + \langle g, f \rangle + \langle f, -g \rangle + \langle g, -g \rangle \\
&= \langle f, f \rangle + \langle f, g \rangle - \langle f, g \rangle - \langle g, g \rangle \\
&= \langle f, f \rangle - \langle g, g \rangle.
\end{aligned}$$

This much is true regardless of how f and g are chosen. If it should happen that $f^2(x) = g^2(x)$, for example, $f(x) = x$, $g(x) = |x|$, then $\langle f, f \rangle = \langle g, g \rangle$, so we conclude that $\langle f + g, f - g \rangle = 0$. In other words, $f + g$ and $f - g$ are orthogonal. The conclusion is of course true also whenever $\langle f, f \rangle = \langle g, g \rangle$.

To see the importance of orthonormal sequences in general, we consider the following problem: Let $\langle f, g \rangle$ be an inner product, and let $\{\phi_k\}_{k=1,2,...}$ be a sequence of elements, orthonormal with respect to the inner product. Using the **norm** defined by $\|f\| = \langle f, f \rangle^{1/2}$, we try to determine coefficients c_k, $k = 1, 2, \ldots, N$, such that

$$\left\| g - \sum_{k=1}^{N} c_k \phi_k \right\|$$

is minimized for given g and N.

■ 1.4 THEOREM

Let $\{\phi_k\}_{k=1,2,...}$ be an orthonormal sequence in a space with an inner product in which numerical multiplication and addition are defined. Then, given an element g of the space, the distance

$$d_N = \left\| g - \sum_{k=1}^{N} c_k \phi_k \right\|$$

is minimized for $N = 1, 2, \ldots$ by taking c_k to be the kth Fourier coefficient: $c_k = \langle g, \phi_k \rangle$.

Proof. The fact that the sequence ϕ_k is orthonormal makes the solution very simple; we get

$$0 \le \left\| g - \sum_{k=1}^{N} c_k \phi_k \right\|^2 = \left\langle g - \sum_{k=1}^{N} c_k \phi_k, g - \sum_{k=1}^{N} c_k \phi_k \right\rangle$$

$$= \|g\|^2 - 2 \sum_{k=1}^{N} c_k \langle g, \phi_k \rangle + \sum_{k=1}^{N} c_k^2$$

$$= \|g\| - \sum_{k=1}^{N} \langle g, \phi_k \rangle^2 + \sum_{k=1}^{N} [\langle g, \phi_k \rangle^2 - 2c_k \langle g, \phi_k \rangle + c_k^2]$$

$$= \|g\|^2 - \sum_{k=1}^{N} \langle g, \phi_k \rangle^2 + \sum_{k=1}^{N} [\langle g, \phi_k \rangle - c_k]^2. \tag{1}$$

We have added and subtracted $\sum_{k=1}^{N} \langle g, \phi_k \rangle^2$ at the next-to-last step. But the first two terms in the last expression are independent of the choice of the c_k, and the last sum is then minimized by taking $c_k = \langle g, \phi_k \rangle$.   ■

The numbers $c_k = \langle g, \phi_k \rangle$ are called the **Fourier coefficients** of g with respect to the orthonormal sequence $\{\phi_k\}$. An important remark about the conclusion of Theorem 1.4 is that the c_k are uniquely determined, independently of N. In other words, if we wanted to improve the closeness of the approximation to g by increasing N, then Theorem 1.4 says that the c_k already computed are to be left unchanged, and it is only necessary to compute additional coefficients $c_k = \langle g, \phi_k \rangle$, $k = N + 1, \ldots$.

As a by-product of the proof of Theorem 1.4, we have the following.

■■ **1.5 BESSEL'S INEQUALITY** $\|g\|^2 \geq \sum\limits_{k=1}^{\infty} \langle g, \phi_k \rangle^2.$

Proof. The inequality $0 \leq \|g\|^2 - \sum_{k=1}^{N} \langle g, \phi_k \rangle^2 + \sum_{k=1}^{N} [\langle g, \phi_k \rangle - c_k]^2$ was established in the last step of equation (1). On taking $c_k = \langle g, \phi_k \rangle$, the inequality becomes

$$0 \leq \|g\|^2 - \sum_{k=1}^{N} \langle g, \phi_k \rangle^2.$$

Bessel's inequality follows by letting N tend to infinity.   ■

If the distance d_N in Theorem 1.4 always tends to zero as $N \to \infty$, the orthonormal system $\{\phi_k\}$ is said to be **complete,** and the inequality in Bessel's inequality 1.5 becomes an equation, called the **Parseval equation:**

$$\|g\|^2 = \sum_{k=1}^{\infty} \langle g, \phi_k \rangle^2.$$

(The proof, assuming completeness, is left as an exercise.) Taken in the proper context, all the orthonormal sequences considered in this book turn out to be complete, and we'll assume that about each of them. Unfortunately, the proofs of these facts lie considerably deeper than we're prepared to go here.

 **example** 2

The approximation to g by a sum $\sum_{k=1}^{N} c_k \phi_k$ has been measured by a norm in Theorem 1.4. To see what this means for approximation by trigonometric polynomials, we use the inner product given by Equation 1.1 on the space $C[-\pi, \pi]$ of continuous functions on $[-\pi, \pi]$. Given the orthonormal sequence

$$\phi_1(x) = \frac{1}{\sqrt{2\pi}}, \qquad \phi_{2n}(x) = \frac{\cos nx}{\sqrt{\pi}}, \qquad \phi_{2n+1}(x) = \frac{\sin nx}{\sqrt{\pi}},$$

we try to minimize, for given g in $C[-\pi, \pi]$, the norm

$$\left\| g - \sum_{k=1}^{2N+1} c_k \phi_k \right\|.$$

We have seen that this is done by taking $c_k = \langle g, \phi_k \rangle$. By the definition of the inner product, we can take out constant factors to get

$$\langle g, \phi_k \rangle = \begin{cases} (1/\sqrt{2\pi}) \int_{-\pi}^{\pi} g(t)\, dt, & k = 1, \\ (1/\sqrt{\pi}) \int_{-\pi}^{\pi} g(t) \cos nt\, dt, & k = 2n, \\ (1/\sqrt{\pi}) \int_{-\pi}^{\pi} g(t) \sin nt\, dt, & k = 2n+1. \end{cases}$$

Hence the terms $c_k \phi_k$ become

$$\langle g, \phi_1 \rangle \phi_1(x) = \frac{1}{2\pi} \int_{-\pi}^{\pi} g(t)\, dt,$$

$$\langle g, \phi_{2n} \rangle \phi_{2n}(x) = \left[\frac{1}{\pi} \int_{-\pi}^{\pi} g(t) \cos nt\, dt \right] \cos nx,$$

$$\langle g, \phi_{2n+1} \rangle \phi_{2n+1}(x) = \left[\frac{1}{\pi} \int_{-\pi}^{\pi} g(t) \sin nt\, dt \right] \sin nx.$$

Then

$$\sum_{k=1}^{2N+1} c_k \phi_k(x) = \frac{a_0}{2} + \sum_{k=1}^{N} (a_k \cos kx + b_k \sin kx),$$

where a_k and b_k are the trigonometric Fourier coefficients as defined in Section 2 of Chapter 10. The square of the norm to be minimized takes the form

$$\int_{-\pi}^{\pi} \left[g(x) - \frac{a_0}{2} - \sum_{k=1}^{N} (a_k \cos kx + b_k \sin kx) \right]^2 dx,$$

and Theorem 1.4 says that the minimum will be attained for any fixed N by taking a_k and b_k to be the Fourier coefficients of g. Since the trigonometric system is complete, this minimum tends to zero as N tends to infinity, and the Parseval equation for the trigonometric system becomes, after division by the number π,

$$\frac{1}{\pi} \int_{-\pi}^{\pi} g(x)^2 \, dx = \tfrac{1}{2}a_0^2 + \sum_{k=0}^{\infty} (a_k^2 + b_k^2).$$

This equation has an interesting interpretation in terms of sound intensity, as shown in the exercises.

The minimization of an integral of the form

$$\int_{a}^{b} \left[g(x) - \sum_{k=1}^{N} c_k \phi_k(x) \right]^2 dx$$

is called a best **mean-square approximation** to g. In this sense we can say that the Fourier approximation provides the best mean-square approximation by a trigonometric polynomial.

EXERCISES

1. **(a)** Verify that

$$\langle f, g \rangle = \int_{a}^{b} f(x)g(x) \, dx$$

defines an inner product on the set $C[a, b]$ of continuous functions defined on $[a, b]$; that is, verify Equations 1.2 of the text.

(b) What condition is required of a continuous function $r(x)$ defined on $[a, b]$ in order that

$$\langle f, g \rangle = \int_{a}^{b} f(x)g(x)r(x) \, dx$$

satisfy $\langle f, f \rangle \geq 0$?

2. Define $\langle \mathbf{x}, \mathbf{y} \rangle = x_1 y_1 + x_2 y_2 + x_3 y_3$ for   vectors $\mathbf{x} = (x_1, \ x_2, \ x_2)$, $\mathbf{y} = (y_1, \ y_2, \ y_3)$ in $\mathcal{R}^3$.

(a) Verify that $\langle \mathbf{x}, \mathbf{y} \rangle$ is equal to the dot product $\mathbf{x} \cdot \mathbf{y}$ on $\mathcal{R}^3$.

(b) Verify that $\langle \mathbf{x}, \mathbf{y} \rangle$ as defined above satisfies the four properties of a general inner product listed under Equations 1.2.

3. Let $\{\phi_k\}$ be an orthonormal sequence of functions in $C[a, b]$, and let g be in $C[a, b]$. Show that if the real-valued function

$$\Delta(c_1, \ldots, c_N) = \int_a^b \left[g(x) - \sum_{k=1}^N c_k \phi_k(x) \right]^2 dx$$

has a local minimum as a function of $(c_1, \ldots, c_N)$, then

$$c_k = \int_a^b g(x) \phi_k(x) \, dx.$$

[*Hint:* Differentiate the formula for Δ under the integral sign. This is justified under the assumptions made here.]

4. Let $\{\phi_k\}$ be an orthonormal sequence of functions in $C[a, b]$ and let g be in $C[a, b]$. Use Bessel's inequality to show the following:

(a) $\sum_{k=1}^\infty [\int_a^b g(x)\phi_k(x) \, dx]^2$ converges.

(b) If c_k is the kth Fourier coefficient of g with respect to $\{\phi_k\}$, then $\lim_{k\to\infty} c_k = 0$.

(c) Find an example of a function whose trigonometric Fourier series has coefficients a_k and b_k such that $\sum_{k=1}^\infty (a_k^2 + b_k^2)$ converges, but such that $\sum_{k=1}^\infty a_k$ or $\sum_{k=1}^\infty b_k$ does not converge.

5. Define $\langle \mathbf{x}, \mathbf{y} \rangle = r_1 x_1 y_1 + r_2 x_2 y_2 + r_3 x_3 y_3$, where $\mathbf{x} = (x_1, \ x_2, \ x_3)$, $\mathbf{y} = (y_1, \ y_2, \ y_3)$, for positive r_k. Verify that this definition of $\langle \mathbf{x}, \mathbf{y} \rangle$ satisfies the relations 1.2 with x replacing f and y replacing g. Is this still true if $r_1 = 0$?

6. Use the properties 1.2 of a general inner product, together with its relation to length, to prove the following relations. Can you give a geometric interpretation for (b) and (c)?

(a) $\|f - g\|^2 = \|f\|^2 + \|g\|^2 - 2\langle f, g \rangle$.

(b) $\|f + g\|^2 + \|f - g\|^2 = 2\|f\|^2 + 2\|g\|^2$.

(c) $\|f - g\|^2 = \|f - h\|^2 + \|h - g\|^2$ if and only if $\langle f - h, h - g \rangle = 0$.

7. Use the displayed equations in the proof of Theorem 1.4 to show that the **Parseval equation** $\|g\|^2 = \sum_{k=1}^\infty \langle g, \phi_k \rangle^2$ holds for an orthonormal system $\{\phi_k\}$ if and only if the system is **complete,** that is, if and only if $\|g - \sum_{k=1}^N \langle g, \phi_k \rangle \phi_k \| \to 0$ as $N \to \infty$.

8. Periodic sound sources are perceived by the human ear as pressure variations $g(t)$ representable by Fourier series:

$$g(t) = \tfrac{1}{2}a_0 + \sum_{k=1}^\infty \left(a_k \cos \frac{k\pi t}{p} + b_k \sin \frac{k\pi t}{p} \right).$$

The values $g(t)$ are the deviations, plus or minus, from ambient atmospheric pressure at times t. The **intensity** of a pressure function of period $2p$ is defined to be the average value $(1/2p) \int_0^{2p} g^2(t) \, dt$.

(a) Show that $g(t) = a \cos (k\pi t/p) + b \sin (k\pi t/p)$ has intensity $\tfrac{1}{2}(a^2 + b^2)$.

(b) Assuming the validity of Parseval's equation for the general pressure function $g(t)$, show that the intensity of g equals the sum of the intensities of the individual terms in the Fourier expansion of g.

9. The inner product on continuous functions in $C[0, 1]$ given by

$$\langle f, g \rangle = \int_0^1 f(x)g(x)x \, dx$$

arises naturally in problems involving Bessel functions. Verify the conditions listed in Equations 1.2 of the text, paying particularly careful attention to the positivity condition.

10. Which of the following definitions for the weight function $r(x)$ makes

$$\langle f, g \rangle = \int_{-1}^1 f(x)g(x)r(x) \, dx$$

a genuine inner product on $C[-1, 1]$, the continuous real functions on $-1 \leq x \leq 1$? Look carefully at the conditions given by Equations 1.2, and for those choices of $r(x)$ that don't define inner products, state why.

(a) $r(x) = x^2$.

(b) $r(x) = x$.

(c) $r(x) = (1 - x^2)^{-1/2}$.

(d) $r(x) = x, x \geq 0, r(x) = 0, x < 0$.

(e) $r(x) = |x|$.

(f) $r(x) = x + 1$.

11. The nth **Chebychev polynomial** is defined by

$$T_n(x) = \cos(n \arccos x), \qquad -1 \leq x \leq 1.$$

(a) Compute $T_n(x)$ for $n = 0, 1, 2, 3$.

(b) Show that $|T_n(x)| \leq 1$ for $-1 \leq x \leq 1$.

(c) Verify that the relation

$$\langle f, g \rangle = \int_{-1}^1 f(x)g(x)(1 - x^2)^{-1/2} \, dx$$

defines an inner product on $C[-1, 1]$; that is, verify properties 1.2.

(d) Prove the orthogonality relations

$$\int_{-1}^1 T_k(x)T_l(x)(1 - x^2)^{-1/2} \, dx = \begin{cases} 0, & k \neq l, \\ \pi, & k = l \neq 0. \end{cases}$$

[*Hint:* Make the change of variable $x = \cos \theta$.]

(e) Show that $T_n(x)$ is always a polynomial by setting $\arccos x = \theta$ and expanding the right side of $\cos n\theta = \operatorname{Re} e^{in\theta} = \operatorname{Re}(\cos \theta + i \sin \theta)^n$ by the binomial theorem.

12. For continuous functions $f(x, y)$ and $g(x, y)$ defined on the rectangle $0 \leq x \leq p, 0 \leq y \leq q$ in $\mathcal{R}^2$, define

$$\langle f, g \rangle = \int_0^p \int_0^q f(x, y)g(x, y) \, dx \, dy.$$

(a) Show that $\langle f, g \rangle$ is an inner product on the set of continuous functions on the rectangle.

(b) Show that the functions

$$\phi_{jk}(x, y) = \frac{4}{pq} \sin \frac{j\pi x}{p} \sin \frac{k\pi y}{q}, \qquad j, k = 1, 2, 3, \ldots,$$

form an orthonormal set relative to the above inner product:

$$\langle \phi_{jk}, \phi_{lm} \rangle = \begin{cases} 0, & (j, k) \neq (l, m) \\ 1, & (j, k) = (l, m). \end{cases}$$

13. **Heat equation in a rectangular plate.** The heat equation in rectangular xy coordinates is $a^2(u_{xx} + u_{yy}) = u_t$.

(a) Assuming product solutions of the form $u(x, y, t) = X(x)Y(y)T(t)$, derive the solutions

$$u_{mn}(x, y, t) = e^{\pi^2 a^2(m^2 + n^2)t} \sin \frac{m\pi x}{p} \sin \frac{n\pi y}{q}$$

with boundary values identically zero on the edges of the rectangle with corners at $(0, 0)$, $(p, 0)$, (p, q), $(0, q)$.

(b) Use the orthonormal system of the previous exercise to derive the formal expansion

$$u(x, y, t) = \sum_{m,n=1}^{\infty} b_{jk} e^{\pi^2 a^2(m^2 + n^2)t} \sin \frac{m\pi x}{p} \sin \frac{n\pi y}{q},$$

$$b_{mn} = \frac{4}{pq} \int_0^p \int_0^q f(x, y) \sin \frac{m\pi x}{p} \sin \frac{n\pi y}{q} \, dx \, dy$$

for the zero-boundary problem with initial temperature function $f(x, y)$ on the rectangle.

(c) Suppose $w(x, y)$ solves the Laplace equation $u_{xx} + u_{yy} = 0$ on the rectangle with given boundary values on the edges of the rectangle. Verify that if $f(x, y)$ is replaced by $f(x, y) - w(x, y)$ in part (b) to produce solution $U(x, y, t)$, then $u(x, y, t) = w(x, y) + U(x, y, t)$ solves the heat equation subject to the same boundary conditions and with initial temperature function $f(x, y)$.

2 EIGENFUNCTIONS AND SYMMETRY

It is easy to check that the functions $\cos kx$ and $\sin kx$ are solutions of the harmonic oscillator equation $y'' = -k^2 y$. Standard terminology associated with these relations is that each of $\cos kx$ and $\sin kx$ is an **eigenfunction** of the differential operator d^2/dx^2, corresponding to the **eigenvalue** $-k^2$. More generally, let L be a linear differential operator. We say that a function $y = y_\lambda(x)$, not identically zero, is an **eigenfunction** of L corresponding to the **eigenvalue** λ if $Ly_\lambda = \lambda y_\lambda$ for some number λ. If you've found an eigenfunction y corresponding to an eigenvalue λ, then cy is also an eigenfunction for each constant $c \neq 0$; the reason is that because L is linear, $L(cy) = cLy = c(\lambda y) = \lambda(cy)$.

To understand the connection between eigenfunctions and orthogonal sets of functions, we need one more definition. Suppose that $\mathcal{F}$ is a set of functions with an inner product $\langle f, g \rangle$. (We assume that sums and numerical multiples of functions in $\mathscr{F}$ are also in $\mathscr{F}$.) Let L be a linear operator from $\mathscr{F}$ to $\mathscr{F}$. Then L is **symmetric** with respect to the inner product if $\langle Lf, g \rangle = \langle f, Lg \rangle$ for all functions f and g for which Lf and Lg are defined as elements of $\mathscr{F}$.

In the following examples we denote the set of continuous functions on $[a, b]$ by $C[a, b]$. Eventually we'll introduce a generalization of the Fourier inner product 1.1, defined in Section 1, to the **weighted inner product** on $C[a,b]$ with positive **weight function** $r(x)$:

■ 2.1
$$\langle f, g \rangle_r = \int_a^b f(x)g(x)r(x)\, dx.$$

At first we'll continue to concentrate on the special case $r(x) = 1$.

2A   *Fourier Inner Product*

We begin with the simplest example of a symmetric operator on $C[0, \pi]$. In all examples it's important to understand that symmetry of an operator depends not just on a formula for a differential operator but also on a precise specification of the functions on which the differential operator acts. This specification is usually made by imposing certain boundary conditions on the functions, and suitable conditions for a particular operator are called **symmetric boundary conditions.**

e x a m p l e 1

Let $\langle f, g \rangle$ be the Fourier inner product defined by Equation 1.1 of Section 1. If we let $Lf = d^2f/dx^2$, then it is clear that Lf is in $C[0, \pi]$ only for those f in $C[0, \pi]$ that happen to have continuous second derivatives. Thus, for L to be symmetric, we must have $\langle Lf, g \rangle = \langle f, Lg \rangle$ for all twice continuously differentiable f and g on $[0, \pi]$. Equivalently, we must have, because of the definition of the inner product,

$$\int_0^\pi f''(x)g(x)\, dx = \int_0^\pi f(x)g''(x)\, dx.$$

This follows from integration by parts twice:

$$\int_0^\pi f''(x)g(x)\, dx = [f'(x)g(x)]_0^\pi - \int_0^\pi f'(x)g'(x)\, dx$$

$$= [f'(x)g(x) - f(x)g'(x)]_0^\pi + \int_0^\pi f(x)g''(x)\, dx.$$

Hence to make L symmetric we restrict its domain to some subset of $C[0, \pi]$ for which the bracketed terms will always add up to zero. This can be done in several ways. For example, we may restrict L to the subset consisting of those functions f in $C[0, \pi]$ for which $f(0) = f(\pi) = 0$, or to the subset for which $f'(0) = f'(\pi) = 0$. With either of these restrictions, L becomes symmetric. Notice that while a restriction of the required type is a boundary condition in that it specifies the values of f at the endpoints 0 and π of its domain of definition, the symmetry follows from making rather special choices for these "symmetric" conditions.

The connection between orthogonal functions and symmetric operators is as follows.

■ 2.2 THEOREM

Let L be a symmetric linear operator defined on a set of functions with an inner product. If y_1 and y_2 are eigenfunctions of L corresponding to distinct eigenvalues λ_1 and λ_2, then y_1 and y_2 are orthogonal.

Proof. We assume that

$$Ly_1 = \lambda_1 y_1, \qquad Ly_2 = \lambda y_2,$$

and prove that $\langle y_1, y_2 \rangle = 0$. We have

$$\langle Ly_1, y_2 \rangle = \langle \lambda y_1, y_2 \rangle = \lambda_1 \langle y_1, y_2 \rangle,$$

and

$$\langle y_1, Ly_2 \rangle = \langle y_1, \lambda_2 y_2 \rangle = \lambda_2 \langle y_1, y_2 \rangle.$$

Because L is symmetric, $\langle Ly_1, y_2 \rangle = \langle y_1, Ly_2 \rangle$, so $\lambda_1 \langle y_1, y_2 \rangle = \lambda_2 \langle y_1, y_2 \rangle$, equivalently $(\lambda_1 - \lambda_2)\langle y_1, y_2 \rangle = 0$. Since λ_1 and λ_2 are not equal, we must have $\langle y_1, y_2 \rangle = 0$. ■

A differential operator of **Sturm-Liouville form** is one that can be written

■ 2.3 $$Ly = (py')' + qy,$$

where $p(x)$ is a continuously differentiable function and $q(x)$ is assumed only continuous. (The operator d^2/dx^2 is a special case if we set $p(x) = 1$, $q(x) = 0$.) We want first to see what boundary conditions should be imposed on the domain of L in order to make L symmetric with respect to the inner product

$$\langle f, g \rangle = \int_a^b f(x)g(x)\, dx.$$

The following identity reduces the problem to its essence.

■ 2.4 LAGRANGE IDENTITY

If L is given by $Ly = (py')' + qy$ on an interval $[a, b]$, then

$$\langle y_1, Ly_2 \rangle - \langle Ly_1, y_2 \rangle = [p(x)(y_1(x)y_2'(x) - y_1(x)'y_2(x))]_a^b.$$

This identity shows that any conditions on the coefficient $p(x)$, or on the set $\mathcal{F}$ containing y_1 and y_2, that makes

$$p(a)[y_1(a)y_2'(a) - y_1'(a)y_2(a)] = p(b)[y_1(b)y_2'(b) - y_1'(b)y_2(b)]$$

are symmetric conditions that will make L symmetric on $\mathcal{F}$.

Proof. Starting with the definition of L we rearrange $y_1(Ly_2) - (Ly_1)y_2$ as follows:

$$\begin{aligned}
y_1(Ly_2) - (Ly_1)y_2 &= y_1[(py_2')' + qy_2] - y_2[(py_1')' + qy_1] \\
&= y_1(py_2')' - y_2(py_1')' \\
&= y_1[py_2'' + p'y_2'] - y_2[py_1'' + p'y_1'] \\
&= p'[y_1y_2' - y_1'y_2] + p[y_1y_2'' - y_1''y_2] \\
&= [p(y_1y_2' - y_1'y_2)]'.
\end{aligned}$$

Integrating both sides from a to b gives the Lagrange identity. ■

e x a m p l e 2

The operator defined by $Ly(x) = y''(x)$ is in Sturm-Liouville form, where we take $p(x) = 1$ and $q(x) = 0$. Initially we'll consider L to be operating on twice continuously differentiable functions defined on $[-\pi, \pi]$. To ensure that L is symmetric we'll further restrict the domain of L by requiring that the right side of the Lagrange identity 2.4 be zero. This we do by restricting attention to those functions $y(x)$ for which $y(0) = y(\pi) = 0$, or else to those for which $y'(0) = y'(\pi) = 0$. Solving the eigenvalue problem $y'' = \lambda y$, we find

$$y(x) = c_1 e^{\sqrt{\lambda}x} + c_2 e^{-\sqrt{\lambda}x}.$$

Fixing attention on the boundary conditions $y(0) = y(\pi) = 0$, we see that the coefficients c_1 and c_2 must satisfy

$$c_1 + c_2 = 0, \qquad c_1 e^{\sqrt{\lambda}\pi} + c_2 e^{-\sqrt{\lambda}\pi} = 0.$$

Solving for c_1 and c_2 shows that they are zero unless $e^{2\sqrt{\lambda}\pi} = 1$, in which case the only restriction is that $c_2 = -c_1$. The trivial choice $c_1 = c_2 = 0$ is conveniently ruled out as a part of the definition of eigenfunction. The possibility that $e^{2\sqrt{\lambda}\pi} = 1$ is satisfied by just those values of λ for which $2\sqrt{\lambda}\pi = 2k\pi i$, where k is an integer; that is, $\sqrt{\lambda} = ki$, so eigenvalues $\lambda = -k^2$. Since $c_2 = -c_1$, the corresponding eigenfunctions are

$$y_k(x) = c_1 e^{kix} - c_1 e^{-kix} = 2ic_1 \sin kx.$$

Taking $c_1 = -\frac{1}{2}i$, we get a sequence of eigenfunctions $y_k(x) = \sin kx$, $k = 1, 2, 3, \ldots$, with corresponding eigenvalues $\lambda_k = -k^2$. Orthogonality now follows from Theorem 2.2:

$$\langle y_k, y_l \rangle = \int_0^\pi \sin kx \sin lx \, dx = 0, \qquad k \neq l.$$

e x a m p l e 3

A computation similar to the one in Example 2 identifies the eigenfunctions of d^2/dx^2 that satisfy the symmetric boundary conditions $y'(0) = y'(\pi) = 0$; they are $z_k(x) = \cos kx$, $\lambda_k = -k^2$, $k = 0, 1, 2, \ldots$. Working through the details is left as an exercise.

e x a m p l e 4

The operator defined by $Ly(x) = (1 - x^2)y''(x) - 2xy'(x)$, is in Sturm-Liouville form where we set $p(x) = (1 - x^2)$ and $q(x) = 0$. We'll consider L to be operating on twice continuously differentiable functions defined on $[-1, 1]$. Normally we would make L symmetric by ensuring that the right side of the Lagrange formula 2.4 is always zero for $a = -1$, $b = 1$. But since $p(x) = (1 - x^2)$, we find that $p(-1) = p(1) = 0$. Hence, L is symmetric on $C[-1, 1]$ without further restriction, and its domain consists of all twice continuously differentiable functions in $C[-1, 1]$.

The symmetric operator Lf defined in Example 2 is usually associated with the differential equation

$$(1 - x^2)y'' - 2xy' + n(n + 1)y = 0.$$

This is called the **Legendre equation** of index n, and it is satisfied by the nth **Legendre polynomial** defined by

2.5
$$P_n(x) = \frac{1}{2^n n!} \frac{d^n}{dx^n} (x^2-1)^n, \qquad n = 0, 1, 2, \ldots.$$

That P_n satisfies the Legendre equation can be verified by repeated differentiation (see Exercise 4). The significance of the fact that P_n satisfies the Legendre equation comes from writing the equation in the form $Ly = -n(n + 1)y$, where L is the symmetric operator $Ly = (1 - x^2)y'' - 2xy'$ on $C[-1, 1]$. Then P_n can be looked at as an eigenfunction of L, corresponding to the eigenvalue $-n(n + 1)$. Hence, by Theorem 2.2, the Legendre polynomials are orthogonal; that is,

$$\int_{-1}^{1} P_n(x)P_m(x)\, dx = 0, \qquad n \neq m.$$

Furthermore, a fairly complicated calculation (see Exercise 6) shows that

$$\int_{-1}^{1} P_n^2(x)\, dx = \frac{2}{2n + 1}, \qquad n = 0, 1, 2, \ldots$$

Therefore, the normalized sequence $\{(\sqrt{2n + \frac{1}{2}})P^n(x)\}$ $n = 0, 1, 2, \ldots$ is an orthonormal sequence in $C[-1, 1]$. The graphs of the first four Legendre polynomials on $[-1, 1]$, shown in Figure 11.1, make it plausible that these functions are orthogonal.

The Nth **Fourier-Legendre approximation** to a function g in $C[-1, 1]$ is the finite sum

$$\sum_{k=0}^{N} c_k P_k(x),$$

where

2.6
$$c_k = \frac{2k + 1}{2} \int_{-1}^{1} g(x)P_k(x)\, dx.$$

Theorem 1.4 in Section 1 then implies that the best mean-square approximation to g by a linear combination of Legendre polynomials is given by the Fourier-

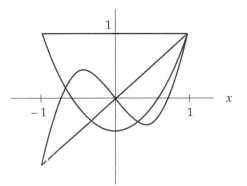

Figure 11.1 $P_n(x)$ for $-1 \leq x \leq 1$, $n = 0, 1, 2, 3$.

Legendre approximation. In other words,

$$\int_{-1}^{1} \left[g(x) - \sum_{k=0}^{N} c_k P_k(x) \right]^2 dx$$

is minimized by computing c_k by Equation 2.6. The precise details under which convergence takes place for particular values of x is generally of less interest than for trigonometric series; however, it's worth mentioning that Fourier-Legendre series converge if $g(t)$ is piecewise-monotone on $[-1, 1]$.

2B Weighted Inner Products

By introducing the more general weighted inner product 1.2, we can extend the scope of the ideas in Section 2A considerably. In particular, we can include the following ordinary differential equations that arise from separating variables in partial differential equations. Each equation is followed by an equivalent equation in Sturm-Liouville form.

Parametric Bessel equation:

$$x^2 y'' + x y' + (\lambda^2 x^2 - n^2) y = 0; \qquad (x y')' + \left(\lambda^2 x - \frac{n^2}{x} \right) y = 0.$$

Chebychev equation:

$$(1 - x^2) y'' - x y' + \lambda^2 y = 0; \qquad (\sqrt{1 - x^2} y')' + \frac{\lambda^2}{\sqrt{1 - x^2}} y = 0.$$

Hermite equation:

$$y'' - x y' + \lambda y = 0; \qquad (e^{-x^2/2} y')' + \lambda e^{-x^2/2} y = 0.$$

Laguerre equation:

$$x y'' + (1 - x) y' + \lambda y = 0; \qquad (x e^{-x} y')' + \lambda e^{-x} y = 0.$$

As is, the first equation in each pair is not in Sturm-Liouville form, but a simple modification allows us to put it in that form so we can apply the theory of Section 2A. The general idea is first to normalize the general eigenvalue-eigenfunction equation

$$a_0(x) y'' + a_1(x) y' + a_2(x) y = \lambda y \qquad \text{to} \qquad y'' + a(x) y' + b(x)(x) y = \lambda c(x) y$$

by dividing by $a_0(x)$. At this point we can apply the exponential integrating factor device of Chapter 1, Section 5A, to the first two terms:

$$\text{If} \qquad p(x) = e^{\int a(x)\, dx}, \qquad \text{then} \qquad p(x) y'' + p(x) a(x) y' = \frac{d}{dx}(p(x) y').$$

After multiplication by $p(x)$, the normalized differential equation can then be written

$$\frac{d}{dx}(p(x) y') + q(x) y = \lambda r(x) y, \qquad \text{where } q(x) = p(x) b(x) \text{ and } r(x) = p(x) c(x).$$

This last equation is in Sturm-Liouville form, and we can apply the Lagrange identity 2.4 to it in proving the next theorem. This theorem subsumes the related discussion in Section 2A.

e x a m p l e 5

Following the routine outlined above to put the parametric Bessel equation
$$x^2 y'' + xy' + (\lambda^2 x^2 - n^2)y = 0$$
in Sturm-Liouville form, we first normalize the differential equation to get $y'' + (1/x)y' + (\lambda - n^2/x^2)y = 0$. Setting $p(x) = e^{\int (1/x)\,dx} = x$, we multiply this equation by $p(x) = x$ to get

$$xy'' + y' + \left(\lambda^2 x - \frac{n^2}{x}\right)y = (xy')' + \left(\lambda^2 x - \frac{n^2}{x}\right)y = 0.$$

Note that $r(x) = p(x) = x$ so that $p(0)) = 0$. To achieve symmetry an arbitrary boundary condition can be imposed at $x = 0$ along with any condition at another point $x = b$ that makes the Wronskian $w[y_1, y_2](b) = 0$.

■ 2.7 THEOREM

The Sturm-Liouville equation
$$(p(x)y')' + q(x)y = \lambda r(x)y$$
is assumed to have $p(x)$ continuously differentiable and $q(x)$ and $r(x)$ continuous on an interval (a, b), which may be infinite. In addition $r(x)$ is to be positive and integrable on (a, b). Suppose $y_1(x)$ and $y_2(x)$ are eigenfunctions that both satisfy the same symmetric boundary conditions. If the corresponding eigenvalues λ_1, λ_2 are distinct, then y_1 and y_2 are orthogonal with respect to the $r(x)$-weighted inner product:

$$\langle y_1, y_2 \rangle_r = \int_a^b y_1(x)y_2(x)r(x)\,dx = 0.$$

Alternatively, the functions $w_1(x) = \sqrt{r(x)}\,y_1(x)$, $w_2(x) = \sqrt{r(x)}\,y_2(x)$ are orthogonal with respect to the Fourier inner product: $\langle w_1, w_2 \rangle = 0$.

Proof. Define the linear operator L_r acting on twice continuously differentiable functions in $C[a, b]$ by $L_r y = r^{-1}(p(x)y')' + q(x)y$. Note that the factors r and r^{-1} cancel in

$$\langle L_r y_1, y_2 \rangle_r = \int_a^b (L_r y_1(x))y_2(x)r(x)\,dx = \langle Ly_1, y_2 \rangle,$$

where $Ly = (p(x)y')' + q(x)y$. Since L is a Sturm-Liouville operator, imposing symmetric boundary conditions and using the Lagrange identity makes L a symmetric operator relative to the Fourier inner product. Since L and L_r have the same eigenvalues-eigenvector pairs, our conclusion follows from Theorem 2.2 on orthogonality. The final statement follows just because $r \geq 0$ and $(\sqrt{r})^2 = r$.   ■

e x a m p l e 6

The eigenfunction problem for the parametric Bessel equation requires us to solve the parametric Bessel equation $x^2 y'' + xy' + (\lambda^2 x^2 - n^2)y = 0$. To do this in terms of Bessel functions, we assume first that $\lambda \neq 0$ and change variable by $N = \pi/\lambda$. Thus $d/dx = (dz/dx)(d/dz) = \lambda\,d/dz$ and $d^2/dx^2 = \lambda^2\,d^2/dz^2$. The differential equation transforms into the Bessel equation

$$z^2 \bar{y}'' + z\bar{y}' + (z^2 - n^2)\bar{y} = 0, \qquad \text{where } \bar{y}(z) = y\left(\frac{z}{\lambda}\right).$$

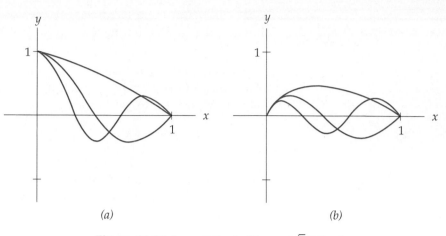

Figure 11.2 (*a*) $y = J_0(\lambda_k x)$. (*b*) $y = \sqrt{x}\, J_0(\lambda_k x)$.

By the results of Chapter 9, Section 7B, a solution that is finite at $z = 0$ is a constant multiple of $\bar{y}(z) = J_n(z)$, so $y(x) = J_n(\lambda x)$. In a typical application, boundary conditions consisting of finiteness at zero and $y(b) = 0$, $b > 0$, are required, so we want $y(b) = J_n(\lambda b) = 0$. In other words, the eigenvalues λ for the problem are those for which λb is a zero of $J_n(x)$. For example, if $n = 0$ and $b = 1$, the first three eigenvalues in increasing order are $\lambda_1 \approx 2.4$, $\lambda_2 \approx 5.5$, $\lambda_3 \approx 8.7$. The graphs of the first three corresponding eigenfunctions are shown in Figure 11.2*a* on the interval $0 \le x \le 1$. These are orthogonal with respect to the weight function $r(x) = x$. This orthogonality is hard to visualize, so the graphs in Figure 11.2*b* are used to show the eigenfunctions modified by the factor $\sqrt{r(x)} = \sqrt{x}$. The modified functions are orthogonal with respect to the unweighted Fourier inner product, and the graphs are the analogs for the Bessel equation of the harmonic-oscillator eigenfunctions $\sin \pi x$, $\sin 2\pi x$, and $\sin 3\pi x$.

EXERCISES

1. Find all eigenfunctions and corresponding eigenvalues of the differential operator d^2/dx^2 satisfying each of the following sets of boundary conditions. That is, solve $y'' = \lambda y$, subject to the following boundary conditions:

(a) $y(0) = y(\pi) = 0$.   **(d)** $y(0) = y'(\pi/2) = 0$.

(b) $y'(0) = y'(\pi)$.   **(e)** $y(0) + y'(0) = y(1) = 0$.

(c) $y(0) = y'(\pi) = 0$.   **(f)** $y'(0) = y(\pi) + y'(\pi) = 0$.

2. Find all eigenfunctions and corresponding eigenvalues of the differential operator

$$L = \frac{d^2}{dx^2} + 2\frac{d}{dx}$$

that satisfy each of the following sets of boundary conditions. That is, solve $L(y) = \lambda y$ subject to the following boundary conditions:

(a) $y(0) = y(\pi) = 0$.   **(c)** $y'(0) = y'(\pi) = 0$.

(b) $y(0) = y(1) = 0$.

3. Show that the Lagrange condition $[p(x)(f_1'(x)f_2(x) - f_1(x)f_2'(x))]_a^b = 0$ of identity 2.4 can be written $p(a)w(a) = p(b)w(b)$, where $w(x)$ is the Wronskian determinant of $f_1(x)$ and $f_2(x)$.

4. **Periodic boundary conditions** Boundary conditions of the form $y(a) = y(b)$, $y'(a) = y'(b)$, $a \neq b$ are called **periodic**.

   (a) Show that a Sturm-Liouville operator $Ly = (p(x)y')' + q(x)$ is symmetric if it is restricted by periodic boundary conditions and $p(a) = p(b)$.

   (b) Suppose a function $y = y(x)$ satisfies periodic boundary conditions at a and b and is continuously differentiable for $a \leq x \leq b$. Show that $y(x)$ can be extended to be continuously differentiable and periodic with period $b - a$ for all real x. (This result is the origin of the term "periodic" in the present context.)

5. Show that the differential operator d^2/dx^2 is symmetric with respect to the inner product

$$\langle y_1, y_2 \rangle = \int_{-1}^{1} y_1(x)y_2(x)\, dx,$$

   and with each of the following sets of boundary conditions.

   (a) $y(-1) = y(1) = 0.$ **(b)** $y'(1) = y'(-1) = 0.$

   (c) $c_1 y(-1) + d_1 y'(-1) = c_2 y(1) + d_2 y'(1) = 0$, where c_k and d_k aren't both zero.

6. Verify that the Legendre polynomial P_n defined by Formula 2.5 satisfies the Legendre equation: $(1 - x^2)y'' - 2xy' + n(n + 1)y = 0$. To do this let $u = (x^2 - 1)^n$. Then $(x^2 - 1)u' = 2nxu$. Differentiate both sides $(n + 1)$ times with respect to x. Use the Leibnitz rule for the derivative of a product:

$$(fg)^{(n)} = \sum_{k=0}^{N} \binom{n}{k} f^{(k)} g^{(n-k)}.$$

7. Compute the Legendre polynomials $P_0(x)$, $P_1(x)$, $P_2(x)$, and $P_3(x)$ from Equation 2.5. Then verify directly by integration that P_1 and P_3 are orthogonal with respect to the Fourier inner product over the interval $[-1, 1]$.

8. **(a)** By using Formula 2.5 and repeated integration by parts, show that

$$\int_{-1}^{1} P_n^2(x)\, dx = \frac{(2n)!}{2^{2n}(n!)^2} \int_{-1}^{1} (1 - x^2)^n\, dx.$$

   (b) Show that $\displaystyle\int_{-1}^{1} (1 - x^2)^n\, dx = 2 \int_{0}^{\pi/2} \sin^{2n+1} \theta\, d\theta.$

   (c) Show that $\displaystyle\int_{0}^{\pi/2} \sin^{2n+1} \theta\, d\theta = \frac{2 \cdot 4 \cdot 6 \cdots (2n)}{1 \cdot 3 \cdot 5 \cdots (2n + 1)}.$

   (d) Show that $\displaystyle\int_{-1}^{1} P_n^2(x)\, dx = \frac{2}{2n + 1}.$

9. Prove that P_n, the nth Legendre polynomial, has n distinct roots in the interval $[-1, 1]$. [*Hint:* Use Formula 2.5 and Rolle's theorem.]

10. **Chebychev equation.** Show that $(1 - x^2)y'' - xy' + \lambda^2 y = 0$ in Sturm-Liouville form is $(\sqrt{1 - x^2}y')' + (\lambda^2/\sqrt{1 - x^2})y = 0.$

11. **Hermite equation.** Show that $y'' - xy' + \lambda y = 0$ in Sturm-Liouville form is $(e^{-x^2/2}y')' + \lambda e^{-x^2/2}y = 0$.

12. **Laguerre equation.** Show that $xy'' + (1 - x)y' + \lambda y = 0$ in Sturm-Liouville form is $(xe^{-x}y')' + \lambda e^{-x}y = 0$.

13. The 3-dimensional Laplace equation in spherical coordinates (r, ϕ, θ) has the form

$$\frac{\partial}{\partial r}\left(r^2\frac{\partial u}{\partial r}\right) + \frac{1}{\sin\phi}\frac{\partial}{\partial\phi}\left(\sin\phi\frac{\partial u}{\partial\phi}\right) + \frac{1}{\sin^2\phi}\frac{\partial^2 u}{\partial\theta^2} = 0.$$

(a) Show that for solutions $u(r, \phi, \theta) = v(r, \phi)$ that are independent of θ, the equation has the form

$$r\frac{\partial^2}{\partial r^2}(rv) + \frac{1}{\sin\phi}\frac{\partial}{\partial\phi}\left(\sin\phi\frac{\partial v}{\partial\phi}\right) = 0.$$

(b) Show that the method of separation of variables applied to the equation of part (a) leads to the two ordinary differential equations

$$r^2R'' + 2rR' = \lambda R,$$

$$\frac{1}{\sin\phi}\frac{d}{d\phi}\left(\sin\phi\frac{d\Phi}{d\phi}\right) = -\lambda\Phi.$$

(c) Show that the equation of Euler type for $R(r)$ has solutions

$$r^{-(1/2)+\sqrt{\lambda+(1/4)}} \qquad \text{and} \qquad r^{-(1/2)-\sqrt{\lambda+(1/4)}}.$$

[*Hint:* Let $R = r^\mu$.]

(d) Show that the equation for Φ can be put in the form of the Legendre equation

$$(1 - x^2)y'' - 2xy' + \lambda y = 0.$$

[*Hint:* Let $x = \cos\phi$, and use the chain rule.]

(e) By setting $\lambda = n(n + 1)$, find a sequence of solutions to the partial differential equation of part (a).

14. Show that in the case of the orthonormal sequence derived from the Legendre polynomials, the general expansion $\sum_{k=0}^{N}\langle g, \phi_k\rangle\phi_k$ reduces to $\sum_{k=0}^{N}c_kP_k(x)$, where c_k is given by Equation 2.6 of the text.

15. Let $a_0(x)$, $a_1(x)$, and $a_2(x)$ be continuous on an interval, with $a_0(x) \neq 0$. Show that the expression $a_0(x)y'' + a_1(x)y' + a_2(x)y$ can be written in the Sturm-Liouville form $(py')' + qy$ after multiplication by $m(x)/a_0(x)$, where $m(x) = \exp\left(\int(a_1(x)/a_0(x))\,dx\right)$. What are $p(x)$ and $q(x)$ in terms of $a_0(x)$, $a_1(x)$, and $a_2(x)$?

16. Write out a formal proof that if y is an eigenfunction of a linear operator L with eigenvalue λ, then so is cy for constant $c \neq 0$.

17. If λ is an eigenvalue of the linear operator L, the set of all linear combinations of corresponding eigenfunctions is called the **eigenspace** of λ for L.

(a) Let L be the linear differential operator d^2/dx^2 restricted to twice continuously differentiable functions $y = y(x)$ satisfying $y(0) = y(\pi) = 0$. Show that the eigenvalues are all of the form $-n^2$, where n is a positive integer, and show that the eigenspace of L is 1-dimensional, consisting of all nonzero constant multiples $cy(x)$ of a single function.

(b) Let L be the linear differential operator d^2/dx^2 restricted to twice continuously differentiable functions $y = y(x)$ satisfying $y(0) = y(2\pi) = 0$ and $y'(0) = y'(2\pi)$. Show that the eigenvalues are all of the form $-n^2$, where n is a positive integer, and show that the eigenspace of L is 2-dimensional, consisting of all nonzero linear combinations $c_1 y_1(x) + c_2 y_2(x)$ of two linearly independent functions.

3   VIBRATIONAL MODES

The purpose of this section is to apply the ideas in Sections 1 and 2 to interpreting solutions of boundary and initial-value problems for higher-dimensional wave equations. Because its solutions are constructed using elementary trigonometric functions, we begin with a problem where separation of variables involves **double Fourier sine series** of an integrable function $f(x, y)$:

◼ 3.1
$$\sum_{j,k=1}^{\infty} b_{jk} \sin \frac{j\pi x}{p} \sin \frac{k\pi y}{q},$$

where
$$b_{jk} = \frac{4}{pq} \int_0^p \int_0^q f(x, y) \sin \frac{j\pi x}{p} \sin \frac{k\pi x}{q} \, dx \, dy.$$

We'll be especially concerned with the geometry of the individual terms in such expansions, in particular with regions in the xy plane in which their graphs lie above or below the plane, and with the **nodal curves** that separate these regions.

3A   Rectangular Membrane

An ideal drumhead is a homogeneous 2-dimensional membrane that stays motionless along a boundary curve while the interior points move vertically in response to initial excitation. We consider small vertical displacements $u(x, y, t)$ of a membrane, depending on time t and position (x, y) in a plane rectangular region $R: 0 \le x \le p, 0 \le y \le q$. For boundary conditions the displacement u is held at value zero all around the edge of the rectangle, and natural initial conditions are $u(x, y, 0) = f(x, y)$ and $u_t(x, y, 0) = g(x, y)$, where $f(x, y)$ and $g(x, y)$ represent initial displacement and initial velocity, respectively, at (x, y). The appropriate partial differential equation for u turns out to be the **2-dimensional wave equation**

$$a^2(u_{xx} + u_{yy}) = u_{tt},$$

where a^2 is a positive constant depending on the uniform density and tension of the membrane. For functions $u(x, y, t) = u(x, t)$ that are independent of y, we have $u_{yy} = 0$, so the differential equation reduces to the 1-dimensional wave equation of Chapter 10, Section 4C. Recalling from Chapter 10, Section 5, the definition of the 2-dimensional Laplace operator by $\Delta u = u_{xx} + u_{yy}$, the 2-dimensional wave equation can be written more concisely as $a^2 \Delta u = u_{tt}$.

We begin by finding product solutions $u(x, y, t) = X(x)Y(y)T(t)$ to the boundary problem, requiring $a^2 \Delta u = u_{tt}$ with $u(x, 0, t)$, $u(p, y, t)$, $u(x, q, t)$, and

$u(0, y, t)$ all to be zero on the boundary of the rectangle. These conditions are equivalent to requiring the factors in the product solution to satisfy $X(0) = X(p) = 0$ and $Y(0) = Y(q) = 0$. Substitution into the differential equation gives

$$a^2(X''YT + XY''T) = XYT'' \quad \text{or} \quad a^2\left(\frac{X''}{X} + \frac{Y''}{Y}\right) = \frac{T''}{T} = -\lambda^2,$$

where $-\lambda^2$ is just a convenient way to write the arbitrary separation constant. Indeed, the resulting equation $T'' + \lambda^2 T = 0$ has the oscillatory solutions $T(t) = A \cos \lambda t + B \sin \lambda t$ that we associate with vibration. (The apparent prejudice here is purely formal; $-\lambda^2$ can assume positive values if λ is allowed to be imaginary.) The equation for X and Y is $X''/X + Y''/Y = -\lambda^2$. Separating variables again gives

$$\frac{X''}{X} = -\frac{Y''}{Y} - \lambda^2 = -l^2,$$

where $-l^2$ is constant because it can't vary with either x or y. The resulting ordinary differential equations are

$$X'' + l^2 X = 0 \quad \text{and} \quad Y'' + (\lambda^2 - l^2)Y = 0.$$

Mindful that the first of these equations has boundary conditions $X(0) = X(p) = 0$, we find solutions $X(x) = \sin j\pi x/p$, so $l^2 = j^2\pi^2/p^2$. Similarly the boundary problem for Y has solutions $Y(y) = \sin(k\pi y/q)$, so

$$\lambda^2 - l^2 = \lambda^2 - \frac{j^2\pi^2}{p^2} = \frac{k^2\pi^2}{q^2}.$$

Note that as a result $\lambda^2 = \pi^2(j^2/p^2 + k^2/q^2)$. Thus our product solutions take the form

$$u_{jk}(x, y, t) = (A_{jk} \cos \lambda_{jk} t + B_{jk} \sin \lambda_{jk} t) \sin \frac{j\pi x}{p} \sin \frac{k\pi y}{q},$$

where $\lambda_{jk} = \pi\sqrt{j^2/p^2 + k^2/q^2}$.

We're now able to consider initial displacement $u(x, y, 0)$ and velocity $u_t(x, y, 0)$. To accommodate a significant amount of generality for these conditions, we use the linearity of the differential equation to try to put together solutions using double Fourier series:

■ **3.2** $$u(x, y, t) = \sum_{j,k=1}^{\infty} \sin \frac{j\pi x}{p} \sin \frac{k\pi y}{q} (A_{jk} \cos \lambda_{jk} t + B_{jk} \sin \lambda_{jk} t).$$

Since we want $u(x, y, 0) = \sum_{j,k=1}^{\infty} A_{jk} \sin(j\pi x/p) \sin(k\pi y/q)$, we determine the A_{jk} as double Fourier coefficients given in general form by Equation 3.1:

■ **3.3** $$A_{jk} = \frac{4}{pq} \int_0^p \int_0^q u(x, y, 0) \sin \frac{j\pi x}{p} \sin \frac{k\pi x}{q} \, dx \, dy.$$

Similarly, if $u_t(x, y, 0) = \sum_{j,k=1}^{\infty} \lambda_{jk} B_{jk} \sin(j\pi x/p) \sin(k\pi y/q)$, we compute B_{jk} by

■ **3.4** $$B_{jk} = \frac{4}{pq\lambda_{jk}} \int_0^p \int_0^q u_t(x, y, 0) \sin \frac{j\pi x}{p} \sin \frac{k\pi x}{q} \, dx \, dy.$$

The convergence of the expansion in Equation 3.2 is quite complicated in general. We get some insight into this just by examining the multiplicity of individual terms. As in the 1-dimensional case, the t-dependent factors are just sinusoidal oscillations with respective circular frequencies $\lambda_{jk} = \pi\sqrt{j^2/p^2 + k^2/q^2}$, and these frequencies are essentially pitches of certain vibrational "modes." Each factor $v_{jk}(x, y) = \sin(j\pi x/p)\sin(k\pi y/q)$ determines a basic **mode** of vibration. It is easy to verify directly, and implicit in the derivation given above, that each v_{jk} is an eigenfunction of the Laplace operator Δ corresponding to eigenvalue $-\lambda_{jk}^2 = -\pi^2(j^2/p^2 + k^2/q^2)$. Hence, in contrast with the case of one space dimension, there is frequently more than one independent eigenfunction for each eigenvalue, since an integer can sometimes be represented as a sum of squares in more than one way. Beyond that, it's easy to verify that a linear combination of eigenfunctions with the same eigenvalue is also an eigenfunction, so the variety among eigenfunctions with a given eigenvalue can be considerable.

example 1

We continue to assume boundary values identically zero. If $p = q = \pi$, then $\lambda^2 = j^2 + k^2$, so the possible eigenvalues are the negatives of integers that can be represented as a sum of squares of two positive integers. For example, eigenvalue $-2 = -1^2 - 1^2$ has the single eigenfunction $v_{11}(x, y) = \sin x \sin y$. But $-5 = -1^2 - 2^2$ has the independent eigenfunctions $v_{12}(x, y) = \sin x \sin 2y$ and $v_{21}(x, y) = \sin 2x \sin y$. The eigenvalue -50 has three independent eigenfunctions: v_{17}, v_{71}, and v_{55}, so, for example, $v_{17} - v_{71} + 2v_{55}$ is also an eigenfunction with eigenvalue -50. Each of these eigenfunctions is made to oscillate with the appropriate circular frequency $\lambda_{jk} = \sqrt{j^2 + k^2}$ in the sense that the expansion in Equation 3.2 will contain, for example, the terms

$$(A_{17}\sin x \sin 7y + A_{71}\sin 7x \sin y + A_{55}\sin 5x \sin 5y)\cos \pi\sqrt{50}\,t,$$

where the constants A_{jk} depend on the initial displacement $u(x, y, 0)$ as in Equation 3.3.

Figure 11.3 shows the graphs of three of the infinitely many different eigenfunctions with eigenvalue -50. A glance at Figure 11.3b suggests that there are curves along which the displacement is always zero; such a curve is called a **nodal curve** and is analogous to a *nodal point*, or **node**, which also remains fixed during oscillation in one space dimension. Nodal curves are significant because they mark the boundaries of regions on which displacement actually

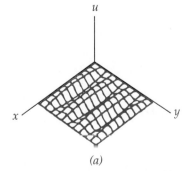

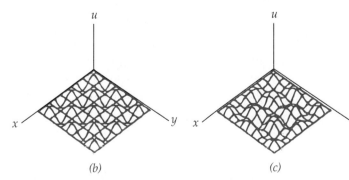

(a)　(b)　(c)

Figure 11.3 (a) $\frac{1}{3}v_{17}$. (b) $\frac{1}{3}v_{55}$. (c) $\frac{1}{6}v_{17} + \frac{1}{4}v_{71} + \frac{1}{4}v_{55}$.

$$u_{13} + \alpha u_{31} = 0$$

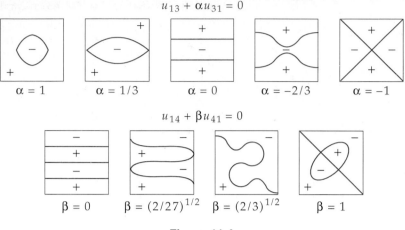

$$u_{14} + \beta u_{41} = 0$$

Figure 11.4

takes place. Figure 11.4 shows a selection of nodal curves associated with the eigenvalues -10 and -17. Each of these drawings represents the level set at level zero of the corresponding eigenfunction, restricted in these examples to a π by π square. Each level set can be realized as the result of a gradual deformation of the adjacent ones. The signs show the regions in which the eigenfunction is positive or negative.

The orthogonality of the eigenfunctions $v_{jk}(x, y) = \sin(j\pi x/p) \sin(k\pi y/p)$ over the p by q square follows easily by elementary integration. It's left as an exercise to show that the Laplace operator Δ is symmetric with respect to the zero boundary values that we've quite naturally imposed. Orthogonality then follows from Theorem 2.2 without any additional computation.

3B   *Circular Membrane*

We can think of a conventional drumhead as a homogeneous membrane bounded by a circular frame. The relevant wave equation is still $a^2 \Delta u = u_{tt}$, but to adapt the method of Section 3A to a circular domain we need to use the chain rule to express the **Laplace operator in polar coordinates:**

$$\Delta u = u_{rr} + \frac{1}{r} u_r + \frac{1}{r^2} u_{\theta\theta}.$$

(The polar coordinate transformation is singular at the origin, so $r = 0$ is a singular point for the transformed operator.) The wave equation for a circular region becomes $a^2(u_{rr} + r^{-1}u_r + r^{-2}u_{\theta\theta}) = u_{tt}$. Looking for product solutions $u(r, \theta, t) = R(r)\Theta(\theta)T(T)$, we find

■ **3.5** $$R''\Theta T + \frac{1}{r} R'\Theta T + \frac{1}{r^2} R\Theta''T = \frac{1}{a^2} R\Theta T''.$$

Division by $R\Theta T$ gives

$$\frac{R''}{R} + \frac{1}{r}\frac{R'}{R} + \frac{1}{r^2}\frac{\Theta''}{\Theta} = \frac{1}{a^2}\frac{T''}{T}.$$

Initial conditions specify $u(r, 0)$ and $u_t(r, 0)$ as initial displacement and velocity. One boundary condition for a membrane of radius b is $u(b, t) = 0$. The singularity at $r = 0$ gives rise to a new boundary point, where we impose just the condition that the solution remain bounded. We first look at solutions that are independent of θ; these can be generated by initial conditions that are independent of θ. (In practice this might be done by striking a drum at its precise center.) A solution independent of θ is called a **radial function,** because position in the plane of equilibrium is uniquely determined by specifying r.

example 2

For solutions independent of θ, we have $\Theta'' = 0$, so radial solutions to the circular drum problem satisfy

$$\frac{R''}{R} + \frac{1}{r}\frac{R'}{R} = \frac{1}{a^2}\frac{T''}{T}.$$

Both sides of the equation must be constant, which we call $-\lambda^2$ for convenience. The two separated equations are

$$rR'' + R' + \lambda^2 rR = 0 \quad \text{and} \quad T'' + \lambda^2 a^2 T = 0.$$

The first of these equations is the Bessel equation of order 0, with solution $R(r) - J_0(\lambda r)$ bounded at $r = 0$. (We reject all solutions independent of this one, since they are guaranteed to be unbounded by Theorem 7.5 of Chapter 9, Section 7B.) The boundary condition at $r = b$ requires $R(b) = J_0(\lambda b) = 0$, so λ will be chosen from among the increasing sequence of positive values λ_j for which $J_0(\lambda_j b) = 0$. The key λ's are then r_j/b, where the r_j are just the positive zeros of J_0, the first five of which are located approximately at $r_1 = 2.4, r_2 = 5.5, r_3 = 8.7, r_4 = 11.8, r_5 = 14.9$. The corresponding solutions to the equation for T are the usual sinusoidal ones, so the θ-independent problem has solutions

$$u(r, t) = \sum_{j=0}^{\infty} J_0(\lambda_j r)(A_j \cos \lambda_j at + B_j \sin \lambda_j at).$$

The A's and B's are determined by the initial conditions as follows. In Section 2B we saw that the functions $\sqrt{r} J_0(\lambda_j r)$ are orthogonal on the interval $0 \leq r \leq 1$ so that

$$\int_0^b r J_0(\lambda_j r) J_0(\lambda_k r) \, dr = 0, \quad j \neq k.$$

Making a Fourier-Bessel expansion requires that the eigenfunctions be normalized by factors designed to make this integral equal 1 when $j = k$. This comes about if we multiply $J_0(\lambda_j r)$ by $N_j = (\int_0^b r J_0^2(\lambda_j r) \, dr)^{-1/2}$. Hence we can compute the coefficients by

$$A_j = N_j^2 \int_0^b r J_0(\lambda_j r) u(r, 0) \, dr, \quad B_j = \frac{N_j^2}{a \lambda_j} \int_0^b r J_0(\lambda_j r) u_t(r, 0) \, dr.$$

The coefficient formulas can be simplified somewhat for computing purposes, but such computations are complicated at best. We'll concentrate here on studying the eigenfunctions and their associated vibrational modes. Figure

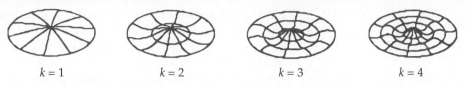

$k = 1$ $k = 2$ $k = 3$ $k = 4$

Figure 11.5 Graphs of $J_0(x_k r)$, $r = \sqrt{x^2 + y^2} \leq 1$ with $J_0(x_k) = 0$.

11.5 shows in perspective the first four modes of the radially symmetric circular membrane of radius 1, starting with the "fundamental." The circles are the stationary nodal curves of an associated oscillation such as $J_0(r_k r) \sin a r_k t$, where γ_k is the kth zero of J_0. The radii are, respectively, $\rho_l = r_l / r_k$, $l = 1, 2, \ldots,$ k. Note that the kth mode oscillates with circular frequency $a r_k$. Since none of these frequencies is a rational multiple of another, a linear combination of these oscillations will never be periodic, so a conventional drum will not sound "harmonious" by itself the way some musical instruments do. With some notable exceptions, the same remark applies to the rectangular drum.

---

example 3

Allowing dependence on the polar angle θ in the wave equation, we can first separate out the time dependence in Equation 3.5 for product solutions like this:

$$\frac{R''}{R} + \frac{1}{r}\frac{R'}{R} + \frac{1}{r^2}\frac{\Theta''}{\Theta} = \frac{1}{a^2}\frac{T''}{T} = -\lambda^2.$$

Of the two resulting equations

$$T'' + (\lambda a)^2 T = 0 \qquad \text{and} \qquad \frac{R''}{R} + \frac{1}{r}\frac{R'}{R} + \frac{1}{r^2}\frac{\Theta''}{\Theta} = -\lambda^2,$$

the second separates with an additional constant k^2 to give

$$r^2 \frac{R''}{R} + r\frac{R'}{R} + \lambda^2 r^2 = -\frac{\Theta''}{\Theta} = k^2.$$

Thus we have $\Theta'' + k^2 \Theta = 0$ and $r^2 R'' + r R' + (\lambda^2 r^2 - k^2) R = 0$. To be well-defined as a function of position in a plane, $\Theta(\theta)$ must have period 2π as a function of the angle θ. Hence k must be an integer in the general solution formula $\Theta_k(\theta) = \alpha_k \cos k\theta + \beta_k \sin k\theta$. The equation for $R(r)$ is a parametric Bessel equation with bounded solutions $R_k(r) = J_k(\lambda r)$. (Recall that by Theorem 7.5 of Chapter 9, Section 7B, the solutions independent of these are all unbounded near $r = 0$.) To make all solutions be zero on the boundary of a disk of radius b, we require $R_k(b) = J_k(\lambda b) = 0$; thus λb must be a zero r_{kj} of the Bessel function J_k. We define λ_{kj} by $\lambda_{kj} = r_{kj}/b$. Using these values for λ in the sinusoidal solutions $\cos \lambda a t$ and $\sin \lambda t$ of $T'' + (\lambda/a)^2 T = 0$, we find four types of product solution to the differential equation:

$$J_k(\lambda_{kj} r) \cos k\theta \cos \lambda_{kj} a t, \qquad J_k(\lambda_{kj} r) \cos k\theta \sin \lambda_{kj} a t,$$
$$J_k(\lambda_{kj} r) \sin k\theta \cos \lambda_{kj} a t, \qquad J_k(\lambda_{kj} r) \sin k\theta \sin \lambda_{kj} a t.$$

When $k = 0$, we get only the two terms $J_0(\lambda_{j0} r) \cos \lambda_{j0} a t$ and $J_0(\lambda_{j0} r) \sin \lambda_{j0} a t$ that appeared in the θ-independent solutions of Example 2.

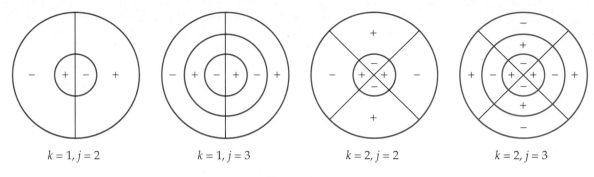

$k = 1, j = 2$ $k = 1, j = 3$ $k = 2, j = 2$ $k = 2, j = 3$

Figure 11.6

Associated with each circular frequency $\lambda_{kj}a$ is a vibrational mode of the form

$$A_{kj}J_k(\lambda_{kj}r) \cos k\theta + B_{kj}J_k(\lambda_{kj}r) \sin k\theta = C_{kj}J_k(\lambda_{kj}r) \cos k(\theta - \phi_{kj}).$$

The A's and B's, or alternatively the C's and ϕ's, are determined as usual by initial displacement and velocity. It follows that the typical nodal curve tied to a given frequency has a fairly simple structure consisting of concentric circles, where the Bessel factor is zero, together with evenly spaced radial lines along which the trigonometric factor is zero. Figure 11.6 shows some nodal curves for some $u_{kj}(r,\theta) = J_k(\lambda_{kj}r) \cos k\theta$; comparison with Figure 11.4 shows that rectangular drums offer more variety in the way of regions that oscillate at a given circular frequenc

---

When an ideal drumhead of whatever shape is set in motion, the oscillatory modes are eigenfunctions $u_\lambda(x, y)$ of the Laplace operator: $\Delta u_\lambda = \lambda u_\lambda$. The only thing about these sets of eigenfunctions that distinguishes one drum shape from another is that they must all take on the value zero on the boundary of the drum. The corresponding eigenvalues determine the oscillation frequencies of the various modes, and it is these frequencies that determine the quality of the sound produced by the drum. Can two differently shaped drums have exactly the same eigenvalues? (Roughly speaking, can you hear the shape of a drum by detecting these frequencies?) This question was much discussed but remained unanswered for about 40 years. It turns out that you can't hear the shape of a drum in this way; mathematicians Carolyn Gordon, David Webb, and Scott Wolpert collaborating in 1991 managed to find distinct nonconvex drum shapes that generate identical sequences of eigenvalues. See *American Scientist*, vol. 84, no. 1, 1996, pp. 46–55, for an elementary exposition.

EXERCISES

1. Make the analogs of the sketches in Figure 11.4 for the eigenfunctions $u_{12}(x, y) = \sin x \sin 2y$ and $u_{21}(x, y) = \sin 2x \sin y$

2. Make the analogs of the sketches in Figure 11.4 for the eigenfunctions $u_{13}(x, y) = \sin x \sin 3y$ and $u_{31}(x, y) = \sin 3x \sin y$.

3. Make the analog of a sketch in Figure 11.4 for the eigenfunction $u_{12}(x, y) + u_{21}(x, y) = \sin x \sin 2y + \sin 2x \sin y$. [*Hint:* Aside from the boundary square, show that the level set satisfies $\cos x + \cos y = 0$.]

4. Make the analogs of the sketches in Figure 11.6 for the eigenfunctions $u_{11}(r, \theta) = J_1(r_{11}r) \cos \theta$ and $u_{21}(r, \theta) = J_2(r_{21}r) \cos 2\theta$. Here r_{k1} is the smallest positive zero of J_k.

5. Example 1 of the text shows that for a π by π membrane the mode of vibration linked to a given frequency can in general be decomposed into a number n of different sinusoidal products.

   (a) For $n = 65$ there are four such products. What are they?

   (b) How many such products are there when $n = 325$, and what are they?

6. The 2-dimensional heat equation in rectangular coordinates is $a^2(u_{xx} + u_{yy}) = u_t$. Find the product solutions that are identically zero on the boundary of the p by q rectangle with opposite corners at $(0, 0)$ and (p, q).

7. The 2-dimensional heat equation in polar coordinates is $a^2(u_{rr} + r^{-1}u_r + r^{-2}u_{\theta\theta}) = u_t$. Find the product solutions that are identically zero on the circle of radius b centered at $(0, 0)$.

8. Consider the following vibrational modes of a rectangular π by π membrane:

$$u_{jk}(x, y, t) = \sin jx \sin ky \sin \sqrt{j^2 + k^2}\, t.$$

   (a) Show that two modes u_{jk} and u_{lm} have a common time period of oscillation when the ratio $(j^2 + k^2)/(l^2 + m^2)$ is the square of a rational number.

   (b) Give some examples of modes that have a common period. [*Hint:* Consider integer pairs $j = 2ab$, $k = a^2 - b^2$, where $a > b$.]

   (c) Give some examples of two modes that don't have a common period, explaining why they don't.

9. **Green's theorem** for a plane region R with piecewise-smooth boundary curve ∂R asserts the validity of an equation relating an area integral for a gradient field $\nabla f = (f_x, f_y)$ to an arc-length integral of a directional derivative: $\int_R (f_{xx} + f_{yy})\, dx\, dy = \int_{\partial R} (\partial f / \partial \mathbf{n})\, ds$, where $\mathbf{n}$ is the outward-pointing unit normal vector on ∂R and f is continuously differentiable.

   (a) Use Green's theorem to derive **Green's first identity** for the Laplace operator Δ:

$$\int_R f \Delta g \, dx \, dy + \int_R \nabla f \cdot \nabla g \, dx \, dy = \int_{\partial R} f \frac{\partial g}{\partial \mathbf{n}} \, ds.$$

   (b) Use part (a) to prove **Green's second identity:**

$$\int_R (f \Delta g - g \Delta f) \, dx \, dy = \int_{\partial R} \left(f \frac{\partial g}{\partial \mathbf{n}} - g \frac{\partial f}{\partial \mathbf{n}} \right) ds.$$

10. Define an inner product $\langle f, g \rangle = \int_R fg \, dx \, dy$.

   (a) Use Green's second identity of the previous exercise to show that the Laplace operator Δ is symmetric when acting on twice continuously differentiable functions on R with boundary values zero: $\langle \Delta f, g \rangle = \langle f, \Delta g \rangle$.

(b) Show that the symmetry result of part (a) holds if the condition that f and g be zero on the boundary is replaced by the condition that their normal derivatives be zero there.

11. The Bessel equation $x^2y'' + xy' + (x^2 - n^2)y = 0$ with solutions $J_n(x)$ derived in Chapter 9, Section 7B, is somewhat less general than the parametric Bessel equation $x^2y'' + xy' + (\lambda^2x^2 - n^2)y = 0$ that we need for solving the wave equation. Show that the more general parametric equation is satisfied by $y(x) = J_n(\lambda x)$.

12. Show that the Fourier-Bessel coefficients in Example 2 of the text can be computed by

$$A_k = \frac{1}{M_k} \int_0^1 xJ_0(x_k x)u(bx, 0) \, dx, \qquad B_k = \frac{b}{ax_k M_k} \int_0^1 xJ_0(x_k x)u_t(bx, 0) \, dx,$$

where x_k is the kth positive zero of $J_0(x)$ and $M_k = \int_0^1 xJ_0^2(x_k x) \, dx$.

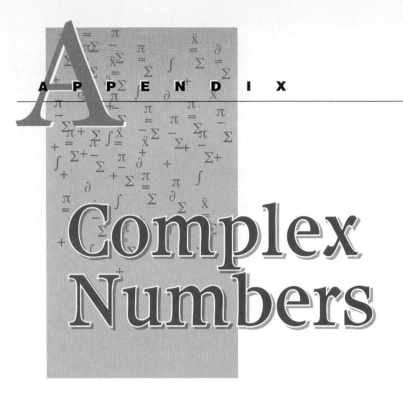

Complex Numbers

Complex numbers arise naturally when we try to extend our method for solving linear constant-coefficient differential equations of order two or more.

example 1

The differential equation $d^2y/dx^2 + y = 0$ has characteristic equation $r^2 + 1 = 0$, or $r^2 = -1$. Since there are no solutions r to this equation among the real numbers, the thing to do is work in the larger complex number field that contains solutions $r = \pm i$ having the property that $i^2 = -1$. (This is somewhat analogous to what we do when faced with the problem of finding a solution to the equation $r^2 = 2$ among the rational numbers; since it can be proved that there are no rational solutions, we enlarge the number system to include some irrational numbers such as $\pm\sqrt{2}$.) Following the routine that we use for real roots of a characteristic equation, we would expect solutions to the differential equation $d^2y/dx^2 + y = 0$ to look like

$$y(x) = c_1 e^{ix} + c_2 e^{-ix}.$$

Interpreting the formula at the end of Example 1 in a useful way is the subject matter of Section 2 of Chapter 2. The rest of this appendix is taken up with the arithmetic of complex numbers, that is, the addition, multiplication, and division of numbers such as i, $-2i$, and $1 - \sqrt{3}\,i$. A **complex number** $z = x + iy$, with **real part** $\operatorname{Re}(z) = x$ and **imaginary part** $\operatorname{Im}(z) = y$, can be identified with the point, or vector, (x, y) in $\mathcal{R}^2$. This identification not only gives us a useful pictorial representation for z but makes $z = x + iy$ a clearly defined mathematical entity. In particular, the **purely real** complex numbers $x + i0$ are

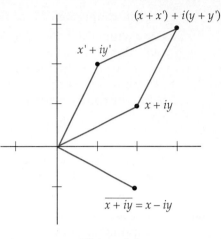

Figure A.1

identified with the points $(x, 0)$ on the **real axis,** and are written x, and the **purely imaginary** complex numbers $0 + iy$ are identified with the points $(0, y)$ on the **imaginary axis,** and are written iy. Of course, to justify using the term "number," we have to know how to operate arithmetically with complex numbers.

Complex addition corresponds to addition of vectors in $\mathfrak{R}^2$ and is defined by applying the usual rules of real number arithmetic to the real and imaginary parts of the numbers as well as to the symbol i:

$$(x + iy) + (x' + iy') = (x + x') + i(y + y').$$

Given our identification of $z = x + iy$ and $w = x' + iy'$ with the points (x, y) and (x', y'), it follows that the sum $z + w$ is identified with the point $(x + x', y + y')$. Thus the location of the point for the sum can be arrived at by vector addition using the parallelogram law. See Figure A.1.

It is customary simply to write x for $x + i0$, iy for $0 + iy$, and 0 for $0 + i0$, as well as $-(x + iy)$ for $-x + i(-y)$.

Complex multiplication is defined by using the usual rules of real number arithmetic except that i^2 can be replaced by -1:

$$(x + iy)(x' + iy') = (xx' - yy') + i(xy' + yx').$$

It follows that multiplication is commutative; that is, $(x + iy)(x' + iy') = (x' + iy')(x + iy)$. In particular $iy = yi$, and it's customary to write, for example, $1 - 2i$ instead of $1 + i(-2)$.

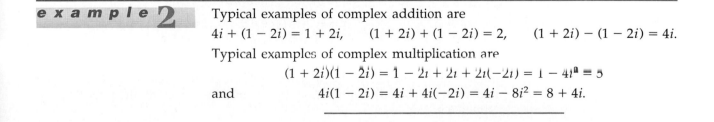

e x a m p l e 2

Typical examples of complex addition are

$$4i + (1 - 2i) = 1 + 2i, \qquad (1 + 2i) + (1 - 2i) = 2, \qquad (1 + 2i) - (1 - 2i) = 4i.$$

Typical examples of complex multiplication are

$$(1 + 2i)(1 - 2i) = 1 - 2i + 2i + 2i(-2i) = 1 - 4i^2 = 5$$

and

$$4i(1 - 2i) = 4i + 4i(-2i) = 4i - 8i^2 = 8 + 4i.$$

The **conjugate** of a complex number $z = x + iy$ is the number $\bar{z} = x - iy$. For example, $\overline{2 + 3i} = 2 - 3i$, $\overline{2 - i} = 2 + i$, and $\overline{-i} = i$. Geometrically, $\bar{z}$ is the reflection of z in the real number line, as indicated in Figure A.1.

To see how the **quotient** of two complex numbers should be defined, we can formally multiply the numerator and the *nonzero* denominator by the conjugate of the denominator:

$$\frac{x' + iy'}{x + iy} = \frac{(x' + iy')(x - iy)}{(x + iy)(x - iy)} = \frac{(xx' + yy') + i(xy' - yx')}{x^2 + y^2}.$$

Showing that this procedure always results in a uniquely defined quotient when $x + iy \neq 0$ is left as an exercise.

e x a m p l e 3

Computing a quotient is mostly multiplication:

$$\frac{2 + 3i}{2 - i} = \frac{(2 + 3i)(2 + i)}{(2 - i)(2 + i)} = \frac{(2 \cdot 2 - 1 \cdot 3) + i(2 \cdot 3 + 1 \cdot 2)}{2^2 + 1^2}$$

$$= \frac{1 + 8i}{5} = \tfrac{1}{5} + \tfrac{8}{5}i.$$

Note that the only "division" in the conventional sense is by a real number in the last step.

The **absolute value** of $z = x + iy$ is defined to be $|z| = \sqrt{x^2 + y^2}$. When we identify $x + iy$ with the point (x, y) in the plane, we see that $|x + iy|$ is just the distance from (x, y) to $(0, 0)$, that is, to the complex number zero.

e x a m p l e 4

For purely real complex numbers, the absolute value is the same as for real numbers, for example, $|-3| = 3$. Otherwise, we have, for example,

$$|3i| = \sqrt{0^2 + 3^2} = 3,$$

$$|\sqrt{2} - i\sqrt{2}| = \sqrt{(\sqrt{2})^2 + (\sqrt{2})^2} = \sqrt{2 + 2} = 2,$$

and

$$\left| \frac{1}{2} + \frac{i\sqrt{3}}{2} \right| = \sqrt{\left(\frac{1}{2}\right)^2 + \left(\frac{\sqrt{3}}{2}\right)^2} = \sqrt{\tfrac{1}{4} + \tfrac{3}{4}} = 1.$$

Interpreting multiplication and division of complex numbers is done best by writing nonzero complex numbers $z = x + iy$ in **polar form,** with $x = |z| \cos \theta$ and $y = |z| \sin \theta$, illustrated in Figure A.2:

$$x + iy = |z| \cos \theta + i|z| \sin \theta$$
$$= |z|(\cos \theta + i \sin \theta).$$

Here $|z| = \sqrt{x^2 + y^2}$ is the radial distance of z from the origin and θ is an angle that the line joining z to 0 makes with the positive real axis. Because $\cos \theta$ and $\sin \theta$ are periodic with period 2π, the number z has infinitely many polar angles $\theta + 2\pi k$.

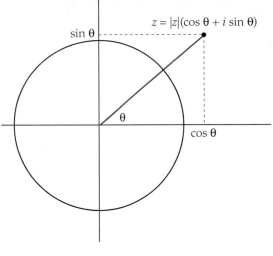

Figure A.2

---

example 5

(a) The angle that the complex number $3i$ makes with the positive real axis is $\pi/2$, and $|3i| = 3$, so $3i = 3(\cos{(\pi/2)} + i\sin{(\pi/2)})$.

(b) The angle that the number $\sqrt{2} \;-\; i\sqrt{2}$ makes with the positive real axis is $-\pi/4$, and $|\sqrt{2} - i\sqrt{2}| = 2$, so

$$\sqrt{2} - i\sqrt{2} = 2\left(\cos\left(-\frac{\pi}{4}\right) + i\sin\left(-\frac{\pi}{4}\right)\right) = 2\left(\cos\frac{\pi}{4} - i\sin\frac{\pi}{4}\right).$$

Here we've used the fact that $\cos\theta$ is an even function and $\sin\theta$ is an odd function.

(c) The angle that the number $\frac{1}{2} + i\sqrt{3}/2$ makes with the positive real axis is $\pi/3$, and $|\frac{1}{2} + i\sqrt{3}/2| = 1$, so

$$\frac{1}{2} + \frac{i\sqrt{3}}{2} = 1\left(\cos\frac{\pi}{3} + i\sin\frac{\pi}{3}\right).$$

As a reward for using the relatively complicated looking polar form, we gain an advantage in interpreting products of complex numbers. If z and z' are complex numbers with polar angles θ and θ', respectively, their product can be written

$$zz' = |z|(\cos\theta + i\sin\theta)|z'|(\cos\theta' + i\sin\theta')$$
$$= |z||z'|((\cos\theta\cos\theta' - \sin\theta\sin\theta') + i(\cos\theta\sin\theta' + \sin\theta\cos\theta'))$$
$$= |z||z'|(\cos{(\theta + \theta')} + i\sin{(\theta + \theta')}).$$

The last step follows from the addition formulas for the cosine and sine functions. What we arrive at is a complex number zz' in polar form having absolute value $|z||z'|$ and polar angle $\theta + \theta'$. That is, if $\theta(w)$ stands for the polar angle of w, then

$$|zz'| = |z||z'| \qquad \text{and} \qquad \theta(zz') = \theta(z) + \theta(z').$$

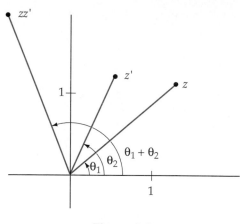

Figure A.3

These equations are illustrated in Figure A.3 and can be stated in words as follows.

1. **The absolute value of a product of complex numbers is the product of the absolute values of the factors.**
2. **The polar angle of a product of complex numbers is the sum of the polar angles of the factors.**

These rules tell us geometrically how to locate a product in the plane when we have already plotted the individual factors.

---

example 6

Where is the product of $\sqrt{2} - i\sqrt{2}$ and $\frac{1}{2} + i\sqrt{3}/2$ located in the complex plane? The absolute values of the two numbers are 2 and 1, respectively, and their polar angles are $-\pi/4$ $(-45°)$ and $\pi/3$ $(60°)$ according to Example 5. Hence the absolute value of the product is 2, and the polar angle is $-\pi/4 + \pi/3 = \pi/12$, or $15°$.

Polynomial roots have their existence guaranteed by the fundamental theorem of algebra, which states that an nth-degree polynomial equation with complex coefficients,

$$z^n + a_{n-1}z^{n-1} + \cdots + a_1 z + a_0 = 0, \qquad n \geq 1,$$

has a complex solution $z = r_1$. It then follows that there is a polynomial factorization

$$z^n + a_{n-1}z^{n-1} + \cdots + a_1 z + a_0 = (z - r_1)(z - r_2) \cdots (z - r_n)$$

corresponding to roots $r_1, \ldots, r_n$. The number of times a root r_k appears in the factorization is called the **multiplicity** of that root. Finding the roots, or equivalently the polynomial factorization, is immediate if $n = 1$ and quite simple in case $n = 2$ from the quadratic formula for roots r_1, r_2 of $z^2 + az + b$, namely, $r = \frac{1}{2}(-a \pm \sqrt{a^2 - 4b})$. (See Exercise 23 for the derivation.) Analogous, but fairly complicated, formulas exist for polynomials of degree 3 and 4, but it has been proved that no such general formulas exist for degree 5 or more. Certain particular equations can be solved quite easily, as the following examples show.

---

example 7

The 5th-degree polynomial $z^5 - z$ can be factored as follows:
$$z^5 - z = z(z^4 - 1) = z(z^2 - 1)(z^2 + 1) = z(z - 1)(z + 1)(z - i)(z + 1).$$
Thus the solutions of the equation $z^5 - z = 0$ are evidently 0, 1, -1, i, and $-i$.

---

example 8

If the 4th-degree equation $z^4 + z^2 - 6 = 0$ is regarded as quadratic in z^2, we find by the factorization $(z^2 - 2)(z^2 + 3)$, or by the quadratic formula, that either $z^2 = 2$ or $z^2 = -3$. Thus the four solutions to the equation are $\pm\sqrt{2}$ and $\pm\sqrt{3}i$.

EXERCISES

Locate the following complex numbers in a sketch showing real and imaginary axes.

1. $-2 + 2i$.

2. $(2 + i) + (-1 - 2i)$.

3. $i(1 + 2i)$.

4. $2(\cos(\pi/4) + i\sin(\pi/4))$.

5. $(1 + i)/i$.

6. $(2 + i)/(-1 - 2i)$.

7. $i(1 + 2i)^2$.

8. $2(\cos(\pi/6) - i\sin(\pi/6))$.

For each of the following complex numbers z, find $\bar{z}$ and $1/z$ in the form $x + iy$. Then find $|z|$, and write z in polar form $|z|(\cos\theta + i\sin\theta)$.

9. $1 + i$.

10. $-1 + 2i$.

11. $2i$.

12. $(2 + i)/i$.

13. $i(\cos t + i\sin t)$.

14. $(1 + 2i)(3 - i)$.

15. $(1 + i)(1 - i)$.

16. $(2 - i)(1 + i^{-1})$.

17. $(1 + i)/(1 - i)$.

18. Let z and w be complex numbers, $z \neq 0$. Show that the equation
$$\frac{w}{z} = \frac{1}{|z|^2}\, w\bar{z}$$
defines a *unique* quotient w/z such that $(w/z)z = w$. [*Hint:* Suppose that $qz = w$. Multiply by $\bar{z}$; then solve for q.]

19. Show that if $|z| = 1$, then $1/z = \bar{z}$.

20. For a given real number θ prove the following:

(a) $(\cos\theta + i\sin\theta)^2 = \cos 2\theta + i\sin 2\theta$.

(b) $(\cos\theta + i\sin\theta)^n = \cos n\theta + i\sin n\theta$ for integers $n \geq 0$ (**De Moivre's formula**).

(c) The formula in part (b) for $n = 1$.

(d) The formula in part (b) for integers $n < 0$.

21. Prove the following:

    (a) The equation $x^2 = -1$ has no solutions x in the real numbers $\mathcal{R}$.

    (b) The equation $x^2 = 2$ has no solutions x in the rational numbers $\mathcal{Q}$, where $\mathcal{Q}$ is the set of all quotients m/n of integers. [*Hint:* Suppose $(m/n)^2 = 2$ and that all common factors in m and n have been canceled; then arrive at a contradiction.]

22. Verify for arbitrary complex numbers the commutative, associative, and distributive laws listed below using the analogous properties for real numbers.

    (i) $z_1 z_2 = z_2 z_1$ and $z_1 + z_2 = z_2 + z_1$.
    (ii) $z_1(z_2 z_3) = (z_1 z_2)z_3$ and $z_1 + (z_2 + z_3) = (z_1 + z_2) + z_3$.
    (iii) $z_1(z_2 + z_3) = z_1 z_2 + z_1 z_3$.

23. (a) Show the equation $z^2 + az + b = 0$, where a, b are complex constants, can be rewritten in the form $(z + a/2)^2 + (4b - a^2)/4 = 0$. Hence show that all roots of the equation are given by the **quadratic formula** $z = (-a \pm \sqrt{a^2 - 4b})/2$.

    (b) Show that for real a and b, the roots are complex conjugates if $a^2 < 4b$.

    (c) If a and b are allowed to be complex, the expression $a^2 - 4b$ inside the square-root symbol in the quadratic formula will itself in general be a complex number $u + iv$. To find the two distinct square roots of a non-zero complex number $u + iv$, let

    $$x = \sqrt{\frac{\sqrt{u^2 + v^2} + u}{2}}, \qquad y = \sqrt{\frac{\sqrt{u^2 + v^2} - u}{2}}.$$

    Show that the square roots of $u + iv$ are $\pm(x + iy)$ if $v \geq 0$ and $\pm(x - iy)$ if $v \leq 0$.

    (d) Solve $z^2 + z + i = 0$.

24. There are n distinct solutions of $z^n = w$ if $w \neq 0$, that is, nth roots of a complex number $\omega \neq 0$. These roots can be found by first expressing ω in multiple polar form by allowing for the 2π periodicity of cosine and sine as follows:

    $$\omega = |\omega|(\cos(\theta + 2k\pi) + i\sin(\theta + 2k\pi)), \qquad k = 0, 1, \ldots, n - 1.$$

    The desired roots are then

    $$z_k = |\omega|^{1/n}\left(\cos\left(\frac{\theta}{n} + \frac{2k\pi}{n}\right) + i\sin\left(\frac{\theta}{n} + \frac{2k\pi}{n}\right)\right), \qquad k = 0, 1, \ldots, n - 1.$$

    (a) Verify that the numbers z_k satisfy $z_k^n = \omega$, $k = 0, 1, \ldots, n - 1$.

    (b) Find the cube roots of i, first in polar form, then in $x + iy$ form.

    (c) Find the fourth roots of $-i$, first in polar form, then in $x + iy$ form.

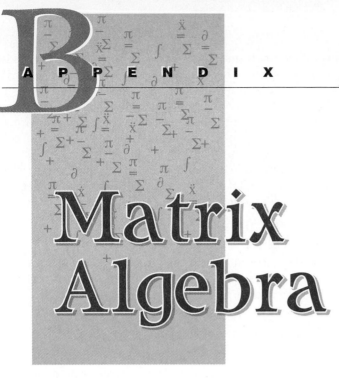

Matrix Algebra

Matrices are important in many areas of mathematics, but their use in this book is for solving systems of linear differential equations and systems of linear algebraic equations. The systems

$$\frac{dx}{dt} = 2x + 5y \qquad\qquad 5x + y - z = 0$$
$$\text{and}$$
$$\frac{dy}{dt} = 2x - y \qquad\qquad x + 2y + 3z = 0$$

are examples of the two kinds of system we are concerned with. Once we know the general form of these two systems, it is clear that they are completely determined by the respective rectangular arrays of numbers

$$\begin{pmatrix} 2 & 5 \\ 2 & -1 \end{pmatrix} \quad \text{and} \quad \begin{pmatrix} 5 & 1 & -1 \\ 1 & 2 & 3 \end{pmatrix}.$$

Any such rectangular array is called a **matrix.** Thus

$$\begin{pmatrix} 1 & 0 \\ 2 & 1 \\ 1 & 1 \end{pmatrix}, \quad \begin{pmatrix} 1 & 2 & 1 & 7 \\ 0 & 0 & 0 & 1 \\ 0 & 1 & 2 & 5 \end{pmatrix}, \quad \begin{pmatrix} 0 \\ 4 \\ 8 \end{pmatrix}, \quad \begin{pmatrix} 0.111 & 0.325 & 5.002 \\ 0.200 & 0.007 & 3.125 \\ 3.001 & 0.555 & 1.007 \end{pmatrix}$$

are examples of matrices. We'll sometimes want the entries in a matrix to be functions rather than constants, but since the underlying algebra is the same in both instances, we'll restrict ourselves at first to the simpler-looking case of constant entries. A horizontal line of entries in a matrix is called a **row,** and a vertical line of entries is called a **column.** The number of rows and columns in a matrix determine its **dimensions.** Thus the four previous examples have dimensions 3 by 2, 3 by 4, 3 by 1, and 3 by 3. Note that the number of rows is always listed before the number of columns. A 1 by n matrix is called an n-**dimensional row vector,** and an n by 1 matrix is called an n-**dimensional column vector.** A matrix is **square** if it has the same number of rows as columns.

Two matrices are equal if they have the same dimensions and corresponding entries are equal. We'll use capital letters to denote matrices and corresponding lowercase letters to denote a typical entry. Thus we may write

$$A = \begin{pmatrix} a_{11} & a_{12} & a_{13} \\ a_{21} & a_{22} & a_{23} \end{pmatrix} \quad \text{or} \quad A = (a_{ij}),$$

where a_{ij}, the ijth entry, is the entry in the ith row and jth column. If there is only one row or one column, we usually omit the unnecessary subscript and use a boldface letter to denote such a row or column. For example,

$$\mathbf{x} = \begin{pmatrix} x_1 \\ x_2 \\ x_3 \end{pmatrix}, \quad \mathbf{a} = (a_1 \quad a_2 \quad a_3),$$

The set of all n-tuples, whether written vertically or horizontally, is often denoted by $\mathscr{R}^n$.

SUM AND SCALAR MULTIPLE

Addition and multiplication by a number can be defined for vectors and matrices. If A and B have the same dimensions, then the **sum** $A + B$ is defined to be the matrix having the same dimensions as A and B and with the ijth entry equal to $a_{ij} + b_{ij}$, where $A = (a_{ij})$ and $B = (b_{ij})$.

---

example **1**

$$\begin{pmatrix} 1 & 1 \\ 0 & 2 \end{pmatrix} + \begin{pmatrix} -1 & 1 \\ 1 & 2 \end{pmatrix} = \begin{pmatrix} 0 & 2 \\ 1 & 4 \end{pmatrix}.$$

$$\begin{pmatrix} 1 & 2 & 1 \\ -1 & 1 & 0 \end{pmatrix} + \begin{pmatrix} -1 & -2 & -1 \\ 1 & -1 & 0 \end{pmatrix} = \begin{pmatrix} 0 & 0 & 0 \\ 0 & 0 & 0 \end{pmatrix}.$$

Addition is not defined for matrices with different dimensions, so we can't add

$$\begin{pmatrix} 1 & 1 \\ 0 & 2 \end{pmatrix} + \begin{pmatrix} 1 & 2 & 1 \\ -1 & 1 & 0 \end{pmatrix}.$$

For any matrix A and any number r, the **scalar multiple** rA is defined to be the matrix with the same dimensions as A and with ijth the entry ra_{ij}.

example 2

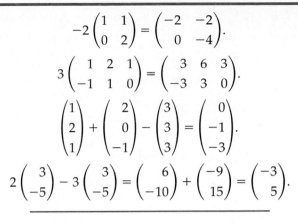

$$-2\begin{pmatrix} 1 & 1 \\ 0 & 2 \end{pmatrix} = \begin{pmatrix} -2 & -2 \\ 0 & -4 \end{pmatrix}.$$

$$3\begin{pmatrix} 1 & 2 & 1 \\ -1 & 1 & 0 \end{pmatrix} = \begin{pmatrix} 3 & 6 & 3 \\ -3 & 3 & 0 \end{pmatrix}.$$

$$\begin{pmatrix} 1 \\ 2 \\ 1 \end{pmatrix} + \begin{pmatrix} 2 \\ 0 \\ -1 \end{pmatrix} - \begin{pmatrix} 3 \\ 3 \\ 3 \end{pmatrix} = \begin{pmatrix} 0 \\ -1 \\ -3 \end{pmatrix}.$$

$$2\begin{pmatrix} 3 \\ -5 \end{pmatrix} - 3\begin{pmatrix} 3 \\ -5 \end{pmatrix} = \begin{pmatrix} 6 \\ -10 \end{pmatrix} + \begin{pmatrix} -9 \\ 15 \end{pmatrix} = \begin{pmatrix} -3 \\ 5 \end{pmatrix}.$$

A sum of scalar multiples of matrices is called a **linear combination** of matrices. Because of the way the matrix operations of addition and scalar multiplication are defined, a linear combination can of course always be written as a single matrix. However, it often happens that we want to expand a matrix into a linear combination. For example, if

$$\mathbf{e}_1 = \begin{pmatrix} 1 \\ 0 \\ 0 \end{pmatrix}, \qquad \mathbf{e}_2 = \begin{pmatrix} 0 \\ 1 \\ 0 \end{pmatrix}, \qquad \mathbf{e}_3 = \begin{pmatrix} 0 \\ 0 \\ 1 \end{pmatrix},$$

then any 3-dimensional column can be written as a linear combination of $\mathbf{e}_1$, $\mathbf{e}_2$, $\mathbf{e}_3$:

$$\begin{pmatrix} c_1 \\ c_2 \\ c_3 \end{pmatrix} = c_1 \begin{pmatrix} 1 \\ 0 \\ 0 \end{pmatrix} + c_2 \begin{pmatrix} 0 \\ 1 \\ 0 \end{pmatrix} + c_3 \begin{pmatrix} 0 \\ 0 \\ 1 \end{pmatrix}$$

$$= c_1\mathbf{e}_1 + c_2\mathbf{e}_2 + c_3\mathbf{e}_3.$$

Once the dimension you are working in has been identified, you can use the **e** notation without ambiguity. Let $\mathbf{e}_k$ be the n-dimensional column with 1 in the kth entry and 0 elsewhere. Then any n-dimensional column can be represented uniquely as a linear combination of $\mathbf{e}_1, \ldots, \mathbf{e}_n$. For this reason, the set $\{\mathbf{e}_1, \ldots, \mathbf{e}_n\}$ is called a **basis** for the n-dimensional vectors.

example 3

With $\mathbf{e}_1$, $\mathbf{e}_2$ of dimension 2, we have

$$\begin{pmatrix} 3 \\ 7 \end{pmatrix} = 3\begin{pmatrix} 1 \\ 0 \end{pmatrix} + 7\begin{pmatrix} 0 \\ 1 \end{pmatrix}$$

$$= 3\mathbf{e}_1 + 7\mathbf{e}_2.$$

example 4

A vector function can have a basis representation. For example,

$$\begin{pmatrix} t \\ e^t \\ e^{-t} \end{pmatrix} = t \begin{pmatrix} 1 \\ 0 \\ 0 \end{pmatrix} + e^t \begin{pmatrix} 0 \\ 1 \\ 0 \end{pmatrix} + e^{-t} \begin{pmatrix} 0 \\ 0 \\ 1 \end{pmatrix}$$

$$= t\mathbf{e}_1 + e^t\mathbf{e}_2 + e^{-t}\mathbf{e}_3.$$

Using both addition and scalar multiplication, we can write a linear combination of matrices with the same dimensions. For example, using 2 by 2 matrices,

$$2\begin{pmatrix} 1 & -1 \\ 2 & 3 \end{pmatrix} - 3\begin{pmatrix} 0 & 1 \\ 2 & 1 \end{pmatrix} = \begin{pmatrix} 2 & -2 \\ 4 & 6 \end{pmatrix} + \begin{pmatrix} 0 & -3 \\ -6 & -3 \end{pmatrix}$$

$$= \begin{pmatrix} 2 & -5 \\ -2 & 3 \end{pmatrix}.$$

We write $-A$ for $(-1)A$ and $A - B$ for $A + (-1)B$. Also, for any shape n by m, there is a **zero matrix,** denoted by 0, with all entries equal to zero, such that $A + 0 = A$.

Notational Warning. When just a single 0 is used to denote the zero matrix, it must be made clear from the context what dimension is intended. For example, if $A = \begin{pmatrix} 1 & 2 \\ 4 & 3 \end{pmatrix}$, then the 0 in $A + 0$ must mean $\begin{pmatrix} 0 & 0 \\ 0 & 0 \end{pmatrix}$, because no other dimensions would allow addition to a 2 by 2 matrix.

As a practical matter, it's important that if vectors $\mathbf{v}_1, \ldots, \mathbf{v}_n$ are used to express a vector $\mathbf{x}$ as a linear combination

$$\mathbf{x} = c_1\mathbf{v}_1 + \cdots + c_n\mathbf{v}_n,$$

then the coefficients $c_1, \ldots, c_n$ should be uniquely determined by the vector $\mathbf{x}$. In particular, the zero vector $\mathbf{x} = 0$ should be uniquely expressible, in which case $c_1 = c_2 = \cdots = c_n = 0$. It's easy to see that uniqueness in general follows from this special case, because

$$c_1\mathbf{v}_1 + \cdots + c_n\mathbf{v}_n = d_1\mathbf{v}_1 + \cdots + d_n\mathbf{v}_n$$

is equivalent to $(c_1 - d_1)\mathbf{v}_1 + \cdots + (c_n - d_n)\mathbf{v}_n = 0$; thus $c_k - d_k = 0$ for $k = 1, 2, \ldots, n$. Another way of expressing the uniqueness condition on $\mathbf{v}_1, \ldots, \mathbf{v}_n$ is to say that no one of them should be expressible as a linear combination of the remaining ones. If the vectors $\mathbf{v}_1, \ldots, \mathbf{v}_n$ satisfy any one of these equivalent conditions, then the vectors are said to be **linearly independent;** otherwise they are **linearly dependent.**

example 5

The vectors $\mathbf{e}_1, \mathbf{e}_2, \mathbf{e}_3$ in $\mathcal{R}^3$ are linearly independent, because

$$c_1 \begin{pmatrix} 1 \\ 0 \\ 0 \end{pmatrix} + c_2 \begin{pmatrix} 0 \\ 1 \\ 0 \end{pmatrix} + c_3 \begin{pmatrix} 0 \\ 0 \\ 1 \end{pmatrix} = \begin{pmatrix} 0 \\ 0 \\ 0 \end{pmatrix}$$

implies $c_1 = c_2 = c_3 = 0$. On the other hand,

$$2\begin{pmatrix} 1 \\ 1 \\ 1 \end{pmatrix} + 3\begin{pmatrix} 0 \\ 1 \\ 2 \end{pmatrix} - \begin{pmatrix} 2 \\ 5 \\ 8 \end{pmatrix} = \begin{pmatrix} 0 \\ 0 \\ 0 \end{pmatrix},$$

so the three vectors on the left are linearly dependent; the third one is, in fact, 2 times the first plus 3 times the second.

MATRIX PRODUCTS

First consider an m by n matrix A and n-dimensional column $\mathbf{x}$. The **product** $A\mathbf{x}$ is defined to be the m-dimensional column vector equal to the linear combination of the columns of A with coefficients from $\mathbf{x}$, taken in the same order. For example,

$$\begin{pmatrix} a_{11} & a_{12} \\ a_{21} & a_{22} \end{pmatrix}\begin{pmatrix} x_1 \\ x_2 \end{pmatrix} = x_1\begin{pmatrix} a_{11} \\ a_{21} \end{pmatrix} + x_2\begin{pmatrix} a_{12} \\ a_{22} \end{pmatrix}$$

$$= \begin{pmatrix} a_{11}x_1 & a_{12}x_2 \\ a_{21}x_1 & a_{22}x_2 \end{pmatrix}.$$

Thus the product $A\mathbf{x} = b$ represents the system of linear equations

$$a_{11}x_1 + a_{12}x_2 = b_1,$$
$$a_{21}x_1 + a_{22}x_2 = b_2,$$

where b_1, b_2 are the entries in the column vector $\mathbf{b}$. Note also that if $\mathbf{x} = \mathbf{e}_j$, with 1 in the jth entry and 0 elsewhere, then $A\mathbf{e}_j$ is just the jth column of A:

$$A\mathbf{e}_j = \begin{pmatrix} a_{1j} \\ \vdots \\ a_{mj} \end{pmatrix}, \qquad \text{where } A = (a_{ij}).$$

If B is an n by p matrix, the matrix product AB is the m by p matrix whose columns are the respective products of A with the columns of B. For example,

$$\begin{pmatrix} a_{11} & a_{12} \\ a_{21} & a_{22} \end{pmatrix}\begin{pmatrix} b_{11} & b_{12} \\ b_{21} & b_{22} \end{pmatrix} = \begin{pmatrix} a_{11}b_{11} + a_{12}b_{21} & a_{11}b_{21} + a_{12}b_{22} \\ a_{21}b_{11} + a_{22}b_{21} & a_{21}b_{12} + a_{22}b_{22} \end{pmatrix}.$$

---

example 6

If
$$A = \begin{pmatrix} 4 & 3 \\ -1 & 2 \\ 1 & -1 \end{pmatrix} \qquad \text{and} \qquad B = \begin{pmatrix} b_{11} & b_{12} \\ b_{21} & b_{22} \end{pmatrix},$$

then

$$AB = \begin{pmatrix} 4 & 3 \\ -1 & 2 \\ 1 & -1 \end{pmatrix}\begin{pmatrix} b_{11} & b_{12} \\ b_{21} & b_{22} \end{pmatrix} = \begin{pmatrix} 4b_{11} + 3b_{21} & 4b_{12} + 3b_{22} \\ -b_{11} + 2b_{21} & -b_{12} + 2b_{22} \\ b_{11} - b_{21} & b_{12} - b_{22} \end{pmatrix}.$$

If
$$C = \begin{pmatrix} c_{11} & c_{12} \\ c_{21} & c_{22} \end{pmatrix} \qquad \text{and} \qquad D = \begin{pmatrix} 1 & 2 \\ 4 & 5 \end{pmatrix},$$

then
$$CD = \begin{pmatrix} c_{11} & c_{12} \\ c_{21} & c_{22} \end{pmatrix} \begin{pmatrix} 1 & 2 \\ 4 & 5 \end{pmatrix} = \begin{pmatrix} c_{11} + 4c_{12} & 2c_{11} + 5c_{12} \\ c_{21} + 4c_{22} & 2c_{21} + 5c_{22} \end{pmatrix}.$$

Note that the only requirement on the dimensions of matrices in a product is that the number of columns in the first matrix be the same as the number of rows in the second. Hence we can form the product

$$AD = \begin{pmatrix} 4 & 3 \\ -1 & 2 \\ 1 & -1 \end{pmatrix} \begin{pmatrix} 1 & 2 \\ 4 & 5 \end{pmatrix} = \begin{pmatrix} 16 & 23 \\ 7 & 8 \\ -3 & -3 \end{pmatrix}.$$

The entry in the first row and second column of AD is the dot product $(4)(2) + (3)(5) = 23$ of the first row of A and the second column of D. Matrix multiplication is sometimes called **row-by-column multiplication.**

---

e x a m p l e 7

The system of linear algebraic equations
$$\begin{array}{c} 3x + 7y = 1, \\ 2x + 5y = 2, \end{array} \quad \text{or} \quad \begin{pmatrix} 3 & 7 \\ 2 & 5 \end{pmatrix}\begin{pmatrix} x \\ y \end{pmatrix} = \begin{pmatrix} 1 \\ 2 \end{pmatrix}$$
can be written $A\mathbf{x} = \mathbf{c}$, where
$$A = \begin{pmatrix} 3 & 7 \\ 2 & 5 \end{pmatrix}, \quad \mathbf{x} = \begin{pmatrix} x \\ y \end{pmatrix}, \quad \text{and} \quad \mathbf{c} = \begin{pmatrix} 1 \\ 2 \end{pmatrix}.$$
Similarly, the system of differential equations
$$\begin{array}{c} \dfrac{dx}{dt} = 3x + 7y - 1, \\[2mm] \dfrac{dy}{dt} = 2x + 5y - 2, \end{array} \quad \text{or} \quad \begin{pmatrix} \dot{x} \\ \dot{y} \end{pmatrix} = \begin{pmatrix} 3 & 7 \\ 2 & 5 \end{pmatrix}\begin{pmatrix} x \\ y \end{pmatrix} - \begin{pmatrix} 1 \\ 2 \end{pmatrix},$$
can be written $d\mathbf{x}/dt = A\mathbf{x} - \mathbf{c}$, since the vector derivative is computed in terms of coordinates by
$$\frac{d\mathbf{x}}{dt} = \frac{d}{dt}\begin{pmatrix} x \\ y \end{pmatrix} = \begin{pmatrix} \dot{x} \\ \dot{y} \end{pmatrix}.$$

It is worth remarking that the algebraic system becomes a special case of the differential system if we make the restriction that the solutions be constant, in which case $d\mathbf{x}/dt = 0$. It is easy to see that the algebraic system has solution $x = -9$, $y = 4$. It follows that the differential system has those constant solutions also. Of course, it has many nonconstant solutions in addition. (The constant solutions are called **equilibrium solutions,** and they are important in the discussion of stability in Section 8 of Chapter 6.)

For general questions about matrix products, it is sometimes useful to write the entries in the product of $A = (a_{ij})$ *and* $B = (b_{ij})$ by using the summation notation. The ijth entry in AB is

$$\sum_{k=1}^{n} a_{ik} b_{ik}$$

Note that the summation index k runs over the column index of A (that is, along a row) and over the row index of B (that is, along the column).

The next theorem lists the basic properties of matrix products. Of course, we assume that the various matrices have the proper shapes so that the combinations are well-defined. The **derivative of a matrix,** and in particular of a vector, is computed entry by entry.

■ B.1 THEOREM

1. $(A + B)C = AC + BC$.

2. $C(A + B) = CA + CB$.

3. $(rA)B = r(AB) = A(rB)$.

4. $A(BC) = (AB)C$.

5. $\dfrac{d(AB)}{dt} = A\,\dfrac{dB}{dt} + \dfrac{dA}{dt}\,B$.

Proof. We prove only property 4 since it is the most complicated; the others are left as exercises. We let A be m by n, let B be n by p, and let C be p by q. Then the ijth entry of BC is $\sum_{k=1}^{p} b_{ik}c_{kj}$, so the ijth entry of $A(BC)$ is

$$\sum_{l=1}^{n} a_{il}\left(\sum_{k=1}^{p} b_{lk}c_{kj}\right) = \sum_{l=1}^{n}\sum_{k=1}^{p} a_{il}b_{lk}c_{kj}.$$

Similarly, the ijth entry of $(AB)C$ is

$$\sum_{k=1}^{p}\left(\sum_{l=1}^{n} a_{il}b_{lk}\right)c_{kj} = \sum_{k=1}^{p}\sum_{l=1}^{n} a_{il}b_{lk}c_{kj}.$$

The two double sums on the right consist of the same terms added in different orders, so the ijth entries in $A(BC)$ and $(AB)C$ are the same.   ■

Formulas 1 and 2 in Theorem B.1 are called the **distributive laws** for matrix multiplication. Two of them are needed because **matrix multiplication is not in general commutative;** that is, there are square matrices A and B for which AB and BA both make sense but such that $AB \neq BA$. Exercise 7 gives an example. Formula 4 is the **associative law** for matrix multiplication, and it shows that it makes sense to write the product ABC, since the result is independent of the order in which the products are formed. Properties 1, 2, and 3 of Theorem B.1 express the **linearity** of matrix multiplication; in particular, if **x** and **y** are n-dimensional column vectors and A is an m by n matrix, the two equations

$$A(\mathbf{x} + \mathbf{y}) = A\mathbf{x} + A\mathbf{y}, \qquad A(r\mathbf{x}) = rA\mathbf{x}$$

express the fact that A acts as a linear function from $\mathcal{R}^n$ to $\mathcal{R}^m$.

example 8

We verify here that $(A + B)C = AC + BC$ for

$$A = \begin{pmatrix} 1 & 2 \\ -1 & 2 \end{pmatrix}, \qquad B = \begin{pmatrix} 0 & 2 \\ 2 & 1 \end{pmatrix}, \qquad C = \begin{pmatrix} -1 & 1 & 2 \\ 1 & 2 & 1 \end{pmatrix}.$$

We have

$$\left(\begin{pmatrix} 1 & 2 \\ -1 & 2 \end{pmatrix} + \begin{pmatrix} 0 & 2 \\ 2 & 1 \end{pmatrix}\right)\begin{pmatrix} -1 & 1 & 2 \\ 1 & 2 & 1 \end{pmatrix} = \begin{pmatrix} 1 & 4 \\ 1 & 3 \end{pmatrix}\begin{pmatrix} -1 & 1 & 2 \\ 1 & 2 & 1 \end{pmatrix}$$

$$= \begin{pmatrix} 3 & 9 & 6 \\ 2 & 7 & 5 \end{pmatrix}.$$

Also,

$$\begin{pmatrix} 1 & 2 \\ -1 & 2 \end{pmatrix}\begin{pmatrix} -1 & 1 & 2 \\ 1 & 2 & 1 \end{pmatrix} + \begin{pmatrix} 0 & 2 \\ 2 & 1 \end{pmatrix}\begin{pmatrix} -1 & 1 & 2 \\ 1 & 2 & 1 \end{pmatrix} = \begin{pmatrix} 1 & 5 & 4 \\ 3 & 3 & 0 \end{pmatrix} + \begin{pmatrix} 2 & 4 & 2 \\ -1 & 4 & 5 \end{pmatrix}$$

$$= \begin{pmatrix} 3 & 9 & 6 \\ 2 & 7 & 5 \end{pmatrix}.$$

IDENTITY MATRICES

A square matrix of the form

$$I = \begin{pmatrix} 1 & 0 \\ 0 & 1 \end{pmatrix} \quad \text{or} \quad I = \begin{pmatrix} 1 & 0 & 0 \\ 0 & 1 & 0 \\ 0 & 0 & 1 \end{pmatrix} \quad \text{or} \quad I = \begin{pmatrix} 1 & 0 & \cdots & 0 & 0 \\ 0 & 1 & \cdots & 0 & 0 \\ \vdots & \vdots & \ddots & \vdots & \vdots \\ 0 & 0 & \cdots & 0 & 1 \end{pmatrix}$$

that has 1's on its **main diagonal** and zeros elsewhere is called an **identity matrix.** It has the property that

$$IA = A, \qquad BI = B$$

for any matrices A, B such that the products are defined. Thus it is an identity element for matrix multiplication somewhat as the number 1 is an identity for multiplication of numbers. There is an n by n identity matrix for every value of n, but as with the zero matrices, it must be made clear from the context what the dimension of an identity matrix is.

e x a m p l e 9   You can check the following.

$$\begin{pmatrix} 1 & 0 \\ 0 & 1 \end{pmatrix}\begin{pmatrix} 1 & 2 & 3 \\ 4 & 5 & 6 \end{pmatrix} = \begin{pmatrix} 1 & 2 & 3 \\ 4 & 5 & 6 \end{pmatrix}\begin{pmatrix} 1 & 0 & 0 \\ 0 & 1 & 0 \\ 0 & 0 & 1 \end{pmatrix} = \begin{pmatrix} 1 & 2 & 3 \\ 4 & 5 & 6 \end{pmatrix}.$$

$$\begin{pmatrix} 1 & 0 & 0 \\ 0 & 1 & 0 \\ 0 & 0 & 1 \end{pmatrix}\begin{pmatrix} 1 & 2 \\ 3 & 4 \\ 5 & 6 \end{pmatrix} = \begin{pmatrix} 1 & 2 \\ 3 & 4 \\ 5 & 6 \end{pmatrix}\begin{pmatrix} 1 & 0 \\ 0 & 1 \end{pmatrix} = \begin{pmatrix} 1 & 2 \\ 3 & 4 \\ 5 & 6 \end{pmatrix}.$$

Notice that while we use the letter I to denote the identity matrix when it occurs on either side of a given matrix A, these identity matrices will be of different sizes unless A itself is a square matrix. However, we do have, for example,

$$\begin{pmatrix} 1 & 0 \\ 0 & 1 \end{pmatrix}\begin{pmatrix} a & b \\ c & d \end{pmatrix} = \begin{pmatrix} a & b \\ c & d \end{pmatrix}\begin{pmatrix} 1 & 0 \\ 0 & 1 \end{pmatrix} = \begin{pmatrix} a & b \\ c & d \end{pmatrix}.$$

If x is an n-dimensional column vector, then $I\mathbf{x} = \mathbf{x}$ looks like this when $n = 3$:

$$\begin{pmatrix} 1 & 0 & 0 \\ 0 & 1 & 0 \\ 0 & 0 & 1 \end{pmatrix}\begin{pmatrix} x_1 \\ x_2 \\ x_3 \end{pmatrix} = \begin{pmatrix} x_1 \\ x_2 \\ x_3 \end{pmatrix}.$$

INVERTIBILITY

If A is a given square matrix and B is a square matrix of the same size such that

$$AB = BA = I,$$

then A is said to be an **invertible** matrix and B is an **inverse** of A. In fact, Exercise 18c shows that a matrix A can have at most one inverse, so we can speak of the inverse of A and denote it by A^{-1}. Thus, if A is invertible,

$$AA^{-1} = A^{-1}A = I.$$

---

example 10

If $A = \begin{pmatrix} 1 & 2 \\ 3 & 7 \end{pmatrix}$, then $A^{-1} = \begin{pmatrix} 7 & -2 \\ -3 & 1 \end{pmatrix}$, because

$$AA^{-1} = \begin{pmatrix} 1 & 2 \\ 3 & 7 \end{pmatrix}\begin{pmatrix} 7 & -2 \\ -3 & 1 \end{pmatrix} = \begin{pmatrix} 1 & 0 \\ 0 & 1 \end{pmatrix}$$

and

$$A^{-1}A = \begin{pmatrix} 7 & -2 \\ -3 & 1 \end{pmatrix}\begin{pmatrix} 1 & 2 \\ 3 & 7 \end{pmatrix} = \begin{pmatrix} 1 & 0 \\ 0 & 1 \end{pmatrix}.$$

In general, any 2 by 2 matrix $\begin{pmatrix} a & b \\ c & d \end{pmatrix}$ is invertible if $ad - bc \neq 0$. The real-valued function $\det A = ad - bc$ of the 2 by 2 matrix A is called the **determinant** of the matrix. In that case

■ **B.2**
$$\begin{pmatrix} a & b \\ c & d \end{pmatrix}^{-1} = \frac{1}{ad - bc}\begin{pmatrix} d & -b \\ -c & a \end{pmatrix}.$$

In particular,

$$\begin{pmatrix} 1 & 3 \\ -1 & 2 \end{pmatrix}^{-1} = \frac{1}{5}\begin{pmatrix} 2 & -3 \\ 1 & 1 \end{pmatrix} = \begin{pmatrix} \frac{2}{5} & -\frac{3}{5} \\ \frac{1}{5} & \frac{1}{5} \end{pmatrix}.$$

Formula 1.2 is worth remembering, and we leave it as an exercise to show that the formula is correct.

If A is an n by n matrix, then the matrix equation $A\mathbf{x} = \mathbf{b}$ is equivalent to a system of n linear equations in n unknowns. If A happens to be an invertible matrix with inverse A^{-1}, then we can solve the system in matrix form by multiplying both sides on the left by A^{-1} to get

$$A^{-1}A\mathbf{x} = A^{-1}\mathbf{b}.$$

Since $A^{-1}A = I$, we have $A^{-1}A\mathbf{x} = I\mathbf{x} = \mathbf{x}$, so the equation becomes

$$\mathbf{x} = A^{-1}\mathbf{b}.$$

In other words, $\mathbf{x} = A^{-1}\mathbf{b}$ is a solution to the given equation and, in fact, is the only solution, because $\mathbf{x}$ could have been an arbitrary vector satisfying $A\mathbf{x} = \mathbf{b}$.

---

example 11

The system

$$\begin{aligned} x + 2y &= 3 \\ 3x + 7y &= -4 \end{aligned} \quad \text{is equivalent to} \quad \begin{pmatrix} 1 & 2 \\ 3 & 7 \end{pmatrix}\begin{pmatrix} x \\ y \end{pmatrix} = \begin{pmatrix} 3 \\ -4 \end{pmatrix}.$$

By Formula 1.2,

$$\begin{pmatrix} 1 & 2 \\ 3 & 7 \end{pmatrix}^{-1} = \begin{pmatrix} 7 & -2 \\ -3 & 1 \end{pmatrix},$$

so we multiply the left side of the equation by the inverse to get

$$\begin{pmatrix} 7 & -2 \\ -3 & 1 \end{pmatrix}\begin{pmatrix} 1 & 2 \\ 3 & 7 \end{pmatrix}\begin{pmatrix} x \\ y \end{pmatrix} = \begin{pmatrix} 1 & 0 \\ 0 & 1 \end{pmatrix}\begin{pmatrix} x \\ y \end{pmatrix} = \begin{pmatrix} x \\ y \end{pmatrix}.$$

Hence multiplying also on the right gives

$$\begin{pmatrix} x \\ y \end{pmatrix} = \begin{pmatrix} 7 & -2 \\ -3 & 1 \end{pmatrix}\begin{pmatrix} 3 \\ -4 \end{pmatrix} = \begin{pmatrix} 29 \\ -13 \end{pmatrix}.$$

Thus $(x, y) = (29, -13)$ is the unique solution.

POWERS OF A SQUARE MATRIX

If A is a square matrix, it can be multiplied by itself repeatedly, and we define

$$A^2 = AA, \ A^3 = AAA = AA^2, \dots, A^n = AA^{n-1}.$$

These powers of A all have the same dimension as A, but note that if B is a nonsquare matrix, then B^2 never makes sense.

---

example 12

If $A = \begin{pmatrix} 2 & 1 \\ 0 & 3 \end{pmatrix}$, then

$$A^2 = \begin{pmatrix} 2 & 1 \\ 0 & 3 \end{pmatrix}\begin{pmatrix} 2 & 1 \\ 0 & 3 \end{pmatrix} = \begin{pmatrix} 4 & 5 \\ 0 & 9 \end{pmatrix},$$

$$A^3 = \begin{pmatrix} 2 & 1 \\ 0 & 3 \end{pmatrix}\begin{pmatrix} 4 & 5 \\ 0 & 9 \end{pmatrix} = \begin{pmatrix} 8 & 19 \\ 0 & 27 \end{pmatrix}, \text{ etc.}$$

EXERCISES

1. Given that

$$A = \begin{pmatrix} -1 & 2 \\ 0 & 1 \end{pmatrix}, \qquad B = \begin{pmatrix} 1 & 4 \\ 1 & 1 \end{pmatrix}, \qquad \mathbf{x} = \begin{pmatrix} 1 \\ 2 \end{pmatrix}, \qquad \mathbf{b} = \begin{pmatrix} 1 \\ 2 \end{pmatrix},$$

$$C = \begin{pmatrix} 1 & 1 & 1 \\ 0 & 1 & 1 \\ 0 & 0 & 1 \end{pmatrix}, \qquad D = \begin{pmatrix} 1 & 0 & 1 \\ 1 & 0 & 1 \\ 2 & 2 & 2 \end{pmatrix}, \qquad \mathbf{y} = \begin{pmatrix} 1 \\ 2 \\ 3 \end{pmatrix}, \qquad \mathbf{d} = \begin{pmatrix} 1 \\ 1 \\ 1 \end{pmatrix},$$

compute

(a) AB.

(b) $A\mathbf{x} + \mathbf{b}$.

(c) $BA + B^2$.

(d) CD.

(e) $C\mathbf{y} + \mathbf{d}$.

(f) $DC + C^2$.

2. For each of the following systems find a matrix A, a column vector $\mathbf{x}$, and a column vector $\mathbf{b}$ such that the system can be written in the form $A\mathbf{x} = \mathbf{b}$.

(a) $x - y = 1$,
    $x + y = 2$.

(b) $2x + 3y = 0$,
    $x + 3y = 0$.

(c) $x + y = 1$,
    $y = 1$.

(d) $x + y + z = 0$,
    $x + y - z = 1$,
    $x - y - z = 0$.

(e) $x + y = 0$,
    $y + z = 0$,
    $x - z = 1$.

(f) $u + v - w = 1$,
    $u - v + w = 2$.

3. For each of the following systems of differential equations, find a matrix A, a column vector $\mathbf{x}$, and a column vector $\mathbf{b}$ such that the system can be written in the form $d\mathbf{x}/dt = A\mathbf{x} + \mathbf{b}$. Note that in general the entries in A, $\mathbf{b}$, and $\mathbf{x}$ will depend on t.

(a) $dx/dt = 2x + 3y$,
    $dy/dt = 2x - 4y$.

(b) $dx/dt = x + y + t$,
    $dy/dt = x - y - t$.

(c) $dx/dt = x + y - z$,
    $dy/dt = x - y + z$,
    $dz/dt = x + ty$.

(d) $du/dt = tu - tv + w$,
    $dv/dt = u + v$,
    $dw/dt = u + v + t$.

4. Find matrices A and vectors $\mathbf{b}$ such that each of the following systems of equations can be written in the form $\dot{\mathbf{x}} = A\mathbf{x} + \mathbf{b}$, where $\mathbf{x}$ is an n by 1 column vector of the appropriate dimension.

(a) $\dot{x} - 2y = 1$,
    $x + 3\dot{y} = 2$.

(b) $\dot{x} = 1$,
    $\dot{y} = 2$.

(c) $\dot{x} - \dot{y} = 0$,
    $\dot{x} + \dot{y} = 0$.

(d) $\dot{x} - y + z = 1$,
    $x - \dot{y} - z = 0$
    $x + y + \dot{z} = 0$.

(e) $\dot{x} - 2y = 0$,
    $\dot{y} + z = 0$,
    $x + \dot{z} = 1$.

5. Express each of the following vector functions as a linear combination of $\mathbf{e}_1$ and $\mathbf{e}_2$ in the case of dimension 2 or $\mathbf{e}_1$, $\mathbf{e}_2$, and $\mathbf{e}_3$ in the case of dimension 3.

(a) $f(t) = \begin{pmatrix} e^t \\ e^t \\ t^2 \end{pmatrix}$.

(b) $f(t) = \begin{pmatrix} t^2 \\ t^3 \\ t^4 \end{pmatrix}$.

(c) $f(t) = \begin{pmatrix} t - 1 \\ t - 1 \\ t + 2 \end{pmatrix}$.

(d) $f(t) = \begin{pmatrix} 1 \\ t \end{pmatrix}$.

(e) $f(t) = \begin{pmatrix} \cos t \\ \sin t \end{pmatrix}$.

(f) $f(t) = \begin{pmatrix} 1 \\ 1 \end{pmatrix}$.

6. Recall that the derivative of a vector-valued function is computed by elementwise differentiation of the entries. The same is true for matrices, for example,

$$\frac{d}{dt}\begin{pmatrix} 1 & t \\ t & 2 \end{pmatrix} = \begin{pmatrix} 0 & 1 \\ 1 & 0 \end{pmatrix}.$$

Compute the derivative $dA(t)/dt$ for each of the following matrices $A(t)$.

(a) $A(t) = \begin{pmatrix} e^t & te^t \\ 0 & e^t \end{pmatrix}.$

(b) $A(t) = \begin{pmatrix} t^2 & 2t \\ 3 & 4 \end{pmatrix}\begin{pmatrix} t & 2t \\ t & 3t \end{pmatrix}.$

(c) $A(t) = \begin{pmatrix} e^t & 0 & 0 \\ 0 & e^{2t} & 0 \\ 0 & 0 & e^{3t} \end{pmatrix}.$

(d) $A(t) = \begin{pmatrix} t & t^2 & t^3 \\ 1 & t^3 & t^4 \\ 1 & 1 & t^5 \end{pmatrix}.$

7. Let $U = \begin{pmatrix} -1 & 2 \\ 2 & -4 \end{pmatrix}$, $V = \begin{pmatrix} 2 & 6 \\ 1 & 3 \end{pmatrix}$. Compute UV and VU. Are they the same? Is it possible for the product of two matrices to be zero without either factor being the zero matrix?

8. (a) Using the matrix C of Exercise 1, compute the vectors $C\mathbf{e}_1$, $C\mathbf{e}_2$, $C\mathbf{e}_3$, where the $\mathbf{e}_k$ are the natural basis vectors in $\mathcal{R}^3$.

(b) Show that if A is an m by n matrix, then $A\mathbf{e}_j$ is the jth column of A, where the $\mathbf{e}_k$ are the natural basis vectors in $\mathcal{R}^n$.

9. Compute the products.

(a) $\begin{pmatrix} 1 & 2 & 3 \\ 4 & 5 & 6 \\ 7 & 8 & 9 \end{pmatrix}\begin{pmatrix} 0 \\ 1 \\ 0 \end{pmatrix}.$

(b) $\begin{pmatrix} 0 & 1 & 1 \\ 1 & 0 & 1 \\ 1 & 1 & 0 \end{pmatrix}\begin{pmatrix} 2 \\ 1 \\ 3 \end{pmatrix}.$

(c) $(2 \quad 1 \quad 4)\begin{pmatrix} 3 \\ 5 \\ 7 \end{pmatrix}.$

(d) $\begin{pmatrix} 3 \\ 5 \\ 7 \end{pmatrix}(2 \quad 1 \quad 4).$

(e) $\begin{pmatrix} 2 & 1 \\ 5 & 6 \\ 3 & 4 \end{pmatrix}\begin{pmatrix} 1 & -1 & 1 \\ -1 & 1 & -1 \end{pmatrix}.$

(f) $\begin{pmatrix} 2 & 0 & 0 \\ 0 & 4 & 0 \\ 0 & 0 & 5 \end{pmatrix}\begin{pmatrix} -1 \\ 1 \\ -1 \end{pmatrix}.$

10. Prove Equations 1, 2, 3, and 5 of Theorem B.1.

11. If A is a square matrix, it can be multiplied by itself, and we can define $A^2 = AA$, $A^3 = AAA = A^2A$, $A^n = AA\cdots A$ (n factors). These powers of A all have the same dimension as A. Find A^2 and A^3 if:

(a) $A = \begin{pmatrix} 2 & 1 \\ 0 & 1 \end{pmatrix}.$

(b) $A = \begin{pmatrix} 1 & 0 & -1 \\ -1 & 0 & 1 \\ 2 & 1 & -1 \end{pmatrix}.$

(c) $A = \begin{pmatrix} 2 & 0 \\ 1 & 1 \end{pmatrix}.$

(d) $A = \begin{pmatrix} 2 & 0 \\ 0 & 3 \end{pmatrix}.$

[Note that 0 is the only number whose cube is 0. Part (b) of this problem thus illustrates another difference between the arithmetic of numbers and of matrices.]

12. The numerical equation $a^2 = 1$ has $a = 1$ and $a = -1$ as its only solutions.

(a) Show that if $A = I$ or $-I$, then $A^2 = I$, where I is an identity matrix of any dimension.

(b) Show that $\begin{pmatrix} a & b \\ c & -a \end{pmatrix}^2 = \begin{pmatrix} 1 & 0 \\ 0 & 1 \end{pmatrix}$ if $a^2 + bc = 1$; so the equation $A^2 = I$ has infinitely many different solutions in the set of 2 by 2 matrices.

(c) Show that every 2 by 2 matrix A for which $A^2 = I$ is either I, $-I$, or one of the matrices described in part (b).

13. (a) Express the matrix $\begin{pmatrix} 1 & -1 \\ 1 & 0 \end{pmatrix}$ as a linear combination of the matrices $\begin{pmatrix} 2 & -1 \\ 0 & 1 \end{pmatrix}$ and $\begin{pmatrix} 5 & -3 \\ 1 & 2 \end{pmatrix}$.

(b) Can an arbitrary 2 by 2 matrix be expressed as a linear combination of the last two matrices in part (a)?

14. Verify the matrix multiplications in Example 9 of the text.

15. Show that if A is an n by n matrix and I is the n by n identity matrix, then:

(a) $(A - I)(A + I) = A^2 - I$.   (b) $(A + I)^2 = A^2 + 2A + I$.

(c) Show that, if B is also n by n, it is not true in general that $(A - B)(A + B) = A^2 - B^2$ or that $(A + B)^2 = A^2 + 2AB + B^2$.

16. Find inverses for those of the following 2 by 2 matrices that have inverses.

(a) $\begin{pmatrix} 1 & 1 \\ 1 & 2 \end{pmatrix}$.   (c) $\begin{pmatrix} \frac{1}{2} & \frac{1}{4} \\ \frac{1}{4} & \frac{1}{5} \end{pmatrix}$.

(b) $\begin{pmatrix} 3 & 6 \\ 2 & 4 \end{pmatrix}$.   (d) $\begin{pmatrix} -7 & -5 \\ 12 & 9 \end{pmatrix}$.

17. Solve the matrix equation $Ax = b$ by multiplying both sides by A^{-1} with

(a) $A = \begin{pmatrix} 2 & -1 \\ 3 & 4 \end{pmatrix}$; $b = \begin{pmatrix} 1 \\ 1 \end{pmatrix}$.   (b) $A = \begin{pmatrix} 7 & 2 \\ 1 & 1 \end{pmatrix}$; $b = \begin{pmatrix} -2 \\ 4 \end{pmatrix}$.

18. (a) Show that if A and B are invertible matrices of the same dimension, then AB is invertible and $(AB)^{-1} = B^{-1}A^{-1}$.

(b) Show that if $A_1, \ldots, A_n$ are invertible matrices with the same dimension, then the matrix product $A_1 A_2 \cdots A_n$ is invertible and $(A_1 A_2 \cdots A_n)^{-1} = A_n^{-1} \cdots A_2^{-1} A_1^{-1}$.

(c) Show that if $AB = BA = I$ and $AC = CA = I$, then $B = C$.

DETERMINANTS

For a square matrix A, the determinant is a numerical-valued function written $\det A$. However, for a displayed matrix it is also customary just to replace the parentheses by vertical bars. Thus the notations

$$\begin{vmatrix} 1 & 4 & 5 \\ 6 & 7 & -3 \\ -2 & 1 & 0 \end{vmatrix}, \qquad \begin{vmatrix} a & b \\ c & d \end{vmatrix}$$

mean the same as

$$\det \begin{pmatrix} 1 & 4 & 5 \\ 6 & 7 & -3 \\ -2 & 1 & 0 \end{pmatrix}, \qquad \det \begin{pmatrix} a & b \\ c & d \end{pmatrix}.$$

Our definition of determinant will be inductive; that is, we will define det A first for 1 by 1 matrices, and then for each n define the determinant of an n by n matrix in terms of determinants of certain $(n-1)$ by $(n-1)$ matrices called *minors*. For any matrix A, the matrix obtained by deleting the ith row and jth column of A is called the ijth **minor** of A and is denoted by A_{ij}. Recall that we use the lowercase letter a to denote the ijth entry of a matrix A. Thus the ijth minor A_{ij} corresponds to the entry a_{ij} in a natural way, because the minor is obtained by deleting the row and column containing a_{ij}.

---

e x a m p l e 13   Let

$$A = \begin{pmatrix} -5 & -6 & 7 \\ 8 & -9 & 0 \\ -3 & 4 & 2 \end{pmatrix}, \qquad B = \begin{pmatrix} 1 & 2 \\ 3 & 4 \end{pmatrix}.$$

Then some examples of entries and corresponding minors are

$$a_{11} = -5, \qquad A_{11} = \begin{pmatrix} -9 & 0 \\ 4 & 2 \end{pmatrix},$$

$$a_{23} = 0, \qquad A_{23} = \begin{pmatrix} -5 & -6 \\ -3 & 4 \end{pmatrix},$$

$$b_{11} = 1, \qquad B_{11} = (4),$$

$$b_{12} = 2, \qquad B_{12} = (3).$$

We can now give the definition of **determinant.** For a 1 by 1 matrix, $A = (a)$, we define

$$\det A = a.$$

For an n by n matrix, $A = (a_{ij})$, $i, j = 1, \ldots, n$, we define

■ **B.3** $\det A = a_{11} \det A_{11} - a_{12} \det A_{12} + \cdots - (-1)^n a_{1n} \det A_{1n}.$

The definition is inductive in the sense that defining the determinant of an n by n matrix A requires us to know the determinants of the $(n-1)$ by $(n-1)$ minors A_{ij}. But the simple definition for the 1 by 1 case allows us to go on to 2 by 2, then 3 by 3, and so on. In words, the formula says that det A is the sum, with alternating signs, of the elements of the first row of A, each multiplied by the determinant of its corresponding minor. For this reason, the numbers

$$\det A_{11}, \; -\det A_{12}, \ldots, (-1)^{n+1} \det A_{1n}$$

are called the *cofactors* of the corresponding elements of the first row of A. In general, the **cofactor** of the entry a_{ij} in A is defined to be $(-1)^{i+j}$, det A_{ij}. Thus, in Example 13 the entry $a_{23} = 0$ in the matrix A has cofactor

$$(-1)^{2+3} \det \begin{pmatrix} -5 & -6 \\ -3 & 4 \end{pmatrix} = 38.$$

The factor $(-1)^{i+j}$ associates plus and minus signs with det A_{ij} according to the pattern

$$\begin{pmatrix} + & - & + & - & \cdots \\ - & + & - & + & \cdots \\ + & - & + & - & \cdots \\ - & + & - & + & \cdots \\ \vdots & \vdots & \vdots & \vdots & \end{pmatrix}.$$

e x a m p l e 14

(a) $\det \begin{pmatrix} 1 & 2 \\ 3 & 4 \end{pmatrix} = (1)(4) - (2)(3) = 4 - 6 = -2.$

(b) $\det \begin{pmatrix} -5 & -6 & 7 \\ 8 & -9 & 0 \\ -3 & 4 & 2 \end{pmatrix} = -5 \det \begin{pmatrix} -9 & 0 \\ 4 & 2 \end{pmatrix} - (-6) \det \begin{pmatrix} 8 & 0 \\ -3 & 2 \end{pmatrix}$

$$+ 7 \det \begin{pmatrix} 8 & -9 \\ -3 & 4 \end{pmatrix}$$

$$= (-5)(-18 - 0) + (6)(16 - 0) + 7(32 - 27)$$
$$= 90 + 96 + 35 = 221.$$

(c) $\det \begin{pmatrix} a & b \\ c & d \end{pmatrix} = ad - bc.$

The result of Example 14c is worth remembering as a general rule of calculation. **The determinant of a 2 by 2 matrix is the product of the entries on the main diagonal minus the product of the other two entries.** Thus 2 by 2 determinants can often be computed mentally, and 3 by 3 determinants in one or two lines.

It is an important fact, whose proof we omit, that if in Equation B.3 the elements and cofactors of the first row are replaced by the elements and cofactors of any other row, or of any column, then the expansion is still valid. Formally, the statement can be expressed as follows.

■ B.4 THEOREM

If A is a square matrix, then

$$\det A = \sum_{j=1}^{n} (-1)^{i+j} a_{ij} \det A_{ij}, \qquad \textbf{expansion by } i\textbf{th row,}$$

and $\qquad \det A = \sum_{i=1}^{n} (-1)^{i+j} a_{ij} \det A_{ij}, \qquad \textbf{expansion by } j\textbf{th column.}$

Notice that Equation B.3, which we used to define determinant, appears as a special case of the first equation in the theorem when we let $i = 1$. The alternating pattern of cofactor signs applies to all expansions by row or column.

example 15

The determinant computation in part (b) of Example 14 can be done also by using the elements and cofactors of the second row:

$$\det \begin{pmatrix} -5 & -6 & 7 \\ 8 & -9 & 0 \\ -3 & 4 & 2 \end{pmatrix}$$

$$= -(8) \det \begin{pmatrix} -6 & 7 \\ 4 & 2 \end{pmatrix} + (-9) \det \begin{pmatrix} -5 & 7 \\ -3 & 2 \end{pmatrix} - (0) \det \begin{pmatrix} -5 & -6 \\ -3 & 4 \end{pmatrix}$$

$$= -8(-12 - 28) - 9(-10 + 21)$$

$$= 221.$$

Computing the same determinant using the elements and cofactors of the third column gives

$$\det \begin{pmatrix} -5 & -6 & 7 \\ 8 & -9 & 0 \\ -3 & 4 & 2 \end{pmatrix}$$

$$= 7 \det \begin{pmatrix} 8 & -9 \\ -3 & 4 \end{pmatrix} - (0) \det \begin{pmatrix} -5 & -6 \\ -3 & 4 \end{pmatrix} + 2 \det \begin{pmatrix} -5 & -6 \\ 8 & -9 \end{pmatrix}$$

$$= 7(32 - 27) + 2(45 + 48)$$

$$= 221.$$

The following theorem shows the effect on det A of the row operations we use to solve linear systems. The proof is an easy consequence of the definition of det A (see Exercise 16).

■ B.5 THEOREM

Let A be a square matrix. Then:

(a) Multiplying a row, or a column, of A by a number r multiplies det A by r.

(b) Adding a multiple of one row to another leaves det A unchanged; likewise for columns.

(c) Interchanging two rows or two columns changes the sign of det A.

Thus, in putting a matrix in another form B, we just need to keep track of the row multipliers r and the sign changes that occur when rows are interchanged. Then k det A = det B, where k is plus or minus the product of the row multipliers. A little thought shows that the row operations just listed can be used to alter a matrix so that the first nonzero entry in each row is 1 and so that a column containing a 1 has all its other entries equal to zero. Such a matrix R is said to be in **reduced form,** and det R is either 0, 1, or -1.

e x a m p l e 16 Let

$$A = \begin{pmatrix} 1 & 3 & -2 \\ 2 & -4 & 1 \\ 3 & 5 & -2 \end{pmatrix}, \quad C = \begin{pmatrix} 1 & 3 & 0 \\ 2 & -4 & 5 \\ 3 & 5 & 4 \end{pmatrix}.$$

The third column of C is equal to the third column of A plus 2 times the first column. We compute

$$\det C = (1)(-16 - 25) - (3)(8 - 15) + (0)(10 + 12) = -20.$$

It follows that $\det A = -20$ also.

e x a m p l e 17 Let

$$A = \begin{pmatrix} 2 & 4 & -1 & 0 \\ 3 & 0 & 2 & 3 \\ -1 & 2 & 3 & 1 \\ 0 & 1 & -2 & -1 \end{pmatrix}.$$

By adding 2 times column 3 to column 1, and 4 times column 3 to column 2, we obtain

$$B = \begin{pmatrix} 0 & 0 & -1 & 0 \\ 7 & 8 & 2 & 3 \\ 5 & 14 & 3 & 1 \\ -4 & -7 & -2 & -1 \end{pmatrix},$$

and by Theorem B.5b, $\det A = \det B$. The expansion of $\det B$ by the first row has only one nonzero term, and we get

$$\det B = (-1) \det \begin{pmatrix} 7 & 8 & 3 \\ 5 & 14 & 1 \\ -4 & -7 & -1 \end{pmatrix}$$

$$= -\det \begin{pmatrix} 7 & 1 & 3 \\ 5 & 9 & 1 \\ -4 & -3 & -1 \end{pmatrix}$$

$$= -\det \begin{pmatrix} 0 & 1 & 0 \\ -58 & 9 & -26 \\ 17 & -3 & 8 \end{pmatrix}$$

Then $\det B = -(-1)((-58)(8) - (17)(-26)) = -22.$

The product of two square matrices is again a square matrix. It is a remarkable fact that the determinant of the product equals the product of the determinants of the individual matrices. The theorem is called the **product rule for determinants**.

The proof is fairly long, and we omit it.

■ B.6 PRODUCT RULE

If A and B are square matrices of the same size, then

$$\det (AB) = (\det A)(\det B).$$

A consequence of the product rule is that for an invertible matrix A,

$$\det (A^{-1}) = \frac{1}{\det A}.$$

There is a determinant formula for the inverse of an invertible matrix that generalizes the very simple Equation B.2 that holds for 2 by 2 matrices. The formula is reasonably efficient for simple 3 by 3 or even 4 by 4 matrices. For large matrices, it is usually preferable to use row operations to compute a determinant. However, the inversion formula allows us to see some general facts about determinants that are awkward to derive in other ways. We first prove a theorem from which the facts about inverses follow easily.

■ B.7 THEOREM

For an n by n matrix A,

$$\sum_{i=1}^{n} (-1)^{i+j} a_{ik} \det A_{ij} = \begin{cases} \det A & \text{if } k = j, \\ 0 & \text{if } k \neq j, \end{cases}$$

and

$$\sum_{i=1}^{n} (-1)^{i+j} a_{kj} \det A_{ij} = \begin{cases} \det A & \text{if } k = i, \\ 0 & \text{if } k \neq i. \end{cases}$$

Proof. Consider the expression

$$\sum_{i=1}^{n} (-1)^{i+j} x_i \det A_{ij},$$

where $x_1, \ldots, x_n$ may be any n numbers. We see that this is a certain determinant; in fact, it is the expansion, using the jth column, of the matrix obtained from A by replacing the jth column by $x_1, \ldots, x_n$. Now consider what happens if we take $x_1 = a_{1k}, x_2 = a_{2k}, \ldots, x_n = a_{nk}$, in other words, if we enter the elements from the kth column in the jth column. If $k = j$, we just get $\det A$. If $k \neq j$, we have the determinant of a matrix with two columns (the jth and kth) identical, so the result is zero. (Exercise 14.) This proves the first equation; the second is proved similarly by reversing the roles of row and column. ■

At this point we need the idea of transposing a square matrix, that is, reflecting it across its main diagonal. The matrix A and its **transpose** A^t are related as shown:

$$A = \begin{pmatrix} a_{11} & a_{12} & \cdots \\ a_{21} & a_{22} & \cdots \\ \vdots & \vdots & \end{pmatrix}; \qquad A^t = \begin{pmatrix} a_{11} & a_{21} & \cdots \\ a_{12} & a_{22} & \cdots \\ \vdots & \vdots & \end{pmatrix}.$$

Thus forming the transpose interchanges rows and columns.

Now the number $(-1)^{i+j} \det A_{ij}$ is the ijth cofactor of A; we'll abbreviate the cofactor $\tilde{a}_{ij}$ and write $\tilde{A}$ for the matrix with entries $\tilde{a}_{ij}$. Using this notation, we can interpret the previous theorem as a statement about the matrix products $\tilde{A}^t A$ and $A\tilde{A}^t$. In fact, we have

■■ **B.8** $\tilde{A}^t A = A\tilde{A}^t = (\det A)I,$

because the jkth entry in the product $\tilde{A}^t A$ is the product of the jth row of $\tilde{A}^t$ (i.e., the jth column of $\tilde{A}$) and the kth column of A, in other words, the sum $\Sigma_{i=1}^n \tilde{a}_{ij} a_{ik}$. But by Theorem B.7 this sum is equal to $\det A$ if $k = j$ and 0 otherwise. Hence $\tilde{A}^t A$ is equal to $(\det A)I$, a scalar multiple of the identity matrix. A similar calculation using the second equation in Theorem B.7 shows that $A\tilde{A}^t = (\det A)I$ also. Thus we have proved Formula B.8.

We can now easily write down a formula for A^{-1} when $\det A \neq 0$. A somewhat elaborated version of this theorem, described in Exercise 17, is called **Cramer's rule.**

■■ **B.9 THEOREM**

If $\det A \neq 0$, then A is invertible, and

$$A^{-1} = \frac{1}{\det A} \tilde{A}^t,$$

where $\tilde{A}^t$ is the transpose of the matrix of cofactors of A. Conversely, if A is invertible, then $\det A \neq 0$.

Proof. If $\det A \neq 0$, then Formula B.8 can be written

$$\left(\frac{1}{\det A} \tilde{A}^t \right) A = A \left(\frac{1}{\det A} \tilde{A}^t \right) = I.$$

Hence A^{-1} exists and equals $(\det A)^{-1}\tilde{A}^t$. Conversely, if A is invertible, then there is a matrix B such that $AB = BA = I$. By the product rule for determinants,

$$(\det A)(\det B) = \det I = 1,$$

so neither $\det A$ nor $\det B$ can be zero.   ■

---

example 18

We can use Theorem B.9 to compute the inverse of the matrix

$$A = \begin{pmatrix} 2 & 3 & 4 \\ 5 & 6 & 7 \\ 8 & 9 & 0 \end{pmatrix}.$$

The matrix with entries $\det A_{ij}$ is easily computed to be

$$\begin{pmatrix} -63 & -56 & -3 \\ -36 & -32 & -6 \\ -3 & -6 & -3 \end{pmatrix}.$$

To get the matrix of cofactors, we insert the factor $(-1)^{i+j}$. This changes the sign of every second entry, giving

$$\tilde{A} = \begin{pmatrix} 63 & 56 & 3 \\ 36 & -32 & 6 \\ -3 & -6 & -3 \end{pmatrix}.$$

The transposed matrix is

$$\widetilde{A}^t = \begin{pmatrix} -63 & 36 & -3 \\ 56 & -32 & 6 \\ -3 & 6 & -3 \end{pmatrix}.$$

The cofactors in $\widetilde{A}$ can be used to expand $\det A$ using the last column. We expand $\det A$ by its last column to get $\det A = 4(-3) + 7(6) + 0(-3) = 30$. Finally,

$$A^{-1} = \frac{1}{\det A} \widetilde{A}^t = \begin{pmatrix} -\frac{63}{30} & \frac{36}{30} & -\frac{3}{30} \\ \frac{56}{30} & -\frac{32}{30} & \frac{6}{30} \\ -\frac{3}{30} & \frac{6}{30} & -\frac{3}{30} \end{pmatrix}.$$

The following remarkable theorem is used in the discussion of the exponential matrix in Chapter 7, Section 4.

■ B.10 CAYLEY-HAMILTON THEOREM

Let

$$P(\lambda) = \det(A - \lambda I) = (-1)^n \lambda^n + d_{n-1}\lambda^{n-1} + \cdots + d_1\lambda + d_0$$

be the **characteristic polynomial** of the n by n matrix A. Then $P(A)$ is equal to the zero matrix; that is, $(-1)^n A^n + d_{n-1}A^{n-1} + \cdots + d_1 A + d_0 I = 0$.

Proof. Apply Theorem B.9 to the matrix $A - \lambda I$. It follows that if $P(\lambda) \neq 0$ then

$$P(\lambda)I = (A - \lambda I)(\widetilde{A - \lambda I})^t,$$

where the entries in the cofactor matrix are polynomials of degree at most $n - 1$ in λ. Note that $(\widetilde{A - \lambda I})^t$ can be written as a polynomial in λ with matrix coefficients B_k so that

$$P(\lambda)I = (A - \lambda I)(B_{n-1}\lambda^{n-1} + \cdots + B_1\lambda + B_0),$$

where the n by n matrices B_k have constant entries constructed from the entries of A. Now

$P(A) - P(\lambda)I$

$$= [(-1)^n A^n + d_{n-1}A^{n-1} + \cdots + d_0 I] - [(-1)^n\lambda^n I + d_{n-1}\lambda^{n-1}I + \cdots + d_0 I]$$

$$= (-1)^n(A^n - \lambda^n I) + d_{n-1}(A^{n-1} - \lambda^{n-1}I) + \cdots + d_1(A - \lambda I) = (A - \lambda I)Q(A).$$

We can factor out $(A - \lambda I)$ using the usual difference of powers factorization, because the powers of A commute. Now add our expressions above for $P(\lambda)I$ to both sides to get

$$P(A) = (A - \lambda I)[Q(A) + (\widetilde{A - \lambda I})^t].$$

On the right we have a polynomial with matrix coefficients of degree at least 1 in λ, unless the factor in brackets is zero. Since the left side doesn't contain λ, the factor in brackets must be the zero matrix, so $P(A)$ is the zero matrix. ■

It's worth pointing out that if a square matrix A is invertible, then the Cayley-Hamilton theorem can be used to write A^{-1} as a polynomial in A. For if

$$a_0 I + a_1 A + \cdots + a_{n-1}A^{n-1} \pm A^n = 0$$

is the characteristic equation of A with λ replaced by A, we can multiply by A^{-1} to get

$$a_0 A^{-1} + a_1 I + \cdots + a_{n-1} A^{n-2} \pm A^{n-1} = 0.$$

Now solve for A^{-1}, noting that $a_0 = \det A \neq 0$.

---

example 19

The matrix $A = \begin{pmatrix} 2 & 3 & 1 \\ 0 & 2 & 2 \\ 0 & 0 & 2 \end{pmatrix}$ has characteristic equation

$$(2 - \lambda)^3 = 8 - 12\lambda + 6\lambda^2 - \lambda^3 = 0,$$

so $8A^{-1} - 12I + 6A - A^2 = 0$,   or $8A^{-1} = 12I - 6A + A^2 = 0$.

Hence we need only divide by 8 after computing

$$8A^{-1} = 12 \begin{pmatrix} 1 & 0 & 0 \\ 0 & 1 & 0 \\ 0 & 0 & 1 \end{pmatrix} - 6 \begin{pmatrix} 2 & 3 & 1 \\ 0 & 2 & 2 \\ 0 & 0 & 2 \end{pmatrix} + \begin{pmatrix} 4 & 12 & 10 \\ 0 & 4 & 8 \\ 0 & 0 & 4 \end{pmatrix} = \begin{pmatrix} 4 & -6 & 4 \\ 0 & 4 & -4 \\ 0 & 0 & 4 \end{pmatrix}.$$

EXERCISES

1. Find AB, BA, and the determinants of A, B, AB, and BA when

   (a) $A = \begin{pmatrix} 1 & -2 \\ 3 & 1 \end{pmatrix}$, $B = \begin{pmatrix} 0 & 1 \\ 2 & -3 \end{pmatrix}$.

   (b) $A = \begin{pmatrix} 2 & 0 & 0 \\ 0 & 3 & 0 \\ 0 & 0 & 4 \end{pmatrix}$, $B = \begin{pmatrix} -1 & 0 & 1 \\ 2 & -1 & -3 \\ 0 & 3 & 5 \end{pmatrix}$.

2. For the matrices in Exercise 1b, what are $\det (2A)$ and $\det (2B)$?

3. A **diagonal matrix** D is a square matrix such that only the main diagonal entries $d_i = d_{ii}$ are allowed to be nonzero. We sometimes write $D = \text{diag } (d_1, \ldots, d_n)$. Show that $\det D = d_1 d_2 \cdots d_n$. In particular, $\det I = 1$.

4. What is the relation between $\det A$ and $\det (-A)$?

5. Verify the product rule for the pairs of matrices (a) and (b) in Exercise 1.

6. Apply the product rule to show that if A is invertible, then $\det A \neq 0$ and $(\det A^{-1}) = (\det A)^{-1}$.

7. Let A be an m by m matrix and B an n by n matrix. Consider the $(m + n)$ by $(m + n)$ matrix $\begin{pmatrix} A & 0 \\ 0 & B \end{pmatrix}$, which has A in the upper left corner, B in the lower right corner, and zeros elsewhere. Show that its determinant is equal to $(\det A)(\det B)$. [*Hint:* Consider the cases $A = I$ and $B = I$. Then use the product rule.]

8. Use the method of Example 17 of the text to evaluate

(a) $\det \begin{pmatrix} -1 & 0 & 1 & 2 \\ 0 & 1 & 2 & -1 \\ 1 & 2 & -1 & 0 \\ 2 & -1 & 0 & 1 \end{pmatrix}$.

(b) $\det \begin{pmatrix} 1 & 1 & 1 & 1 \\ 1 & 2 & 4 & 8 \\ 1 & 3 & 9 & 27 \\ 1 & 4 & 16 & 64 \end{pmatrix}$.

9. (a) Compute

$$\begin{pmatrix} 1 & 2 & 3 & 4 \\ 0 & -1 & 5 & 6 \\ 0 & 0 & 3 & -1 \\ 0 & 0 & 0 & 4 \end{pmatrix}.$$

(b) A matrix A, like the one in part (a), in which every element below the diagonal is 0, is said to be **upper triangular.** Show that if A is any triangular matrix, then $\det A$ is equal to the product of the diagonal elements.

10. Let A be the 3 by 3 matrix

$$\begin{pmatrix} 1 & 4 & 1 \\ 2 & 5 & 6 \\ -1 & 3 & 7 \end{pmatrix}.$$

(a) What is A^t?

(b) Compute $\det A$ and $\det A^t$.

11. Show that for an n by n matrix A

$$\det (A\tilde{A}^t) = \det (\tilde{A}^t A) = (\det A)^n.$$

12. (a) Show that if A, B, and C are n by n matrices such that

$$AB = CA = I$$

then $B = C$. [*Hint:* Multiply $CA = I$ by B on the right.]

(b) Show that if A and B are n by n matrices such that $AB = I$, then A and B are both invertible, with $A^{-1} = B$ and $B^{-1} = A$. [*Hint:* Use the product rule for determinants, together with part (a).]

13. Use Theorem B.9 to determine which of the following matrices have inverses and then use either Theorem B.9 or the remark following the Cayley-Hamilton theorem to find the inverses of the ones that are invertible.

(a) $\begin{pmatrix} 1 & 0 & 0 \\ 3 & 1 & 5 \\ -2 & 0 & 1 \end{pmatrix}$.

(d) $\begin{pmatrix} t & 0 & 0 \\ 0 & 2 & 0 \\ 0 & 0 & 1 \end{pmatrix}$, t real.

(b) $\begin{pmatrix} 1 & 2 & 3 \\ -1 & 1 & 0 \\ 0 & 3 & 3 \end{pmatrix}$.

(e) $\begin{pmatrix} 1 & 2 & 1 \\ 0 & 0 & 1 \\ 0 & 0 & 3 \end{pmatrix}$.

(c) $\begin{pmatrix} 2 & 4 & 8 \\ 1 & 0 & 0 \\ 1 & -3 & -7 \end{pmatrix}$.

(f) $\begin{pmatrix} 1 & -1 & 1 \\ 0 & -1 & 1 \\ 0 & 0 & 1 \end{pmatrix}$.

(g) $\begin{pmatrix} 1 & 2 & -1 & 3 \\ 0 & 2 & 0 & 1 \\ 0 & 0 & 1 & 1 \\ 0 & 0 & 0 & 4 \end{pmatrix}$.

(i) $\begin{pmatrix} 1 & 0 & 0 \\ 0 & e_t & te^t \\ 0 & 0 & e^t \end{pmatrix}$, t real.

(h) $\begin{pmatrix} 1 & 0 & 1 & 0 \\ 0 & 2 & 0 & 0 \\ 0 & 0 & 3 & 0 \\ 0 & 0 & 0 & 4 \end{pmatrix}$.

14. Show that if a square matrix A has two rows proportional, then $\det A = 0$. Show that the same result holds for columns.

15. Let A and E be identical square matrices except in some one row or column. Let C also be the same as A and B except that in that one row or column its entries are the sums of the corresponding entries in A and B. Show that $\det C = \det A + \det B$.

16. Prove (a), (b), and (c) of Theorem B.5. The first two follow immediately from expansion by minors of the distinguished row. The third is immediate for 2 by 2 matrices and can be proved by induction using expansion by minors of a row different from the two to be interchanged.

17. **Cramer's rule** says that if M is a square matrix with nonzero determinant, then the solution of the system $M\mathbf{x} = \mathbf{c}$ has its kth coordinate given by $x_k = \det M_k / \det M$, where the matrix M_k is obtained by replacing the entries in the kth column of M by the corresponding entries in the vector $\mathbf{c}$.

(a) Use Cramer's rule to solve $M\mathbf{x} = \mathbf{c}$, where

$$M = \begin{pmatrix} 1 & 3 & -2 \\ 2 & -4 & 1 \\ 3 & 5 & -2 \end{pmatrix} \quad \text{and} \quad \mathbf{c} = \begin{pmatrix} 1 \\ 2 \\ -2 \end{pmatrix}.$$

(b) Show that Cramer's rule is just a written-out version of Theorem B.9.

18. Find coefficients α and β, depending on a, b, c, and d, that express the inverse of the 2 by 2 matrix $A = \begin{pmatrix} a & b \\ c & d \end{pmatrix}$ in the form $A^{-1} = \alpha I + \beta A$. [*Hint:* By Theorem B.10, A satisfies its characteristic equation.]

Existence of Solutions

The Picard method described at the beginning of the first section below is based on a fundamental, and very useful, idea, the iterative solution of equations. The general idea behind the method has important applications in many branches of pure and applied mathematics. Proofs of the theorems stated below are quite technical, and in that way they exceed the demands made by the rest of the book. Instead of proving the theorems, the accompanying discussion aims simply at explaining in concrete terms the process of constructing solutions.

PICARD METHOD

In this section we take up an iterative method for finding approximate solutions to the vector differential equation with initial condition:

$$\frac{d\mathbf{x}}{dt} = \mathbf{F}(t, \mathbf{x}), \qquad \mathbf{x}(t_0) = \mathbf{x}_0.$$

The method has the advantage that it can be used to prove the existence of solutions under certain hypotheses.

The first step in applying the Picard method to the preceding vector differential equation is to replace it by an equivalent vector **integral equation:**

$$\mathbf{x}(t) = \mathbf{x}_0 + \int_{t_0}^{t} \mathbf{F}(u, \mathbf{x}(u)) \, du.$$

It is easy to check (see Exercise 8) that if $\mathbf{x} = \mathbf{x}(t)$ is a solution of either equation, then it is a solution of the other, in particular because the initial condition

$\mathbf{x}(t_0) = \mathbf{x}_0$ is satisfied by the integral equation. The vector integral is, of course, computed by integrating the individual coordinate functions of the integrand. Next we compute a sequence of approximations to the desired solution by substituting each approximation into the next integrand:

$$\mathbf{x}_1(t) = \mathbf{x}_0 + \int_{t_0}^{t} \mathbf{F}(u, \mathbf{x}_0)\, du,$$

$$\mathbf{x}_2(t) = \mathbf{x}_0 + \int_{t_0}^{t} \mathbf{F}(u, \mathbf{x}_1(u))\, du,$$

$$\vdots$$

$$\mathbf{x}_{k+1}(t) = \mathbf{x}_0 + \int_{t_0}^{t} \mathbf{F}(u, \mathbf{x}_k(u))\, du.$$

To get some feeling for the kind of approximation that the method produces, we take up an example for which a solution formula can be found by ordinary methods.

---

e x a m p l e 1

The real-valued differential equation $dx/dt = x + 1$ is easily seen to have the solution $x(t) = -1 + e^t$, satisfying the initial condition $x(0) = 0$. Let's arrive at the solution using the Picard method. Since $t_0 = 0$, $x_0 = 0$, and $F(t, x) = 1 + x$, we find

$$x_1(t) = \int_0^t 1\, du = t,$$

$$x_2(t) = \int_0^t (1 + u)\, du = t + \frac{t^2}{2},$$

$$x_3(t) = \int_0^t \left(1 + u + \frac{u^2}{2}\right) du = t + \frac{t^2}{2} + \frac{t^3}{3 \cdot 2},$$

$$\vdots$$

$$x_k(t) = \int_0^t \left(1 + u + \frac{u^2}{2} + \cdots + \frac{u^{k-1}}{(k-1)!}\right) du = t + \frac{t^2}{2} + \cdots + \frac{t^k}{k!}.$$

If we let $k \to \infty$, we get

$$\lim_{k \to \infty} x_k(t) = \lim_{k \to \infty} \left(t + \frac{t^2}{2} + \cdots + \frac{t^k}{k!}\right)$$

$$= \sum_{k=1}^{\infty} \frac{t^k}{k!} = -1 + e^t.$$

Thus, the kth Picard approximation happens in this example to be just the first k terms in the power series expansion of the solution.

---

e x a m p l e 2

The nonlinear system with initial conditions

$$\dot{x} = xy, \qquad x(0) = 1,$$

$$\dot{y} = x + y, \qquad y(0) = 1,$$

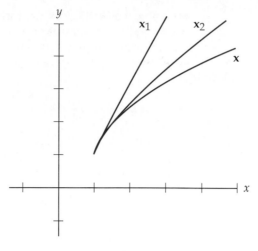

Figure C.1

has Picard approximations given by the formulas

$$x_{k+1}(t) = 1 + \int_0^t x_k(u)y_k(u)\ du,$$

$$y_{k+1}(t) = 1 + \int_0^t (x_k(u) + y_k(u))\ du.$$

The first approximation, made using $(x_0, y_0) = (1, 1)$, is

$$x_1(t) = 1 + \int_0^t 1\ du = 1 + t,$$

$$y_1(t) = 1 + \int_0^t (1 + 1)\ du = 1 + 2t.$$

The next step gives

$$x_2(t) = 1 + \int_0^t (1 + u)(1 + 2u)\ du = 1 + t + \frac{3t^2}{2} + \frac{2t^3}{3},$$

$$y_2(t) = 1 + \int_0^t (1 + u) + (1 + 2u)\ du = 1 + 2t + \frac{3t^2}{2}.$$

We sketch the *trajectories* $\mathbf{x}_k(t) = (x_k(t), y_k(t))$ of the two approximations, which are shown in Figure C.1 along with a computer approximation to the solution itself.

The Picard method can of course be applied to differential equations of order higher than one by first reducing such equations to first-order systems. Thus solving an initial-value problem for the system of Example 2 is equivalent to solving an initial-value problem for the single second-order equation $\ddot{y} - \dot{y} - xy = 0$. The solutions to the nonlinear system are not available in terms of elementary functions. However, the Picard approximations to the associated initial-value problems converge to a solution, thus showing that solutions do exist. Proving this convergence is the standard method for proving existence in the first two theorems of the next section.

EXISTENCE THEOREMS

Without some restriction on the nature of the function $\mathbf{F}(t, \mathbf{x})$, we cannot prove that the Picard iterates $\mathbf{x}_k(t)$ converge to a solution $\mathbf{x}(t)$ of $d\mathbf{x}/dt = \mathbf{F}(t, \mathbf{x})$. All such restrictions that have been used have the common feature that they prevent the values of $\mathbf{F}$ from changing too fast in response to changes in $\mathbf{x}$. The most widely used restriction is the assumption that the coordinate functions F_j have continuous partial derivatives with respect to the real coordinates x_k in $\mathbf{x}$. Under such a condition the Picard iterates converge to a unique solution of the differential equation with prescribed initial value. The precise statement follows.

■ C.1 THEOREM

Suppose that $\mathbf{F}(t, \mathbf{x})$ and its partial derivatives with respect to the variables x_k in $\mathbf{x}$ are continuous on an open rectangle in $\mathscr{R}^{n+1}$ containing $(t_0, \mathbf{x}_0)$. Then the initial-value problem

$$\dot{\mathbf{x}} = \mathbf{F}(t, \mathbf{x}), \qquad \mathbf{x}(t_0) = \mathbf{x}_0,$$

has a unique solution defined on some t interval containing t_0.

example 3

If F and x are real-valued, with $F(t, x) = 1 + x^2$, then $F_x(t, x) = 2x$. If we take $x(t) = \tan t$, we see that we get a solution for $-\pi/2 < t < \pi/2$, but in no larger interval containing $t = 0$.

example 4

If $\dot{\mathbf{x}} = \mathbf{F}(t, \mathbf{x})$ is linear, then $\mathbf{F}$ has the form

$$\mathbf{F}(t, \mathbf{x}) = A(t)\mathbf{x} + \mathbf{b}(t),$$

with coordinate functions

$$F_j(t, x_1, \ldots, x_n) = \sum_{k=1}^{n} a_{jk}(t)x_k + b_j(t).$$

The partial derivatives in question are $(\partial F_j/\partial x_k)(t, \mathbf{x}) = a_{jk}(t)$, so checking the hypothesis of Theorem C.1 is just a matter of verifying that the entries in the matrix $A(t)$ are continuous on some common interval; if these entries are constant, then Chapter 7 shows more, namely, that all solutions are expressible using elementary formulas.

For linear systems, the conclusion of Theorem C.1 can be improved so that the domain of the solutions coincides with the domain of continuity of the coefficients, rather than shrinking as may happen for nonlinear systems. The result is fundamentally important if the coefficient matrix has variable coefficients.

■ C.2 THEOREM

If the matrix $A(t)$ and vector function $\mathbf{b}(t)$ have continuous entries on a common interval $a < t < b$, then the initial-value problem

$$\dot{\mathbf{x}} = A(t)\mathbf{x} + \mathbf{b}(t), \qquad \mathbf{x}(t_0) = \mathbf{x}_0,$$

has for each t_0 in the interval a unique solution defined on the entire interval $a < t < b$.

---

example 5

The result of Theorem C.1 appears very concretely in the case of dimension $n = 1$. The exponential integrating factor $M(t)$ of Chapter 1, Section 5, provides an explicit solution formula. The fundamental matrix $X(t)$ of Chapter 7, Section 4C, serves as a substitute in higher dimensions to $M(t)$ but is usually too hard to compute.

An **Euler polygon** for an initial-value problem is a concatenation of line segments joining successive Euler approximation points. See Figure 2.8 in Chapter 2 for dimension 1. The definition applies to first-order normal-form systems of any dimension. An analysis replacing Picard approximations by Euler polygons shows that the assumption about $\mathbf{F}(t, \mathbf{x})$ can be weakened to continuity alone if all we want is existence of solutions on some restricted interval. Thus it might seem that the continuity requirement on partial derivatives of $\mathbf{F}$ is somewhat artificial. However, the stronger assumption is usually appropriate, because some such condition is necessary anyway to guarantee uniqueness.

It would be nice if continuity of $\mathbf{F}$ implied the convergence of Euler polygons to a solution graph of $\dot{\mathbf{x}} = \mathbf{F}(t, \mathbf{x})$, $\mathbf{x}(t_0) = \mathbf{x}_0$ as the time difference between approximation points tends to zero. The truth is less pleasant, but in a way more interesting, as the following theorem shows.

■ C.3 PEANO'S THEOREM

Let $\mathbf{F}(t, \mathbf{x})$ be continuous for t in an interval containing $t_0 \leq t \leq b$ and all $\mathbf{x}$ in some neighborhood of $\mathbf{x}_0$. Let $\mathbf{x}_k(t)$ be as sequence of Euler polygons for the initial-value problem

$$\dot{\mathbf{x}} = \mathbf{F}(t, \mathbf{x}), \qquad \mathbf{x}(t_0) = \mathbf{x}_0,$$

with maximum time difference tending to zero. Then a *subsequence* of $\mathbf{x}_k(t)$ converges uniformly to a solution for $t_0 \leq t \leq b$.

EXERCISES

1. For each initial-value problem, write an equivalent integral equation.

(a) $\dot{x} = tx + t^2$, $x(0) = 1$.

(b) $\dot{x} = 1$, $x(0) = 2$.

(c) $\dot{x} = x - t$, $x(0) = 0$.

2. For each equation given in Exercise 1, find the Picard approximate solutions $x_1(t)$ and $x_2(t)$. Start with $x_0(t) \equiv x(0)$ in each case.

3. **(a)** Find a vector integral equation equivalent to the system
$$\dot{x} = ty, \qquad x(0) = 1,$$
$$\dot{y} = tx, \qquad y(0) = 2.$$

(b) Find the Picard approximate solution $(x_1(t), y_1(t))$ by starting with $(x(0), y(0)) = (1, 2)$.

(c) Find the Picard approximate solution $(x_2(t), y_2(t))$ from the approximation found in part (b).

(d) Sketch the trajectory of the approximate solution found in part (c).

4. **(a)** The second-order differential equation
$$\frac{d^2y}{dt^2} + ty = 0, \qquad y(0) = 0, \qquad \dot{y}(0) = 1$$

is equivalent to a first-order system of differential equations. What is that system?

(b) Find a pair of integral equations equivalent to the system found in part (a).

(c) Starting with the constant initial solution, find the next two Picard approximate solutions and pick out the relevant part as an approximate solution to the second-order differential equation in part (a).

5. Which of the following problems have solutions? Justify your answers.

(a) $\dot{x} - F(t, x) = t^2 + x^2, 0 \le t \le 1, 0 \le x < 1.$

(b) $\dot{x} = x^2, -\infty < t < \infty, -\infty < x < \infty.$

(c) $\dot{x} = tx + y^2, \dot{y} = x + y, 0 \le t \le 1, 0 \le x, 0 \le y \le 1.$

6. Let A be an n by n constant matrix. Prove that the Picard iterates $x_n(t)$ for the differential equation $dx/dt = Ax$ are just the partial sums of the power series expansion of the matrix exponential e^{tA} applied to the initial vector x_0.

7. Sketch the exact solution graph along with the graphs of the first three Picard approximations found in Example 1 of the text.

8. Verify that the initial-value problem and integral equation
$$\dot{\mathbf{x}} = \mathbf{F}(t, \mathbf{x}), \qquad \mathbf{x}(t_0) = \mathbf{x}_0 \qquad \text{and} \qquad \mathbf{x}(t) = \mathbf{x}_0 + \int_{t_0}^{t} \mathbf{F}(u, \mathbf{x}(u)) \, du$$

have the same solutions if $\mathbf{F}$ is continuous:

(a) Show by using one version of the fundamental theorem of calculus that a solution of the initial-value problem is also a solution of the integral equation.

(b) Show by using the other version of the fundamental theorem of calculus that a solution of the integral equation is necessarily differentiable and solves the initial-value problem.

9. Just by looking at the direction field of $dx/dt = \sqrt{1 - x^3}$ it's hard to guess whether the graph actually touches the horizontal line $x = 1$ or is only asymptotic to it.

(a) Show that, for a given value of x, a solution $x(t)$ of $dx/dt = \sqrt{1 - x^3}$ always increases at least as fast as a solution to $dx/dt = \sqrt{1 - x}$. [*Hint:* Note that $1 - x^3 = (1 + x + x^2)(1 - x)$.]

(b) Show that $dx/dt = \sqrt{1 - x}$ has solutions $x(t) = 1 - (t - c)^2/4$.

(c) Show that all solution graphs of $dx/dt = \sqrt{1 - x^3}$ have a point on the line $x = 1$. [*Hint:* Compare, without computing, the integrals of $(1 - x)^{-1/2}$ and $(1 - x^3)^{-1/2}$.]

*10. It's easy to verify that the initial-value problem

$$\frac{dx}{dt} = \sqrt{x}, \qquad x(0) = 0$$

has two different solutions, $x(t) = \frac{1}{4}t^2$ and $x(t) = 0$ for all t and that Picard iteration sequences are generated, for various choices of the initial estimate $x_0(t)$, by

$$x_{n+1}(t) = \int_0^t \sqrt{x_n(u)} \, du.$$

So far we've considered starting the iteration only with the constant $x_0(t) = x(0)$, but here we'll make another choice in parts (b) through (f).

(a) Show that starting with $x_0(t) = 0$ for all t makes $x_n(t) = 0$ for all t and $n = 1, 2, 3, \ldots$ so that the limit of this sequence is the identically zero solution.

(b) Show that starting with $x_0(t) = t$ for all $t \geq 0$ produces a sequence of the form $x_n(t) = c_n t^{r_n}$.

(c) Show that the exponents in part (b) are $r_n = (2^{n+1} - 1)/2^n$ and that $\lim_{n \to \infty} r_n = 2$.

(d) Show that the coefficients c_n in part (b) satisfy

$$c_n = \frac{2^n \sqrt{c_{n-1}}}{2^{n+1} - 1}$$

(e) Show that the coefficients c_n in part (b) satisfy $\frac{1}{4} \leq c_n \leq c_{n-1}$, $n = 2, 3, 4, \ldots$, and so conclude that $\lim_{n \to \infty} c_n$ exists and is at least $\frac{1}{4}$. [*Hint:* Use induction and part (d).]

(f) Show that the sequence c_n has limit $\frac{1}{4}$ and hence that the sequence $c_n t^{r_n}$ tends to the corresponding value of the nontrivial solution $\frac{1}{4}t^2$. [*Hint:* Let n tend to infinity in the equation of part (d).]

*11. Carry out the analog of parts (b) through (f) of the previous problem for the initial choice $x_0(t) = t^2$. All other assumptions are to remain the same. (This problem is really easier than the previous one, except that you're not given as much detailed guidance here.)

12. It can be shown that the 1-dimensional problem $\dot{x} = x/|x|^{3/4} + t \sin(\pi/t)$, $x(0) = 0$, has Euler polygons that fail to converge as the time steps tend to zero. (See chapter 1, exercise 12, in E. A. Coddington and N. Levinson, *Theory of Ordinary Differential Equations*, Krieger, 1984.)

(a) Show that Theorem C.3 implies the existence of a solution in a neighborhood of $t = 0$.

(b) Show that Theorem C.1 fails to apply.

D

Indefinite Integrals

The table at the end of this appendix lists some frequently occurring integrals. As a supplement to the table it's often useful to use a symbolic calculator that provides some indefinite integrals. If you don't see how to compute an indefinite integral directly and don't find it in a table or from a calculator, you can try one of the following techniques. Integration constants are omitted, since they're not the main issue here.

IDENTITY SUBSTITUTIONS

Rewriting the integrand using an algebraic, trigonometric, exponential, or logarithmic identity will sometimes convert an apparently intractable integrand into an amenable one.

***

example 1   The integral $\int (e^x + e^{3x})^2\, dx$ can be rewritten by squaring out the binomial to get

$$\int (e^x + e^{3x})^2\, dx = \int (e^{2x} + 2e^{4x} + e^{6x})\, dx$$

$$= \tfrac{1}{2}e^{2x} + \tfrac{1}{2}e^{4x} + \tfrac{1}{6}e^{6x}.$$

669

example 2

To integrate $\cos^2 x$ recall the trigonometric identity $\cos 2x = 2 \cos^2 x - 1$, which is equivalent to $\cos^2 x = \frac{1}{2}(1 + \cos 2x)$. Thus formula 34 in the table follows from

$$\int \cos^2 x \, dx = \frac{1}{2} \int (1 + \cos 2x) \, dx = \frac{1}{2}x + \frac{1}{4} \sin 2x.$$

Identities to facilitate the integration of rational functions $P(x)/Q(x)$, where $P(x)$ and $Q(x)$ are polynomials, are derived by **partial-fraction decomposition,** described in detail in Chapter 8, Section 1, on Laplace transforms.

example 3

A partial-fraction decomposition is used to compute formula 6 in the table:

$$\int \frac{1}{(x - a)(x - b)} \, dx = \int \left(\frac{1}{(a - b)(x - a)} - \frac{1}{(a - b)(x - b)} \right) dx$$

$$= \frac{\ln |x - a|}{a - b} - \frac{\ln |x - b|}{a - b} = \frac{1}{a - b} \ln \left| \frac{x - a}{x - b} \right|.$$

SUBSTITUTION FOR THE INTEGRATION VARIABLE

Awkwardness in an integrand can sometimes be circumvented by a substitution. If the given integration variable is x, a substitution $x = g(u)$, $dx = g'(u) \, du$ may simplify the integrand enough that the corresponding integral in u can be computed. Then replace u by its equivalent in terms of x using the inverse relation $u = h(x)$, where $g(h(x)) = x$.

example 4

An awkward occurrence of $\sqrt{x}$ can be circumvented by letting $x = u^2$, $dx = 2u \, du$. For example,

$$\int \frac{dx}{\sqrt{x} + 1} = \int \frac{2u \, du}{u + 1} = 2 \int \left(1 - \frac{1}{u + 1} \right) du.$$

The last step comes from division of u by $u + 1$. Now integrate with respect to u and reintroduce x using the inverse relation $u = \sqrt{x}$ to get

$$\int \frac{dx}{\sqrt{x} + 1} = 2(u - \ln (u + 1)) = 2(\sqrt{x} - \ln (\sqrt{x} + 1)).$$

example 5

To integrate $\sqrt{1 - x^2}$, set $x = \sin u$, $dx = \cos u \, du$. Then

$$\int \sqrt{1 - x^2} \, dx = \int \sqrt{1 - \sin^2 u} \cos u \, du$$

$$= \int \cos^2 u \, du = \frac{1}{2}u + \frac{1}{4} \sin 2u, \qquad \text{by Example 2.}$$

Since $\sin 2u = 2 \sin u \cos u$, we find $\int \sqrt{1 - x^2} \, dx = \frac{1}{2} \arcsin x + \frac{1}{2}x\sqrt{1 - x^2}.$

SUBSTITUTION FOR A PART OF THE INTEGRAND

Here we try to write a given integral in the form $\int f(g(x))g'(x)\,dx$ and set $u = g(x)$, $du = g'(x)\,dx$ in the hope that we can compute the indefinite integral $F(u) = \int f(u)\,du$. If $F'(u) = f(u)$, the result is

$$\int f(g(x))g'(x)\,dx = \int f(u)\,du = F(g(x)).$$

e x a m p l e 6

In the integral $\int \cos^2 x \sin x \,dx$ we note the square of a function, namely, $g(x) = \cos x$, multiplied by a function, $\sin x$, which is easily modified to be the derivative $g'(x) = -\sin x$. By including the constant factor -1 in the integrand and compensating with a minus sign before the integral, we rewrite the integral as

$$\int \cos^2 x \sin x\,dx = -\int (\cos x)^2(-\sin x)\,dx.$$

It's now natural to think of substituting u for $g(x) = \cos x$ and du for $g'(x) = (-\sin x)\,dx$ to get

$$\int \cos^2 x \sin x\,dx = -\int u^2\,du$$

$$= -\tfrac{1}{3}u^3 = -\tfrac{1}{3}\cos^3 x.$$

INTEGRATION BY PARTS

This technique is one of the most important because of its frequent use in deriving other general formulas; it is embodied in the formula

$$\int f(x)g'(x)\,dx = f(x)g(x) - \int f'(x)g(x)\,dx,$$

which follows from the product rule for differentiation. To apply the method, you need to recognize the integrand of a given integral as a product of two functions; one of them, $f(x)$, you differentiate and the other one, $g'(x)$, you try to identify as a function you can integrate easily. If one choice for $f(x)$ and $g'(x)$ fails to work, you may want to try another. Formulas 21, 23, and 27 in the table can be computed by a single application of integration by parts, and formulas 25, 28, and 33 by repeated integration by parts.

e x a m p l e 7

In $\int x \sin x \,dx$ a good choice is $f(x) = x$ and $g'(x) = \sin x$, because $f'(x) = 1$ simplifies the remaining integration, and $g(x) = -\cos x$ is easy to integrate. Thus

$$\int x \sin x\,dx = (x)(-\cos x) - \int (1)(-\cos x)\,dx$$

$$= -x \cos x + \sin x.$$

INTEGRAL TABLE

1. $\int (ax + b)^n\, dx = \dfrac{1}{a(n+1)}\, (ax+b)^{n+1}, \qquad n \neq -1$

2. $\int \dfrac{dx}{ax+b} = \dfrac{1}{a}\, \ln|ax+b|$

3. $\int \dfrac{x\, dx}{ax+b} = \dfrac{x}{a} - \dfrac{b}{a^2}\, \ln|ax+b|$

4. $\int \dfrac{x\, dx}{(ax+b)^2} = \dfrac{b}{a^2(ax+b)} + \dfrac{1}{a^2}\, \ln|ax+b|$

5. $\int x(ax+b)^n\, dx = \dfrac{1}{a^2(n+2)}\, (ax+b)^{n+2} - \dfrac{b}{a^2(n+1)}\, (ax+b)^{n+1},$
$$n \neq -1,\ -2$$

6. $\int \dfrac{1}{(x-a)(x-b)}\, dx = \dfrac{1}{a-b}\, \ln\left|\dfrac{x-a}{x-b}\right|$

7. $\int \dfrac{dx}{(ax+b)(cx+d)} = \dfrac{1}{ad-bc}\, \ln\left|\dfrac{ax+b}{cx+d}\right|, \qquad ad-bc \neq 0$

8. $\int \dfrac{x\, dx}{(ax+b)^3} = \dfrac{b}{2a^2(ax+b)^2} - \dfrac{1}{a^2(ax+b)}$

9. $\int \dfrac{dx}{(x^2-a^2)(x^2-b^2)} = \dfrac{1}{2b(b^2-a^2)}\, \ln\left|\dfrac{x-b}{x+b}\right| + \dfrac{1}{2a(a^2-b^2)}\, \ln\left|\dfrac{x-a}{x+a}\right|,$
$$a \neq b,\ a,\ b \neq 0$$

10. $\int \dfrac{dx}{x^2(x^2-a^2)} = \dfrac{1}{a^2 x} + \dfrac{1}{2a^3}\, \ln\left|\dfrac{x-a}{x+a}\right|, \qquad a \neq 0$

11. $\int x\sqrt{ax+b}\, dx = \dfrac{2(3ax-2b)}{15a^2}\, (ax+b)^{3/2}$

12. $\int \dfrac{x\, dx}{\sqrt{ax+b}} = \dfrac{2(ax-2b)}{3a^2}\, \sqrt{ax+b}$

13. $\int \dfrac{dx}{(x+a)^2+b^2} = \dfrac{1}{b}\, \arctan \dfrac{x+a}{b}$

14. $\int \dfrac{dx}{a^2-x^2} = \dfrac{1}{2a}\, \ln\left|\dfrac{a+x}{a-x}\right|$

15. $\int \dfrac{dx}{x(ax^2+b)} = \dfrac{1}{2b}\, \ln\left|\dfrac{x^2}{ax^2+b}\right|$

16. $\int \sqrt{x^2 \pm p^2}\, dx = \tfrac{1}{2}x\sqrt{x^2 \pm p^2} \pm \tfrac{1}{2}p^2\, \ln|x + \sqrt{x^2 \pm p^2}|$

17. $\int \sqrt{p^2 - x^2}\, dx = \tfrac{1}{2}x\sqrt{p^2 - x^2} + \tfrac{1}{2}p^2\, \arcsin \dfrac{x}{p}$

18. $\int \dfrac{dx}{\sqrt{x^2 \pm p^2}} = \ln|x + \sqrt{x^2 \pm p^2}|$

19. $\displaystyle\int \frac{dx}{\sqrt{p^2 - x^2}} = \arcsin \frac{x}{p}$

20. $\displaystyle\int e^{ax}\, dx = \frac{1}{a}\, e^{ax}$

21. $\displaystyle\int xe^{ax}\, dx = \frac{1}{a^2}\, (ax - 1)e^{ax}$

22. $\displaystyle\int x^2 e^{ax}\, dx = \frac{1}{a^3}\, (a^2 x^2 - 2ax + 2)e^{ax}$

23. $\displaystyle\int \ln ax\, dx = x \ln ax - x$

24. $\displaystyle\int x \ln ax\, dx = \tfrac{1}{2}x^2 \ln ax - \tfrac{1}{4}x^2$

25. $\displaystyle\int x^2 \ln ax\, dx = \tfrac{1}{3}x^3 \ln ax - \tfrac{1}{9}x^3$

26. $\displaystyle\int \sin ax\, dx = -\frac{1}{a}\cos ax$

27. $\displaystyle\int x \sin ax\, dx = \frac{1}{a^2}\sin ax - \frac{1}{a}x \cos ax$

28. $\displaystyle\int x^2 \sin ax\, dx = \frac{2}{a^2}x \sin ax + \frac{2}{a^3}\cos ax - \frac{1}{a}x^2 \cos ax$

29. $\displaystyle\int \sin^2 ax\, dx = \frac{x}{2} - \frac{\sin 2ax}{4a}$

30. $\displaystyle\int \sin^3 ax\, dx = -\frac{1}{a}\cos ax + \frac{\cos^3 ax}{3a}$

31. $\displaystyle\int \cos ax\, dx = \frac{1}{a}\sin ax$

32. $\displaystyle\int x \cos ax\, dx = \frac{1}{a^2}\cos ax + \frac{1}{a}x \sin ax$

33. $\displaystyle\int x^2 \cos ax\, dx = \frac{2}{a^2}x \cos ax - \frac{2}{a^3}\sin ax + \frac{1}{a}x^2 \sin ax$

34. $\displaystyle\int \cos^2 ax\, dx = \frac{x}{2} + \frac{\sin 2ax}{4a}$

35. $\displaystyle\int \cos^3 ax\, dx = \frac{1}{a}\sin ax - \frac{\sin^3 ax}{3a}$

36. $\displaystyle\int \sin ax \sin bx\, dx = \frac{\sin (a - b)x}{2(a - b)} - \frac{\sin (a + b)x}{2(a + b)}, \qquad |a| \neq |b|$

37. $\displaystyle\int \cos ax \cos bx\, dx = \frac{\sin (a - b)x}{2(a - b)} + \frac{\sin (a + b)x}{2(a + b)}, \qquad |a| \neq |b|$

38. $\displaystyle\int \sin ax \cos bx\, dx = -\frac{\cos (a - b)x}{2(a - b)} - \frac{\cos (a + b)x}{2(a + b)}, \qquad |a| \neq |b|$

39. $\displaystyle\int \tan ax\, dx = -\frac{1}{a} \ln \cos ax$

40. $\displaystyle\int \tan^2 ax\, dx = \frac{1}{a} \tan ax - x$

41. $\displaystyle\int \tan^3 ax\, dx = \frac{1}{2a} \tan^2 ax + \frac{1}{a} \ln \cos ax$

42. $\displaystyle\int \sec ax\, dx = \frac{1}{a} \ln \tan \left(\frac{ax}{2} + \frac{\pi}{4} \right)$

43. $\displaystyle\int \sec^2 ax\, dx = \frac{1}{a} \tan ax$

44. $\displaystyle\int \sec^3 ax\, dx = \frac{1}{2a} \tan ax \sec ax + \frac{1}{2a} \ln \tan \left(\frac{ax}{2} + \frac{\pi}{4} \right)$

45. $\displaystyle\int \tan ax \sec ax\, dx = \frac{1}{a} \sec ax$

46. $\displaystyle\int e^{ax} \sin bx\, dx = \frac{e^{ax}}{a^2 + b^2} (a \sin bx - b \cos bx)$

47. $\displaystyle\int e^{ax} \cos bx\, dx = \frac{e^{ax}}{a^2 + b^2} (a \cos bx + b \sin bx)$

ANSWERS TO SELECTED EXERCISES

Answers to exercises are provided as a spot check on the correctness and accuracy of work, and not as definitive validations. It's good to maintain the habit of checking results both for general reasonableness and for precision; the latter can usually be done routinely after solving a differential equation by substitution into the equation, while the former depends on the development of sound reasoning, particularly when it comes to the interpretation of a formal solution. Many exercises in this book include as integral parts of their statements what amount to answers; for many others, brief answers are provided here as the outcome of more elaborate work. For numbered exercises with several similar parts, the general, though not invariable, practice is to give an answer for part (a) and then for alternate parts thereafter.

Chapter 1

Section 1

1. **(a)** Yes   **(c)** Yes   **(e)** Yes   **(g)** Yes   **(i)** Yes   **(k)** Yes   **(m)** Yes   **(o)** Yes
2. **(b)** $x(t) = 0$
3. **(f)** $y = \tan(x/4)$ **(g)** $y = -\sqrt{2 - x^2}$ **(h)** $y = \sqrt{(x^2 - 1)/3}$ **(i)** $y = -\sqrt{3x}$ **(j)** $y = e^{1/x}$
4. **(f)** $y' = (1 + y^2)(\arctan y)/x$ **(g)** $y' = x/y$ **(h)** $y' = xy/(x^2 - 1)$ **(i)** $y' = y/(2x)$
   (j) $y' = (-y \ln y)/x$
12. **(a)** $b = -(a/A)\sqrt{g/2}, \ c = \sqrt{h(0)}$ **(b)** $(A/a)\sqrt{2h_0/g}$ time units; about 7.4 min

Sections 2A, B

10. **(a)** Always concave up   **(c)** Concave up when $y > -x - 1$
    (e) Concave up when $k\pi < y < (k + \frac{1}{2})\pi$, k integer   **(g)** Concave up when $\cos x > 0$ **(h)** Straight lines
13. **(c)** $y = -x - 1$ **(d)** $y = x; \ y = -x$
17. **(a)** $y' = y/x$ **(c)** $y' = -y/x$ **(e)** $y' = 3y/x$ **(g)** $y' = y \cot x$ **(i)** $y' = -\sin x$ **(k)** $y' = -y/x$

Section 2C

1. **(a)** $R = S$ **(c)** S is union of first and third quadrants; for R, delete the axes   **(e)** $R = S$
   (g) $R = S = \mathcal{R}^2$ **(i)** $R = S = \mathcal{R}^2$
4. **(b)** $y_1 = \begin{cases} (x - x_0)^2/4, & x > x_0, \\ 0, & x \le x_0, \end{cases} y_2 = 0$ 13. **(d)** Yes; Theorem 2.1 applies to the entire xy plane

Section 3A

1. **(a)** $y = x^2/2 - x^3/3 + 1$ **(c)** $y = -\frac{1}{2}\ln(1 - x^2) + 1$ **(e)** $y = -\sin x + 2x + 1$ **(g)** $z = (t - 1)e^t + 2$
   (i) $x = e^t - t$
8. $y_{\max} \approx 388{,}199$ ft at about $t = 155$ s   9. $v \approx 567$ ft/s at about $t = 18$ s
10. **(a)** 4984 ft after 1 s   **(b)** $v_0 = 401$ ft/s   12. About 80.25 ft/s   13. About 60 ft/s   15. -50 ft/s^2
16. **(a)** About 2.9 s   **(b)** About 129 ft   **(c)** About 25 ft/s^2; 155 ft
18. **(a)** $v_0 = \sqrt{g} \approx 5.67$ ft/s   **(b)** About 3.1 ft   19. **(a)** About 88.89 ft/s^2 **(b)** About 133 ft/s
21. **(a)** $66\frac{2}{3}$ ft/s^2 **(b)** About 621 ft   30. **(c)** $x = (3 \pm \sqrt{3})L/6$

Section 3B

1. $y = \sqrt{\frac{2}{3}x^3 - \frac{13}{3}}$ 3. $y = \sqrt{2e^{-x} - 1}$ 5. $v = \arccos(1 - \sin u), 0 < u < \pi$ 18. $16a^2t^2/A^2$ feet
19. **(c)** $(A/(ka))\sqrt{2h_0/g}$ time units; 7.4 min   **(d)** $k \approx 0.74$
20. **(a)** $ka(t) = 100/\sqrt{2g(60 - t)}; ka(0) \approx 1.6$ ft^2; $ka(30) \approx 2.28$ ft^2 23. **(a)** $A = \pi(2h - h^2)$ **(b)** 33.6 s
24. **(c)** $3/3200 \approx 0.0009$ in^2 **(d)** $r = 3\sqrt{6/\pi} \approx 4.1$ in   25. $r = \sqrt[3]{3Ct/(4\pi)}$ 28. **(a)** $z = z_0/\sqrt{1 + 2z_0^2 t}$
29. **(a)** $y(x)$ a; graph increasing   **(b)** $y(x) > x^2/2$; graph increasing and concave up
   (c) $y \ge e^x, x \ge 0; y \le e^x, x < 0$ **(d)** $x^3/6 + x \le y(x) \le x^3/3 + x$ **(e)** $y(x) \ge e^x, x \ge 0; 0 < y \le e^x, x < 0$
   (f) Graph concave down, and $y(x) \le x$
31. **(a)** $y = (-1 + Ke^{2x})/(1 + Ke^{2x}), K \ne 0$

Chapter 1

Section 3C

1. $y = ke^x - 2x - 5$ **3.** $\arctan(2y/x - 1)/\sqrt{3} = (\sqrt{3}/2)\ln|x| + C$ **5.** $\ln(x^7 + y^2) + 2\arctan(y/x) + C$
7. **(a)** $\ln|x - 1 + \sqrt{2}(y + 2)| - \ln|x - 1 - \sqrt{2}(y + 2)| - \sqrt{2}\ln|(x - 1)^2 - 2(y + 2)^2| = C$
(c) $2\arctan[(y + \frac{1}{2})/(x + \frac{1}{2})] = \ln((x + \frac{1}{2})^2 + (y + \frac{1}{2})^2) + C$ **(e)** $7y^2 - 4x^2 - 6xy - 8y + 14x = K$
8. **(b)** $(x + 2y + 1)^2 - 6y = C$ **9.** $x = \pm(1/k)e^{(n+1/2)\pi}$, n an integer
19. $y = x + 1$, $x < 1$, $y(x) = x^2 + 1$, $x \geq 1$

Section 4

1. **(a)** $xy + x^2y^2 = 2$ **(c)** $x(y + e^y) = 1$ **(e)** Not exact, but has solution $y = 0$
(f) $y + (y - x + 1)e^x = 1 + e$
2. **(a)** $k = 1$, $xy + y^3/3 + y = C$ **(c)** $k = 0$, $x = K$; $k = 1$, $(x + y)e^x = K \neq 0$
5. **(a)** $x + x/y = C$ **(b)** $2xe^{x-y} + y^2 = C$ **(c)** $x^4 + 2x^3y - x^2y^2 = C$ **(d)** $y + \arctan(y/x) = C$
7. $xy + y^2/2 = C$ **9.** $y\cos x = C$ **11.** $x = Cy^2$ **13.** $x^2y^3 = C$ **15.** $y = Cx$ **17.** $x^2/y + y = C$
19. $x\sin y = C$ **21.** $\ln(x^2 + y^2) + 2\arctan(y/x) = C$

Section 5A

1. **(a)** $M(x) = e^{2x}$ **(c)** $M(x) = x^2$ **(e)** $M(x) = e^{-x - e^{-2x}/2}$
2. **(a)** $s = 1 - e^{-t^2/2}$ **(c)** $y = 0$ **(e)** $x = 1$ **(g)** $y = (e^x - e^{-x})/2$
3. $y = x - 1 + ce^x$, $0 < x < 1$; $y = x^2/3 + (ce - \frac{1}{3})/x$, $1 \leq x$ **5.** $y = x - c_1e^{-x} + c_2$ **7.** $y = x\ln|x| + c_1x + c_2$
8. $y = x + c_1\int e^{-x^2/2}\,dx + c_2$ **9.** $y = c_1e^{-x} + c_2x + c_3$ **11.** $y = c_1x^2 + c_2x + c_3$ **12.** $y = c_1 + c_2x + c_3|x|^{1/2}$

Section 5B

1. $s = 10e^{-t/50}$ **5.** $u_{min} = 47.5°$ at about 1 h 23 min
6. **(a)** About 10.4 lb   **(b)** After about 54 min   **(c)** About 73.2 more hours   **7.** $100/e \approx 36.8$ lb
8. **(c)** $x = t - t^2/100$, $0 \leq t \leq 100$ **(d)** $y = \frac{1}{50}(50 - t)^2\ln(50 - t) + (50 - t) - \frac{1}{50}(1 + \ln 50)(50 - t)^2$, $0 \leq t \leq 50$
9. About 14.35 lb   **10.** 75 lb   **11.** **(a)** 155 lb   **(b)** 200 lb; about 29.4 min   **12.** 25 lb
13. $50\ln 2 \approx 34.66$ lb
14. **(a)** $u = 10(1 - e^{-2t})$; $b = 10$ **(b)** $u = 20(e^t - e^{-2t})$; $b = 5$
(c) $u = 10$ if $0 \leq t \leq 5$, $u = 20 - 10e^{10-2t}$ if $5 < t$; $b = 20$ **(d)** $u = 10 - 20e^{-t} + 10e^{-2t}$; $b = 10$
(e) $u = 10 - 5/(t + 1)$; $b = 10$
15. **(a)** $50 + 5 - e^{-t}$ **(b)** $50(t + 2)e^{-t}$ **(c)** $100e^{-t}$ **(d)** $25(\sin t - \cos t) + 25e^{-t}$
(e) $200 - 50e^{-2t} - 50e^{-t}$
16. $t \approx 1.19$ **17.** **(a)** About 0.0924   **(b)** After an additional 7.5 min
18. **(a)** About 53°   **(b)** About 54°   **20.** $k \approx 0.02$; 144°
21. **(a)** $dx/dt = rx/(l + rt)$ **(b)** $x = x_0(l + rt/l)$
22. **(a)** Bug arrives after $e^{1/b} - 1$ min   **(b)** Bug arrives after $1(e^{r(1-x_0/l)} - 1)/r$ min

Section 5C

1. $y' = y$ **2.** $y' = xy$ **3.** $y' = 2y$ **4.** $y' = 2xy + (1 - \pi/2)x$ **5.** $y' = -y$
6. $y' = -(y + 1 - \pi/4)/\sqrt{2}$ **7.** $y' = -4x^2(y + 1)$ **8.** $y' = -xy + 2x$
9. **(c)** $y_1 = 6/(2 - 3x^2)$, $y_2 = \frac{3}{2}(1 + e^{3x^2})$ **10.** **(a)** $dv/dt = 132 - 2v$ **(b)** $v(t) = 66 + 34e^{-2t}$
12. **(a)** $dv/dt = -2\sqrt{kg/m}\,v + 2g$ **(b)** $v(t) = \sqrt{gm/k}$
13. **(a)** $dv/dt = \alpha g(1 - \sqrt[\alpha]{k/gm}\,v)$ **(b)** $v(t) = \sqrt[\alpha]{gm/k}$
14. **(a)** $m\,dv/dt = -2kv_0v + gm + kv_0^2$ **(b)** $v = \frac{1}{2}(v_0 - gm/(kv_0))e^{-2(kv_0/m)t} + \frac{1}{2}(v_0 + gm/(kv_0))$
19. $y = (1 + 7e^{-3x})^{-1/3}$ **20.** $y = x/(1 - \ln x)$; $y = 0$ **21.** $(y + y_0)^2v^2 = \frac{2}{3}((y + y_0)^3 - y_0^3)$

Chapter Review and Supplement

1. $xy - x^2/2 = C$ **3.** $x = (t + C)e^t$ **5.** $\ln(y^4 + 1) = 4e^x + C$
7. $y = x^3(Ce^{-2x} + \frac{1}{2})$; C may differ from $x < 0$ to $x > 0$ **9.** $x = (t^3 - 12/t^2)/5$ **11.** $x = (3t + C)^{-1}$, also $x = 0$
13. $x = -t + \tan(t + C)$ **19.** $200(1 - e^{-t/100})$; $200 - 150e^{-t/100}$
22. **(a)** $h_0 = \sqrt{5} - 1$, about $1\frac{1}{4}$ h before midnight   **(b)** About $3\frac{2}{3}$ mi   **(c)** $k = 3/\ln(\sqrt{5} + 2) \approx 2.078$

Chapter 2

Sections 1, 2

1. $P(-10) = \frac{400}{9}$ **3.** $T = \ln 3/k$ **5.** $P(t) = 72,250,000$ **7.** $a = k,\ b = k/P_\infty$

9. **(c)** $\ln (P/P_0) - (1 + c)\ln [(P_\infty - P)/(P_\infty - P_0)] = kt$ **11.** $T(1) \approx 0.85$ g, $T(60) \approx 0.00005$ g, $Pb_{206}(2) \approx 0.28$ g

13. 5.75 g/m   **15.** $U(10,000) \approx 0.972 U_0$

17. $Pb(t) = 1 - e^{-k_1 t} + k_1 p(e^{-k_2 t} - e^{-k_1 t})/(k_2 - k_1) + k_1(1 - p)(e^{-k_3 t} - e^{-k_1 t})/(k_3 - k_1)$

Section 3

1. **(a)** 99.5 ft/s   **(b)** 40,000 ft   **2.** **(a)** 0.29   **(b)** 0.28   **3.** **(a)** 0.36   **(b)** 10^{-4}

4. **(a)** $v = (m/k)(g - Ce^{-kt/m})$ **(b)** $v = (mg/k) - (mg/k - v_0)e^{-kt/m}$ **6.** **(b)** $v(t) = mv_0/(m + v_0 kt)$

7. **(a)** $v = 4v_0/(2 + 10^{-3}\sqrt{v_0}\ t)^2$

9. **(a)** $10 \ln \frac{3}{2} \approx 4.05$ s   **(b)** 304.4 ft   **(c)** 4.7 s, 121 ft/s

(e) Max. height 402.5 ft after 5 s; final velocity equals initial velocity after another 5 s

10. **(a)** $(m/k)(mg/k + v_0)(1 - e^{-kt/m}) - (mg/k)t$ **14.** **(b)** $t_{max} = \sqrt{m/(kg)}\ \arctan (v_0\sqrt{k/(mg)})$

15. **(b)** $v = \sqrt{gm/k}\ (ce^{2bt} - 1)/(ce^{2bt} + 1)$, where $b = \sqrt{kg/m}$

Section 4

1. **(a)** $v = \alpha v_e/k + cm^{k/\alpha}$ **3.** **(b)** $m_r/m_0 = e^{-c/v_e}$ **(c)** $10^{-37,000}$

4. **(a)** $10 \ln \frac{3}{2} \approx 4.05465$ ft/s   **(b)** $(200/g)(1 - (\frac{2}{3})^{g/20})$ ft/s

12. **(a)** $v = v_e \ln (m_0/m)$ **(b)** $v = c(1 - (m/m_0)^{2v_e/c})/(1 + (m/m_0)^{2v_e/c})$

13. $v = c((c + v_0) - (c - v_0)(m/m_0)^{2v_e/c})/((c + v_0) + (c - v_0)(m/m_0)^{2v_e/c})$

14. **(b)** Difference is $v_e \ln (1 + m_f/(2m_r + m_f))$ **18.** **(b)** $m(t) = m(0)e^{(v(0)-v(t))/v_e} + (1/v_e)e^{-v(t)/v_e}\int_0^t e^{v(t)/v_e}F(t)\ dt$

19. $(m_0 - m_1)(1 - e^{(v_2-v_1)/v_e})$ **20.** **(a)** $x(t) = (m_0 v_0/r)\ln (1 + rt/m_0)$ **23.** **(b)** $s_0 = (\pi/2)v_0\sqrt{m_0/a}$

24. **(a)** 240 ft/s   **(b)** 234 ft/s

25. **(a)** $16t(20 - t)/(10 - t)$ **(b)** $(3200/19)((1 - t/10)^{-0.9} - (1 - t/10))$ **(c)** 485 ft/s for (a), 447 ft/s for (b)

Section 5

1. $y_e = 0$; all solutions approach y_e as $t \to \infty$ **3.** $y_e = 0$; no solutions approach y_e as $t \to \infty$ **5.** $y_e = \pm 1$

6. No equilibrium solutions   **10.** **(b)** $\sqrt[3]{ma/k}$ **13.** $v_\infty = e^{gm/k} - 1$ **14.** **(a)** $v_\infty = (1/\alpha)\arctan (mg/k)$

15. **(b)** $v_\infty = gm/k$ if $m \leq k/g$; $v_\infty = \sqrt{(2gm/k)} - 1$ if $m > k/g$ **18.** **(a)** $0 < h < \frac{1}{4}$ **(b)** $(1 \pm \sqrt{1 - 4h})/2$

19. **(a)** $0 < h < kc^2/4$ **(b)** $(c \pm \sqrt{c^2 - 4h/k})/2$ **21.** **(b)** $0 < h < kc^{\alpha+\beta}\alpha^\alpha\beta^\beta/(\alpha + \beta)^{\alpha+\beta}$

Sections 6A, B

1. **(a)** $y = Ce^x$; with $y(0) = 1,\ y = e^x$ **(b)** $y = Ce^x - x - 1$; with $y(0) = 0,\ y = e^x - x - 1$

17. **(a)** $400((k + t/50) - (1 + t/50)^{-3})$ **18.** **(a)** $400((1 - t/50)^{-3} - (1 - t/50))$

19. **(b)** 101.99 ft/s   **(c)** 99.45 ft/s   **21.** **(c)** $S(35.5) \approx 48.4$ **22.** 15.6 s   **23.** 18 s

Section 6C

1. 23.50393   **3.** 0.56714   **5.** Apply ln; $x = 1$ **7.** $x \approx -0.72449,\ x \approx 1.22074$ **9.** $x \approx 1 - 3098$

11. $x \approx 0.40556$ **14.** 4.98 s

Chapter 3

Section 1

1. **(a)** $-e^{-2x}$ **(c)** $27e^{3x}$ **(e)** $-2 \sin x$

2. **(a)** $y = c_1 e^{-3x} + c_2 e^{2x}$ **(c)** $y = c_1 e^{-x} + c_2 x e^{-x}$ **(e)** $y = c_1 + c_2 e^x$ **(g)** $c_1 e^{x/2} + c_2 e^x$

3. **(a)** $c_1 = \frac{2}{5},\ c_2 = \frac{8}{5}$ **(c)** $c_1 = 1,\ c_2 = 3$ **(e)** $c_1 = e/(e - 1),\ c_2 = 1/(1 - e)$ **(g)** $c_1 = c_2 = 0$

6. **(a)** $y = c_1 + c_2 e^{3x}$ **(c)** $y = -1 + c_1 e^x + c_2 e^{-x}$ **7.** **(d)** $y = c_1 \ln |x| + c_2$ **11.** $(1 - 2x)e^x,\ (1 - x)e^x,\ e^x$

13. **(c)** $y_0 \cosh \sqrt{g/l}\ t + v_0\sqrt{l/g} \sinh \sqrt{g/l}\ t$ **14.** **(c)** $y = (e^x + e^{-x})/20$

Sections 2A, B

1. **(a)** $x = \pi/2$ **(c)** $x = -3\pi/4$ **2.** **(a)** $c_1 = c_2 = \frac{1}{2}$ **(c)** $c_1 = -\pi i/2,\ c_2 = -c_1$

6. **(a)** $y = 1 + d_1 \cos x + d_2 \sin x$ **(c)** $y = d_1 \cos \sqrt{2}x + d_2 \sin \sqrt{2}x$

17. **(a)** $y'' + 4y = 0$ **(c)** $y'' - 2y' + 5y = 0$ **(e)** $4y'' + y = 0$ **(h)** $y'' = 0$

Chapter 3

Section 2C

1. **(a)** $y = c_1 e^x + e^{-x/2}(c_2 \cos (\sqrt{3}/2)x + c_3 \sin (\sqrt{3}/2)x)$ **(c)** $y = c_1 + c_2 e^{\sqrt{2}x} + c_3 e^{-\sqrt{2}x}$
   (f) $y = c_1 e^{\sqrt{2}x} + c_2 x e^{\sqrt{2}x} + c_3 e^{-\sqrt{2}x} + c_4 x e^{-\sqrt{2}x}$ **(j)** $y = c_1 e^x + c_2 e^{-x} + c_3 \cos x + c_4 \sin x$
2. **(a)** $y'' + 16y = 0$ **(c)** $y^{(6)} + 48y^{(4)} + 768y'' + 4096y = 0$ **(e)** $y^{(4)} - 4y''' + 8y'' - 8y' + 4y = 0$
   (g) $y^{(6)} = 0$ **(j)** $y^{(6)} + 6y^{(4)} + 9y'' + 4y = 0$
3. **(a)** $y''' - y' = 0$ **(c)** $y''' - y'' + y' - y = 0$ **(e)** $y''' = 0$ **(g)** $y^{(4)} - y = 0$ **(j)** $y^{(4)} + 10y'' + 9y = 0$
4. **(a)** $c_1 = 0, c_2 = c_3 = 1$ **(b)** $c_1 = c_2 = c_3 = c_4 = 1$ **(c)** $c_1 = c_3 = 1, c_2 = -1$ **(d)** $c_1 = c_2 = c_3 = c_4 = 1$
   (e) $c_1 = c_2 = c_3 = 1$ **(f)** $c_1 = c_2 = 1, c_3 = 0$ **(g)** $c_1 = c_3 = 1, c_2 = c_4 = -1$ **(h)** $c_1 = c_2 = c_3 = 1$
5. **(a)** $y = c_1 + c_2 x + c_3 x^2 + c_4 x^3 + c_5 e^x + c_6 e^{-x}$ **(c)** $y = c_1 + c_2 e^x + c_3 e^{-x} + c_4 \cos x + c_5 \sin x$
   (e) $y = c_1 e^{2x} + c_2 e^{-2x} + c_3 e^{4x} + c_4 e^{-4x}$
7. **(a)** $y(x) = c(1 - \cos 2\pi x)$ 8. **(a)** $y(x) = c \sin \pi x$ 9. **(b)** $y(x) = c \sin n\pi x$
12. $y = (R/24)(x^4 + L^3 x - 2Lx^3)$

Section 2D

1. $p = 2\pi/3, \nu = 3/(2\pi), A = 1$ 3. $p = 2\pi/3, \nu = 3/(2\pi), A = \sqrt{5}$ 6. $p = \pi/2, \nu = 2/\pi, A = \sqrt{17}$
7. $y'' + y = 0$ 9. $y'' + 4y = 0$ 11. $y'' + 2y' + 2y = 0$ 13. No second-order equation
15. $2 \cos (x - \pi/3)$ 16. $2 \cos (x - \pi/4)$ 17. $3 \cos (2x - \pi/2)$ 18. $4 \cos (5x + \pi/4)$ 19. $\cos (4x - \pi)$
20. $4 \cos (x - \pi/6)$ 21. $6 \cos (2x - 2\pi/3)$ 22. $\cos (x - \pi/2)$ 23. $y'' + 4y = 0; y(0) = 0, y'(0) = 6$
24. $y'' + 9y = 0; y(0) = 0, y'(0) = 3$ 25. $y'' + y = 0; y(0) = 0, y'(0) = -2\sqrt{3}$
26. $y'' + 2y = 0; y(0) = 1/\sqrt{2}, y'(0) = -1$ 27. $y'' + 2y' + 5y = 0; y(0) = -1, y'(0) = 1$
28. $y'' + 4y' + 13y = 0; y(0) = -\frac{1}{2}, y'(0) = 1 + 3\sqrt{3}/2$ 29. $y' - 2y' + 2y = 0; y(0) = 0, y'(0) = 1$
30. $y'' - 4y' + 8y = 0; y(0) = 1/\sqrt{2}, y'(0) = 2\sqrt{2}$

Sections 3A, B

2. **(a)** Linear   **(c)** Nonlinear   3. **(a)** $y = x + c_1 \cos x + c_2 \sin x$ **(c)** $y = e^x + c_1 e^{x/\sqrt{2}} + c_2 e^{-x/\sqrt{2}}$
4. **(a)** $D^2 - 2D + 2$ **(c)** D^2 **(e)** $D^4 + 18D^2 + 81$ **(g)** $D^4 - 4D^3 + 8D^2 - 6D + 4$
5. **(a)** $A \cos x + B \sin x$ **(c)** Axe^x **(e)** $Ax^2 e^x + Bx^3 e^x$
6. **(a)** $y = \frac{1}{3}e^{2x} + c_1 e^x + c_2 e^{-x}$ **(c)** $y = \frac{1}{4}e^x + c_1 e^{-x} + c_2 x e^{-x}$ **(e)** $y = (x/2)e^x - x + c_1 e^x + c_2 e^{-x}$
   (g) $\sin x + c_1 \cos x + c_2 \sin x$ **(j)** $y = -(x/3)e^x + (x^2/6)e^x + c_1 e^x + e^{-x/2}(c_1 \cos (\sqrt{3}/2)x + c_2 \sin (\sqrt{3}/2)x$
   (l) $y = \frac{2}{7}x e^{4x} + c_1 e^{4x} + c_3 e^{-3x}$ **(n)** $y = -\frac{1}{2}x^2 + c_1 + c_2 e^x + c_2 e^{-x}$
7. **(a)** $y = -\frac{1}{2}\cos x + c_1 e^x + c_2 e^{-x}$ **(c)** $y = \frac{1}{2}x e^x + c_1 e^x + c_2 e^{-x}$ **(e)** $y = \frac{1}{6}x^3 e^x + c_1 e^x + c_2 x e^x$
8. **(a)** $y = -\frac{1}{2}\cos x + \frac{3}{4}e^x - \frac{1}{4}e^{-x}$ **(c)** $(x/2)e^x + \frac{1}{4}e^x - \frac{1}{4}e^{-x}$ **(e)** $y = (x^3/6)e^x + xe^x$

Section 3C

1. **(a)** $g(x - t) = (e^{x-t} - e^{-(x-t)})/2$ **(b)** $g(x - t) = \sin (x - t)$ **(c)** same as (a)   **(e)** $g(x - t) = (x - t)e^{x-t}$
2. **(a)** $y_p = -x + (e^x - e^{-x})/2$ **(c)** $y_p = -\frac{1}{9}e^{-x} - (x/3)e^{-x} + \frac{1}{9}e^{2x}$ **(e)** $y_p = -\frac{1}{3}e^x + \frac{1}{4}e^{2x+1} + \frac{1}{12}e^{-2x-3}$
3. **(a)** $y_p = \begin{cases} -\frac{1}{4} + (e^{2x} + e^{-2x})/8, & x \geq 0 \\ 0, & x < 0 \end{cases}$ **(c)** $y_p = \begin{cases} -\frac{1}{4} + (e^{2x} + e^{-2x})/8, & x \geq 0 \\ \frac{1}{4} - (e^{2x} + e^{-2x})/8, & x < 0 \end{cases}$
   (e) $y_p = \begin{cases} \frac{1}{4} - x/4 + (e^{2x-2} - e^{-2x+2})/16, & x \geq 1 \\ 0, & x < 1 \end{cases}$
5. $(e^{-x} + e^{-2x}) \ln (1 + e^x) - (1 + \ln 2)e^{-x} + (1 - \ln 2)e^{-2x}$
6. **(a)** $y_p(x) = e^x(1 - x + x \ln x)$ **(b)** $x_p(x) = e^x(-1 - x + x \ln (-x))$

Section 4A

1. **(a)** $\ddot{y} = y - 1$ **(b)** $\ddot{y} = \dot{y}$ **(c)** $\ddot{y} = y$ **(d)** $\ddot{y} = y + 1$ **(e)** $-\ddot{y} = \frac{1}{4}y + 6$ **(f)** $\ddot{y} = \dot{y} + 2ky = k$
2. **(a)** $t_1 \approx 1.5$ **(b)** $t_1 \approx 0.5$ **(c)** $t_1 \approx 0.4$ **(d)** $t_1 \approx 0.6$ **(e)** $t_1 \approx 2.1$

Sections 4B, C

1. **(a)** $y = c_1 t^2 + c_2$ **(c)** $y = \ln |t + c_1| + c_2$ **(e)** $y = t^4/16 + c_1 \ln |t| + c_2$
2. **(a)** $y = c_2 e^{c_1 t}$ **(c)** $y = c_1 \pm \sqrt{c_2 - 2t}$ **(e)** $y = \ln |c_1 e^t + c_2 e^{-t}|$
3. **(c)** $y = \sqrt{2} \tan (t/\sqrt{2})$ **(d)** $y = \tanh t$ 5. **(a)** $y(x) = (1/k) \cosh kx$
9. **(a)** $y = a, y = a \tan (at + b), y = -1/(t + a), y = -a \coth (at + b), y = -a \tanh (at + b), a, b$ const
10. **(a)** $y(t) = (3 - \cosh (\sqrt{2} t))/2$ **(c)** $t \approx 1.11$ 12. **(c)** $v = \sqrt{2gy/7}$ **(d)** $y(t) = gt^2/14$

Chapter 3

Chapter Review and Supplement

1. $y = (x^2/2)e^{-x} + \frac{3}{4}e^x + c_1 e^x + c_2 x e^{-x}$ **3.** $y = -\frac{1}{2}\sin x + c_1 e^x + c_2 e^{-x}$ **5.** $y = \frac{1}{3} + e^{-x}(c_1\cos\sqrt{2}x + c_2\sin\sqrt{2}x)$

7. $y = -(x/6)\cos 3x + \cos 3x + \frac{1}{18}\sin 3x$ **9.** $y = -\frac{1}{5}\cos 3x + c_1\cos 2x + c_2\sin 2x$ **11.** $y = -\cos x + c_2\sin x$

13. $y_p = Ax^2 e^{2x} + Bx^3 e^{2x}$ **15.** $y_p = Axe^{2x} + Bx^2 e^{2x} + Cxe^{3x}$ **17.** $y_p = Axe^{2x} + B\cos x + C\sin x$

19. $y_p = Ax + Bx^2 + Cx^3 + Dxe^x$ **21.** $y_p = Ax^3 + Bx^4 + Cx^5 + Dx^6$

Chapter 4

Section 1

2. **(d)** 322 ft/s   **(e)** 2.65 s   **3.** **(c)** $y(t) = (mg/k)t - (m^2 g/k^2 - mv_0/k)(1 - e^{-kt/m})$ **5.** **(b)** About 964 ft

6. **(b)** Nonlinear $v_\infty = \sqrt{mg/k}$; linear $v_\infty = mg/k$ **7.** **(a)** $25/8 \approx 3.1$ s, 156.25 ft   **(b)** 2.7 s, 129.8 ft

8. **(b)** $t_{max} = (m/k)\ln(1 + (kv_0/mg))$, $y_{max} = mv_0/k - (m^2 g/k^2)\ln(1 + (kv_0/mg))$ **11.** $k = 5$

12. **(a)** 5.75 s; about 4518 ft   **(b)** 4.6 min   **13.** 7.9 s; 5.6 s   **16.** $k = 3.125$ **18.** $\sqrt[\alpha]{gm/k}$

19. **(a)** $0, \ln 2$ **(b)** $0, \ln 3$ **(c)** $\ln 2$

20. **(a)** $\ddot{y} = \frac{16}{25}y$ **(b)** $y = 5e^{4t/5} + 5e^{-4t/5}$ **(c)** $\dot{y} = 4e^{4t/5} - 4e^{-4t/5}$ **(d)** 2.6 s   **(e)** 31 ft/s

     (f) 25.6 ft/s^2

22. **(b)** 273 mi

Section 2

2. 3.26 ft   **3.** **(b)** $k^2 < 4gm^2/l$ **9.** **(a)** $\theta(t) = 0.1\cos 4t$ **(b)** $\pi/8 \approx 0.4$ s   **(c)** 0.4 rad/s

10. $\frac{1}{16}$ rad or about 3.6°   **11.** $P = \pi/4$ s, $A \approx 0.12$, $\theta_{max} \approx 0.96$ rad/s   **13.** 1.57 s   **14.** 0.7 ft/s

15. About $9\frac{3}{4}$ in   **16.** $c_1 = \theta_0$, $c_2\sqrt{l/g}\,\psi_0$, $\omega = \sqrt{g/l}$ **17.** About 26.4 ft/s^2 **22.** **(a)** 433 mi

Section 3

1. **(a)** Critically damped   **(b)** Underdamped   **(c)** Harmonic   **(d)** Overdamped   **(e)** Underdamped

     (f) Underdamped

3. **(a)** $\ddot{x} + 4x = 0$ **(b)** $\ddot{x} + 9x = 0$ **(e)** $\ddot{x} + 6\dot{x} + 8x = 0$

4. **(b)** $t \approx 5.01$ **6.** **(a)** $h = 6$ **(b)** $h \approx 8.9$ **(c)** 80 lb

7. **(a)** $\omega = \sqrt{7}/2$ **(b)** $h = \frac{21}{5}$ **(c)** $\omega = \sqrt{2}$ **(d)** $100 < h < 100.25$

8. **(a)** $8h > k^2$ **(b)** $m > \frac{1}{4}$ **(c)** $k = \sqrt{3}$ **(d)** $k < 2$

9. **(a)** $\omega = 0, \omega = \pm 1$ **(b)** $h = m$ **(c)** $k = \frac{1}{2}$ **(d)** $4m^2\omega^2 + k^2 = 4mh$

15. **(a)** $k = 2mV_0/(eE)$, $h = mV_0^2/(eE)^2$ **16.** **(a)** $h = \frac{5}{2}$; $\omega/(2\pi) \approx 0.64$ per second   **(b)** Reduced by 64%

22. **(b)** 0.5 ft; $\frac{2}{5}\sqrt{\pi/5}$ s   **23.** 2546 lb

28. **(a)** $k = 6, h = 45$ **(b)** $m = \frac{1}{26}, k = \frac{1}{13}$ **(c)** $m = 1, h = \frac{1}{2}$ **(d)** No such m exists

29. $m_0 = \frac{9}{8}$ **30.** **(a)** $v = \sqrt{2gl}$

Section 4

1. **(b)** $Q(t) = e^{-5t}(\cos 80t + 16\sin 80t)$ **(c)** $Q(t) = (5/6416)e^{-t} + (1/6416)e^{-5t}(6411\cos 80t + (1603/4)\sin 80t)$

     (d) $Q(t) = (290{,}725)^{-1}(8\cos 20t + 241\sin 20t + 290{,}717 e^{-5t}\cos 80t + 289{,}753 e^{-5t}\sin 80t)$

5. **(c)** $\dfrac{CE_0}{\sqrt{1 + R^2 C^2 \omega^2}}\left(\cos(\omega t - \alpha) - \dfrac{e^{-t/RC}}{\sqrt{1 + R^2 C^2 \omega^2}}\right)$, $\alpha = \arctan/RC\omega$

Sections 5A, B

1. Circle of radius 1 centered at $(1, 0)$ **2.** Hyperbola opening right and left   **3.** Parabola opening right

5. Single point $(2, 0)$ **7.** Circles centered at $(1, 0)$ **9.** Parabolas opening right

11. Single points on the y axis and half-lines of slope -1 approaching these points

13. Two families of parallel lines with slopes 1 and -1 **15.** $0, 1$ **16.** $0, \pm 1$

31. **(e)** $y = 2^{-1/3}(4 - 3t)^{2/3}$ **37.** **(b)** $y = t^3 = 3t_0^2 t + 2t_0^3$, $z = 3t^2 - 3t_0^2$

51. **(a)** $C = \frac{1}{2}z_0^2 - (g/l)\cos y_0$ **(b)** $|z_0| > \sqrt{2(g/l)(1 + \cos y_0)}$ **52.** $y = \frac{2}{3}\sqrt{g} \approx 3.8$ rad/s

Sections 6A, B

3. $E/(mg)$ **6.** **(b)** $\sqrt{E}$; $\sqrt{2E}$ **16.** $y = \pm\sqrt{C}\sin(t + b)$ **19.** **(a)** $(0, 0)$ unstable; $U(y) = -y^6/6$

21. **(b)** $\theta^* = 1.018$ **24.** $x = gat^2/(2 + 2a^2) + c_1 t + c_2$ **25.** **(d)** $y = 0$ is stable

27. $y = 1, y = -1$ are stable; $y = 0$ is unstable

Chapter 4

Sections 7A, B

1. **(a)** $y(t) \approx 0$ if $t = 0$, 2.67, 4.35, 5.75, and five more   2. $y(t) \approx 0$ if $t = 2.92$, 6.04, 9.17, 12.3, and eight more
11. **(a)** $k = \frac{8}{9}$ **(b)** 9.2 s   **(c)** $k \approx 0.6212$ **(d)** 8.4 s   12. $\dot{x}(0) \approx 3.67$

Chapter 5

Sections 1A–C

1. **(a)** $x = 2e^t - 1$, $y = 2e^t$ **(c)** $x = 0$, $y = e^{t/2}$, $z = -e^{t/3}$
2. **(a)** $\mathbf{F}(t, \mathbf{x}) = (x + 1, y)$, $v = \sqrt{(x+1)^2 + y^2}$ **(c)** $\mathbf{F}(t, x) = (x, y/2, z/3)$, $v = \sqrt{x^2 y^2/4 + z^2/9}$
8. **(a)** $\dot{x} = -x - y$, $x(0) = 1$; $\dot{y} = x$, $y(0) = 1$ 9. **(a)** $y(t) = e^{-t/2}(\cos(\sqrt{3}t/2) + \sqrt{3}\sin(\sqrt{3}t/2))$, $x = \dot{y}$
10. **(a)** $\dot{x} = y$, $\dot{y} = x^2 - y^2 + e^t$ **(c)** $\dot{x} = y$, $\dot{y} = z$, $\dot{z} = z^2 - xy - t$
11. **(a)** $dx/dt = (t + y)/2$, $dy/dt = (t - y)/2$ **(c)** $dx/dt = x - 3y + t$, $dy/dt = -3x + y - t$
13. **(a)** Trajectories: $y = ce^x - x - 1$; constant solutions are points (x, y) where $x = y$
    (c) Trajectories: $x^2 - y^2 = c$; constant solutions: $x = y = 0$
15. **(a)** $x = (z_0/k)(1 - e^{-kt})$, $y = -gt/k + (kw_0 + g)(1 - e^{-kt})/k^2$
19. **(b)** $u = (c_0 + p(u_0 - v_0)e^{-(p+q)t})/(p + q)$ **(c)** $u = (c_0 - q(u_0 - v_0)e^{-(p+q)t})/(p + q)$ **(d)** $t = \ln 2/(p + q)$

Section 1D

1. $(-\frac{1}{5}, \frac{1}{5})$ 2. $(0, 0)$, $(1, 1)$ 3. $(0, a)$, a arbitrary   4. $(k\pi, l\pi)$, k, l integers   5. $(0, 0, 0)$, $(-\frac{1}{2}, \frac{1}{2}, -\frac{1}{2})$
6. $(a, a, 0)$, $(0, 0, b)$, a, b arbitrary   7. $\pm(c/\sqrt{1 + c^2}, 1/\sqrt{1 + c^2})$; all lie on $x^2 + y^2 = 1$ 9. $(0, 0)$, $(d/c, a/b)$
11. $(0, 0, 0)$, $(3, 3, 0)$, $(0, 2, 3)$ 14. **(a)** $x^2 + y^2 = 1$ or $x + y = 0$, circle or line

Section 1E

1. $x = 1/(1 - at)$, $t < 1/a$

Sections 2A–C

1. **(a)** Nonlinear   **(b), (c)** Linear
2. **(a)** $x = c_1 e^{-2t} + c_2 e^{2t}$, $y = -c_1 e^{-2t} - \frac{1}{2}c_2 e^{2t}$ **(c)** $x = c_1 e^{(1+\sqrt{2})t} + c_2 e^{(1-\sqrt{2})t} - 1t + 4$
$\qquad\qquad y = (1/\sqrt{2})c_1 e^{(1+\sqrt{2})t} - (1/\sqrt{2})c_2 e^{(1-\sqrt{2})t} + t - 3$
3. $x = -c_1 t^3 + (2c_1 - c_2)t$, $y = c_1 t^2 + c_2$
4. **(a)** $dx/dt = x/2 + t/2$, $dy/dt = -x/2 + t/2$ **(c)** $dx/dt = x - 3y + t$, $dy/dt = -3x + y - t$
5. **(a)** $x = 3e^{t/2} - t - 2$, $y = -6e^{t/2} + t^2/2 + t + 5$ **(c)** $x = -(7e^{4t} + 8e^{-2t} + 4t + 1)/16$
$\qquad\qquad y = (7e^{4t} - 8e^{-2t} + 4t + 1)/16$
6. **(a)** $x = e^t - t - 1$, $y = t$ **(c)** $x = 2 - t + t^2/2 - 2e^{-t}$, $y = -2 + t - t^2/2 + 2e^{-t}$
7. **(a)** $\dot{x} = u$, $x(0) = 0$; $\dot{u} = x - v - y$, $u(0) = 0$; $\dot{y} = v$, $g(0) = 0$, $\dot{v} = x - y - y$, $v(0) = 1$
    (c) No first-order normal form
11. **(b)** $\ddot{z} = 2w$, $\ddot{w} = 2z$ 12. **(d)** $\mathbf{v} = (u, v + \omega R)$

Chapter Review and Supplement

1. $x = \tan(t + \pi/4)$, $y = 2e^t$ 3. $x = 3(e^t - e^{-t})/2$, $y = -1 + 3(e^t + e^{-t})/2$ 5. $x = e^{-5t}$, $y = 2e^{-5t}$
7. $x = e^{2t}$, $y = 2e^{2t} - t$ 9. $x = 2\cos t + \cos 2t$, $y = 4\cos t - \cos 2t$ 11. $x = -\sin t$, $y = \cos t$, $z = -e^t$
14. **(b)** $x = \cos t$, $y = \sin t$, $z = t$

Chapter 6

Section 1

1. **(a)** $dy/dt = -y/25 + z/25$, $dx/dt = y/100 - z/25$ **(b)** $y = 25e^{-t/50} - 15e^{-3t/50}$, $z = \frac{25}{2}e^{-t/50} + \frac{15}{2}e^{-3t/50}$
2. **(a)** $c_1 = -\frac{48}{5}$, $c_2 = -\frac{32}{5}$ 3. **(a)** $\dot{x} = -ax + ay$, $\dot{y} = ax - ay$ 4. **(a)** $\dot{x} = -ax + az$, $\dot{y} = ax - ay$, $\dot{z} = ay - az$
5. **(b)** $\dot{x} = -2x/100 + 2y/(200 - t)$, $x(0) = 0$, $\dot{y} = x/100 - 2y/(200 - t)$, $y(0) = 10$
6. **(b)** $\dot{x} = -x/(50 + t) + 2y/(100 - t)$, $x(0) = 0$, $\dot{y} = x/(50 + t) - 2y/(100 - t)$, $y(0) = 10$
7. **(b)** $\dot{x} = -2x/(50 - t) + y/(100 + t)$, $x(0) = 0$, $\dot{y} = 2x/(50 - t) - y/(100 + t)$, $y(0) = 10$
8. **(a)** $\dot{x} = -x/50 + y/50$, $\dot{y} = x/50 - y/25 + z/50$, $\dot{z} = y/50 - z/50$
    (c) $(x, y, z) = (\frac{1}{6}(e^{-3t/50} - 3e^{-t/50} + 2), \frac{1}{3}(1 - e^{-3t/50}), \frac{1}{6}(e^{-3t/50} + 3e^{-t/50} + 2))$

Chapter 6

Section 1
9. (c) $x = (100 - t) - (100 - t)^2/100$, $y = t(100 - t)/100 - (100 - t)^2/100 + (100 - t)^3/10^4$
11. (d) $x = 18 - 8(1 - t/100)^5$, $y = 12 + 8(1 - t/100)^5$ 12. (d) $x(100) \approx 1.23$, $y(100) \approx 8.77$

Section 3
1. (a) $I_1 = 4$, $I_2 = 2$, $I_3 = 6$ (b) Currents reverse direction   (c) $I_1 = 22$, $I_2 = 11$, $I_3 = 33$ (d) $I_1 = \infty$
3. (a) $\dot{I}_1 = 10E - 200I_1 - 150I_2$, $\dot{I}_2 = -6E + 120I_1 + 80I_2$, $I_3 = I_1 + I_2$
$\quad$ (b) $I_1 = \frac{1}{20}E + c_1 e^{-20t} - \frac{3}{2}c_2 e^{-100t}$
$\qquad I_2 = \qquad\quad -\frac{6}{5}c_1 e^{-20t} + c_2 e^{-100t}$
$\qquad I_3 = \frac{1}{20}E - \frac{1}{5}c_1 e^{-20t} - \frac{1}{2}c_2 e^{-100t}$
$\quad$ (c) $c_1 = 1225/16$, $c_2 = 751/8$
5. $I_1 = I_2 + I_5$, $I_2 = I_3 + I_6$, $I_4 = I_5 + I_6$, $R_2 I_2 + L_3 \dot{I}_3 + R_1 I_1 = E$,
$L_5 \ddot{I}_5 + (1/c_4)I_4 + R_1 \dot{I}_1 = 0$,
$R_2 \dot{I}_2 + L_2 \ddot{I}_6 + (1/c_2)I_6 - L_5 \ddot{I}_5 = 0$

Sections 4A, B
1. (a) $c_1 = c_2 = c_3 = \frac{1}{2}$, $c_4 = -\sqrt{3}/6$ (b) $c_1 = (x_0 + y_0)/2$, $c_3 = (y_0 - x_0)/2$, $c_2 = (u_0 - v_0)/2$, $c_4 = \sqrt{3}(v_0 - u_0)/6$
3. $\sqrt{(5 \pm \sqrt{5})/2}$ 6. (b) $\sqrt{h_1/m_1}$, $\sqrt{h_2/m_2}$ 7. $\sqrt{g(2 \pm \sqrt{2})/2}$
8. (a) $\frac{1}{2}(h_1 + h_2)x^2 - h_2 xy + \frac{1}{2}(h_2 + h_3)y^2$ 9. $x_e = (l + 2l_1 - l_2 - l_3)/3$, $y_e = (2l + l_1 + l_2 - 2l_3)/3$
10. (b) $x_e = l/3$, $y_e = 2l/3$ 11. (b) $x_e = 2h/5$, $y_e = 3h/5$ (c) $\sqrt{h}$, $\sqrt{5h}$
14. (b) $m(\ddot{x}, \ddot{y}, \ddot{z}) = -h_1(1 - l_1(x^2 + y^2 + z^2)^{-1/2})(x, y, z) - h_2(1 - l_2((x - b)^2 + y^2 + z^2)^{-1/2})(x - b, y, z)$
15. $m\ddot{\mathbf{x}} = \dfrac{h_1(|\mathbf{a} - \mathbf{x}| - l_1)(\mathbf{a} - \mathbf{x})}{|\mathbf{a} - \mathbf{x}|} + \dfrac{h_2(|\mathbf{b} - \mathbf{x}| - l_2)(\mathbf{b} - \mathbf{x})}{|\mathbf{b} - \mathbf{x}|} + \dfrac{h_3(|\mathbf{c} - \mathbf{x}| - l_3)(\mathbf{c} - \mathbf{x})}{|\mathbf{c} - \mathbf{x}|}$

Section 5
2. 1.108 m/s, 1.109 m/s   9. (a) $G \approx 6.67 \cdot 10^{-11}$ (b) 9.805 m/s²   12. $x_{max} = x_0 + v_0^2/(2g)$
13. (b) $v = \sqrt{G(m_1 + m_2)/a}$ 14. (d) About 27.35 days   16. (a) $\omega = \sqrt{G(m_1 + m_2)/a^3}$
24. (b) 0.9 cm; 3 km   26. $\dot{x} = \sqrt{2k(1/x_0 - 1/x) + z_0^2}$ 27. $v = \frac{1}{2}\sqrt{GM/r}$

Section 6
1. (a) $E = 2$ (b) $E = 2$ (c) $E = 2$ (d) $E = 3$
3. (a) $(4, -3)$ (b) $x - y = 2k\pi$, k integer   (c) x, y, z axes
$\quad$ (d) $(0, 0, 0)$ and sphere of radius 1 centered there
7. $(\dot{x}^2 - \dot{x}_0^2)/2 + \displaystyle\int_{x_0}^{x} f(t)\, dt$ 8. $U(x, y) = -(ax^2/2 + bxy + dy^2/2)$
14. (b) $U = -gl \cos \theta$, $T = l^2(\dot{\theta}^2 + \dot{\phi}^2 \sin^2 \theta)/2$

Sections 7A, B
1. (b) $x = \frac{3}{2}e^t - \frac{1}{2}e^{-t}$, $y = \frac{3}{2}e^t + \frac{1}{2}e^{-t}$ 10. (b) $x_1(25.3) \approx 50.2$, $x_2(100) \approx 88.3$ 11. (b) $x_1(8.6) \approx 7.45$
22. (a) 4.20   (b) 4.16   (c) 3.72   24. (a) 3.404   (b) 2.9485   (c) 3.0405

Section 8A
1. (a) $\lambda = \pm 1$; saddle   (c) $\lambda = 2 \pm i$; unstable spiral   (e) $\lambda = \pm\sqrt{2}i$; stable center
$\quad$ (g) $\lambda = 0, 1$; degenerate
2. (a) $\lambda = \pm 1$; saddle   (c) $\lambda = \pm i$; stable center   10. (a) C   (b) B   (c) A   (d) D

Section 8B
1. (a) $\lambda = \pm 1$; saddle   (c) $\lambda = \pm i$; center   (e) $\mathbf{x}_0 = (0, 0)$: $\lambda = 0, 1$; degenerate
$\quad$ (f) $\mathbf{x}_0 = (1, 1)$. λ 2, 1) asymptotically stable node   (g) $\mathbf{x}_0 = (0, 0)$, $\lambda = 1 = \pm i$; unstable spiral
9. Equilibrium points are $(0, 0)$ and each individual point on the circle $x^2 + y^2 = 1$.   10. $(0, 0), (1, 1), (2, 0)$
11. $(0, 0)$ unstable; $(\pm 1, 0)$ asymptotically stable
12. (a) $y = y_0 e^{-t}/\sqrt{1 + y_0^2 - y_0^2 e^{-2t}}$ (b) $x = x_0 e^t/\sqrt{x_0^2 e^{2t} + 1 - x_0^2}$
20. (a) $(\frac{1}{2}, 0), (\frac{3}{2}, 0), (-2, 0)$

Chapter 6

Section 8C

1. **(a)** Strict if $a, b > 0$ **(c)** Strict if $a, b > 0$ **(e)** Nonstrict if $a, b > 0$
2. **(a)** Strict if $a, b, c > 0$ **(c)** Not Liapunov   **(e)** Not strict if $a, b, c > 0$ **5.** **(a)** All points $(x, y, 0)$
6. **(a)** All points $(x, y, 0)$

Section 9

1. **(a)** $f(t) = 1$ minimizes   **(b)** $f(t) = 1 + t$ minimizes   **2.** **(a)** $\ddot{x} = t/2$ **(c)** $1 + \dot{x}^2 = k$ **(e)** $2t\dot{x} = t + c$
3. **(a)** $x = t^3/12 + c_1 t + c_2$ **(c)** $x = c_1 t + c^2$ **(e)** $x = t/2 + k \ln t + c$
4. **(a)** $x^2 + y^2 = k\sqrt{x^2 + y^2 + \dot{x}^2 + \dot{y}^2}$ **(c)** $\dot{x} = k_1, \dot{y} = k_2$ **(e)** $\dot{y} = \dot{y}, \dot{x} = \dot{x}$, hence no information
5. **(a)** $x^2 + y^2 = c$ **(c)** $x = k_1 t + c_1, y = k_2 t + c_2$
   (e) Any choice for $x(t), y(t)$ will suffice that meets the boundary conditions
8. $x = c \cosh(t/c + k)$ **13.** $x = c_1 t + c_2, y = k_1 t + k_2, z = g_1 t + g_2$

Chapter 7

Section 1

1. If $\lambda = -5$, (b) is an eigenvector. If $\lambda = 7$, (a) and (c) are eigenvectors   **2.** **(a)** $\lambda = -1, 3$ **(c)** $\lambda = 0, 4$

3. **(a)** $\lambda = -1, \mathbf{u} = \begin{pmatrix} -2 \\ 1 \end{pmatrix}; \lambda = 3, \mathbf{u} = \begin{pmatrix} 2 \\ 1 \end{pmatrix}$ **(c)** $\lambda = 0, \mathbf{u} = \begin{pmatrix} -2 \\ 1 \end{pmatrix}; \lambda = 4, \mathbf{u} = \begin{pmatrix} 2 \\ 1 \end{pmatrix}$

4. **(a)** $\lambda_1 = 1, \mathbf{u}_1 = (1, 2); \lambda_2 = -1, \mathbf{u}_2 = (1, 1)$ **(b)** $\lambda_1 = 3, \mathbf{u}_1 = (1, 0); \lambda_2 = 2, \mathbf{u}_2 = (0, 1)$
   (c) $\lambda_1 = 5, \mathbf{u}_1 = (1, 1); \lambda_2 = 1, \mathbf{u}_2 = (1, 0)$ **(d)** $\lambda_1 = 2 + i, \mathbf{u}_1 = (i, 1); \lambda_2 = 2 - i, \mathbf{u}_2 = (i, -1)$

5. **(a)** $\mathbf{x} = c_1 e^{-t} \begin{pmatrix} 1 \\ 1 \end{pmatrix} + c_2 e^t \begin{pmatrix} 1 \\ 2 \end{pmatrix} = \begin{pmatrix} c_1 e^{-t} + c_2 e^t \\ c_1 e^{-t} + 2c_2 e^t \end{pmatrix}$ **(c)** $\mathbf{x} = c_1 e^t \begin{pmatrix} 1 \\ 0 \end{pmatrix} + c_2 e^{5t} \begin{pmatrix} 1 \\ 1 \end{pmatrix} = \begin{pmatrix} c_1 e^t + c_2 e^{5t} \\ c_2 e^{5t} \end{pmatrix}$

6. **(a)** $\begin{pmatrix} -e^{-t} + 2e^t \\ -e^{-t} + 4e^t \end{pmatrix}$ **(c)** $\begin{pmatrix} e^{5t} \\ e^{5t} \end{pmatrix}$ **7.** **(c)** $x = -e^{-2t}/2 + 2e^t + 2e^{3t} - \frac{5}{2}$
   $$y = e^{-2t}/2 - 8e^t + 4e^{3t} + \frac{9}{2}$$
   $$z = e^{-2t}/2 - 2e^t + 2e^{3t} + \frac{3}{2}$$

13. **(a)** Solutions oscillate and tend to 0   **(c)** Solutions oscillate and tend to ∞ **18.** **(a)** $\begin{pmatrix} -2 & 2 \\ -10 & 7 \end{pmatrix}$

Sections 2A, B

1. **(a)** $\begin{pmatrix} e^{-t} & 0 \\ 0 & e^t \end{pmatrix}$ **(c)** $\begin{pmatrix} \cos t + i \sin t & 0 \\ 0 & \cos t - i \sin t \end{pmatrix}$ **5.** **(b)** $e^{5t} \begin{pmatrix} 1 + 4t & -4t \\ 4t & 1 - 4t \end{pmatrix}$

10. **(a)** $\begin{pmatrix} 2e^{-t} - e^t & -e^{-t} + e^t \\ 2e^{-t} - 2e^t & -e^{-t} + 2e^t \end{pmatrix}$ **(c)** $\begin{pmatrix} e^t & e^{5t} - e^t \\ 0 & e^{5t} \end{pmatrix}$ **11.** **(c)** $\frac{1}{2} \begin{pmatrix} e^{2t} + 1 & e^{2t} - 1 \\ e^{2t} - 1 & e^{2t} + 1 \end{pmatrix}$

Section 2C

1. **(a)** $c_1 \begin{pmatrix} 1 \\ 0 \end{pmatrix} + c_2 \begin{pmatrix} 0 \\ e^t \end{pmatrix}$ **2.** **(a)** $A = \begin{pmatrix} 0 & 0 \\ 0 & 1 \end{pmatrix}$ **3.** **(a)** $M(0) \neq I$

Section 3

1. **(a)** $\begin{pmatrix} -5e^t + 6e^{3t} & 3e^{2t} - 3e^{3t} \\ -10e^{2t} + 10e^{3t} & 6e^{2t} - 5e^{3t} \end{pmatrix}$ **(c)** $\begin{pmatrix} 1 & t & e^t + t - 1 \\ 0 & 1 & 1 - e^t \\ 0 & 0 & e^t \end{pmatrix}$

(e) $\frac{1}{2} \begin{pmatrix} 1 + e^{2t} & -1 + e^{2t} & -1 + e^{2t} & 0 \\ -1 + 2e^t - e^{2t} & 1 + 2e^t - e^{2t} & 1 - e^{2t} & 0 \\ -2e^t + 2e^{2t} & -2e^t + 2e^t & 2e^{2t} & 0 \\ -1 + 2e^t(1 + t) - e^{2t} & 1 + 2te^t - e^{2t} & 1 - e^{2t} & e^{2t} \end{pmatrix}$

3. $e^{tA} = \begin{pmatrix} e^{2t} & -\frac{1}{2}e^t \cos t + \frac{1}{2}e^{2t}(1 - t) & te^{2t} & -\frac{1}{2}e^t \sin t + \frac{1}{2}te^{2t} \\ 0 & e^t \cos t & 0 & e^t \sin t \\ 0 & -\frac{1}{2}e^{2t} + \frac{1}{2}e^t(\cos t + \sin t) & e^{2t} & \frac{1}{2}e^{2t} + \frac{1}{2}e^t(\sin t - \cos t) \\ 0 & -e^t \sin t & 0 & e^t \cos t \end{pmatrix}$

Chapter 7

Section 3

5. (a) If $\alpha \neq \beta$, $e^{tA} = \begin{pmatrix} e^{\alpha t} & (e^{\alpha t} - e^{\beta t})/(\alpha - \beta) \\ 0 & e^{\beta t} \end{pmatrix}$ 7. (b) $\begin{pmatrix} e^{\alpha t} & \beta t e^{\alpha t} \\ 0 & e^{\alpha t} \end{pmatrix}$

9. (a) $\begin{pmatrix} 1 & 0 & 0 \\ -13 & 1 & -5 \\ 2 & 0 & 1 \end{pmatrix}$ (c) $\dfrac{1}{4}\begin{pmatrix} 0 & 4 & 0 \\ 7 & -22 & 8 \\ -3 & 10 & -4 \end{pmatrix}$ (e) $\begin{pmatrix} 1 & -1 & 0 \\ 0 & -1 & 1 \\ 0 & 0 & 1 \end{pmatrix}$

   (g) $\dfrac{1}{8}\begin{pmatrix} 8 & -8 & 8 & -6 \\ 0 & 4 & 0 & -1 \\ 0 & 0 & 8 & -2 \\ 0 & 0 & 0 & 2 \end{pmatrix}$ (i) $\begin{pmatrix} 1 & 0 & 0 \\ 0 & e^{-t} & -te^{-t} \\ 0 & 0 & e^{-t} \end{pmatrix}$

Section 4

1. (a) $x = c_1(e^{-t} + 2e^{-t}) + c_2(e^t - e^{-t}) - 2 - e^{2t}/3$ (c) $x = c_1 e^t + c_2(e^{5t} - e^t) - 1 - te^t - e^t/4;\; y = c_2 e^{5t} - e^t/4$
    $y = c_1(-2e^t + 2e^{-t}) + c_2(2e^t - e^{-t}) - 3 - 4e^{2t}/3$

6. (a) $\mathbf{x}_p = \begin{pmatrix} t^4/6 \\ 7t^3/6 \end{pmatrix}$

Sections 5A, B

1. (a) Unstable   (b) Unstable   (c) Unstable   (d) Stable
2. (a) $(-1, 0)$, $(1, \pm\sqrt{2})$, all unstable   (b) $\pm(1/\sqrt{2}, 1/\sqrt{2})$, both unstable   (c) $(0, 0)$, unstable
    (d) $(0, 0)$, unstable
3. (a) $(0, 0, 0)$, stable   (b) $\pm(1/\sqrt{2}, 1/\sqrt{2}, 0)$, both unstable   4. $-(\lambda + 1)^3 - 1$; -2. $(-1 \pm \sqrt{3}i)/2$; stable
6. (b) $(0, 0)$, $(\frac{1}{2}, \pm\sqrt{2})$ 7. (b) $(0, 0)$, $(0, 1)$, $(\pm\frac{1}{2}\sqrt{3}, -\frac{1}{2})$

Chapter 8

Section 1

3. (a) $4s/(s^2 + 4)^2$ (b) $(s + 2)/(s^2 + 1)$ (c) $(-s^2 + 2s + 2)/s^3$ (d) $(s \cos a - \sin a)/(s^2 + 1)$
    (e) $(s - 1)/(s - 3)^2$ (f) $2s/(s^2 - 1)$
4. (a) $(e^t - e^{-t})/2$ (b) $(e^{2t} + e^{-2t})/2$ (c) $\sin 2t$ (d) $t \sin 2t$ (e) $\frac{1}{3}e^{2t} \sin 3t$ (f) $t - te^t$
5. (a) $3e^t - t - 1$ (b) $(e^{-2t} + 1)/2$ (c) $-\frac{3}{13}e^{-3t} + \frac{3}{13}\cos 2t + \frac{2}{13}\sin 2t$ (d) $-\frac{5}{2}\cos t + \frac{3}{2}\sin t + \frac{1}{2}e^{-t} + 1$
    (e) $\frac{456}{111}e^{t/2} - \frac{12}{111}\cos 3t - \frac{2}{111}\sin 3t - 4$ (k) $x = \frac{4}{3}e^{-t} - \frac{10}{3}e^{t/2} + 2e^t + 2t,\; y = -\frac{4}{3}e^{-t} - \frac{5}{3}e^{t/2} + 2e^t + \frac{1}{2}t^2 - t + 1$
6. (d) $\frac{1}{2}(t - a)^2 H_a(t) + 1$

Section 2

1. (a) $e^{-t} + t - 1$ (b) $(t^5 + 10t^3)/30$ (c) $t^2/2$ 2. (a) $e^{-t} + t - 1$ (b) $e^{2t} - e^t$ (c) $\frac{1}{2}(t - 2)^2 H_2(t)$
3. (a) $\int_0^t ue^{-3u}\,du = (-\frac{1}{3})te^{-3t} - (\frac{1}{9})e^{-3t} + \frac{1}{9}$ (b) $\frac{1}{4}(1 - \cos 2(t - 2))H_2(t)$ (c) $e^{-t}\sin t$ (d) $\sin t$
    (e) $1 + H_1(t) = 1,\; 0 \le t \le 1,\; 0$ otherwise   (f) $\cos\sqrt{10}t$
4. (a) $\frac{17}{10}e^t + \frac{3}{10}e^{-t} - \frac{1}{5}\sin 2t - 1$ (b) $(\sqrt{2}/4)\sin\sqrt{2}t + t/2$ (c) $5 - 3e^{-t} - te^{-t} - t + t^2/2$
5. $y(t) = -1 + 2e^{-t/2}\cosh(\sqrt{5}t/2)$

Section 3

1. (a)

4. (a) $0,\; 0 \le t \le t_0$; $(t - t_0)/h,\; t_0 \le t \le t_0 + h$; $1,\; t_0 + h \le t$ (c) $\lim_{h \to 0+} y(t) = H_{t_0}(t)$
5. $y(t) = (t - t_0)H_{t_0}(t),\; t_0 > 0$
6. (a) $y_\pi = \sin(t - \pi)H_\pi(t);\; y_{2\pi} = \sin t\, H_{2\pi}(t)$ (c) $c \sin t(H_{2\pi}(t) - H_\pi(t))$

Chapter 8

Section 3

7. (a) $(1 - e^{1-t})H_1(t)$ (c) $0, 0 \le t < 2; e^{t-2} - e^{2-t}, 2 \le t$ 8. $0, 0 \le t < 1; (t-1)e^{1-t}, 1 \le t$

11. (a) $(1/s)(e^{-s} - e^{-2s})$ (c) $(1/s)(1 + e^{-s} - 2e^{-2s})$ (e) $(1/s^2)(1 - 2(s+1)e^{-s} + (2s+1)e^{-2s})$

Chapter 9

Section 1

1. (a), (c) Independent   (e) Dependent   3. (a) xe^{2x} (c) x (e) $x^2 + 1$ 4. (a) $c_1 \int e^{-x^2/2}\, dx + c_2$

5. (a) $xe^{\alpha x}$ 6. (a) $1 + (x/2) \ln |(x-1)/(x+1)|$ 7. $\sum_{k=0}^{\infty} x^{2k}/(k!(2k-1))$

12. (a) $c_1 x + c_2/x$ (c) $c_1/x + c_2 (\ln x)/x$ 14. (a) $x = c_1 t + c_2 e^t, y = c_1$ 15. (c) $(x-1)y'' - xy' + y = 0$

17. (b) $(\sin x)/\sqrt{x}, (\cos x)/\sqrt{x}$

Section 2

1. (c) $\frac{1}{3}x^4 + c_1 x^3 + c_2 x$ (d) $x^3 e^x + c_1 x^2 e^x + c_2 e^x$

2. (a) $y = c_1 e^x + c_2 e^{-2x} + e^{2x}/4$ (c) $y = c_1 \cos x + c_2 \sin x + (\cos x) \ln (\cos x) + x \sin x, |x| < \pi/2$

   (e) $y = c_1 x + c_2 x^2 + \frac{1}{2}$

3. (a) $x^2 e^x/2$ (c) $(e^{-x} + e^{-2x})(\ln (1 + e^x) - 1)$ (e) $x^3/3$

4. (a) $G(x, t) = e^{t-x} - e^{2(t-x)}$ (c) $G(x, t) = t(\ln x - \ln t)$

5. (a) $y = 4e^{-x} - 3e^{-2x} + \begin{cases} 4e^{-x} - 3e^{-2x}, & x < 1 \\ 4e^{-x} - 3e^{-2x} + \frac{1}{2} + \frac{1}{2}e^{2-2x} - e^{1-x}, & 1 \le x \end{cases}$ (e) $y = -1 + x + \ln x$

8. See Chapter 3, Section 3C   9. (a) $x^3/2 - x_0 x^2 + x_0^2 x/2$ (c) $-x + x^2/x_0 - x \ln |x/x_0|, x_0 \ne 0$

Section 3

1. (a) Unique   (c) Infinitely many

2. (a) (a, b), where $b - a = k\pi/\sqrt{2}$ (b) No conjugate pairs   (c) (a, b), where $b - a = 2k\pi/\sqrt{3}$

   (d) No conjugate pairs

3. (a) $G(x, t) = \begin{cases} -(1/\sqrt{2}) \cos \sqrt{2}x \sin \sqrt{2}t, & 0 \le t \le x \\ -(1/\sqrt{2}) \sin \sqrt{2}x \cos \sqrt{2}t, & x \le t \le \pi/(2\sqrt{2}) \end{cases}$

   (c) $G(x, t) = \begin{cases} -(2/\sqrt{3})e^{(t-x)/2} \cos (\sqrt{3}x/2) \sin (\sqrt{3}t/2), & 0 \le t \le x \\ -(2/\sqrt{3})e^{(t-x)/2} \sin (\sqrt{3}x/2) \cos (\sqrt{3}t/2), & x \le t \le \pi/\sqrt{3} \end{cases}$

4. (a) $y_p = x/2 - (\sqrt{2}\pi/8) \sin \sqrt{2}x$ (c) $y_p = \sin x + \frac{1}{2}e^{-(x-\pi/\sqrt{3})/2} [\cos \sqrt{3}(\pi/3 + x/2) - \cos \sqrt{3}(\pi/3 - x/2)]$

5. (a) $y = \cos (\sqrt{2}x) + x/1 - (\sqrt{2}\pi/8) \sin (\sqrt{2}x)$ (c) $y = -e^{-x/2} \cos (\sqrt{3}x/2) + y_p(x)$

7. $y_k(x) = \sin (k\pi(x-a)/(b-a))$ 11. (a) $y = (R/24)(L^3 x - 2Lx^3 + x^4)$ (c) $y = (Rx^2/24)(6L^2 - 4Lx + x^2)$

13. (a) $y = (\sinh \sqrt{k}x)/\sinh \sqrt{k}$ 14. (a) $y = (e^{px} - 1)/(e - 1)$

Section 4

1. (a), (b) $y_3 = (e + 1)^{-1}(e^x + e^{-x+1}) - 1$ 2. (b) $y = -\cos x + c \sin x + 1, c$ const   4. (a) $y'(0) \approx 0.6577$

10. $y'(1) = -2.17356$ 11. $y'(1) = 1.60693$

Section 5

1. (a) $y = 1 + 2x + 3x^2 + \cdots$ (b) $y = -1 + x - \frac{1}{2}x^2 + \cdots$ (c) $y = 2 + x^2 + \frac{1}{4}x^4 + \cdots$

   (d) $y = 1 + \frac{1}{6}x^3 + \frac{1}{180}x^6 + \cdots$

2. $y = 1 + x + \frac{1}{2}x^2 + \frac{1}{3}x^3 + \frac{5}{24}x^4 + \cdots$ 3. $y = 2 + \frac{1}{2}(x-1)^2 + \frac{1}{3}(x-1)^3 + \frac{1}{120}(x-1)^5 + \cdots$ 6. $x^5/20$

7. $y = \exp (e^x - 1); y = 1 + x + x^2 + \cdots$ 8. $y(x) = \sum_{k=0}^{\infty} x^k/k! = e^x$ 9. $y(x) = \sum_{k=0}^{\infty} x^k/k! = e^x$

Section 6

3. (a) $y = c_0 \sum_{k=0}^{\infty} (-1)^k x^{2k}/k!$ 4. (a) $y = c_0 + c_1 \sum_{k=0}^{\infty} 2^{-k} x^{2k+1}/((2k+1)k!)$

5. (a) $c_0 \sum_{k=0}^{\infty} \frac{(-1)^k x^{2k}}{2^k k!} + c_1 \sum_{k=0}^{\infty} \frac{(-1)^k 2^k k! x^{2k+1}}{(2k+1)!}$ 6. (a) $y = x/2$ 8. (b) $y(x) = \sum_{k=0}^{\infty} (-1)^k 2^{-2k}(k!)^{-2} x^{2k}$

10. (a) $\sum_{k=0}^{\infty} \frac{(x-1)^{3k}}{k!}$ 11. (b) $r = \frac{1 \pm \sqrt{5}}{2}$

Chapter 9

Section 7A
1. **(a)** $x = 0$, regular   **(c)** $x = 1$, regular   **(e)** No singular points   **(g)** No singular points
2. **(a)** $\cos(\ln x)$, $\sin(\ln x)$ **(c)** $x^{1/2}\cos((\sqrt{3}/2)\ln x)$, $x^{1/2}\sin((\sqrt{3}/2)\ln x)$ **(e)** x, x^{-5}
   (g) $x^{(5+\sqrt{21})/2}$, $x^{(5-\sqrt{21})/2}$
3. e^{nx}, n integer

Section 7B
6. **(a)** 1.4048, 5.5201, and three more   **(b)** 3.8317, 7.0156, and three more   **10.** $y(x) = 3J_1(x)/J_1(2)$

11. **(b)** $y(x) = \dfrac{4}{J_0(6)}J_0(2x)$ **12.** **(b)** $\dfrac{J_1(5x)}{J_1(10)}$

Chapter 10

Section 1
2. **(c)** $u(x, y) = xy + f(x) + g(y)$, f, g differentiable   **3.** $u(x, y) = y + f(y)e^{-x} + g(y)xe^{-x}$, f, g differentiable
5. **(a)** Elliptic   **(c)** Hyperbolic   **7.** $(r + 1)^2 = s^2$ **9.** $u(x) = c_1 e^{\sqrt{c/a}\,x} + c_2 e^{-\sqrt{c/a}\,x}$

Section 2
1. **(a)** $\displaystyle\sum_{k=1}^{\infty} \frac{2(-1)^{k+1}}{k}\sin kx$ **(c)** $\dfrac{\pi^2}{3} + \displaystyle\sum_{k=1}^{\infty}\frac{4(-1)^k}{k^2}\cos kx$ **(e)** $\dfrac{1}{2} + \dfrac{2}{\pi}\displaystyle\sum_{k=0}^{\infty}\frac{1}{2k+1}\sin(2k+1)x$

   (g) $\displaystyle\sum_{k=0}^{\infty}\frac{4}{2k+1}\sin(2k+1)x$ **(i)** $-\dfrac{\pi}{2} + \dfrac{4}{\pi}\displaystyle\sum_{k=0}^{\infty}\frac{1}{(2k+1)^2}\cos(2k+1)x$
4. **(a)** $\sin^2 x = \frac{1}{2} - \frac{1}{2}\cos 2x$ **(c)** $\sin 2x\cos x = \frac{1}{2}\sin x + \frac{1}{2}\sin 3x$ **9.** Yes

Section 3
1. $\dfrac{4}{\pi}\displaystyle\sum_{k=1}^{\infty}\frac{(-1)^k}{k}\sin\frac{k\pi x}{2}$

3. **(a)** $\dfrac{4}{\pi}\displaystyle\sum_{k=0}^{\infty}\frac{1}{2k+1}\sin(2k+1)x$ **(c)** $\dfrac{8}{\pi}\displaystyle\sum_{k=1}^{\infty}\frac{k}{4k^2-1}\sin 2kx$ **(e)** $\displaystyle\sum_{k=1}^{\infty}\left(\frac{2(-1)^{k+1}}{k\pi} - \frac{6}{k^2\pi^2}\sin\frac{2k\pi}{3}\right)\sin\frac{k\pi x}{3}$

4. **(a)** 1   **(b)** $\dfrac{2}{\pi} + \dfrac{4}{\pi}\displaystyle\sum_{k=1}^{\infty}\frac{1}{1-4k^2}\cos 2kx$ **(e)** $\dfrac{1}{3} + \dfrac{2}{\pi}\displaystyle\sum_{k=1}^{\infty}\frac{1}{k}\sin\frac{k\pi}{3}\cos\frac{k\pi x}{3}$

Sections 4A, B
1. **(a)** $u(x, t) = e^{-(\pi^2 a^2/p^2)t}\sin(\pi x/p)$ **(c)** $u(x, t) = 2/\pi\sum_{k=1}^{\infty}(1/k)e^{-k^2\pi^2 a^2 t}\sin k\pi x$
   (e) $u(x, t) = e^{-a^2 t}\sin x + \frac{1}{2}e^{-4a^2 t}\sin 2x$
2. **(a)** $v(x) = -1 + x$ **(c)** $v(x) = 1 + x$
3. **(a)** $\left(1 - \dfrac{8}{\pi}\right)e^{-(\pi^2 a^2/p)t}\sin\dfrac{\pi x}{p} + \dfrac{2}{\pi}e^{-4(\pi^2 a^2/p^2)t}\sin\dfrac{2\pi x}{p} - \dfrac{8}{3\pi}e^{-9(\pi^2 a^2/p^2)t}\sin\dfrac{3\pi x}{p} + \dfrac{1}{\pi}e^{-16(\pi^2 a^2/p^2)t}\sin\dfrac{4\pi x}{p} - \cdots$
   (c) $u(x, t) = \sum_{k=1}^{\infty}(4/2\pi k)e^{-4k^2\pi^2 a^2 t}\sin(2k\pi x)$
5. **(a)** $u(x, t) = e^{\lambda t}(c_1 e^{r_1 x} + c_2 e^{r_2 x})$, r_1, $r_2 = -\frac{1}{2}(1 \pm \sqrt{1 + 4\lambda})$ **(c)** $u(x, t) = Cx^\lambda e^{\lambda t/2}$
7. **(a)** $v(x) = x^2 + 1$ **(c)** $v(x) = x^3/6 + 5x/6$

9. $\dfrac{1}{2} - \dfrac{4}{\pi^2}\displaystyle\sum_{k=0}^{\infty}\frac{1}{(2k+1)^2}e^{-(2k+1)^2\pi^2 a^2 t}\cos(2k+1)\pi x$ **11.** $u(x, t) = e^{-a^4 t}\cos x$
13. **(b)** $C\exp(-\lambda^2 t^2/2)\cos\lambda(x - \alpha)$ **14.** $Cx^{-\lambda}\exp(-\lambda t^2/2)$

Chapter 10

Section 4C

1. **(a)** $u(x, t) = \cos (at) \sin x$

   (c) $u(x, t) = \dfrac{2}{\pi a} \sin (at) \sin x - \dfrac{1}{\pi a} \sin (2at) \sin 2x + \dfrac{2}{9\pi a} \sin (3at) \sin 3x + \dfrac{2}{25\pi a} \sin (5at) \sin 5x -$

   $\dfrac{1}{9\pi a} \sin (6at) \sin 6x + \cdots$

2. **(a)** $v(x) = (g/2a^2)x^2 + \beta x + \alpha$ 7. $u(x, t) = (c_1 e^{\sqrt{2}x} + c_2 e^{-\sqrt{2}x})(c_3 e^{r_1 t} + c_4 e^{r_2 t}),\ r_1, r_2 = -\tfrac{1}{2}(k \pm \sqrt{k^2 + 4a^2\lambda})$

Section 5

3. **(a)** $u(x, y) = 0$

4. **(a)** $u(x, y) = \displaystyle\sum_{j=0}^{\infty} \dfrac{4 \sin ((2j + 1)\pi x)}{(2j + 1)\pi \sinh ((2j + 1)\pi)} [\sinh ((2j + i)\pi y) + \sinh ((2j + 1)\pi(1 - y))]$

   (c) Same as part (a) after insertion of factor 2 after the first square bracket; uses the work in part (a)

   (e) $u = u_0 + u_2 + v_0 + v_1$, where

   $$u_0(x, y) = \sum_{j=0}^{\infty} \dfrac{4 \sin ((2j + 1)\pi x)}{(2j + 1)\pi \sinh (2\pi(2j + 1))} \sinh ((2j + 1)\pi(2 - y))$$

   $$u_2(x, y) = \sum_{j=0}^{\infty} \dfrac{4 \sin ((2j + 1)\pi x)}{(2j + 1)\pi \sinh (2(2j + 1)\pi)} \sinh ((2j + 1)\pi y)$$

   $$v_0(x, y) = \sum_{k=1}^{\infty} \dfrac{4(-1)^{k+1}}{k\pi \sinh (k\pi/2)} \sinh \dfrac{k\pi(1 - x)}{2} \sin \dfrac{k\pi y}{2}$$

   $$v_1(x, y) = \sum_{k=1}^{\infty} \dfrac{4(-1)^{k+1}}{k\pi \sinh \dfrac{k\pi}{2}} \sinh \dfrac{k\pi x}{2} \sin \dfrac{k\pi y}{2}$$

Chapter Review and Supplement

1. **(a)** $r^2 + 2r = s^2$ **(c)** $r^2 + s^2 = 1$ 6. $X'' + 5X' + \lambda X = 0,\ T' + \lambda T = 0$

7. **(a)** $A_2 = 2,\ A_6 = 0$ otherwise   9.   Top to bottom: E, D, A, B, C

10. **(a)** $X'' + \lambda^2 X = 0,\ T'' + (\lambda/k)^2 T = 0$ **(c)** $X'' + \lambda^2 X = 0,\ T' + \lambda^2 T = 0$

11. $y(x, t) = \dfrac{16}{\pi^3 a} \displaystyle\sum_{k=1}^{\infty} \dfrac{(-1)^{k-1}}{(2k - 1)^3} \sin \dfrac{(2k - 1)\pi a t}{2} \sin \dfrac{(2k - 1)\pi x}{2}$

Chapter 11

Section 1

4. **(c)** $f(x) = \begin{cases} 1, & 0 \le x \le \pi \\ -1, & -\pi \le x < 0 \end{cases}$ 10.   **(a)** Yes   **(c)** Yes   **(d)** No

Section 2

1. **(a)** $\lambda = n^2,\ y_n(x) = \sin nx,\ n = 1, 2, \ldots$ **(c)** $\lambda(n + \tfrac{1}{2})^2,\ y_n = \sin (n + \tfrac{1}{2})x,\ n = 0, 1, 2, \ldots$

   (e) λ such that $\sqrt{\lambda} = \tan \sqrt{\lambda},\ y_n = \sin \sqrt{\lambda}(x - 1)$

2. **(a)** $\lambda = n^2 - 1,\ y_n(x) = e^{-x} \sin nx,\ n = 1, 2, \ldots$ **(c)** $\lambda = n^2 - 1,\ y_n(x) = e^{-x} (\sin nx + n \cos nx),\ n = 1, 2, \ldots$

7. $P_0(x) = 1,\ P_1(x) = x,\ P_2(x) = (3x^2 - 1)/2,\ P_3(x) = (5x^3 - 3x)/2$

Section 3

5. **(a)** Solutions go with pairs $(j, k) = (1, 8), (8, 1), (4, 7), (7, 4)$

   (b) Solutions go with pairs $(j, k) = (1, 18), (18, 1), (6, 17), (17, 6), (10, 15), (15, 10)$

6. $u_{jk}(x, y, t) = \exp \left[-a^2\pi^2 \left(\dfrac{j^2}{p^2} + \dfrac{k^2}{q^2} \right) t \right] \sin \dfrac{j\pi x}{p} \sin \dfrac{k\pi y}{q}$ 7. $\exp (-a^2\lambda_{kj}^2 t)J_k(\lambda_{kj}r) \begin{cases} \cos kx, \\ \sin kx, \end{cases} \lambda_{kj} = r_{kj}/b$

8. **(b)** $(j, k) = (6, 8),\ (l, m) = (3, 4);\ (j, k) = (4, 2),\ (l, m) = (10, 5)$ **(c)** $(j, k) = (3, 4),\ (l, m) = (5, 3)$

INDEX